Epigenomics
From Chromatin Biology to Therapeutics

Understanding mechanisms of gene regulation that are independent of the DNA sequence itself – epigenetics – has the potential to overthrow long-held views on central topics in biology, such as the biology of disease or the evolution of species. High-throughput technologies reveal epigenetic mechanisms at a genome-wide level, giving rise to epigenomics as a new discipline with a distinct set of research questions and methods. Leading experts from academia and from the biotechnology and pharmaceutical industries explain the role of epigenomics in a wide range of contexts, covering basic chromatin biology, imprinting at a genome-wide level, and epigenomics in disease biology and epidemiology. Details on assays and sequencing technology serve as an up-to-date overview of the available technological tool kit. A reliable guide for newcomers to the field as well as experienced scientists, this is a unique resource for anyone interested in applying the power of twenty-first-century genomics to epigenetic studies.

KRISHNARAO APPASANI is the Founder and Chief Executive Officer of GeneExpression Systems, a conference-producing organization focusing on biomedical and physical sciences. He is editor of *MicroRNAs: From Basic Science to Disease Biology* (2007) and *RNA Interference: From Basic Science to Drug Development* (2005), also published by Cambridge University Press.

Epigenomics

FROM CHROMATIN BIOLOGY TO THERAPEUTICS

Edited by

Krishnarao Appasani
GeneExpression Systems, Inc.

Foreword by

Azim Surani
University of Cambridge

CAMBRIDGE
UNIVERSITY PRESS

CAMBRIDGE
UNIVERSITY PRESS

University Printing House, Cambridge CB2 8BS, United Kingdom

Published in the United States of America by Cambridge University Press, New York

Cambridge University Press is part of the University of Cambridge.

It furthers the University's mission by disseminating knowledge in the pursuit of education, learning and research at the highest international levels of excellence.

www.cambridge.org
Information on this title: www.cambridge.org/9781107697836

© Cambridge University Press 2012

First published 2012
First paperback edition 2014

A catalogue record for this publication is available from the British Library

Library of Congress Cataloguing in Publication data
Epigenomics, from chromatin biology to therapeutics / edited by Krishnarao Appasani.
     p.   cm.
Includes bibliographical references and index.
ISBN 978-1-107-00382-8 (alk. paper)
1. Epigenetics.   2. Gene expression.   I. Appasani, Krishnarao, 1959–
QH450.E66   2012
572.8'65–dc23

                                                              2012013324

ISBN 978-1-107-00382-8 Hardback
ISBN 978-1-107-69783-6 Paperback

In memory of:

*My mentor, Late **Har Gobind Khorana***
Massachusetts Institute of Technology, USA
Winner of the Nobel Prize in Physiology or Medicine, 1968

and

*My friend, Late **Marshall Nirenberg***
National Institutes of Health, USA
Winner of the Nobel Prize in Physiology or Medicine, 1968

Contents

Contents

The colour plates are to be found between pages 162 and 163

Contributors

Luis G. Acevedo
Active Motif, Inc.
1914 Palomar Oaks Way, Suite 150
Carlsbad, CA 92008, USA

Schahram Akbarian
Brudnick Neuropsychiatric Research Institute
Department of Psychiatry
University of Massachusetts Medical School
Worcester, MA 01604, USA

Ioanna Andreou
QIAGEN GmbH
QIAGEN Strasse 1
Hilden, D-40724, Germany

Krishnarao Appasani
GeneExpression Systems, Inc.
PO Box 540170
Waltham, MA 02454, USA

Raghu K. Appasani
GeneExpression Systems, Inc.
PO Box 540170
Waltham, MA 02454, USA

Julia Arand
Saarland University
FR 8.3 Biosciences
Laboratory of EpiGenetics

Campus Building A2.4
Postbox 151150
Saarbrücken, D-66041, Germany

David M. Ashley
Andrew Love Cancer Centre
Deakin University
Geelong, Victoria 3220, Australia

Alexander R. Ball, Jr.
Department of Biological Chemistry
School of Medicine
University of California
Irvine, CA 92697, USA

Yehudit Bergman
Department of Developmental Biology and Cancer Research
Institute for Medical Research Israel-Canada
The Hebrew University Medical School
Jerusalem 91120, Israel

Marina Bibikova
Illumina, Inc.
9885 Towne Centre Drive
San Diego, CA 92121, USA

Angela Bithell
Department of Molecular and Cellular Neurobiology
King's College London
Institute of Psychiatry
125 Coldharbour Lane
London, SE5 9NU, UK

Francesca Bonafè
Department of Biochemistry "G. Moruzzi"
University of Bologna
via Irnerio, 48
Bologna, I-40126, Italy

Eric E. Bouhassira
Department of Cell Biology
Albert Einstein College of Medicine
1300 Morris Park Ave–Ullmann 903
Bronx, NY 10461, USA

Victoria L. Boyd
Applied Biosystems
850 Lincoln Centre Drive
Foster City, CA 94404, USA

Noel J. Buckley
King's College London
Institute of Psychiatry
125 Coldharbour Lane
London, SE5 9NU, UK

Lars Olov Bygren
Novum Karolinska Institutet
Unit of Preventive Nutrition
Department of Biosciences and Nutrition
Hälsovägen 7
Huddinge, SE-141 57, Sweden

Claudio M. Caldarera
National Institute for Cardiovascular Research
Bologna, I-40126, Italy

Gemma Carvill
Division of Genetic Medicine
University of Washington
Seattle, WA 98195, USA

James W. F. Catto
Academic Urology Unit
Institute for Cancer Studies
The Medical School
University of Sheffield
Beech Hill Road,
Sheffield, S10 2RX, UK

Sarah Derks
Department of Internal Medicine
VU University Medical Center
1007 MB, Amsterdam, The Netherlands

Ewa Dudziec
Academic Urology Unit
Institute for Cancer Studies
The Medical School
University of Sheffield

Beech Hill Road,
Sheffield, S10 2RX, UK

Jeffrey D. Falk
Aviva Systems Biology
5754 Pacific Center Blvd., Suite 201
San Diego, CA 92121, USA

Presently at: RiboMed Biotechnologies, Inc.
2736 Loker Avenue West, Suite C
Carlsbad, CA 92010, USA

Jian-Bing Fan
Illumina, Inc.
9885 Towne Centre Drive
San Diego, CA 92121, USA

Joseph M. Fernandez
Active Motif, Inc.
1914 Palomar Oaks Way, Suite 150
Carlsbad, CA 92008, USA

David E. Fisher
Massachusetts General Hospital
Harvard Medical School
55 Fruit Street, Bartlett Hall 622
Boston, MA 02114, USA

Emanuela Fiumana
Department of Biochemistry "G. Moruzzi"
University of Bologna
via Irnerio, 48
Bologna, I-40126, Italy

Tamara B. Franklin
Brain Research Institute
University of Zurich/ETH Zurich
Winterthurerstrasse 190
Zürich, CH-8057, Switzerland

Presently at: EMBL Monterotondo
Adriano Buzzati-Traverso Campus
Via Ramarini 32
00015 Monterotondo, Italy

Fei Gao
Beijing Genomics Institute
Main Building, Beishan Industrial Zone
Yantian District
Shenzhen 518083, P. R. China

Arkadiusz Gertych
Translational Cytomics Group
Department of Surgery
Cedars-Sinai Medical Center
Los Angeles, CA 90048, USA

Emanuele Giordano
University of Bologna
Laboratory of Cellular and Molecular Engineering
via Venezia, 52
Cesena, I-43027, Italy
and
Dipartimento di Biochimica "G. Moruzzi"
via Irnerio, 48
Bologna, I-40126, Italy

David Goldman
Laboratory of Neurogenetics
National Institute on Alcohol Abuse and Alcoholism
National Institutes of Health
5625 Fisher's Lane
Rockville, MD 20852, USA

Markus Grammel
Laboratory of Chemical Biology
1230 York Avenue, Box 250
The Rockefeller University
New York, NY 10065, USA

Carlo Guarnieri
Department of Biochemistry "G. Moruzzi"
University of Bologna
via Irnerio, 48
Bologna, I-40126, Italy

Kevin L. Gunderson
Illumina, Inc.
9885 Towne Centre Drive
San Diego, CA 92121, USA

Victoria (Fatemeh) G. Haghighi
Columbia University Department of Psychiatry
New York State Psychiatric Institute
1051 Riverside Drive, Box 42
New York, NY 10032, USA

Xu Han
Beijing Genomics Institute
Main Building, Beishan Industrial Zone
Yantian District
Shenzhen 518083, P. R. China

Yong-Mahn Han
Department of Biological Sciences
KAIST
335 Gwahangno, Yuseong
Daejeon 305-701, Korea

Howard C. Hang
Laboratory of Chemical Biology
1230 York Avenue, Box 250
The Rockefeller University
New York, NY 10065, USA

Aditi Hazra
Harvard School of Public Health
Channing Laboratory
181 Longwood Avenue, Room 355
Boston, MA 02115, USA

Laura B. K. Herzing
Northwestern University Feinberg School of Medicine
Children's Memorial Research Center
2300 Children's Plaza Box 211
Chicago, IL 60614, USA

Norbert Hochstein
QIAGEN GmbH
QIAGEN Strasse 1
Hilden, D-40724, Germany

Robin Holliday
Australian Academy of Sciences
12 Roma Court
West Pennant Hills, NSW 2125, Australia

Dorothee Honsel
QIAGEN GmbH
QIAGEN Strasse 1
Hilden, D-40724, Germany

Mary A. Jelinek
Active Motif, Inc.
1914 Palomar Oaks Way, Suite 150
Carlsbad, CA 92008, USA

Guanyu Ji
Beijing Genomics Institute
Main Building, Beishan Industrial Zone
Yantian District
Shenzhen 518083, P. R. China

Yan Jiang
Brudnick Neuropsychiatric Research Institute
Department of Psychiatry
University of Massachusetts Medical School
Worcester, MA 01604, USA

Atsushi Kaneda
Genome Science Division, RCAST
The University of Tokyo
4-6-1 Komaba, Meguro-ku
Tokyo 153-8904, Japan

Richard A. Katz
Epigenetics and Progenitor Cells Program
Fox Chase Cancer Center, Room R422
333 Cottman Avenue
Philadelphia, PA 19111, USA

Hyemin Kim
Department of Biological Sciences and Center for Stem Cell Differentiation
KAIST
335 Gwahangno, Yuseong
Daejeon 305-701, Korea

Richard Kroon
QIAGEN GmbH
QIAGEN Strasse 1
Hilden, D-40724, Germany

Tapas K. Kundu
Molecular Biology and Genetics Unit
Jawaharlal Nehru Centre for Advanced Scientific Research
Jakkur, P.O. Bangalore-560 064
Karnataka, India

Benoit Labonté
Department of Psychiatry
McGill University
Douglas Mental Health Institute
6875 LaSalle Boulevard
Montreal, QC H4H 1R3, Canada

Daeyoup Lee
Department of Biological Sciences
KAIST
335 Gwahangno, Yuseong
Daejeon 305-701, Korea

Konstantin Lepikhov
Saarland University
FR 8.3 Biosciences
Laboratory of EpiGenetics
Campus Building A2.4
Postbox 151150
Saarbrücken, D-66041, Germany

Andrea Linnemann-Florl
QIAGEN GmbH
QIAGEN Strasse 1
Hilden, D-40724, Germany

Dirk Loeffert
QIAGEN GmbH
QIAGEN Strasse 1
Hilden, D-40724, Germany

Dylan Maixner
Laboratory of Neurogenetics
National Institute on Alcohol Abuse and Alcoholism
National Institutes of Health
5625 Fisher's Lane
Rockville, MD 20852, USA

Isabelle M. Mansuy
Brain Research Institute
University of Zürich/Swiss Federal Institute of Technology Zürich
Winterthurerstrasse 190
Zürich, CH-8057, Switzerland

Andreas Missel
QIAGEN GmbH
QIAGEN Strasse 1
Hilden, D-40724, Germany

D. V. Mohankrishna
Molecular Biology and Genetics Unit
Jawaharlal Nehru Centre for Advanced Scientific Research
Jakkur, P.O. Bangalore-560 064
Karnataka, India

Joana Carvalho Moreira de Mello
Department of Genetics and Evolutionary Biology
Institute of Bioscience
University of São Paulo
Rua do Matão, 277 sala 350
São Paulo, SP 05508-900, Brazil

Paolo G. Morselli
School of Plastic Surgery
University of Bologna
Bologna, I-40126, Italy

Rituparna Mukhopadhyay
Department of Cell Biology
Albert Einstein College of Medicine
1300 Morris Park Avenue, Ullmann 903
Bronx, NY 10461, USA

Claudio Muscari
Department of Biochemistry "G. Moruzzi"
University of Bologna
via Irnerio, 48
Bologna, I-40126, Italy

Takashi Nagano
Nuclear Dynamics Programme
The Babraham Institute
Babraham Research Campus
Cambridge, CB22 3AT, UK

Frank Narz
QIAGEN GmbH
QIAGEN Strasse 1
Hilden, D-40724, Germany

Shuji Ogino
Department of Pathology
Brigham and Women's Hospital, Harvard Medical School
75 Francis Street
Boston, MA 02115, USA

Carlo M. Oranges
School of Plastic Surgery
University of Bologna
Bologna, I-40126, Italy

Shari Orlanski
Developmental Biology and Cancer Research
Medical School
Hebrew University of Jerusalem
Jerusalem, Israel

Alice Pasini
Laboratory of Cellular and Molecular Engineering
University of Bologna, Campus of Cesena
via Venezia, 52
Cesena, I-43027, Italy
and
Department of Biochemistry "G. Moruzzi"
University of Bologna
via Irnerio, 48
Bologna , I-40126, Italy

Ralf Peist
QIAGEN GmbH
QIAGEN Strasse 1
Hilden, D-40724, Germany

Lygia V. Pereira
Departimento di Genética e Biologia Evolutiva
Instituto de Biociências, USP
Rua do Matão, 277 sala 350
São Paulo, SP 05508-900, Brazil

Andrey Poleshko
Fox Chase Cancer Center
333 Cottman Avenue
Philadelphia, PA 19111, USA

Claire Rougeulle
UMR 7216 Epigenetics and Cell Fate
CNRS/Université Paris Diderot
Bâtiment Lamarck, Case 7042
35 rue Hélène Brion
75013 Paris Cedex 13, France

Thea Rütjes
QIAGEN GmbH
QIAGEN Strasse 1
Hilden, D-40724, Germany

Ana Sanz
Active Motif, Inc.
1914 Palomar Oaks Way, Suite 150
Carlsbad, CA 92008, USA

Benjamin G. Schroeder
NUGEN Technologies Inc.
201 Industrial Road, Suite 310
San Carlos, CA 94070, USA

Gerald Schock
Epigenetics and Whole Genome Amplification
QIAGEN GmbH
QIAGEN Strasse 1
Hilden, D-40724, Germany

Kornel Schuebel
Laboratory of Neurogenetics
National Institute on Alcohol Abuse and Alcoholism
National Institutes of Health
5625 Fisher's Lane
Rockville, MD 20852, USA

B. Ruthrotha Selvi
Molecular Biology and Genetics Unit
Jawaharlal Nehru Centre for Advanced Scientific Research
Jakkur, P.O. Bangalore-560 064
Karnataka, India

Hogyu Seo
Department of Biological Sciences
KAIST
335 Gwahangno, Yuseong
Daejeon 305-701, Korea

Natalia Shalginskikh
Fox Chase Cancer Center
333 Cottman Avenue
Philadelphia, PA 19111, USA

Andrew Sharp
Division of Human Genetics
Department of Clinical Laboratory Sciences
University of Cape Town Medical School
Cape Town, South Africa

Presently at: Department of Genetics and Genomic Sciences
Mount Sinai School of Medicine
New York, NY 10002, USA

Jun S. Song
Institute for Human Genetics
Department of Epidemiology and Biostatistics
University of California–San Francisco
San Francisco, CA 94107, USA

Lennart Suckau
QIAGEN GmbH
QIAGEN Strasse 1
Hilden, D-40724, Germany

Azim Surani
The Gurdon Institute
University of Cambridge
Tennis Court Road
Cambridge, CB2 1QN, UK

Jian Tajbakhsh
Group Translational Cytomics
Department of Surgery
Cedars-Sinai Medical Center
Los Angeles, CA 90048, USA

Gustavo Turecki
McGill Group for Suicide Studies
Department of Psychiatry
McGill University
Douglas Mental Health Institute
6875 LaSalle Boulevard
Montreal, QC H4H 1R3, Canada

Céline Vallot
UMR 7216 Epigenetics and Cell Fate
CNRS/Université Paris Diderot
Bâtiment Lamarck, Case 7042
35 rue Hélène Brion
75013 Paris Cedex 13, France

Manon van Engeland
GROW-School for Oncology and Developmental Biology
Maastricht University Medical Center
6221 LK, Maastricht, The Netherlands

Jörn Walter
Saarland University
Laboratory of Epigenetics
Campus Building A2.4
Postbox 151150
Saarbrücken, D-66041, Germany

Nicholas C. Wong
Postdoctoral Fellow
Developmental Epigenetics, Early Development and Disease
Murdoch Childrens Research Institute
The Royal Children's Hospital
Flemington Road
Parkville, Victoria 3052, Australia

Mark Wossidlo
Saarland University
FR 8.3 Biosciences
Laboratory of Epigenetics
Campus Building A2.4
Postbox 151150
Saarbrücken, D-66041, Germany

Honglong Wu
Beijing Genomics Institute
Main Building, Beishan Industrial Zone
Yantian District
Shenzhen 518083, P. R. China

Yurong Xin
Columbia University Department of Psychiatry
New York State Psychiatric Institute
1051 Riverside Drive, Box 42
New York, NY 10032, USA

Zhixiang Yan
Beijing Genomics Institute
Main Building, Beishan Industrial Zone
Yantian District
Shenzhen 518083, P. R. China

Yu-Ying Yang
Laboratory of Chemical Biology
1230 York Avenue, Box 250
The Rockefeller University
New York, NY 10065, USA

Mingzhi Ye
Beijing Genomics Institute
Main Building, Beishan Industrial Zone
Yantian District
Shenzhen 518083, P. R. China

Kyoko Yokomori
Department of Biological Chemistry
School of Medicine
University of California–Irvine
Irvine, CA 92697, USA

Sephorah Zaman
Columbia University Department of Psychiatry
New York State Psychiatric Institute
1051 Riverside Drive, Box 42
New York, NY 10032, USA

Weihua Zeng
Department of Biological Chemistry
School of Medicine

University of California–Irvine
Irvine, CA 92697, USA

Gerald Zon
Formerly at: Life Technologies
850 Lincoln Center Drive
Foster City, CA 94404, USA

Foreword

Each cell type in an individual organism has a unique epigenome, although they share identical genetic information. Epigenomes are established by heritable but reversible modifications of DNA and histones that affect gene expression without altering the DNA sequence. Epigenetics is currently one of the most exciting areas of biological research. These advances in research teach us how the genetic information or the code is interpreted to generate diversity of cells and tissues.

Epigenomes can change in response to environmental factors such as signaling molecules to elicit an appropriate response during development and specification of cell fates. Unlike genetic mutations, epimutations are reversible, which indicates that epigenetic regulation of gene expression can be manipulated. This provides one of the reasons why advances in basic mechanisms will lead to advances in biomedicine, through discovery of agents that can affect and alter the epigenome and gene transcription in diseased tissues, including those aberrant modifications that accumulate in the course of aging.

In this book, Krishnarao Appasani has assembled a stellar group of researchers to provide their expert views on the current status of the field and on some of the likely advances in the future. Among the contributors to the book is Robin Holliday, a pioneer in the field of epigenetics through his work on DNA methylation, a heritable but reversible DNA modification of great importance in development and disease. He writes about epigenetics from a historical perspective. At the same time, Jörn Walter and colleagues present aspects on how DNA methylation can be erased. This is currently a major area of research, especially in the context of enzymes that convert 5-methylcytosine to 5-hydroxymethylcytosine. There is also a discussion on the emerging field of long non-coding RNAs that are likely to play a significant role in inducing epigenetic changes.

Epigenomics: From Chromatin Biology to Therapeutics provides expert reviews on many other diverse areas, which will serve as a valuable compendium. This book is very timely, as it captures some of the spectacular advances in epigenetics, which have led to unraveling the mechanistic basis for some of the phenomena such as genomic imprinting. The book also covers advances in chromatin biology, new techniques and methods that can be used to carry out genome scale analysis. These advances are relevant to many areas of developmental biology, regenerative medicine, and stem cell research. What are the mechanisms that ensure self-renewal of stem cells, and how do epigenetic mechanisms contribute to the differentiation of diverse cell types from undifferentiated cells? Conversely, what is the role of epigenetic mechanisms in the reprogramming of somatic cells to pluripotent stem cells, which is an area of great potential for progress in regenerative medicine? This is also critical for understanding the basis of some of human diseases, including cancers and neurodegenerative disorders.

These advances will also ultimately lead to a better understanding of epigenetic modifications induced by environmental factors, and their apparent subsequent inheritance through the germline to affect subsequent generations. Such transgenerational epigenetic inheritance phenomena, which could be of great importance for human diseases, are currently not well understood. This book provides a wealth of information for those who are already in the field, and for others who aspire to join them and make their contributions to shape the future advances in research on epigenomics.

Azim Surani Ph.D. F.Med.Sci. F.R.S.
Marshall-Walton Professor and Head of Wellcome Laboratories
Wellcome Trust Cancer Research UK Gurdon Institute
University of Cambridge

Preface

It is not enough to discover and prove a useful truth previously unknown, but it is necessary also to be able to propagate it and get it recognized.

Jean-Baptiste Lamarck, French naturalist (1744–1829) *Philosophie Zoologique* (1809), vol. 2, p. 450

Epigenetics: a Greek term meaning "above and beyond the gene." The late Conrad Waddington, the last Renaissance evolutionary biologist from the University of Edinburgh, Scotland, coined the word *epigenetics* in the 1940s for a phenomenon: *"that phenotypes arises from genotype through programmed change."* Taking the past two decades of work on epigenetics into consideration Adrian Bird of the University of Edinburgh provided a unifying definition of epigenetics as: *"the structural adaptation of chromosomal regions so as to register, signal or perpetuate altered activity states."* According to Andrew Feinberg of the Johns Hopkins University, Baltimore, USA, the modern definition of epigenetics is *"information heritable during cell division other than the DNA sequence itself."* Epigenetics is one of the fundamental mechanisms that is involved in embryo development and differentiation of cell types. One of the most exciting frontiers in both epigenetics and genomics (genome sciences) is the new field of *epigenomics* which can be defined as: *"the study of epigenetic marks in a given cell type using genomics technologies."* This new discipline promises novel insights into the genome because of its potential to detect quantitative alterations, multiplex modifications, and regulatory sequences outside of genes.

Epigenomics: From Chromatin Biology to Therapeutics is mainly intended for readers in the genomics, biotechnology, and molecular medicine fields. There are quite a number of books already available covering epigenetics/epigenomics.[1] For example, the book by Jablonka and Lamb (1995) emphasizes the importance of "epigenetic inheritance and evolution" specially focusing on the Lamarckian approach, whereas the book by Allis *et al.* (2007) nicely provides the principles of epigenetics. Three recent books, by Esteller (2009), Ferguson-Smith *et al.* (2009), and Tost (2010), have covered the importance of DNA methylation in

---

[1] Jablonka, E., and Lamb, M. (1995) *Epigenetic Inheritance and Evolution: The Lamarckian Dimension*, New York: Oxford University Press; Allis, C. D., *et al.* (2007) *Epigenetics*, Cold Spring Harbor Laboratory Press; Esteller, M. (2009) *Epigenetics in Biology and Medicine*, Boca Raton, FL: CRC Press; Ferguson-Smith, A.C., *et al.* (2009) *Epigenomics*, New York: Springer; Tost, J. (2010) *DNA Methylation: Methods and Protocols*, New York: Springer.

development and cancer. The present book differs, in that it is the first text completely devoted to the new field of *epigenomics* covering the basic biological, biochemical, molecular, and genomics aspects of epigenetics and their importance in health and disease biology. This book focuses on the history, biology, and biochemistry of chromatin, epigenomic imprinting, assay technology platforms, and cancer biology. Particularly, this book highlights the importance of epigenomics in variation (includes polymorphism), epidemiology (environment and nutrition), and neurodegenerative diseases. The goal is to have this book serve as a reference for graduate students, post-doctoral researchers, and teachers and as an explanatory analysis for executives and scientists in biotechnology and pharmaceutical companies. Our hope is that this volume will serve as a prologue to the field for both new comers and those already active in the field. This book is differentiated from others due to its careful integration of relevant emerging chromatin immunoprecipitation (ChIP) and sequencing platforms that have been continuously used today by various scientists to decipher the code of "epi-epigenomes" in normal and diseased cells. We carefully chose the chapters written by experts in the field from academia and industry and made appropriate sections to maintain the theme expressed in the subtitle of this book: *From Chromatin Biology to Therapeutics.*

Three epigenetic systems, i.e., X-chromosome inactivation, genetic imprinting, and epigenetic modifications, are the building blocks of the field of epigenomics. *X-chromosome inactivation* is the fundamental and common type of epigenetic marking that occurs during embryogenesis in female mammals. This mechanism was experimentally demonstrated in 1961 by Mary Lyon of the Medical Research Council's Mammalian Genetics Unit in Oxfordshire, UK; thus it is referred to as *lyonization*. In 1975 Robin Holliday of the Medical Research Council's National Institute for Medical Research, London, UK, and Arthur Riggs of the Beckman Research Institute of the City of Hope, Duarte, USA independently proposed a molecular model for *somatic cell inheritance* emphasizing that DNA methylation could be an important mechanism for the control of gene expression in higher organisms.

A second form of epigenetic inheritance is *genomic imprinting*, in which "stamping" of the genetic information occurs according to whether it is inherited from the mother or the father. This has been independently shown by Azim Surani of the University of Cambridge, UK, Bruce Cattanach of the Medical Research Council's Mammalian Genetics Unit, and Davor Solter of the Max-Planck Institute for Immunobiology, Freiburg, Germany. Genomic imprinting is the prime example of "transgenerational epigenetic inheritance" because the imprint that is established in the germ-line of a parent is passed on to the offspring where it is "read" in the next generation. This discovery was a harbinger of the exciting (and currently flourishing) research area of epigenetics and epigenomics. DNA methylation, histone modifications, and nucleosome positioning are the important parts of the machinery of the epigenome. Each part involves a battery of enzymes and protein complexes, including methyltransferases, acetyl transferases, histone deactylases, and several signaling molecules. The biochemistry of these modifications is the subject of this book. Understanding chromatin biology and the dynamics of the epigenome in development, aging, and disease states will help us to better understand the "histone code," and to develop models for screening environmental compounds and chemicals.

It is increasingly being recognized that *epigenetic modifications* are critical to disease pathogenesis. The first example of a human disease with an epigenetic mechanism was shown in cancer by Andrew Feinberg and Bert Vogelstein of the Johns Hopkins University, in 1983, but today it can be seen in several human diseases such as cardiovascular, diabetes, neurodegenerative and inflammatory diseases. Much of the recent growth in the field can be attributed to the technology-enabled ability to survey epigenetic modifications on a genome-wide scale. Large-scale epigenomic mapping projects have the potential to provide global, integrated views of different cellular states. Modern biomedical science is finally bringing together the intellectual forces of international academic researchers, industry scientists, and clinicians to map all the *methylomes* and/or *epigenomes*. Such collaborations have been initiated, and are relevant for the emerging science of epigenetics and epigenomics.

Epidemiology is the study of factors affecting the health and illness of human populations. Some of those factors include nutrition (diet), chemical exposure, behavior, and environment. It turns out that nutrition affects the way our genetic code is expressed: this was shown in the 1980s by Lars Olov Bygren of Umeå University, Sweden by analyzing agricultural records of the children growing up in Norrbotten in the nineteenth century. Effects of nurture (environment) on a species' nature (genes) were not supposed to happen so quickly and should take place over many generations and through millions of years of natural selection as proposed by evolutionary biologist Charles Darwin in his famous *On the Origin of Species*. But after all that, now Bygren and other scientists have amassed historical evidence as a trump card to play against Darwin! The present *environmental/epigenetic inheritance* principles are similar to those proposed earlier by evolutionist Jean-Baptiste Lamarck (1744–1829), who believed that the environment plays an important role in an organism's acquisition of evolutionary characteristics. Recent neo-Lamarckian researchers now believe that the environment plays a key role in a species acquiring inherited characteristics that drive variation and evolution. Decades later, many of Lamarck's theories are now being shown to be surprisingly correct.

Randy Jirtle of Duke University, Research Triangle, North Carolina, USA has shown for the first time that chemical exposure alters the physical characteristics of an organism by reducing the DNA methylation, and affects the next generations. Behavior (childhood abuse and suicidal nature) affects gene expression and passes to the next generation, as shown recently by Gustavo Turecki, a medical scientist from McGill University, Montreal, Canada. Bruno Reversade, a developmental biologist at Singapore's Institute of Medical Biology, is studying the effect of environment and genetic variant(s) on the *biology of monozygous twins*. However, no common allele or environmental factors or epigenetic factors have yet been identified behind the twinning process. In conclusion, it is now quite evident that chemical exposures and nutrition play a significant role at the epigenetic level and affect gene expression and regulation.

Many people have contributed to making our involvement in this project possible. We thank our teachers for their excellent teaching, guidance, and mentorship, which helped us to bring about this educational enterprise. We are extremely grateful to all of the contributors to this book, without whose commitment this book would not have been possible. Many people have had a hand in the preparation of this book. Each chapter has been passed back and forth between the authors for criticism and revision; hence each chapter represents a joint composition. We thank our readers, who have made our hours putting together this volume worthwhile. We are indebted to the staff of the Cambridge University Press, and in particular Katrina Halliday for her generosity and efficiency throughout the editing of this book; she truly understands the urgency and need of this volume. We also extend our appreciation to Hans Zauner for his excellent cooperation during the development of this volume. We thank English molecular biologist Robin Holliday (presently in Australia) and Swedish epidemiologist Lars Olov Bygren, men of encyclopedic interests and knowledge, for their understanding and support in writing on historical and epidemiological aspects respectively in this book. We want to thank Professor Azim Surani, an English developmental molecular biologist, and one of the pioneers in the field of epigenetics, for his kindness in writing the Foreword to this book. Last, but not least, we thank Shyamala Appasani for her understanding and cooperation during the development of this interesting volume.

This book is the second joint project of father and son. A portion of the royalties will be contributed to the Dr. Appasani Foundation (a non-profit organization devoted to bringing social change through the education of youth in developing nations) and MINDS Foundation (**M**ental **I**llness and **N**eurological **D**iseases), which is committed to taking a grassroots approach in eliminating stigma and providing educational, financial, medical, and moral support for patients suffering from mental illness in developing countries.

Krishnarao Appasani
Raghu K. Appasani

Part I

Basics of chromatin biology
and biochemistry

1 Introduction to epigenomics

Krishnarao Appasani and Raghu K. Appasani

A scientific man ought to have no wishes, no affections, – a mere heart of stone.
        Charles R. Darwin, English naturalist (1809–82)

The present "epigenetic soft inheritance" principles are somewhat related to those proposed earlier by evolutionist Jean-Baptiste Lamarck (1744–1829), who believed that the environment plays an important role in organisms' acquisition of evolutionary characteristics. Recent neo-Lamarckian researchers now similarly believe that the environment plays a key role in a species acquiring inherited characteristics that drive variation and evolution. Many of Lamarck's theories are now being shown to be surprisingly correct. Most of the non-heritable signals reside in and are controlled by chromatin, which is the complex of DNA and protein that makes up chromosomes. Chromatin is easily visualized by cytological stains, hence its name, which literally means "colored material" and was coined by German cytogeneticist Walther Flemming in 1879 (Flemming, 1879). Cytological stains distinguish the chromatin mass into euchromatin and heterochromatin. Chromatin that presents in a tightly condensed form is also known as heterochromatin (biologically inactive), and the non-condensed or extended loop form is called euchromatin which is the transcriptionally active form.

To date several epigenetic systems in humans have been described.

X-chromosome inactivation (dosage compensation)

This is one common type of epigenetic marking which occurs during embryogenesis in female mammals. The epigenetic mark "turns off" one of the X chromosomes that the female has inherited from her parents. The purpose of inactivating one of the X chromosomes is dosage compensation which is seen in all mammals. In 1949, two Canadian scientists, Murray Barr and Ewart Bertram, identified a highly condensed structure in the interphase nuclei of somatic cells in female cats but not in male cats. Later, this structure became

Epigenomics: From Chromatin Biology to Therapeutics, ed. K. Appasani. Published by Cambridge University Press. © Cambridge University Press 2012.

known as the Barr body, which was identified as a highly condensed X chromo-some (Barr and Betram, 1949). Almost a decade later, Susumu Ohno of the Beckman Institute City of Hope, Duarte, USA, proposed X-chromosome inacti-vation. However, in 1961, Mary Lyon (an English geneticist at the Medical Research Council's Mammalian Genetics Unit in Oxfordshire, UK) clearly described the dosage-compensation phenomenon in mammals that occurs by the inactivation of a single X chromosome in females (Lyon, 1961). At the same time Liane B. Russell of the Oak Ridge National Laboratory of the USA proposed the same theory. Today, the mechanism of X-inactivation is referred to as the *Lyon hypothesis* or *lyonization*.

Genetic imprinting

A second form of epigenetic inheritance is genomic imprinting, in which the "stamping" (expression) of the genetic information occurs according to whether it is inherited from the mother or the father. This is termed *monoallelic expression*. In other words, it is a form of gene regulation whereby some genes are silenced when inherited from the father and some are silenced when inherited from the mother. Almost a quarter of a century after the discovery of X-chromosome inactivation by Mary Lyon, three groups (Azim Surani of the University of Cambridge, UK, Bruce Cattanach of the Medical Research Council's Mammalian Genetics Unit, and Davor Solter of the Max-Planck Institute for Immunobiology, Freiburg, Germany) independently showed that the maternal and paternal genomes play different roles during early development, which they named *genomic imprinting*.

Genetic imprinting is an inheritance process that follows non-Mendelian inher-itance. Forms of genomic imprinting have been demonstrated in insects, mam-mals, and flowering plants. At the cellular level, imprinting is an epigenetic process that can be divided into three stages: (1) establishment of the imprint during gametogenesis; (2) maintenance of the imprint during embryogenesis and in the adult somatic cells; and (3) erasure and re-establishment of the imprint in the germ cells. Genomic imprinting is an epigenetic process that involves meth-ylation and histone modifications. These epigenetic markers are established in the germ-line and are maintained throughout all the somatic cells of an organism. Genomic imprinting involves a chemical marking process called methylation in which a methyl (-CH$_3$) group is added to cytosines in the DNA. Genomic imprint-ing is the prime example of transgenerational epigenetic inheritance because the imprint that is established in the germ-line of a parent is passed on to the offspring where it is "read" in the next generation.

Epigetic modifications and their mechanisms

DNA methylation The most widely studied epigenetic modification in humans is cytosine methylation. DNA methylation occurs almost exclusively in the context of CpG dinucleotides. The CpG dinucleotides tend to cluster in regions called CpG islands, defined as regions of more than 200 bases with G+C content of at least 50% and a ratio of observed to statistically expected CpG frequencies of at least 0.6 (Esteller, 2008).

Histone modifications Histones are key players in the epigenetics field, and the *nucleosome* is the fundamental unit of chromatin structure, which comprises a core of eight histones (H2A, H2B, H3, and H4 grouped into two H2A–H2B dimers and one H3–H4 tetramer) around which 147 base pairs of DNA are wrapped in 1.65 spherical turns. Histone H1 is called the linker histone. It does not form part of the nucleosome but binds to the linker DNA, sealing off the nucleosome at the location where DNA enters and leaves (Kouzarides, 2007). Histones influence every aspect of DNA function. The functions of chromatin are to package DNA into a smaller volume to fit into the cell, to strengthen the DNA to allow mitosis and meiosis, and to serve as a mechanism to control gene expression. Histones that are present in nucleosomes undergo two biochemical modifications: methylation and acetylation.

Nucleosome positioning Nucleosomes are a barrier to transcription that block access of activators and transcription factors to their sites on DNA; at the same time they inhibit the elongation of the transcripts by engaging polymerases. The packaging of DNA into nucleosomes appears to affect all stages of transcription, thereby regulating gene expression (Schones *et al.*, 2008).

All of the above steps involve a battery of enzymes and protein complexes, including histone methyltransferases (HMTs), histone acetyl transferases (HATs), histone deactylases (HDACs), histone demethylases (KDMs), histone deacety-lases, DNA methyltransferases (DNMTs), DNA demethylases, and nucleosome remodeling complexes. The activity of all of these proteins could be regulated by signaling molecules. The structure and function of chromatin varies consid-erably as the cell progresses through the cell cycle, by participating in DNA replication, repair, and damage. In the past three Nobel Prizes have been awarded (to Thomas Morgan, 1933, Aaron Klug, 1982, and Roger Kornberg, 2006) in the field of the elucidation of the function and structure of chromatin. However, its involvement in development, differentiation, disease, and genomic imprinting is still unclear, and a mystery to biologists.

Scope of this book

This text consists of 35 chapters, grouped into six parts, and many of the afore-mentioned applications are described within the various sections of the book, which are summarized as follows.

Part I: Basics of chromatin biology and biochemistry

This section consists of five chapters. In the early 1940s Conrad Waddington had published important research in embryology and genetics; he realized that there should be a synergistic relationship between these two fields (Waddington, 1939). The new field was later termed *epigenetics* (Waddington, 1942). In effect, epigenetics was extinguished in the resulting explosion of molecular genetics. The historical details and the development of epigenetics field have been well described in Chapter 2 by Robin Holliday, a pioneer molecular biologist and discoverer of DNA modification mechanisms and gene activity during development (Holliday and Pugh, 1975). Richard Katz and his colleagues in Chapter 3 describe the use of the green fluorescent protein reporter model system and siRNA screens to understand the functional networks of human epigenetic factors; this has resulted in detection of the interruption of epigenetic silencing functions in human cells. The recent upsurge of discoveries in epigenetics has revealed numerous tissue-specific DNA methylation and histone modification patterns that distinguish different parts of the human genome. It is believed that these modifications can induce global chromatin compaction or decondensation. Some functions of these modifications may be to control nucleosome stability and positioning in promoters and enhancers which critically regulate transcriptional activities. Nucleosome positioning is found among different cell types, and the interplay among transcription factors, chromatin remodelers, non-coding RNAs, and histones may play crucial roles in establishing and maintaining cell-type-specific chromatin structure. In Chapter 4, David Fisher and his colleagues summarize the results of high-throughput genome-wide mapping of nucleosomes with respect to their positioning and transcriptional regulation.

Posttranslational modifications, namely protein acetylation and methylation, are crucial in protein regulation and regulators of various epigenetic *histone code* phenomena (Allis *et al.*, 2007; Kouzarides, 2007). Proteins can be posttranslationally acetylated at their N-termini as well as at serine, threonine, and lysine residues; however, lysine acetylation is the most prevalent type of acetylation. Protein acetylation is a reversible posttranslational modification that is regulated by lysine acetyltransferases and deacetylases. Chemical reporter and mass-spectrometry-based proteomics strategies are the two major technological developments in recent years towards the analysis of posttranslational modifications. Therefore, to understand the protein acetylation and methylation mechanisms in the context of epigenetic regulation, Howard Hang and his colleagues used chemical reporter and mass-spectrometry methods as described in Chapter 5 of this book.

In Chapter 6, Takashi Nagano summarizes recent findings on the role of long non-coding RNAs in epigenetic silencing. Our knowledge of RNA has continuously expanded since then through identification and analyses of various distinct RNAs (including ribosomal RNA, transfer RNA, small nuclear RNAs, small nucleolar RNAs, and microRNAs). Each of these RNA classes has distinct features of its own, and the members of each class are thought to share similar function. In the 1990s several long non-coding RNAs (lncRNAs) were identified including: *Xist* and *Airn* and *Kcnq1ot1* that play essential roles in regulating transcriptional silencing of other genes. *Xist* is essential for X-chromosome inactivation in mammalian female cells (Plath *et al.*, 2003).

Part II: Epigenomic imprinting and stem cells

This section consists of seven chapters detailing the molecular processes that are involved in the genomic imprinting, the fundamental phenomenon occurring in embryos. In the early hours of life, the parental genomes of highly specialized germ cells get reprogrammed in order to comply with the totipotency state of the resulting zygote, which later on upon further development gives rise to all the cells of a multicellular organism. Two of the striking events happening in the developing zygote are the processes of DNA methylation and demethylation. Although the biological significance and mechanism of DNA methylation has been explored by experiments in transgenic mice, the mechanism of DNA demethylation is still not understood. DNA demethylation is not the only player in epigenetic reprogramming – histone modifications, histone variants, and small RNAs also contribute in leading mammals through proper development. In Chapter 7, Jorn Walter and his colleagues emphasize the importance of DNA demethylation in the zygote in order to better understand epigenetic reprogramming. Histone modifications largely take place on histone N-termini, regulating access to the underlying DNA. Histone proteins and their associated covalent modifications can alter chromatin structure, and determine how and when the DNA packaged in the nucleosomes is accessed, leading to the histone code hypothesis. In Chapter 8 Kim *et al.* describe the importance of histone modifications of lineage-specific genes in embryonic stem cells during differentiation.

The status of the X chromosome reflects major epigenetic instability of human embryonic stem cells (hESCs). While most studies on the epigenetic stability of hESCs have focused on imprinted loci and X-inactivation status, some have addressed the epigenetic variation at gene promoters in the rest of the genome, especially CpG-rich promoters. Normally, human pluripotent stem cells do not stand in a defined epigenetic state. Multiple variations can occur in the epigenome of hESCs along passages and within different cell lines specially reflecting a tremendous epigenetic plasticity of hESCs. Female pluripotent stem cells can harbor various X-chromosome patterns which are not stably maintained throughout numerous cell divisions. Human pluripotent stem cells (hPSCs) potentially stand as promising therapeutic tools for degenerative diseases. Through their capacity to differentiate to any cell type, they offer the possibility of a renewable source of replacement cells to treat various diseases including Parkinson's and Alzheimer's diseases. In Chapter 9, Vallot and Rougeulle elegantly describe the epigenetic stability of hPSCs.

Stem cells are undifferentiated cells with self-renewal and differentiation potential. Among stem cells, embryonic are considered as the best for cardiac regeneration. In the last decade another group of somatic stem cells, derived from adipose tissue, has been studied. In Chapter 10, Pasini *et al.* describe the features of adipose-derived stem cells, how to isolate them from lipoaspirates, and how they differentiate into cardiomyocytes. This chapter focuses especially on the epigenetic modifications (impact of CpG methylation) that influence the cells' commitment towards a cardiac phenotype. Complex gene regulatory networks control the acquisition and maintenance of cellular phenotype and function. In order to identify the molecular hallmarks of "stemness" it is a prerequisite to understand their transcriptional regulatory pathways or "epigenomic signatures." Epigenetic modifiers are recruited by transcription factors, and the transcription factor position is determined by the epigenetic status of the target gene chromatin. Therefore, it is important to identify factors that act to coordinate regulation of the transcriptome and epigenome. In Chapter 11, Bithell and Buckley discuss the involvement of transcription factors in the regulation of the epigenome, and provide an example of repressor element 1 silencing transcription factor (REST), which is a key neural regulator that balances stem-cell maintenance and differentiation.

As we mentioned earlier, many epigenetic elements such as DNA methylation, histone modifications, and microRNAs play crucial roles in gene regulation. MicroRNAs (miRNAs) are short sequences of RNA, 21–4 nucleotides in length, which play an integral role in the regulation of protein expression. Several such tiny microRNAs have been identified in stem cells. These tiny molecules play crucial roles in the gene network(s) involved in embryonic stem-cell maintenance, proliferation, differentiation, self-renewal, and pluripotency. In Chapter 12, Orlanski and Bergman summarize the importance of these tiny molecules in stem-cell biology and how these small molecules will likely have great impact on the use of stem-cell therapeutics. DNA replication is a process fundamental to cell proliferation, and occurs only once per cell cycle in eukaryotic cells. Additionally, after each DNA replication, the epigenetic information (such as DNA methylation and histone modifications) that constitutes the memory and the identity of the cells also is replicated. Well-regulated DNA replication is essential for normal development of an organism because the entire genome must faithfully be duplicated during a single cell cycle. Failure to do so may result in over-replication or under-replication of the genome giving rise to various abnormalities. Likely, these functional constraints might have caused eukaryotes to evolve complex tissue-specific replication programs. Thus, the timing during S phase at which each DNA segment replicates is critical for the transmission of *epigenetic memory*. The timing of replication affects gene expression or the probability of gene activation. In Chapter 13, Mukhopadhyay and Bouhassira emphasize the importance of the timing of replication in the context of epigenetic memory using genome-wide *epigenetic profiling* studies.

Part III: Epigenomic assays and sequencing technology

This section consists of seven chapters that emphasize the technological platform used to study epigenetic phenomena. A number of technologies (such as methylation-specific restriction enzyme digestion, methylation-dependent fragment separation, bisulfite DNA sequencing, pyrosequencing, and MALDI mass spectrometry) have evolved in recent years that have enhanced our abilities to identify and characterize epigenetic modifications on a genome-wide scale, and these developments have facilitated comparative studies to elucidate the functional relationship between epigenetic modifications and gene expression. The most widely studied epigenetic modification in humans is cytosine methylation. DNA methylation occurs almost exclusively in the context of CpG dinucleotides.

The conversion from the unmethylated to the methylated state has been coined an *epigenetic transition*. There are about 29 million CpG dinucleotides present in the mammalian genome. Broadly speaking there are two categories of CpG dinucleotides – clustered and unclustered. Clustered CpG dinucleotides are found primarily within and near gene loci in the mammalian genome – termed *CpG islands*. These CpG islands mediate a variety of biological processes such as gene expression, X-chromosome inactivation, imprinting, cellular differentiation, aging, and chromatin structure.

Technically, it is a quite challenge to discriminate accurately between the unmethylated and methylated states. One of such techniques, the protein-based affinity-capture method, is discussed by Acevedo *et al.* in Chapter 14 of this book. The greatest potential of this technique lies in its application in the clinical setting where it can be used to detect the methylation status of clinically significant targets that will help in the development of diagnostic or therapeutic assays. During the past decade the development of chromatin immunoprecipitation (ChIP) technologies and more recently the next-generation sequencing (ChIP–seq) technologies have emerged as the predominant methodologies for high-throughput epigenome-wide mapping. In order to provide complete genome coverage using ChIP–chip studies, typically 10^8 cells of starting material are required (which is difficult to get from stem cells or cancer tissues), and large sets of microarrays. To overcome this obstacle, a novel ChIP–chip technology, ChIP–DSL (ChIP–DNA selection and ligation) has been developed, which is discussed in Chapter 15 by Falk. ChIP–DSL technology provides increased sensitivity and utilizes a single microarray, made it adaptable to achieve rapid global profiling of epigenetic modifications.

Aberrant CpG methylation correlates in mammals with many diseases like cancer and developmental disorders. Therefore, there is a need to develop potential diagnostic marker assays to quantify DNA methylation especially at the individual CpG positions. Sanger sequencing has been a valuable technology to elucidate DNA methylation patterns; however, it is cumbersome and expensive. On the other hand, pyrosequencing, a real-time sequencing method (Ronaghi *et al.*, 1998), is cost-effective and sensitive for quantifying minor differences in modified CpG dinucleotides. Chapter 16, written by Löffert *et al.*, provides an overview of the development of high-resolution CpG methylation assays using such a pyrosequencing platform. Scientists at Illumina have developed a low-cost, high-throughput, and genome-wide DNA methylation profiling technology using the Illumina BeadChip platform. Nevertheless, in Chapter 17, Bibikova *et al.* have shown that universal arrays can access individual CpG sites across both CpG islands and genomic regions with low CpG density, and therefore give a good overview of epigenetic profiles on a genome-wide scale. Capillary electrophoresis (CE) can also be used for DNA methylation analysis. Scientists from Applied Biosystems have developed CE-based DNA methylation analysis for global analysis. Using labeled primers in the polymerase chain reaction (PCR) followed by separation on capillaries Schroeder *et al.* have described how to differentiate methylated and unmethylated gDNA, as discussed in Chapter 18.

In Chapter 19, Mingzhi Ye's team from the Beijing Genomics Institute evaluate five high-throughput sequencing methods (including BS–seq, MeDIP–seq, MBD–seq, MeDIP–BS, and ChIP–BS) in the analysis of DNA methylation at the chromatin level, and conclude that ChIP–BS method works efficiently. Very recent methods such as single molecular DNA sequencing technology will guide us into a new direction to unravel the secrets of epigenome/methylomes. In Chapter 20 Tajbakhsh and Gertych introduce a novel cytometric approach termed "Three-dimensional quantitative DNA methylation imaging." This method extracts fluorescence signals from three-dimensional images of chromatin texture in the nuclei of thousands of cells in parallel.

Part IV: Epigenomics in disease biology

This section consists of six chapters. It is increasingly being recognized that epigenetic abnormalities are critical to disease pathogenesis. Studies of epigenetic changes associated with different conditions can not only improve our understanding of the biology of the diseases and hold great promise for improving their management, but also be invaluable for providing insights into basic aspects of epigenetic regulation. The first example of a human disease with an epigenetic mechanism was cancer. In 1983, widespread loss of DNA methylation was observed in colorectal cancers compared with matched normal mucosa from the same patients (Feinberg and Vogelstein, 1983). Gene silencing is a major epigenetic gene-inactivation mechanism, by DNA methylation of the promoter region, and is involved in the initiation and progression of cancer. At present, cancer is by far the best-studied disorder with respect to epigenetic abnormalities. Several books on epigenetics by others have detailed its prominence in cancer biology. This book provides recent glimpses on the epigenome-wide studies on a few other cancers along with colorectal cancer.

For classification of cancer cases using DNA methylation data, a subset of colorectal cancer was found to show accumulation of CpG island methylation, the so-called "CpG island methylator phenotype"; this is described in detail by Hazra and Ogino in Chapter 34. Genome-wide approaches to searching for aberrantly methylated regions in cancer have been developed since 1993. in Chapter 21 Kaneda summarized cancer classification and genome-wide approaches in order to identify novel tumor-suppressor/inactivated genes and methylation markers. Colorectal cancer is the third leading cause of cancer death in both the USA and Europe. This cancer is thought to arise from pluripotent stem cells located in intestinal crypts which can develop into aberrant crypt foci and premalignant adenomas of which about 5–6% will develop into a carcinoma with invasive and metastatic potential. Nowadays colorectal cancer is one of the best-studied malignancies and provides an excellent model for the study of the complexity of epigenetics and its driving role in colorectal carcinogenesis. In Chapter 22 Derks and van Engeland outline the current knowledge of promoter CpG-island hypermethylation in colorectal cancer and the promising role of this biomarker for early disease detection, and the prediction of prognosis and response to therapy.

Urothelial carcinoma of the bladder is the third most prevalent cancer and one of the most expensive to treat. The etiology of the disease is acquired carcinogen exposure with little or no familial component. In general, low-grade tumors are characterized by infrequent aberrant hypermethylation, DNA hypomethylation, and the down-regulation of microRNAs, whereas high-grade tumors have extensive promoter hypermethylation and upregulation of many microRNAs. Numerous epigenetic alterations have been observed in this cancer, and these represent potential biomarkers or therapeutic targets due to their reversible nature. In Chapter 23, Dudziec and Catto provided an overview of this cancer. A number of animal studies linking environmental exposure to changes in epigenetic modifications such as DNA methylation suggest that changes in phenotype can arise from the environment via changes in the epigenome. The link between environment and epigenetics is becoming clearer in cancer; one important area is childhood cancer. Genome-scale analysis offers an unbiased approach in cataloging DNA methylation changes associated with disease and could be readily applied to childhood cancer. The clear differences between adult and childhood cancers preclude direct extrapolation of findings in adult studies to their childhood counterparts. Therefore there is a need to investigate the epigenomes of childhood cancer in addition to current efforts. In Chapter 24, Wong and Ashley emphasize the importance of genome-wide studies in childhood cancer.

Epigenetic chromatin regulation is crucial for myogenesis and muscle regeneration. Muscular dystrophies are a group of hereditary muscle diseases marked by muscle weakness and loss of muscle tissue. Facioscapulohumeral muscular dystrophy is one of the most common muscular dystrophies; identification of disease-specific genes for this disease is a quite challenge. Development of therapies that control the epigenetic chromatin alterations, both to stimulate muscle regeneration and to alleviate the pathological changes, is an important direction of research. In addition, therapeutic epigenetic manipulation may be of clinical value in the treatment of muscular dystrophies. Yokomori and his colleagues in Chapter 25 discuss their discovery of epigenetic changes in this condition, in the muscle and name it as an "epigenetic abnormality" disease.

The protein lysine acetylation is one of the important covalent modifications, which was first observed on histones and later on non-histone proteins. Although histone acetylation was originally reported by Vincent Allfrey in 1964 (Allfrey et al., 1964), the first histone acetyltransferase (GCN5) was identified from Tetrahymena by David Allis's group in the mid-1990s (Brownell and Allis, 1995). This was followed by the identification of several acetyltransferases, most of which are positive regulators of transcription. Nucleosomal histone acetylation has been considered to be an indispensable component of transcriptionally active chromatin. This "acetylation" is an integral component of a code or the "epigenetic language" rather than a mere "mark." The small molecule modulators of acetyltransferases could be useful both as biological probes to elucidate the epigenetic language of cellular functions as well as therapeutic tools to target disease conditions. In Chapter 26 Selvi et al. summarize the roles of histone

acetyltransferases and all the possible methods available to develop histone acetylation drugs and therapeutics.

Epigenetic changes often precede disease pathology, making them valuable
diagnostic indicators for disease risk or prognostic indicators for disease progression. Several inhibitors of histone deacetylation or DNA methylation are
approved for hematological cancers by the US Food and Drug Administration
(FDA) and have been in clinical use for several years. More recently, histone
methylation and microRNA expression have gained attention as potential therapeutic targets. A key challenge for future epigenetic therapies will be to develop
inhibitors with specificity to particular regions of chromosomes, thereby reducing
side effects. The identification of histone demethylase enzymes has opened a new
frontier in the study of dynamic epigenetic regulation. Recently, it has become
clear that histone methylation contributes to maintaining the undifferentiated
state of embryonic stem cells and to the epigenetic landscape during early development. The involvement of histone demethylase enzymes in disease also provides a unique opportunity for pharmacological intervention by designing small-
molecule inhibitors that exploit the structure and enzymatic reaction mechanisms of these newly discovered enzymes to counteract their function. Indeed,
some small-molecule inhibitors have already been identified. Hopefully, these
will help to explore how dynamic histone methylation contributes to normal
biological functions and disease.

Part V: Epigenomics in neurodegenerative diseases

This section consists of five chapters. The central nervous system is one of the
most complex systems in humans. Recent studies have shed some light on the
relationship between epigenetic alterations and neurodegenerative and/or neurological diseases such as: Rett syndrome, Rubinstein–Taybi syndrome, ATRX syndrome, fragile X syndrome, amyotrophic lateral sclerosis, Alzheimer's disease,
multiple sclerosis, Parkinson's and Huntington's diseases, and congenital myotonic dystrophy. Epigenetic alterations are likely to be found in other disorders,
e.g., autoimmune, cardiovascular, and metabolic diseases. As noted by Andrew
Feinberg of the Johns Hopkins University, *"epigenomics provides the context for
understanding the function of genome sequence, analogous to the functional anatomy
of the human body provided by Vesalius a half-millennium ago"* (Feinberg, 2010).

Neurodevelopmental diseases (such as autism, schizophrenia, depression, and
Alzheimer's disease) are complex disorders that likely arise from the interaction of
alleles at multiple loci with environmental factors. However, additional information that affects phenotype is encoded in the distribution of epigenetic markers,
including DNA methylation and histone modifications. It is also the fact that
epigenetic phenomena have been reported in dynamic regulation of DNA methylation within differentiated neurons of the human cerebral cortex throughout
development, maturation, and aging. Epigenetic alterations at selected genomic
loci may affect social cognition, learning and memory, and stress-related behaviors and contribute to aberrant gene expression in a range of neurodevelopmental
disorders. In Chapter 27, Haghighi *et al.* outline stable histone methylation

markers in the postmortem brain, and provide strategies to be considered in the design of epigenetic studies in brain-based disorders. At both the single-gene and genome-wide levels, epigenetics influences phenotypic outcome, and its disruption is a key component of multiple disorders including neurodevelopmental and autism-spectrum disorder syndromes. The study of epigenetics and its (dys)regulation continues to provide much insight into normal processes and the etiologies of many disorders. In Chapter 28, Herzing focuses on the role of epigenetics in neurodevelopmental disorders with overlapping autism spectrum and seizure phenotypes (including Rett and Angelman syndromes) with special emphasis on her group's work on the chromosome 15q11–q13 locus.

There is also early evidence for potential epigenetic targets in the treatment of psychosis spectrum disorders. There is ample evidence that some of the most frequently prescribed psychopharmacological agents are associated with chromatin remodeling events and DNA and histone modification changes in brain cells. But whether these effects are required for therapeutic action or, alternatively, reflect some sort of epiphenomenon is presently unclear. Most of the antipsychotic drugs available today in the marketplace are aimed at potential epigenetic targets. These drugs mainly act at dopaminergic and/or serotonergic receptor systems and exert therapeutic effects only on psychotic symptoms but do not work for cognitive impairment. In an elegant study by Eric Nestler and his colleagues from the Mount Sinai School of Medicine, it was observed that conventional antidepressant therapies work partially by targeting monoamine metabolism and reuptake mechanisms at the terminals of (serotonergic, noradrenergic, and dopaminergic) neurons. In Chapter 29, Akbarian nicely introduces the concept of epigenetics as it pertains to the neurosciences and psychiatry and outlines the epigenetic targets and novel treatments for cognitive/emotional brain disorders.

Mental retardation is a common congenital disorder; patients afflicted with this condition exhibit impaired development of adaptive and cognitive abilities. Mental retardation is a genetically and clinically heterogeneous disorder with mutations in over 300 genes that disrupt the organization of neuronal cells into complex networks as well as the ability of these networks to remodel in response to learning and experience. Thus, understanding the neurobiology that governs this pathophysiology requires knowledge of the molecular functions and pathways which are affected in the pathogenesis of mental retardation. In Chapter 30, Carvill and Sharp provide an insight into how genome-wide DNA methylation studies helped them to detect important genes involved in the pathogenesis of mental retardation.

Long-term memory formation is a complex process requiring *de novo* protein synthesis, which is often preceded by increased gene transcription rates. Recent evidence suggests that changes in chromatin structure, also known to affect gene transcription, are likely to play a critical part in the establishment and maintenance of long-term memory. The regulation of both acetylation and phosphorylation of histones has been implicated in long-term memory processes. In particular, protein kinases and protein phosphatases have been suggested to play a role in chromatin remodeling. In Chapter 31, Franklin and Mansuy summarize the role of protein kinases and protein phosphatases in chromatin remodeling during long-term memory formation and maintenance.

Part VI: Epigenetic variation, polymorphism, and epidemiological perspectives
This section consists of four chapters highlighting the epigenomic diversity, polymorphism, and epidemiological aspects of epigenetics. Epidemiology is the study of factors affecting the health and illness of human populations. Some of those factors include: environment, nutrition (diet), chemical exposure, and behavior. After all, *"What you're getting fed in the womb influences your phenotype; physical and physiological attributes."* It turns out that nutrition affects the way our genetic code is expressed (Richards, 2006). In the 1980s, Lars Olov Bygren, a preventive-health specialist who is now at the prestigious Karolinska Institute in Stockholm, Sweden (contributor of the last chapter, Chapter 35), began to wonder what long-term effects the feast and famine years might have had on children growing up in Norrbotten in the nineteenth century. By analyzing agricultural records, Bygren and two colleagues determined how much food had been available to the parents and grandparents of these children when they were young. Around the time he started collecting the data, Bygren had become fascinated with research showing that conditions in the womb could affect an individual's health not only when as a fetus but well into adulthood. In 1986, for example, *The Lancet* published the first of two groundbreaking papers showing that if a pregnant woman ate poorly, her child would be at significantly higher-than-average risk for cardiovascular disease as an adult. Bygren wondered whether that effect could start even before pregnancy: could parents' experiences early in their lives somehow change the traits they passed to their offspring? Effects of nurture (environment) on a species' nature (genes) were not supposed to happen so quickly and should take place over many generations and through millions of years of natural selection as proposed by the great English evolutionary biologist Charles Darwin, in his famous *On the Origin of Species*. But Bygren and other scientists have now amassed historical evidence suggesting that powerful environmental conditions (near death from starvation, for instance) can somehow leave an imprint on the genetic material in eggs and sperm. These genetic imprints can short-circuit evolution and pass along new traits in a single generation.

More recently, however, researchers have begun to realize that epigenetics can also help explain certain scientific mysteries that traditional genetics never could: for instance, why one member of a pair of identical twins can develop bipolar disorder or asthma even though the other is fine. Or why autism strikes boys four times as often as girls. Or why extreme changes in diet over a short period in Norrbotten could lead to extreme changes in longevity. In these cases, the genes may be the same, but their patterns of expression have clearly been tweaked. Biologists offer this analogy as an explanation: if the genome is the hardware, then the epigenome is the software. "I can load Windows, if I want, on my Mac," says Joseph Ecker, a Salk Institute biologist and leading epigenetic scientist. "You're going to have the same chip in there, the same genome, but different software. And the outcome is a different cell type" (Cloud, 2010).

Randy Jirtle and Robert Waterland of Duke University, Research Triangle, North Carolina, USA have shown for the first time that chemical exposure alters the physical characteristics of an organism by reducing DNA methylation, and affects the next generations. The environment in which we live, especially the early-life environment, shapes our behavior. Adversity during early life is strongly associated with problems in behavioral regulation and psychopathology in adulthood. Until recently, the mechanisms responsible for behavioral changes induced by early-life adversity were not clear. Childhood abuse and suicidal behavior affects gene expression and passes to the next generation. Gustavo Turecki, a Canadian scientist, is conducting studies to understand better the characteristics of individuals with major depression, focusing on issues such as personality traits and other possible psychiatric disorders. He has provided molecular clues as to why do some people who become depressed commit suicide while others who have the same illness do not. Labonte and Turecki summarize their findings in Chapter 32, suggesting that epigenetic changes are induced by the early environment and impact on the regulation of gene expression in the brain. Together, these findings suggest that epigenetics may act as a mechanism whereby environmental factors act on the modulation of long-term behavioral responses.

In mammals dosage compensation of X-linked products between males and females is achieved by an extreme epigenetic process: the transcriptional inactivation of one of the two X chromosomes in female cells, a process called X-chromosome inactivation. X-chromosome inactivation is primarily dependent on the expression of the *XIST/Xist* (inactive X-specific transcript) gene from the future inactive X, and co-localization in *cis* of its non-coding RNA. The interaction of *Xist* RNA with the future inactive X triggers a series of epigenetic changes in its chromatin, mainly DNA methylation and histone modifications, which determine its transcriptional silenced state. Pereira and Carvalho Moreira de Mello in Chapter 33 discuss how X-chromosome inactivation has been traditionally studied in humans, and how the use of contemporary genetic tools such as single-nucleotide polymorphism (SNP) DNA/RNA genotyping helps to analyze allele-specific gene expression.

Colorectal cancer is the third most common cancer and the third leading cause of cancer mortality in the USA. Colorectal cancer has a complex etiology resulting from germ-line and tumoral genetic variation, epigenetic changes, lifestyle risk factors, and clinical alterations. Further epigenetic epidemiologic studies are needed to link dietary factors related to one-carbon metabolism, cellular epigenetic alterations, and hypothesized pathogenic mechanisms. Hazra and Ogino summarize the details of dietary factors with respect to epidemiology in Chapter 34.

To some degree, the influence of epigenetic factors – environmental factors outside the gene such as a cell's exposure to chemical, physical, or biological agents modifying a gene, e.g., by methylation or phosphorylation – can be transmitted to an individual's descendants even if these factors are not present after fertilization or while programming takes place during gametogenesis (Petronis, 2010). We call this phenomenon *transgenerational epigenetic hereditability* or *gametic epigenetic inheritance*. Bygren in Chapter 35 discusses insights into complex etiologies, mainly of cardiovascular disease, related to epigenetic epidemiology.

Current issues in epigenomics

Identical twins

Globally, only 1 in every 250 to 300 births is identical twins. Bruno Reversade, a developmental biologist from the Singapore's Institute of Medical Biology, is studying the biology of monozygous twins by collecting samples around the globe. Generally, we do know that assisted reproduction (in vitro fertilization) causes higher rates of monozygotic twinning because of ovarian stimulation with drugs. This phenomenon is quite rare in nature but exists in the following three examples:

(1) Mohammad Pur Umri, a small village situated where Rivers Ganga and Yamuna meet near Allahabad City in the northeast of India, has roughly 1 in 10 live births of homozygous twins. There must be something in the soil or environment and/or genetic factors that are stimulating the ovaries to produce such eggs.

(2) Another example of monozygotic twinning comes from Jordan where a family has 15 pairs of monozygotic twins.

(3) Linha São Pedro, a town in Brazil, is inhabited predominantly by blond-haired, blue-eyed people descended from German immigrants. In the 1990s, 10% of the births there were twins, and almost half of those were monozygotic.

Probably, a particular allele (genetic variant) could be shared by most of these monozygous twins in this extraordinary phenomenon. However, no such common allele nor environmental factor nor epigenetic factor has yet been identified behind this twinning process. Arturas Petronis, from the University of Toronto, Canada is also studying identical twins that differ in appearance, personality, and their propensity to develop disease.

One possibility is that epigenetic differences might even start accumulating from day one of development (even the first four cells in a mouse embryo sometimes exhibit differences in histone methylation). According to Petronis diverging epigenetic profiles might be what drives the cells to split into twins in the first place. Although monozygotic twinning is more prevalent than any genetic disease, its cause remains enigmatic (Cyranosky, 2009).

Future developments in epigenomics

Epigenetic modifications provide a cellular memory for transcriptional control in all cell types via four mechanisms, i.e., RNA interference, histone code, nucleosomal remodeling, and CpG methylation. Scientific understanding of the DNA "hardware" of the human genome is well established, but the epigenomic "software" has not yet been systematically investigated at a genome-wide level (Satterlee *et al.*, 2010). The main hurdle for such an endeavor is the large number of epigenomes present even within an individual. Each of us has essentially one genome; however, each cell type in each individual is believed to have a distinct epigenome that reflects its developmental state.

Many groups are joining forces toward developing an organized *Human Epigenome Project* to exploit the new technologies to understand better the basis of normal development and human disease. We hope that epigenome maps will reveal new principles in the regulation of genome structure and function. Large-scale epigenomic projects around the globe include: the Asian Project (Korea, Japan, China, and Singapore) and the European Epigenomics Project (which comprises several European Union countries as members). Canada and Australia have initiated their own epigenetics alliance. The prominent ones in the USA are: the National Institutes of Health Roadmap Epigenomics Program and the International Human Epigenome Consortium (Bernstein *et al.*, 2010). After all, the goal of these projects (including the *Encyclopedia of DNA Elements*: ENCODE, 2007) is to gather the experts in the field and share the data and crack the secrets of the epigenome(s).

The field of epigenomics has an impact on modern biomedicine and the commercial enterprise of development of new theranostics (therapeutics and diagnostics) for several human diseases (Appasani, 2007). The recognition of epigenetics/epigenomics as a significant contributor to normal development and disease has opened new avenues for drug discovery and therapeutics, with a range of prospects that continues to expand as our knowledge of epigenetic regulation advances.

REFERENCES

Allfrey, V. G., Faulkner, R., and Mirsky, A. E. (1964). Acetylation and methylation of histones and their possible role in the regulation of RNA synthesis. *Proceedings of the National Academy of Sciences USA*, **51**, 786–794.

Allis, C. D., Jenuwein, T., Reinberg, D., and Caparros, M. L. (2007). *Epigenetics*. Cold Spring Harbor, NY: Cold Spring Harbor Laboratory Press.

Appasani, K. (2007). Epigenomics and sequencing: an intertwined and emerging big science of the next decade. *Pharmacogenomics*, **8**, 1109–1113.

Barr, M. L. and Bertram, E. G. (1949). A morphological distinction between neurons of the male and female, and the behaviour of the nucleolar satellite during accelerated nucleoprotein synthesis. *Nature*, **163**, 676–677.

Bernstein, B. E., Stamatoyannopoulos, J. A., Costello, J. F., *et al.* (2010). The NIH Roadmap epigenomics mapping consortium. *Nature Biotechnology*, **28**, 1045–1048.

Bird, A. (2007). Perceptions of epigenetics. *Nature*, **447**, 396–398.

Brownell, J. E. and Allis, C. D. (1995). An activity gel assay detects a single, catalytically active histone acetyltransferase subunit in *Tetrahymena* macronuclei. *Proceedings of the National Academy of Sciences USA*, **92**, 6364–6368.

Callinan, P. A. and Feinberg, A. P. (2006). The emerging science of epigenomics. *Human Molecular Genetics*, **15**, R95–R101.

Cloud, J. (2010). Why your DNA isn't your destiny. *Time Magazine*, January 06.

Cyranoski, D. (2009). Two by two. *Nature*, **458**, 826–829.

ENCODE Project Consortium (2007). Identification and analysis of functional elements in 1% of the human genome by the ENCODE pilot project. *Nature*, **447**, 799–816.

Esteller, M. (2008). Epigenetics in evolution and disease. *Lancet*, **372**, S90–S96.

Esteller, M. (2009). *Epigenetics in Biology and Medicine*. Boca Raton, FL: CRC Press.

Feinberg, A. P. (2010). Epigenomics reveals a functional genome anatomy and a new approach to common disease. *Nature Biotechnology*, **28**, 1049–1052.

Feinberg, A. P. and Vogelstein, B. (1983). Hypomethylation distinguishes genes of some human cancers from their normal counterparts. *Nature*, **301**, 89–92.

Ferguson-Smith, A. C., Greally, J. M., and Martienssen, R. A. (2009). *Epigenomics*. New York: Springer.

Flemming, W. (1879). Beitrage zur Kenntniss der Zelle und ihrer Lebenserscheinungen. *Archives für mikroskopische Anatomie*, **16**, 302–436.

Holliday, R. and Pugh, J. E. (1975). DNA modification mechanisms and gene activity during development. *Science*, **187**, 226–232.

Jablonka, E. and Lamb, M. (1995). *Epigenetic Inheritance and Evolution: The Lamarckian Dimension*. New York: Oxford University Press.

Kouzarides, T. (2007). Chromatin modifications and their function. *Cell*, **128**, 693–705.

Lyon, M. (1961). Gene action in the X-chromosome of the mouse (*Mus musculus* L). *Nature*, **190**, 372–373.

Petronis, A. (2010). Epigenetics as a unifying principle in the aetiology of complex traits and diseases. *Nature*, **465**, 721–727.

Plath, K., Fang, J., Mlynarczyk-Evans, S. K., *et al.* (2003). Role of histone H3 lysine 27 methylation in X inactivation. *Science*, **300**, 131–135.

Richards, E. J. (2006). Opinion: Inherited epigenetic variation – revisiting soft inheritance. *Nature Reviews Genetics*, **7**, 395–401.

Riggs, A. D. (1975). X-inactivation, differentiation and DNA methylation. *Cytogenetics and Cell Genetics*, **14**, 9–25.

Ronaghi, M., Uhlen, M., and Nyren, P. (1998). A sequencing method based on real-time pyrophosphate detection. *Science*, **281**, 363–365.

Satterlee, J. S., Schubeler, D., and Ng, H. H. (2010). Tackling the epigenome: challenges and opportunities for collaboration. *Nature Biotechnology*, **28**, 1039–1044.

Schones, D. E., Cui, K., Cuddapah, S., *et al.* (2008). Dynamic regulation of nucleosome positioning in the human genome. *Cell*, **132**, 887–898.

Tost, J. (2010). *DNA Methylation: Methods and Protocols*. New York: Springer.

Waddington, C. H. (1939). *Introduction to Modern Genetics*. London: Macmillan.

Waddington, C. H. (1942). The epigenotype. *Endeavour*, **1**, 18–20.

2 Epigenetics and its historical perspectives

Robin Holliday

2.1 Introduction

In the nineteenth century the leading biologists regarded development and inheritance as one and the same problem, and this is well exemplified by the classic text by Wilson (1896), *The Cell in Development and Inheritance*. It has been convincingly argued that this viewpoint was the main reason for the neglect of Mendel's discovery of the laws of inheritance in 1865 (Sandler and Sandler, 1985). Mendel's genius was to realize that genetics could be successfully studied without taking account of development. It is a remarkable fact that the 35-year neglect of Mendel's discoveries had a very strong influence on future fields of embryology and development, and also genetics. The rediscovery of Mendel's work at the turn of the century led to many further studies of inheritance in animals and plants, and the launch of the new science of genetics. Thereafter genetics proceeded quite independently of development. It was remarkable that the most prominent leader of this new science, using *Drosophila melanogaster*, was T. H. Morgan, who was himself an embryologist. Similarly, developmental biologists and embryologists pursued their research with little regard for genetics.

This situation persisted for the first 40 years of the twentieth century, that is, until C. H. Waddington championed the new field of epigenetics (Waddington, 1939). Waddington had published important research in both embryology and genetics, and he realized that there should be a synergistic relationship between these two fields. Not many other scientists were influenced by him; instead, studies of genetics in animals and plants continued, and there were quite separate studies of development. Then Beadle and Tatum initiated the biochemical genetics of *Neurospora*, and microbial genetics became the dominant theme. This included bacteria and their bacteriophages, and then yeast and other fungi. At the same time the structure of DNA was determined, and this provided a molecular basis for genetic replication, mutation, and the genetic code. The unraveling of the triplet nucleotide code was the major achievement of molecular biology.

Epigenomics: From Chromatin Biology to Therapeutics, ed. K. Appasani. Published by Cambridge University Press. © Cambridge University Press 2012.

Yet more was to come with the discovery of restriction enzymes, together with the cloning and sequencing of DNA. Now, the sequencing of DNA revealed the amino acid sequence of innumerable proteins. In effect, epigenetics was extinguished in the resulting explosion of molecular genetics.

In the late 1960s several leading molecular biologists turned their attention to animals, including the nematode *Caenorhabditis elegans*, the fly *Drosophila melanogaster*, the zebrafish, and mice, and the plant *Arabidopsis*. In addition, cultured cells, including human cells, became a major component of cell biology and molecular biology.

2.2   Inheritance in somatic cells

It has long been known that differentiated cells in complex organisms retain their phenotype through cell division. These cells can be said to have "housekeeping" proteins, common to all cells, and "luxury" proteins, which occur in specialized cells, such as fibroblasts, lymphocytes, or epithelial cells. When these cells divide the set of luxury proteins that characterize the particular cell type is stably maintained. Also, the genes coding for luxury proteins of a different cell type are silent. There are also stem cells which divide to form one differentiated cell type and another stem cell. Thus there are switches in gene activity during differentiation and development. The mechanisms that determine the activity or inactivity of specific genes were unknown. In all these cases the DNA in the cells is the same, so what could be changing DNA activities?

Another example of somatic inheritance is the X chromosome of female eutherian mammals. During embryonic development, one X chromosomes is inactivated, whereas the other is active. The switching off of most of the genes on each X chromosome is random, and once it has occurred it is not reversed. The DNA in the two X chromosomes is essentially the same, and they are in the same nuclear environment, yet the genes in one are active and in the other are inactive. X-chromosome inactivation is an example of dosage compensation which produces female cells equivalent to those in the male cells which have one X chromosome.

In 1975 two publications independently proposed a molecular model for somatic cell inheritance (Holliday and Pugh, 1975; Riggs 1975). This was based on the specific methylation of cytosines in DNA, and the presence of a maintenance methylase. The substrate for this enzyme is hemi-methylated DNA formed in newly replicated DNA, and it would not act on non-methylated DNA. Thus, the pattern of methylated cytosines in DNA would be inherited through sequential cell divisions. It was further proposed that the presence or absence of DNA methylation would determine whether a particular gene was active or inactive. There would be other sequence specific *de novo* methylases which could act on unmethylated DNA and bring about the switches in gene activity that occur during development and differentiation. It should be noted that the methylation of CpG doublets would be enormously stable in the presence of a maintenance methylase, since the loss of one methyl group, by whatever mechanism, would lead to its immediate replacement by the enzyme.

There was no direct evidence for this molecular model, therefore gene activity and gene inactivity could depend on either the presence or absence of DNA methylation in particular genes. However, there is an inactivation center in the X chromosome, and the spreading of inactivation from this suggested that methylation inactivated genes. The significance of the model was the fact that it provided for the first time a molecular switch at the DNA level which might explain the on/off activity of genes coding for luxury proteins, the somatic segregation of gene activities, and also chromosome inactivation. Neither publication used the term epigenetic, but as evidence for the model slowly began to accumulate, epigenetics and epigenetic became standard terminology. Thus, epigenetics can be defined as the study of changes in DNA activity or function that are not due to any change in DNA sequence.

It soon became apparent that the methylation of DNA was associated with gene inactivity or silencing. In particular there were CpG islands in the promoter region of many structural genes. These regions were not depleted in CpG as in the rest of genomic DNA. When CpG islands are methylated, then the gene is inactive, and when unmethylated transcription occurs. There were examples of the frequent reactivation of silent genes by 5-azacytidine, a potent inhibitor of DNA methylases. It also could induce differentiation (for example, myocytes and adipocytes) in cultured cells. The presence or absence of methylation in a given stretch of DNA could in many cases be determined by the use of methyl-sensitive restriction enzymes (reviewed in Russo *et al.*, 1996). These identify only a subset of methylated cytosines, so new methods were developed to determine the methylation status of all cytosines in DNA. In particular, the bisulfite sequencing method reveals the sequence of all five bases in DNA (Frommer *et al.*, 1992; Millar *et al.*, 2003). The sequencing of the human genome did not distinguish between cytosine and 5-methylcytosine, but the epigenome project will provide this information in many different cell types (Beck and Olek, 2003). It is an enormous undertaking.

2.3   Epimutations and dual inheritance

There was a long-standing controversy concerning the genetics of cultured mammalian cells. Some maintained that they could be studied on much the same way as micro-organisms. For example, using Chinese hamster ovary (CHO) cells, Siminovitch's laboratory isolated a large number of recessive mutant strains even though the initial isolate was diploid or pseudo-diploid. It later became apparent that many genes had become functionally hemizygous, because one of the genes had been silenced by DNA methylation (Holliday and Ho, 1990). Therefore standard mutations could be isolated in the single active gene. Since a silent gene is also present, such a strain can be reactivated to wild-type by 5-azacytidine, which is a powerful inhibitor of DNA methylases. Thus in such studies with mammalian cells there are two types of inheritance: standard classical mutations and genes silenced by DNA methylation. It became appropriate to refer to dual inheritance. Also, an epimutation could silence a gene by the *de novo* methylation of a promoter sequence. Reactivation by 5-azacytidine is also due to an epimutation (Holliday, 1991). Direct evidence for dual inheritance was obtained using the APRT (adenine phosphoribosyl transferase) gene of CHO cells. In this study bisulfite sequencing showed that the silenced APRT gene had heavily methylated CpG island promoter sequences, and reactivated genes had no methylation (Paulin *et al.*, 1998). It had also been demonstrated that genes could be silenced by the uptake of 5-methyl deoxycytidine monophosphate (methyl-dCMP) into permeabilized cells (Holliday and Ho, 1991; Nyce, 1991). We isolated a strain of CHO which had a very high frequency of spontaneous gene silencing, so it could be called an *epimutator* (Holliday and Ho, 1998). We showed that it had a low level of the enzyme methyl-dCMP deaminase, and also that H^3 methyl-dCMP was incorporated into DNA, with less being converted to thymidine mono-phosphate. We also isolated another strain that had a very low level of gene silencing, and a greater amount of methyl-dCMP deaminase. It can be regarded as an *anti*-epimutator strain.

It is unfortunate that these studies on *de novo* methylation in mammalian cells and its removal have not been followed up using the same or a similar exper-imental system. One important problem is the origin of the multiple methylated cytosines in CpG islands. The evidence is that a low level of methylated CpG is followed by a spreading of methylation to other nearby CpG doublets. A solution to the problem is particularly important in cancer cells, which will be discussed below.

2.4   Cancer and *de novo* methylation

For many years it was believed that mutations in tumor suppressor genes, such as that coding for p53, were one of the main causes of cancer. When it was discov-ered that genes could be silenced by DNA methylation, there was an explosion in cancer research, starting at the end of the twentieth century. The methylation status of tumor suppressors, oncogenes, and other genes of interest was examined, and it was found that many were hypermethylated, particularly in CpG island promoters. In a review published in 2003, nearly 60 examples are listed (Table 1.1 in Millar *et al.*, 2003).

The important problem is the origin of this hypermethylation. Most of these discoveries are descriptive, in the sense that the presence or absence of methyl-ation in cancer cells and controls is documented. In very few, if any, cases can an experimental analysis be done, such as that described in CHO cells in the previous section. It is well known that general metabolism in cancer cells is perturbed, although this is no longer a fashionable field of research. It may well be that initial *de novo* methylation is not due to a methylating enzyme, but to the incorporation of 5-methyl dCTP which is the phosphorylation product of 5-methyl dCMP (5-methyldeoxycytidine mono- and triphosphates). This metabolite will be present in cell nuclei because DNA is continually subject to repair. In normal cells it is very important that 5-methyl dCMP is not incorporated at random into DNA, so there is a specific mechanism for its removal by the enzyme 5-methyl dCMP deaminase to form dTMP. In cancer cells, the uptake of 5-methyl dCTP could only be a trickle, and could not possibly produce the hypermethylation that is observed. So the likelihood of specific methylation sites that trigger further methylation should be examined in detail. (It should be noted that in X-chromosome inactivation, DNA methylation of CpG islands spreads along the chromosome from the initial *Xist* inactivation site.) There is a real need for new experimental approaches to unravel the problem of DNA hypermethylation of CpG islands in cancer cells.

2.5 Chromatin remodeling

So far the discussion has been largely based on the chemical modification of cytosine in DNA to form methyl cytosine. Yet it has long been known that DNA exists in an open form which can be transcribed to RNA, and a closed form – usually referred to as heterochromatin – which is not transcribed. The DNA interacts with a group of histone proteins to form nucleosomes, which are the constituents of chromatin. The activity or inactivity of chromatin is closed related to the modification of histones by acetylation, methylation, phosphorylation, and ubiquitination. In the switching of gene activity it has sometimes been argued that the primary changes occur in histones and that DNA methylation provides a memory of the chromatin state that is heritable. Alternatively, DNA methylation or the removal of methyl groups may be the primary events that trigger all the subsequent changes required for chromatin remodeling.

5-Azacytidine is an inhibitor of DNA methylation which can reactivate silent genes at high frequency (see previous discussion). This suggests that DNA methylation is the primary event in gene silencing, and that its removal activates the gene. These experiments therefore indicate that chromatin modeling by histone modification is a secondary event. Another important issue is the heritability of histone modification, irrespective of DNA methylation. Is this a supposition, or based on direct evidence? Much discussion of chromatin modeling and remodeling will be found elsewhere in this book.

2.6 Differences between genetic and epigenetic systems

Much less is known about the epigenetic inheritance system than traditional well-established genetics. Genetics is based on cell lineages and clonal inheritance. Gametogenesis produces haploid cells that fuse to form a diploid zygote. This zygote produces a clone of cells with same genotype, but very different phenotypes. A mutation or chromosomal change can also be clonally inherited within the organism. Epigenetic changes are responsible for the different phenotypes in any multicellular differentiated organism. Often epigenetic events can occur in groups of cells, for example the induction of mesoderm or muscle tissue. This is due to an external signal (a morphogen, growth factor, or hormone) which interacts with surface receptors, and may alter the methylation status of the DNA that is heritable. Some epigenetic events are clonal, and X-chromosome inactivation is an excellent example. Genetic changes are rarely reversed, whereas epigenetic events

Table 2.1. Differences between classical genetics and epigenetic inheritance

Genetic system	Epigenetic system
Based on mutations which are heritable changes in DNA sequence	Based on heritable changes in DNA methylation, or other heritable chromatin changes
Very stable	Stable or unstable
Not subject to environmental influences (except mutagens or DNA damage)	Subject to environmental influences
Inheritance based on cell lineages (clonal)	Often polyclonal; sometimes clonal (X-chromosome inactivation)
Transmitted unchanged through mitosis and meiosis	Normally reversed or reprogrammed during gametogenesis

are often reversed. Genomic imprinting is epigenetic, because male and female gametes have different information superimposed on DNA. This information may be due to differences in DNA methylation, which may persist or be lost during development. If they persist they are reversed prior to or during gametogenesis. Environmental influences do not affect the genotype, apart from mutagens, and there is no inheritance of acquired characteristics. Epigenetics is quite different because during development there are continual interactions between cells that produce different phenotypes, which may or may not be heritable. It can be said that different cell types such as fibroblasts and lymphocytes have the same genotype, but different *epigenotypes*. The environment of a cell may be all important in determining its properties and its fate during development. The major differences between genetic and epigenetic systems are listed in Table 2.1.

This raises the question of epigenetic transgenerational effects, but it is not at all clear to what extent these are an *adaptive* response. There are now many examples of environmental influences that appear to affect the subsequent generation, or generations (Jablonka and Lamb, 1995, 2005). This in turn suggests that evolution itself may be influenced by the epigenetic transmission of environmental influences. The importance and significance of this has often been overstated, possibly because it might undermine Darwinian natural selection as the driving force of evolution. In fact, evolutionary biologists have now sequenced so much DNA that evolutionary trees can be constructed which can verify or improve the existing taxonomic classification of animals or plants. Epigenetic transmission may be based on differences in DNA methylation. It would seem that there is little likelihood of permanent epigenetic changes embedded in the genome of different species. However, such epigenetic changes might be recognizable if closely related species had fewer changes in DNA sequence than might be expected.

One of the challenges for the future is the reprogramming of epigenetic information in the germ-line, gametogenesis, zygote, or developing egg. It is most likely that pattern of DNA methylation is the most important component of this, and there is already evidence that massive changes in methylation occur during these stages of the life cycle. In addition to reprogramming, there is the imposition of imprinting in male and female gametes. This may have the effect of making particular genes hemizygous, that is, haploid with regard to the activity of single genes. Such haploidy might be important because switching of gene activities in early development is easier if there is only one gene copy rather than two (Holliday, 1990).

2.7 An epigenetic component in long-term memory?

The first suggestion that DNA methylation has an important biological role was made by Griffiths and Mahler (1969), who proposed that it could provide a basis for memory in the brain. Much later on Crick (1984) pointed out that neural circuits are not likely to have the stability to explain long-term memory over many decades. He proposed a protein model that would provide long-term chemical stability, which has affinities with the maintenance of DNA methylation. The suggestion was that modified protein dimers would be maintained indefinitely if a half-modified dimer was immediately fully modified by an enzyme that did not act on monomers or unmodified dimers. Another possibility is that there is a stable epigenetic component in this memory (Holliday, 1999). In the simplest instance there could be a signal to a receptive neuron which has the effect of methylating a particular gene, that was previously unmethylated. This would be completely stable in the presence of a maintenance methylase and absence of any demethylating mechanism.

In effect, the presence or absence of methylation is like a single item of a zero–one code. The complexity of the code can rise exponentially with each additional methylation site. Thus, two sites produces four code items, 10 sites produces about a thousand, 20 sites produces a million, 30 sites produces a billion, and so on. Thus, a relatively small number of methylation sites could encode a huge amount of information. This information would be at the DNA level, but the recognition of such a code would be at the level of the axon or synapse in many different neurons. The sites in DNA could be close together or dispersed, and it may be inappropriate to assign them to genes. Almost nothing is known about the epigenetics of brain function, and the possibility of uncovering an epigenetic code is far in the future. Nevertheless, the discussion of possible mechanisms for memory storage and other functions is not out of place (see, for example, Meaney and Ferguson-Smith, 2010).

2.8 A wider definition of epigenetics

A widespread definition of epigenetics is that it is the study of information in DNA that is not based on changes in base sequence. Waddington clearly had a broad view of the field, and to him epigenetics was the study of all those mechanisms that are involved in the unfolding of the genetic program for development, and to that we can add the maintenance of the adult. This broad definition clearly includes a number of mechanisms based on RNA function. For example, the alternative splicing of RNA transcripts provides the means of producing not one but many related proteins from one gene. A given pattern of alternative splicing will be appropriate for one cell type, and a different pattern for another cell type, and so on. This surely is an epigenetic mechanism which must be accurately controlled. It has been suggested that the accuracy of RNA splicing is achieved by small RNA molecules that hybridize to the ends of adjacent exons (Murray and Holliday 1979a, 1979b; Burnett, 1982; Holliday and Murray, 1994). The intron can then loop out and be removed, by a mechanism akin to DNA repair. This provides one function for small non-coding RNA. There are now known to be a huge number of small RNAs in cell nuclei, and also large non-coding sequences. This provides functions for much of the DNA that many had referred to as "junk DNA." It can be convincingly argued that many of the functions of DNA are to set up cell regulatory mechanisms, based on RNA, many of which can be regarded as epigenetic (Mattick and Makunin, 2006; Mattick *et al.*, 2009). An exciting possibility is that small RNAs can provide signals by moving from one cell to another. Another epigenetic mechanism is provided by the existence of large non-coding RNAs in the oocyte, egg, or early embryo, which could have an essential spatial, positional, or structural role. These may provide a structural framework that is essential for the three-dimensional distribution of proteins (Holliday, 1989; Kloc *et al.*, 2005). As well as RNA there are proteins that may have an epigenetic function, such as the polycomb group of proteins. Many of these mechanisms are reviewed elsewhere in this book.

Another specialized epigenetic mechanism is required for the creation of antibody diversity. It is now known that there are specific mechanisms for producing a high frequency of mutations in the DNA coding for antibodies. In particular, the enzyme cytosine deaminase converts cytosine to uracil, or conversion of 5-methylcytosine to thymine, and thus changes a G–C base pair into an A–T. These are mutations produced at high frequency by enzymes (Petersen-Mahrt, 2005). It is possible that they are also very important in pluripotent cells (Morgan *et al.*, 2004).

2.9   The chronology of epigenetics

Epigenetics is derived from the Greek word *epigenesis*, which is the theory that an embryo develops progressively from an undifferentiated egg. Obviously this is correct, so Waddington coined the word *epigenetics* in the early 1940s, which is the study of the unfolding of the genetic program for development. For many decades the word remained largely unused, or used to label any inherited event that was not explained by genetics. The proposal in 1975 that methylation of DNA could provide a mechanism for the control of gene activity during development became the first tangible epigenetic model. Evidence that DNA methylation was an important control mechanism accumulated in the 1980s, but it was not commonly referred to as an epigenetic mechanism. It has been said that the paper "The inheritance of epigenetic defects" (Holliday, 1987) *"was the critical paper that lit the fuse for the explosion in the use of 'epigenetic' in the 1990s"* (Haig, 2004). It could be said that the word gained full respectability and became part of the scientific establishment in 2004, when the 69th Cold Spring Harbor Symposium on

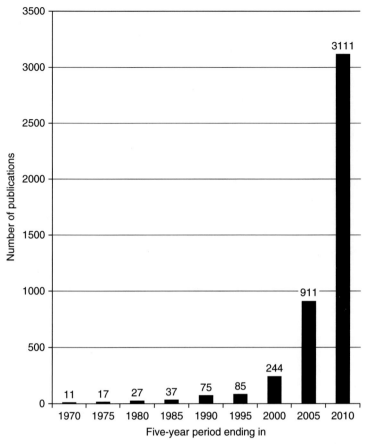

Figure 2.1 Number of publications with "epigenetic(s)" in their title (from PubMed).

Quantitative Biology had the simple title *Epigenetics* (Stillman and Stewart, 2004). Since then the number of publications with epigenetics in their title has increased enormously as shown in Figure 2.1. One might well ask why the number of publications remained fairly low up to the mid-1990s, and thereafter skyrocketed. It is probable that the merging of studies in molecular biology, genetics, cell biology, and development required some new simple terminology, and the science of epigenetics was finally established. This was nearly 60 years after Waddington's initial proposal, and about 20 years after the DNA methylation model was published. A wealth of data will come from the epigenome project which will further advance research and lead to innumerable insights. Epigenetics and the epigenome project will together continue to be a dominant force in the biological literature.

REFERENCES

Beck, S. and Olek, A. (eds.) (2003). *The Epigenome: Molecular Hide and Seek*. Weinheim: Wiley VCH.

Burnett, L. (1982). A model for the mechanism and control of eukaryote gene splicing. *Journal of Theoretical Biology*, **97**, 351–366.

Crick, F. H. C. (1984). Memory and molecular turnover. *Nature*, **312**, 101.

Frommer, M., MacDonald, L. E., Millar, D. S., Collis, C. M., Watt, F., Grigg, G. W., Molloy, P. L., and Paul, C. I. (1992). A genomic sequencing protocol which yields a positive display of 5-methyl cytosine residues in individual strands. *Proceedings of the National Academy of Sciences USA*, **89**, 1827–1831.

Griffith, J. S. and Mahler, H. R. (1969). DNA ticketing theory of memory. *Nature* **223**, 580–582.

Haig, D. (2004). The (dual) origin of epigenetics. *Cold Spring Harbor Symposia on Quantitative Biology*, **LXIX**, 1–4.

Holliday, R. (1987). The inheritance of epigenetic defects. *Science*, **238**, 163–170.

Holliday, R. (1989). A molecular approach to the problem of positional information in eggs and early embryos. *New Biologist*, **1**, 336–343.

Holliday, R. (1990). Genomic imprinting and allelic exclusion. In *Genomic Imprinting*, eds. Monk, M. and Surani, A., pp. 125–129. Cambridge: Company of Biologists.

Holliday, R. (1991). Mutations and epimutations in mammalian cells. *Mutation Research*, **250**, 345–363.

Holliday, R. (1999). Is there an epigenetic component in long-term memory? *Journal of Theoretical Biology*, **200**, 339–341.

Holliday, R. and Ho, T. (1990). Evidence for allelic exclusion in Chinese hamster ovary cells. *New Biologist*, **2**, 719–726.

Holliday, R. and Ho, T. (1991). Gene silencing in mammalian cells by uptake of 5-methyldeoxycytidine 5′ phosphate. *Somatic Cell and Molecular Genetics*, **17**, 537–542.

Holliday, R. and Ho, T. (1998). Evidence for gene silencing by endogenous DNA methylation. *Proceedings of the National Academy of Sciences USA*, **95**, 8727–8732.

Holliday, R. and Murray, V. (1994). Specificity in splicing. *BioEssays*, **16**, 771–774.

Holliday, R. and Pugh, J. E. (1975). DNA modification mechanisms and gene activity during development. *Science*, **187**, 226–232.

Jablonka, E. and Lamb, M. (1995). *Epigenetic Inheritance and Evolution: The Lamarckian Dimension*. Oxford: Oxford University Press.

Jablonka, E. and Lamb, M. (2005). *Evolution in Four Dimensions*. Boston, MA: MIT Press.

Kloc, M., Wilk, K., Vargas, D., *et al*. (2005). Potential structural role of non-coding and coding RNAs in the organization of the cytoskeleton at the vegetal cortex of *Xenopus* oocytes. *Development*, **132**, 3445–3457.

Mattick, J. S. and Makunin, I. V. (2006). Non-coding RNA. *Human Molecular Genetics*, **15**, R17–R29.

Mattick, J. S., Amaral, P. P., Dinger, M. E., Mercer, T. R., and Mehler, M. F. (2009). RNA regulation of epigenetic processes. *BioEssays*, **31**, 51–59.

Meaney, M. J. and Ferguson-Smith, A. C. (2010). Epigenetic regulation of the neural transcriptome. *Nature Neuroscience*, **13**, 1313–1318.

Millar, D. S., Holliday, R., and Grigg, G. W. (2003) Five not four: history and significance of the fifth base. In *The Epigenome: Molecular Hide and Seek*, eds. Beck, S. and Olec, A. pp. 3–38. Weinheim: Wiley VCH.

Morgan, H. D., Dean, W., Coker, H. A., Reik, W., and Petersen-Mahrt, S. K. (2004). Activation induced cytidine deaminase deaminates 5 methyl cytosine in DNA and is expressed in pluripotent tissues: implications for epigenetic reprogramming. *Journal of Biological Chemistry*, **279**, 52 353–52 360.

Murray, V. and Holliday, R. (1979a). A mechanism for RNA–RNA splicing and a model for the control of gene expression. *Genetical Research*, **34**, 173–188

Murray, V. and Holliday, R. (1979b). Mechanism for the splicing of gene transcripts. *FEBS Letters*, **106**, 5–7.

Nyce, J. (1991). Gene silencing in mammalian cells by direct incorporation of electroporated 5-methyl-2′ deoxycytidine 5′phosphate. *Somatic Cell and Molecular Genetics*, **17**, 543–550.

Paulin, R. P., Ho, T., Balzer, H. J., and Holliday, R. (1998). Gene silencing by DNA methylation and dual inheritance of Chinese hamster ovary cells. *Genetics*, **149**, 1081–1088.

Petersen-Mahrt, S. (2005). DNA deamination in immunity. *Immunological Reviews*, **203**, 80–97.

Riggs, A. D. (1975). X-inactivation, differentiation and DNA methylation. *Cytogenetics and Cell Genetics*, **14**, 9–25.

Russo, V. E. A., Riggs, A. D., and Martienssen, R. (eds.) (1996). *Epigenetics: Mechanisms of Gene Regulation*. New York: Cold Spring Harbor Laboratory Press.

Sandler, I. and Sandler, L. (1985). A conceptual ambiguity that contributed to the neglect of Mendel's paper. *History and Philosophy of Life Sciences*, **7**, 3–70.

Stillman, B. and Stewart, D. (eds.) (2004). Epigenetics. *Cold Spring Harbor Symposia on Quantitative Biology*, **LXIX**.

Waddington, C. H. (1939). *Introduction to Modern Genetics*. London: Allen & Unwin.

Wilson, E. B. (1896). *The Cell in Development and Inheritance*. New York: Macmillan.

3 Functional networks of human epigenetic factors

Andrey Poleshko, Natalia Shalginskikh, and Richard A. Katz*

3.1 Introduction

The term "epigenetic" was originally invoked to describe the cell type diversification that is revealed during development, or adult tissue maintenance. A long-standing question is how the nearly identical DNA blueprint in each body cell is parsed functionally to give rise to distinct cell types. We now know that selective utilization of the DNA blueprint is controlled by chromatin modifications: DNA methylation (primarily 5meCpG) and nucleosomal histone modifications (e.g., acetylation, methylation), as well as chromatin remodeling factors and microRNAs. Based on the early evidence for a role of DNA methylation in the control of gene activity, a modern definition of epigenetics emerged, to describe situations in which changes in chromatin modifications, not DNA sequence, alter gene activity (Holliday, 2006). The term "epigenetic" is now used to describe control mechanisms mediated by the complex array of "heritable" histone tail modifications, and DNA methylation (Klose and Bird, 2006; Kouzarides, 2007). Precise temporal and spatial control of these modifications must be maintained during normal development and tissue maintenance, and misregulation of epigenetic events can lead to cancer and other diseases (Feinberg, 2007).

A general understanding of how chromatin modifications can regulate gene expression is in place. For example, histone modification marks are "written" and "erased" by modifying enzymes (Taverna *et al.*, 2007), the marks are recognized by "readers" via specific binding domains, and the readers finally recruit "effectors" that modulate gene expression. Similarly 5meC DNA modifications can be recognized by methyl-CpG binding domain proteins (MBDs), leading to repression (Klose and Bird, 2006). However, it has become apparent that there are enormous complexities in such processes. For example, the positions and types of histone tail marks are vast, and their roles can be context-dependent, combinatorial, or

---

* Author to whom correspondence should be addressed.

Epigenomics: From Chromatin Biology to Therapeutics, ed. K. Appasani. Published by Cambridge University Press. © Cambridge University Press 2012.

dynamic with respect to gene activity (Berger, 2007; Grewal and Jia, 2007). Furthermore, there is evidence for histone–histone modification crosstalk, as well as histone–DNA methylation crosstalk (Cedar and Bergman, 2009). Although epigenetic gene expression states are generally binary (on/off switches), genes that are in an "off state" may display marks of active genes, signifying a "poised state" (Bernstein *et al.*, 2007). It is therefore not always clear which epigenetic modifications, and corresponding factors, are critical "drivers" for establishing the gene expression state, and which are ancillary, redundant, or have other roles, such as in DNA repair. Recent genome-wide analyses of histone marks and DNA methylation patterns (comprising the "epigenome") have begun to reveal the distribution of epigenetic mark signatures, and such information may now be integrated with DNA sequence data (Hawkins *et al.*, 2010).

The genome-wide technologies are now bringing into view the physical and functional organization of normal and diseased epigenomes, with the hope of developing disease treatments, and to potentially influence cellular identity for stem-cell-based regenerative therapy. Indeed, the ability to manipulate the epigenome has already provided therapeutic approaches, primarily using compounds that inhibit repressive epigenetic enzymes: DNA methyltransferases (DNMTs) and histone deacetylases (HDACs). HDAC inhibitors (HDACis) (e.g., trichosatin A, TSA) were identified in the 1970s through their ability to drive the differentiation of cancer cells to normal counterparts, without the knowledge of their molecular targets (Marks and Breslow, 2007). It was later realized that removal of acetylation marks on histones by HDACs could repress gene expression, and that the inhibition of HDACs leads to increased histone acetylation and reactivation of silent genes. Following these early studies, decades of research have led to treatment approaches for epigenetic diseases including cancer, with a focus on restoring the functions of epigenetically silent genes. In addition to case-by-case analyses of critical genes (e.g., silenced tumor suppressor genes), more recent findings indicate that epigenetic drugs can promote, or accelerate, the "re-setting" of cellular identity (Huangfu *et al.*, 2008; Kim *et al.*, 2010). It is hoped the complete characterization of normal and diseased epigenomes will provide a more detailed understanding of the specific roles of epigenetic marks. However, such studies are correlative, and it will therefore be critical to accelerate functional studies on the hundreds of epigenetic factors that deposit these marks, to identify causal roles in establishing and maintaining critical epigenomic patterns.

The abilities of epigenetic drugs to either promote cancer cell differentiation or influence cellular epigenetic memory indicate that many cell-type-specific genes are not permanently silenced, and are vulnerable to manipulation. Because the reversal of gene silencing appears to contribute to the mechanisms of the current epigenetic therapy drugs, we set out to obtain a broad view of human epigenetic factors that can participate functionally in epigenetic silencing networks. Such factors may be "druggable" or "targetable" for disruption, leading to reactivation of epigenetically silenced genes. Our approach has been to use a reporter model system in which widely distributed, epigenetically silent green fluorescence protein (GFP) genes provide a readout to detect interruption of epigenetic silencing functions. As described below, development of this approach required purification of cells that exhibit reversible epigenetic silencing of the introduced reporter genes. Furthermore, our studies indicate that the inserted genes can be epigenetically repressed, largely independent of their chromosomal position, allowing for measurements of epigenome-wide effects. This approach thereby allows robust, high-throughput analyses of epigenome-wide changes in epigenetic silencing following disruption or inhibition of individual silencing factors. We have been able to use this system to functionally interrogate individual factor roles in epigenetic silencing, using gene-by-gene small interfering RNA (siRNA) screening (Poleshko *et al.*, 2010), whereby a gene "hit" is indicated by reactivation of the silent reporter genes. This readout is interpreted to mean that a key, targetable silencing factor has been depleted. Based on the high degree of epigenetic crosstalk, and potential redundancies, it was not anticipated that knockdown of single factors would be sufficient to promote reactivation. However, as described (Poleshko *et al.*, 2010) and summarized below, we found that, indeed, many individual epigenetic factors are "targetable."

3.2   Methodology

3.2.1   Design of reporter cell systems

The strategies for isolation and analysis of GFP-silent reporter human cell populations and clones have been described previously (Katz *et al.*, 2007) (Figure 3.1A). This approach is based on the principle that epigenetic gene silencing is heritable through cell division, and can be reversible. Briefly, HeLa cells were infected with a retroviral vector encoding the GFP gene under control of several alternative promoters (Poleshko *et al.*, 2008, 2010). To purify cells harboring silent GFP genes, GFP-negative [GFP(−)] cells were first sorted at 7 to 10 days postinfection. To identify cells in which the GFP gene had undergone silencing, the sorted GFP(−) cultures were treated with epigenetic drugs known to reactivate silent genes. TSA or 5-aza-C, HDAC and DNMT inhibitors (HDACis, DNMTis) respectively, were used at concentrations of 0.5 μM (TSA) or 5 μM (5-aza-C). Cells in which the GFP reporter gene was reactivated were then isolated by cell sorting and were passaged for about 2 weeks during which time the GFP gene was re-silenced in a fraction of cells. The resulting re-silenced GFP(−) cells were then sorted. These enriched GFP(−) cells could be passaged for months without significant spontaneous GFP reactivation. Treatment of these GFP(−) cells with TSA or 5-aza-C resulted in reactivation of the GFP gene, as expected (Figure 3.1B). The percentage of GFP-positive cells [GFP(+)], and their intensities were quantitated by flow cytometry after 48 to 72 hrs. Cell populations isolated on the basis of GFP reactivation in response to either TSA or 5-aza-C were denoted TI or AI, respectively.

3.2.2   SiRNA-based knockdown analysis

A gene-by-gene siRNA knockdown approach (Boutros and Ahringer, 2008; Mohr *et al.*, 2010) was implemented to uncover roles for specific silencing factors and

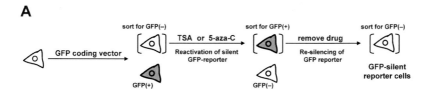

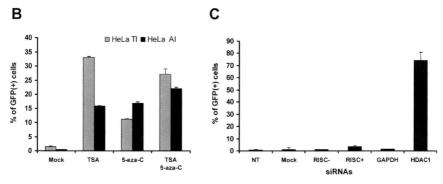

Figure 3.1 Design and use of GFP-silent reporter cell lines. (A) Sorting strategy for isolation of cell populations and clones containing silent GFP-reporter genes. (B) GFP-silent reporter cell populations selected using either an HDAC inhibitor TSA (HeLa TI) or a DNMT inhibitor 5-aza-C (HeLa AI) were treated with TSA, 5-aza-C, or a combination. Percentage of GFP-positive cells was measured by flow cytometry. Error bars indicated standard deviation ($n = 3$). (C) Response of HeLa TI cell population following treatment with the indicated siRNAs. Negative control siRNAs RISC−, RISC+, and GAPDH and the positive control siRNA HDAC1 were analyzed. Percentage of GFP-positive cells was measured by flow cytometry. Error bars indicated standard deviation ($n = 3$).

networks using GFP-silent reporter cells (Poleshko *et al.*, 2010). The strategy was based on the principle that knockdown of critical epigenetic factors would result in reactivation of the silent GFP genes. The screen was performed using siRNAs according to the manufacturer's protocol (QIAGEN, Inc. or Dharmacon Corp.). The transfection volumes and cell number were chosen according to the multi-well plate format using the manufacturer's protocol, with a final concentration of 50 nM siRNA. SiRNA and the DharmaFECT transfection reagent were mixed in Hank's Buffered Salt Solution (HBSS) and then incubated for 20 minutes. The siRNA transfection mix was added to a cell suspension (wet reverse transfection method) and plated. After 48 hrs, the transfection medium was replaced with complete growth medium, and cells were incubated for an additional 48 hrs. The percentage and intensity of GFP-positive cells were then measured by flow cytometry (Figure 3.1C). HDAC1 had been identified in a proof-of-concept study as an siRNA target leading to reactivation of the silent GFP gene, and was used as a positive control in subsequent screening assays (Poleshko *et al.*, 2008, 2010) (Figure 3.1C).

3.2.3 High-throughput siRNA screening

A well-by-well, gene-by-gene, high-throughput approach was designed for siRNA-based screening. The high-throughput siRNA screen was designed using a robotic pipetting system for siRNA transfection, optimized in a 96-well format, with multiple readout options for measuring GFP expression. Initially, HDAC1 siRNAs and GAPDH siRNAs were used as positive and negative controls, respectively, to assess the robustness and quality of responses in a 96-well format. Using these control siRNAs, the efficiency of the screening protocol, the Z'-factor (Zhang *et al.*, 1999), was calculated. The Z'-factor was measured in three independent experiments using the HeLa TI reporter cell population and yielded an average value of 0.79, indicating high reproducibility and sensitivity.

Screening was performed using the QIAGEN Inc. human "epigenetics set" siRNA library, which targets approximately 200 human genes (Poleshko *et al.*, 2010). For the primary screen, HeLa TI reporter cells harboring the silent GFP genes were treated with two independent siRNAs for each of the 200 gene targets, using the wet reverse transfection protocol that was optimized for HeLa cells in a 96-well plate format. A transfection mix containing siRNAs and DharmaFECT transfection reagent was first dispensed into 96-well plates, followed by addition of the reporter cell suspension in complete medium without antibiotics. The final concentration of siRNA was 50 nM in 100 μl of the transfection solution, with a cell number of 5000 per well. Cells were incubated with transfection medium for 48 hrs, followed by complete growth media. After a total of 96 hrs post-transfection, the percentage and intensity of GFP-positive cells were measured using a 96-well Millipore-Guava EasyCyte Plus flow cytometer. The data were scored as percent of GFP-positive cells. A secondary "validation" screen was then performed using four independent siRNAs against statistically verified targets identified in the primary screen.

3.2.4   Considerations for design and analysis of the siRNA screen

For the primary screen, a low stringency criterion for scoring was applied to reduce potential false-negative responses by any individual siRNA. For example, a target gene was scored positive if at least one of two individual siRNAs produced 15% of GFP-positive cells or higher. A total of 28 gene targets (out of about 200 tested) passed this primary criterion and were chosen for validation. The validation process was carried out as a secondary screen with four individual siRNAs per target gene. Raw values (percent GFP positive cells) for each single siRNA were averaged, and the fold-difference relative to the negative control, together with the p-value, was determined. An individual siRNA was scored positive if the fold-difference above the control value was higher than, or equal to 10, and p-value was less than 0.0001. The gene was validated as a "hit" if at least two out of four individual siRNAs fulfilled these criteria. Overall, 15 out of 28 primary hits were validated. As in any siRNA screen, there is a potential for false negatives; consequently, it is difficult to draw conclusions based on lack of a response to a particular target gene, but extensive follow up analyses can be carried out with multiple siRNAs to address this issue (Poleshko *et al.*, 2010).

3.3 Results and discussion

3.3.1 Proof-of-concept study to identify factors and networks that participate in epigenetic gene silencing

As noted above, epigenetic aberrations are implicated in the initiation and progression of many human diseases. As epigenetic changes are typically reversible, there is intense interest in the development of so-called "epigenetic therapies." In addition to disease targets, it may also be possible to manipulate cellular identity by targeting epigenetic processes. As an entry point to identify "targetable" factors that participate in epigenetic processes, we chose to study "epigenetic gene silencing," initially using a human cancer cell line, HeLa. HDACis and DNMTis (Bolden *et al.*, 2006; Yoo and Jones, 2006; Marks and Breslow, 2007; Fandy, 2009), known therapeutic compounds, relieve epigenetic gene silencing by inhibiting enzymes that modify chromatin, leading to reactivation of silent genes (Esteller, 2006; Jones and Baylin, 2007). Therefore, siRNA-based knockdown of key regulators of epigenetic silencing would be expected to phenocopy such effects.

To facilitate detection of factors that modulate epigenetic gene silencing, we implemented the GFP-silent reporter-based approach that provides a robust readout (Figure 3.1A). The populations of reporter cells harbor epigenetically silent GFP-reporter genes at essentially random locations, and the genes can be reactivated after treatment with epigenetic drugs or siRNA-based knockdown of epigenetic silencing factors (Figure 3.1). Compared to other detection methods for gene activity, this system allows analysis of reporter gene expression in individual cells, and as well as a measure of the fraction of cells in which the GFP gene is reactivated. HeLa cells were used as a representative human cell line with which to establish the GFP-silent reporter system, although specific perturbations of epigenetic pathways likely preexist in these cancer cells.

HeLa cells were retrovirally transduced with GFP genes and a three-step cell sorting method was performed to obtain a population of cells that harbor silent GFP-reporter genes that could be reactivated in response to epigenetic drugs (Figure 3.1A). We obtained a cell population that contains silent GFP-reporter genes deposited at dispersed chromosomal location (Narezkina *et al.*, 2004). To examine potential position-specific effects, we also isolated individual cell clones that harbor GFP insertions at specific locations (Katz *et al.*, 2007). The GFP-silent reporter provided a one-step, quantifiable assay and serves as a sensor to monitor the effects of epigenome-wide depletion of epigenetic regulators.

Two epigenetic drugs, TSA and 5-aza-C, were used for transient GFP reactivation and sorting to reveal GFP-silent cell populations, denoted HeLa TI and HeLa AI, respectively (Figure 3.1A). To investigate possible biases or differences introduced by the use of a specific epigenetic drug in the selection process, HeLa TI and HeLa AI cell populations were treated with TSA or 5-aza-C, and GFP-positive cells were measured by flow cytometry (Figure 3.1B). Treatment with either drug resulted in reactivation of the GFP-reporter gene in both populations, suggesting DNA methylation–histone modification crosstalk (Cedar and Bergman, 2009).

To identify factors that participate in silencing, we developed an siRNA-based gene-by-gene knockdown approach. Using this GFP-reporter gene system, siRNA-based knockdown of epigenetic repressors resulted in reactivation of GFP-reporter genes (Figure 3.1C). Previously, we identified HDAC1, among the family of HDACs, as being essential for maintenance of epigenetic silencing in this system (Poleshko *et al.*, 2008).

To validate the use of siRNA technology for identification of epigenetic factor roles, reporter cells were transfected with variety of siRNA controls (Figure 3.1C). As there are many non-specific cellular responses to siRNA transfection (Birmingham *et al.*, 2006; Cullen, 2006; Fedorov *et al.*, 2006), it is important to challenge the system with exhaustive siRNA controls. As shown in Figure 3.1C, among the challenges were tests to detect any non-specific effects of siRNA transfection including transfection reagent alone, a control siRNA modified to avoid interaction with the RNA-induced silencing complex (RISC) (denoted RISC−), a control siRNA with no complementarity to human mRNAs (RISC+), and an siRNA that targets a housekeeping gene (GAPDH). We observed no significant reactivation of reporter genes with any of these challenges, but robust reactivation after knockdown of HDAC1. These findings confirmed that our reporter gene system was an appropriate tool for development of high-throughput siRNA-based screening approaches to identify cellular epigenetic silencing factors.

3.3.2 Design of siRNA library

The latest advances in RNA interference (RNAi)-based knockdown technologies and automated screening systems allow a variety of high-throughput screening methods and readout options. Recent RNAi-based approaches include target gene knockdown using siRNA duplexes, or vector-based shRNA constructs. A variety of RNAi libraries are available commercially (Iorns *et al.*, 2007; Kassner, 2008; Paddison, 2008). RNAi library approaches can be divided into two types: (1) gene-by-gene siRNA/shRNA libraries, containing individual siRNAs or shRNAs plasmids/vectors, or (2) introduction of pooled shRNA vector libraries, followed by a selection for cell phenotypes of interest and identification of gene-specific shRNAs by sequencing-based or barcode analyses. Synthetic siRNA libraries are most useful for transient analyses, while shRNA libraries delivered by viral vectors can provide stable gene silencing and an ability to select for outgrowth of specific phenotypes (Echeverri and Perrimon, 2006).

We initiated our screening approach for identification of cellular factors and networks that are involved in maintenance of epigenetic silencing using a biased siRNA library that targets about 200 functionally diverse epigenetic regulators, including activators, chromatin remodelers, and silencing factors (QIAGEN, Inc.). The original library was formulated with two independent siRNAs against each target gene, with up to eight siRNAs available for follow-up analyses. We chose a gene-by-gene siRNA approach in order to monitor both cell viability and morphological changes in response to individual siRNAs, thereby increasing the information content of the data output. In contrast, when using pooled shRNA-based selection approaches, non-viable cells are lost.

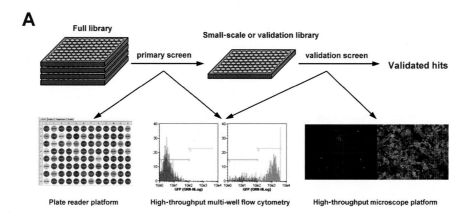

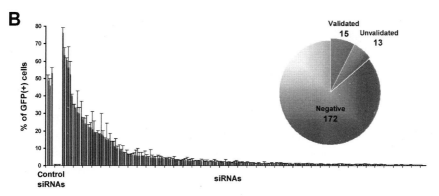

Figure 3.2 SiRNA-based high-throughput functional screen for detection of epigenetic silencing factors. (A) A schematic representation of the siRNA screening procedure using the GFP-silent reporter cell population, showing results with different GFP detection methods. General representations of readout results with plate reader format, multi-well flow cytometry, and high-content microscopy are indicated. Flow cytometry profiles and microscope images represent samples with GAPDH and HDAC1 siRNA, as negative and positive controls, respectively. (B) Representative results of primary siRNA screen (Poleshko *et al.*, 2010). Screening results are scored as percentage of GFP-positive cells measured by 96-well flow cytometry. Triplicate siRNAs for HDAC1 and GAPDH served as positive and negative control, respectively. Error bars indicated range ($n = 2$) determined with duplicate plates. Pie chart illustrates the siRNA screen hit rate. See plate section for color version.

3.3.3 SiRNA high-throughput screening

We have implemented three general methods for identification and quantification of GFP reactivation and the effects of siRNA knockdown on cell viability and morphology (Figure 3.2A):

(1) A plate reader platform (Envision, Perkin-Elmer) was designed for large, or genome-wide siRNA screens. This system provides a rapid quantification of GFP-fluorescence within a single well and is capable of detecting low GFP signals. However, the readout is the total GFP signal and does not report the number of GFP-positive cells, or their individual intensities (Carpenter and Sabatini, 2004; Echeverri and Perrimon, 2006) (Figure 3.2A). Nevertheless, this system is essential for large-scale screens.

(2) Multi-well flow cytometry systems (Millipore-Guava) provide a quantitative analysis of the number of GFP-positive cells within a population, as well their individual GFP intensities (Figure 3.2A). At the same time, flow cytometry systems can be used for detecting secondary cellular phenotypes produced by siRNA target gene knockdown, such as cell viability, apoptosis, cell cycle effects, and others. The only disadvantage of this system is the relatively slow speed of sample analysis as compared to a plate reader. This instrumentation is therefore less then adequate for genome-wide screening (e.g., 20 000 targets).

(3) Automated fluorescent microscopy provides an image-based analysis that is divided into two stages, image acquisition and computational analysis. Fluorescent microscopy allows a comprehensive analysis of GFP signal and morphology of individual cells (Figure 3.2A) (Sacher *et al.*, 2008; Conrad and Gerlich, 2010). This method can be applied for simple quantification of the number of GFP-positive cells in small-scale screens, or for comprehensive analysis (e.g., cell morphology changes) for validated gene hits.

For screening the "epigenetics set" siRNA library we used a flow cytometry readout system for both the primary and secondary screening steps (Poleshko *et al.*, 2010). During the primary screen, HeLa TI cells were transfected with two independent siRNAs for each of the 200 gene targets and the percentage of GFP-positive cells was measured (Figure 3.2B). The distribution of GFP intensities of GFP-positive cells was also monitored in the primary screen. As shown, the read-out of GFP-reporter gene expression is not binary, as might be expected, due to the varying half-life of target gene proteins or the particular role of the factor in epigenetic silencing. From the primary screen, a set of 28 gene targets were selected for validation, based on criteria described in the "Methodology" section.

3.3.4   Identities of screen hits

As siRNAs can produce non-specific or off-target effects, it was crucial to confirm the observed GFP-reactivation phenotypes (Cullen, 2006; Mohr *et al.*, 2010). Using stringent technical validation criteria described above, we identified 15 target genes (Figure 3.2B).

Gene hits included diverse epigenetic regulators, with significant enrichment for repressive factors (Figure 3.3). Four chromatin modifiers were identified. Two factors, KMT1E/SETDB1 and KMT5C/Suv420H2, catalyze the placement of repressive H3K9me3 and H4K20me3 chromatin marks respectively, while other factors, KDM2A/FBXL11 and KDM4A/JMJD2A, remove "active" H3K36me3 marks (Allis *et al.*, 2007; Kouzarides, 2007; Frescas *et al.*, 2008). The reversal of epigenetic silencing by knockdown of these individual modifying enzymes suggests that a specific constellation of methylation states of histones H3 and H4 is required for maintenance of epigenetic silencing. The screen also identified roles for Polycomb repressive factors RING1 and HPH2. These proteins have been identified as components of the human silencing maintenance complex (hPRC1L), which places a repressive monoubiquitin mark on histone H2A at lysine K119 (Wang *et al.*, 2004).

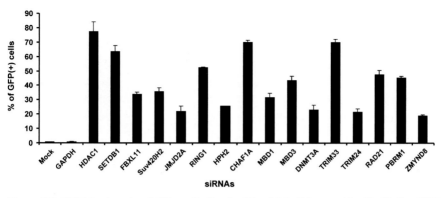

Figure 3.3 SiRNA treatment and GFP reactivation for 15 validated hit genes identified in the siRNA screen shown in Figure 3.2B. HDAC1 and GAPDH siRNA were used as positive and negative controls, respectively. Percentage of GFP-positive cells was measured by flow cytometry. Error bars indicate standard deviation ($n = 3$).

A role for the histone chaperone CHAF1A/CAF-1 p150, a non-enzymatic factor, was also detected. CAF-1 p150 is a subunit of the Chromatin Assembly Factor-1 complex (CAF-1) (composed of p150, p60, and p48 subunits), and it had been reported that the p150 subunit forms a specific complex responsible for maintenance of repressive histone marks during S-phase chromatin duplication (Sarraf and Stancheva, 2004). SiRNA knockdown of CHAF1A/CAF-1 p150 resulted in robust reactivation of the GFP-reporter gene (Figure 3.3) in agreement with such an essential role in silencing maintenance. A specific role for CAF-1 p150 in S-phase was tested. Cycling and arrested reporter cells were treated with several siRNAs that promote GFP-reporter reactivation, including CHAF1A/CAF-1 p150. Compared to other factors, the ability of CHAF1A/CAF-1 p150 to promote reactivation was highly dependent on cell cycling, consistent with a role in S-phase (Poleshko et al., 2010). We also confirmed that CHAF1A/CAF-1 p150 siRNA treatment resulted in the loss of repressive histone marks, H3K9me3 and H4K20me3, at the GFP-reporter promoter suggesting a role for CHAF1A in inheritance of epigenetic marks through chromatin replication in S-phase (Poleshko et al., 2010). These findings are also in line with an earlier report showing that CAF-1 p150 is required for maintenance of these repressive marks in murine pluripotent embryonic cells (Houlard et al., 2006).

Consistent with the ability of DNA methyltransferase inhibitors to reactivate GFP, screening detected a role for a DNMT, namely DNMT3A. This enzyme is considered to be a *de novo* DNMT, in contrast to the "maintenance" enzyme, DNMT1. A role for a *de novo* DNMT was unexpected; however such mechanistic distinctions between the DNMTs are being reassessed (Jones and Liang, 2009). The screen also detected two CpG-methyl-binding domain proteins (MBDs), MBD1 and MBD3, which maintain silencing through binding to methylated DNA (Klose and Bird, 2006). Taken together, these results support a role for the DNA methylation machinery and associated proteins in maintenance of epigenetic silencing in this system, as an example of epigenetic 'crosstalk' between the DNA methylation and histone modification complexes.

Of the remaining factors identified in the screen, TRIM24 (TIF1α) and TRIM33 (TIF1γ) are interacting members of the TIF1 family of transcriptional regulators that can function as repressors (Peng *et al.*, 2002), and RAD21 is a member of the cohesin family that has been implicated in transcriptional control (Peters *et al.*, 2008). Two other factors were identified, ZMYND8 (RACK7), a putative transcriptional regulator, and PBRM1/BAF180, a component of the human SWI/SNF chromatin remodeling complex PBAF.

3.3.5 Evaluating position-specific versus epigenome-wide roles for silencing factors

The GFP-silent HeLa reporter cells were isolated as population with widely dispersed insertions of the GFP-reporter gene (Narezkina *et al.*, 2004), a strategy designed to reduce potential position-specific effects of silencing factors. The siRNA screen was performed using this population of reporter cells in an effort to identify epigenome-wide networks that maintain epigenetic silencing. The robust GFP reactivation in a large fraction of reporter cells within the population upon treatment with several siRNAs (Figure 3.3) indeed suggested that the identified factors function independently of the chromosomal location of the silent GFP genes. However, more modest reactivation in response to a particular siRNA could indicate that only a fraction of the silent GFP loci are regulated by the factor, possibly due to position-specific effects. To distinguish between these possibilities, the response of GFP-reporter cell populations was compared with those of three independent cell clones (Figure 3.4A). The relative levels of silent GFP-reporter gene reactivation between the reporter cell population and individual clones indicate that the factor set is highly independent of the chromosomal position of the reporter gene. However, some differences in the degree of response within individual clones suggest that the location can modulate the GFP reactivation.

3.3.6 Evidence for crosstalk between DNA methylation and histone modifications

The experiments described above were performed using GFP-silent reporter cells that were selected following transient reactivation with the HDACi TSA (HeLa TI). Screening performed with these cells identified roles for DNMT3A, MBD1, and MBD, indicating a contribution of the DNA methylation machinery to reporter-gene silencing. To determine if the strategy for deriving the reporter cells might contribute to the factor repertoire detected by screening, a second reporter-cell population was derived using the DNMTi 5-aza-C. This cell population (HeLa AI) was then challenged with a subset of siRNAs corresponding to the targets identified in the screen of HeLa TI reporter cells, including HDAC1, SETDB1, DNMT3A, and MBD3. Knockdown of each of the tested factors resulted in reactivation of the silent GFP gene in HeLa AI cells (Figure 3.4B). These findings indicate that the identified factor set does not reflect the reporter cell selection procedure. Furthermore, the results support a role for functional crosstalk between the histone-modifying and DNA-methylation machineries. Numerous examples of

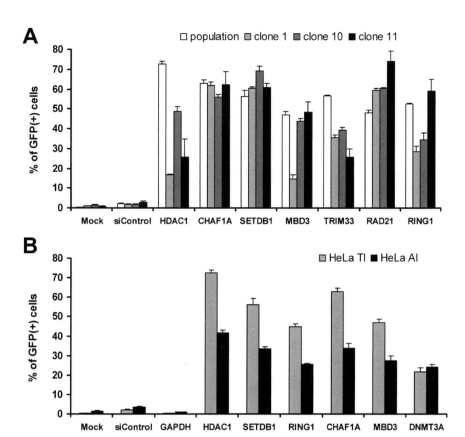

Figure 3.4 SiRNA treatment and GFP reactivation using different GFP-silent reporter cell populations and clones. Percentage of GFP-positive cells was measured by flow cytometry. SiControl is Dharmacon non-targeting siRNA control (RISC–). Error bars indicated standard deviation ($n = 3$). (A) GFP-silent HeLa TI cell population, and individual clones were analyzed with the indicated siRNAs, and reactivation was measured as percent GFP-positive cells. (B) Comparison of HeLa GFP-silent reporter cell populations prepared using the HDAC inhibitor TSA (HeLa TI) or a DNMT inhibitor 5-aza-C (HeLa AI) (see Figure 3.1A). The reporter cell populations were transfected with the indicated siRNAs, and reactivation was measured as percent GFP positive cells.

such bidirectional crosstalk between have been described (Fischle, 2008; Vaissiere *et al.*, 2008; Cedar and Bergman, 2009), and such interactions may reflect a self-reinforcing silencing mechanism (Feinberg and Tycko, 2004).

3.3.7 Interpreting output from epigenetic factor screens

It is clear that reactivation of epigenetically silent genes, in this case a reporter gene, represents a relevant response with pronounced biological implications. Figure 3.5A provides an overview of the potential of siRNA screens, as well some limitations. Reactivation by siRNA-based factor knockdown can be direct or indirect, and must be monitored case by case (Poleshko *et al.*, 2010). As mentioned, the ability to detect reactivation after knockdown of individual factors indicates a surprising lack of redundancies. In contrast, redundant functions may preclude the detection of factor roles. Lack of detection by siRNA knockdown

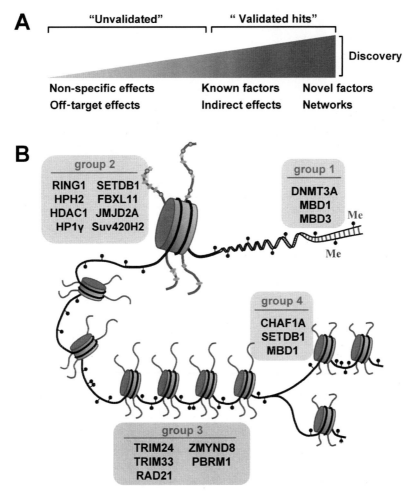

Figure 3.5 (A) Prospects and limitations of siRNA screening for identification of epigenetic factors. (B) Summary of factors identified, displayed as a model, based on predicted functions. The factors are grouped according to general functions: Group 1, DNMT and 5MeCpG recognition factors, MBDs; Group 2, histone modifiers and a recognition factor, HP1γ. Group 3, transcriptional roles; Group 4, inheritance of epigenetic marks through chromatin replication (Sarraf and Stancheva, 2004; Poleshko *et al.*, 2010).

may also reflect an inability to deplete the factor due to a long protein half-life, or a requirement for prolonged cell division for effects of the knockdown to be manifested. The screen readout is a temporal snapshot and not optimized for interrogation of each factor. As such, factor roles cannot be ruled out on the basis of lack of response to siRNA knockdown.

The detection of multiple factors within a complex reinforces factor roles, as we have observed with the Polycomb Group hPRC1L factors (RING1 and HPH2). However, the lack of hits among all of the components of a complex (e.g., hPRCIL) may indicate either a technical inability to deplete factors, or perhaps specialized roles for a subset of factors outside of the complex. Another consideration is that depletion of a single factor may destabilize other proteins in the complex (Fritsch *et al.*, 2010) and such instability of partner factors can amplify the effect of knocking down a single factor.

Lastly, scanning for functional roles of silencing factors provides a rich approach for generating or reinforcing hypotheses, in particular regarding causal versus correlative roles of epigenetic marks. For example, we provisionally interpret our results to indicate that H3K36 demethylases, KDM2A/FBXL11 and KDM4A/JMJD2A, are required to remove H3K36me3 marks in an active gene body for silencing to occur. That is, the H3K36me3 mark is not simply a consequence of active transcription, but is a key controlling mark. Such screen results can thus provide a roadmap for detailed mechanistic follow up experiments.

3.3.8 Identifying networks of "targetable" epigenetic factors

To effectively manipulate epigenetic processes for cancer therapy, or to potentially reprogram cellular identity, a global view of epigenetic silencing factor functions is essential. The reporter gene system described here appears to provide a sensor for epigenome-wide silencing factor function. The identified epigenetic silencing factor roles appear to be largely position and promoter independent. In support of this interpretation, we have found that the *Oct4* gene in HeLa cells is reactivated in parallel with the silent GFP after treatment with epigenetic drugs (data not shown).

Earlier work, using a different screening method, had shown that several tumor suppressor genes at unique chromosomal locations were silenced by a distinct set of epigenetic factors (Gazin *et al.*, 2007). As in our study, depletion of single factors could promote reactivation of dispersed silent genes, thereby defining an elaborate factor network or pathway.

Our findings have uncovered significant interactions and interdependence of factors in maintenance of epigenetic silencing, as well as the stated unexpected lack of redundancy. Both features may be explained by the fact that many factors provide enzymatic, as well as scaffolding roles (Feinberg and Tycko, 2004). Thus, each of the identified factors that may participate in this network may play either a critical enzymatic role in the maintenance of epigenetic silencing, or be essential for assembly of repressive complexes. The high level of functional interdependence among factors was consistent with physical findings demonstrating that the depletion of one factor leads to loss of multiple repressive marks (Poleshko *et al.* 2010). These results indicate the potential for identifying specific targets for epigenetic therapy to manipulate cellular identity or reprogram diseased cells.

3.4 Summary

We have designed and implemented a model to study epigenetic gene silencing using a reporter gene system. The approach is based on isolation of human cells harboring silent GFP-reporter genes, followed by a gene-by-gene siRNA-based knockdown screen, with reactivation of the silent GFP genes providing a functional readout for loss of epigenetic silencing functions. Using this system we identified a factor network (Figure 3.5B) and determined that factor roles are largely position- and selection-independent. Our findings from the use of a small, biased siRNA screen are not only informative, but also provided the framework for a now completed genome-wide functional siRNA screen to discover novel silencing factors, and more extensive networks (Poleshko *et al.*, unpublished data). A reporter-gene approach of this type will be essential to begin to identify targetable factors for reversal of epigenetic gene silencing. Furthermore, the approach may be applicable to a variety of cell types, enabling the functional identification of general as well as cell-type-specific factors.

ACKNOWLEDGMENTS

This work was supported by National Institutes of Health grant R01 DK082498 (R.A.K.) and A.P. is a recipient of an American Association for Cancer Research Centennial Predoctoral Fellowship in Cancer Research. We thank Anna Marie Skalka for critical comments and Marie Estes for assistance in preparing the chapter manuscript.

REFERENCES

Allis, C. D., Berger, S. L., Cote, J., *et al.* (2007). New nomenclature for chromatin-modifying enzymes. *Cell*, **131**, 633–636.

Berger, S. L. (2007). The complex language of chromatin regulation during transcription. *Nature*, **447**, 407–412.

Bernstein, B. E., Meissner, A., and Lander, E. S. (2007). The mammalian epigenome. *Cell*, **128**, 669–681.

Birmingham, A., Anderson, E. M., Reynolds, A., *et al.* (2006). 3′ UTR seed matches, but not overall identity, are associated with RNAi off-targets. *Nature Methods*, **3**, 199–204.

Bolden, J. E., Peart, M. J., and Johnstone, R. W. (2006). Anticancer activities of histone deacetylase inhibitors. *Nature Reviews in Drug Discovery*, **5**, 769–784.

Boutros, M. and Ahringer, J. (2008). The art and design of genetic screens: RNA interference. *Nature Reviews in Genetics*, **9**, 554–566.

Carpenter, A. E. and Sabatini, D. M. (2004). Systematic genome-wide screens of gene function. *Nature Reviews in Genetics*, **5**, 11–22.

Cedar, H. and Bergman, Y. (2009). Linking DNA methylation and histone modification: patterns and paradigms. *Nature Reviews in Genetics*, **10**, 295–304.

Conrad, C. and Gerlich, D. W. (2010). Automated microscopy for high-content RNAi screening. *Journal of Cell Biology*, **188**, 453–461.

Cullen, B. R. (2006). Enhancing and confirming the specificity of RNAi experiments. *Nature Methods*, **3**, 677–681.

Echeverri, C. J. and Perrimon, N. (2006). High-throughput RNAi screening in cultured cells: a user's guide. *Nature Reviews in Genetics*, **7**, 373–384.

Esteller, M. (2006). Epigenetics provides a new generation of oncogenes and tumour-suppressor genes. *British Journal of Cancer*, **94**, 179–183.

Fandy, T. E. (2009). Development of DNA methyltransferase inhibitors for the treatment of neoplastic diseases. *Current Medical Chemistry*, **16**, 2075–2085.

Fedorov, Y., Anderson, E. M., Birmingham, A., *et al.* (2006). Off-target effects by siRNA can induce toxic phenotype. *RNA*, **12**, 1188–1196.

Feinberg, A. P. (2007). Phenotypic plasticity and the epigenetics of human disease. *Nature*, **447**, 433–440.

Feinberg, A. P. and Tycko, B. (2004). The history of cancer epigenetics. *Nature Reviews in Cancer*, **4**, 143–153.

Fischle, W. (2008). Talk is cheap: cross-talk in establishment, maintenance, and readout of chromatin modifications. *Genes and Development*, **22**, 3375–3382.

Frescas, D., Guardavaccaro, D., Kuchay, S. M., *et al.* (2008). KDM2A represses transcription of centromeric satellite repeats and maintains the heterochromatic state. *Cell Cycle*, **7**, 3539–3547.

Fritsch, L., Robin, P., Mathieu, J. R., *et al.* (2010). A subset of the histone H3 lysine 9 methyltransferases Suv39h1, G9a, GLP, and SETDB1 participate in a multimeric complex. *Molecular Cell*, **37**, 46–56.

Gazin, C., Wajapeyee, N., Gobeil, S., Virbasius, C. M., and Green, M. R. (2007). An elaborate pathway required for Ras-mediated epigenetic silencing. *Nature*, **449**, 1073–1077.

Grewal, S. I., and Jia, S. (2007). Heterochromatin revisited. *Nature Reviews in Genetics*, **8**, 35–46.

Hawkins, R. D., Hon, G. C., and Ren, B. (2010). Next-generation genomics: an integrative approach. *Nature Reviews in Genetics*, **11**, 476–486.

Holliday, R. (2006). Epigenetics: a historical overview. *Epigenetics*, **1**, 76–80.

Houlard, M., Berlivet, S., Probst, A. V., *et al.* (2006). CAF-1 is essential for heterochromatin organization in pluripotent embryonic cells. *PLoS Genetics*, **2**, 1686–1696.

Huangfu, D., Osafune, K., Maehr, R., *et al.* (2008). Induction of pluripotent stem cells from primary human fibroblasts with only Oct4 and Sox2. *Nature Biotechnology*, **26**, 1269–1275.

Iorns, E., Lord, C. J., Turner, N., and Ashworth, A. (2007). Utilizing RNA interference to enhance cancer drug discovery. *Nature Reviews in Drug Discovery*, **6**, 556–568.

Jones, P. A. and Baylin, S. B. (2007). The epigenomics of cancer. *Cell*, **128**, 683–692.

Jones, P. A. and Liang, G. (2009). Rethinking how DNA methylation patterns are maintained. *Nature Reviews in Genetics*, **10**, 805–811.

Kassner, P. D. (2008). Discovery of novel targets with high-throughput RNA interference screening. *Combinatorial Chemistry High Throughput Screening*, **11**, 175–184.

Katz, R. A., Jack-Scott, E., Narezkina, A., *et al.* (2007). High-frequency epigenetic repression and silencing of retroviruses can be antagonized by histone deacetylase inhibitors and transcriptional activators, but uniform reactivation in cell clones is restricted by additional mechanisms. *Journal of Virology*, **81**, 2592–2604.

Kim, K., Doi, A., Wen, B., *et al.* (2010). Epigenetic memory in induced pluripotent stem cells. *Nature*, **467**, 285–290.

Klose, R. J. and Bird, A. P. (2006). Genomic DNA methylation: the mark and its mediators. *Trends in Biochemical Sciences*, **31**, 89–97.

Kouzarides, T. (2007). Chromatin modifications and their function. *Cell*, **128**, 693–705.

Marks, P. A. and Breslow, R. (2007). Dimethyl sulfoxide to vorinostat: development of this histone deacetylase inhibitor as an anticancer drug. *Nature Biotechnology*, **25**, 84–90.

Mohr, S., Bakal, C., and Perrimon, N. (2010). Genomic screening with RNAi: results and challenges. *Annual Reviews of Biochemistry*, **79**, 37–64.

Narezkina, A., Taganov, K. D., Litwin, S., *et al.* (2004). Genome-wide analyses of avian sarcoma virus integration sites. *Journal of Virology*, **78**, 11 656–11 663.

Paddison, P. J. (2008). RNA interference in mammalian cell systems. *Current Topics in Microbiology and Immunology*, **320**, 1–19.

Peng, H., Feldman, I., and Rauscher, F. J., III (2002). Hetero-oligomerization among the TIF family of RBCC/TRIM domain-containing nuclear cofactors: a potential mechanism for regulating the switch between coactivation and corepression. *Journal of Molecular Biology*, **320**, 629–644.

Peters, J. M., Tedeschi, A., and Schmitz, J. (2008). The cohesin complex and its roles in chromosome biology. *Genes and Development*, **22**, 3089–3114.

Poleshko, A., Palagin, I., Zhang, R., *et al.* (2008). Identification of cellular proteins that maintain retroviral epigenetic silencing: evidence for an antiviral response. *Journal of Virology*, **82**, 2313–2323.

46   **A. Poleshko *et al.***

Poleshko, A., Einarson, M. B., Shalginskikh, N., *et al.* (2010). Identification of a functional network of human epigenetic silencing factors. *Journal of Biological Chemistry*, **285**, 422–433.

Sacher, R., Stergiou, L., and Pelkmans, L. (2008). Lessons from genetics: interpreting complex phenotypes in RNAi screens. *Current Opinions in Cell Biology*, **20**, 483–489.

Sarraf, S. A. and Stancheva, I. (2004). Methyl-CpG binding protein MBD1 couples histone H3 methylation at lysine 9 by SETDB1 to DNA replication and chromatin assembly. *Molecular Cell*, **15**, 595–605.

Taverna, S. D., Li, H., Ruthenburg, A. J., Allis, C. D., and Patel, D. J. (2007). How chromatin-binding modules interpret histone modifications: lessons from professional pocket pickers. *Nature Structural and Molecular Biology*, **14**, 1025–1040.

Vaissiere, T., Sawan, C., and Herceg, Z. (2008). Epigenetic interplay between histone modifications and DNA methylation in gene silencing. *Mutation Research*, **659**, 40–48.

Wang, H., Wang, L., Erdjument-Bromage, H., *et al.* (2004). Role of histone H2A ubiquitination in Polycomb silencing. *Nature*, **431**, 873–878.

Yoo, C. B. and Jones, P. A. (2006). Epigenetic therapy of cancer: past, present and future. *Nature Reviews in Drug Discovery*, **5**, 37–50.

Zhang, J. H., Chung, T. D., and Oldenburg, K. R. (1999). A simple statistical parameter for use in evaluation and validation of high-throughput screening assays. *Journal of Biomolecular Screening*, **4**, 67–73.

4 Nucleosome positioning in promoters: significance and open questions

Jun S. Song* and David E. Fisher

4.1 Introduction

The structure and function of the human genome are so intricately intertwined that understanding gene regulation necessitates viewing the genome as a dynamic three-dimensional entity which controls its own information content. The spatial organization of DNA into chromatin provides a critical layer of tissue-specific instructions for regulating the accessibility and distal interactions of genomic loci. Even though recent high-throughput studies have begun to unravel how chromatin structure affects transcriptional regulation, our understanding still remains very rudimentary. Several puzzling issues persist in the literature, and different interpretations have now gained strong supporters who propound conflicting views. This chapter highlights some of the computational and biological issues that may prove beneficial for scientists to bear in mind when entering the study of epigenetics.

The fundamental repeating subunits of chromatin are the nucleosomes which consist of 146 base pairs (bp) of DNA wrapping around histone octamers in 1.65 left-handed superhelical turns (Luger and Richmond, 1998). Chromatin is organized in a hierarchical order, the first level of which is represented by a linear array of nucleosomes, with roughly one nucleosome every 200 bp. The beads-on-a-string configuration is believed to be folded into a fiber of 30 nm diameter in transcriptionally repressed heterochromatin regions (Robinson and Rhodes, 2006; Routh et al., 2008), and the fiber further folds into a highly condensed form during metaphase. The precise structure of the 30-nm chromatin fiber and beyond still remains unknown. It is believed that global high-order chromatin structure requires covalent modifications of histone N-terminal tails, such as H3K9 and H3K27 trimethylation; proteins that recognize such modifications and that interact with each other may then facilitate the dense compaction of chromatin.

* Author to whom correspondence should be addressed.

Epigenomics: From Chromatin Biology to Therapeutics, ed. K. Appasani. Published by Cambridge University Press. © Cambridge University Press 2012.

This chapter focuses on the first level of local chromatin organization and discusses the recently discovered relations between nucleosome positioning and transcriptional regulation. Nucleosomes in the gene body hinder, but may not actually prevent, transcription initiation and elongation (Hodges *et al.*, 2009); RNA polymerases are able to transcribe nucleosomal DNA and translocate the histone octamer, or some parts of the complex, behind the polymerases (Hodges *et al.*, 2009; Kulaeva *et al.*, 2009). Nucleosomes in promoters and enhancers may block important regulatory sites from being accessed by transcription factors and thus repress transcription initiation. Given the vast size of eukaryotic genomes, however, nucleosomes can also facilitate transcription by bringing distal regulatory sites into proximity (Thomas and Elgin, 1988; Jackson and Benyajati, 1993; Lupien *et al.*, 2008; He *et al.*, 2010). In addition, histone depletion has been shown to reduce the expression of 10% of genes in yeast, indicating that functional nucleosomes may be needed to maintain transcription at those genomic loci (Wyrick *et al.*, 1999). Nucleosome positioning may thus have either a repressive or an activating effect on transcription.

There are two aspects of nucleosome positioning: rotational and translational. Rotational positioning involves the axial orientation of DNA which defines the contact points of the minor grooves with the histone surface, so that AT-rich sequences preferentially face inward towards the histones while GC-rich sequences face outward (Satchwell *et al.*, 1986). Since one complete helical turn of nucleosomal DNA contains 10.0–10.7 bp (Hayes *et al.*, 1991), various periodic dinucleotides which facilitate the rotational setting of bendable DNA have been used to predict translational positioning, which describes genomic locations of nucleosomes (Ioshikhes *et al.*, 1996; Levitsky, 2004; Segal *et al.*, 2006; Yuan and Liu, 2008). A lot of confusion now exists regarding what "periodic" means and how predictive the models of sequence-dependent nucleosome positioning really are. High-throughput genome-wide mapping of nucleosomes is currently providing a wealth of data to help us infer the rules which guide nucleosome positioning and thereby regulate transcription. This chapter will discuss several technical aspects involved in utilizing the technology and analyzing the resulting data.

In addition to positioning, nucleosome stability may also have an important role in regulating the rate of transcription. Recent studies suggest that the histone variants H3.3 and H2A.Z may together significantly decrease the stability of nucleosomes in transcriptionally active promoters and enhancers (Jin *et al.*, 2009). Despite the common belief that neutralization of H3K4 via acetylation weakens the electrostatic interaction with the negatively charged DNA backbone, it is not clear whether histone modifications themselves have any effect on nucleosome stability (Bresnick *et al.*, 1991). This chapter will not discuss these issues further.

4.1.1 Micrococcal nuclease bias

Several different experimental protocols can be found in previous publications (Yuan *et al.*, 2005; Dennis *et al.*, 2007; Lee *et al.*, 2007; Ozsolak *et al.*, 2007; Mavrich *et al.*, 2008). The main idea is to digest chromatin with micrococcal nuclease

Table 4.1 Sequence bias of MNase cleavage sites

Seq	Frequency	Seq	Frequency	Seq	Frequency	Seq	Frequency
AA	0.19 (0.1)	AT	0.08 (0.08)	AG	0.02 (0.07)	AC	0.01 (0.05)
CA	0.13 (0.07)	CT	0.07 (0.07)	CG	0.002 (0.01)	CC	0.007 (0.05)
GA	0.10 (0.06)	GT	0.05 (0.05)	GG	0.02 (0.05)	GC	0.01 (0.04)
TA	0.16 (0.07)	TT	0.10 (0.10)	TG	0.02 (0.07)	TC	0.01 (0.06)

Note: Dinucleotides containing MNase cleavage sites in the middle were aligned. The numbers in parentheses indicate background frequencies in the human genome. AA, CA, GA and TA are overrepresented by almost two-fold in MNase cleavage sites compared to genomic background, while CG and CC are underrepresented by ten-fold. The data are from short-read sequencing of mononucleosomal DNA on chromosomes 21 and 22 in CD4$^+$ T-cells (Schones *et al.*, 2008).

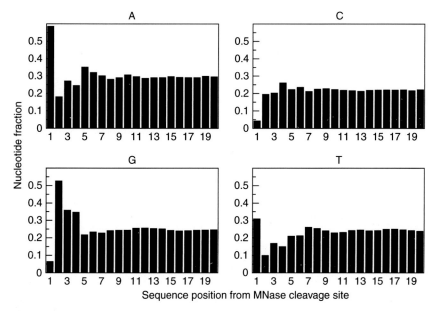

Figure 4.1 Nucleotide distribution of MNase-cleaved DNA. MNase preferentially cuts between NA while avoiding SC dinucleotides. The G nucleotide also occurs often immediately adjacent to the cleavage sites. These biases should be taken into account when inferring the precise nucleosome locations and modeling sequence-dependent DNA structural properties.

(MNase), which preferentially cleaves unprotected linker DNA, and to gel extract mononucleosome bands of size 100–200 bp. The obtained DNA can be either hybridized onto a high-density tiling microarray or sequenced. Even though MNase may preferentially cut linker DNA instead of nucleosomal DNA, it has been known that the enzyme has a sequence bias towards AT-rich sites (Horz and Altenburger, 1981). Our analysis of MNase cleavage sites from high-throughput sequencing of mono-nucleosomal DNA confirms this bias, as shown in Table 4.1 and Figure 4.1. Figure 4.1 shows that the cleavage sites also tend to have adjacent G-rich regions, consistent with the previous observation that GC-rich flanking sequences can increase the cutting probability (Horz and Altenburger, 1981). These results suggest that the sequenced DNA ends may not necessarily represent the exact locations of nucleosome boundaries and that caution should be taken when inferring nucleosome locations from short-read sequencing data. Paired-end sequencing would provide a better sense of the distribution of MNase cleavage sites with respect to individual nucleosomal DNA.

4.1.2   To sequence or not to sequence?

The next-generation short-read sequencing technology has rapidly changed the landscape of genomics research. As the price continues to drop, it has become a common practice to perform ChIP-seq in order to discover histone modification regions and transcription factor binding sites. Despite all the advantages of sequencing, one must keep in mind that 80–90% of sequencing data are still background noise from ChIP or repeat sequences that cannot be mapped uniquely to the reference genome. Some researchers have also sequenced the mononucleosomal DNA in model organisms (Mavrich *et al.*, 2008; Kaplan *et al.*, 2009; He *et al.*, 2010) and in humans (Schones *et al.*, 2008). A simple calculation shows that each human cell contains roughly 30 million nucleosomes and that it would require 250 times higher sequencing depth to get the same coverage of all nucleosomes in human as in yeast. Genome-wide profiling of nucleosomes in mammals thus still remains beyond the reach of most researchers. No good approach is also available to sequence nucleosomal DNA from only a small subset of the genome. We thus believe that custom high-density tiling microarrays may still provide a useful alternative to studying nucleosome positioning in targeted genomic loci.

4.2   Factors potentially affecting nucleosome positioning

4.2.1   Pitfalls of sequence analysis

Most computational methods for predicting nucleosome locations are based on the average dinucleotide content or GC-content of *aligned* nucleosomal DNA sequences. Such approaches, although quite successful in many instances, nevertheless have limitations and have also propagated several misconceptions:

(1) Dinucleotides may be insufficient to fully describe the physical properties of DNA; for example, poly-A tracks must have a minimum length of four in order to take a non-canonical form (Delcourt and Blake, 1991), and high-order Markov models may thus be needed to adequately capture the sequence dependence of nucleosome positioning.

(2) Fourier transform is a mathematically rigorous way of analyzing the frequency spectrum of a signal by decomposing the signal as a linear combination of sinusoidal harmonics. Because DNA sequence is a categorical series, i.e., a string in four letters, and not a real-valued function, how we represent the nucleotides as real numbers can significantly influence the spectral analysis; with only few exceptions (Segal, 2008), this important point has been largely ignored to date.

(3) Periodic average occurrences in aligned sequences do not imply periodicity within individual sequences. That is, in the mathematical language of

harmonic analysis, the Fourier spectral density of average dinucleotide frequencies does not summarize the periodicity of dinucleotides in individual sequences.

(4) There may be many-to-one correspondences between DNA sequence motifs and physical properties, and it may thus be necessary to discover periodic physical properties instead of periodic dinucleotides. This observation may explain why previous dinucleotide-based algorithms fitted to a particular genome are species-specific and why the estimated parameters do not generalize well to other species (Lantermann *et al.*, 2010).

4.2.2 Periodicity or artifact?

The relation between DNA sequence and nucleosome formation is indirect. Histones interact with DNA via the electrostatic forces between their amino acid side chains and the negatively charged DNA backbone, so histones do not interact directly with nucleotide bases (Mirzabekov and Rich, 1979). It is thus the sequence-dependent physical properties of DNA, and not the specific sequences themselves, that influence nucleosome positioning. It has been known for many years that certain di- and trinucleotides that occur in ~10 bp periods (corresponding to one helical turn of DNA) may be highly bendable and facilitate the formation of nucleosomes in vitro. The observations have led to the hypothesis that periodic dinucleotides may provide a significant signal that positions nucleosomes genome-wide in vivo (Godde *et al.*, 1996; Godde and Wolffe, 1996; Ioshikhes *et al.*, 1996; Lowary and Widom, 1997, 1998; Cao *et al.*, 1998; Levitsky, 2004; Segal *et al.*, 2006; Gupta *et al.*, 2008; Segal, 2008; Balasubramanian *et al.*, 2009).

This hypothesis currently remains controversial and is largely based on the observed periodic *average* occurrence of nucleotides in *aligned* sequences. However, such periodicities were absent in individual nucleosomal DNA, suggesting that the periodic occurrences may be mere artifacts of alignment (Ozsolak *et al.*, 2007; Segal, 2008; Zhang *et al.*, 2009). For example, one can devise an alignment algorithm that counts dinucleotide locations modulo 10 and that shifts individual sequences by the mode of the distribution of locations in order to produce artificial periodicity in aligned sequences; Figure 4.2A illustrates the spurious periodic occurrence of amino acids in the alignment of 3000 random sequences. In order to resolve the existing controversy, more rigorous methods need to be developed for analyzing DNA sequences in the frequency domain. A successful method should accommodate the possibility that several polynucleotides may be exchangeable in nucleosomal DNA and thus that they should be assigned to the same real number in the spectral analysis.

4.2.3 DNA bendability and nucleosome positioning

To date, several studies have examined the dinucleotide content of nucleosomal and linker DNA (Segal *et al.*, 2006; Peckham *et al.*, 2007; Gupta *et al.*, 2008; Segal, 2008; Yuan and Liu, 2008; Kaplan *et al.*, 2009; Reynolds *et al.*, 2009a, 2009b, 2010; Segal and Widom, 2009), but the motivation behind their models is statistical, not

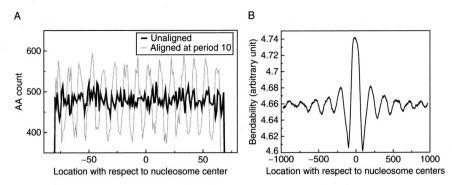

Figure 4.2 Periodic sequences and DNA bendability. (A) Alignment of sequences can produce spurious periodicity of dinucleotides. Aligning 3000 random sequences by shifting to the left or right by at most 5 bp can artificially create periodic amino acid dinucleotide counts (gray line). (B) Average DNA bendability at experimentally detected nucleosomes in *Saccharomyces cerevisiae*. Each peak corresponds to a nucleosome. It is seen that nucleosomal DNA is more flexible than linker DNA.

biophysical, and the results are thus often difficult to interpret. Detailed biochemical assays were performed almost two decades ago to determine the sequence-dependent flexibility of DNA by measuring the frequency of DNase I cleavage (Brukner *et al.*, 1990, 1995; Lahm and Suck, 1991; Weston *et al.*, 1992; Goodsell and Dickerson, 1994; Suck, 1994; Gabrielian and Pongor, 1996). DNase I interacts with a 6-bp-long region in the minor groove and bends the surface towards the major groove; thus, if a DNA segment were highly bendable, then it would be preferentially cut by DNase I. Given a hexamer sequence s, the sequence-dependent cleavage frequency $F(s)$ was modeled as $\log F(s) = \Sigma_i f(s_i)$, where s_i are the four trinucleotide substrings of s and the bendability function $f(s_i)$ is estimated by using maximum likelihood (Brukner *et al.*, 1990, 1995). The estimates were later improved (Gabrielian and Pongor, 1996) by incorporating the sequence spectrum in chicken nucleosomal DNA (Satchwell *et al.*, 1986).

Our analysis shown in Figure 4.2B demonstrates that the bendability estimates correlate very well with the experimentally determined nucleosome occupancy in both yeast (Lee *et al.*, 2007; Mavrich *et al.*, 2008; Lantermann *et al.*, 2009) and human (Ozsolak *et al.*, 2007, 2008; Schones *et al.*, 2008), suggesting that this model is not species-specific and may actually reflect the biophysical properties of nucleosomal DNA universal in all species. Surprisingly, discrepancies between our calculation and experimental nucleosome maps were observed in proximal gene promoters, where nucleosome positioning is highly regulated by chromatin remodeling proteins. It can be thus hypothesized that the physical properties of DNA define a thermodynamic potential energy landscape that needs to be locally minimized for preferential nucleosome formation throughout the genome, but this potential energy is easily overcome by ATP-dependent chromatin remodeling proteins.

4.2.4 Nucleosome positioning, histone modifications, and DNA methylation

There exists a fascinating interplay among DNA methylation, histone modifications, and nucleosome positioning, although the dynamics and regulation of their interactions remain largely unknown. The enzyme DNMT3L can bind unmethylated H3K4 and recruit the DNA methyltransferase DNMT3A, thus inducing *de novo* DNA methylation (Hu *et al.*, 2009). Consistent with the finding that nucleosomes may facilitate *de novo* DNA methylation, a recent study shows that DNA methylation in *Arabidopsis thaliana* and human occurs preferentially in nucleosomal DNA and in 10-bp periods (Chodavarapu *et al.*, 2010), perhaps corresponding to CpG facing outward from the histone surface; however, it is not clear whether the periodicity is actually present in individual methylated nucleosomal DNA, because their Fourier analysis computed the spectral density of an average methylation profile, instead of taking the average of spectral densities of individual methylation levels.

DNA methylation, in turn, may influence nucleosome positioning, perhaps by increasing the stiffness or some other physical property of DNA (Davey *et al.*, 2004). Wrapping DNA around histones creates non-homogeneous stress along the DNA, so that different parts experience different superhelical twists, destabilization forces, and bending (Hayes *et al.*, 1991). In particular, the locations 1.5 helical turns away from the dyad axis have a strong bent or kink (Hogan *et al.*, 1987); it has been shown that DNA methylation of short CpG tandem repeats at these locations significantly decreases the stability of the associated nucleosome and can lead to nucleosome repositioning (Davey *et al.*, 2004). About 20% of human *RefSeq* genes have $(CpG)_3$ within 100 bp of transcription start sites, and their known functions are highly enriched in DNA repair and cell cycle regulation. It would be interesting to see whether some of those promoters have aberrant methylation in cancer and whether the methylation leads to repositioning of nucleosomes to block transcription initiation.

Finally, covalently modified histone tails provide recognition sites for nucleosome remodeling SWI/SNF complexes that can reposition or disassemble nucleosomes. There is evidence that the genomic regions in mammals with the highest level of discrepancy between computational predictions and experimentally detected nucleosome locations occur in promoters and enhancers, where the enzymatic activities of the ATP-dependent chromatin remodelers are high. Since it is these regulatory regions that play critical roles in gene expression, sequence-dependent positioning of nucleosomes appears to have only a secondary effect on gene regulation in general (Peckham *et al.*, 2007; Mavrich *et al.*, 2008; Zhang *et al.*, 2009). We will come back to this point in the next section.

4.3   Hypotheses

4.3.1   Transcription-coupled nucleosome positioning

In vitro reconstituted nucleosomes on genomic DNA have been recently shown to differ significantly from in vivo nucleosome locations in yeast (Zhang *et al.*, 2009). In particular, the phasing of +1 nucleosomes, i.e. the nucleosomes found immediately downstream of transcription initiation sites, disappeared in vitro. The results strongly support the idea that transcription machineries, and not the sequence-dependent structural properties of DNA, may be the dominant forces that define the local chromatin structure of promoters. It remains unclear how or whether certain proteins block nucleosome sliding, but it has been observed that transcription factor (TF) binding sites preferentially localize in nucleosome-free linker DNA (Yuan *et al.*, 2005; Ozsolak *et al.*, 2007).

Many TFs directly interact with chromatin remodelers and guide site-specific nucleosome remodeling or disassembly. For example, the master regulator of melanocyte differentiation MITF interacts with different SWI/SNF complexes to regulate distinct sets of differentiation and proliferation genes (Keenen *et al.*, 2010). SWI/SNF uses the energy released from ATP hydrolysis to reposition nucleosomes (Zofall *et al.*, 2006) and can also disassemble adjacent nucleosomes (Dechassa *et al.*, 2010). Another intriguing story that is beginning to emerge is that various non-coding RNAs may interact with chromatin modifying enzymes and lead to site-specific histone modifications (Guttman *et al.*, 2009; Khalil *et al.*, 2009; Huarte *et al.*, 2010), which may then be recognized by chromatin remodelers. Locally transcribed non-coding RNAs could also influence nucleosome positioning by interacting with repressor complexes. For example, the histone-remodeling complex NoRC interacts with promoter-associated RNA from rDNA loci, and its transient release from the RNA allows its remodeling activity of nucleosomes at the transcription start site of rDNA (Zhou *et al.*, 2009).

Thus, TFs and non-coding RNAs crucially interact with local chromatin structure. What is cause and what is effect remains to be seen, but it is likely that they mutually influence each other in a complicated feedback loop.

4.3.2   Chromatin structure as a cancer lineage classifier

Even though different cell types from an individual have the identical DNA sequence, gene expression levels vary greatly across cell types. One way of establishing cell-type-specific expression programs is via the interplay between chromatin structure and TFs. On the one hand, chromatin controls the tissue-specific accessibility of TFs to DNA and thus guides tissue-specific gene expression. In turn, TFs affect local chromatin states by recruiting chromatin modifiers. These considerations further highlight the fact that sequence-dependent positioning of nucleosomes cannot account for tissue-specific variations in local chromatin structure.

In the melanoma cell line MALME, our nucleosome mapping showed that 47 out of 57 (83%) promoters bound by the oncogenic transcription factor MITF had the recognition motif E-box (CACGTG or CATGTG) in nucleosome-free linker DNA (p-value $= 5 \times 10^{-8}$) (Ozsolak *et al.*, 2007). Scanning for E-boxes in linker DNA within microRNA promoters also allowed us to find MITF-regulated microRNAs (Ozsolak *et al.*, 2008). We have further shown that evolutionarily conserved sequences are preferentially found within linker DNA (p-value $= 3.6 \times 10^{-11}$) (Ozsolak *et al.*, 2007). Sequence analyses restricted to nucleosome-free regions may thus greatly facilitate the discovery of functional TF motifs and eliminate many false positives. These findings strongly support that nucleosome positioning is highly regulated near functional DNA elements. Importantly, different cell types have distinct patterns of accessible DNA, implying that the tissue of origin of cancers may be classified based on nucleosome positioning in promoters. Local chromatin structure may thus help define the gene expression patterns that ultimately specify cell lineages (Ozsolak *et al.*, 2007). These findings strongly suggest that cell-lineage-specific chromatin structure is partially preserved during malignant transformation of cells and suggests its application as a diagnostic tool in the future.

4.3.3 Retrotransposons and nucleosome positioning signals

Roughly half of the mammalian genomes are repeat sequences, which arose from retrotransposons that can expand by a copy-and-paste mechanism involving RNA intermediates. The enormous number of retrotransposons in the human genome indicates that should these repeat elements carry special nucleosome positioning signals, they would significantly affect the distribution of nucleosomes in the genome by constraining the locations of nearby nucleosomes. Examining the sequence structure of Alu retrotransposons shows that it is partitioned into two distinct GC-rich blocks separated by a poly-A stretch. The GC-rich blocks may provide the Alu repeats with a strong preference for positioning two adjacent nucleosomes (Peckham *et al.*, 2007). In support, Figure 4.3 shows distinct bimodal nucleosome occupancy probabilities around 10 000 Alu sequences aligned at the poly-A center, where the occupancy was both computationally predicted using an available algorithm (Segal *et al.*, 2006) and determined based on the recent sequencing data in CD4$^+$ T-cells (Schones *et al.*, 2008). Retrotransposons could thus influence how DNA winds around histones throughout the genome. Since

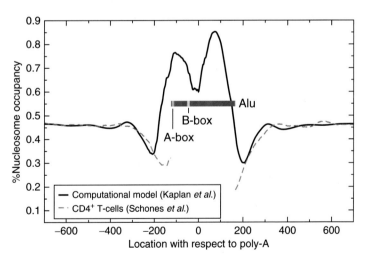

Figure 4.3 Alu retrotransposons may provide strong nucleosome positioning signals. The solid line shows the predicted probability of finding a nucleosome at each position in Alu retrotransposons in the human genome. The first nucleosome preferentially blocks the transcription initiation sequences A-box and B-box, indicating that the default nucleosome positioning may inhibit transcription. The dotted line shows the average occupancy from published experimental data of nucleosome locations in CD4$^+$ T-cells (Schones *et al.*, 2008) and shows significant nucleosome depletion flanking the Alu, consistent with the computational prediction; the central region is missing, because the short-reads could not be mapped uniquely to the reference genome.

most retrotransposons have DNA methylation, it would be also interesting to study whether the modification has any effect on nucleosome positioning.

4.4   Concluding remarks

The recent upsurge of discoveries in epigenetics has revealed numerous tissue-specific DNA methylation and histone modification patterns that distinguish different parts of the human genome. It is believed that these modifications can induce global chromatin compaction or decondensation. At least locally, the ultimate functions of the modifications may be to control nucleosome stability and positioning in promoters and enhancers which critically regulate transcriptional activities. It is conceivable that therapeutic drugs targeting the DNA minor groove and affecting nucleosome positioning at transcription initiation sites may be developed in the future to alter gene regulation in vivo.

The field of epigenetics is vast, and our understanding continues to evolve. In the midst of these rapid developments, misconceptions and preconceptions are easy to form and hard to rectify. Nucleosome positioning in regulatory sites is highly dynamic and reflects the transcriptional activities of genes. In these regions, the sequence-dependent structural properties of DNA cannot explain the differences in nucleosome positioning found among different cell types, and the interplay among TFs, chromatin remodelers, non-coding RNAs, and histones may play crucial roles in establishing and maintaining cell-type-specific chromatin structure. It is, however, still possible that DNA sequence plays an important role in defining a broad potential energy landscape which nucleosomes tend to minimize throughout the human genome. It is certainly not a question of "either–or", and where ignorance hovers, it would be best to remain open-minded.

REFERENCES

Balasubramanian, S., Xu, F., and Olson, W. K. (2009). DNA sequence-directed organization of chromatin: structure-based computational analysis of nucleosome-binding sequences. *Biophysics Journal*, **96**, 2245–2260.

Bresnick, E. H., John, S., and Hager, G. L. (1991). Histone hyperacetylation does not alter the positioning or stability of phased nucleosomes on the mouse mammary tumor virus long terminal repeat. *Biochemistry*, **30**, 3490–3497.

Brukner, I., Jurukovski, V., and Savic, A. (1990). Sequence-dependent structural variations of DNA revealed by DNase I. *Nucleic Acids Research*, **18**, 891–894.

Brukner, I., Sanchez, R., Suck, D., and Pongor, S. (1995). Sequence-dependent bending propensity of DNA as revealed by DNase I: parameters for trinucleotides. *EMBO Journal*, **14**, 1812–1818.

Cao, H., Widlund, H. R., Simonsson, T., and Kubista, M. (1998). TGGA repeats impair nucleosome formation. *Journal of Molecular Biology*, **281**, 253–260.

Chodavarapu, R. K., Feng, S., Bernatavichute, Y. V., *et al.* (2010). Relationship between nucleosome positioning and DNA methylation. *Nature*, **466**, 388–392.

Davey, C. S., Pennings, S., Reilly, C., Meehan, R. R., and Allan, J. (2004). A determining influence for CpG dinucleotides on nucleosome positioning in vitro. *Nucleic Acids Research*, **32**, 4322–4331.

Dechassa, M. L., Sabri, A., Pondugula, S., *et al.* (2010). SWI/SNF has intrinsic nucleosome disassembly activity that is dependent on adjacent nucleosomes. *Molecular Cell*, **38**, 590–602.

Delcourt, S. G. and Blake, R. D. (1991). Stacking energies in DNA. *Journal of Biological Chemistry*, **266**, 15 160–15 169.

Dennis, J. H., Fan, H. Y., Reynolds, S. M., *et al.* (2007). Independent and complementary methods for large-scale structural analysis of mammalian chromatin. *Genome Research*, **17**, 928–939.

Gabrielian, A. and Pongor, S. (1996). Correlation of intrinsic DNA curvature with DNA property periodicity. *FEBS Letters*, **393**, 65–68.

Godde, J. S. and Wolffe, A. P. (1996). Nucleosome assembly on CTG triplet repeats. *Journal of Biological Chemistry*, **271**, 15 222–15 229.

Godde, J. S., Kass, S. U., Hirst, M. C., and Wolffe, A. P. (1996). Nucleosome assembly on methylated CGG triplet repeats in the fragile X mental retardation gene 1 promoter. *Journal of Biological Chemistry*, **271**, 24 325–24 328.

Goodsell, D. S. and Dickerson, R. E. (1994). Bending and curvature calculations in B-DNA. *Nucleic Acids Research*, **22**, 5497–5503.

Gupta, S., Dennis, J., Thurman, R. E., *et al.* (2008). Predicting human nucleosome occupancy from primary sequence. *PLoS Computational Biology*, **4**, e1000134.

Guttman, M., Amit, I., Garber, M., *et al.* (2009). Chromatin signature reveals over a thousand highly conserved large non-coding RNAs in mammals. *Nature*, **458**, 223–227.

Hayes, J. J., Clark, D. J., and Wolffe, A. P. (1991). Histone contributions to the structure of DNA in the nucleosome. *Proceedings of the National Academy of Sciences USA*, **88**, 6829–6833.

He, H. H., Meyer, C. A., Shin, H., *et al.* (2010). Nucleosome dynamics define transcriptional enhancers. *Nature Genetics*, **42**, 343–347.

Hodges, C., Bintu, L., Lubkowska, L., Kashlev, M., and Bustamante, C. (2009). Nucleosomal fluctuations govern the transcription dynamics of RNA polymerase II. *Science*, **325**, 626–628.

Hogan, M. E., Rooney, T. F., and Austin, R. H. (1987). Evidence for kinks in DNA folding in the nucleosome. *Nature*, **328**, 554–557.

Horz, W. and Altenburger, W. (1981). Sequence specific cleavage of DNA by micrococcal nuclease. *Nucleic Acids Research*, **9**, 2643–2658.

Hu, J. L., Zhou, B. O., Zhang, R. R., *et al.* (2009). The N-terminus of histone H3 is required for de novo DNA methylation in chromatin. *Proceedings of the National Academy of Sciences USA*, **106**, 22187–22192.

Huarte, M., Guttman, M., Feldser, D., *et al.* (2010). A large intergenic noncoding RNA induced by p53 mediates global gene repression in the p53 response. *Cell*, **142**, 409–419.

Ioshikhes, I., Bolshoy, A., Derenshteyn, K., Borodovsky, M., and Trifonov, E. N. (1996). Nucleosome DNA sequence pattern revealed by multiple alignment of experimentally mapped sequences. *Journal of Molecular Biology*, **262**, 129–139.

Jackson, J. R. and Benyajati, C. (1993). DNA-histone interactions are sufficient to position a single nucleosome juxtaposing *Drosophila* Adh adult enhancer and distal promoter. *Nucleic Acids Research*, **21**, 957–967.

Jin, C., Zang, C., Wei, G., *et al.* (2009). H3.3/H2A.Z double variant-containing nucleosomes mark 'nucleosome-free regions' of active promoters and other regulatory regions. *Nature Genetics*, **41**, 941–945.

Kaplan, N., Moore, I. K., Fondufe-Mittendorf, Y., *et al.* (2009). The DNA-encoded nucleosome organization of a eukaryotic genome. *Nature*, **458**, 362–366.

Keenen, B., Qi, H., Saladi, S. V., Yeung, M., and de la Serna, I. L. (2010). Heterogeneous SWI/SNF chromatin remodeling complexes promote expression of microphthalmia-associated transcription factor target genes in melanoma. *Oncogene*, **29**, 81–92.

Khalil, A. M., Guttman, M., Huarte, M., *et al.* (2009). Many human large intergenic non-coding RNAs associate with chromatin-modifying complexes and affect gene expression. *Proceedings of the National Academy of Sciences USA*, **106**, 11 667–11 672.

Kulaeva, O. I., Gaykalova, D. A., Pestov, N. A., *et al.* (2009). Mechanism of chromatin remodeling and recovery during passage of RNA polymerase II. *Nature Structural and Molecular Biology*, **16**, 1272–1278.

Lahm, A. and Suck, D. (1991). DNase I-induced DNA conformation. II. A structure of a DNase I-octamer complex. *Journal of Molecular Biology*, **222**, 645–667.

Lantermann, A., Stralfors, A., Fagerstrom-Billai, F., Korber, P., and Ekwall, K. (2009). Genome-wide mapping of nucleosome positions in *Schizosaccharomyces pombe*. *Methods*, **48**, 218–225.

Lantermann, A. B., Straub, T., Stralfors, A., *et al.* (2010). *Schizosaccharomyces pombe* genome-wide nucleosome mapping reveals positioning mechanisms distinct from those of *Saccharomyces cerevisiae*. *Nature Structural and Molecular Biology*, **17**, 251–257.

Lee, W., Tillo, D., Bray, N., *et al.* (2007). A high-resolution atlas of nucleosome occupancy in yeast. *Nature Genetics*, **39**, 1235–1244.

Levitsky, V. G. (2004). RECON: a program for prediction of nucleosome formation potential. *Nucleic Acids Research*, **32**, W346–349.

Lowary, P. T., and Widom, J. (1997). Nucleosome packaging and nucleosome positioning of genomic DNA. *Proceedings of the National Academy of Sciences USA*, **94**, 1183–1188.

Lowary, P. T. and Widom, J. (1998). New DNA sequence rules for high-affinity binding to histone octamer and sequence-directed nucleosome positioning. *Journal of Molecular Biology*, **276**, 19–42.

Luger, K. and Richmond, T. J. (1998). DNA binding within the nucleosome core. *Current Opinion in Structural Biology*, **8**, 33–40.

Lupien, M., Eeckhoute, J., Meyer, C. A., *et al.* (2008). FoxA1 translates epigenetic signatures into enhancer-driven lineage-specific transcription. *Cell*, **132**, 958–970.

Mavrich, T. N., Ioshikhes, I. P., Venters, B. J., *et al.* (2008). A barrier nucleosome model for statistical positioning of nucleosomes throughout the yeast genome. *Genome Research*, **18**, 1073–1083.

Mirzabekov, A. D. and Rich, A. (1979). Asymmetric lateral distribution of unshielded phosphate groups in nucleosomal DNA and its role in DNA bending. *Proceedings of the National Academy of Sciences USA*, **76**, 1118–1121.

Ozsolak, F., Song, J. S., Liu, X. S., and Fisher, D. E. (2007). High-throughput mapping of the chromatin structure of human promoters. *Nature Biotechnology*, **25**, 244–248.

Ozsolak, F., Poling, L. L., Wang, Z., *et al.* (2008). Chromatin structure analyses identify miRNA promoters. *Genes & Development*, **22**, 3172–3183.

Peckham, H. E., Thurman, R. E., Fu, Y., *et al.* (2007). Nucleosome positioning signals in genomic DNA. *Genome Research*, **17**, 1170–1177.

Reynolds, S., Bilmes, J., and Noble, W. (2009a). Low frequency oscillations in single-nucleotide content play a role in nucleosome positioning in *H. sapiens*. In *Proceedings of the 13th International Conference on Research in Computational Molecular Biology (RECOMB)*.

Reynolds, S., Bilmes, J., and Noble, W. (2009b). On the relationship between DNA periodicity and local chromatin structure. In *Proceedings of the 13th International Conference on Research in Computational Molecular Biology (RECOMB)*.

Reynolds, S., Bilmes, J., and Noble, W. (2010). Predicting nucleosome positioning using multiple evidence tracks. In *Proceedings of the 14th International Conference on Research in Computational Molecular Biology (RECOMB)*.

Robinson, P. J. and Rhodes, D. (2006). Structure of the '30 nm' chromatin fibre: a key role for the linker histone. *Current Opinion in Structural Biology*, **16**, 336–343.

Routh, A., Sandin, S., and Rhodes, D. (2008). Nucleosome repeat length and linker histone stoichiometry determine chromatin fiber structure. *Proceedings of the National Academy of Sciences USA*, **105**, 8872–8877.

Satchwell, S. C., Drew, H. R., and Travers, A. A. (1986). Sequence periodicities in chicken nucleosome core DNA. *Journal of Molecular Biology*, **191**, 659–675.

Schones, D. E., Cui, K., Cuddapah, S., *et al.* (2008). Dynamic regulation of nucleosome positioning in the human genome. *Cell*, **132**, 887–898.

Segal, E. and Widom, J. (2009). Poly(dA:dT) tracts: major determinants of nucleosome organization. *Current Opinion in Structural Biology*, **19**, 65–71.

Segal, E., Fondufe-Mittendorf, Y., Chen, L., *et al.* (2006). A genomic code for nucleosome positioning. *Nature*, **442**, 772–778.

Segal, M. R. (2008). Re-cracking the nucleosome positioning code. *Statistical Applications in Genetics and Molecular Biology*, **7**, Article 14.

Suck, D. (1994). DNA recognition by DNase I. *Journal of Molecular Recognition*, **7**, 65–70.

Thomas, G. H. and Elgin, S. C. (1988). Protein/DNA architecture of the DNase I hypersensitive region of the *Drosophila* hsp26 promoter. *EMBO Journal*, **7**, 2191–2201.

Weston, S. A., Lahm, A., and Suck, D. (1992). X-ray structure of the DNase I-d(GGTATACC)2 complex at 2.3 Å resolution. *Journal of Molecular Biology*, **226**, 1237–1256.

Wyrick, J. J., Holstege, F. C., Jennings, E. G., *et al.* (1999). Chromosomal landscape of nucleosome-dependent gene expression and silencing in yeast. *Nature*, **402**, 418–421.

Yuan, G. C. and Liu, J. S. (2008). Genomic sequence is highly predictive of local nucleosome depletion. *PLoS Computational Biology*, **4**, e13.

Yuan, G. C., Liu, Y. J., Dion, M. F., *et al.* (2005). Genome-scale identification of nucleosome positions in *S. cerevisiae*. *Science*, **309**, 626–630.

Zhang, Y., Moqtaderi, Z., Rattner, B. P., *et al.* (2009). Intrinsic histone-DNA interactions are not the major determinant of nucleosome positions in vivo. *Nature Structural and Molecular Biology*, **16**, 847–852.

Zhou, Y., Schmitz, K. M., Mayer, C., *et al.* (2009). Reversible acetylation of the chromatin remodelling complex NoRC is required for non-coding RNA-dependent silencing. *Nature Cell Biology*, **11**, 1010–1016.

Zofall, M., Persinger, J., Kassabov, S. R., and Bartholomew, B. (2006). Chromatin remodeling by ISW2 and SWI/SNF requires DNA translocation inside the nucleosome. *Nature Structural and Molecular Biology*, **13**, 339–346.

5 Chemical reporters of protein methylation and acetylation

Markus Grammel, Yu-Ying Yang, and Howard C. Hang*

5.1 Introduction

Posttranslational modifications (PTMs) are of crucial importance in protein regulation and their impact on virtually every aspect of biology has fostered the development of numerous experimental methods for their analysis. In particular, the application of sophisticated chemical approaches for the detection of PTMs has been of great value. Here we summarize our efforts to establish a chemical-reporter-based system for two PTMs, namely protein acetylation and methylation. These two covalent protein modifications are fundamental regulators of various epigenetic phenomena (Kouzarides, 2007) and have taken center stage in the current interpretation of what is commonly referred to as the "histone code" (Strahl and Allis, 2000).

Proteins can be posttranslationally acetylated at their N-termini as well as at serine, threonine, and lysine residues; however, lysine acetylation is the most prevalent type of acetylation (Allfrey et al., 1964). Lysine acetylation of histones, particularly at N-terminal tails, is a key mechanism of transcriptional regulation. In addition to the acetylation of histone tails hundreds of acetylation substrates have been identified in eukaryotic and prokaryotic cells (Choudhary et al., 2009; Wang et al., 2010; Zhao et al., 2010). Protein acetylation is a reversible PTM that is regulated by lysine acetyltransferases (KATs) and deacetylases (KDACs); KATs catalyze the transfer of the acetyl group from acetyl-CoA (an activated thioester) (Figure 5.1C) to their cognate substrate proteins, while KDACs remove the acetyl group from the lysine residue (Figure 5.1E). Lysine acetylation neutralizes the positive charge on the lysine residue and can lead to structural changes of the protein itself or influence its interaction with other proteins. Furthermore, the acetylated residue can serve to recruit other proteins that selectively recognize

---

* Author to whom correspondence should be addressed.

Epigenomics: From Chromatin Biology to Therapeutics, ed. K. Appasani. Published by Cambridge University Press. © Cambridge University Press 2012.

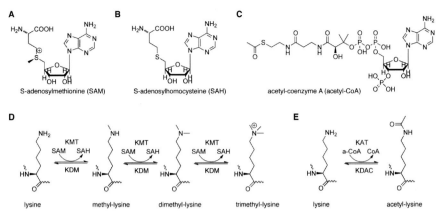

Figure 5.1 Protein acetylation and methylation. (A) The cofactor for protein methylation S-adenosylmethionine (SAM). (B) The cofactor product of the methylation reaction, S-adenosylhomocysteine (SAH). (C) The cofactor for protein acetylation, acetyl-coenzyme A (acetyl-CoA). (D) Protein lysine residues can be modified with one, two, or three methyl groups. Methyl groups are transferred by lysine methyltransferases (KMTs) from the methylation cofactor SAM onto the protein substrate. This reaction also forms SAH. The reverse reaction is catalyzed by lysine demethylases (KDMs). (E) Acetylation of lysine side chains is catalyzed by lysine acetyltransferases (KATs), which transfer an acetyl group from the activated thioester of acetyl-CoA onto the lysine substrate residue. Lysine deacylases (KDACs) catalyze the removal of acetyl groups from lysine residues.

the modified residue based on the acetyl-lysine-recognizing bromodomain (Sanchez and Zhou, 2009).

Lysine methylation of proteins is characterized by the attachment of one, two, or three methyl groups to the ε-amino group on lysine residues, which are installed by a family of lysine-specific methyltransferases (KMTs) (Figure 5.1D). The KMTs are classified by the presence of the catalytic SET-domain and can be grouped into seven subfamilies and a few orphan members, which together comprise more than 150 proteins in humans (Dillon *et al.*, 2005). The KMTs utilize the cofactor S-adenosyl-L-methionine (SAM) as the methyl donor (Figure 5.1A). Contrary to the classical view of protein methylation as an irreversible protein modification, various enzymes with lysine demethylation activity have been identified in recent years (Shi, 2007). Protein methylation and acetylation differ in the way they confer function on proteins. While acetylation changes the iso-electric point and hence protein conformation and interaction, methylation is thought to act mostly through the recruitment of other protein effectors. Several protein domains have been shown to facilitate protein binding to methylated lysine residues: the chromodomain, tudor domain, WD40-repeat domain, and PHD finger domain (Ruthenburg *et al.*, 2007). These domains mediate the interaction with different methylated motifs and are believed to translate the information stored in the "histone code" into a transcriptional response (Strahl and Allis, 2000).

Two major technological developments in recent years have facilitated the analysis of PTMs, namely the development of chemical reporters and the

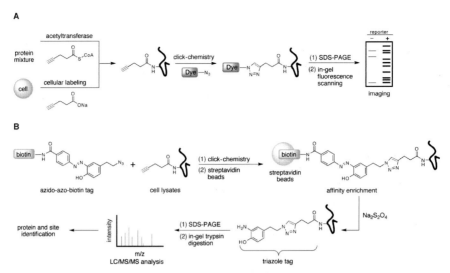

Figure 5.2 General strategy for the analysis of protein acetylation with alkyne-based chemical reporters. (A) Proteins labeled by 4-pentynoyl groups in vitro or in cells can be reacted with azido-functionalized dyes via click-chemistry to enable the in-gel fluorescence imaging of labeled proteins. (B) 4-Pentynoylated proteins in cell lysates can be reacted with a cleavable azido-azo-functionalized biotin tag and retrieved from protein mixtures on streptavidin beads. Bound proteins can be selectively eluted from the beads by selective reduction of the azo-bond using 25 mM $Na_2S_2O_4$, separated on SDS-PAGE and then trypsin-digested for mass spectrometric analysis to enable protein and site identification. See plate section for color version.

improvements in mass-spectrometry-based proteomics. To satisfy increasing interest in protein acetylation and methylation as regulated and reversible PTMs, particularly in the context of epigenetic regulation, the use of chemical reporters for the detection and analysis of these PTMs has been investigated by our groups and others (Yu *et al.*, 2006; Peters *et al.*, 2010; Yang *et al.*, 2010). The application of chemical reporters is based on a two-step strategy (Prescher and Bertozzi, 2005). In the first step the reporter is enzymatically installed onto target proteins in vitro or metabolically incorporated into the proteome by the endogenous enzymatic machinery (Figure 5.2A). The reporter replaces the natural PTM with a chemically similar moiety, which differs only in the presence of a functional chemical handle. In the second step, the introduced chemical handle allows modification of all labeled proteins with the desired analytical tag via bioorthogonal chemical ligation, thereby selectively forming a covalent bond between the functional handle and the chemical tag (Figure 5.2A). Bioorthogonal ligation reactions allow the selective modification of the incorporated chemical handle with a plethora of analytical tags, customized for the individual experiment (van Swieten *et al.*, 2005). Various ligation reactions have been used in this context over the years, but the copper(I)-catalyzed alkyne–azide cycloaddition reaction (CuAAC), also termed click-chemistry, has emerged as the predominantly used reaction with superior selectivity and kinetics (Sletten and Bertozzi, 2009). Click-chemistry is based on the ability of copper(I) to catalyze the formation of a triazole between a terminal alkyne and an azide, resulting in a covalent bond between the two functionalities (Figure 5.2A). Various PTMs have been studied using metabolically incorporated chemical reporters based on alkyne or azide analogs. Our laboratory has successfully used alkynyl fatty acid analogs to investigate protein palmitoylation and lipoproteins in immune cells and bacteria, respectively, as well as isoprenoid analogs to analyze protein prenylation (Ivanov *et al.* 2010; Rangan *et al.*, 2010; Yount *et al.*, 2010). This chapter highlights the application of chemical reporters for the analysis of protein acetylation and methylation and their potential application to epigenetics.

5.2   Methodology

This section will give a brief overview of the techniques used and direct the interested reader to the relevant literature (all experimental details for protein acetylation chemical reporters are described in Yang *et al.*, 2010).

5.2.1   Synthesis of chemical reporters

Acetyl-CoA analogs for in vitro transfer experiments were generated with a terminal alkyne group at the ω-position of the fatty acid chain (Figure 5.3A). Commercially available 3-butynoic acid, 4-pentynoic acid, or 5-hexynoic acid was dissolved in anhydrous dichloromethane and the anhydride was formed by addition of N,N′-dicyclohexylcarbodiimide (DCC). The reaction mixture was concentrated, dissolved in dimethylformamide and the thioester with coenzyme-A hydrate was formed in the presence of triethylamine. All alkynyl-CoA analogs were purified by high-performance liquid chromatography (HPLC). Sodium salts of alkynyl-acetate analogs for cellular labeling were formed from commercially available 3-butynoic acid, 4-pentynoic acid, or 5-hexynoic acid in the presence of sodium hydroxide.

The SAM analog propargyl S-adenosylhomocysteine (PHC) was formed by acidic alkylation of S-adenosylhomocysteine (SAH) (Figure 5.1B) with propargyl-

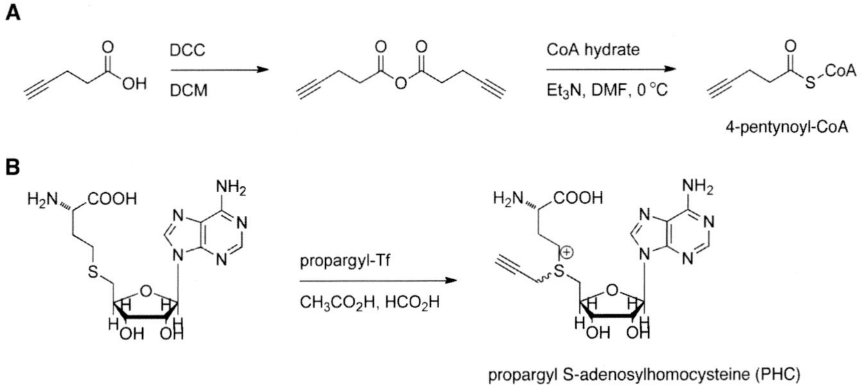

Figure 5.3 Synthesis of 4-pentynoyl-CoA and propargyl S-adenosylhomocysteine (PHC). (A) Synthetic scheme for the in vitro acetylation reporter, 4-pentynoyl-CoA. (B) Synthetic scheme for propargyl S-adenosylhomocysteine (PHC).

trifluoromethanesulfonate (propargyl-Tf) (Figure 5.3B) (Dalhoff *et al.*, 2006). Propargyl-Tf was freshly generated before the alkylation reaction due to its highly reactive nature. Propargyl alcohol, dissolved in anhydrous dichloromethane, was slowly added to a cooled (0 °C) suspension of trifluoromethanesulfonic anhydride and the non-nucleophilic base poly(4-vinylpyridine). After basic work-up the organic solvent was carefully evaporated (highly reactive triflate). Propargyl-Tf was dissolved in a 1:1 mixture of formic acid and acetic acid and added to a cooled (0 °C) solution of S-adenosylhomocysteine in the same solvent. The reaction was allowed to proceed at room temperature, was quenched with water, and washed with diethylether. The resulting aqueous solution was lyophilized and the precipitate suspended in 1 N sulfuric acid. PHC was precipitated by addition of 15 volumes of ice-cold ethanol.

5.2.2   In vitro transfer assays

As an in vitro model system to test the functionality of our acetylation chemical reporters we used the KAT p300. Using published in vitro conditions, two types of in vitro experiments were conducted (An and Roeder, 2002). First, we used H3 peptide (amino acids 1–20) as a substrate for p300. Possible covalent modifications on H3 peptide were monitored using MALDI-TOF-MS analysis. Additionally, histone H3 protein was used as a substrate for p300 and its modification was monitored by click-chemistry-mediated tagging with azido-rhodamine and in-gel fluorescent scanning (see Section 5.2.5.).

5.2.3   Labeling of cell lysates or living cells

To investigate protein acetylation in Jurkat T-cells, the alkynyl-acetate sodium salt analogs were added to the growth medium for different periods of time and at different concentrations. Unlike labeling studies using the acetylation reporters, the methylation reporter PHC required labeling of cell lysates instead of living cells due to its highly charged character and the resulting poor uptake into living cells (McMillan *et al.*, 2005).

5.2.4   Determining the specificity of acetylation and methylation reporters

To determine the specificity of the chemical reporters used, various control experiments were conducted. Inhibition of protein synthesis with cyclohexi-mide was used to reveal the posttranslational character of the observed modification. Small molecule inhibitors of KATs and KDACs were used to show the dependence of the observed labeling on the endogenous acetylation machinery. For this purpose, cells were treated with suberoylanilide hydroxamic acid (SAHA) (Richon *et al.*, 1998), a KDAC inhibitor, and curcumin, a p300 inhibitor (Marcu *et al.*, 2006). In the case of the methylation reporter PHC, we employed competition experiments with SAM or S-adenosylhomocysteine (SAH), a product inhibitor of KMTs. Furthermore, we assessed unspecific chemical reactivity of PHC by heat inactivation (10 minutes, 80 °C) of cell lysates prior to labeling with the reporter.

5.2.5 Detection of chemical reporters by click-chemistry

Click-chemistry was conducted as previously published to allow the rapid detection of incorporated chemical reporters (Grammel *et al.*, 2010; Rangan *et al.*, 2010; Yang *et al.*, 2010). For this purpose cell lysates were reacted with azido-rhodamine fluorescence dye and separated by SDS-PAGE. After protein separation the gel was analyzed by in-gel fluorescence scanning, which was conducted with a Typhoon 9400 (Amersham Biosciences) at 532 nm excitation and 580 nm emission.

5.2.6 Affinity enrichment of labeled proteins with immobilized streptavidin

Click-chemistry was also used to tag labeled proteins with an azido-azo-biotin affinity tag for affinity enrichment and subsequent protein identification (Figure 5.2B). For large-scale click-chemistry preparations, slightly longer incubation times were used, but all other parameters were kept the same. After click-chemistry proteins were precipitated and resulting pellets were resuspended in 4% SDS buffer, diluted to lower SDS concentrations to allow binding and incubated with streptavidin beads. On-bead captured proteins were washed multiple times followed by on-bead reduction and alkylation of cysteine residues. Selectively bound proteins were eluted by reductive cleavage of the azobenzene bond with 25 mM sodium dithionite. Upon desalting of the recovered protein samples, they were separated by SDS-PAGE and subsequently digested with trypsin. The resulting peptides were extracted from the gel matrix and analyzed by LC-MS/MS on an LTQ-Orbitrap (Thermo Fisher Scientific Inc.) mass spectrometer run in standard data-dependent mode.

5.3 Results and discussion

5.3.1 Monitoring protein acetylation in vitro

To assess the utility of alkynyl acetyl-CoA derivatives as chemical reporters for protein acetylation we established an in vitro acetylation assay. All three synthesized analogs 3-butynoyl-CoA, 4-pentynoyl-CoA, and 5-hexynoyl-CoA were tested for their ability to serve as chemical reporters for p300-mediated H3 peptide acetylation. Transfer of the chemical reporter groups from the acetyl-CoA analogs onto the H3 peptide substrate was monitored by MALDI-MS. MS analysis revealed that 4-pentynoyl-CoA is readily used by p300 as a substrate (Figure 5.4A), while 5-hexynoyl-CoA is a less efficient reporter. 3-Butynoyl-CoA was not used at all by p300 and decomposed rapidly under the reaction conditions used, which we attributed to the highly acidic propargylic protons adjacent to the thioester bond. To test whether 4-pentynoyl-CoA is also transferred onto a protein substrate we conducted an in-vitro assay with complete H3 protein as substrate. In this case, transfer was analyzed by click-chemistry and in-gel fluorescent scanning. Incubation of H3 with the chemical reporter alone resulted only in minimal fluorescent labeling, while concomitant addition of p300 and chemical reporter gave an intense and defined fluorescent signal (Figure 5.4B). The observed fluorescent signal was time- and concentration-dependent. Also, this in-vitro assay was successfully used to determine the Michaelis constant (K_M, 0.69 ± 0.12 μM)

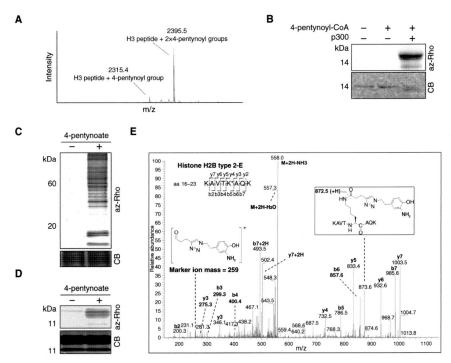

Figure 5.4 Chemical reporter allows detection of protein acetylation. (A) MALDI-TOF analysis of H3 peptide (amino acids 1–20) incubated in vitro with 4-pentynoyl-CoA and p300 KAT. (B) H3 histone protein was incubated in the presence or absence of p300 KAT and/or 4-pentynoyl-CoA. Histone H3 was subsequently reacted with azido-rhodamine (az-Rho) by click-chemistry (CuAAC), separated by SDS-PAGE, and visualized by in-gel fluorescence scanning. (C) Jurkat T-cells were labeled with 4-pentynoate (5 mM). Cells were lysed, proteins were tagged with az-Rho by CuAAC, separated by SDS-PAGE, and visualized by in-gel fluorescence scanning. (D) Jurkat T-cells were labeled with 4-pentynoate (5 mM). Cells were lysed and histones were isolated. Isolated histones were tagged with az-Rho by CuAAC, separated by SDS-PAGE, and visualized by in-gel fluorescence scanning. (E) Figure shows MS/MS spectrum of 4-pentynoate labeled peptide. The protein was isolated from total cell lysates. It was tagged with azido-azo-biotin and eluted by reduction with sodium dithionite, leaving a characteristic fragment ion (m/z 872) to determine the site of modification.

and the turnover number (k_{cat}, 2.10 ± 0.06 s^{-1}) of p300 with 4-pentynoyl-CoA as substrate. Importantly, the determined values compared well with the previously published values for acetyl-CoA (Thompson *et al.*, 2001).

5.3.2 Metabolic labeling of cells with chemical reporters for protein acetylation

While it was crucial to prove the chemical transfer onto a well-defined substrate in vitro and develop the assay for future applications, the essence of chemical reporters lies in their use in living cells or organisms. In contrast to the acetyl-CoA analogs used for in-vitro assays, we hypothesized that 4-pentynoate would be utilized by the endogenous cellular metabolic machinery to form 4-pentynoyl-CoA as a substrate for protein acetylation in vivo (we also tested 3-butynoate and 5-hexynoate). Therefore we labeled Jurkat T-cells with increasing concentrations of 4-pentynoate by addition to the growth medium for different lengths of time. The labeling profile displayed time and concentration dependence and an optimal dose and time window of 5 mM and 6 hours was established (Figure 5.3C). Most crucially, it had to be demonstrated that the observed labeling truly indicated an acetylation surrogate modification of protein lysine residues. We addressed this question by interrogating the dependence of the labeling signal on the enzymatic activity of KATs and KDACs as well as analyzing the incorporation of the chemical reporter onto core histones, a prominent known acetylation substrate.

Inhibition of protein synthesis with cycloheximide did not alter the labeling profile, indicating that the observed incorporation is a posttranslational process. We further used the KDAC inhibitor SAHA to show that inhibition of KDACs leads to a decrease in signal incorporation. We attribute this interference with chemical reporter labeling to a reduced number of possible acetylation substrates due to a hyperacetylated state of the cell. In addition, we could also show that direct inhibition of endogenous p300 with the inhibitor curcumin reduced the labeling of known p300 substrates. Most convincingly, we could prove the incorporation of the chemical reporter onto histones. Metabolic labeling of Jurkat T cells and subsequent extraction of core histones revealed chemical reporter specific labeling (Figure 5.4D). Most importantly, we could also prove direct chemical modification of endogenous histone proteins with the chemical reporter by mass spectrometry. For this purpose cells were labeled with 4-pentynoate, proteins were tagged with azido-azo-biotin, and cell lysates were digested with trypsin in solution. Tagged peptides were enriched with streptavidin beads, selectively eluted, and analyzed by mass spectrometry. Amongst the identified protein was histone H2B which showed, as all other identified peptides, the characteristic fragment ion (mass = 259 Da), indicative of the modified lysine residue (Figure 5.4E). These experiments show that 4-pentynoate can be used in vivo as a chemical reporter for acetylation, which is incorporated in a KAT-dependent manner onto well-known acetylation substrates. Most importantly it serves as a robust reporter for histone acetylation, as proven by various modes of analysis.

5.3.3   Proteomic identification of acetylated proteins

Proteomic analysis of PTMs by MS has become a popular approach over the past decade, which allows the identification of hundreds of modified proteins in one single experiment (Mann and Jensen, 2003). As indicated in the previous chapters, the use of chemical reporters provides a unique way to selectively enrich for proteins carrying a particular PTM. Based on the data discussed above we selected 4-pentynoate to analyze the acetylome of Jurkat T-cells. Upon labeling with 4-pentynoate Jurkat T-cells were lysed and resulting lysates were tagged with azido-azo-biotin by click-chemistry for selective enrichment and elution of labeled proteins (Figure 5.2B). Eluted proteins were identified using a standard GeLC-MS protocol. In three independent experiments we identified approximately 194 4-pentynoate labeled, putatively acetylated, proteins, of which 86% were also identified by other proteomic studies using anti-acetyl-Lys antibodies for enrichment (Choudhary *et al.*, 2009; Wang *et al.*, 2010). We also verified the enrichment of a number of identified proteins by Western blotting (e.g., cofilin, Hsp90). The results gained in our validation experiments (e.g., KDAC inhibition) and the overlap of identified proteins with other orthogonal proteomics and enrichment approaches for acetylated proteins strongly support the use of 4-pentynoate as a general chemical reporter for protein acetylation.

5.3.4   Progress towards a general chemical reporter for protein methylation

Protein lysine methylation is certainly one of the subtlest PTMs, comprising the addition of as little as one methyl group. As mentioned above, this modification does not change the charge state of the residue and only introduces a subtle hydrophobic modification to proteins. Due to these constraints there are no enrichment methods available that are based on the changed physicochemical properties of the methylated protein. Furthermore, available pan-selective lysine-methyl antibodies still display sequence preferences and hence bind only subsets of methylated proteins. These challenges make the development of a general chemical reporter for protein methylation especially valuable. However, the small size of the chemical group to be transferred as well as the high reactivity of the sulfonium center of SAM makes any functionalization for click-chemistry purposes difficult. In this paragraph we would like to share our (unpublished) attempts to develop an alkynyl chemical reporter for protein methylation.

Based on our previous success with alkynyl chemical reporters we conceived a propargyl S-adenosylhomocysteine (PHC, Figure 5.3B) derivative of SAM, which provides the shortest possible terminal alkyne moiety for derivatization of SAM's sulfonium center. Dalhoff and colleagues had reported similar compounds for the enzymatic transfer of different unsaturated alkyl chains from SAM analogs onto DNA by DNA-methyltranferases (Dalhoff *et al.*, 2005). The authors argued that the π-orbital of alkynes or olefins would stabilize the transition state during the S_N2 transfer reaction.

Due to its highly charged character we expected poor cell permeability qualities of PHC, therefore all labeling experiments were conducted with cell lysates in lieu of living cells. Addition of PHC to fresh cell lysates resulted in a multitude of labeled proteins. As described above for the acetylation reporter, it was essential to determine the chemical selectivity of the reporter. Adenosine dialdehyde (AdOx) has been widely used as a general indirect methyltransferase inhibitor (Chen, 2004). AdOx inhibits SAH hydrolase and leads to increased levels of SAH, which in turn serves as a product inhibitor of methyltransferases. Hence, treatment of cells with AdOx should lead to a hypomethylated substrate population and to increased incorporation of PHC (observed with ^{3}H-SAM), provided it serves as a functional SAM analog. Therefore we pre-incubated cells with AdOx and subsequently incubated the resulting cell lysates with PHC (Figure 5.5). In contrast to the impact AdOx treatment had on ^{3}H-SAM labeled cell lysates, we observed only one differential band appearing when incubated with PHC. As another means of testing for enzymatic transfer instead of unspecific chemical reactivity, we heat inactivated cell lysates for 10 minutes at 80 °C before incubation with PHC. The heat inactivated samples still displayed labeling of equivalent overall intensity, yet slightly changed in its labeling quality. Finally, we also co-incubated cell

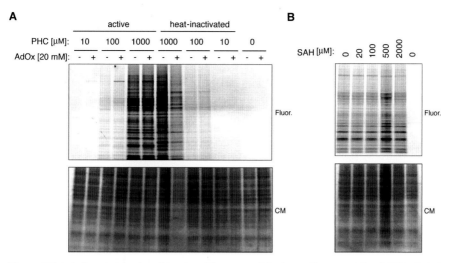

Figure 5.5 Labeling proteins in cell lysates with propargyl S-adenosylhomocysteine (PHC). Labeled proteins were tagged with azido-rhodamine using click-chemistry, separated by SDS-PAGE, and visualized by in-gel fluorescence scanning. (A) Fresh cell lysates from adenosine dialdehyde (AdOx; 20 mM) treated or untreated HeLa cells, or heat-inactivated (10 minutes, 80 °C) lysates of the same cells, were labeled with different concentrations of propargyl S-adenosylhomocysteine (PHC). (B) HeLa cell lysates were labeled with 500 μM PHC and increasing concentrations of SAH. (Fluor., in-gel fluorescence scanning; CM, coomassie.)

lysates with PHC and increasing concentrations of SAH, or SAM, which however did not reduce the fluorescence intensity. Taking into consideration these data, we concluded that PHC possesses too much intrinsic chemical reactivity and unselectively covalently modifies proteins. In addition, we also conducted in-vitro methyltransferase assays with PHC, in which we could not observe transfer of the propargyl group onto the provided peptide substrate.

5.4 Outlook

Alkyne- and azide-based chemical reporters have now made the leap from the early stages of development to be a robust research tool, not only accessible to chemistry laboratories but also to the wider research community (commercially available kits are sold by different vendors). We expect that the chemical reporters for acetylation described in this chapter will join the growing repertoire of chemical reporters and will especially find fruitful applications in epigenomics research. One aspect of growing importance in the epigenomics field, also emphasized in this volume, is the development of drugs targeting epigenetic mechanisms (Spannhoff et al., 2009). Considerable effort has been invested in the development of HDAC inhibitors for therapeutic purposes, especially as anti-cancer drugs, since aberrant epigenetic acetylation patterns have been associated with various neoplasms (Bolden et al., 2006). We envision that robust assays for histone acetylation, as well as general protein acetylation, will be of great importance in the future for different drug-development purposes. Since the use of radioactivity is a laborious and time-consuming process, there should be great interest in flexible and rapid alternatives – as provided by click-chemistry-based chemical reporters. One challenge still remains, which is that, due to the presence of multiple KATs and thousands of substrates, it is very difficult to identify specific enzyme–substrate pairs. An elegant method to solve this complex puzzle is the so-called "bump-and-hole" approach, which can be combined with a chemical reporter strategy. Shokat and colleagues have successfully applied this strategy to identify protein kinase substrates (Shogren-Knaak *et al.*, 2001). This approach is based on an allele-specific chemical reporter, or inhibitor, which provides the required "bump" to fit the "hole" that is provided by an enzyme mutant, similar to the "lock and key" analogy. This method could possibly facilitate the development of orthogonal chemical reporter systems for individual KATs. Finally, maybe no other technology has shaped the field of epigenetic research as much as chromatin immunoprecipitation (ChIP) and its derivatives ChIP-chip and ChIP-seq (Park, 2009). We envisage that chemical reporters could be used as an alternative for antibody-based capture for the enrichment of a chromatin subset which carries a particular histone modification.

Protein methylation is still lacking a general chemical reporter at this time, even though two independent recent publications have presented data on in-vitro transfer of an alkyne chemical reporter by KMTs (Binda *et al.*, 2010; Peters *et al.*, 2010). Peters and colleagues have introduced the first chemical reporter (AdoEnYn) that is transferred onto a peptide or protein substrate by a KMT (Peters *et al.*, 2010). However, all experiments were conducted with isolated components in vitro and it remains to be seen whether the same reporter will be accepted by multiple KMTs in vivo. Interestingly, the authors also observed the highly unstable nature of PHC, which led them to use a seemingly more stable pent-2-en-4-ynyl analog, emphasizing that it will be crucial to find the right balance between the inherent chemical reactivity and the steric perturbation of the reporter group. In spite of the unstable nature of PHC, Binda and colleagues were able to show in-vitro transfer of the terminal alkyne group of PHC onto histone H3 protein and a H3 peptide GST fusion by the KMT SETDB1 (Binda *et al.*, 2010). Still, no other KMT tested by the authors accepted PHC for in vitro transfer. Therefore PHC does not seem to be suitable as a general reporter for protein methylation, due to its high selectivity for individual KMTs as well as its unspecific reactivity in cell lysates. While the development of a general chemical reporter for protein methylation is an evidently challenging task, it bears great potential for the epigenetic research community since histone methylation is now recognized as a major dynamic regulator of chromatin biology whose tight control is essential for proper gene expression. Aberrant histone methylation has been implicated in many diseases, especially various forms of cancer. The success of epigenetic therapeutic interventions with HDAC inhibitors and DNA demethylating agents has therefore spurred the development of therapies targeting the histone methylation machinery (Spannhoff *et al.*, 2009). For these reasons, histone and protein methylation will continue to be a growing field in basic and applied biological research that demands rapid and robust assays for KMT activity analysis – WANTED: A chemical reporter for protein methylation.

REFERENCES

Allfrey, V. G., Faulkner, R., and Mirsky, A. E. (1964). Acetylation and methylation of histones and their possible role in the regulation of RNA synthesis. *Proceedings of the National Academy of Sciences USA*, **51**, 786–794.

An, W. and Roeder, R. (2002). Direct association of p300 with unmodified H3 and H4 N-termini modulates p300-dependent acetylation and transcription of nucleosomal templates. *Journal of Biological Chemistry*, **278**, 1504–1510.

Bolden, J. E., Peart, M. J., and Johnstone, R. W. (2006). Anticancer activities of histone deacetylase inhibitors. *Nature Reviews Drug Discovery*, **5**, 769–784.

Chen, D. (2004). Effects of adenosine dialdehyde treatment on in vitro and in vivo stable protein methylation in HeLa cells. *Journal of Biochemistry*, **136**, 371–376.

Choudhary, C., Kumar, C., Gnad, F., *et al.* (2009). Lysine acetylation targets protein complexes and co-regulates major cellular functions. *Science*, **325**, 834–840.

Dalhoff, C., Lukinavičius, G., Klimašauskas, S., and Weinhold, E. (2005). Direct transfer of extended groups from synthetic cofactors by DNA methyltransferases. *Nature Chemical Biology*, **2**, 31–32.

Dalhoff, C., Lukinavičius, G., Klimašauskas, S., and Weinhold, E. (2006). Synthesis of S-adenosyl-L-methionine analogs and their use for sequence-specific transalkylation of DNA by methyltransferases. *Nature Protocols*, **1**, 1879–1886.

Dillon, S. C., Zhang, X., Trievel, R. C., and Cheng, X. (2005). The SET-domain protein superfamily: protein lysine methyltransferases. *Genome Biology*, **6**, 227.

Grammel, M., Zhang, M. M., and Hang, H. C. (2010). Orthogonal alkynyl amino acid reporter for selective labeling of bacterial proteomes during infection. *Angewandte Chemie (International Edition)*, **49**, 5970–5974.

Ivanov, S. S., Charron, G., Hang, H. C., and Roy, C. R. (2010). Lipidation by the host prenyltransferase machinery facilitates membrane localization of *Legionella pneumophila* effector proteins. *Journal of Biological Chemistry*, **285**, 34 686–34 698.

Kouzarides, T. (2007). Chromatin modifications and their function. *Cell*, **128**, 693–705.

Mann, M. and Jensen, O. N. (2003). Proteomic analysis of posttranslational modifications. *Nature Biotechnology*, **21**, 255–261.

Marcu, M. G., Jung, Y., Lee, S., *et al.* (2006). Curcumin is an inhibitor of p300 histone acetyltransferase. *Medicinal Chemistry (Shāriqah [United Arab Emirates])*, **2**, 169–174.

McMillan, J. M., Walle, U. K., and Walle, T. (2005). S-adenosyl-L-methionine: transcellular transport and uptake by Caco-2 cells and hepatocytes. *Journal of Pharmacy and Pharmacology*, **57**, 599–605.

Park, P. J. (2009). ChIP-seq: advantages and challenges of a maturing technology. *Nature Reviews Genetics*, **10**, 669–680.

Peters, W., Willnow, S., Duisken, M., *et al.* (2010). Enzymatic site-specific functionalization of protein methyltransferase substrates with alkynes for click labeling. *Angewandte Chemie (International Edition)*, **49**, 5170–5173.

Prescher, J. A. and Bertozzi, C. R. (2005). Chemistry in living systems. *Nature Chemical Biology*, **1**, 13–21.

Rangan, K. J., Yang, Y., Charron, G., and Hang, H. C. (2010). Rapid visualization and large-scale profiling of bacterial lipoproteins with chemical reporters. *Journal of the American Chemical Society*, **132**, 10 628–10 629.

Richon, V. M., Emiliani, S., Verdin, E., *et al.* (1998). A class of hybrid polar inducers of transformed cell differentiation inhibits histone deacetylases. *Proceedings of the National Academy of Sciences USA*, **95**, 3003–3007.

Ruthenburg, A. J., Li, H., Patel, D. J., and Allis, C. D. (2007). Multivalent engagement of chromatin modifications by linked binding modules. *Nature Reviews Molecular Cell Biology*, **8**, 983–994.

Sanchez, R. and Zhou, M. M. (2009). The role of human bromodomains in chromatin biology and gene transcription. *Current Opinion in Drug Discovery and Development*, **12**, 659.

Shi, Y. (2007). Histone lysine demethylases: emerging roles in development, physiology and disease. *Nature Reviews Genetics*, **8**, 829–833.

Shogren-Knaak, M. A., Alaimo, P. J., and Shokat, K. M. (2001). Recent advances in chemical approaches to the study of biological systems. *Annual Review of Cell and Developmental Biology*, **17**, 405–433.

Sletten, E. and Bertozzi, C. (2009). Bioorthogonal chemistry: fishing for selectivity in a sea of functionality. *Angewandte Chemie (International Edition)*, **48**, 6974–6998.

Spannhoff, A., Hauser, A., Heinke, R., Sippl, W., and Jung, M. (2009). The emerging therapeutic potential of histone methyltransferase and demethylase inhibitors. *ChemMedChem*, **4**, 1568–1582.

Strahl, B. D. and Allis, C. D. (2000). The language of covalent histone modifications. *Nature*, **403**, 41–45.

van Swieten, P. F., Leeuwenburgh, M. A., Kessler, B. M., and Overkleeft, H. S. (2005). Bioorthogonal organic chemistry in living cells: novel strategies for labeling biomolecules. *Organic and Biomolecular Chemistry*, **3**, 20–27.

Thompson, P. R., Kurooka, H., Nakatani, Y., and Cole, P. A. (2001). Transcriptional coactivator protein p300: kinetic characterization of its histone acetyltransferase activity. *Journal of Biological Chemistry*, **276**, 33 721–33 729.

Wang, Q., Zhang, Y., Yang, C., *et al.* (2010). Acetylation of metabolic enzymes coordinates carbon source utilization and metabolic flux. *Science*, **327**, 1004–1007.

Yang, Y., Ascano, J. M., and Hang, H. C. (2010). Bioorthogonal chemical reporters for monitoring protein acetylation. *Journal of the American Chemical Society*, **132**, 3640–3641.

Yount, J. S., Moltedo, B., Yang, Y., *et al.* (2010). Palmitoylome profiling reveals S-palmitoylation-dependent antiviral activity of IFITM3. *Nature Chemical Biology*, **6**, 610–614.

Yu, M., de Carvalho, L. P. S., Sun, G., and Blanchard, J. S. (2006). Activity-based substrate profiling for Gcn5-related N-acetyltransferases: the use of chloroacetyl-coenzyme A to identify protein substrates. *Journal of the American Chemical Society*, **128**, 15 356–15 357.

Zhao, S., Xu, W., Jiang, W., *et al.* (2010). Regulation of cellular metabolism by protein lysine acetylation. *Science*, **327**, 1000–1004.

6 Long non-coding RNA in epigenetic gene silencing

Takashi Nagano

6.1 Introduction

The term non-coding RNA (ncRNA) refers to RNA that does not code for protein. This classification derives from the RNA research before 1970s, where the function of RNA had been studied in the dominant context of how genomic DNA prescribes for protein synthesis, having protein-coding messenger RNA (mRNA) at its core. Our knowledge on RNA has continuously expanded thereafter through identification and analyses of various distinct RNAs lacking protein-coding potential, including ribosomal RNA, transfer RNA, small nuclear RNAs, small nucleolar RNAs, microRNAs, etc. Each of these RNA classes (including mRNA) has distinct features of its own, and the members of each class are thought to share similar function.

After 1990s, several key discoveries have been made that shed light on some "enigmatic" ncRNAs that had not been classified as above. The ncRNAs *Xist* (Penny *et al.*, 1996; Marahrens *et al.*, 1997) and *Airn* (previously referred to as *Air*; Sleutels *et al.*, 2002), which lack known features as functional ncRNA unlike various established ncRNA classes listed above, play essential roles in regulating transcriptional silencing of other genes. These findings had a great impact on the research on epigenetic gene silencing in higher mammals, and many studies carried out later based on these discoveries have contributed to the prosperity of epigenetics research today. On the other hand, we now know, through recent progress in transcriptome analyses, that a huge number of ncRNAs are transcribed from most regions of the genome in human and mouse (FANTOM Consortium *et al.*, 2005; ENCODE Project Consortium, 2007). The recent data that a substantial population among them shows evolutionary conservation and regulated expression (Guttman *et al.*, 2009) suggest that they likely include as yet unidentified functional ncRNAs similar to *Xist* and *Airn* (some of them are actually emerging as reviewed below). Therefore the ncRNAs that do not have an

Epigenomics: From Chromatin Biology to Therapeutics, ed. K. Appasani. Published by Cambridge University Press. © Cambridge University Press 2012.

appropriate criterion to form a specific ncRNA class now attract intense scientific interest as a possible mine of epigenetic regulators. The unclassifiable, enigmatic ncRNAs are collectively called long ncRNAs (lncRNAs) or macro RNAs, simply because they are relatively long. The term "mRNA-like ncRNAs" can also be applicable to many of the lncRNAs, since they share several features with protein-coding mRNA such as modifications through 5′ capping, splicing, and polyadenylation tailing. In contrast to the small ncRNAs that comprise distinct classes, there is neither a clear definition nor established subclasses for lncRNA. It is a collection of miscellaneous ncRNAs longer than around 100 nucleotides or so.

Among the lncRNAs, *Xist*, *Airn*, and *Kcnq1ot1* are the typical and pioneering examples which suggested important regulatory roles well before the lncRNA drew much attention as today. *Xist* is essential for X-chromosome inactivation (X inactivation) in mammalian female cells (Plath *et al.*, 2002). Based on the microscopic observation, *Xist* has been described as coating or painting the entire X chromosome in one of the earliest steps in X inactivation (Clemson *et al.*, 1996). Although this characteristic observation allows us to speculate how *Xist* works on chromatin at the molecular level, there are many basic questions still left unanswered 15 years after the first observation. For example, are there specific regions on the future inactive X chromosome that serve as interaction sites for *Xist*, or does *Xist* associate with the future inactive X chromosome uniformly? How does *Xist* specifically interact with the X chromosome? What molecular role does *Xist* play in the inactivation of genes on the future inactive X chromosome after associating with chromatin? These important but still open questions about *Xist* are more or less common to *Airn* and *Kcnq1ot1*, which are involved in gene silencing through epigenetic regulation and share functional similarities with *Xist* (Umlauf *et al.*, 2008).

Below, the important findings on *Airn* and *Kcnq1ot1* will be reviewed in comparison with *Xist*, to highlight a general mechanism of lncRNA-mediated epigenetic gene silencing. In addition, the latest progress on the other but similar lncRNAs is summarized to show how the commonality is expanding into the lncRNA in general.

6.2   Overview of *Airn*- and *Kcnq1ot1*-dependent gene silencing

The *Airn* transcription unit initiates in the second intron of the mouse *Igf2r* gene located on chromosome 17, in an antisense direction to *Igf2r* (Figure 6.1A). *Airn* promoter DNA is methylated on the maternal allele before fertilization and *Airn* is expressed only from the paternal allele. In mouse embryonic day 11.5 (E11.5) placenta, a cluster of several nearby protein coding genes *Igf2r*, *Slc22a2*, and *Slc22a3* are silenced and expressed only from the maternal allele (Zwart *et al.*, 2001). Deletion of the *Airn* promoter on the paternal allele results in aberrant activation of paternal *Igf2r*, *Slc22a2*, and *Slc22a3* genes (Wutz *et al.*, 2001; Zwart *et al.*, 2001). Similar derepression of the three paternal genes is caused when *Airn* RNA is truncated by inserting a polyadenylation cassette just downstream to the paternal *Airn* transcription start site (Sleutels *et al.*, 2002). Based on these results,

A *Airn* gene cluster in E11.5 placenta

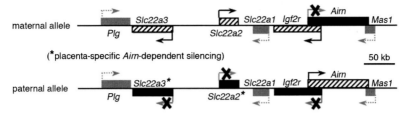

B *Kcnq1ot1* gene cluster in placenta

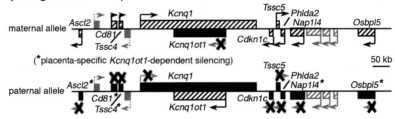

Figure 6.1 Schematic drawing of the *Airn* and *Kcnq1ot1* gene clusters where two typical lncRNAs are expressed and involved in silencing of neighboring genes in *cis*. Black rectangles, allele-specific silenced genes; hatched rectangles, transcribed genes; gray rectangles, inactive genes. (A) *Airn* non-coding and neighbor protein-coding genes at the proximal region on mouse chromosome 17. In embryonic day 11.5 (E11.5) placenta, *Airn* is expressed only from paternal allele, and *Igf2r*, *Slc22a2*, and *Slc22a3* genes are expressed only from maternal allele. Silencing of paternal *Igf2r*, *Slc22a2*, and *Slc22a3* genes is *Airn*-dependent in E11.5 placenta, while *Slc22a3* is expressed from both alleles in E15.5 placenta. In embryo proper and adult tissues, expression of paternal *Slc22a2* and *Slc22a3* (marked with *) is independent from *Airn*. (B) *Kcnq1ot1* non-coding RNA is expressed from the paternal allele and responsible for silencing of neighbor protein-coding genes in *cis* at the distal region on mouse chromosome 7. In embryo proper and adult tissues, expression of several paternal alleles (marked with *) is independent from *Kcnq1ot1*.

Airn from the paternal allele is responsible for silencing the paternal alleles of *Igf2r*, *Slc22a2*, and *Slc22a3* in *cis*. In addition, the allele-specific silencing of the *Slc22a3* gene is dynamically regulated and it shows bi-allelic expression in E15.5 placenta in contrast to E11.5 placenta (Zwart *et al.*, 2001). On the other hand, the expression status of the gene cluster in embryo proper and adult tissue is slightly different, where only *Igf2r*, but not *Slc22a2* and *Slc22a3*, shows *Airn*-dependent silencing in *cis* (Zwart *et al.*, 2001).

Kcnq1ot1 shares several features with *Airn*. As shown in Figure 6.1B, *Kcnq1ot1* is also expressed from the paternal allele, and is responsible for silencing a cluster of genes in *cis* in placenta. *Kcnq1ot1* is transcribed in an antisense direction from intron 10 of one of these genes (*Kcnq1*), and the others are located distantly within a 780-kb domain. Deletion of the *Kcnq1ot1* promoter or truncation of the RNA results in derepression of *Kcnq1ot1*-dependent genes (Fitzpatrick *et al.*, 2002; Mancini-Dinardo *et al.*, 2006; Shin *et al.*, 2008). Similarly to the *Airn* gene cluster, several genes in the *Kcnq1ot1* gene cluster show different expression patterns between placenta and embryo proper. Among *Kcnq1ot1*-dependent genes in placenta, only a subset shows *Kcnq1ot1* dependency in embryo proper as well (Lewis *et al.*, 2006).

Taken together, these results suggest that either the lncRNA molecules or transcription per se through the lncRNA transcription units is essential for repressing not only genes overlapping with lncRNAs (i.e., *Igf2r* and *Kcnq1*) but also distant genes in *cis*. In addition, the mechanisms working in the embryo proper and adult tissues appear to be different from placenta.

6.3   lncRNA interacts with chromatin

One possible molecular mechanism of lncRNA-dependent gene silencing is that the lncRNA associates directly with target genes and recruits factors necessary for silencing. This idea originates from the famous microscopic observations of Clemson *et al.* (1996) who found that RNA fluorescence in situ hybridization (FISH) signals for *XIST* appear to cover the entire X chromosome. Comparable observations for *Airn* and *Kcnq1ot1* have been reported recently. Simultaneous detection of *Airn* RNA FISH signals and *Slc22a3* DNA FISH signals in mouse placenta reveal that *Airn* covers the paternal *Slc22a3* gene in *cis* in correlation with allelic silencing (Nagano *et al.*, 2008) (Figure 6.2A). Similar conclusions were drawn from measurements between *Kcnq1ot1* RNA FISH signals and DNA FISH signals for genes silenced by the lncRNA (Terranova *et al.*, 2008; Redrup *et al.*, 2009). In all three cases above, silenced genes tended to be more closely associated

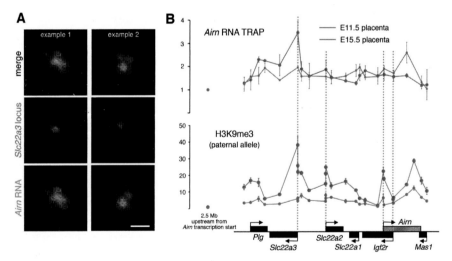

Figure 6.2 *Airn* accumulates specifically at the promoter of the *Slc22a3* gene in dynamic correlation with allele-specific gene silencing and repressive histone modification. (A) Simultaneous detection of *Airn* RNA FISH signals (bottom panels) and *Slc22a3* DNA FISH signals (middle panels). *Slc22a3* loci densely covered by *Airn* RNA (left column) are dominant in E11.5 placenta where paternal *Slc22a3* gene is silenced in *Airn*-dependent manner, whereas *Slc22a3* loci with little overlap with *Airn* RNA (right column) are observed more often in E15.5 placenta where paternal *Slc22a3* gene is relieved from silencing. Scale bar = 1 μm. (B) Association of *Airn* RNA with chromatin analyzed by RNA TRAP (tagging and recovery of associated proteins) (upper panel) and distribution of trimethylation of H3K9 on the paternal allele (lower panel), each normalized to 2.5 Mb upstream from *Airn* transcription start site, in E11.5 (pink lines) and E15.5 (blue lines) placentas. See plate section for color version.

or covered by the lncRNA signals. These observations suggest a common or similar mechanism involving an interaction of lncRNA with chromatin.

The association between the *Xist* RNA and inactive X chromosome has been known under the microscope for over a decade, and several lines of evidence also show that RNA (especially ncRNA) is integrated in (or associates with) chromatin (Rodriguez-Campos and Azorin, 2007; Mondal *et al.*, 2010). However, biochemical evidence describing the interaction of the lncRNAs and chromatin at specific loci has long been lacking. Two techniques, RNA TRAP (tagging and recovery of associated proteins) and ChOP (chromatin oligo-affinity precipitation), have recently been used to analyze lncRNA interactions with chromatin at specific loci at the molecular level. RNA TRAP employs RNA FISH signal-dependent chromatin labeling followed by recovery and quantification of the labeled chromatin (Carter *et al.*, 2002). The ChOP assay starts with solubilized, cross-linked chromatin and captures RNA-associated chromatin by hybridizing a tagged oligonucleotide with the target RNA followed by pull-down of the tag and associated chromatin (Mariner *et al.*, 2008). In short, RNA TRAP and ChOP can identify genomic sites relatively enriched with a specific RNA. RNA TRAP was used to quantify *Airn* association at 20 genomic locations across the entire *Airn*-regulated cluster in E11.5 placenta (where *Slc22a3* shows allele-specific expression) and E15.5 placenta (where *Slc22a3* is expressed bi-allelically) (Nagano *et al.*, 2008) (Figure 6.2B). The results showed that *Airn* accumulates specifically at the promoter of the distal *Slc22a3* gene in dynamic correlation with allele-specific silencing. Using a similar method, evidence for association of *KCNQ1OT1* (Murakami *et al.*, 2007) and *Xist* (Murakami *et al.*, 2009) with some of its target gene bodies and promoters is found, although the extent of association was not clear due to non-quantitative analysis of a limited number of data points. In support of these findings, Pandey *et al.* (2008) employed ChOP to show association of *Kcnq1ot1* with promoters of some target genes in mouse placenta. Although it is still unclear whether *Kcnq1ot1* and *Xist* associate with their target genes in a promoter-specific manner or by uniformly "coating" the entire cluster, these results suggest that chromatin association is a common feature of *Airn*-, *Kcnq1ot1*- and *Xist*-dependent gene silencing.

6.4 Epigenetic modifications that correlate with lncRNA association

Several groups have reported analysis of epigenetic modifications across the *Airn* and *Kcnq1ot1* clusters and the inactive X chromosome, and in some cases correlated the presence of repressive modifications with lncRNA association.

Nagano *et al.* (2008) reported histone modifications in the *Airn* gene cluster in placenta where *Airn* regulates all three genes in the cluster. They found *Airn* association correlated strongly with paternal allele-specific presence of trimethylation of histone H3 lysine 9 (H3K9me3) at the promoter of the silenced *Slc22a3* gene (Figure 6.2B). Trimethylation of histone H3 lysine 27 (H3K27me3) and histone H3 lysine 4 (H3K4me3) were also found at the *Slc22a3* promoter but were not paternal-allele-specific and/or did not correlate with developmentally dynamic pattern of gene silencing. On the other hand, DNA methylation is found at the maternal allele of *Airn* promoter and paternal allele of *Igf2r* promoter. The DNA methylation at maternal *Airn* promoter is established during oogenesis (before fertilization: Stoger *et al.*, 1993), and this is essential for allele-specific expression of *Airn* (Seidl *et al.*, 2006). In contrast, DNA methylation on the paternal *Igf2r* promoter is established in *Airn*-dependent manner (Sleutels *et al.*, 2002) but not necessary for silencing the paternal *Igf2r* allele (Li *et al.*, 1993).

In the *Kcnq1ot1* gene cluster, Umlauf *et al.* (2004) and Lewis *et al.* (2004) found paternal-allele-specific enrichment of H3K27me3 and H3K9me3 at promoter regions of *Kcnq1ot1*-silenced genes. Using a *Kcnq1ot1* truncation mutant, Pandey *et al.* (2008) showed that the presence of the H3K27me3 modification was dependent on *Kcnq1ot1* lncRNA. Consistent with these results, Terranova *et al.* (2008) found that *Kcnq1ot1* RNA FISH signals often colocalize with immuno-staining for H3K9me3 and H3K27me3 in trophectodermal stem (TS) cells. In addition, *Kcnq1ot1* RNA FISH signals also colocalized with immunostaining for monoubiquitination of histone H2A lysine 119 (H2AK119u1) (Terranova *et al.*, 2008), which is associated with prevention of transcription elongation (Stock *et al.*, 2007). On the other hand, DNA methylation in the *Kcnq1ot1* gene cluster is found at maternal allele of *Kcnq1ot1* promoter and paternal alleles of *Cdkn1c* and *Phlda2* promoters (Lewis *et al.*, 2004). The DNA methylation at maternal *Kcnq1ot1* promoter is established during oogenesis and is essential for allele-specific expression of *Kcnq1ot1* (Lewis *et al.*, 2004), while DNA methylation at the paternal *Cdkn1c* promoters is established in a *Kcnq1ot1*-dependent manner but not necessary for silencing paternal *Cdkn1c* allele at least in several embryonic tissues like kidney, liver, and lung (Shin *et al.*, 2008). Such a hierarchy of DNA methylation, lncRNA expression, and allele-specific gene silencing in the *Kcnq1ot1* gene cluster is similar to the *Airn* gene cluster described above.

The X chromosome destined to be inactivated and be coated with *Xist* first loses euchromatin-associated histone modifications such as acetylation of histone H3 lysine 9 (H3K9ac), dimethylation and trimethylation of histone H3 lysine 4 (H3K4me2 and H3K4me3), followed by accumulation of repressive histone marks like dimethylation of histone H3 lysine 9 (H3K9me2), H3K27me3, and H2AK119u1; deposition of macroH2A and introduction of DNA methylation come later (Heard, 2005).

These results that the lncRNA association correlates with different sets of histone marks (H3K9me3 but not H3K27me3 for *Airn* gene cluster, H3K9me3, H3K27me3, and H2AK119u1 for *Kcnq1ot1*, and H3K9me2, H3K27me3, and H2AK119u1 for *Xist*) suggest possible differences in the silencing mechanism for each.

6.5   lncRNA recruits chromatin modifiers to chromatin

Wagschal *et al.* (2008) provided the first indication that a specific histone modifier G9a, a H3K9 histone methyltransferase, was involved in *Kcnq1ot1* silencing. A G9a knockout resulted in reduced H3K9me2 and H3K9me3 in the paternal

Kcnq1ot1 cluster; however, derepression of *Kcnq1ot1*-regulated genes in placenta was highly variable. In contrast, Nagano *et al.* (2008) showed consistent derepression of the paternal *Slc22a3* allele in the same G9a knockout. *Airn* co-immunoprecipitated with G9a in placenta and G9a enrichment at the *Slc22a3* promoter was found to be dependent on full-length *Airn* RNA (Nagano *et al.*, 2008). Taken together, these results support a clear functional link between the *Airn* RNA molecule and G9a at the paternal *Slc22a3* promoter.

In contrast, *Kcnq1ot1* co-immunoprecipitates with G9a, Ezh2, and Suz12 (Pandey *et al.*, 2008), the latter two of which are the members of Polycomb repressive complex 2 (PRC2) and mediate H3K27 methylation. Further, G9a and PRC members have been found at some promoters of repressed paternal alleles (Umlauf *et al.*, 2004; Wagschal *et al.*, 2008), and G9a, Ezh2, and Eed (another member of PRC2) are required for *Kcnq1ot1*-dependent silencing of several genes (Mager *et al.*, 2003; Terranova *et al.*, 2008; Wagschal *et al.*, 2008), although the affected genes in the absence of each one of the three are not necessarily the same. Since H3K27me3 enrichment at most genes in the *Kcnq1ot1* cluster is reduced in the absence of *Kcnq1ot1* (Pandey *et al.*, 2008), it is likely that *Kcnq1ot1* recruits PRC2 to target genes to achieve silencing. On the other hand, Terranova *et al.* (2008) showed by RNA immunoFISH that Rnf2 (Ring1b), an ubiquitin E3 ligase that is a member of Polycomb repressive complex 1 (PRC1) and responsble for H2AK119u1, often associates with *Kcnq1ot1* RNA FISH signals in TS cells and preimplantation embryos. This is consistent with the finding that PRC1 is recruited through the affinity for H3K27me3 (Simon and Kingston, 2009). H2AK119u1 is thought to act adversely to gene expression by pausing transcription (Simon and Kingston, 2009), and it has further been shown that Rnf2 is essential for allele-specific silencing of the *Cdkn1c* gene (Terranova *et al.*, 2008). In addition, Mohammad *et al.* (2010) recently reported that *Kcnq1ot1* also binds with DNA methyltransferase Dnmt1 and is responsible for Dnmt1 occupancy and allele-specific DNA methylation at the *Cdkn1c* and *Slc22a18* promoter regions, although the experimental perturbation of the binding actually abolishes allele-specific gene expression only in limited tissues.

Zhao *et al.* (2008) showed that *Xist* also has an ability to bind with PRC2 through an important motif within *Xist* called A-repeat, which is required for silencing (Wutz *et al.*, 2002). They also showed that A-repeat is transcribed as 1.6-kb-long *RepA* RNA distinctly from *Xist*, and *RepA* as well as *Xist* associates with PRC2. However, the significance of A-repeat to recruit PRC2 in vivo is elusive, since PRC2 can be recruited to the inactive X chromosome in the absence of A-repeat (Plath *et al.*, 2003).

6.6   lncRNAs and nuclear compartmentalization

Though it is clear that there are specific association between lncRNAs and their target genes, it is not entirely clear if the lncRNAs are recruited to them with chromatin modifiers as shown in Figure 6.3A. Recent results on *Xist* and X inactivation suggest that the process should not be oversimplified. A more

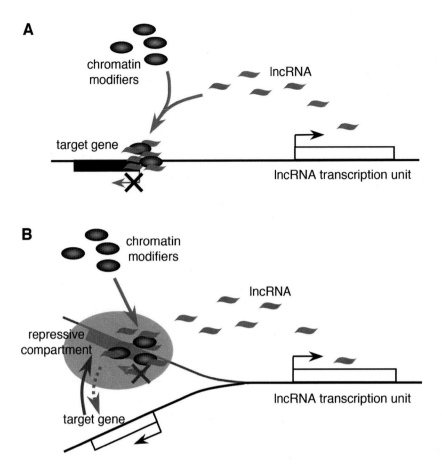

Figure 6.3 Schematic drawing of two possible lncRNA-dependent gene silencing mechanisms. Open rectangles represent active genes, whereas solid rectangles show silenced genes. (A) The lncRNA first accumulates at target gene, then (or at the same time) chromatin modifiers are recruited to silence the target gene. (B) The lncRNA forms a repressive compartment possibly in conjunction with chromatin modifiers and/or scaffold proteins. The target gene then associates with the repressive compartment in conjunction with silencing. This association could be either dynamic (e.g., *Slc22a3* gene in placenta) or static (e.g., X inactivation).

probable explanation based on X inactivation research is that a lncRNA compartment is formed away from the target genes, then the target genes are recruited to the compartment and silenced, and finally the silencing is stably maintained in some cases (Figure 6.3B).

During X inactivation, *Xist* initially forms a structure encapsulating mainly non-genic repetitive sequences of the future inactive X chromosome (Clemson *et al.*, 2006). The "*Xist* compartment" mostly excludes RNAPII and the repetitive sequences inside are silenced. The repressive histone marks such as H3K27me3 are accumulated inside the compartment following the RNAPII exclusion (Chaumeil *et al.*, 2006). However, X-linked gene silencing is not complete at this stage and active genes remain on the surface or outside the *Xist* compartment. Interestingly, these events in earlier stages are separable from pan-chromosomal silencing, not only temporally but also functionally. The A-repeat on the *Xist* RNA, which is essential for silencing (Wutz *et al.*, 2002), is not required to form the *Xist* compartment (Chaumeil *et al.*, 2006). In other words, a mutant *Xist* lacking A-repeat can form the *Xist* compartment with repressive chromatin marks on the X chromosome, but cannot organize pan-chromosomal gene silencing.

The X-linked genes destined for inactivation are subsequently drawn inside the *Xist* compartment coincident with silencing, clearly later than the *Xist* compartment formation (Chaumeil *et al.*, 2006). As mentioned above, A-repeat on *Xist* RNA is essential for this step, although it is still not clear enough how A-repeat works at this stage. One clue reported recently is that the ongoing transcription of full-length young LINE-1 elements, which is A-repeat-dependent, helps silencing spread in remaining active regions on the X chromosome (Chow *et al.*, 2010). In addition to A-repeat, Agrelo *et al.* (2009) showed that nuclear scaffold proteins SATB1 and SATB2 (special AT-rich binding proteins) are crucial in silencing, possibly by holding genes to be silenced within the repressive compartment.

Once X inactivation is established, then DNA methylation and histone H4 hypoacetylation play key roles to maintain it, even without *Xist* RNA (Csankovszki *et al.*, 2001). Several lines of evidence support that this step is also separable from former steps in X inactivation. Smchd1 (structural maintenance of chromosomes hinge domain containing 1) is required for DNA methylation on the inactive X, but dispensable for initial events such as *Xist* compartment formation and H3K27me3 modification (Blewitt *et al.*, 2008). Also, the mutant *Xist* without A-repeat can create stable hypoacetylated histone H4 mark throughout the entire X chromosome in *cis* (Pullirsch *et al.*, 2010).

Although little is known about compartment formation with *Airn* and *Kcnq1ot1* lncRNAs compared to *Xist*, both *Airn* and *Kcnq1ot1* associate with (or form) similar repressive compartments excluding RNAPII (Terranova *et al.*, 2008; T. Nagano and P. Fraser, unpublished observation). In the *Airn* cluster, actively transcribed *Slc22a3* alleles are predominantly found outside the *Airn* compartment (Nagano *et al.*, 2008), suggesting dynamic oscillation between the silencing compartment and transcription factories (Figure 6.3B). It is noteworthy that *Airn* RNA associates with a H3K27me3-enriched nuclear compartment in trophectoderm cells of the blastocyst (Terranova *et al.*, 2008), in which *Airn*-dependent gene silencing has not yet been established (Szabó and Mann, 1995). Thus, like *Xist*, formation of other lncRNA repressive compartments may precede recruitment of genes to be silenced. Given the similarities, it is an interesting idea that a similar nuclear scaffold protein such as SATB1 contributes to *Airn*- and *Kcnq1ot1*-dependent gene silencing.

6.7   Not a single silencing mechanism per single lncRNA?

As reviewed above, *Airn* is required for silencing the paternal allele of the *Slc22a3* gene by recruiting G9a to the promoter region in placenta. Then one of the remaining questions is if the *Igf2r* gene is also silenced by the same *Airn*-dependent mechanism as *Slc22a3* in placenta, embryo proper, and adult tissues. The results by Nagano *et al.* (2008) show that *Igf2r* silencing is maintained without G9a in placenta in contrast to *Slc22a3*. In addition, not only G9a and H3K9me3 but also *Airn* were not enriched at the *Igf2r* promoter (Nagano *et al.*, 2008), even though the *Igf2r* promoter is in the *Airn* transcription unit. This indicates that, even in placenta, *Igf2r* silencing occurs via a mechanism distinct from *Slc22a3*, which does not involve *Airn* accumulation at least once silencing is established. Likewise, the ChOP assay detected *Kcnq1ot1* at some but not all promoter regions of *Kcnq1ot1*-dependent genes (Pandey *et al.*, 2008).

Nagano *et al.* (2008) analyzed *Airn* accumulation in placenta only after E11.5, when allele-specific silencing of *Igf2r* is established, rather than in the peri-implantation stage when *Igf2r* silencing is first established (Szabó and Mann, 1995; Lerchner and Barlow, 1997). This leaves open the possibility that *Airn* accumulation at *Igf2r* is required only for the establishment of silencing. Once silencing is established, repressive chromatin modifications set during the establishment may be inherited through cell division to maintain silencing without the continuous presence of the lncRNA.

In contrast, another possibility is that lncRNA transcription per se, rather than the lncRNA molecule itself, is important for gene silencing. A variety of mechanisms for this possibility have been discussed (Seidl *et al.* 2006; Pauler *et al.*, 2007). For example, it has been hypothesized that transcription through putative regulatory DNA elements within the lncRNA transcription unit could negatively affect expression of lncRNA-dependent genes in *cis* (Pauler *et al.*, 2007), although no direct evidence to support specific regulatory elements has been reported thus far. Alternately, a negative effect of lncRNA transcription on neighboring gene expression may be another hypothetical transcription-based silencing mechanism. This effect is generally referred to as transcriptional interference, in which transcription of one gene typically prevents recruitment of the initiation complex or elongation on neighboring genes (Mazo *et al.*, 2007). Transcriptional interference appears to be more robust when one promoter is stronger than the other (Callen *et al.*, 2004). The finding that bi-allelic expression of *Igf2r* is induced when the *Airn* promoter is replaced by an exogenous weaker promoter supports this possibility (Stricker *et al.*, 2008).

In summary, while *Airn* accumulation at the *Slc22a3* promoter region in placenta correlates with gene silencing well after the silencing is established, this correlation does not apply to the *Igf2r* gene silencing in placenta. This indicates that a single lncRNA employs multiple modes of gene silencing simultaneously in the same tissue.

6.8   Recent expansion of the commonality

Based on the recent findings about the three "pioneering" lncRNAs, *Xist*, *Airn*, and *Kcnq1ot1*, it seems a general concept that lncRNA molecules bind with a repressive chromatin modification complex/factor and recruit it to chromatin. Indeed, there are a number of additional examples of similar lncRNAs that were identified more recently. They work not only in *cis*, but some act in *trans*, and some even on multiple targets at a genome-wide scale.

HOTAIR is a lncRNA transcribed from the *HOXC* locus, which is responsible for PRC2 occupancy and H3K27me3 of *HOXD* locus on different chromosome (Rinn *et al.*, 2007). However, the target loci where *HOTAIR* regulates PRC2 accumulation are not only *HOXD*. A recent report showed that breast cancer cells highly express *HOTAIR*, which likely induces genome-wide retargeting of PRC2 and confers increased invasiveness on the cancer cells through an altered gene expression pattern (Gupta *et al.*, 2010). Moreover, *HOTAIR* can act as a scaffold for two distinct histone modification complexes, PRC2 and LSD1-containing complex (Tsai *et al.*, 2010). Since LSD1 is a demethylase that mediates enzymatic demethylation of active histone modification H3K4me2, *HOTAIR* can coordinate two different histone modifications.

ANRIL is a lncRNA antisense to the *p15/CDKN2B/INK4B* tumor suppressor gene. This lncRNA is transcribed from the region deleted in hereditary cutaneous malignant melanoma patients in the *INK4B/ARF/INK4A* tumor suppressor gene cluster (Pasmant *et al.*, 2007). *ANRIL* is also known as a genetic susceptibility locus associated with various diseases, such as coronary disease, intracranial aneurysm, and type 2 diabetes (Pasmant *et al.*, 2011). Yap *et al.* (2010) recently reported that *ANRIL* associates with CBX7, a member of PRC1, and this association is important for CBX7 to repress the *INK4B/ARF/INK4A* locus.

A lncRNA named *lincRNA-p21* acts as a repressor in p53-dependent transcriptional responses (Huarte *et al.*, 2010). *LincRNA-p21* expression is directly activated by p53 in response to DNA damage, and *lincRNA-p21* regulates multiple target genes at a genome-wide scale, which are involved in either apoptosis or cell cycle regulation. *LincRNA-p21* associates with and recruits heterogeneous nuclear ribonucleoprotein K (hnRNP-K; a component of a repressor complex that acts in the p53 pathway) to regulate target loci.

Moreover, Khalil *et al.* (2009) found over 3000 loci in the human genome which encode lncRNA based on chromatin-modification signatures in several kinds of cells. Of these, they showed that nearly 40% physically associate with repressive chromatin modifiers such as PRC2 and/or CoREST (corepressor for RE1 silencing transcription factor), and are likely to repress target genes in *trans* such as *HOTAIR*.

Thus, as anticipated when we came to know that human and mouse cells express huge numbers of lncRNA, the lncRNAs that associate with chromatin modification factor/complex account for a substantial population among the entire lncRNAs. Some of them regulate multiple functionally linked target genes simultaneously to orchestrate gene expression on a genome-wide scale. Given the large number of such lncRNAs and the important roles in normal development and homeostasis that they play, they constitute a large source of therapeutic targets for many diseases as well.

6.9   Closing remarks

As reviewed above, we now know that many lncRNAs involved in epigenetic gene silencing employ the similar mechanism in which chromatin modifiers are bound and recruited to target loci by the lncRNA. Apart from small differences in detail among individual lncRNAs, there is still a big unanswered question about how the lncRNA complex specifically find its target loci. In this context, the correlation of local LINE density and its transcription with gene silencing on the inactive X chromosome (Chow *et al.*, 2010) may provide possible clues. When we find the answer, we will be one step closer to applying the amazing potential of lncRNA that can regulate multiple specific genes at the right timing, to overcome intractable diseases.

REFERENCES

Agrelo, R., Souabni, A., Novatchkova, M., *et al.* (2009). SATB1 defines the developmental context for gene silencing by *Xist* in lymphoma and embryonic cells. *Developmental Cell*, **16**, 507–516.

Blewitt, M. E., Gendrel, A. V., Pang, Z., *et al.* (2008). SmcHD1, containing a structural-maintenance-of-chromosomes hinge domain, has a critical role in X inactivation. *Nature Genetics*, **40**, 663–669.

Callen, B. P., Shearwin, K. E., and Egan, J. B. (2004). Transcriptional interference between convergent promoters caused by elongation over the promoter. *Molecular Cell*, **14**, 647–656.

Carter, D., Chakalova, L., Osborne, C. S., Dai, Y. F., and Fraser, P. (2002). Long-range chromatin regulatory interactions in vivo. *Nature Genetics*, **32**, 623–626.

Chaumeil, J., Le Baccon, P., Wutz, A., and Heard, E. (2006). A novel role for *Xist* RNA in the formation of a repressive nuclear compartment into which genes are recruited when silenced. *Genes & Development*, **20**, 2223–2237.

Chow, J. C., Ciaudo, C., Fazzari, M. J., *et al.* (2010). LINE-1 activity in facultative heterochromatin formation during X chromosome inactivation. *Cell*, **141**, 956–969.

Clemson, C. M., McNeil, J. A., Willard, H. F., and Lawrence, J. B. (1996). *XIST* RNA paints the inactive X chromosome at interphase: evidence for a novel RNA involved in nuclear/chromosome structure. *Journal of Cell Biology*, **132**, 259–275.

Clemson, C. M., Hall, L. L., Byron, M., McNeil, J., and Lawrence, J. B. (2006). The X chromosome is organized into a gene-rich outer rim and an internal core containing silenced nongenic sequences. *Proceedings of the National Academy of Sciences USA*, **103**, 7688–7693.

Csankovszki, G., Nagy, A., and Jaenisch, R. (2001). Synergism of *Xist* RNA, DNA methylation, and histone hypoacetylation in maintaining X chromosome inactivation. *Journal of Cell Biology*, **153**, 773–784.

ENCODE Project Consortium (2007). Identification and analysis of functional elements in 1% of the human genome by the ENCODE pilot project. *Nature*, **447**, 799–816.

FANTOM Consortium and RIKEN Genome Exploration Research Group and Genome Science Group (Genome Network Project Core Group) (2005). The transcriptional landscape of the mammalian genome. *Science*, **309**, 1559–1563.

Fitzpatrick, G. V., Soloway, P. D., and Higgins, M. J. (2002). Regional loss of imprinting and growth deficiency in mice with a targeted deletion of KvDMR1. *Nature Genetics*, **32**, 426–431.

Gupta, R. A., Shah, N., Wang, K. C., *et al.* (2010). Long non-coding RNA *HOTAIR* reprograms chromatin state to promote cancer metastasis. *Nature*, **464**, 1071–1076.

Guttman, M., Amit, I., Garber, M., *et al.* (2009). Chromatin signature reveals over a thousand highly conserved large non-coding RNAs in mammals. *Nature*, **458**, 223–227.

Heard, E. (2005). Delving into the diversity of facultative heterochromatin: the epigenetics of the inactive X chromosome. *Current Opinion in Genetics and Development*, **15**, 482–489.

Huarte, M., Guttman, M., Feldser, D., *et al.* (2010). A large intergenic noncoding RNA induced by p53 mediates global gene repression in the p53 response. *Cell*, **142**, 409–419.

Khalil, A. M., Guttman, M., Huarte, M., *et al.* (2009). Many human large intergenic non-coding RNAs associate with chromatin-modifying complexes and affect gene expression. *Proceedings of the National Academy of Sciences USA*, **106**, 11 667–11 672.

Lerchner, W. and Barlow, D. P. (1997). Paternal repression of the imprinted mouse *Igf2r* locus occurs during implantation and is stable in all tissues of the post-implantation mouse embryo. *Mechanisms of Development*, **61**, 141–149.

Lewis, A., Mitsuya, K., Umlauf, D., *et al.* (2004). Imprinting on distal chromosome 7 in the placenta involves repressive histone methylation independent of DNA methylation. *Nature Genetics*, **36**, 1291–1295.

Lewis, A., Green, K., Dawson, C., *et al.* (2006). Epigenetic dynamics of the *Kcnq1* imprinted domain in the early embryo. *Development*, **133**, 4203–4210.

Li, E., Beard, C., and Jaenisch, R. (1993). Role for DNA methylation in genomic imprinting. *Nature*, **366**, 362–365.

Mager, J., Montgomery, N. D., de Villena, F. P., and Magnuson, T. (2003). Genome imprinting regulated by the mouse Polycomb group protein Eed. *Nature Genetics*, **33**, 502–507.

Mancini-Dinardo, D., Steele, S. J., Levorse, J. M., Ingram, R. S., and Tilghman, S. M. (2006). Elongation of the *Kcnq1ot1* transcript is required for genomic imprinting of neighboring genes. *Genes & Development*, **20**, 1268–1282.

Marahrens, Y., Panning, B., Dausman, J., Strauss, W., and Jaenisch, R. (1997). *Xist*-deficient mice are defective in dosage compensation but not spermatogenesis. *Genes & Development*, **11**, 156–166.

Mariner, P. D., Walters, R. D., Espinoza, C. A., *et al.* (2008). Human Alu RNA is a modular transacting repressor of mRNA transcription during heat shock. *Molecular Cell*, **29**, 499–509.

Mazo, A., Hodgson, J. W., Petruk, S., Sedkov, Y., and Brock, H. W. (2007). Transcriptional interference: an unexpected layer of complexity in gene regulation. *Journal of Cell Science*, **120**, 2755–2761.

Mohammad, F., Mondal, T., Guseva, N., Pandey, G. K., and Kanduri, C. (2010). *Kcnq1ot1* noncoding RNA mediates transcriptional gene silencing by interacting with Dnmt1. *Development*, **137**, 2493–2499.

Mondal, T., Rasmussen, M., Pandey, G. K., Isaksson, A., and Kanduri, C. (2010). Characterization of the RNA content of chromatin. *Genome Research*, **20**, 899–907.

Murakami, K., Oshimura, M., and Kugoh, H. (2007). Suggestive evidence for chromosomal localization of non-coding RNA from imprinted LIT1. *Journal of Human Genetics*, **52**, 926–933.

Murakami, K., Ohhira, T., Oshiro, E., *et al.* (2009). Identification of the chromatin regions coated by non-coding *Xist* RNA. *Cytogenetic and Genome Research*, **125**, 19–25.

Nagano, T., Mitchell, J. A., Sanz, L. A., *et al.* (2008). The *Air* noncoding RNA epigenetically silences transcription by targeting G9a to chromatin. *Science*, **322**, 1717–1720.

Pandey, R. R., Mondal, T., Mohammad, F., *et al.* (2008). *Kcnq1ot1* antisense noncoding RNA mediates lineage-specific transcriptional silencing through chromatin-level regulation. *Molecular Cell*, **32**, 232–246.

Pasmant, E., Laurendeau, I., Héron, D., *et al.* (2007). Characterization of a germ-line deletion, including the entire *INK4/ARF* locus, in a melanoma-neural system tumor family: identification of *ANRIL*, an antisense noncoding RNA whose expression coclusters with *ARF*. *Cancer Research*, **67**, 3963–3969.

Pasmant, E., Sabbagh, A., Vidaud, M., and Bièche, I. (2011). *ANRIL*, a long, noncoding RNA, is an unexpected major hotspot in GWAS. *FASEB Journal*, **25**, 444–448.

Pauler, F. M., Koerner, M. V., and Barlow, D. P. (2007). Silencing by imprinted noncoding RNAs: is transcription the answer? *Trends in Genetics*, **23**, 284–292.

Penny, G. D., Kay, G. F., Sheardown, S. A., Rastan, S., and Brockdorff, N. (1996). Requirement for *Xist* in X chromosome inactivation. *Nature*, **379**, 131–137.

Plath, K., Mlynarczyk-Evans, S., Nusinow, D. A., and Panning, B. (2002). *Xist* RNA and the mechanism of X chromosome inactivation. *Annual Review of Genetics*, **36**, 233–278.

Plath, K., Fang, J., Mlynarczyk-Evans, S. K., *et al.* (2003). Role of histone H3 lysine 27 methylation in X inactivation. *Science*, **300**, 131–135.

Pullirsch, D., Härtel, R., Kishimoto, H., *et al.* (2010). The Trithorax group proteins Ash2l and Saf-A are recruited to the inactive X chromosome at the onset of stable X inactivation. *Development*, **137**, 935–943.

Redrup, L., Branco, M. R., Perdeaux, E. R., *et al.* (2009). The long noncoding RNA *Kcnq1ot1* organises a lineage-specific nuclear domain for epigenetic gene silencing. *Development*, **136**, 525–530.

Rinn, J. L., Kertesz, M., Wang, J. K., *et al.* (2007). Functional demarcation of active and silent chromatin domains in human *HOX* loci by noncoding RNAs. *Cell*, **129**, 1311–1323.

Rodriguez-Campos, A. and Azorin, F. (2007). RNA is an integral component of chromatin that contributes to its structural organization. *PLoS ONE*, **2**, e1182.

Seidl, C. I., Stricker, S. H., and Barlow, D. P. (2006). The imprinted Air ncRNA is an atypical RNAPII transcript that evades splicing and escapes nuclear export. *EMBO Journal*, **25**, 3565–3575.

Shin, J. Y., Fitzpatrick, G. V., and Higgins, M. J. (2008). Two distinct mechanisms of silencing by the *KvDMR1* imprinting control region. *EMBO Journal*, **27**, 168–178.

Simon, J. A. and Kingston, R. E. (2009). Mechanisms of polycomb gene silencing: knowns and unknowns. *Nature Reviews Molecular Cell Biology*, **10**, 697–708.

Sleutels, F., Zwart, R., and Barlow, D. P. (2002). The non-coding *Air* RNA is required for silencing autosomal imprinted genes. *Nature*, **415**, 810–813.

Stock, J. K., Giadrossi, S., Casanova, M., *et al.* (2007). Ring1-mediated ubiquitination of H2A restrains poised RNA polymerase II at bivalent genes in mouse ES cells. *Nature Cell Biology*, **9**, 1428–1435.

Stoger, R., Kubicka, P., Liu, C. G., *et al.* (1993). Maternal-specific methylation of the imprinted mouse *Igf2r* locus identifies the expressed locus as carrying the imprinting signal. *Cell*, **73**, 61–71.

Stricker, S. H., Steenpass, L., Pauler, F. M., *et al.* (2008). Silencing and transcriptional properties of the imprinted *Airn* ncRNA are independent of the endogenous promoter. *EMBO Journal*, **27**, 3116–3128.

Szabó, P. E. and Mann, J. R. (1995). Allele-specific expression and total expression levels of imprinted genes during early mouse development: implications for imprinting mechanisms. *Genes & Development*, **9**, 3097–3108.

Terranova, R., Yokobayashi, S., Stadler, M. B., *et al.* (2008). Polycomb group proteins Ezh2 and Rnf2 direct genomic contraction and imprinted repression in early mouse embryos. *Developmental Cell*, **15**, 668–679.

Tsai, M. C., Manor, O., Wan, Y., *et al.* (2010). Long noncoding RNA as modular scaffold of histone modification complexes. *Science*, **329**, 689–693.

Umlauf, D., Goto, Y., Cao, R., *et al.* (2004). Imprinting along the *Kcnq1* domain on mouse chromosome 7 involves repressive histone methylation and recruitment of Polycomb group complexes. *Nature Genetics*, **36**, 1296–1300.

Umlauf, D., Fraser, P., and Nagano, T. (2008). The role of long non-coding RNAs in chromatin structure and gene regulation: variations on a theme. *Biological Chemistry*, **389**, 323–331.

Wagschal, A., Sutherland, H. G., Woodfine, K., *et al.* (2008). G9a histone methyltransferase contributes to imprinting in the mouse placenta. *Molecular and Cellular Biology*, **28**, 1104–1113.

Wutz, A., Theussl, H. C., Dausman, J., *et al.* (2001). Non-imprinted *Igf2r* expression decreases growth and rescues the Tme mutation in mice. *Development*, **128**, 1881–1887.

Wutz, A., Rasmussen, T. P., and Jaenisch, R. (2002). Chromosomal silencing and localization are mediated by different domains of *Xist* RNA. *Nature Genetics*, **30**, 167–174.

Yap, K. L., Li, S., Muñoz-Cabello, A. M., *et al.* (2010). Molecular interplay of the noncoding RNA *ANRIL* and methylated histone H3 lysine 27 by polycomb CBX7 in transcriptional silencing of INK4a. *Molecular Cell*, **38**, 662–674.

Zhao, J., Sun, B. K., Erwin, J. A., Song, J. J., and Lee, J. T. (2008). Polycomb proteins targeted by a short repeat RNA to the mouse X chromosome. *Science*, **322**, 750–756.

Zwart, R., Sleutels, F., Wutz, A., Schinkel, A. H., and Barlow, D. P. (2001). Bidirectional action of the *Igf2r* imprint control element on upstream and downstream imprinted genes. *Genes & Development*, **15**, 2361–2366.

Part II

Epigenomic imprinting
and stem cells

7 Active DNA demethylation: the enigma starts in the zygote

Julia Arand, Konstantin Lepikhov, Mark Wossidlo, and Jörn Walter*

7.1 Specificity and distribution of DNA methylation in mammals

In mammals, DNA methylation occurs exclusively at C5 positions of cytosines, predominantly in a CpG dinucleotide context. The CpG dinucleotides are actually underrepresented in mammalian genomes and unequally distributed across the genomes (Varriale and Bernardi 2010). CpG dinucleotides are enriched in CpG islands, short genomic regions of up to 3 kb in size, which are often located close to or overlap with promoters (Bird *et al.*, 1985; Bird, 1986; Saxonov *et al.*, 2006). CpG islands tend to be largely unmethylated. In contrast, CpG regions with medium or low CpG content tend to be largely hypermethylated. These methylated CpGs comprise about 70–80% of all CpGs (Ehrlich *et al.*, 1982). At promotors with low or medium CpG content, methylation can be frequently observed in a cell-specific manner. Such methylation, in general, imposes silencing of the related genes. Furthermore, DNA methylation can also be guided to particular genes, chromosomal regions and even entire chromosomes to cause silencing of only one allele, as can be seen for imprinted genes and the X chromosome. For imprinted genes the allele-specific marking by DNA methylation occurs in the late germ cells (Reik and Walter, 2001).

A considerable amount of CpGs can also be found in repetitive DNA sequences such as transposons and retrotransposon-like elements. These sequences make up a large fraction (about 40–50%) of the mammalian genome and have usually low CpG content, with CpGs being highly methylated (Ehrlich *et al.*, 1982; Gama-Sosa *et al.*, 1983). It is believed that this DNA methylation is necessary to silence the transcriptional activity of retrotransposable elements and hence to maintain the genome stability (Bourc'his and Bestor, 2004; Sasaki and Matsui, 2008; Ooi *et al.*, 2009).

* Author to whom correspondence should be addressed.

Epigenomics: From Chromatin Biology to Therapeutics, ed. K. Appasani. Published by Cambridge University Press. © Cambridge University Press 2012.

Several reports have suggested that methylation can also be found outside of the CpG context, particularly in mammalian embryonic stem (ES) cells. A recent genome-wide analysis of human ES cells suggested a substantial amount of non-CpG methylation predominantly in the sequence context CpA (Lister *et al.*, 2009). The cellular distribution and enzymatic control of such non-symmetrical methylation is still rather unclear in mammals, unlike in plants, where the presence and enzymatic control of such non-CpG methylation is quite well understood (Finnegan and Kovac, 2000; Feng *et al.*, 2010).

Enzymes responsible for setting and maintaining the DNA methylation are called DNA methyltransferases (DNMTs) and can be divided into two functional groups. The *de novo* DNMTs, DNMT3a and DNMT3b, establish new methylation marks at unmethylated sequences. The maintenance DNMT, named DNMT1, prefers hemi-methylated sites and is responsible for the maintenance of methylation marks during the cell cycle – i.e., after replication, the newly synthesized strand receives the same methylation pattern as the parental strand. In this way, DNA methylation can be stably passed from one cell to the daughter cells over generations. Due to this mechanism, DNA methylation is the most stable epigenetic modification and is often used to maintain a silent status of the genome and it functions in concert with repressive histone modifications or non-coding RNAs (e.g., X inactivation).

Besides 5-methylcytosine (5mC), 5-hydroxymethylcytosine (5hmC) was also recently found in mammalian DNA (Kriaucionis and Heintz, 2009; Tahiliani *et al.*, 2009). This modification and its impact on biological processes are far from being understood. Hydroxylation of methylated cytosines could act as a masking of the methyl group and inhibit the effect of methylation. In fact, hydroxymethylated sites could be recognized as unmethylated sites and so the related genes can be expressed, since methyl-binding proteins cannot bind any more (Valinluck *et al.*, 2004), but still the sites retain the modification for further recognitions. Hydroxymethylation is generated by ten–eleven translocation (Tet) enzymes (Tet1, 2, and 3) (Tahiliani *et al.*, 2009; Ito *et al.*, 2010). These enzymes use 5mC as a substrate for hydroxylation. In mouse, 0.3% of all cytosines are hydroxymethylated in ES cells and up to 1.2% in differentiated tissues (Szwagierczak *et al.*, 2010). In differentiated tissues 5mC accounts for 4–5% of all cytosine residues (Gama-Sosa *et al.*, 1983). That means, in differentiated tissues, 5mC is 4–5 times more abundant than 5hmC. Unfortunately, the discrimination of 5hmC and 5mC by conventional detection methods is not so simple. 5mC and an unmodified cytosine can be very well discriminated by different methods. The gold standard for single base resolution is so far bisulfite sequencing, which allows the conversion of cytosine to uracil, but leaves the 5mC unchanged. 5hmC behaves the same as 5mC in this method and therefore cannot be distinguished. So far, for the discrimination of 5hmC and unmodified cytosines antibodies or mass spectrometry can be used. Glycosylation of 5hmC also makes it possible to distinguish 5hmC from 5mC by the glycosylation itself or by further discrimination by restriction enzymes (Szwagierczak *et al.*, 2010).

DNA methylation is crucial for mammalian life, since deletion of DNMTs is embryonic lethal (Okano *et al.*, 1999). However, in mammalian development, there are phases where this stable mark is globally removed. Afterwards new methylation patterns can be set to govern new transcriptional permissive or repressive landscapes, i.e, the DNA methylation pattern has to be reprogrammed. Yet, the DNA demethylation mechanism is still an enigma.

7.2   DNA demethylation during mouse development

Two global DNA demethylation phases can be observed in mammalian development; one during the generation of the germ cells, after migration of the primordial germ cells to the genital ridge (germ cell generation) and one after fertilization of the oocyte in preimplantation development (Figure 7.1). Associated with this DNA demethylation, changes in histone modification patterns and the activation of pluripotency associated genes can be observed.

The global demethylation processes can be monitored by a decreasing immunofluorescence signal of the 5mC antibody and methylation sensitive HpaII digestion of whole DNA preparations (Monk *et al.*, 1987; Santos *et al.*, 2002; Seki *et al.*, 2005). Furthermore, bisulfite sequencing or HpaII digestion of selected sequences also reveals demethylation (Kafri *et al.*, 1992; Hajkova *et al.*, 2002;

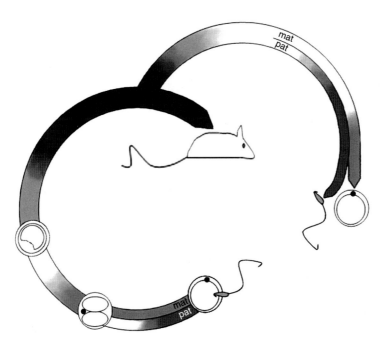

Figure 7.1 DNA methylation reprogramming in a life cycle. There are two waves of DNA demethylation in mammalian development. After fertilization of the oocyte, the paternal genome is very quickly demethylated in zygotes. Further global demethylation can be observed up to the blastocyst stage. This is the time point when differentiation, accompanied or facilitated by remethylation, starts to create all the different cell types that are needed to build up the multicellular organism. During the generation of the germ cells, there is a second round of global DNA demethylation. Afterwards methylation marks are set in a gender-specific manner and the cycle of life can start again.

Lane *et al.*, 2003; Kim *et al.*, 2004), although a global sequence-specific methyl-ation analysis for cell types that are characterized by demethylation process is still lacking. Moreover, sequence-specific methylation analysis revealed differences between both waves of demethylation, especially for imprinted genes. While during germ cell generation the inherited methylation marks at imprinted genes are erased, so that they can be set to be gender-specific, in preimplantation development after fertilization, the marks at imprinted loci are preserved (Reik and Walter, 2001). During germ cell generation the demethylation starts at 8 days post conception (dpc) and continues up to 12.5–13.5 dpc (Hajkova *et al.*, 2002; Lane *et al.*, 2003; Seki *et al.*, 2005). In preimplantation development, the demethylation wave starts directly after fertilization and continues up to the morula stage. During the morula stage cells acquire *de novo* methylation and start to differentiate to trophectoderm, which forms a part of the placenta, and the more methylated inner cell mass, which gives rise to the embryo (Santos *et al.*, 2002). In germ cell development the remethylation starts at 14.5 dpc, but it is delayed for imprints and not that extensive in the female.

Nevertheless, the most remarkable DNA demethylation process can be observed in very early preimplantation development, in the mouse zygote, shortly after fertilization.

7.3   DNA demethylation in the mouse zygote

The life of a mammal starts with the fusion of an oocyte and a sperm. Both cells are highly specialized. The metaphase II arrested oocyte has a large amount of cyto-plasm, containing probably all the factors that are needed for early post-fertilization events. The sperm cell is very small and the DNA is tightly packed with protamines. Still some minor amounts of histones are left in sperm (about 1% in mouse and up to 10% in human) (Brykczynska *et al.*, 2010). These nucleo-somal structures are retained at regions of developmental importance (Hammoud *et al.*, 2009). Also methylation patterns are very dissimilar between the two gametes. Compared to oocyte DNA, sperm DNA is hypermethylated (Monk *et al.*, 1987), but still less methylated than DNA in differentiated tissues (Gama-Sosa *et al.*, 1983). Especially pronounced in sperm is the hypomethylation at promoters of genes important for embryonic development (Hammoud *et al.*, 2009).

Directly after fertilization, protamines have to be replaced by histones and the maternal DNA has to accomplish meiosis. Subsequently, the maternal and pater-nal pronuclei are formed and during further development, they expand until they fuse at syngamy phase. During this development, a rapid loss of DNA methylation immunofluorescence signal is remarkably observed exclusively in the paternal pronucleus (Mayer *et al.*, 2000; Santos *et al.*, 2002). Bisulfite analysis on one-cell embryos confirmed this observation for several repetitive elements (LINE-1, ETn, major satellites), single-copy genes (e.g., *Igf2*), and foreign DNA insertions (Oswald *et al.*, 2000; Lane *et al.*, 2003; Kim *et al.*, 2004; Wossidlo *et al.*, 2010). However, besides the maternal DNA, the methylation level for intracisternal A particles (IAPs) (LTR retrotransposon) and imprinted loci (H19) is preserved (Warnecke *et al.*, 1998; Lane *et al.*, 2003). This points to a directed demethylation or a specific protection against it. One of the factors involved in the demethylation protection is PGC7/Stella (Nakamura *et al.*, 2007). The depletion or mislocation of PGC7/Stella from zygotic pronuclei leads to demethylation of the maternal pronucleus and hypomethylation of maternal imprints (Nakamura *et al.*, 2007).

DNA demethylation of the paternal pronucleus is not unique to mouse zygotes, but can also be observed in rat, pig, bovine, sheep, rabbit, and human zygotes, although it is not so clearly pronounced as in mouse (Dean *et al.*, 2001; Fulka *et al.*, 2004; Zaitseva *et al.*, 2007; Hou *et al.*, 2008; Lepikhov *et al.*, 2008).

This fast, conserved amongst many mammals, temporally and locally very precisely confined demethylation has been the subject for numerous DNA demethylation studies, although the amount of material available for the analysis is very limited if mammalian zygotes are studied. Nevertheless, for immunofluorescence analysis, using the zygote is also an advantage. In fact, both processes, demethylation and the maintenance of methylation (in the paternal and maternal pronucleus, respectively), occur within one cell, but are locally compartmentalized. Hence, by manipulating zygotic development, it is easy to compare the influence on the methylation or demethylation by comparing the signals in both pronuclei.

7.4 Passive DNA demethylation

There are two possibilities for how the DNA demethylation can occur:

(1) Passive demethylation – the reduction of methylation after DNA replication in the absence of maintenance methyltransferase activity. In this case, each round of replication would reduce methylation levels twofold due to the semiconservative nature of DNA replication.

(2) Active demethylation – a process which involves various enzymatic activities and is replication independent, i.e., is not caused by mechanistic dilution of 5mC through the replication cycles (Figure 7.2A).

Passive demethylation would be the most obvious and easiest way to generate global DNA demethylation waves during development. In that way, the modification is lost by a missing maintenance methylation over several cell divisions. Indeed, in preimplantation development and germ cell generation, demethylation includes several replications, and furthermore earlier studies have shown an absence of DNMT1o, the oocyte-specific maintenance methyltransferase, from the nucleus during preimplantation development up to the 8-cell stage, where it transiently enters the nucleus (Ratnam *et al.*, 2002). However, more recent studies have revised this observation and shown that the somatic form of DNMT1, DNMT1s, is actually found in the nuclei of cleavage stage preimplantation embryos (Cirio *et al.*, 2008; Kurihara *et al.*, 2008). Furthermore, demethylation of several sequences methylated in sperm and

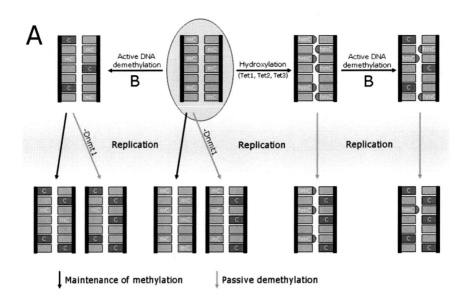

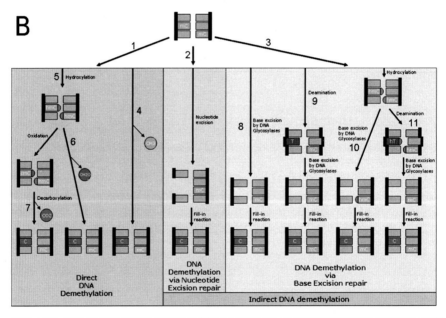

Figure 7.2 DNA demethylation pathways. (A) Demethylation of a methylated DNA molecule could include different pathways. It can be either an active mechanism (arrow left and right and 1B) or passive (red arrows) in the absence of Dnmt1 or passive due to a modification of the methylated cytosine on the parental strand (e.g., hydroxylation), which Dnmt1 cannot read as a hemi-methylated site. DNA demethylation might also include passive and active processes together. (B) Active demethylation can be performed directly (1) or indirectly, by nucleotide (2) or base (3) excision repair. Direct demethylation includes removal of the methyl group (3), or hydroxylation of the methyl group (4) followed by removal of the hydroxymethyl group (5) or further oxidation resulting in a final decarboxylation of the hydroxymethylgroup (6). The base excision repair (BER) pathway requires the recognition of the methylated cytosine by DNA glycosylases. Either DNA glycosylases directly recognize methylcytosine (7), or DNA deaminases could deaminate 5mC to T (8) and then the T–G mismatch is recognized and repaired. An alternative BER-mediated demethylation pathway includes the hydroxylation of 5mC, and a recognition of 5hmC by a DNA glycosylase (9) or a deamination (10) followed by BER. See plate section for color version.

oocytes only arises from the 8-cell stage onwards (Kafri *et al.*, 1992). Also, germ cell generation during DNMT1 can be found in the nuclei by immunofluorescence, and there have been up till now no arguments for passive methylation in those cells (Hajkova *et al.*, 2002). However, an argument for passive demethylation in preimplantation development is the observation, that at the 4-cell stage, inhibition of the replication prevents the DNA from further demethylation (Howlett and Reik, 1991).

In the zygote, where demethylation can be observed within one cell cycle, the 5mC immunofluorescence signal visibly decreases in the paternal pronucleus before replication starts (Wossidlo *et al.*, 2010). But more precise bisulfite methylation analysis in relation to the replication timing did not show what was expected from the immunofluorescence signal. LINE-1 elements show an unexpectedly mild decrease of methylation marks before replication. Further demethylation to the level maintained through the rest of preimplantation development was accomplished only after replication (Wossidlo *et al.*, 2010). Nevertheless, the mosaic methylation pattern and absence of fully methylated sequences in bisulfite sequencing analysis in late zygotes for LINE-1 elements points towards an active demethylation process. Also the observation that *Igf2* is demethylated in presence of a replication blocker shows that replication is probably not directly involved or just a part of the demethylation process in the zygote (Oswald *et al.*, 2000; Lane *et al.*, 2003; Wossidlo *et al.*, 2010).

However, the modification of 5mC to 5hmC by Tet enzymes could, besides the depletion of DNMT1, also lead to a passive demethylation. At least in vitro, DNMT1 is not able to recognize hemihydroxymethylated CpG sites, which occur after replication, and therefore the newly synthesized strand will not contain 5hmC or 5mC (Valinluck and Sowers, 2007) (Figure 7.2A).

7.5   Active DNA demethylation mechanisms

In mammals, active DNA demethylation is observed in many different cell types – for example in the immune system (Bruniquel and Schwartz, 2003), but, except for the two demethylation waves during development, it is not regarded as global – it just affects specific genomic regions.

Active DNA demethylation can be achieved in two different ways: either by removing the methyl group from cytosine (direct demethylation) or by removing the whole base or nucleotide (indirect demethylation) (Morgan *et al.*, 2005) (Figure 7.2B).

The ability to perform the direct demethylation was shown for MBD2, but these experiments could not be reproduced by others, which makes such possibility and the consideration of the cleavage of very stable C–C bonds rather doubtful (Bhattacharya *et al.*, 1999; Ng *et al.*, 1999). However, 5hmC, produced from 5mC by hydroxylation catalyzed by Tet enzymes, can potentially serve as an intermediate in the demethylation process, which provides more pathways for direct demethylation. Either the hydroxymethyl group can be removed directly, e.g., by DNMTs – this has been shown to occur in vitro (Liutkeviciute *et al.*, 2009), or further hydroxylation and decarboxylation steps might lead to an unmodified cytosine.

Indirect demethylation pathways, where the methylated base or nucleotide is removed, involve DNA repair mechanisms. In the nucleotide excision repair (NER) pathway several nucleotides are cut out and replaced, while in the base excision repair (BER) pathway just the base is replaced.

Nucleotide excision repair is a very versatile repair pathway, which can withdraw defective DNA caused by chemicals or ultraviolet light irradiation. In the whole repair process more than 30 proteins are involved. In short, at the defective sequence 24–32 nucleotides are released, resulting in a single-strand gap, which has to be filled by DNA polymerases (Liu *et al.*, 2010). In this fill-in reaction, previously methylated cytosines would be replaced by unmethylated cytosines. This pathway has been proposed as a DNA demethylation pathway, because the knockdown of Gadd45α (growth arrest and DNA-damage-inducible protein 45 alpha) was found to increase overall methylation levels in cell culture experiments. Furthermore, Gadd45α overexpression can lead to T-cell autoreactivity and excessive B-cell stimulation by inducing DNA hypomethylation (Barreto *et al.*, 2007; Li *et al.*, 2010). But, in contrast to that, other research groups were unable to reproduce the DNA demethylation activity of Gadd45a, and also the study of Gadd45a knockout mice did not reveal any differences in methylation level (Jin *et al.*, 2008; Engel *et al.*, 2009).

Base excision repair handles small chemical alterations of DNA bases, which might include methylation of cytosine or further modifications of 5mC. This pathway involves four to five steps starting with base excision by a DNA glycosylase, specific for the damaged base. In the following steps the arising gap in the DNA strand is prepared for DNA polymerase synthesis and in the end the nick is sealed by a DNA ligase (Hegde *et al.*, 2008). Besides mammals, DNA demethylation can also be found in plants and here the involvement of DNA glycosylases in this process has been shown (Feng *et al.*, 2010).

7.6   Base excision repair in the zygote

Recent papers have provided clues for the involvement of the BER pathway in active DNA demethylation in the zygote. Wossidlo *et al.* have shown the appearance of γH2A.X foci starting at the time point of active DNA demethylation and these foci are associated with PARP1 (poly ADP-ribose polymerase) (Wossidlo *et al.*, 2010). Here, γH2A.X is a modification of a histone variant, which has been shown to be recruited to DNA strand breaks. DNA single-strand breaks arise in BER after cleavage of the base with DNA glycosylases and following endonuclease activity. Furthermore, Hajkova *et al.* have also shown the appearance of chromatin-bound XRCC1 (x-ray repair complementing defective repair in Chinese hamster cells 1) in the paternal pronucleus (Hajkova *et al.*, 2010). PARP1 and XRCC1 are both synergistically involved in BER and serve as single-strand break sensor proteins (Hegde *et al.*, 2008). Interestingly, also during replication, single-strand breaks are more prominent in the paternal pronucleus (Wossidlo *et al.*, 2010).

7.7   Initiation of base excision repair

Having these results and also the downstream process of the DNA demethylation in mind, the remaining puzzle is which enzyme or combination of enzymes is involved in initiating the BER pathway for DNA demethylation. In the easiest scenario, 5mC is directly recognized by a 5mC-specific DNA glycosylase. The existence of such enzymes in plants has actually been shown. Here, the DNA glycosylases Demeter (DME), Repressor of silencing 1 (ROS1) and Demeter-Like 1 and 2 (DML1, DML2) fulfill the removal of 5mC. Of these, DME is necessary for global DNA demethylation and the demethylation of genomic imprints in endosperm (Hsieh *et al.*, 2009), while ROS1, DML1, and DML2 are involved in regulation of the methylation pattern of genic regions (Penterman *et al.*, 2007). In animals, no homologs for these enzymes have been found. But in vitro experiments show that the human and chicken DNA glycosylase MBD4 and the thymine DNA glycosylase TDG, in specific conditions, are able to demethylate 5mC (Zhu *et al.*, 2000a, 2000b). However, MBD4 knockout mice still show the demethylation of the paternal pronucleus in the zygote (Santos and Dean, 2004). Perhaps MBD4 could be involved in more specialized DNA demethylation processes (Kim *et al.*, 2009).

Instead of direct 5mC recognition, this nucleotide could also be further modified, for example deaminated or hydroxylated. Deamination of 5mC results in thymine, which will be replaced by unmethylated cytosine by T:G mismatch repair during BER. There are two enzyme families that have been discussed as being possibly involved in deaminating 5mC: DNA deaminases or DNMTs. For the DNA deaminase activation-induced deaminase (AID) influence on the methylation level in primordial germ cells was shown, and AID knockout mice reveal higher DNA methylation (Popp *et al.*, 2010), suggesting a defective demethylation process. But AID-deficient mice are viable without demonstrating any phenotypic defects, pointing to the fact that AID is probably not the only player in demethylation. Further, DNMT3A and DNMT3B were shown to deaminate 5mC in the absence of S-adenosylmethionine (SAM), the donor of methyl groups for DNMTs (Metivier *et al.*, 2008). So far it is not known if SAM concentration is locally reduced in the paternal pronucleus. Since Elp3, a component of the elongator complex, is needed for the demethylation of the paternal pronucleus (Okada *et al.*, 2010), it could influence SAM levels through its radical SAM domain activity. This could be deduced from the fact that an overexpression of Elp3 with a mutated SAM domain hinders the paternal DNA demethylation (Okada *et al.*, 2010).

Hydroxylation of 5mC, which is also proposed as an intermediate in the direct DNA demethylation or passive demethylation, could also be an intermediate in the active DNA demethylation mediated by BER. Alternatively, 5hmC could be recognized by a specific DNA glycosylase. Such a DNA glycosylase was long ago reported to be found in calf thymus (Cannon *et al.*, 1988). Instead of direct removal of 5hmC, this base could also be first deaminated, resulting in 5-hydroxymethyluracil (5hmU) (or 5hT) and then subsequently recognized by a DNA glycosylase. There are DNA glycosylases in mammalian tissues, specific for such oxidized bases, e.g., the DNA glycosylases NEIL and SMUG1. An hypothesis from Hegde *et al.* proposed that whereas SMUG1 cleaves hydroxymethylated bases in non-cycling genomes, NEIL1 acts during replication on single-stranded DNA or DNA bubbles (see review about BER: Hegde *et al.*, 2008). Such a replication association of NEIL1 would be very interesting, because of the observation that occurrence of DNA single-strand breaks during replication in the paternal pronucleus is much higher than in the maternal. It probably means that there are more repair processes going on in the paternal pronucleus. Moreover, the demethylation of LINE-1 was shown to proceed through replication (Wossidlo *et al.*, 2010).

7.8   Conclusion

The life of a mammal starts with the fertilization of an oocyte by a sperm. In the early hours of life, the parental genomes of highly specialized germ cells get reprogrammed in order to comply with the totipotency state of the resulting zygote, which later on upon further development gives rise to all the cells of a multicellular organism. One of the striking events that happens in the developing zygote is DNA demethylation of the paternal genome. Although the mechanism of DNA demethylation is still not understood, many facts very strongly point to an active demethylation process mediated by BER. Nevertheless, the scenario of passive demethylation has also to be considered. And, potentially, both act in concert.

The biological significance of DNA methylation for embryo development was shown by knocking out DNMTs (Li *et al.*, 1992; Okano *et al.*, 1999). The significance of the reverse process – global DNA demethylation – is still controversial. The complication arises from the fact that the enzymes involved in global DNA demethylation are not known yet; therefore it is not feasible at the moment to abolish DNA demethylation by switching off the key players in the same way as it was done for DNMTs. Furthermore, there is also the possibility that there are many enzymes involved in the process, with one enzyme (or pathway) being able to compensate the loss of another. The latter was corroborated by Popp *et al.*, who found an influence of AID knockout on global DNA demethylation in primordial germ cells. The depletion of AID still does not completely prevent demethylation, suggesting that other enzymes are also involved in the process (Popp *et al.*, 2010) or points towards the existence of several back-up pathways to ensure proper epigenetic reprogramming. However, at least reprogramming of specific sequences is crucial for life, since the defects in reprogramming of imprints in primordial germ cells lead to severe complications (McGrath and Solter, 1984). Similarly, activation of pluripotency genes in preimplantation development is crucial for normal development.

To better understand the process of global DNA demethylation, the epigenetic role of DNA hydroxymethylation has to be clarified. The recent discovery of the importance of Tet1 for ES cells to express the pluripotency gene *Nanog* (Ito *et al.*, 2010) makes it very likely that 5hmC is an intermediate for active (and/or passive) demethylation in the two global demethylation phases during the development of a mammal.

Solving the enigma of the DNA demethylation process might give hints for the improvement of in vitro reprogramming, which is so far very inefficient.

However, DNA demethylation is not the only player in epigenetic reprogramming – histone modifications, histone variants, and small RNAs also contribute in leading animals through proper development. So far, the cooperation between different epigenetic mechanisms is not clear and requires further investigations. Better understanding of epigenetic reprogramming would help to improve the efficiency of therapeutic cloning and induced pluripotent stem cell (iPS) technologies, which have great potential in fighting against many diseases.

REFERENCES

Barreto, G., Schafer, A., Marhold, J., et al. (2007). Gadd45a promotes epigenetic gene activation by repair-mediated DNA demethylation. Nature, **445**, 671–675.

Bhattacharya, S. K., Ramchandani, S., Cervoni, N., et al. (1999). A mammalian protein with specific demethylase activity for mCpG DNA. Nature, **397**, 579–583.

Bird, A. (1986). CpG-rich islands and the function of DNA methylation. Nature, **321**, 209–213.

Bird, A., Taggart, M., Frommer, M., et al. (1985). A fraction of the mouse genome that is derived from islands of nonmethylated, CpG-rich DNA. Cell, **40**, 91–99.

Bourc'his, D. and Bestor, T. H. (2004). Dnmt3L and the establishment of maternal genomic imprints. Nature, **431**, 96–99.

Bruniquel, D. and Schwartz, R. H. (2003). Selective, stable demethylation of the interleukin-2 gene enhances transcription by an active process. Nature Immunology, **4**, 235–240.

Brykczynska, U., Hisano, M., Erkek, S., et al. (2010). Repressive and active histone methylation mark distinct promoters in human and mouse spermatozoa. Nature Structural and Molecular Biology, **17**, 679–687.

Cannon, S. V., Cummings, A., and Teebor, G. W. (1988). 5-Hydroxymethylcytosine DNA glycosylase activity in mammalian tissue. Biochemical and Biophysical Research Communications, **151**, 1173–1179.

Cirio, M. C., Ratnam, S., Ding, F., et al. (2008). Preimplantation expression of the somatic form of Dnmt1 suggests a role in the inheritance of genomic imprints. BMC Developmental Biology, **8**, 9.

Dean, W., Santos, F., Stojkovic, M., et al. (2001). Conservation of methylation reprogramming in mammalian development: aberrant reprogramming in cloned embryos. Proceedings of the National Academy of Sciences USA, **98**, 13 734–13 738.

Ehrlich, M., Gama-Sosa, M. A., Huang, L. H., et al. (1982). Amount and distribution of 5-methylcytosine in human DNA from different types of tissues and cells. Nucleic Acids Research, **10**, 2709–2721.

Engel, N., Tront, J. S., Erinle, T., et al. (2009). Conserved DNA methylation in Gadd45a(–/–) mice. Epigenetics, **4**, 98–99.

Feng, S., Jacobsen, S. E., and Reik, W. (2010). Epigenetic reprogramming in plant and animal development. Science, **330**, 622–627.

Finnegan, E. J. and Kovac, K. A. (2000). Plant DNA methyltransferases. Plant Molecular Biology, **43**, 189–201.

Fulka, H., Mrazek, M., Tepla, O., et al. (2004). DNA methylation pattern in human zygotes and developing embryos. Reproduction, **128**, 703–708.

Gama-Sosa, M. A., Midgett, R. M., Slagel, V. A., et al. (1983). Tissue-specific differences in DNA methylation in various mammals. Biochimica et Biophysica Acta, **740**, 212–219.

Hajkova, P., Erhardt, S., Lane, N., et al. (2002). Epigenetic reprogramming in mouse primordial germ cells. Mechanisms of Development, **117**, 15–23.

Hajkova, P., Jeffries, S. J., Lee, C., et al. (2010). Genome-wide reprogramming in the mouse germ line entails the base excision repair pathway. Science, **329**, 78–82.

Hammoud, S. S., Nix, D. A., Zhang, H., et al. (2009). Distinctive chromatin in human sperm packages genes for embryo development. Nature, **460**, 473–478.

Hegde, M. L., Hazra, T. K., and Mitra, S. (2008). Early steps in the DNA base excision/single-strand interruption repair pathway in mammalian cells. Cell Research, **18**, 27–47.

Hou, J., Liu, L., Zhang, J., *et al.* (2008). Epigenetic modification of histone 3 at lysine 9 in sheep zygotes and its relationship with DNA methylation. *BMC Developmental Biology*, **8**, 60.

Howlett, S. K. and Reik, W. (1991). Methylation levels of maternal and paternal genomes during preimplantation development. *Development*, **113**, 119–127.

Hsieh, T. F., Ibarra, C. A., Silva, P., *et al.* (2009). Genome-wide demethylation of *Arabidopsis* endosperm. *Science*, **324**, 1451–1454.

Ito, S., D'Alessio, A. C., Taranova, O. V., *et al.* (2010). Role of Tet proteins in 5mC to 5hmC conversion, ES-cell self-renewal and inner cell mass specification. *Nature*, **466**, 1129–1133.

Jin, S. G., Guo, C., and Pfeifer, G. P. (2008). GADD45A does not promote DNA demethylation. *PLoS Genetics*, **4**, e1000013.

Kafri, T., Ariel, M., Brandeis, M., *et al.* (1992). Developmental pattern of gene-specific DNA methylation in the mouse embryo and germ line. *Genes and Development*, **6**, 705–714.

Kim, M. S., Kondo, T., Takada, I., *et al.* (2009). DNA demethylation in hormone-induced transcriptional derepression. *Nature*, **461**, 1007–1012.

Kim, S. H., Kang, Y. K., Koo, D. B., *et al.* (2004). Differential DNA methylation reprogramming of various repetitive sequences in mouse preimplantation embryos. *Biochemical and Biophysical Research Communications*, **324**, 58–63.

Kriaucionis, S. and Heintz, N. (2009). The nuclear DNA base 5-hydroxymethylcytosine is present in Purkinje neurons and the brain. *Science*, **324**, 929–930.

Kurihara, Y., Kawamura, Y., Uchijima, Y., *et al.* (2008). Maintenance of genomic methylation patterns during preimplantation development requires the somatic form of DNA methyltransferase 1. *Developmental Biology*, **313**, 335–346.

Lane, N., Dean, W., Erhardt, S., *et al.* (2003). Resistance of IAPs to methylation reprogramming may provide a mechanism for epigenetic inheritance in the mouse. *Genesis*, **35**, 88–93.

Lepikhov, K., Zakhartchenko, V., Hao, R., *et al.* (2008). Evidence for conserved DNA and histone H3 methylation reprogramming in mouse, bovine and rabbit zygotes. *Epigenetics and Chromatin*, **1**, 8.

Li, E., Bestor, T. H., and Jaenisch, R. (1992). Targeted mutation of the DNA methyltransferase gene results in embryonic lethality. *Cell*, **69**, 915–926.

Li, Y., Zhao, M., Yin, H., *et al.* (2010). Overexpression of the growth arrest and DNA damage-induced 45α gene contributes to autoimmunity by promoting DNA demethylation in lupus T cells. *Arthritis and Rheumatism*, **62**, 1438–1447.

Lister, R., Pelizzola, M., Dowen, R. H., *et al.* (2009). Human DNA methylomes at base resolution show widespread epigenomic differences. *Nature*, **462**, 315–322.

Liu, L., Lee, J., and Zhou, P. (2010). Navigating the nucleotide excision repair threshold. *Journal of Cell Physiology*, **224**, 585–589.

Liutkeviciute, Z., Lukinavicius, G., Masevicius, V., *et al.* (2009). Cytosine-5-methyltransferases add aldehydes to DNA. *Nature Chemical Biology*, **5**, 400–402.

Mayer, W., Niveleau, A., Walter, J., *et al.* (2000). Demethylation of the zygotic paternal genome. *Nature*, **403**, 501–502.

McGrath, J. and Solter, D. (1984). Completion of mouse embryogenesis requires both the maternal and paternal genomes. *Cell*, **37**, 179–183.

Metivier, R., Gallais, R., Tiffoche, C., *et al.* (2008). Cyclical DNA methylation of a transcriptionally active promoter. *Nature*, **452**, 45–50.

Monk, M., Boubelik, M., and Lehnert, S. (1987). Temporal and regional changes in DNA methylation in the embryonic, extraembryonic and germ cell lineages during mouse embryo development. *Development*, **99**, 371–382.

Morgan, H. D., Santos, F., Green, K., *et al.* (2005). Epigenetic reprogramming in mammals. *Human Molecular Genetics*, **14**, R47–58.

Nakamura, T., Arai, Y., Umehara, H., *et al.* (2007). PGC7/Stella protects against DNA demethylation in early embryogenesis. *Nature Cell Biology*, **9**, 64–71.

Ng, H. H., Zhang, Y., Hendrich, B., *et al.* (1999). MBD2 is a transcriptional repressor belonging to the MeCP1 histone deacetylase complex. *Nature Genetics*, **23**, 58–61.

Okada, Y., Yamagata, K., Hong, K., *et al.* (2010). A role for the elongator complex in zygotic paternal genome demethylation. *Nature*, **463**, 554–558.

Okano, M., Bell, D. W., Haber, D. A., *et al.* (1999). DNA methyltransferases Dnmt3a and Dnmt3b are essential for de novo methylation and mammalian development. *Cell*, **99**, 247–257.

Ooi, S. K., O'Donnell, A. H., and Bestor, T. H. (2009). Mammalian cytosine methylation at a glance. *Journal of Cell Science*, **122**, 2787–2791.

Oswald, J., Engemann, S., Lane, N., *et al.* (2000). Active demethylation of the paternal genome in the mouse zygote. *Current Biology*, **10**, 475–478.

Penterman, J., Zilberman, D., Huh, J. H., *et al.* (2007). DNA demethylation in the *Arabidopsis* genome. *Proceedings of the National Academy of Sciences USA*, **104**, 6752–6757.

Popp, C., Dean, W., Feng, S., *et al.* (2010). Genome-wide erasure of DNA methylation in mouse primordial germ cells is affected by AID deficiency. *Nature*, **463**, 1101–1105.

Ratnam, S., Mertineit, C., Ding, F., *et al.* (2002). Dynamics of Dnmt1 methyltransferase expression and intracellular localization during oogenesis and preimplantation development. *Developmental Biology*, **245**, 304–314.

Reik, W. and Walter, J. (2001). Genomic imprinting: parental influence on the genome. *Nature Reviews Genetics*, **2**, 21–32.

Santos, F. and Dean, W. (2004). Epigenetic reprogramming during early development in mammals. *Reproduction*, **127**, 643–651.

Santos, F., Hendrich, B., Reik, W., *et al.* (2002). Dynamic reprogramming of DNA methylation in the early mouse embryo. *Developmental Biology*, **241**, 172–182.

Sasaki, H. and Matsui, Y. (2008). Epigenetic events in mammalian germ-cell development: reprogramming and beyond. *Nature Reviews Genetics*, **9**, 129–140.

Saxonov, S., Berg, P., and Brutlag, D. L. (2006). A genome-wide analysis of CpG dinucleotides in the human genome distinguishes two distinct classes of promoters. *Proceedings of the National Academy of Sciences USA*, **103**, 1412–1417.

Seki, Y., Hayashi, K., Itoh, K., *et al.* (2005). Extensive and orderly reprogramming of genome-wide chromatin modifications associated with specification and early development of germ cells in mice. *Developmental Biology*, **278**, 440–458.

Szwagierczak, A., Bultmann, S., Schmidt, C. S., *et al.* (2010). Sensitive enzymatic quantification of 5-hydroxymethylcytosine in genomic DNA. *Nucleic Acids Research*, **38**, e181.

Tahiliani, M., Koh, K. P., Shen, Y., *et al.* (2009). Conversion of 5-methylcytosine to 5-hydroxymethylcytosine in mammalian DNA by MLL partner TET1. *Science*, **324**, 930–935.

Valinluck, V. and Sowers, L. C. (2007). Endogenous cytosine damage products alter the site selectivity of human DNA maintenance methyltransferase DNMT1. *Cancer Research*, **67**, 946–950.

Valinluck, V., Tsai, H. H., Rogstad, D. K., *et al.* (2004). Oxidative damage to methyl-CpG sequences inhibits the binding of the methyl-CpG binding domain (MBD) of methyl-CpG binding protein 2 (MeCP2). *Nucleic Acids Research*, **32**, 4100–4108.

Varriale, A. and Bernardi, G. (2010). Distribution of DNA methylation, CpGs, and CpG islands in human isochores. *Genomics*, **95**, 25–28.

Warnecke, P. M., Mann, J. R., Frommer, M., *et al.* (1998). Bisulfite sequencing in preimplantation embryos: DNA methylation profile of the upstream region of the mouse imprinted H19 gene. *Genomics*, **51**, 182–190.

Wossidlo, M., Arand, J., Sebastiano, V., *et al.* (2010). Dynamic link of DNA demethylation, DNA strand breaks and repair in mouse zygotes. *EMBO Journal*, **29**, 1877–1888.

Zaitseva, I., Zaitsev, S., Alenina, N., *et al.* (2007). Dynamics of DNA-demethylation in early mouse and rat embryos developed in vivo and in vitro. *Molecular Reproduction and Development*, **74**, 1255–1261.

Zhu, B., Zheng, Y., Angliker, H., *et al.* (2000a). 5-Methylcytosine DNA glycosylase activity is also present in the human MBD4 (G/T mismatch glycosylase) and in a related avian sequence. *Nucleic Acids Research*, **28**, 4157–4165.

Zhu, B., Zheng, Y., Hess, D., *et al.* (2000b). 5-Methylcytosine-DNA glycosylase activity is present in a cloned G/T mismatch DNA glycosylase associated with the chicken embryo DNA demethylation complex. *Proceedings of the National Academy of Sciences USA*, **97**, 5135–5139.

8 Histone modifications of lineage-specific genes in human embryonic stem cells during in vitro differentiation

Hyemin Kim, Hogyu Seo, Daeyoup Lee, and Yong-Mahn Han*

8.1 Introduction

In the nuclei of all eukaryotic cells, there exists genomic DNA which is highly folded, constrained, and compacted by histone and non-histone proteins in a dynamic polymer called chromatin. As a physiological template of genetic information, chromatin is subject to a diverse array of posttranslational modifications. The modifications largely take place on histone N-termini, regulating the access to the underlying DNA. Histone proteins and their associated covalent modifications can alter chromatin structure, and determine how and when the DNA packaged in the nucleosomes is accessed, leading to the Histone Code Hypothesis. For example, trimethylation of Histone 3 at lysine 9 (H3K9me3) attracts the heterochromatin-specific protein HP1, which induces a spreading wave of further H3K9me3 followed by further HP1 binding (Bannister *et al.*, 2001).

In the context of human embryonic stem cells (hESCs), the histone modifications are also found to be essential in the regulation of gene expression or repression, in association with the combination of histone modification, or histone code. In ESCs, alike somatic cells, trimethylation of H3 at lysine 4 (H3K4me3) and acetylation of H3 at lysine 9 (H3K9ac) are found in the promoter regions of actively transcribed genes on nucleosomes (Bernstein *et al.*, 2002; Santos-Rosa *et al.*, 2002; Liang *et al.*, 2004; Pokholok *et al.*, 2005). Dimethylation of H3 at lysine 36 (H3K36me2) is also a candidate of transcription activation (Rao *et al.*, 2005). On the other side, repressive histone marks such as trimethylation of H3 at lysine 27 (H3K27me3) and trimethylation of H3 at lysine 9 (H3K9me3) are responsible for transcriptional repression at promoter regions (Meshorer *et al.*, 2006; Wen *et al.*, 2009; Hawkins *et al.*, 2010). Whole-genome

---

* Author to whom correspondence should be addressed.

Epigenomics: From Chromatin Biology to Therapeutics, ed. K. Appasani. Published by Cambridge University Press. © Cambridge University Press 2012.

mapping for H3K4me3 and H3K27me3 shows that ES marker genes are changed from a chromatin state modified by H3K4me3 alone to the presence of both modifications during hESC differentiation after treatment with retinoic acid (RA) (Pan *et al.*, 2007). This study presents data suggesting that silencing of ES marker genes in hESC derivatives is predominantly governed by repressive histone marks.

In ESCs, H3K4me3 and H3K27me3, named as bivalent domains, are known to play a crucial role in maintaining pluripotency (Azuara *et al.*, 2006; Bernstein *et al.*, 2006; Mikkelsen *et al.*, 2007). The bivalent domains are also known to be essential for differentiation of hESCs to neural stem cells as well as β-III tubulin-positive neurons (Golebiewska *et al.*, 2009). In comparison with lineage-committed cells, ES cells exhibit an open and decondensed chromatin that is more transcriptionally accessible, with a high level of histone acetylation and H3K4 methylation to sustain pluripotency (Azuara *et al.*, 2006; Meshorer *et al.*, 2006). Recent works show that lineage-committed cells are characterized by significantly expanded repressive chromatin domains which selectively affect pluripotency and developmental genes (Wen *et al.*, 2009; Hawkins *et al.*, 2010), and this supports the observation of the increase in transcriptionally inactive heterochromatin regions with H3K4 and H3K36 methylation in differentiating ES cells (Meshorer *et al.*, 2006; Wen *et al.*, 2009; Hawkins *et al.*, 2010). Also, the epigenetic marks are modified in the promoters of ES marker genes such as *OCT4*, *SOX2*, and *NANOG* to regulate the gene expression. These transcription factors play essential roles in the determination and maintenance of the pluripotency. For example, loss of OCT4 directs early embryos into the trophectoderm lineage without pluripotent inner cell mass (Nichols *et al.*, 1998) whereas overexpression of OCT4 results in differentiation towards the extra-embryonic endoderm and mesoderm lineages (Niwa *et al.*, 2000). Therefore, it can be concluded that the epigenetic controls exert a significant influence in the regulation of the genes necessary for maintaining the pluripotency, and thereby mastermind the fate of ES cells in essence.

Here, we investigated epigenetic behaviors of lineage-specific genes in terms of the histone modifications by comparison with transcriptional expression profiles in hESCs during differentiation to hepatocytes. Our findings demonstrate that temporal expression of lineage-specific genes is modulated by unique epigenetic modifications during differentiation of hESCs into a specialized cell type.

8.2 Methodology

8.2.1 Differentiation of hESCs into hepatocytes

Human ESCs (CHA-hES4 cell line) were cultured in DMEM/F12 medium (Invitrogen, Carlsbad, CA, USA) containing 20% knockout serum replacement (Invitrogen), 1% non-essential amino acids (Invitrogen), 0.1 mM β-mercaptoethanol (Sigma-Aldrich, St. Louis, MO, USA), 100 U/ml penicillin–streptomycin (Invitrogen), and 4 ng/ml basic fibroblast growth factor (bFGF) (Invitrogen) on

mitomycin C (Sigma-Aldrich)-treated mouse embryonic fibroblast (MEF) feeders at
37 °C, 5% CO_2 in air. Differentiation of hESCs into hepatocytes was carried out as
previously described (Cai *et al.*, 2007). Briefly, hESCs were cultured to confluency in
MEF-conditioned media (CM) using a feeder-free system for 3 days. Then, hESCs
were incubated in RPMI-1640 (Hyclone, Logan, UT, USA) containing 0.5 mg/ml
albumin fraction V (Sigma-Aldrich) and 50 ng/ml Activin A (Peprotech, Rocky
Hill, NJ, USA) for 1 day, and further cultured in the same RPMI medium supplemented with 1% insulin–transferrin–selenium (ITS, Sigma-Aldrich) for 4 days. After
treatment with Activin A, the differentiated cells were cultured in hepatocyte culture medium (HCM, Lonza, Baltimore, MD, USA) containing 30 ng/ml fibroblast
growth factor 4 (FGF4, Peprotech) and 20 ng/ml bone morphogenetic protein 2
(BMP2, Peprotech) for 5 days, and further cultured in HCM supplemented 20 ng/ml
hepatocyte growth factor (HGF, Peprotech) for 5 days. Maturation of hESC-derived
hepatocytes was induced by culturing in HCM supplemented with 10 ng/ml oncostatin M (R&D Systems, Minneapolis, MN, USA) and 0.1 μM dexamethasone (Sigma-
Aldrich) for 5 days. The culture media were changed daily.

8.2.2   Real-time reverse transcription

Total RNA was isolated from cells using TRIzol Reagent (Invitrogen) and reverse-
transcribed using M-MLV Reverse Transcriptase (Enzynomics, Seoul, Korea)
according to the manufacturer's protocol. The primers used in this study are listed
in Table 8.1. The expression level of respective genes was measured by real-time
reverse transcription (RT) PCR using 2 × Prime Q-Master Mix (GENET BIO, Seoul,
Korea). Relative expression level was analyzed by using an iCycler iQ5 Real-Time
detection system (Bio-Rad Laboratories, Hercules, CA, USA). The reaction parameters for real-time RT-PCR analysis were 95 °C for 10 minutes followed by 40 cycles
of 95 °C for 30 seconds, 60 °C for 30 seconds, and 72 °C for 30 seconds, and a final
elongation step at 72 °C for 5 minutes. All reactions were repeated more than two
times. For comparative quantification, the expression levels of genes of interest
were normalized to that of *GAPDH* and expressed as fold-change relative to the
expression level of undifferentiated hESCs. The sample ΔCt (SΔCt) value was
calculated as the difference between the Ct values of *GAPDH* and the target. The
ΔCt value of undifferentiated hESCs was used as a control ΔCt (CΔCt) value. The
relative gene expression levels between the sample and the control were determined using the formula $2^{-(SΔCt-CΔCt)}$.

8.2.3   Immunofluorescence analysis

Cells were fixed in 4% formaldehyde (Sigma-Aldrich) for 30 minutes at room
temperature, rinsed three times in PBS containing 0.1% Tween 20 (PBST) for
10 minutes, permeabilized in PBS containing 0.1% Triton X-100 (Sigma-Aldrich)
for 15 minutes, and blocked for 1 hour in PBS containing 5% FBS (Hyclone). Cells
were incubated with primary antibodies diluted in PBS containing 5% FBS overnight at 4 °C and rinsed six times in PBST. The primary antibodies against human
OCT4, SOX2 (Santa Cruz Biotechnology, Santa Cruz, CA, USA), and SOX17 (R&D
Systems) were diluted 1 : 200, and antibodies against human α-fetoprotein (AFP,

Table 8.1 Primers used for RT-PCR analysis of specific marker genes in the differentiated cells and detection of genomic DNA in chromatin immunopreciptation samples (ChIP)

Gene	Real-time PCR primer sequence	Chromatin immunoprecipitation	
		Primer sequence	Genomic region
OCT4	F: tcggggtgggagagcaact	F: accattgccaccaccattag	−140 ~ −36
	R: gggtgatcctcttctgcttc	R: ccactagccttgacctctgg	
SOX2	F: accagctcgcagacctacat	F: ctgcgagaggggatacaaag	−243 ~ −98
	R: tggagtgggaggaagaggta	R: cgggtttttgcatgaaagg	
NANOG	F: tgatttgtgggcctgaaga	F: ggttctgttgctcggttttc	−174 ~ −49
	R: gttgtttgcctttgggactg	R: actgacccacccttgtgaat	
SOX17	F: cagaatccagacctgcacaa	F: gaatggacgctcggtatgtt	−331 ~ −239
	R: gcggccggtacttgtagtt	R: gagactcgaaaagccgtctg	
GATA4	F: tccaaaccagaaaacggaag	F: gagaaatattggaagcgccttt	−208 ~ −62
	R: ctgtgcccgtagtgagatga	R: acaagggttggagaatgtgc	
FOXA2	F: ctgagcgagatctaccagtgga	F: gagcctccacatccaaacac	−366 ~ −267
	R: agtcgttgaaggagagcgagt	R: cagcagctcttgggttcaa	
CXCR4	F: ggtggtctatgttggcgtct	F: ggagaaccagcggttaccat	−319 ~ −208
	R: tggagtgtgacagcttggag	R: ggctgcgctctaagttcaa	
ALB	F: cgattggtgagaccagaggt	F: cattgacaaggtcttgtggaga	−163 ~ −53
	R: tggagactggcacacttgag	R: gctgccaaccgattacaaa	
HNF4A	F: cgagcagatccagttcatca	F: aacccaggttggactctcac	−382 ~ −289
	R: tcacacatctgtccgttgct	R: aacccagagccaggtgtatg	
AFP	F: agcttggtggtggatgaa	F: ccgctatgctgttaattattgg	−137 ~ −35
	R: gcaatgcctgttaactagtaacctt	R: tctgcaatgacagcctcaag	
TDO2	F: caaatcctctgggagttgga	F: tggcctctgttgattcattg	−253 ~ −164
	R: gtccaaggctgtcatcgtct	R: gtgagtgatctgccaaattgag	
GAPDH	F: gaaggtgaaggtcggagtc	F: cggctactagcggttttacg	−122 ~ +67
	R: gaagatggtgatgggatttc	R: aagaagatgcggctgactgt	

Dako, Glostrup, DK, Denmark) and α-1-antitrypsin (AAT, Invitrogen) were diluted 1:100. Alexa Fluor 488- or 594-conjugated secondary antibody (Invitrogen) diluted 1:200 was incubated at room temperature for 1 hour and washed six times in PBST. Then, cells were mounted in VECTASHIELD Mounting Medium containing DAPI (4'-6-diamidino-2-phenylindole) (Vector Laboratories, Burlingame, CA, USA).

8.2.4 Chromatin immunoprecipitation analysis

The chromatin immunoprecipitation (ChIP) assay was performed as previously described (Park *et al.*, 2007). Briefly, approximately 1×10^6 cells were incubated in cell culture medium containing 1.2% formaldehyde at 25 °C for 10 minutes, and quenched by the addition of 0.123 M glycine for 5 minutes at 25 °C. Cells were harvested by scraping, washed twice in phosphate-buffered saline, and three times in ChIP lysis buffer, and resuspended in 200 µl of ChIP lysis buffer containing high salt. Cross-linked chromatin was fragmented by sonication, and pre-cleared with protein A/G PLUS-agarose (Santa Cruz Biotechnology) at 4 °C for 30 minutes. Each primary antibody was incubated overnight with chromatin at 4 °C.

Antibodies (Millipore-Upstate, Temecula, CA, USA) used for the ChIP assay were as follows: normal rabbit IgG (#12–370), anti-trimethyl-H3K4 (#04–745), anti-trimethyl-H3K27 (#07–449), anti-dimethyl-H3K9 (#07–441), anti-trimethyl-H3K9 (#07–442), and anti-acetyl-H3K9 (#06–942). Immunocomplexes were harvested by incubation with protein A/G PLUS-agarose for 2 hours at 4 °C. Immunoprecipitates were washed twice with lysis buffer containing high salt, and rinsed four times with wash buffer. Samples were resuspended in elution buffer and incubated at 67 °C overnight. DNA samples were isolated using phenol/chloroform extraction, precipitated with ethanol, and resuspended in 50 μl of TE buffer. Quantitative PCR was carried out on an iQ5 Real-Time PCR Detection System (Bio-Rad Laboratories). The primer sequences and genomic regions used for PCR are shown in Table 8.1.

8.3   Results and discussion

8.3.1   Differentiation of hESCs to hepatocytes in vitro

Human ESCs can be employed as good materials for exploring genetic and epigenetic mechanisms of human developmental processes because they can differentiate into all cell types. Differentiation of hESCs into a specialized cell type should require dramatic changes of epigenetic states for expression of lineage-determining genes. Here, we investigated epigenetic marks such as histone modifications at the promoters of lineage-specific genes in hESCs during differentiation into hepatocytes in vitro (Figure 8.1A). CHA-hES4 cells specifically expressed ES markers such as OCT4 and SOX2 (Figure 8.1A). hESCs were induced into definitive endoderm (DE) cells by treatment with Activin A for 5 days under serum-free conditions. Activin A-treated cells were stained positively for DE markers SOX17 and GATA4 (Figure 8.1A). hESC-derived DE cells were further differentiated to hepatocytes as described in the section on 'Materials and methods'. The resultant hepatocytes expressed AAT and AFP in the cytoplasm (Figure 8.1.A) and secreted albumin into the culture medium (Figure 8.1B). They were also able to store glycogen and take up acetylated-LDL in the cytoplasm (Figure 8.1B). These results demonstrate that hESC-derived hepatocytes have the cellular and molecular characteristics of adult liver cells.

8.3.2   Histone modifications of ES marker genes in hESCs during differentiation to hepatocytes

This study was carried out to understand epigenetic behaviors such as histone modifications during differentiation of hESCs to hepatocytes. In hESCs, core transcriptional factors such as OCT4, SOX2, and NANOG are essential for maintaining pluripotency and self-renewal by forming a regulatory circuitry that is composed of autoregulatory and feedforward loops (Boyer *et al.*, 2005). Transcription of ES marker genes was enhanced in hESCs and significantly decreased during differentiation into hepatocytes (Figure 8.2A). To detect epigenetic modifications during differentiation of hESCs to hepatocytes, histone modifications were analyzed by chromatin immunoprecipitation. As shown in

A

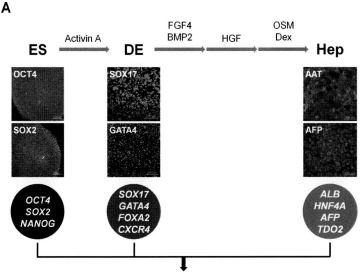

B

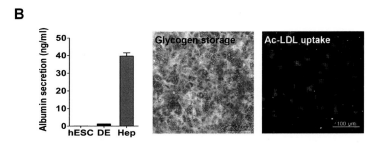

Figure 8.1 (A) Schematic representation of differentiation of hESCs into hepatocytes and analysis of histone modifications at lineage-specific genes in hESCs (ES), DE cells (DE), and hepatocytes (Hep). (B) Immunofluorescence detection of ES (OCT4 and SOX2), DE (SOX17 and GATA4), and hepatocyte (AAT and AFP) markers was performed in hESCs, DE cells, and hepatocytes, respectively. See plate section for color version.

Figure 8.3A, permissive histone marks such as H3K4me3 and H3K9ac were enriched within the promoter of *OCT4* gene in hESCs but gradually decreased in DE cells and hepatocytes. A repressive histone mark H3K27me3 showed a gradual increment during differentiation of hESCs to hepatocytes. Another repressive histone mark H3K9me2 disappeared in DE cells and then abruptly increased in hepatocytes. These results indicate that modest expression of *OCT4* gene in DE cells may be regulated by the competing activities of acetylation and dimethylation at H3K9. Silencing of this gene in hepatocytes appeared to be responsible for occupancies of H3K27me3 and H3K9me2. Modification of H3K9me3 at the *OCT4* promoter was maintained throughout hESC differentiation (Figure 8.3A). These findings demonstrate that transcription of *OCT4* is activated in hESCs by the permissive marks H3K4me3 and H3K9ac, and is suppressed in DE cells and hepatocytes by the repressive marks H3K27me3 and H3K9me2. According to quantitative ChIP-PCR in the *SOX2* promoter region,

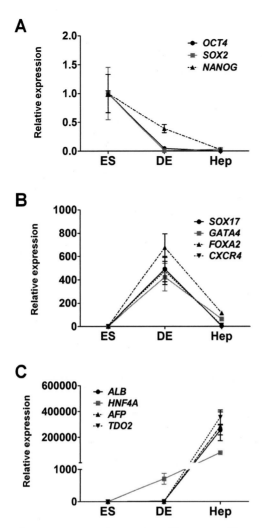

Figure 8.2 Temporal expression of lineage-specific genes during differentiation of hESCs to hepatocytes. The relative expression of ES (A), DE (B), and hepatocyte (C) marker genes during differentiation was determined by real-time RT-PCR. Gene expression was normalized to *GAPDH*. The data are presented as mean ±SEM (*n*=2).

both permissive marks H3K4me3 and H3K9ac were predominant in hESCs, but not detected in DE cells and hepatocytes (Figure 8.3A). During hESC differentiation, levels of H3K27me3 and H3K9me3 were gradually enhanced for complete repression of the *SOX2* gene. We could not observe any significant changes of H3K9me2 level in DE cells and hepatocytes. Thus, we conclude that permissive histone marks such as H3K4me3 and H3K9ac regulate expression of the *SOX2* gene in hESCs, and H3K27me3 and H3K9me3, repressive marks, are associated with suppression of the gene in DE cells and hepatocytes. Permissive histone marks in the promoter region of the *NANOG* gene gradually decreased during differentiation of hESCs to hepatocytes. Repression of the *NANOG* gene was governed in DE cells by H3K27me3 and H3K9me2, and in hepatocytes by H3K9me2

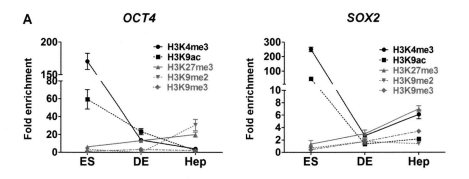

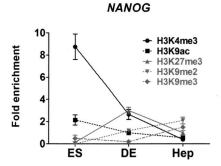

Figure 8.3 Histone modifications in the promoter regions of lineage-specific genes during differentiation of hESCs to hepatocytes. ChIP analysis for histone modifications was performed in the promoter regions of ES (A), DE (B), and hepatocyte (C) marker genes during hESC differentiation. Data validated by real-time PCR are presented as fold enrichment of precipitated DNA associated with a given histone modification relative to a 100-fold dilution of input chromatin. The data are presented as mean ±SEM ($n = 2$).

and H3K9me3 (Figure 8.3A). Thus, repressive histone marks suppress the expression of the *NANOG* gene in hESCs during differentiation to hepatocytes. Collectively, this study represents that the expression of pluripotency marker genes is regulated by a canonical epigenetic regulation in hESCs during differentiation to hepatocytes.

8.3.3 Histone modifications of DE marker genes in hESCs during differentiation to hepatocytes

Next, we analyzed epigenetic changes of DE marker genes during differentiation of hESCs to hepatocytes. A number of developmental genes maintain a bivalent chromatin structure in which H3K4me3 is colocalized with H3K27me3 in mouse ESCs and hESCs (Azuara *et al.*, 2006; Bernstein *et al.*, 2006; Pan *et al.*, 2007; Zhao *et al.*, 2007). Here, we also observed bivalent histone modifications in the promoters of some DE marker genes (*SOX17*, *GATA4*, and *FOXA2*) in hESCs (Figure 8.3B). In contrast, histone modifications of most ES and hepatocyte marker genes are not bivalent in promoter regions in hESCs (Figure 8.3A and C). DE marker genes such as *SOX17* (SRY-box 17), *GATA4* (GATA binding

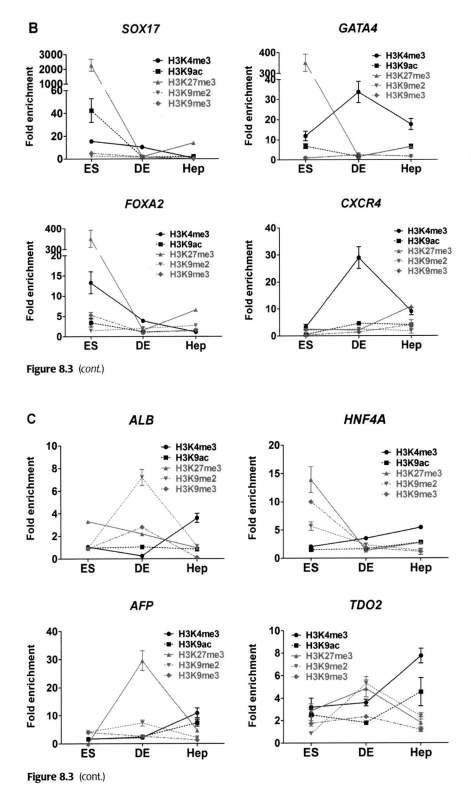

Figure 8.3 (*cont.*)

Figure 8.3 (cont.)

protein 4), *FOXA2* (forkhead box A2), and *CXCR4* (CXC motif receptor 4) were highly expressed in DE cells but only weakly transcribed in hepatocytes (Figure 8.2B). Interestingly, although transcription of the *SOX17* gene was absolutely repressed in hESCs, histone modifications in the promoter region were bivalent in that both permissive and repressive histone marks were enriched (Figure 8.3B). Extreme enrichment of H3K27me3 in hESCs seems to suppress activation of the *SOX17* gene. Furthermore, modifications of both permissive and repressive histone marks were reduced during differentiation of hESCs to hepatocytes. Therefore, the *SOX17* gene in DE cells is likely activated by demethylation of H3K27 rather than trimethylation of H3K4. In addition, H3K9me2 and H3K9me3 did not appear to play an important role in regulation of the *SOX17* gene expression in both DE cells and hepatocytes. Therefore, we suggest that H3K27me3 may function as a key regulator for the expression of the *SOX17* gene during differentiation of hESCs to hepatocytes. The *GATA4* promoter region was modified by H3K4me3, H3K9ac, and H3K27me3 in hESCs (Figure 8.3B). H3K27me3 in the *GATA4* promoter region largely disappeared, whereas H3K4me3 was highly enhanced during the differentiation of hESCs to DE cells. From these results, we conclude that specific expression of *GATA4* gene in DE cells is modulated by trimethylation of H3K4 and demethylation of H3K27. The *GATA4* gene was also weakly transcribed in hepatocytes (Figure 8.2B). It is conceivable that this gene expression in hepatocytes may be mainly activated by acetylation of H3K9. Co-occupancy of permissive (H3K4me3 and H3K9ac) and repressive histone marks (H3K27me3 and H3K9me3) was detected at the promoter region of the *FOXA2* gene in hESCs (Figure 8.3B). This gene is likely to be suppressed in hESCs by H3K27me3 and H3K9me3, despite the presence of permissive histone marks. Levels of H3K4me3 and H3K9ac in DE cells decreased approximately threefold compared to hESCs, and H3K27me3 completely disappeared during differentiation of hESCs to DE cells (Figure 8.3B). Thus, expression of the *FOXA2* gene in DE cells and hepatocytes appears to be modulated by demethylation of H3K27 because permissive histone marks such as H3K4me3 and H3K9ac were present at a lower level in DE cells than in hESCs. Our findings indicate that the expression of *SOX17*, *GATA4*, and *FOXA2* genes may be regulated by the methylation state of histone H3K27 during differentiation of hESCs to DE cells. In the *CXCR4* promoter region, frequencies of H3K4me3 and H3K9ac were enhanced while those of H3K27me3, H3K9me2, and H3K9me3 were not changed in hESCs during early development to DE cells (Figure 8.3B). Thus, expression of the *CXCR4* gene is likely to be modulated in DE cells by abundant permissive histone marks and low levels of repressive histone marks, representing gene regulation by a typical histone modification mode. Unlike other DE marker genes such as *SOX17*, *GATA4*, and *FOXA2*, the promoter of the *CXCR4* gene had a paucity of H3K27me3 in hESCs. Nonetheless, transcription of this gene in hepatocytes was suppressed by H3K27me3 and H3K9me3. Therefore, it is likely that the expression of the *CXCR4* gene in DE cells is regulated by only the permissive histone marks H3K4me3 and H3K9ac. During differentiation into a specialized lineage, the bivalent domains in genes required to maintain identity of a specific cell type might be switched to a more permissive transcriptional mode. However, all promoter loci of lineage-specific genes do not contain bivalent domains in hESCs. In fact, some developmental genes are marked only by the permissive H3K4me3 and H3 acetylation marks in mouse ESCs (Szutorisz *et al.*, 2005), and H3K27me3 is associated with widespread repression of developmental genes in hESCs (Pan *et al.*, 2007). Mechanisms explaining how developmental genes are kept silent in ESCs remain elusive.

8.3.4   Histone modifications of hepatocyte marker genes in hESCs during differentiation to hepatocytes

Figure 8.2C illustrates the transcription of hepatocyte marker genes. Interestingly, expression of *HNF4A* (hepatocyte nuclear factor 4-α) was not detected in hESCs, but was activated in DE cells. Given that *HNF4A* is also a DE marker (Grapin-Botton and Constam, 2007; Seguin *et al.*, 2008), this result implies that hESC-derived DE cells may have a predisposition to develop into hepatocytes. Like HepG2 cancer cells, other hepatocyte marker genes such as α-1-antitrysin (*AAT*), glucose-6-phosphatase (*G6P*), and cytochrome enzymes *CYP2B6* and *CYP3A4* were transcriptionally expressed in hESC-derived hepatocytes (data not shown). During hESC differentiation, H3K4me3 in the *ALB* (albumin) promoter region was increased only in hepatocytes, while levels of H3K4me3 and H3K9ac were similar to those of IgG controls in hESCs and DE cells (Figure 8.3C), indicating that H3K4me3 is a key modulator in the expression of the *ALB* gene. Repression of *ALB* gene in hESCs correlated with the existence of H3K27me3. Three repressive histone marks were enriched in the *ALB* promoter region in DE cells, coinciding with suppression of the gene. A gradual increment of *HNF4A* gene expression was consistent with competitive behaviors between permissive and repressive histone marks (Figure 8.3C). These findings indicate that expression of the *HNF4A* gene is modulated by histone modifications during differentiation of hESCs into hepatocytes. In hESC-derived hepatocytes, the *AFP* promoter region was mainly occupied with permissive histone marks H3K4me3 and H3K9ac (Figure 8.3C). Interestingly, a paucity of H3K27me3 was observed in the promoter region of this gene in hESCs. Instead, di- and trimethylation of H3K9 were responsible for the suppression of this gene in hESCs. H3K27me3 probably acts to repress expression of this gene in DE cells. Also, *TDO2* (tryptophan 2,3-dioxygenase) expression in hepatocytes seemed to be regulated only by permissive histone marks H3K4me3 and H3K9ac (Figure 8.3C). Suppression of the *TDO2* gene in DE cells is likely governed by an enrichment of repressive histone marks. These results demonstrate that expression of hepatocyte marker genes is closely correlated with histone modifications in their promoter regions during differentiation of hESCs to hepatocytes. Repressive histone marks enriched in the promoter of hepatocyte marker genes in hESCs and DE cells are replaced with permissive histone marks in hepatocytes, thereby resulting in the activation of hepatocyte marker genes. In particular, H3K4me3 appears to be a key regulator in the expression of hepatocyte marker genes in hESC-derived hepatocytes. With the exception of the *AFP* gene, repression of hepatocyte marker genes in other cells of this lineage may be governed by H3K27me3. Similarly, gene repression of many neuron-specific genes by H3K27me3 is observed at the progenitor stage during the differentiation of mouse ESCs into neurons (Mohn *et al.*, 2008). H3K27 methylation is critical for repressing expression of lineage-specific genes in ESCs (Azuara *et al.*, 2006; Bracken *et al.*, 2006).

8.4   Concluding remarks

Based on the results obtained in this study, we propose schematic models on behaviors of epigenetic marks for the expression of lineage-specific genes during differentiation of hESCs to hepatocytes (Figure 8.4). A canonical epigenetic regulation is represented in the expression of pluripotency marker genes in hESCs during differentiation to hepatocytes (Figure 8.4A). Expression of pluri-potency marker genes such as *OCT4*, *SOX2*, and *NANOG* is consistent with the epigenetic states of permissive histone marks at the promoters in hESCs. Upon

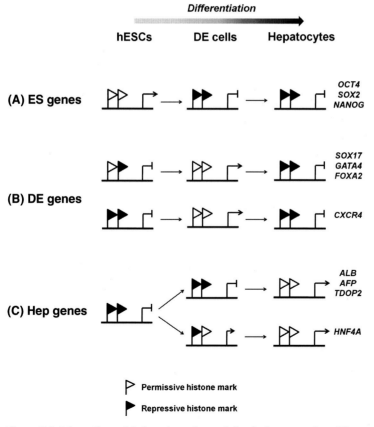

Figure 8.4 Schematic models for epigenetic regulation in the expression of lineage-specific genes in hESCs during differentiation to hepatocytes.

hESC differentiation, permissive histone modifications are replaced with repressive forms gradually increased at the promoters of pluripotency marker genes. Promoters of some DE marker genes such as *SOX17*, *GATA4*, and *FOXA2* retain bivalent domains of histone modifications in hESCs (Figure 8.4B). The bivalent histone modifications of DE marker genes in hESCs are shifted to permissive forms in DE cells and then reset with repressive marks in hepatocytes, whereas the *CXCR4* gene containing non-bivalent domains was transcriptionally activated by the enrichment of permissive marks such as H3K4me3 and H3K9ac. Transcriptional expression of hepatocyte marker genes by histone modifications was similar to that of ES marker genes. Repressive histone marks enriched in the promoter of hepatocyte marker genes in hESCs and DE cells are replaced with permissive histone marks in hepatocytes (Figure 8.4C). Our findings provide an in vitro model of epigenetic mechanisms involved in human hepatocyte development.

REFERENCES

Azuara, V., Perry, P., Sauer, S., *et al.* (2006). Chromatin signatures of pluripotent cell lines. *Nature Cell Biology*, **8**, 532–538.

Bannister, A. J., Zegerman, P., Partridge, J. F., *et al.* (2001). Selective recognition of methylated lysine 9 on histone H3 by the HP1 chromo domain. *Nature*, **410**, 120–124.

Bernstein, B. E., Humphrey, E. L., Erlich, R. L., *et al.* (2002). Methylation of histone H3 Lys 4 in coding regions of active genes. *Proceedings of the National Academy of Sciences USA*, **99**, 8695–8700.

Bernstein, B. E., Mikkelsen, T. S., Xie, X., *et al.* (2006). A bivalent chromatin structure marks key developmental genes in embryonic stem cells. *Cell*, **125**, 315–326.

Boyer, L. A., Lee, T. I., Cole, M. F., *et al.* (2005). Core transcriptional regulatory circuitry in human embryonic stem cells. *Cell*, **122**, 947–956.

Bracken, A. P., Dietrich, N., Pasini, D., Hansen, K. H., and Helin, K. (2006). Genome-wide mapping of Polycomb target genes unravels their roles in cell fate transitions. *Genes and Development*, **20**, 1123–1136.

Cai, J., Zhao, Y., Liu, Y., *et al.* (2007). Directed differentiation of human embryonic stem cells into functional hepatic cells. *Hepatology*, **45**, 1229–1239.

Golebiewska, A., Atkinson, S. P., Lako, M., and Armstrong, L. (2009). Epigenetic landscaping during hESC differentiation to neural cells. *Stem Cells*, **27**, 1298–1308.

Grapin-Botton, A. and Constam, D. (2007). Evolution of the mechanisms and molecular control of endoderm formation. *Mechanisms of Development*, **124**, 253–278.

Hawkins, R. D., Hon, G. C., Lee, L. K., *et al.* (2010). Distinct epigenomic landscapes of pluripotent and lineage-committed human cells. *Cell Stem Cell*, **6**, 479–491.

Liang, G., Lin, J. C., Wei, V., *et al.* (2004). Distinct localization of histone H3 acetylation and H3-K4 methylation to the transcription start sites in the human genome. *Proceedings of the National Academy of Sciences USA*, **101**, 7357–7362.

Meshorer, E., Yellajoshula, D., George, E., *et al.* (2006). Hyperdynamic plasticity of chromatin proteins in pluripotent embryonic stem cells. *Developmental Cell*, **10**, 105–16.

Mikkelsen, T. S., Ku, M., Jaffe, D. B., *et al.* (2007). Genome-wide maps of chromatin state in pluripotent and lineage-committed cells. *Nature*, **448**, 553–560.

Mohn, F., Weber, M., Rebhan, M., *et al.* (2008). Lineage-specific polycomb targets and de novo DNA methylation define restriction and potential of neuronal progenitors. *Molecular Cell*, **30**, 755–66.

Nichols, J., Zevnik, B., Anastassiadis, K., *et al.* (1998). Formation of pluripotent stem cells in the mammalian embryo depends on the POU transcription factor Oct4. *Cell*, **95**, 379–391.

Niwa, H., Miyazaki, J., and Smith, A. G. (2000). Quantitative expression of Oct-3/4 defines differentiation, dedifferentiation or self-renewal of ES cells. *Nature Genetics*, **24**, 372–376.

Pan, G., Tian, S., Nie, J., *et al.* (2007). Whole-genome analysis of histone H3 lysine 4 and lysine 27 methylation in human embryonic stem cells. *Cell Stem Cell*, **1**, 299–312.

Park, S. Y., Kim, J. B., and Han, Y. M. (2007). REST is a key regulator in brain-specific homeobox gene expression during neuronal differentiation. *Journal of Neurochemistry*, **103**, 2565–2574.

Pokholok, D. K., Harbison, C. T., Levine, S., *et al.* (2005). Genome-wide map of nucleosome acetylation and methylation in yeast. *Cell*, **122**, 517–527.

Rao, B., Shibata, Y., Strahl, B. D., and Lieb, J. D. (2005). Dimethylation of histone H3 at lysine 36 demarcates regulatory and nonregulatory chromatin genome-wide. *Molecular Cell Biology*, **25**, 9447–9459.

Santos-Rosa, H., Schneider, R., Bannister, A. J., *et al.* (2002). Active genes are tri-methylated at K4 of histone H3. *Nature*, **419**, 407–411.

Seguin, C. A., Draper, J. S., Nagy, A., and Rossant, J. (2008). Establishment of endoderm progenitors by SOX transcription factor expression in human embryonic stem cells. *Cell Stem Cell*, **3**, 182–195.

Szutorisz, H., Canzonetta, C., Georgiou, A., *et al.* (2005). Formation of an active tissue-specific chromatin domain initiated by epigenetic marking at the embryonic stem cell stage. *Molecular Cell Biology*, **25**, 1804–1820.

Wen, B., Wu, H., Shinkai, Y., Irizarry, R. A., and Feinberg, A. P. (2009). Large histone H3 lysine 9 dimethylated chromatin blocks distinguish differentiated from embryonic stem cells. *Nature Genetics*, **41**, 246–250.

Zhao, X. D., Han, X., Chew, J. L., *et al.* (2007). Whole-genome mapping of histone H3 Lys4 and 27 trimethylations reveals distinct genomic compartments in human embryonic stem cells. *Cell Stem Cell*, **1**, 286–298.

9 Epigenetic stability of human pluripotent stem cells

Céline Vallot and Claire Rougeulle*

9.1 Epigenetic regulation in pluripotent stem cells: does it matter?

Human pluripotent stem cells (hPSC) potentially stand as promising therapeutic tools for degenerative diseases. Through their capacity to differentiate to any cell type, they offer the possibility of a renewable source of replacement cells to treat various diseases including Parkinson's and Alzheimer's diseases. These cells can be obtained by two different means (Figure 9.1). Since 1998 (Thomson *et al.*, 1998), it has been possible to derive human embryonic stem cells (hESCs) from the inner cell mass cells of blastocysts with the potential to self-renew indefinitely but also to give rise to the three embryonic germ layers (endoderm, ectoderm, and mesoderm) when differentiating. Since 2007 (Takahashi *et al.*, 2007; Yu *et al.*, 2007), induced pluripotent stem cells have also been derived from adult differentiated cells by reprogramming; they harbor self-renewing properties and are able to differentiate into cells of all three germ layers.

In order to use hPSC for therapeutic purposes, extensive culture protocols are needed to maintain the population of undifferentiated cells. During these long-lasting culture periods, stem cells are susceptible to various changes, genetic as well as epigenetic, which could jeopardize their medical use. Genetic alterations have been carefully monitored in hESCs (Maitra *et al.*, 2005; Baker *et al.*, 2007) and occur frequently in these cell types: the most recurrent karyotypic changes are a gain of chromosomes 12 and 17 and to a lesser extent of the X chromosome (Baker *et al.*, 2007). Whereas genetic abnormalities are now reasonably counter-selected for in routine hESCs expansion, less is known about epigenetic stability in pluripotent stem cell culture and how it could affect downstream applications. Recent studies have deployed tremendous efforts to unravel the epigenomes of human embryonic stem cells (Meissner, 2010) in terms of histone modifications profiles (Guenther

---

* Author to whom correspondence should be addressed.

Epigenomics: From Chromatin Biology to Therapeutics, ed. K. Appasani. Published by Cambridge University Press. © Cambridge University Press 2012.

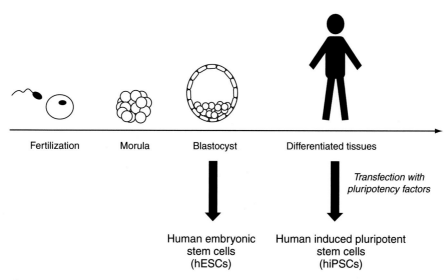

Figure 9.1 Different sources of human pluripotent stem cells. These cells can be obtained by deriving human embryonic stem cells from the inner cell mass cells of blastocysts or by reprogramming adult differentiated cells into induced pluripotent stem cells.

et al., 2007; Pan *et al.*, 2007; Zhao *et al.*, 2007; Xie *et al.*, 2009; Hawkins *et al.*, 2010) as well as DNA methylation distribution (Lister *et al.*, 2009; Laurent *et al.*, 2010). This type of analysis was also conducted for induced pluripotent stem cells mostly in order to compare them with hESCs (Hawkins *et al.*, 2010; Meissner, 2010). However, all these studies do not address the potential stability or instability of these newly well-defined epigenomes and its implication concerning therapeutic issues.

Before addressing the stability of epigenetic marks in pluripotent cells, one important consideration is to understand whether they matter or not in this cell system. Various experiments performed on mouse embryonic stem cells have shown that pluripotent cells can tolerate a global loss of epigenetic marks, such as DNA methylation and H3K27me3. Indeed, the knockout of DNA methyltransferases (Tsumura *et al.*, 2006) or Polycomb repressive complex 2 (PRC2) (Chamberlain *et al.*, 2008) does not alter self-renewing properties or the maintenance of cell pluripotency. These studies suggest that epigenetic mechanisms might not be essential for self-renewal and for early differentiation steps (Christophersen and Helin, 2010). However, these enzymes, as well as the marks they lay down, are essential for embryonic development (Okano *et al.*, 1999; O'Carroll *et al.*, 2001) and especially for embryonic stem cell full differentiation (reviewed in Meissner, 2010). While epigenetic marks may not be the key regulator of embryonic stem cell identity, they control genes involved in lineage commitment as well as specific parts of the genome, such as centromeres, imprinted loci, and one X chromosome in female cells. A deregulation of epigenetic modifications in embryonic stem cells might therefore not have a direct effect in undifferentiated cells but drastic consequences on their differentiated offspring, and that is of major concern for the use of pluripotent stem cells in clinics. In order to pursue the road towards therapeutic applications, these cells have to be extensively monitored in terms of genetic and epigenetic stability. In this chapter, we review the studies that have addressed the epigenetic instability of pluripotent stem cells, in order to apprehend how we could best monitor this shakiness, possibly to circumvent it and counter-select it in the future. To do so, we will classify the studies according to the part of the genome they are analyzing, and finish by investigating whether induced pluripotent stem cells display similar variations.

9.2   Imprinted loci in pluripotent cells

Human embryonic stem cells are derived from embryos conceived by assisted reproductive technology (ART) and several studies have indicated that there might be an increased frequency of ART conceptions among children with imprinting disorders, such as the Beckwith–Wiedemann or Angelman syndromes (Cox et al., 2002; Orstavik et al., 2003). While this link has to be confirmed, it suggests that the stability of imprinting in hESCs requires particular attention. Ex vivo culture could be a major factor of instability; culture of preimplantation embryos for ART prior to hESCs derivation (reviewed in Huntriss and Picton, 2008), as well as long-lasting periods of culture to maintain hESCs might induce imprinting defects. In addition, culture conditions have been shown to artificially influence epigenetic mechanisms in various cell systems, such as cancer cell lines (Smiraglia et al., 2001).

Imprinted genes are loci for which the transcriptional activity of each allele depends on its parental origin. Some imprinted genes are expressed from the paternally inherited allele (e.g., IGF2) and others from the maternally inherited one (e.g., CDKN1C). In some cases however imprinting patterns are more complex and some imprinted genes demonstrate tissue-specific or developmental-stage-specific imprinting (e.g., IGF2R or UBE3A). This parental memory is initiated in the germ-line and maintained throughout somatic development due to epigenetic modifications (DNA methylation, histone marks, etc.). More than 100 imprinted or potentially imprinted genes have been identified in the human genome (Luedi et al., 2007). Most of them (around 80%) are organized in large clusters spanning several megabases. Each cluster is coordinately regulated by an imprinting center (IC) characterized by a differentially methylated region (DMR). Parental specific DNA methylation at these ICs is established in the germ-line and maintained after fertilization throughout development. Other DMRs, outside of ICs, are also found in imprinted regions for which DNA methylation is established during development. Abnormalities in the epigenetic regulation of imprinted regions cause severe syndromes and have been observed in various cancers (Horsthemke and Buiting, 2008). For example, aberrant loss or gain of DNA methylation at the IC leading to aberrant gene expression is the most frequent cause of Beckwith–Wiedemann syndrome (Cooper et al., 2005). Therefore, the epigenetic status of imprinting regions should be closely monitored in hESCs to prevent the appearance of de-regulations which could stand as major limitations for their therapeutic use.

Several studies on the expression status of imprinted genes in hESCs have been conducted since 2005. All the results are summarized in Table 9.1. The first study by Rugg-Gunn and colleagues concerned four hESC lines (Rugg-Gunn et al., 2005)

Table 9.1 Summary of imprinting defects in hESCs. This table summarizes all the imprinted genes whose expression has been studied in hESCs. Genes are classified according to their genomic localization and their expression pattern in hESCs, as reported in the different publications. The expression of a given gene can be monoallelic, variable, and/or biallelic in hESCs depending on the study. For each publication, the indication in brackets represents the number of hESC samples in which the corresponding gene was found biallelically expressed over the total number of samples that were tested

Locus	Gene	Parental expression	Monoallelic	Variable	Biallelic	Known disease
1p36.33	TP73	[a]			Kim 2007 (9/9)	
6q25	IGF2R	[a]			Kim 2007 (12/12)	
7p11	GRB10	[a]				
7q21	PEG10	paternal	Frost 2011(0/1), Sun 2006 (0/1)	Kim 2007 (1/4)	Frost 2011(1/1)	
7q32	CPA4	maternal	Frost 2011 (0/1)	Kim 2007 (1/12)	Frost 2011(2/2)	Silver–Russell syndrome
	MEST IT 1	paternal		Kim 2007 (1/12)		
	MEST iso1	paternal		Kim 2007 (10/12),		
	MEST iso2	paternal		Adewumi 2007 (11/16)		
10q26	INPP5F	paternal	Frost 2011(0/3)			
11p15.5	H19	maternal	Sun 2006 (0/1)	Kim 2007 (2/11), Frost 2011 (1/5),	Kim 2007 (6/6),	Beckwith–Wiedemann
	IGF2	paternal	Frost 2011(0/2),	Rugg-Gunn 2005 (1/4),	Frost 2011 (5/5)	syndrome, Silver–Russell
	KCNQ1	maternal	Rugg-Gunn 2005 (0/4)	Adewumi 2007 (1/20)		syndrome, cancer
	KCNQ1OT1	paternal	Kim 2007 (0/8, Frost 2011	Kim 2007 (9/10),		
	CDKN1C	maternal	(0/2), Sun 2006 (0/1)	Adewumi 2007 (6/20)		
	SLC22A18	[a]	Kim 2007 (0/12), Frost 2011	Adewumi 2007 (8/11)		
	PHLDA2	maternal	(0/2), Rugg-Gunn 2005 (4/4),	Kim 2007 (6/10)		
			Adewumi 2007 (0/23)			
			Kim 2007 (0/3), Frost 2011 (0/1)			
			Rugg-Gunn 2005 (0/4)			
11p3	WT1	[a]				
14q32	DLK1	paternal		Frost 2011 (3/4)		Maternal and paternal
	MEG3	maternal		Kim 2007 (1/10),		uniparental disomy
				Adewumi 2007 (1/7)		chromosome 14

Table 9.1 (cont.)

Locus	Gene	Parental expression	Monoallelic	Variable	Biallelic	Known disease
15q11–q13	IPW	paternal	Rugg-Gunn 2005 (0/4), Adewumi 2007 (0/14)	Kim 2007 (1/11)		Prader–Willi syndrome, Angelman syndrome, cancer
	NDN	paternal	Kim 2007 (0/6), Frost 2011 (0/2)	Kim 2007 (7/11)		
	MAGEL2	paternal	Kim 2007 (0/1), Sun 2006 (0/1)			
	SNRPN	paternal	Kim 2007 (0/10), Frost 2011 (0/2)			
	ATP10A	maternal	Kim 2007 (0/5), Frost 2011 (0/1), Adewumi 2007 (0/3)			
	PEG3	paternal				
19q13.4						
20q13.2	NESP55/GNAS5	maternal	Rugg-Gunn 2005 (0/4), Frost 2011(0/1)	Kim 2007 (1/9), Adewumi 2007 (1/8)	Frost 2011(5/5)	

[a] tissue-specific/polymorphic imprint.

and six genes (*NESP55*, *IPW*, and four genes belonging to the 11p15.5 cluster: *IGF2*, *KCNQ1OT1*, *H19*, *SLC22A18*). All genes were monoallelically expressed in all cell lines even after prolonged passage, except for *H19*, which was biallelically expressed after long-lasting culture in H9 cells. Sun *et al.* confirmed this apparent stability for one hESC line (Sun *et al.*, 2006). As the number of samples described in these studies was low, their conclusion for overall stability of imprinting in hESCs needed further confirmation. In 2007, the analysis of two large cohorts of hESCs (22 and 46 hESCs respectively) (Adewumi *et al.*, 2007; Kim *et al.*, 2007) gave detailed information on the allelic expression of various clusters of imprinted genes in hESCs originating from different laboratories. Kim and colleagues showed that 11 out of the 22 imprinted genes they had examined (50%) displayed variable allelic expression (Kim *et al.*, 2007), clearly demonstrating that imprinting can be disrupted in hESCs. To a lesser extent, the International Stem Cell Initiative also found that three genes out of the ten (30%) they had studied were biallelically expressed in some hESCs (Adewumi *et al.*, 2007). More recently, the analysis of 26 imprinted genes in eight hESCs (Frost *et al.*, 2011) confirmed the imprinting variability in hESCs: 11 out 26 genes (42%) displayed biallelic expression in a fraction of the cell lines.

Altogether, these results demonstrate that hESCs display some levels of imprinting instability which depends on the gene which is considered. Imprinted genes are either frequently deregulated in hESCs, like *IGF2R*, *MEST* (isoform 2), or *ATP10A*, or stable in almost all cell lines studied like *KCNQ1*, *CDKN1C*, or *NDN* for example. This behavior does not seem to be cluster-specific as in one common cluster, e.g., 11p15.5, neighboring genes can display opposite properties, e.g., *CDKN1C* and *SLC22A18*. It has to be noted that some imprinted genes, such as *IGF2R*, are found biallelically expressed in all studied hES cell lines. This could reflect the biparental expression of the gene at the corresponding developmental stage (i.e., blastocyst) rather than epigenomic instability. Further expression analysis will be needed to determine the 'normal' allelic status of these genes in the inner cell mass of the embryo.

This gene-specific pattern for imprinting instability in hESCs raises several questions regarding its origin. Are genes prone to instability different from those protected from variation? First, looking at the parental origin of the expressed allele, out of the 18 genes which can be biallelically expressed in at least one studied hESC, seven are paternally expressed, six are maternally expressed, and the others show a tissue-specific or polymorphic (i.e., which varies among individuals) imprint (*WT1* for example can display biallelic or monoallelic expression from either the paternal or maternal allele in adult tissues); out of the seven genes which are always monoallelically expressed, five are paternally and two are maternally expressed. There is no significant difference of parental origin between the genes prone or resistant to imprinting instability (Fisher's exact test). In other words, the parental origin of the expressed allele does not favor the stability of imprinting in hESCs.

The level and pattern of expression of imprinted genes during early embryonic development might be a factor affecting the stability of the imprinting status in hESCs (Rugg-Gunn *et al.*, 2007). However, analyzing these parameters in the human embryo is challenging as access to early embryos is limited; moreover, the collected embryos need to have a polymorphism in the imprinted genes which are studied in order to conclude on their allelic expression, which decreases the number of informative samples. It is important to note that several studies aimed at addressing this question have limited their analysis to the level of expression of imprinted genes and not their allelism (Salpekar *et al.*, 2001). In contrast, the pattern of the expression of *SNRPN*, which is monoallelically expressed in all 36 studied hESCs (Table 9.1), has been carefully studied in human preimplantation embryos (Huntriss *et al.*, 1998). The authors show that *SNRPN* starts to be monoallelically expressed at a very early stage in development (4-cell stage). Such precocious establishment of stable monoallelic expression could confer imprinting stability in hESC culture. However, further studies of imprinted genes in human embryos are necessary to confirm such a trend. Indeed, *IGF2*, which is also monoallelically expressed early during preimplantation development (at the 8-cell stage: Lighten *et al.*, 1997), is often biallelically expressed in hESCs.

Importantly, the imprinting irregularities observed in hESCs are not equivalent to the ones found in imprinting disorders. Indeed, in most cases, the biallelic expression of genes has not been linked to any DNA methylation defect at related DMRs, as it is the case for Beckwith–Wiedemann syndrome for example. When studying *H19* biallelic expression in H9 cells, Rugg-Gunn and colleagues did not find any loss of gametic imprint (Rugg-Gunn *et al.*, 2005). Similarly, Frost and colleagues did not observe any clear correlation between the allelic expression and the methylation status of the associated DMR (Frost *et al.*, 2011): *SLC22A18*, which is frequently biallelically expressed, always keeps a differentially methylated DMR. Intriguingly, for monoallelically expressed genes, DMRs were sometimes disrupted, without perturbating the normal allelic expression of the surrounding imprinted genes. Altogether, DNA methylation at imprinted loci seems more malleable in hESCs than in differentiated cells, but surprisingly does not systematically influence the allelic expression of corresponding imprinted genes.

In summary, monoallelic expression is not stable in hESCs at all imprinted loci. Some genes are prone to frequent biallelic expression in these pluripotent cells. As these variations in expression do not correspond to modifications of the methylation status of control regions, they might not be stably maintained throughout various cell divisions and during differentiation. However, other epigenetic modifications acting at the level of histones could also be involved in the deregulation of imprinted expression in hESCs. This needs to be carefully examined in order to understand the mechanisms at the basis of imprinting instability in hESCs.

9.3   Looking at the X chromosomes in female hESCs

In mammals, sex chromosome dosage compensation is achieved by silencing one of the two X chromosomes during early female development. The kinetics and mechanism of X-chromosome inactivation (XCI) have mainly been studied in mice, and have just started to be analyzed in human development, especially in hESCs which are far more accessible than human embryos per se. In mice, the paternal X chromosome is first inactivated at the 4-cell stage (Chow and Heard, 2009). This imprinted inactivation persists until the blastocyst stage and continues in extra-embryonic tissues. The paternal X is however reactivated in cells of the inner cell mass of the blastocyst, whereafter one X will be randomly inactivated upon cell differentiation. The non-coding *Xist* RNA is a key player in XCI (Penny *et al.*, 1996): it is essential for the initiation and propagation of silencing of the inactive X (Xi).

Little is known about how XCI takes place during human development. Up to now, no evidence for an imprinted form of X inactivation in preimplantation has been put forward. In contrast to the mouse, random X inactivation takes place in both embryonic and extra-embryonic lineage (Moreira de Mello *et al.*, 2010); however the timing of X-inactivation onset remains unknown. *Xist*, as in mouse, is a major regulator of X-inactivation in human cells (Brown *et al.*, 1991).

In 1997, two studies have measured *Xist* expression levels in human preimplantation embryos (Daniels *et al.*, 1997; Ray *et al.*, 1997). Using reverse transcription (RT) PCR, both detected *Xist* transcription starting from the 4-cell (Daniels *et al.*, 1997) or 5-cell stage (Ray *et al.*, 1997). *Xist* transcripts were detected in both male and female embryos, but levels of transcription were hardly quantifiable and extensive amplification was used. Using a single-cell approach by FISH, van den Berg *et al.* conducted a more thorough analysis of X inactivation status in fresh and cryopreserved embryos fixed at the 8-cell, morula, and blastocyst stages (van den Berg *et al.*, 2009). The authors showed that a *Xist* signal gradually accumulated to a full domain on one of the X chromosomes at the late morula and blastocyst stage in female embryos specifically. In morulas, 49% of cells already displayed a single *Xist* cloud; in blastocysts, 90% of the nuclei showed a full *Xist* domain. In male embryos, *Xist* RNA signals were limited to pinpoints at the morula stage. Further experiments showed that the *Xist* RNA domains in female embryos corresponded to areas where transcription was absent (using Cot-1 RNA-FISH). Single spots of *CHIC1* expression, which is located on the X chromosome, solely on the X without any *Xist* cloud confirmed the silencing of the *Xist*-coated chromosome. This work strongly suggests the occurrence of XCI in female pre-implantation embryos.

In the same way that mouse embryonic stem cells were used as a tool to exhaustively characterize XCI in mice, hESCs stand as a promising model to study the progress of the X-inactivation process during human development. Moreover, the status of X inactivation in hESCs is of major interest regarding their use in regenerative medicine as deregulation of the XCI can lead to various disorders such as fragile X syndrome, and has been observed in cancer (Kawakami *et al.*, 2004).

Several studies have started to analyze the X-inactivation status in hESCs. In 2004, using gene trap, Dhara and colleagues reported that both X chromosomes in female hESCs were active and that XCI occurred in a random fashion as cells differentiated (Dhara and Benvenisty, 2004). However, a minority of undifferentiated hESCs already displayed XCI. Only few other studies have reported the existence of hESCs with two active X chromosomes (Hall *et al.*, 2008; Silva *et al.*, 2008). The majority of hESCs display a heterogeneous pattern for XCI. In 2005, Hoffman and collaborators studied H9 and H7 cell lines, demonstrating that H9 had already completed XCI while remaining pluripotent, and that H7 had also completed XCI but had lost *Xist* expression (Hoffman *et al.*, 2005). The variability of XCI in hESCs was confirmed by the International Stem Cell Initiative (Adewumi *et al.*, 2007), which studied 31 female hESCs and showed that they displayed various *Xist* RNA levels, ranging from negligible male expression level to levels of transcription observed in female differentiated cells. Shen *et al.* conducted a thorough analysis of two female hESCs, H9 and HSF6, which had undergone XCI and in which part of the cells had lost *Xist* expression (Shen *et al.*, 2008). Interestingly, the authors suggest that transient exposure to stress or suboptimal culture conditions might lead to epigenetic silencing of *Xist* by DNA methylation, and to the loss, on the Xi, of classical XCI markers such as H3K27me3, H4K20me1, and macro-H2A. Silva *et al.* categorized hESCs depending on their XCI status (Silva *et al.*, 2008) (Figure 9.2). Class I hESCs correspond to cells with two active X chromosomes (*XaXa*), where *Xist* will be expressed and one X inactivated upon differentiation. Class II hESCs are cells which have already undergone XCI and have an inactive X coated by *Xist* ncRNA (*XaXi*). Class III hESCs correspond to cells with one inactive X, but where *Xist* is not expressed, and will not be upregulated during differentiation. Interestingly, a given cell line can be found in each of these different status: depending on the laboratory and the number of passages, the H9 cell line for example is described either as class III (Silva *et al.*, 2008), class II (Hoffman *et al.*, 2005), or class I (Dhara

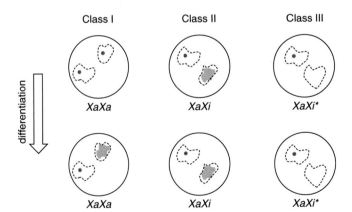

Figure 9.2 X chromosome status in hES cells. Female hES cells can be classified according to their X chromosome status. This can be achieved by RNA-FISH analysis using a probe directed against the ncRNA *Xist* and for an X-linked gene, such as *CHIC1* for example. Here is a schematic vision of an RNA-FISH study: the point probe stands for *CHIC1* mRNA (pinpoint should be smaller), and the gray cloud for *Xist* ncRNA. It must be noted that studying a single X-linked gene by RNA-FISH is not sufficient to fully conclude as to the activity status of the X. The dashed line represents the X chromosome territories in the nucleus. See text for description of the three classes.

and Benvenisty, 2004). Moreover, a given culture of hESCs can also be heterogeneous: Tanasijevic *et al.* showed in the same culture plate co-existing H9 colonies belonging to class II and others to class III (Tanasijevic *et al.*, 2009).

X inactivation in hESCs is highly instable, as the X-inactivation status varies from one cell line to another as well as along passages in a given cell line. Understanding how this process is regulated in hESCs, by either culture conditions or others transcriptional factors, is essential in order to control the development of hESCs for potential therapeutic applications. Few studies have started to analyze how culture conditions could influence and stabilize the X-inactivation status in hESCs. Lengner *et al.* studied how oxidative stress could be linked to X inactivation during hESCs derivation (Lengner *et al.*, 2010). The authors could established *XaXa* hESCs under physiological oxygen concentrations, and showed that even a chronic exposure to atmospheric oxygen concentrations, the standard for cell culture in laboratories, triggered XCI irreversibly. Hanna *et al.* transferred hESCs from the conventional medium, containing β-FGF and knockout serum replacer, to a serum-free '2i' medium supplied with human leukemia inhibitory factor (LIF) and two inhibitors of glycogen synthase kinase 3β (GSK3β) and the mitogen-activated protein kinase (ERK1/2) pathway (Hanna *et al.*, 2010). The transition between the two culture media was achieved by ectopic induction of *OCT4*, *KLF4*, and *KLF2*, which were transiently transfected in hESCs. Previously class II hESCs turned class I in the '2i' complemented media, displaying two active X chromosomes. XCI was recapitulated during differentiation or by switching back to β-FGF-containing media. This artificially generated class I status appears however unstable and these cells eventually died after prolonged culture. Nevertheless this study, which needs to be confirmed using hESCs coming from other laboratories, suggests that X-inactivation status depends on the culture conditions, and more precisely on the signaling pathways that are activated. Taken together, these recent findings indicate that it might be achievable to canalize the instability of X inactivation in hESCs by adapting culture conditions.

9.4   Epigenetic modifications at gene promoters

While most studies on hESCs epigenetic stability have focused on imprinted loci and X-inactivation status, some have addressed the epigenetic variation at gene promoters in the rest of the genome, especially CpG-rich promoters. The first analysis by Maitra *et al.* showed that genes can gain aberrant DNA methylation during long-term cell culture in hESCs (Maitra *et al.*, 2005). Nine hES cell lines at low and late passages and 14 genes that are frequently hypermethylated in cancer were studied. *RASSF1* and to a lesser extent *PTPN6* displayed increased DNA methylation at late passage for seven out of nine and three out of nine hESCs respectively. Bibikova *et al.* led a large-scale analysis of DNA methylation for 371 genes using bead arrays in 14 hESCs and other cancer and normal cell lines which confirmed this culture-dependent variation (Bibikova *et al.*, 2006). For nine hES cell lines, they were able to compare methylation patterns at different passage number, and observed that different hES cell lines showed different changes with time in culture, and that the degree of overall change in methylation was directly linked to the number of passages separating DNA preparations. However the authors did not find a specific common set of modified loci between the nine cell lines, suggesting that the DNA hypermethylation process is not predictable. Allegrucci *et al.* also demonstrated that hESCs expansion led to stable and unpredictable epigenetic changes (Allegrucci *et al.*, 2007). Using restriction landmark genome scanning (RLGS) for four hESCs at different passage number, the authors showed that DNA methylation instability affects 0.33% to 0.79% of the studied loci (for comparison in cancer cells 0.22% to 3.39% of the loci are modified), but no common changes were identified between all cell lines. Moreover, changing the culture method induced additional changes within a few passages. Strikingly, 56% of the identified loci with either aberrant hypo- or hypermethylation occurring during hESCs culture had been reported to change their methylation profile in at least one tumor type. More recently, Calvanese *et al.* reported a class of genes ($n = 20$) that were specifically hypermethylated in hESCs and cancer cell lines compared to normal differentiated tissues (Calvanese *et al.*, 2008). Only some of these loci were de-methylated during differentiation of hESCs. These two studies (Allegrucci *et al.*, 2007; Calvanese *et al.*, 2008) suggest a potential link between epigenetic instability in hESCs and the occurrence of tumor-like epigenetic modifications, which could be a major restraint in their therapeutic use. However, there are two ways of looking at this issue: on one hand, it could be that hESC lines acquire DNA methylation abnormalities providing growth advantages like cancer cells, which would be problematic. On the other hand, the DNA methylation alterations seen in hESCs could also normally occur even in pluripotent cells in vivo and have no consequence, and cancer cells display the same patterns because they have dedif-ferentiated during tumor progression or arisen from tissue stem cells.

Shen *et al.* characterized another type of epigenetic instability in hESCs (Shen *et al.*, 2006): they showed that cells differentiated in vitro from hESCs displayed DNA hypermethylation compared to primary differentiated cells. In this case, the hypermethylated genes were not linked to any known tumor type. Nevertheless, the influence of these abnormalities on the generation of differentiated tissues remains unknown and needs to be deeply investigated.

9.5   Do induced pluripotent stem cells share this variability?

In 2007, Takahashi and colleagues generated induced pluripotent stem (iPS) cells from adult human fibroblasts by overexpressing four factors: Oct4, Sox2, Klf4, and c-Myc (Takahashi *et al.*, 2007). Thus iPS cells stand as a promising tool for regen-erative biology as patient-specific stem cells could be generated, avoiding ethical issues surrounding the use of hES cells.

Very few studies have addressed the epigenetic stability of iPS cells and most analyses have focused on understanding whether iPS cells are comparable to hES cells. In contrast to their mouse counterparts, demonstrating the inherent pluri-potency of both hESC and hiPS is hardly feasible. Indeed, the most stringent assays available for measuring human pluripotency are in vitro differentiation into embryoid bodies and/or in vivo formation of teratomas. Therefore, the comparison of these two cell types mostly stands on molecular terms. Concerning gene expression, several studies showed conflicting results. On one hand, Chin *et al.* argued that iPS cells display a slightly different transcriptome than hESCs (Chin *et al.*, 2009). All iPS cells share a distinct gene expression signature, late-passage iPS cells being closer to hES cells regarding gene expression. Two studies confirmed that iPS and hES cells were transcriptionally distinguishable (Marchetto *et al.*, 2009; Ghosh *et al.*, 2010). On the other hand, Guenther and colleagues showed that the very few differences they could observe between hESCs and iPSCs were random and were smaller than the overall variability between each type of cells (Guenther *et al.*, 2010). Altogether, to conclude on this matter it would be necessary to study two large equivalent cohorts of hESCs and iPSCs: such a comparison would address whether consistent gene expression differences can be observed between those two cell types. Some studies have compared the epigenomes of hESCs and iPSCs. Levels of H3K27me3 and H3K4me3 do not differ between these two types of cells (Chin *et al.*, 2009; Guenther *et al.*, 2010; Hawkins *et al.*, 2010). However, H3K9me3 (Hawkins *et al.*, 2010) as well as DNA methylation (Deng *et al.*, 2009) distribution is not equivalent in both cell types. The consequence of such differences on the ability of iPS cells to mimic hES cells remains to be understood.

So far, only few studies have addressed imprinting stability in hiPS cells. They showed that some methylation imprints are maintained during iPS generation (Chamberlain *et al.*, 2010) and that iPS cells display overall normal and stable imprinted gene expression (Pick *et al.*, 2009). However, abnormal parental imprinting could be detected for some genes and in some iPS clones, and in some cases this coincides with aberrant DNA methylation (Pick *et al.*, 2009). Larger-scale analyses are clearly needed to investigate fully imprinting stability in iPS cells and to which extent this resembles what has been shown in hES cells.

The status of the X chromosome in female iPSCs has also been recently characterized by several groups. Tchieu *et al.* demonstrated that in contrast to the mouse, the reprogramming process in human iPSCs does not seem to involve a reactivation of the somatic Xi, as it resulted in iPSCs that carry one inactive X chromosome expressing *Xist*, like the parental somatic cell and class II hESCs (Tchieu *et al.*, 2010). Extended culture of some iPS cell lines led to a loss of *Xist* expression mimicking the behavior of class III hESCs. However, the authors did observe some changes in Xi chromatin structure in iPSCs compared to the fibroblasts they were derived from: the inactive X in iPSCs was found highly enriched in EZH2, a member of the PRC2 complex responsible for triggering H3K27me3 marks, which is usually lost, or at least difficult to detect on the inactive X in differentiated cells. Lagarkova and colleagues also demonstrated that iPSCs obtained from endothelial cells do not show Xi reactivation, as the cells still express *Xist*, but undergo slight chromatin changes compared to the umbilical cord parental cells and to hES cells (Lagarkova *et al.*, 2010). In some iPSCs for example, the supposed Xi displayed H3K27me3 as well as H3K4me2, whereas in hESCs,

H3K27me3 was more intense and H3K4me2 absent from the Xi. Altogether, these findings suggest partial reprogramming of the Xi chromatin structure during iPSC generation. Strikingly, reactivation of the Xi was reported in iPSCs derived from Rett syndrome patients carrying mutation in the *MECP2* gene (Marchetto *et al.*, 2010). In these cells, normal X inactivation is recapitulated upon neuronal differentiation. The extent to which the observed reprogramming of inactivation is linked to the absence of a functional MECP2 protein, which binds methylated DNA and is involved in epigenetic regulation, is still unclear. Altogether the status of the inactive X chromosome in iPSCs remains to be fully characterized to understand whether the reprogramming of the Xi is limited to chromatin changes or involves a reactivation of the chromosome. The Xi status appears however unstable and prone to evolution in culture like in hESCs, as *Xist* can also be lost in iPSCs. From the few available studies addressing epigenetic stability in iPSCs, both at the level of imprinting and X inactivation, it appears that iPSCs share a common variability with hESCs.

9.6   Conclusion

Human pluripotent stem cells do not stand in a defined epigenetic state. As mentioned above, multiple variations occur in the epigenome of hES cells along passages and within different cell lines. The behavior of imprinted genes attests this variability. Genes such as *IGF2R* are often biallelically expressed in several hES cell lines. Noticeably, these deregulations do not correspond to any alteration in DNA methylation. The status of the X chromosome also reflects major epigenetic instability of hESCs. Female pluripotent stem cells can harbor various X chromosome patterns which are not stably maintained throughout numerous cell divisions; coating of the inactive X chromosome by the *Xist* RNA can be lost and the effect of this phenomenon remains largely unknown. Even isolated promoters can display epigenetic variability and be hypermethylated compared to normal tissues.

At first sight, this epigenetic heterogeneity might be a major drawback for the use of pluripotent stem cells for therapeutic purposes. Indeed, such variations could induce tumor-like alterations. Alternatively, epigenetic variations may only reflect a tremendous epigenetic plasticity of hESCs. In other words, the diverse epigenetic states the hESCs can be found in might all exist in vivo. The observed culture variations could result from the selection of subpopulations of epigenetic variants, perhaps displaying proliferative properties giving them a growth advantage or other beneficial characteristics. These populations of plastic, undifferentiated cells do not develop in vivo as there is no time for selection at this point in development. According to this scenario, the consequences of epigenetic variations for the differentiated offspring could be less dramatic, but remain to be thoroughly investigated. This paradox between plasticity versus abnormality will need to be addressed in the future to definitely validate pluripotent stem cells as a solid therapeutic tool.

REFERENCES

Adewumi, O., Aflatoonian, B., Ahrlund-Richter, L., *et al.* (2007). Characterization of human embryonic stem cell lines by the International Stem Cell Initiative. *Nature Biotechnology*, **25**, 803–816.

Allegrucci, C., Wu, Y. Z., Thurston, A., *et al.* (2007). Restriction landmark genome scanning identifies culture-induced DNA methylation instability in the human embryonic stem cell epigenome. *Human Molecular Genetics*, **16**, 1253–1268.

Baker, D. E., Harrison, N. J., Maltby, E., *et al.* (2007). Adaptation to culture of human embryonic stem cells and oncogenesis *in vivo*. *Nature Biotechnology*, **25**, 207–215.

Bibikova, M., Chudin, E., Wu, B., *et al.* (2006). Human embryonic stem cells have a unique epigenetic signature. *Genome Research*, **16**, 1075–1083.

Brown, C. J., Ballabio, A., Rupert, J. L., *et al.* (1991). A gene from the region of the human X inactivation centre is expressed exclusively from the inactive X chromosome. *Nature*, **349**, 38–44.

Calvanese, V., Horrillo, A., Hmadcha, A., *et al.* (2008). Cancer genes hypermethylated in human embryonic stem cells. *PLoS One*, **3**, e3294.

Chamberlain, S. J., Yee, D., and Magnuson, T. (2008). Polycomb repressive complex 2 is dispensable for maintenance of embryonic stem cell pluripotency. *Stem Cells*, **26**, 1496–1505.

Chamberlain, S. J., Chen, P. F., Ng, K. Y., *et al.* (2010). Induced pluripotent stem cell models of the genomic imprinting disorders Angelman and Prader–Willi syndromes. *Proceedings of the National Academy of Sciences USA*, **107**, 17 668–17 673.

Chin, M. H., Mason, M. J., Xie, W., *et al.* (2009). Induced pluripotent stem cells and embryonic stem cells are distinguished by gene expression signatures. *Cell Stem Cell*, **5**, 111–123.

Chow, J. and Heard, E. (2009). X inactivation and the complexities of silencing a sex chromosome. *Current Opinion in Cell Biology*, **21**, 359–366.

Christophersen, N. S. and Helin, K. (2010). Epigenetic control of embryonic stem cell fate. *Journal of Experimental Medicine*, **207**, 2287–2295.

Cooper, W. N., Luharia, A., Evans, G. A., *et al.* (2005). Molecular subtypes and phenotypic expression of Beckwith–Wiedemann syndrome. *European Journal of Human Genetics*, **13**, 1025–1032.

Cox, G. F., Burger, J., Lip, V., *et al.* (2002). Intracytoplasmic sperm injection may increase the risk of imprinting defects. *American Journal of Human Genetics*, **71**, 162–164.

Daniels, R., Zuccotti, M., Kinis, T., *et al.* (1997). *Xist* expression in human oocytes and preimplantation embryos. *American Journal of Human Genetics*, **61**, 33–39.

Deng, J., Shoemaker, R., Xie, B., *et al.* (2009). Targeted bisulfite sequencing reveals changes in DNA methylation associated with nuclear reprogramming. *Nature Biotechnology*, **27**, 353–360.

Dhara, S. K. and Benvenisty, N. (2004). Gene trap as a tool for genome annotation and analysis of X chromosome inactivation in human embryonic stem cells. *Nucleic Acids Research*, **32**, 3995–4002.

Frost, J. M., Monk, D., Moschidou, D., *et al.* (2011). The effects of culture on genomic imprinting profiles in human embryonic and fetal mesenchymal stem cells. *Epigenetics*, **6**, 52–62.

Ghosh, Z., Wilson, K. D., Wu, Y., *et al.* (2010). Persistent donor cell gene expression among human induced pluripotent stem cells contributes to differences with human embryonic stem cells. *PLoS One*, **5**, e8975.

Guenther, M. G., Levine, S. S., Boyer, L. A., *et al.* (2007). A chromatin landmark and transcription initiation at most promoters in human cells. *Cell*, **130**, 77–88.

Guenther, M. G., Frampton, G. M., Soldner, F., *et al.* (2010). Chromatin structure and gene expression programs of human embryonic and induced pluripotent stem cells. *Cell Stem Cell*, **7**, 249–257.

Hall, L. L., Byron, M., Butler, J., *et al.* (2008). X-inactivation reveals epigenetic anomalies in most hESC but identifies sublines that initiate as expected. *Journal of Cell Physiology*, **216**, 445–452.

Hanna, J., Cheng, A. W., Saha, K., *et al.* (2010). Human embryonic stem cells with biological and epigenetic characteristics similar to those of mouse ESCs. *Proceedings of the National Academy of Sciences USA*, **107**, 9222–9227.

Hawkins, R. D., Hon, G. C., Lee, L. K., *et al.* (2010). Distinct epigenomic landscapes of pluripotent and lineage-committed human cells. *Cell Stem Cell*, **6**, 479–491.

Hoffman, L. M., Hall, L., Batten, J. L., *et al.* (2005). X-inactivation status varies in human embryonic stem cell lines. *Stem Cells*, **23**, 1468–1478.

Horsthemke, B. and Buiting, K. (2008). Genomic imprinting and imprinting defects in humans. *Advances in Genetics*, **61**, 225–246.

Huntriss, J. and Picton, H. M. (2008). Stability of genomic imprinting in embryonic stem cells: lessons from assisted reproductive technology. *Current Stem Cell Research and Therapy*, **3**, 107–16.

Huntriss, J., Daniels, R., Bolton, V., *et al.* (1998). Imprinted expression of SNRPN in human preimplantation embryos. *American Journal of Human Genetics*, **63**, 1009–1014.

Kawakami, T., Zhang, C., Taniguchi, T., *et al.* (2004). Characterization of loss-of-inactive X in Klinefelter syndrome and female-derived cancer cells. *Oncogene*, **23**, 6163–6169.

Kim, K. P., Thurston, A., Mummery, C., *et al.* (2007). Gene-specific vulnerability to imprinting variability in human embryonic stem cell lines. *Genome Research*, **17**, 1731–1742.

Lagarkova, M. A., Shutova, M. V., Bogomazova, A. N., *et al.* (2010). Induction of pluripotency in human endothelial cells resets epigenetic profile on genome scale. *Cell Cycle*, **9**, 937–946.

Laurent, L., Wong, E., Li, G., *et al.* (2010). Dynamic changes in the human methylome during differentiation. *Genome Research*, **20**, 320–331.

Lengner, C. J., Gimelbrant, A. A., Erwin, J. A., *et al.* (2010). Derivation of pre-X inactivation human embryonic stem cells under physiological oxygen concentrations. *Cell*, **141**, 872–883.

Lighten, A. D., Hardy, K., Winston, R. M., *et al.* (1997). IGF2 is parentally imprinted in human preimplantation embryos. *Nature Genetics*, **15**, 122–123.

Lister, R., Pelizzola, M., Dowen, R. H., *et al.* (2009). Human DNA methylomes at base resolution show widespread epigenomic differences. *Nature*, **462**, 315–322.

Luedi, P. P., Dietrich, F. S., Weidman, J. R., *et al.* (2007). Computational and experimental identification of novel human imprinted genes. *Genome Research*, **17**, 1723–1730.

Maitra, A., Arking, D. E., Shivapurkar, N., *et al.* (2005). Genomic alterations in cultured human embryonic stem cells. *Nature Genetics*, **37**, 1099–1103.

Marchetto, M. C., Yeo, G. W., Kainohana, O., *et al.* (2009). Transcriptional signature and memory retention of human-induced pluripotent stem cells. *PLoS One*, **4**, e7076.

Marchetto, M. C., Carromeu, C., Acab, A., *et al.* (2010). A model for neural development and treatment of Rett syndrome using human induced pluripotent stem cells. *Cell*, **143**, 527–539.

Meissner, A. (2010). Epigenetic modifications in pluripotent and differentiated cells. *Nature Biotechnology*, **28**, 1079–1088.

Moreira de Mello, J. C., de Araujo, E. S., Stabellini, R., *et al.* (2010). Random X inactivation and extensive mosaicism in human placenta revealed by analysis of allele-specific gene expression along the X chromosome. *PLoS One*, **5**, e10947.

O'Carroll, D., Erhardt, S., Pagani, M., *et al.* (2001). The Polycomb-group gene *Ezh2* is required for early mouse development. *Molecular Cell Biology*, **21**, 4330–4336.

Okano, M., Bell, D. W., Haber, D. A., *et al.* (1999). DNA methyltransferases Dnmt3a and Dnmt3b are essential for de novo methylation and mammalian development. *Cell*, **99**, 247–257.

Orstavik, K. H., Eiklid, K., van der Hagen, C. B., *et al.* (2003). Another case of imprinting defect in a girl with Angelman syndrome who was conceived by intracytoplasmic semen injection. *American Journal of Human Genetics*, **72**, 218–219.

Pan, G., Tian, S., Nie, J., *et al.* (2007). Whole-genome analysis of histone H3 lysine 4 and lysine 27 methylation in human embryonic stem cells. *Cell Stem Cell*, **1**, 299–312.

Penny, G. D., Kay, G. F., Sheardown, S. A., *et al.* (1996). Requirement for *Xist* in X chromosome inactivation. *Nature*, **379**, 131–137.

Pick, M., Stelzer, Y., Bar-Nur, O., *et al.* (2009). Clone- and gene-specific aberrations of parental imprinting in human induced pluripotent stem cells. *Stem Cells*, **27**, 2686–2690.

Ray, P. F., Winston, R. M., and Handyside, A. H. (1997). *Xist* expression from the maternal X chromosome in human male preimplantation embryos at the blastocyst stage. *Human Molecular Genetics*, **6**, 1323–1327.

Rugg-Gunn, P. J., Ferguson-Smith, A. C., and Pedersen, R. A. (2005). Epigenetic status of human embryonic stem cells. *Nature Genetics*, **37**, 585–587.

Rugg-Gunn, P. J., Ferguson-Smith, A. C., and Pedersen, R. A. (2007). Status of genomic imprinting in human embryonic stem cells as revealed by a large cohort of independently derived and maintained lines. *Human Molecular Genet*, **16** Spec. No. 2, R243–51.

Salpekar, A., Huntriss, J., Bolton, V., *et al.* (2001). The use of amplified cDNA to investigate the expression of seven imprinted genes in human oocytes and preimplantation embryos. *Molecular Human Reproduction*, **7**, 839–844.

Shen, Y., Chow, J., Wang, Z., *et al.* (2006). Abnormal CpG island methylation occurs during in vitro differentiation of human embryonic stem cells. *Human Molecular Genetics*, **15**, 2623–2635.

Shen, Y., Matsuno, Y., Fouse, S. D., *et al.* (2008). X-inactivation in female human embryonic stem cells is in a nonrandom pattern and prone to epigenetic alterations. *Proceedings of the National Academy of Sciences USA*, **105**, 4709–4714.

Silva, S. S., Rowntree, R. K., Mekhoubad, S., *et al.* (2008). X-chromosome inactivation and epigenetic fluidity in human embryonic stem cells. *Proceedings of the National Academy of Sciences USA*, **105**, 4820–4825.

Smiraglia, D. J., Rush, L. J., Fruhwald, M. C., *et al.* (2001). Excessive CpG island hypermethylation in cancer cell lines versus primary human malignancies. *Human Molecular Genetics*, **10**, 1413–1419.

Sun, B. W., Yang, A. C., Feng, Y., *et al.* (2006). Temporal and parental-specific expression of imprinted genes in a newly derived Chinese human embryonic stem cell line and embryoid bodies. *Human Molecular Genetics*, **15**, 65–75.

Takahashi, K., Tanabe, K., Ohnuki, M., *et al.* (2007). Induction of pluripotent stem cells from adult human fibroblasts by defined factors. *Cell*, **131**, 861–872.

Tanasijevic, B., Dai, B., Ezashi, T., *et al.* (2009). Progressive accumulation of epigenetic heterogeneity during human ES cell culture. *Epigenetics*, **4**, 330–338.

Tchieu, J., Kuoy, E., Chin, M. H., *et al.* (2010). Female human iPSCs retain an inactive X chromosome. *Cell Stem Cell*, **7**, 329–342.

Thomson, J. A., Itskovitz-Eldor, J., Shapiro, S. S., *et al.* (1998). Embryonic stem cell lines derived from human blastocysts. *Science*, **282**, 1145–1147.

Tsumura, A., Hayakawa, T., Kumaki, Y., *et al.* (2006). Maintenance of self-renewal ability of mouse embryonic stem cells in the absence of DNA methyltransferases Dnmt1, Dnmt3a and Dnmt3b. *Genes to Cells*, **11**, 805–814.

van den Berg, I. M., Laven, J. S., Stevens, M., *et al.* (2009). X chromosome inactivation is initiated in human preimplantation embryos. *American Journal of Human Genetics*, **84**, 771–779.

Xie, W., Song, C., Young, N. L., *et al.* (2009). Histone H3 lysine 56 acetylation is linked to the core transcriptional network in human embryonic stem cells. *Molecular Cell*, **33**, 417–427.

Yu, J., Vodyanik, M. A., Smuga-Otto, K., *et al.* (2007). Induced pluripotent stem cell lines derived from human somatic cells. *Science*, **318**, 1917–1920.

Zhao, X. D., Han, X., Chew, J. L., *et al.* (2007). Whole-genome mapping of histone H3 Lys4 and 27 trimethylations reveals distinct genomic compartments in human embryonic stem cells. *Cell Stem Cell*, **1**, 286–298.

10 Impact of CpG methylation in addressing adipose-derived stem cell differentiation towards the cardiac phenotype

Alice Pasini, Francesca Bonafè, Emanuela Fiumana, Carlo Guarnieri, Paolo G. Morselli, Carlo M. Oranges, Claudio M. Caldarera, Claudio Muscari, and Emanuele Giordano*

10.1 Introduction

Cardiovascular diseases are a major cause of mortality in industrialized countries. Patients who survive after an acute myocardial infarction (AMI) are prone to ventricular remodeling, resulting from loss of myocardial tissue, and to progressive chronic heart failure (CHF). Heart has just a minimal potential of repair and regeneration, thus the use of new strategies of treatment involving stem cell transplantation and/or endogenous stem cell mobilization is expected to be a promising alternative to standard therapy.

Stem cells are undifferentiated cells with self-renewal and differentiation potential. Among stem cells, embryonic (ESCs) are considered as the best for cardiac regeneration (Nir *et al.*, 2003); on the other hand several issues, including ethical questions, immunorejection, and teratoma formation, limit their practical use. To overcome these restrictions, research interest is focusing on adult stem cells, identified in different tissues and found to be able to differentiate towards the cardiac phenotype. Stem cells obtained from bone marrow contain a subpopulation of hematopoietic stem cells (HSCs) (Goodell *et al.*, 1997), a component of mesenchymal stem cells (MSCs) (Pittenger and Martin, 2004), and multipotent progenitor cells (MAPCs) (Jiang *et al.*, 2002). MSCs derived from bone marrow show some potential of differentiation into beating cardiomyocytes in vitro (Makino *et al.*, 1999; Fukuda, 2001; Hakuno *et al.*, 2002; Toma *et al.*, 2002). Somatic stem cells also include endothelial progenitor cells (EPCs), obtained from peripheral circulation (Badorff *et al.*, 2003), and cells derived from umbilical cord (USSCs). USSCs showed capacity of differentiation towards the cardiac phenotype and to promote angiogenesis (Badorff *et al.*, 2003; Kogler *et al.*, 2004). Resident stem cells, located in cardiac niches, showed a differentiation potential

---

* Author to whom correspondence should be addressed.

Epigenomics: From Chromatin Biology to Therapeutics, ed. K. Appasani. Published by Cambridge University Press. © Cambridge University Press 2012.

towards cardiomyocytes (Bollini *et al.*, 2010). In the last decade another group of somatic stem cells, derived from adipose tissue (ADSCs), has been studied, most of all for their easy mode of extraction, relative abundance, and differentiative capacity towards different lineages.

This chapter will focus on this last family of somatic stem cells. We will describe the features of ADSCs, how to isolate them from lipoaspirates, their cell surface markers, and their differentiative potential. We will also report on ADSCs' ability to differentiate into cardiomyocytes. Finally we will outline the epigenetic signature of ADSCs, to identify whether epigenetic modifications could influence their commitment towards a specific phenotype.

10.2 Adipose-derived stem cells

Adipose derived stem cells (ADSCs) are stromal stem cells isolated from lipoaspirates (Zuk *et al.*, 2001). This cell population, called also processed lipoaspirate (PLA) cells, can be isolated from adipose tissue with a simple and minimal invasive procedure of lipoaspiration, with optimal cell yield given the amount of tissue from which they are derived (~3–5 million cells per 100 ml of lipoaspirate) (Zuk *et al.*, 2001). Therefore ADSCs represent an abundant and easily available pool of stem cells for autologous transplantation. All these features suggest ADSCs as showing promise for regenerative medicine.

10.2.1 Isolation protocol

Several expansion and differentiation protocols for ADSCs have been published in the literature. The protocol universally used for isolating ADSCs is a modified version of the method originally described by Zuk *et al.* (2001): tissue is washed with phosphate-buffered saline (PBS) and treated with collagenase at 37 °C under shaking for about 30 minutes. An equal volume of Dulbecco's modified Eagle's medium (DMEM) supplemented with 10% fetal bovine serum (FBS) is then added to neutralize collagenase activity. This digested solution is then centrifuged for 10 minutes at low speed, the cell pellet resuspended in DMEM/10% FBS, and the resulting solution filtered through a 40- to 200-μm mesh nylon filter. The resulting solution is finally centrifuged again and the cells are resuspended in complete expansion medium (Sanz-Ruiz *et al.*, 2008). We add an intermediate step before resuspension consisting in erythrocyte lysis: the cell pellet is resuspended in lysis buffer (155 mM NH_4Cl, 10 mM $KHCO_3$, 0.1 mM EDTA) and incubated for 5 minutes at room temperature. Lysis is stopped by adding an equal volume of DMEM/10% FBS.

In preclinical and clinical trials ADSCs were used either immediately after tissue digestion with collagenase, or after in vitro expansion during three or four passages (Sanz-Ruiz *et al.*, 2008).

10.2.2 Immunophenotypic characterization

ADSCs show many properties in common with bone-marrow stromal cells (MSCs), including cell-surface phenotype, that are similar but not identical

Table 10.1 Expression of cell-surface markers in ADSCs and MSCs

Cell-surface marker	ADSCs	MSCs
CD9	+	+
CD10	+	+
CD13	+	+
CD29	+	+
CD31	+/− (depending on the study)	−
CD34	−(+ just in early passages)	−
CD44	+	+
CD45	−	−
CD49d	+	−
CD54	+	+
CD55	+	+
CD71	+	+
CD73	+	+
CD90	+	+
CD91	+	+
CD105	+	+
CD106.	−	+
CD146	+	+
CD166	+	−
Stro-1	+/− (depending on the study)	+
SH3	+	+

(Table 10.1). Both ADSCs and MSCs express the cell-surface markers CD9, CD10, CD13, CD29, CD44 (hyaluronate receptor), CD54, CD55, CD71, CD73 (ecto 5′ nucleotidase), CD90 (or Thy-1), and CD91. Both are also positive for endothelial markers CD105/endoglin and CD146. On the other hand, the expression of the hematopoietic lineage marker CD31 is virtually absent (although one study detected a basal expression of CD31 in freshly isolated cells: Boquest *et al.*, 2007) and CD34 is observed only in early passages. CD45 (pan-leucocytes) is lacking. CD49d (α4 integrin) and CD106 (VCAM), a receptor–ligand pair important for hematopoietic stem cells homing and mobilization, show opposite patterns of expression in ADSCs and MSCs. Interestingly ADSCs express just one component of the VCAM receptor, CD49d, but not the component CD106 of the ligand. MSCs only show CD106.

ADSCs tested negative for the following stem cell and hematopoietic markers: CD133, CD117, CD11b (granulocytes, monocytes, natural killer), CD14 (monocytes), and CD19 (lymphocytes). No expression of the histocompatibility locus antigen-DR (HLA-DR) has been shown: the absence of this marker enables these cells for allogenic clinical use because of low immunity reactions (Zuk *et al.*, 2002; Fraser *et al.*, 2004, 2006; Sanz-Ruiz *et al.*, 2008). Stro-1, a marker used for isolating multilineage progenitors from bone marrow, and SH3 were found to be expressed by both ADSCs and MSCs (Zuk *et al.*, 2002). The expression of Stro-1 in ADSCs was not reported by Gronthos *et al.* (2001).

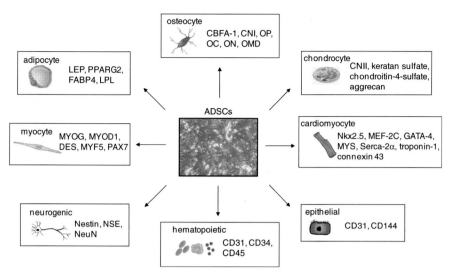

Figure 10.1 Differentiation potential of ADSCs towards multiple cell lineages, with reference to the expression pattern of specific markers.

10.3   Multilineage differentiation potential of ADSCs

Stem cells isolated from adipose tissue show an attractive capacity of differentiation towards many cell lineages (Figure 10.1) depending on culture conditions and media supplementation.

ADSCs can differentiate into the phenotype of the tissue from which they are isolated, therefore into adipocytes. This is highlighted in vitro by large intracellular lipid droplets. Adipogenic differentiation results in upregulation of adipogenic-specific genes: peroxisome proliferator-activated receptor γ2 (*PPARG2*), lipoprotein lipase (*LPL*), leptin (*LEP*), and fatty acid binding protein 4 (*FABP4*) (Zuk *et al.*, 2001, 2002; Noer *et al.*, 2006).

ADSCs can be induced to differentiate towards the osteogenic lineage, using 1,25-dihydroxyvitamin D3 (VD) in culture media. Osteogenic differentiation results in the appearance of alkaline phosphatase (AP) activity, which is involved in matrix calcification, highlighted by functional matrix mineralization. The expression of core-binding factor α-1 (*CBFA-1*), a transcription factor binding to the promoters of different osteogenic genes, is upregulated, as well as several osteogenic markers such as type I collagen (*CNI*), osteopontin (*OP*), osteonectin (*ON*), and osteomodulin (*OMD*), which codify for bone matrix proteins and the specific marker osteocalcin (*OC*) (Zuk *et al.*, 2001, 2002; Boquest *et al.*, 2006).

Chondrogenic differentiation of ADSCs has been obtained under micromass cell culture conditions and results in the formation of spheroid nodules showing the expression of specific chondrogenic genes: keratan sulfate, chodroitin-4-sulfate, cartilaginous type II collagen (*CNII*) and aggrecan (Zuk *et al.*, 2001, 2002; Dragoo *et al.*, 2003).

ADSCs have been induced to myogenic differentiation through culture in myogenic medium. The differentiation process is underlined by the expression of specific myogenic genes: myogenic transcription factor 1 (*MYOD1*), myogenin (*MYOG*), myogenic regulatory factor 5 (*MYF5*), myosin heavy-chain (*MYS*), and desmin (*DES*) (Zuk *et al.*, 2001, 2002; Mizuno *et al.*, 2002; Goudenege *et al.*, 2009).

The differentiation of ADSCs towards cardiac muscle phenotype will be treated in detail in the following section.

Neurogenic differentiation of ADSCs is evaluated by changes in cell morphology towards a neuronal-like phenotype with compact cell bodies and multiple extensions. Neuronal markers *nestin*, *NSE*, and *NeuN* were expressed in induced differentiated cells, but no expression of mature neuron markers *MAP-2* and *NF-70* was observed. Absence of expression of choline acetyltransferase, myelin-basic protein (MBP) and glial fibrillary acidic protein (GFAP) suggests that ADSCs do not assume oligodendrocyte and astrocyte phenotypes (Zuk *et al.*, 2002; Erba *et al.*, 2010).

Although Zuk *et al.* (Zuk, 2010) referred to different studies showing commitment of ADSCs towards other specific phenotypes such as epithelial (Brzoska *et al.*, 2005) and hematopoietic (Cousin *et al.*, 2003), Boquest *et al.* suggested that epigenetic silencing of two endothelial-cell-specific genes, CD31 and CD144, can be associated with poor differentiation potential of ADSCs towards the endothelial phenotype (Boquest *et al.*, 2007).

10.4   Differentiation of ADSCs towards the cardiac phenotype

Different studies have shown evidence of differentiation of ADSCs towards the cardiac phenotype under different conditions.

In vitro cardiac commitment of ADSCs isolated from New Zealand White rabbits subjected to treatment with different concentrations of the demethylating agent 5-azacytidine has been shown (Rangappa *et al.*, 2003). Cells formed a spheroid structure that began to beat spontaneously after 3 weeks. These induced differentiated cells tested positive for the presence of the cardiac specific proteins myosin heavy-chain, α-actinin and troponin-I.

Planat-Bénard *et al.* reported spontaneous in vitro differentiation of mouse ADSCs directly plated in a semisolid methylcellulose medium (Planat-Bénard *et al.*, 2004). After 11 days some round cells started a contractile activity, and tended to assume a myotube-like structure after some days. After 20 to 30 days these cells appeared as a cohesive group of cells beating in synchrony, with branching fibers and tight connections. The contractive clones expressed the cardiac-specific transcription factors *Nkx2.5* and *GATA-4* and the cardiac-specific genes *ANP* and myosin light chain 2a (*MLC-2a*) and 2v (*MLC-2v*). Immunostaining exhibited the presence of multiple cardiomyocyte specific proteins: myocyte enhancer factor 2C (*MEF-2C*), sarcomeric α-actinin, *MLC-2v*, β-myosin heavy chain, connexin 40, and connexin 43.

Cardiomyocyte progenitors derived from brown adipose tissue (BAT) positive for the cell-surface marker CD29 have been directly transplanted into the infarct border zone of an acute myocardial infarction model in rat, leading to the reduction of the infarction area and improvement of left ventricular function (Yamada *et al.*, 2006). Immunohistochemical staining of implanted cells tested positive for cardiomyocyte markers *GATA-4*, *MEF-2C*, cardiac troponin-T, and connexin 43.

Exposure of ADSCs to rat cardiomyocyte extract was shown to induce ADSC differentiation towards the cardiac phenotype. After 3 weeks of treatment, cells changed morphology forming binucleated and striated cells beating spontaneously, and expressed cardiac markers such as sarcomeric α-actinin, desmin, cardiac troponin-T, and connexin 43 (Gaustad *et al.*, 2004).

ADSCs were cultured in a semisolid methylcellulose medium where some clusters changed morphology, becoming elongated and beating. These cells injected into a murine infarct model enhanced myocardial function and promoted angiogenesis and reduction of remodeling (Léobon *et al.*, 2009).

Strem *et al.* directly injected mouse ADSCs into the intraventricular chamber of B6129S recipient mice, immediately following induction of myocardial cryoinjury; after 14 days the injected cells showed the expression of myosin heavy chain, *Nkx2.5*, and troponin-1 (Strem *et al.*, 2005).

All these studies have established that ADSCs can be induced to differentiate towards the cardiac phenotype, enhancing functional cardiac activity when implanted into infarcted animal models. These findings confirm that ADSCs represent a new and attractive strategy for myocardial infarction therapy.

10.5   Epigenetic signature of ADSCs

Epigenetic mechanisms of gene expression modulation are mainly studied in cancer development and progression, represented by transcriptional silencing of tumor suppressor genes via CpG island methylation. However, epigenetics is physiologically involved in development and differentiation, participating in the determination of which gene pathway should be switched on or off (Bernstein *et al.*, 2007).

We pass over the description of epigenetic mechanisms of gene expression regulation that include CpG island DNA methylation, histone modifications, and chromatin remodeling factors, which are treated in other chapters of this book, and we focus on reporting what is known today about the epigenetic profile of ADSCs.

The CpG promoter methylation state of some genes specific for multiple lineages of commitment has been evaluated in undifferentiated and differentiated ADSCs in several studies. The methylation status of most relevant genes is reported in Table 10.2.

Methylation status of genes regulating adipogenesis was first evaluated, since ADSCs are isolated from adipose tissue. Bisulfite sequencing analysis of four adipogenic genes was performed: leptin (*LEP*), peroxisome proliferator-activated receptor γ2 (*PPARG2*), fatty acid binding protein 4 (*FABP4*), and lipoprotein lipase (*LPL*) (Boquest *et al.*, 2006; Noer *et al.*, 2006). The promoters were globally hypomethylated, with 5–30% of methylated sites, depending on the donor; moreover no changes in the methylation status were observed in freshly isolated cells compared to cultured cells (Noer *et al.*, 2006). Induction of differentiation towards

Table 10.2 Promoter methylation status of a pool of genes specific for different lineages of differentiation in ADSCs; CpG methylation status of cardiac genes was tested through methylation-specific PCR

Differentiative lineage	Gene	% CpG methylation
Adipogenic	*LEP*	10–15% methylated
	PPARG2	20–30% methylated
	FABP4	20–30% methylated
	LPL	15–20% methylated
Myogenic	*MYOG*	34% methylated
	MYOD1	0–5% methylated
	PAX7	10% methylated
	MYF5	10% methylated
Endothelial	*CD31*	100% methylated
	CD144	30–40% methylated
Cardiac	*Nkx2.5*	unmethylated
	GATA-4	unmethylated
	MEF-2C	unmethylated
Embryonic	*Nanog*	0–5% methylated
	OCT4	20–30% methylated

osteogenic and endothelial lineages did not cause modifications in the methylation status of adipogenic genes, which remained poorly methylated; however a decrease was expected when the cells were induced to differentiate towards the adipogenic phenotype. This finding suggests that the hypomethylation status of adipogenic genes could represent a constitutive signature of ADSCs.

In previous studies, methylation analysis of non-adipogenic genes was performed, showing promoter hypermethylation of *MYOG* (myogenic), *CD31* and *CD144* (endothelial), and osteoglycine (*OGN*) (osteogenic) (Boquest *et al.*, 2006, 2007). In these studies, despite the highly methylated promoter status, a basal expression of those genes was observed, indicating that DNA methylation does not represent the unique actor of gene expression regulation, where also histone modifications participate, playing a relevant role. An interesting correlation was found between the endothelial gene methylation (*CD31* and *CD144*) and the low differentiation potential of ADSCs towards the endothelial phenotype (Boquest *et al.*, 2007). Despite a basal expression of both genes in undifferentiated cells, *CD31* was found strongly methylated and *CD144* mosaically methylated. Induction of endothelial differentiation determined a marginal upregulation of both genes, without demethylation of promoters. Only CD31+ cells showed consistent demethylation on the 3′ end of the transcription start site of the *CD31* promoter. Induced adipogenic and osteogenic commitment maintained *CD31* and *CD144* methylation patterns. These results suggested that *CD31* and *CD144* hypermethylated status limited the differentiation capacity of ADSCs towards the endothelial phenotype (Boquest *et al.*, 2007).

Following studies confirmed the results regarding *CD31* and analyzed the methylation status of myogenic genes (*MYOD1*, *PAX7*, and *MYF5*), including

MYOG. They showed 34% of CpG sites methylated for *MYOG*, and under 10% for *MYOD1*, *PAX7*, and *MYF5*, suggesting that the methylation status of myogenic genes did not influence the differentiation potential of ADSCs towards the myogenic phenotype (Sørensen *et al.*, 2010a). In this study, Sørensen *et al.* compared also the methylation status of developmental genes such as *OCT4* and *NANOG* in ESCs, ADSCs, and MSCs, showing hypomethylation of the promoter region of developmental genes in each type of stem cells. This common feature is representative for the pluripotency potential of these stem cells. On the other hand ESCs showed hypermethylation of lineage-specific genes, unlike ADSCs and MSCs. None of the studies found in the literature has evaluated the methylation status of cardiac-specific genes. Since several studies indicate the potential commitment of ADSCs towards the cardiac phenotype, it is interesting to evaluate CpG promoter methylation in cardiac genes activated early during cardiac transdifferentiation. Transcription factors indicative for cardiac commitment are *Nkx2.5*, *GATA-4*, and *MEF-2C* (Doevendans and van Bilsen, 1996). We tested by methylation-specific PCR (MSP) analysis their promoter methylation status in undifferentiated ADSCs. The regions amplified by MSP are indicated in Figure 10.2. All the promoters remained unmethylated (Figure 10.3), suggesting that these cells are not prevented from differentiating towards the cardiac phenotype by DNA methylation, as is the case for myogenic genes (Sørensen *et al.*, 2010b).

A genome-wide analysis of hypermethylated genes by immunoprecipitation of methylated DNA (MeDIP) followed by microarray hybridization underlined that ADSCs showed 74% of their genes hypermethylated in common with MSCs derived from bone marrow and 57% with muscle progenitor cells (MPCs), supporting the view of a common mesenchymal origin (Sørensen *et al.*, 2010b). This

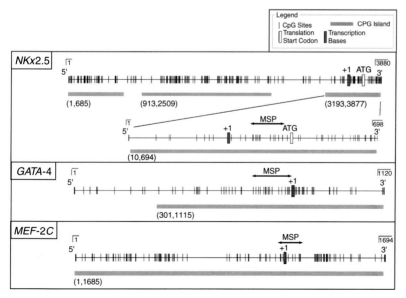

Figure 10.2 CpG island representation of *Nkx2.5*, *GATA-4*, and *MEF-2C* promoters. The arrow indicates the region amplified by methylation-specific PCR (MSP).

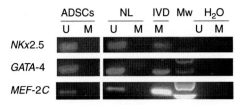

Figure 10.3 Promoter methylation status of cardiac-specific transcription factors *Nkx2.5*, *GATA-4*, and *MEF-2C* by MSP in ADSCs from a representative donor. U, unmethylation specific reaction; M, methylation-specific reaction; NL, normal lymphocytes DNA; IVD, in vitro methylated DNA; Mw, molecular weight marker; H_2O, water negative control.

study also shows that most lineage-specific promoters are hypomethylated and harbor a combination of the bivalent markers, trimethylated H3K4 and H327 (H3K4me3 and H3K27me3), representative of transcriptionally activated and repressed chromatin, respectively. Bivalent chromatin was identified by Bernstein *et al.*, who found the simultaneous presence of both active and repressive chromatin markers in the same domains harboring genes regulating development in ESCs, and suggested that this chromatin configuration keeps genes poised to be rapidly addressed towards either expression or silencing (Bernstein *et al.*, 2006). Also in undifferentiated ADSCs some of the lineage-specific gene promoters (*LPL*, *PPARG2*, and *MYOG*) were shown to be co-enriched in H3K4me3 and H3K27me3 markers, suggesting that these histone modifications are consistent with the inactive state of these promoters in pluripotent cells (Noer *et al.*, 2009). Induction of adipogenic differentiation of ADSCs determined in adipogenic genes (*LPL*, *PPARG2*, and *FABP4*) the demethylation of the repressive marker H3K27me3 to H3K27me2, and the gain of the H3K9 acetylation (H3K9ac) are representative of heterochromatin; the myogenic gene *MYOG* retained the bivalent state as in undifferentiated cells. Culture until senescence resulted in the loss of differentiation capacity in ADSCs, which is associated with (a) the presence of the repressive marker H3K27me3, otherwise lost after the induction of differentiation in early passages, and (b) the maintained active marker H3K4me3: the genes may be transcribed as in early passages but at a strongly reduced level (Noer *et al.*, 2009).

The results obtained to date suggest that presence of hypomethylation of the promoter region of lineage-specific genes in ADSCs does not imply transcriptional activity and does not represent a barrier to differentiation potential towards the specific phenotype. These genes frequently show a bivalent chromatin state in their promoter region, keeping them poised for activation or repression depending on cell program. On the other hand, DNA promoter hypermethylation and repressive histone markers prevent the differentiation capacity of ADSCs.

10.6   Conclusions

The heart has only a minimal capacity of repair and regeneration, consequently the use of new strategies for treatment involving stem cells is becoming a promising alternative to standard therapy.

Among adult stem cells a group of somatic stem cells derived from adipose tissue (ADSCs) has become an attractive candidate for regenerative medicine in recent years. The advantages of ADSCs are identified as their easy means of extraction by liposuction, relative abundance, differentiative capacity towards different lineages, and autologous transplantation.

Several studies have established that ADSCs can be induced to differentiate towards multilineage phenotypes including the cardiac phenotype. Moreover they have been shown to enhance functional heart activity when implanted into infarcted animal models. These findings confirm that ADSCs represent a new and attractive strategy for myocardial infarction therapy.

The epigenetic promoter methylation signature of ADSCs suggests that they are not epigenetically prevented from differentiating towards the cardiac lineage; however more studies are needed to explore the chromatin configuration associated with cardiac gene promoters.

ACKNOWLEDGMENT

The National Institute for Cardiovascular Research (INRC) granted resources to partially cover the fellowships of AP, FB, and EF.

REFERENCES

Badorff, C., Brandes, R. P., Popp, R., *et al.* (2003). Transdifferentiation of blood-derived human adult endothelial progenitor cells into functionally active cardiomyocytes. *Circulation*, **107**, 1024–1032.

Bernstein, B. E., Mikkelsen, T. S., Xie, X., *et al.* (2006). A bivalent chromatin structure marks key developmental genes in embryonic stem cells. *Cell*, **125**, 315–326.

Bernstein, B. E., Meissner, A., and Lander E. S. (2007). The mammalian epigenome. *Cell*, **128**, 669–681.

Bollini, S., Smart, N., and Riley, P. R. (2010). Resident cardiac progenitor cell: at the heart of regeneration. *Journal of Molecular and Cellular Cardiology*, **50**, 296–303.

Boquest, A. C., Noer, A., and Collas. P. (2006). Epigenetic programming of mesenchymal stem cells from human adipose tissue. *Stem Cell Reviews*, **2**, 319–329.

Boquest, A. C., Noer, A., Sørensen, A. L., Vekterud, K., and Collas, P. (2007). CpG methylation profiles of endothelial cell-specific gene promoter regions in adipose tissue stem cells suggest limited differentiation potential toward the endothelial cell lineage. *Stem Cells*, **25**, 852–861.

Brzoska, M., Geiger, H., Gauer, S., and Baer, P. (2005). Epithelial differentiation of human adipose tissue-derived adult stem cells. *Biochemical and Biophysical Research Communications*, **330**, 142–150.

Cousin, B., André, M., Arnaud, E., Pénicaud, L., and Casteilla, L. (2003). Reconstitution of lethally irradiated mice by cells isolated from adipose tissue. *Biochemical and Biophysical Research Communications*, **301**, 1016–1022.

Doevendans, P. A. and van Bilsen, M. (1996). Transcription factors and the cardiac gene programme. *International Journal of Biochemistry and Cell Biology*, **28**, 387–403.

Dragoo, J. L., Samimi, B., Zhu, M., *et al.* (2003). Tissue-engineered cartilage and bone using stem cells from human infrapatellar fat pads. *Journal of Bone and Joint Surgery, British Volume*, **85**, 740–747.

Erba, P., Terenghi, G., and Kingham, P. J. (2010). Neural differentiation and therapeutic potential of adipose tissue derived stem cells. *Current Stem Cell Research and Therapy*, **5**, 153–160.

Fraser, J. K., Schreiber, R. E., Zuk, P. A., and Hedrick, M. H. (2004). Adult stem cell therapy for the heart. *International Journal of Biochemistry and Cell Biology*, **36**, 658–666.

Fraser, J. K., Schreiber, R. E., Strem, B., *et al.* (2006). Plasticity of human adipose stem cells toward endothelial cells and cardiomyocytes. *Nature Clinical Practice Cardiovascular Medicine*, **3** (Suppl. 1), S33–S37.

Fukuda, K. (2001). Development of regenerative cardiomyocytes from mesenchymal stem cells for cardiovascular tissue engineering. *Artificial Organs*, **25**, 187–193.

Gaustad, K. G., Boquest, A. C., Anderson, B. E., Gerdes, A. M., and Collas, P. (2004). Differentiation of human adipose tissue stem cells using extracts of rat cardiomyocytes. *Biochemical and Biophysical Research Communications*, **314**, 420–427.

Goodell, M. A., Rosenzweig, M., Kim, H., *et al.* (1997). Dye efflux studies suggest that hematopoietic stem cells expressing low or undetectable levels of CD34 antigen exist in multiple species. *Nature Medicine*, **3**, 1337–1345.

Goudenege, S., Pisani, D. F., Wdziekonski, B., *et al.* (2009). Enhancement of myogenic and muscle repair capacities of human adipose-derived stem cells with forced expression of MyoD. *Molecular Therapy*, **17**, 1064–1072.

Gronthos, S., Franklin, D. M., Leddy, H. A., *et al.* (2001). Surface protein characterization of human adipose tissue-derived stromal cells. *Journal of Cellular Physiology*, **189**, 54–63.

Hakuno, O., Fukuda, K., Makino, S., *et al.* (2002). Bone marrow-derived regenerated cardiomyocytes (CMG cells) express functional adrenergic and muscarinic receptors. *Circulation*, **105**, 380–386.

Jiang, Y., Vaessen, B., Lenvik, T., *et al.* (2002). Multipotent progenitor cells can be isolated from postnatal murine bone marrow, muscle, and brain. *Experimental Hematology*, **30**, 896–904.

Kögler, G., Sensken, S., Airey, J. A., *et al.* (2004). A new human somatic stem cell from placental cord blood with intrinsic pluripotent differentiation potential. *Journal of Experimental Medicine*, **200**, 123–135.

Léobon, B., Roncalli, J., Joffre, C., *et al.* (2009). Adipose-derived cardiomyogenic cells: in vitro expansion and functional improvement in a mouse model of myocardial infarction. *Cardiovascular Research*, **83**, 757–767.

Makino, S., Fukuda, K., Iyoshi, M. S., *et al.* (1999). Cardiomyocytes can be generated from marrow stromal cells in vitro. *Journal of Clinical Investigation*, **103**, 697–705.

Mizuno, H., Zuk, P. A., Zhu, M., *et al.* (2002). Myogenic differentiation by human processed lipoaspirate cells. *Plastic and Reconstructive Surgery*, **109**, 199–209.

Nir, S. G., David, R., Zaruba, M., Franz, W. M., and Itskovitz-Eldor, J. (2003). Human embryonic stem cells for cardiovascular repair. *Cardiovascular Research*, **58**, 313–323.

Noer, A., Sørensen, A. L., Boquest, A. C., and Collas, P. (2006). Stable CpG hypomethylation of adipogenic promoters in freshly isolated, cultured, and differentiated mesenchymal stem cells from adipose tissue. *Molecular Biology of the Cell*, **17**, 3543–3556.

Noer, A., Lindeman, L. C., and Collas, P. (2009). Histone H3 modifications associated with differentiation and long-term culture of mesenchymal adipose stem cells. *Stem Cells and Development*, **18**, 725–736.

Pittenger, M. F. and Martin, B. J. (2004). Mesenchymal stem cells and their potential as cardiac therapeutics. *Circulation Research*, **95**, 9–20.

Planat-Benard, V., Silvestre, J. S., Cousin, B., *et al.* (2004). Plasticity of human adipose lineage cells toward endothelial cells: physiological and therapeutic perspectives. *Circulation*, **109**, 656–663.

Rangappa, S., Fen, C., Lee, E. H., Bongso, A., and Sim E. K. (2003). Transformation of adult mesenchymal stem cells isolated from the fatty tissue into cardiomyocytes. *Annals of Thoracic Surgery*, **75**, 775–779.

Sanz-Ruiz, R., Santos, M. E., Muñoa, M. D., *et al.* (2008). Adipose tissue-derived stem cells: the friendly side of a classic cardiovascular foe. *Journal of Cardiovascular Translational Research*, **1**, 55–63.

Sørensen, A. L., Timoskainen, S., West, F. D., *et al.* (2010a). Lineage-specific promoter DNA methylation patterns segregate adult progenitor cell types. *Stem Cells and Development*, **19**, 1257–1266.

Sørensen, A. L., Jacobsen, B. M., Reiner, A. H., Andersen, I. S., and Collas, P. (2010b). Promoter DNA methylation patterns of differentiated cells are largely programmed at the progenitor stage. *Molecular Biology of the Cell*, **15**, 2066–2077.

Strem, B. M., Zhu, M., Alfonso, Z., *et al.* (2005). Expression of cardiomyocytic markers on adipose tissue-derived cells in a murine model of acute myocardial injury. *Cytotherapy*, **7**, 282–291.

Toma, C., Pittenger, M. F., Cahill, K. S., Byrne, B. J., and Kessler, P. O. (2002). Human mesenchymal stem cells differentiate to a cardiomyocyte phenotype in the adult murine heart. *Circulation*, **105**, 93–98.

Yamada, Y., Wang, X. D., Yokoyama, S., Fukuda, N., and Takakura, N. (2006). Cardiac progenitor cells in brown adipose tissue repaired damaged myocardium. *Biochemical and Biophysical Research Communications*, **342**, 662–670.

Zuk, P. A. (2010). The adipose-derived stem cell: looking back and looking ahead. *Molecular Biology of the Cell*, **21**, 1783–1787.

Zuk, P. A., Zhu, M., Mizuno, H., *et al.* (2001). Multilineage cells from human adipose tissue: implications for cell-based therapies. *Tissue Engineering*, **7**, 211–228.

Zuk, P. A., Zhu, M., Ashjian, P., *et al.* (2002). Human adipose tissue is a source of multipotent stem cells. *Molecular Biology of the Cell*, **13**, 4279–4295.

11 Regulation of the stem cell epigenome by REST

Angela Bithell* and Noel J. Buckley

11.1 Introduction

Complex gene regulatory networks control the acquisition and maintenance of cellular phenotype and function. This is nowhere more evident than in stem cell maintenance and differentiation. Stem cells are unique in that they can self-renew yet retain the potential to differentiate into specialized cell types in response to signals received during development, injury, or disease. As such they act as a repository of cells that have the potential to be recruited to repair injury or degeneration. This unique plasticity is maintained by signals released from the stem cell niche that coordinately regulate the stem cell transcriptome and epigenome. To a large extent, the hunt for molecular hallmarks of "stemness" has focused on identifying either transcriptional regulatory pathways or epigenomic signatures (Cai *et al.*, 2004; Spivakov and Fisher, 2007). However, it is clear that transcriptional and epigenetic regulation are obverse sides of the same coin, since epigenetic modifiers are recruited by transcription factors and, obversely, transcription factor recruitment is determined by the epigenetic status of the target gene chromatin. Consequently, we need to identify factors that act to coordinate regulation of the transcriptome and epigenome.

Epigenetic marks largely consist of posttranslational modifications on chromatin (largely, but not exclusively, to N-terminal histone tails or methylation of CpG dinucleotides) which act as a layer of information that does not affect DNA sequence but influences the way in which the genome is "read." An ever-increasing list of histone modifications (including acetylation, methylation, phosphorylation, and ubiquitination), and the proteins that "read," "write" (deposit), and remove them, are now known (Kouzarides, 2007; Schones and Zhao, 2008). Much of the recent surge of interest in epigenomics has been fueled

---

* Author to whom correspondence should be addressed.

Epigenomics: From Chromatin Biology to Therapeutics, ed. K. Appasani. Published by Cambridge University Press. © Cambridge University Press 2012.

by the success of next-generation sequencing platforms in generating whole-genome maps of epigenetic marks (Barski *et al.*, 2007; Meissner *et al.*, 2008). An important concept to emerge is the overturning of the simplistic view of active versus repressive chromatin marks, since it is clear that many marks associated with active chromatin coexist with those associated with repressive chromatin, and this is reflected by the co-recruitment of the corresponding histone-modifying activities (Berger, 2007; Meaney and Ferguson-Smith, 2010).

One pivotal reflection of this complexity in relation to development is the concept of bivalency. This idea was born from studies carried out in the laboratories of Amanda Fisher and Eric Lander (Azuara *et al.*, 2006; Bernstein *et al.*, 2006; Mikkelsen *et al.*, 2007), both of whom showed that the promoters of many silent lineage-specific genes in embryonic stem cells (ESCs) carried both "active" epigenetic marks such as H3K4me3 and H3K9ac and also "repressive" marks such as H3K27me3. This was seen as means by which ESCs maintained the chromatin of lineage-specific developmental regulators in a silent but "poised" configuration that was nevertheless readily activatable upon receipt of differentiation cues. Since their original description in ESCs, it has been shown that bivalent promoters are not restricted to ESCs and neither are bivalent signatures restricted to expression of H3K4me3/H3K27me3 marks (Roh *et al.*, 2006; Barski *et al.*, 2007; Mikkelsen *et al.*, 2007). These seminal observations and a host of subsequent studies leave little doubt about the importance of the epigenome in regulation of developmental processes. The epigenome lies at the interface of regulatory signals and control of genomic readout and, as such, acts as a natural developmental regulatory gateway. However, we know relatively little of the transcription factors that are responsible for writing and reading these key epigenomic signatures.

In this chapter we discuss repressor element 1 silencing transcription factor (REST), erstwhile recognized as a key neural transcription factor that acts as a gateway regulator which balances stem cell maintenance and differentiation through coordinated regulation of the epigenome and transcriptome. We will describe studies from several groups, including our own, that have contributed to our current understanding and shape our future goals.

11.1.2 A history of REST

REST, also known as neuron-restrictive silencer factor (NRSF) is a Gli-Kruppel transcription factor originally identified as a transcriptional repressor of a number of neuron-specific genes in non-neuronal cells (Chong *et al.*, 1995; Schoenherr and Anderson, 1995a; Schoenherr *et al.*, 1996). Its identification came about via its recognition of a 21-bp, highly conserved DNA regulatory motif know as repressor element 1 (RE1, or neuron-restrictive silencer element) known to be necessary for repression of the SCG10 and type II sodium channel (NaV1.2) genes in non-neuronal cells (Kraner *et al.*, 1992; Mori *et al.*, 1992). Initially, this led to the perception of REST as a master regulator of neuronal genes (Schoenherr and Anderson, 1995b) though it was later found to also target non-neuronal genes (Schoenherr *et al.*, 1996). It has since become clear that REST indeed plays an important role in neurogenesis and regulation of many aspects of neuronal function. However, it is also apparent that the functional outcomes of REST action are highly context-specific and its diverse repertoire of functions includes tissue-specific roles outside of the nervous system, including in the heart (Kuwahara *et al.*, 2003), the pancreas (Atouf *et al.*, 1997; Martin *et al.*, 2008), and vascular smooth muscle (Cheong *et al.*, 2005). Likewise, it has been implicated in the etiology of many pathological conditions including Huntington's disease (HD, reviewed in Buckley *et al.* [2010]), cancer (Westbrook *et al.*, 2005; Majumder, 2006), Down's syndrome (Canzonetta *et al.*, 2008; Lepagnol-Bestel *et al.*, 2009), cardiac hypertrophy (Kuwahara *et al.*, 2003), and ischemia (Calderone *et al.*, 2003).

REST is widely expressed during development and in the adult organism (Chen *et al.*, 1998; Palm *et al.*, 1998). Despite intensive research to dissect REST function, there remains relatively little known about the factors responsible for regulation of its expression. In the chick, WNT signaling directly regulates REST expression in neural progenitor cells in the spinal cord (Nishihara *et al.*, 2003). Subsequent studies in human and mouse ESCs have identified Nanog, Sox2, Oct4, and REST binding sites in the *Rest* gene and loss of Nanog and Oct4 leads to reduced REST expression (Boyer *et al.*, 2005; Loh *et al.*, 2006; Chen *et al.*, 2008). Human, mouse, and rat *Rest* genes show structural conservation (Koenigsberger *et al.*, 2000) and transcription occurs from three alternate 5′ exons (exons a–c), with a number of splice isoforms generating both full-length REST and several shorter proteins, most notably REST4 (Palm *et al.*, 1998, 1999). Despite multiple promoters, none have been shown to be cell-type-specific and thus fail to account for the complex spatiotemporal expression and functions of REST (Kojima *et al.*, 2001).

REST is essential for mouse embryonic development and constitutive knockouts die mid-gestation, with widespread apoptosis but without evidence of widespread REST target gene derepression although derepression of the neuron-specific gene encoding βIII-tubulin was observed (Chen *et al.*, 1998). This early embryonic death has precluded analysis of REST function at later development stages in vivo and future studies will need to circumvent embryonic lethality using conditional knockout strategies. However, constitutive *Rest*$^{-/-}$ embryos did provide evidence that REST-mediated gene repression is complex and context-specific, given that there was no evidence for widespread derepression of known target genes.

11.2   REST is a transcriptional and epigenetic regulator

11.2.1   REST acts as a hub for recruitment of chromatin-modifying activity

REST acts as a hub for the recruitment of many chromatin-modifying activities (Ooi and Wood, 2007), either by direct interactions or indirectly via recruitment of co-repressor complexes to its N- and C-terminal repressor domains (Figure 11.1). The N-terminal repressor domain acts as a docking site for the co-repressor mSin3 (Naruse *et al.*, 1999; Grimes *et al.*, 2000), which in turn recruits histone deacetylases 1 and 2 (HDAC1/2) (Huang *et al.*, 1999). In cardiac myocytes,

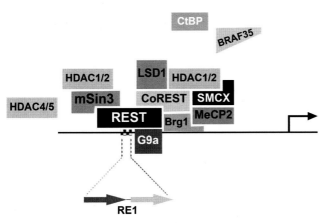

Figure 11.1 REST acts as a hub for recruitment of chromatin-modifying activity. REST binds to RE1 sites in the genome. Its N- and C-terminal co-repressor domains interact with mSin3 and CoREST co-repressor complexes that can recruit a range of histone deacetylases (HDAC1,2,4,5), histone methyltransferase G9a, histone demethylases LSD1 and SMCX (JARID1C), chromatin remodeling factors Brg1 and BRAF35 and the co-repressor CtBP. RE1 motifs are composed of two "half-sites," represented by the two large arrows. LSD1, lysine-specific demethylase 1; RE1, repressor element 1; TSS, transcriptional start site.

HDAC4 and HDAC5 can also form part of the associated mSin3 co-repressor complex (Nakagawa *et al.*, 2006). The C-terminus interacts with the CoREST complex (Andres *et al.*, 1999), which can recruit a range of chromatin-modifying proteins including HDAC1/2 (Ballas *et al.*, 2001), the histone methyltransferase (HMT) G9a (Shi *et al.*, 2003; Roopra *et al.*, 2004), histone demethylases (HDMs) LSD1 (Shi *et al.*, 2004; Lee *et al.*, 2005) and JARID1C/SMCX (Tahiliani *et al.*, 2007), Brg1 (Battaglioli *et al.*, 2002; Ooi *et al.*, 2006), and BRAF35 (Hakimi *et al.*, 2002). CoREST also interacts with the co-repressor CtBP (Garriga-Canut *et al.*, 2006) and methyl-CpG binding protein 2 (MeCP2) (Lunyak *et al.*, 2002; Ballas *et al.*, 2005;) and REST can directly interact with G9a, independent of co-repressor platforms (Roopra *et al.*, 2004).

In concert, the HDAC, HMT, HDM, chromatin-remodeling, and MeCP2 activities recruited by REST and its co-repressors serve to generate a compact or repressive chromatin configuration commonly associated with gene repression or silencing (Figure 11.2). Both Sin3 and CoREST can recruit HDAC activity that can deacetylate H3K9ac/H3K14ac, thus promoting LSD1-mediated demethylation of H3K4me1/me2 (Lee *et al.*, 2005). H3K9 also serves as a substrate for G9a-mediated methylation, and H3K9me2 may recruit HP1 to facilitate chromatin compaction (Roopra *et al.*, 2004). However, the combination of activities recruited at any individual locus is highly context-specific and provides a mechanism by which REST can coordinate a range of different transcriptional and epigenetic outcomes. By way of illustration, Sin3b and CoREST are both recruited to the promoters of *Scn2a2* and *Chrm4* in a NSC line, but the *Scg10* promoter lacks Sin3b (Greenway *et al.*, 2007). Interestingly, all three loci show increased H4ac/H3K9ac and decreased H3K9me2 when REST is inhibited – a common hallmark of REST occupancy on the local chromatin state. In JTC19 cells, the Chrm4 promoter is not occupied by REST

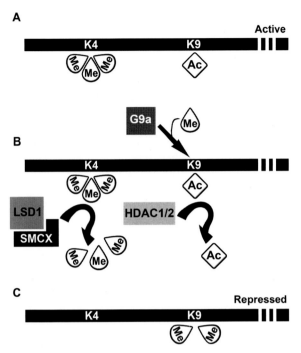

Figure 11.2 A repressive epigenetic signature is established by REST-recruited chromatin modifiers. The figure illustrates some of the changes to post-translational modifications of lysine (K) residues on histone H3 that occur following REST binding to target loci and recruitment of co-repressor platforms. Sites of histone acetylation (Ac) and methylation (Me) are indicated. (A) H3K4me3 and H3K9ac are "active" chromatin marks. (B) LSD1 and SMCX can demethylate H3K4me1/me2/me3 while HDAC1/2 can deacetylate H3K9ac, which can then be methylated by G9a. (C) The resulting chromatin signature is associated with gene repression.

though the *Scn2a2* promoter is, reflecting the silenced chromatin status of the *Chrm4* promoter in these cells (Belyaev *et al.*, 2004). In heart, REST directly represses a number of fetal cardiac genes including *Nppa* and *Nppb* and continued REST occupancy is required for their repression in rat ventricular myocytes (Bingham *et al.*, 2007). However, in a cardiomyocyte cell line it was shown that the mode of action for repression of *Nppa* or *Nppb* is gene-specific; derepression is associated with increased H4ac and H3K4me2 at both promoters, but *Nppa* was repressed through activities recruited by either the N- or C-terminus of REST whilst *Nppa* required both in combination. While in this case REST acts as a dynamic repressor, it can mediate long-term silencing through the interaction of CoREST with MeCP2 and DNA methylation as described during neurogenesis (Ballas *et al.*, 2005). In this scenario, REST clearance from the DNA does not necessarily result in loss of CoREST, which can remain associated with MeCP2 and the DNA.

11.2.2   Genome-wide identification of REST binding sites: functional implications

The unusually long RE1 motif invited use of early bioinformatic approaches to predict REST binding (Schoenherr *et al.*, 1996) and these early, modest attempts identified an additional 22 genes. However, subsequent to the success of the genome-sequencing projects, more sophisticated bioinformatic approaches were applied to identify REST binding sites on a genome-wide basis. The first studies, including our own, used either a modified RE1 consensus or position-specific matrices (Bruce *et al.*, 2004; Johnson *et al.*, 2006; Mortazavi *et al.*, 2006; Wu and Xie, 2006) and led to identification of well over 1000 potential sites. Notably, these were enriched for neuron-specific targets and included many non-coding RNAs (ncRNAs).

Subsequent studies in ESCs (Johnson *et al.*, 2008), NSCs (Johnson *et al.*, 2008; Abrajano *et al.*, 2010), and other neural and non-neural cell types (Abrajano *et al.* 2009a, 2009b; Bruce *et al.*, 2009) have combined bioinformatics with biochemical approaches, most notably genome-wide chromatin immunoprecipitation (ChIP), to enhance binding-site identification. Collectively, they reveal subtleties to the complexity of REST interaction with the genome with additional evidence that inhibition of REST does not lead to widespread derepression of target genes. Interestingly, they also led to discovery of a number of novel non-canonical or atypical RE1 binding motifs that show evolutionary conservation and correlation with cell-type-specific REST binding (Johnson *et al.*, 2007, 2008; Otto *et al.*, 2007; Bruce *et al.*, 2009). Commonly occupied REST target genes in both neural and non-neural cell types bear canonical RE1 motifs and are highly enriched for neuron-specific genes. In contrast, other targets show cell-type-specific REST occupancy (Johnson *et al.*, 2008; Abrajano *et al.*, 2009a, 2009b; Bruce *et al.*, 2009) and this has led to the discovery of potential mechanisms that underlie cell-type-specific REST functions including those involved in maintaining ESC pluripotency (Johnson *et al.*, 2008), neuronal subtype specification (Abrajano *et al.*, 2009b), and glial development and maturation (Abrajano *et al.*, 2009a).

At present, predicted numbers of REST binding sites range from a few hundred into the thousands with further predictions of more than 10 000. It is unclear how much of this variance is due to differences in ChIP platform, threshold of detection, cell type, or species. The threshold of detection may prove interesting since some studies that detect high numbers of binding sites also find a higher proportion with no identifiable RE1 motif. It is possible that some of these sites are not the result of direct REST binding to DNA but recruitment in *cis* or *trans* to remote chromosomal territories via promoter or enhancer interactions. Resolution of these issues awaits employment of a single ChIP platform, a common bioinformatic toolbox, and chromosome capture conformation. Nevertheless, these studies all illustrate cell-type-specific cohorts of REST binding sites that encompass non-neuronal and neuron-specific genes and provide invaluable insights for exploring the role of REST in stem cell populations.

11.3   Investigating the role of REST in pluripotency

Our ability to propagate mouse and human ESCs has greatly facilitated in vitro investigation of developmental programs that generate cellular diversity and has led to unprecedented advances in transgenic mouse technologies. From a regenerative medicine perspective, ESCs have heralded the dawn of research exploring the directed differentiation of in vitro expanded ESCs into a wide range of therapeutically desirable cell types. Thus the key question is what determines pluripotency and the subsequent differentiation of ESCs into cells of the three germ layers?

ESC chromatin is thought to have a unique, "open" configuration relative to more restricted cell-types. The landmark discovery that Sox2, Oct3/4, Klf4, and c-Myc in combination could reprogram somatic cells into an induced pluripotent state (iPS) (Takahashi and Yamanaka, 2006) not only paved the way for the generation of disease-specific iPS cells but has also greatly contributed to our understanding of the importance of the epigenome in defining the pluripotent state. Indeed, the chromatin state of fully reprogrammed iPS cells is similar, albeit not identical, to ESCs and reprogramming protocols can be enhanced by epigenetic modifiers (Wernig et al., 2007; Mikkelsen et al., 2008; Doi et al., 2009).

The cardinal pluripotency transcription factors, Nanog, Sox2, and Oct3/4 are essential for maintenance of ESC pluripotency (Loh et al., 2006; Masui et al., 2007). All of these factors have binding sites in the *Rest* gene and conversely REST targets include many pluripotency genes including *Nanog* and *Rest* itself (Chen et al., 2008; Johnson et al., 2008). Moreover, REST and Oct3/4, Sox2, and Nanog all share a large number of common target genes, which led to the notion that REST is part of the pluripotency network (Johnson et al., 2008) (Figure 11.3). Yet the precise role of REST in ESCs remains poorly understood and its requirement for maintaining pluripotency has been controversial.

In 2008, results from *Rest* $^{+/-}$ ESCs lines and siRNA-mediated REST knockdown in ESCs were reported that suggested that REST was essential in self-renewal and maintenance of pluripotency (Singh et al., 2008). In this study loss of a single

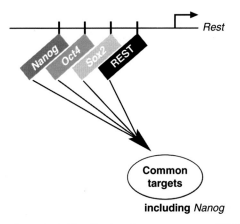

Figure 11.3 Identification of common targets of REST and pluripotency factors. Genome-wide ChIP-PET and ChIPseq studies have revealed that *Rest* is a target of the key pluripotency factors, Nanog, Sox2, and Oct4 as well as REST itself. All four transcription factors have a set of common targets, evidence that REST is part of the pluripotency network. ChIP-PET, ChIP-paired-end tag; ChIPseq, ChIP sequencing.

Rest allele was reported to be sufficient to reduce expression of Nanog, Sox2, Oct4, and c-Myc and affect ESC self-renewal mediated by direct derepression of *miR-21*. However, these findings are inconsistent with earlier and later studies from multiple laboratories including our own. First, REST is dispensable for gastrulation since *Rest*$^{-/-}$ mice develop relatively normally until around E9.25 (Chen *et al.*, 1998) and siRNA-mediated knockdown of REST in ESCs had no effect on morphology or alkaline phosphatase levels (Loh *et al.*, 2006). Subsequent reports failed to replicate the identification of *miR-21* as a REST target (Johnson *et al.*, 2008) or its upregulation in *Rest*$^{-/-}$ ESCs (Jorgensen *et al.*, 2009b) and both *Rest*$^{+/-}$ and *Rest*$^{-/-}$ ESCs retained pluripotency and expression of pluripotency genes including *Nanog*, *Sox2*, and *Oct4* (Buckley *et al.*, 2009; Jorgensen *et al.*, 2009a, 2009b; Yamada *et al.*, 2010; Soldati *et al.*, 2012). Further, upon embryoid body (EB) differentiation, *Rest*$^{-/-}$ ESCs express of markers of all three germ layers and *Rest*$^{-/-}$ ESCs contributed to all three germ layers in chimeric mice and formed teratomas in nude mice (Yamada *et al.*, 2010). Taken together, current data provide strong evidence that REST is not required for maintaining ESC pluripotency.

We have identified over 2500 sites occupied by REST in ESCs, with a much smaller, but largely overlapping number in ESC-derived NSCs (Johnson *et al.*, 2008). Comparison of REST occupancy in ESCs with NSCs identified a set of ESC-specific REST targets enriched for members of the WNT signaling pathway – a key modulator of the pluripotency network. ESC-specific targets also encode proteins involved in chromatin binding, ESC self-renewal, and reprogramming. Interestingly, from the viewpoint of REST function, universal derepression of REST-bound target genes does not occur following loss of REST function, but the individual genes that are dysregulated in ESCs and NSCs are largely different (Johnson *et al.*, 2008; Jorgensen *et al.*, 2009b). REST binding per se does not tell us a lot about REST function but this disparity between binding and transcriptional regulation remains unexplained. Genes that do show significant derepression in ESCs following REST ablation are enriched for genes with neuron-specific functions but key transcriptional regulators that determine commitment to the neural lineage such as *Sox1*, *Mash1*, and *Pax6* do not appear to be direct REST targets (Johnson *et al.*, 2008; Jorgensen *et al.*, 2009b). So given that there are several thousand REST loci occupied by REST in ESCs but that loss of function results neither in loss of self-renewal and pluripotency nor in global derepression of target genes, what is the precise function of REST in terms of transcriptional and epigenetic control?

Sites that are occupied by a transcription factor but appear to be functionally redundant are often referred to as non-functional. This is misleading on at least three counts. First, their functionality, as measured by transcriptional output, is highly cell-specific so a site may be functional in one cell type but not in another. Second, functionality, as measured in cell lines, is a very static snapshot of biology. In reality, cells live in a constantly changing milieu reflecting both maintenance and development and the functionality of any particular transcription factor changes depending not just upon cell type but upon developmental stage. This cell type and stage specificity can be clearly seen in the case of REST binding in ESCs and NSCs; in NSCs, REST occupies a subset of total ESCs sites but, despite this overlap, only a highly restricted and cell-type-specific cohort of targets is transcriptionally regulated in each (Johnson *et al.*, 2008). Third, transcriptional output is not the only measure of activity or functionality of a transcription factor. Recruitment of REST to its binding sites in NSCs can lead to deposition of a local repressive histone signature characterized by low levels of H3K9ac/H4ac and high levels of H3K9me2, which are reversed in the presence of non-functional mutant REST constructs (Greenway *et al.*, 2007). This repressive signature is deposited irrespective of any effect on transcription. It is likely that this holds true for ESCs, supported by our observations of changes to local histone signatures and gene expression in $Rest^{+/+}$ versus $Rest^{-/-}$ ESCs (N. J. Buckley and A. Bithell, unpublished data). We have argued previously that this renders the target gene "poised for repression" such that when an activating signal is received, the gene can maintained in a repressed state and only activated when the repressor is dismissed.

It is likely that REST is necessary for suppression of pluripotency factors and correct establishment of downstream developmental programmes upon ESC differentiation. Although $Rest^{-/-}$ ESCs have no overt loss of pluripotency under proliferative conditions, they reveal a delay in downregulation of pluripotency genes (particularly its direct target, Nanog), and reduced markers of primitive endoderm (Gata-4 and Gata-6) during EB differentiation (Yamada *et al.*, 2010). Meanwhile, REST overexpression results in decreased Nanog, Oct3/4, and Fgf5 and an increased propensity to generate primitive endoderm, in part mediated by Nanog. Thus, it appears that REST is necessary to quickly repress pluripotency factors and permit timely initiation of lineage-specific programs (Soldati *et al.*, 2012). However, the mechanisms by which REST achieves this are not known nor are its parallel effects on the ESC epigenome understood.

Many of REST actions are attributed to its recruitment of HDAC and consequently HDAC inhibitors such as trichostatin A (TSA) are often used as a proxy for inhibition of REST. Undoubtedly many aspects of REST function are due, at least in part, to its recruitment of HDAC1/2 but, as detailed above, REST recruits many other epigenetic modifiers. Differentiation of ESCs (and NSCs) frequently reported in response to global inhibition of HDAC activity stands in marked contrast to the more nuanced effects of REST attenuation or ablation. For example, exposure of ESCs to TSA leads to downregulation of pluripotency genes (including *Nanog*) and upregulation of lineage-specific markers (Karantzali *et al.*, 2008) – an observation not recapitulated in $Rest^{-/-}$ ESCs under proliferative conditions. Interestingly, LSD1 helps to stabilize CoREST complexes and levels of CoREST are reduced and H3K56ac increased in *Lsd1* conditional knockout ESCs without changes to H3K4 methylation (Foster *et al.*, 2010). As for REST, loss of LSD1 does not affect ESC pluripotency, but leads to increased cell death upon differentiation and increased expression of several hundred genes including *Bry* (a direct target of LSD1). Upregulated genes were enriched for unmarked (H3K4/H3K27) and bivalent (H3K4me3/H3K27me3) promoters in ESCs. As part of a CoREST complex, this suggests another mechanism by which REST might selectively modulate the epigenetic signature of key development genes to ensure proper timing of transcriptional programs upon differentiation.

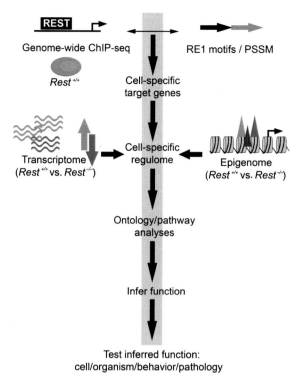

Figure 11.4 A REST 'binding-to-function' pipeline. The flow diagram depicts an integrative approach to investigate the role of REST in ESCs and NSCs. Identification of genome-wide REST binding by genome-wide ChIP sequencing (ChIPseq) and identification of RE1 motifs using a position-weighted matrix (PSSM) generates a set of list target genes. This is then combined with transcriptome data (RNA sequencing [RNA-seq] or microarray) and epigenome data (ChIPseq chromatin state maps) from $Rest^{+/+}$ and $Rest^{-/-}$ to determine ESC- and NSC-specific regulomes. These data are then interrogated with ontological and pathway analysis tools to infer REST function and generate testable hypotheses at the level of the cell, the organism, behavior, or pathological state. Green and red arrows depict up- and downregulated genes. Green/blue/red triangles represent different chromatin marks. See plate section for color version.

In conclusion, REST may act as a gateway regulator between ESC maintenance and differentiation by maintaining a "poised" epigenetic signature at ESC target loci such that cells can respond to differentiation cues by repressing pluripotency determinants and permitting the correct establishment of tissue-specific developmental programs. Our ongoing research in ESCs and NSCs aims to assess genome-wide changes in the epigenome and concomitant changes in gene expression following REST ablation to provide mechanistic insight into the nuanced effects of REST on the stem cell epigenome and transcriptome (Figure 11.4).

11.4   The function of REST in multipotent stem cells: lessons from NSCs

All cells in the nervous system are ultimately derived from NSCs. Understanding the regulatory mechanisms that control NSC maintenance and differentiation remains an outstanding issue in developmental neurobiology and is key to NSC deployment in regenerative medicine strategies to treat injury and neurodegeneration. As with their ESC counterparts, we have only a patchy knowledge of the transcription factors that regulate maintenance and differentiation.

11.4.1   REST and embryonic neurogenesis

REST is expressed in embryonic NSCs and neural progenitor cells (NPCs). The early embryonic lethality of $Rest^{-/-}$ mice means that our insight into REST regulation of neurogenesis has mainly been drawn from studies carried out in cultured cells. These have produced mixed results, due in part to the diversity of cell types employed. One common observation is that downregulation of REST occurs upon neuronal differentiation. However, no such consensus exists for the causative role of REST.

As discussed earlier, unambiguous assessment of the number of REST binding sites is problematic. Different REST occupancy in cultured NSCs and NPCs is likely to be a reflection of restrictions that correspond to different developmental stages. As such, the terms "neural stem cell" and "neuronal progenitor" are operational rather than definitive. For instance ESC-derived NSCs are radial glia-like, tripotent, and express REST protein, while cortical progenitors are more restricted in potential and do not express detectable REST protein, although REST is still detected bound to its target genes (Ballas et al., 2005). This is consistent with the observation that REST remains bound at target genes in cortical progenitors even when the protein is undetectable by Western blot analysis and only disappears upon neuronal differentiation, whereupon gene expression is activated (Ballas et al., 2005). However, for ESC-derived NSCs we found 857 REST-occupied loci genome-wide – many fewer than in ESCs, which may reflect the more restricted potential and different chromatin structure of a multipotent stem cell (Johnson et al., 2008).

Whereas there is no dispute over the downregulation of REST during neural differentiation, there is no clear consensus on the requirement for REST in maintenance or differentiation of NSCs and NPCs. shRNA mediated attenuation of REST in neurospheres led to an decrease in self-renewal and a decrease in neurons, astrocytes, and oligodendrocytes, but an increase in markers of radial glia (Abrajano et al., 2010). Intriguing as these data are, even clonally derived neurospheres are phenotypically heterogeneous so it is difficult to attach these effects to a particular cell type. Inhibition of REST in radial glial-like NSCs or MHP-36 hippocampal NPCs led to changes in neuronal gene expression but not to overt neuronal differentiation (Greenway et al., 2007; Johnson et al., 2008). One caveat is that these attempts to demonstrate REST function have been carried out under proliferative conditions and it is possible that any instructive signal released by downregulation of REST may not be sufficient to overcome the mitogen effects of the medium. This presents a clear parallel to our perception of REST function in ESCs where it is clear that the effects of REST on the initiation of the neurogenic program are only manifest once differentiation has been initiated. Likewise, it is possible that the effects of neurogenesis are also only evident in a permissive neurogenic environment. Indeed, switching REST function from a repressor to an activator by forced expression of REST-VP16 in NSCs leads to overt neuronal differention as evidenced by neurite extension and depolarization-dependent calcium influx (Su *et al.*, 2004). Westbrook *et al.* (2008) did look at the effect of REST ablation under differentiation conditions by introducing REST into *Sox1*:GFP ESCs that allow flow cytometry sorting of early neural precursors followed by culture in differentiation medium. Under these conditions, REST inactivation promoted neuronal differentiation, while constitutive REST expression blocked neural differentiation. At present the mechanism of gene repression and the target genes that link REST directly to neurogenesis remain unidentified.

11.4.2 REST and adult neurogenesis

Although adult NSCs are restricted to two small neurogenic niches in the subventricular zone (SVZ) of the lateral ventricle and subgranular zone (SGZ) of the dentate gyrus, they have attracted a great deal of attention because of their potential for recruitment to repair sites of injury or degeneration. REST is expressed in the presumptive NSCs of both neurogenic niches (Kuwabara *et al.*, 2004; N. J. Buckley and A. Bithell, unpublished observations). Intriguingly, a short dsRNA corresponding to a canonical REST binding (NRSE dsRNA) site is co-expressed with REST in the SGZ (Kuwabara *et al.*, 2004). In an adult hippocampal neural stem cell line, it appears that interaction between the NRSE dsRNA and REST is necessary to convert REST to an activator and can induce neurogenesis, although details of the mechanism of action are unknown. The role of NRSE dsRNA, its interaction with REST, and its role in neurogenesis in vivo are also unknown, but it does seem specific to adult neurogenesis as there is no evidence for NRSE dsRNA in embryonic NSCs or NPCs.

Several epigenetic regulators have been linked to adult NSC maintenance and differentiation. *Bmi1* is a component of the Polycomb group complex and in *Bmi1*-null mice, SVZ NSCs are depleted (Molofsky *et al.*, 2003) whereas overexpression leads to overproduction of NSCs (Fasano *et al.*, 2007, 2009). As yet, no link has been demonstrated between REST and *Bmi1*. *Mbd1*-null mice have impaired neurogenesis in the SGZ (Zhao *et al.*, 2003), an effect that is believed to be mediated by changes in DNA methylation in the promoter of *Fgf2*, a mitogen for SGZ NSCs. Again, although there is no direct link between *Mbd1* and REST in NSCs, in hippocampal neurons, NMDA receptor activation leads to a concomitant release of REST, HDAC1, MeCP2, and Mbd1 from the activity-dependent *Bdnf* promoter (Tian *et al.*, 2009).

When it comes to epigenetic regulation of NSC differentiation there is a potentially more direct connection with REST. The H3K4 HMT MLL1 is required for neuronal differentiation of adult NSCs (Lim *et al.*, 2009) and we have shown that it is a direct target of REST (Johnson *et al.*, 2008). Recruitment of chromatin and histone-modifying activities by REST is thus not the only potential interface between REST and epigenetic regulation. REST targets several genes encoding epigenetic regulatory activities in glia, including *Dnmt1*, *Mbd2*, *Smarcd2*, and

Hdac6 (Abrajano *et al.*, 2009a), but here again, there is no evidence that these genes are regulated by REST in NSCs, or that their activity contributes toward NSC maintenance or differentiation.

11.5   Concluding remarks

The epigenome is critical in acting as a cellular memory and thus determining cell state and potential. What then is the relative contribution of REST to ensuring correct programs of target gene expression through epigenetic modulation? To date, there are relatively few experimental data on the direct role of REST on the stem cell epigenome compared with the transcriptome. What is clear is that changes in REST expression and epigenetic changes both accompany and drive ESC and NSC maintenance and differentiation. Whether these changes occur in concert or independently remains an outstanding question. To understand fully REST action in any cell context, including stem cells, it will be necessary to interrogate a combination of data including gene expression, genome sequence, transcription factor occupancy, and chromatin state maps, in the presence and absence of functional REST. Though this is a highly complex task, these data are beginning to emerge and provide a basis for speculation. Indeed an interesting study in T-cells using a range of extant genome-wide datasets showed correlation of RE1-bound genes with low gene expression and both higher levels of "repres-repressive" chromatin marks (including H3K9me2 and H3K27me3) and lower "active" marks (including H3K9ac, H3K4me3, and H4K8ac) (Zheng *et al.*, 2009). Future experimental comparisons will enable testing of the causal links between REST and the stem cell epigenome.

REFERENCES

Abrajano, J. J., Qureshi, I. A., Gokhan, S., *et al.* (2009a). Differential deployment of REST and CoREST promotes glial subtype specification and oligodendrocyte lineage maturation. *PLoS One*, **4**, e7665.

Abrajano, J. J., Qureshi, I. A., Gokhan, S., *et al.* (2009b). REST and CoREST modulate neuronal subtype specification, maturation and maintenance. *PLoS One*, **4**, e7936.

Abrajano, J. J., Qureshi, I. A., Gokhan, S., *et al.* (2010). Corepressor for element-1-silencing transcription factor preferentially mediates gene networks underlying neural stem cell fate decisions. *Proceedings of the National Academy of Sciences USA*, **107**, 16685–16690.

Andres, M. E., Burger, C., Peral-Rubio, M. J., *et al.* (1999). CoREST: a functional corepressor required for regulation of neural-specific gene expression. *Proceedings of the National Academy of Sciences USA*, **96**, 9873–9878.

Atouf, F., Czernichow, P., and Scharfmann, R. (1997). Expression of neuronal traits in pancreatic beta cells: implication of neuron-restrictive silencing factor/repressor element silencing transcription factor, a neuron-restrictive silencer. *Journal of Biological Chemistry*, **272**, 1929–1934.

Azuara, V., Perry, P., Sauer, S., *et al.* (2006). Chromatin signatures of pluripotent cell lines. *Nat Cell Biology*, **8**, 532–538.

Ballas, N., Battaglioli, E., Atouf, F., *et al.* (2001). Regulation of neuronal traits by a novel transcriptional complex. *Neuron*, **31**, 353–365.

Ballas, N., Grunseich, C., Lu, D. D., Speh, J. C., and Mandel, G. (2005). REST and its co-repressors mediate plasticity of neuronal gene chromatin throughout neurogenesis. *Cell*, **121**, 645–657.

Barski, A., Cuddapah, S., Cui, K., *et al.* (2007). High-resolution profiling of histone methylations in the human genome. *Cell*, **129**, 823–837.

Battaglioli, E., Andres, M. E., Rose, D. W., *et al.* (2002). REST repression of neuronal genes requires components of the hSWI.SNF complex. *Journal of Biological Chemistry*, **277**, 41 038–41 045.

Belyaev, N. D., Wood, I. C., Bruce, A. W., *et al.* (2004). Distinct RE-1 silencing transcription factor-containing complexes interact with different target genes. *Journal of Biological Chemistry*, **279**, 556–561.

Berger, S. L. (2007). The complex language of chromatin regulation during transcription. *Nature*, **447**, 407–412.

Bernstein, B. E., Mikkelsen, T. S., Xie, X., *et al.* (2006) A bivalent chromatin structure marks key developmental genes in embryonic stem cells. *Cell*, **125**, 315–326.

Bingham, A. J., Ooi, L., Kozera, L., White, E., and Wood, I. C. (2007). The repressor element 1-silencing transcription factor regulates heart-specific gene expression using multiple chromatin-modifying complexes. *Molecular and Cellular Biology*, **27**, 4082–4092.

Boyer, L. A., Lee, T. I., Cole, M. F., *et al.* (2005). Core transcriptional regulatory circuitry in human embryonic stem cells. *Cell*, **122**, 947–956.

Bruce, A. W., Donaldson, I. J., Wood, I. C., *et al.* (2004). Genome-wide analysis of repressor element 1 silencing transcription factor/neuron-restrictive silencing factor (REST/NRSF) target genes. *Proceedings of the National Academy of Sciences USA*, **101**, 10 458–10 463.

Bruce, A. W., Lopez-Contreras, A. J., Flicek, P., *et al.* (2009). Functional diversity for REST (NRSF) is defined by in vivo binding affinity hierarchies at the DNA sequence level. *Genome Research*, **19**, 994–1005.

Buckley, N. J., Johnson, R., Sun, Y. M., and Stanton, L. W. (2009). Is REST a regulator of pluripotency? *Nature*, **457**, E5–6; discussion E7.

Buckley, N. J., Johnson, R., Zuccato, C., Bithell, A., and Cattaneo, E. (2010). The role of REST in transcriptional and epigenetic dysregulation in Huntington's disease. *Neurobiology of Disease*, **39**, 28–39.

Cai, J., Weiss, M. L., and Rao, M. S. (2004). In search of "stemness". *Experimental Hematology*, **32**, 585–598.

Calderone, A., Jover, T., Noh, K. M., *et al.* (2003). Ischemic insults derepress the gene silencer REST in neurons destined to die. *Journal of Neuroscience*, **23**, 2112–2121.

Canzonetta, C., Mulligan, C., Deutsch, S., *et al.* (2008). DYRK1A-dosage imbalance perturbs NRSF/REST levels, deregulating pluripotency and embryonic stem cell fate in Down syndrome. *American Journal of Human Genetics*, **83**, 388–400.

Chen, X., Xu, H., Yuan, P., *et al.* (2008). Integration of external signaling pathways with the core transcriptional network in embryonic stem cells. *Cell*, **133**, 1106–1117.

Chen, Z. F., Paquette, A. J., and Anderson, D. J. (1998). NRSF/REST is required in vivo for repression of multiple neuronal target genes during embryogenesis. *Nature Genetics*, **20**, 136–142.

Cheong, A., Bingham, A. J., Li, J., *et al.* (2005). Downregulated REST transcription factor is a switch enabling critical potassium channel expression and cell proliferation. *Molecular Cell*, **20**, 45–52.

Chong, J. A., Tapia-Ramirez, J., Kim, S., *et al.* (1995). REST: a mammalian silencer protein that restricts sodium channel gene expression to neurons. *Cell*, **80**, 949–957.

Doi, A., Park, I. H., Wen, B., *et al.* (2009). Differential methylation of tissue- and cancer-specific CpG island shores distinguishes human induced pluripotent stem cells, embryonic stem cells and fibroblasts. *Nature Genetics*, **41**, 1350–1353.

Fasano, C. A., Dimos, J. T., Ivanova, N. B., *et al.* (2007). shRNA knockdown of *Bmi-1* reveals a critical role for p21-Rb pathway in NSC self-renewal during development. *Cell Stem Cell*, **1**, 87–99.

Fasano, C. A., Phoenix, T. N., Kokovay, E., *et al.* (2009). *Bmi-1* cooperates with *Foxg1* to maintain neural stem cell self-renewal in the forebrain. *Genes and Development*, **23**, 561–574.

Foster, C. T., Dovey, O. M., Lezina, L., *et al.* (2010). Lysine-specific demethylase 1 regulates the embryonic transcriptome and CoREST stability. *Molecular and Cellular Biology*, **30**, 4851–4863.

Garriga-Canut, M., Schoenike, B., Qazi, R., *et al.* (2006). 2-Deoxy-D-glucose reduces epilepsy progression by NRSF-CtBP-dependent metabolic regulation of chromatin structure. *Nature Neuroscience*, **9**, 1382–1387.

Greenway, D. J., Street, M., Jeffries, A., and Buckley, N. J. (2007). RE1 silencing transcription factor maintains a repressive chromatin environment in embryonic hippocampal neural stem cells. *Stem Cells*, **25**, 354–363.

Grimes, J. A., Nielsen, S. J., Battaglioli, E., *et al.* (2000). The co-repressor mSin3A is a functional component of the REST–CoREST repressor complex. *Journal of Biological Chemistry*, **275**, 9461–9467.

Hakimi, M. A., Bochar, D. A., Chenoweth, J., *et al.* (2002). A core-BRAF35 complex containing histone deacetylase mediates repression of neuronal-specific genes. *Proceedings of the National Academy of Sciences USA*, **99**, 7420–7425.

Huang, Y., Myers, S. J., and Dingledine, R. (1999). Transcriptional repression by REST: recruitment of Sin3A and histone deacetylase to neuronal genes. *Nature Neuroscience*, **2**, 867–872.

Johnson, D. S., Mortazavi, A., Myers, R. M., and Wold, B. (2007). Genome-wide mapping of in vivo protein–DNA interactions. *Science*, **316**, 1497–1502.

Johnson, R., Gamblin, R. J., Ooi, L., *et al.* (2006). Identification of the REST regulon reveals extensive transposable element-mediated binding site duplication. *Nucleic Acids Research*, **34**, 3862–3877.

Johnson, R., Teh, C. H., Kunarso, G., *et al.* (2008). REST regulates distinct transcriptional networks in embryonic and neural stem cells. *PLoS Biology*, **6**, e256.

Jorgensen, H. F., Chen, Z. F., Merkenschlager, M., and Fisher, A. G. (2009a). Is REST required for ESC pluripotency? *Nature*, **457**, E4–5; discussion E7.

Jorgensen, H. F., Terry, A., Beretta, C., *et al.* (2009b). REST selectively represses a subset of RE1-containing neuronal genes in mouse embryonic stem cells. *Development*, **136**, 715–721.

Karantzali, E., Schulz, H., Hummel, O., *et al.* (2008). Histone deacetylase inhibition accelerates the early events of stem cell differentiation: transcriptomic and epigenetic analysis. *Genome Biology*, **9**, R65.

Koenigsberger, C., Chicca, J. J., II, Amoureux, M. C., Edelman, G. M., and Jones, F. S. (2000). Differential regulation by multiple promoters of the gene encoding the neuron-restrictive silencer factor. *Proceedings of the National Academy of Sciences USA*, **97**, 2291–2296.

Kojima, T., Murai, K., Naruse, Y., Takahashi, N., and Mori, N. (2001). Cell-type non-selective transcription of mouse and human genes encoding neural-restrictive silencer factor. *Brain Research: Molecular Brain Research*, **90**, 174–186.

Kouzarides, T. (2007). Chromatin modifications and their function. *Cell*, **128**, 693–705.

Kraner, S. D., Chong, J. A., Tsay, H. J., and Mandel, G. (1992). Silencing the type II sodium channel gene: a model for neural-specific gene regulation. *Neuron*, **9**, 37–44.

Kuwabara, T., Hsieh, J., Nakashima, K., Taira, K., and Gage, F. H. (2004). A small modulatory dsRNA specifies the fate of adult neural stem cells. *Cell*, **116**, 779–793.

Kuwahara, K., Saito, Y., Takano, M., *et al.* (2003). NRSF regulates the fetal cardiac gene program and maintains normal cardiac structure and function. *EMBO Journal*, **22**, 6310–6321.

Lee, M. G., Wynder, C., Cooch, N., and Shiekhattar, R. (2005). An essential role for CoREST in nucleosomal histone 3 lysine 4 demethylation. *Nature*, **437**, 432–435.

Lepagnol-Bestel, A. M., Zvara, A., Maussion, G., *et al.* (2009). DYRK1A interacts with the REST/NRSF-SWI/SNF chromatin remodelling complex to deregulate gene clusters involved in the neuronal phenotypic traits of Down syndrome. *Human Molecular Genetics*, **18**, 1405–1414.

Lim, D. A., Huang, Y. C., Swigut, T., *et al.* (2009). Chromatin remodelling factor Mll1 is essential for neurogenesis from postnatal neural stem cells. *Nature*, **458**, 529–533.

Loh, Y. H., Wu, Q., Chew, J. L., *et al.* (2006). The *Oct4* and *Nanog* transcription network regulates pluripotency in mouse embryonic stem cells. *Nature Genetics*, **38**, 431–440.

Lunyak, V. V., Burgess, R., Prefontaine, G. G., *et al.* (2002). Corepressor-dependent silencing of chromosomal regions encoding neuronal genes. *Science*, **298**, 1747–1752.

Majumder, S. (2006). REST in good times and bad: roles in tumor suppressor and oncogenic activities. *Cell Cycle*, **5**, 1929–1935.

Martin, D., Allagnat, F., Chaffard, G., *et al.* (2008). Functional significance of repressor element 1 silencing transcription factor (REST) target genes in pancreatic beta cells. *Diabetologia*, **51**, 1429–1439.

Masui, S., Nakatake, Y., Toyooka, Y., *et al.* (2007). Pluripotency governed by *Sox2* via regulation of *Oct3/4* expression in mouse embryonic stem cells. *Nature Cell Biology*, **9**, 625–635.

Meaney, M. J. and Ferguson-Smith, A. C. (2010). Epigenetic regulation of the neural transcriptome: the meaning of the marks. *Nature Neuroscience*, **13**, 1313–1318.

Meissner, A., Mikkelsen, T. S., Gu, H., *et al.* (2008). Genome-scale DNA methylation maps of pluripotent and differentiated cells. *Nature*, **454**, 766–770.

Mikkelsen, T. S., Ku, M., Jaffe, D. B., *et al.* (2007). Genome-wide maps of chromatin state in pluripotent and lineage-committed cells. *Nature*, **448**, 553–560.

Mikkelsen, T. S., Hanna, J., Zhang, X., *et al.* (2008). Dissecting direct reprogramming through integrative genomic analysis. *Nature*, **454**, 49–55.

Molofsky, A. V., Pardal, R., Iwashita, T., *et al.* (2003). *Bmi-1* dependence distinguishes neural stem cell self-renewal from progenitor proliferation. *Nature*, **425**, 962–967.

Mori, N., Schoenherr, C., Vandenbergh, D. J., and Anderson, D. J. (1992). A common silencer element in the SCG10 and type II Na$^+$ channel genes binds a factor present in non-neuronal cells but not in neuronal cells. *Neuron*, **9**, 45–54.

Mortazavi, A., Leeper Thompson, E. C., Garcia, S. T., Myers, R. M., and Wold, B. (2006). Comparative genomics modeling of the NRSF/REST repressor network: from single conserved sites to genome-wide repertoire. *Genome Research*, **16**, 1208–1221.

Nakagawa, Y., Kuwahara, K., Harada, M., *et al.* (2006). Class II HDACs mediate CaMK-dependent signaling to NRSF in ventricular myocytes. *Journal of Molecular and Cellular Cardiology*, **41**, 1010–1022.

Naruse, Y., Aoki, T., Kojima, T., and Mori, N. (1999). Neural restrictive silencer factor recruits mSin3 and histone deacetylase complex to repress neuron specific target genes. *Proceedings of the National Academy of Sciences USA*, **96**, 13 691–13 696.

Nishihara, S., Tsuda, L., and Ogura, T. (2003). The canonical Wnt pathway directly regulates NRSF/REST expression in chick spinal cord. *Biochemical and Biophysical Research Communications*, **311**, 55–63.

Ooi, L. and Wood, I. C. (2007). Chromatin crosstalk in development and disease: lessons from REST. *Nature Reviews Genetics*, **8**, 544–554.

Ooi, L., Belyaev, N. D., Miyake, K., Wood, I. C., and Buckley, N. J. (2006). BRG1 chromatin remodeling activity is required for efficient chromatin binding by repressor element 1-silencing transcription factor (REST) and facilitates REST-mediated repression. *Journal of Biological Chemistry*, **281**, 38 974–38 980.

Otto, S. J., McCorkle, S. R., Hover, J., *et al.* (2007). A new binding motif for the transcriptional repressor REST uncovers large gene networks devoted to neuronal functions. *Journal of Neuroscience*, **27**, 6729–6739.

Palm, K., Belluardo, N., Metsis, M., and Timmusk, T. (1998). Neuronal expression of zinc finger transcription factor REST/NRSF/XBR gene. *Journal of Neuroscience*, **18**, 1280–1296.

Palm, K., Metsis, M., and Timmusk, T. (1999). Neuron-specific splicing of zinc finger transcription factor REST/NRSF/XBR is frequent in neuroblastomas and conserved in human, mouse and rat. *Brain Research Molecular Brain Research*, **72**, 30–39.

Roh, T. Y., Cuddapah, S., Cui, K., and Zhao, K. (2006). The genomic landscape of histone modifications in human T cells. *Proceedings of the National Academy of Sciences USA*, **103**, 15 782–15 787.

Roopra, A., Qazi, R., Schoenike, B., Daley, T. J., and Morrison, J. F. (2004). Localized domains of G9a-mediated histone methylation are required for silencing of neuronal genes. *Molecular Cell*, **14**, 727–738.

Schoenherr, C. J. and Anderson, D. J. (1995a). The neuron-restrictive silencer factor (NRSF): a coordinate repressor of multiple neuron-specific genes. *Science*, **267**, 1360–1363.

Schoenherr, C. J. and Anderson, D. J. (1995b). Silencing is golden: negative regulation in the control of neuronal gene transcription. *Current Opinion in Neurobiology*, **5**, 566–571.

Schoenherr, C. J., Paquette, A. J., and Anderson, D. J. (1996). Identification of potential target genes for the neuron-restrictive silencer factor. *Proceedings of the National Academy of Sciences USA*, **93**, 9881–9886.

Schones, D. E., and Zhao, K. (2008). Genome-wide approaches to studying chromatin modifications. *Nature Reviews Genetics*, **9**, 179–191.

Shi, Y., Sawada, J., Sui, G., *et al.* (2003). Coordinated histone modifications mediated by a CtBP co-repressor complex. *Nature*, **422**, 735–738.

Shi, Y., Lan, F., Matson, C., *et al.* (2004). Histone demethylation mediated by the nuclear amine oxidase homolog LSD1. *Cell*, **119**, 941–953.

Singh, S. K., Kagalwala, M. N., Parker-Thornburg, J., Adams, H., and Majumder, S. (2008). REST maintains self-renewal and pluripotency of embryonic stem cells. *Nature*, **453**, 223–227.

Spivakov, M. and Fisher, A. G. (2007). Epigenetic signatures of stem-cell identity. *Nature Reviews Genetics*, **8**, 263–271.

Soldati, C., Bithell, A., Johstan, C., *et al.* (2012). Repressor element 1 silencing transcription factor couples loss of pluripotency with neural induction and neural differentiation. *Stem cells*, **30**, 425–434.

Su, X., Kameoka, S., Lentz, S., and Majumder, S. (2004). Activation of REST/NRSF target genes in neural stem cells is sufficient to cause neuronal differentiation. *Molecular and Cellular Biology*, **24**, 8018–8025.

Tahiliani, M., Mei, P., Fang, R., *et al.* (2007). The histone H3K4 demethylase SMCX links REST target genes to X-linked mental retardation. *Nature*, **447**, 601–605.

Takahashi, K. and Yamanaka, S. (2006). Induction of pluripotent stem cells from mouse embryonic and adult fibroblast cultures by defined factors. *Cell*, **126**, 663–676.

Tian, F., Hu, X. Z., Wu, X., *et al.* (2009). Dynamic chromatin remodeling events in hippocampal neurons are associated with NMDA receptor-mediated activation of *Bdnf* gene promoter 1. *Journal of Neurochemistry*, **109**, 1375–1388.

Wernig, M., Meissner, A., Foreman, R., *et al.* (2007). In vitro reprogramming of fibroblasts into a pluripotent ES-cell-like state. *Nature*, **448**, 318–324.

Westbrook, T. F., Martin, E. S., Schlabach, M. R., *et al.* (2005). A genetic screen for candidate tumor suppressors identifies REST. *Cell*, **121**, 837–848.

Westbrook, T. F., Hu, G., Ang, X. L., *et al.* (2008). SCFbeta-TRCP controls oncogenic transformation and neural differentiation through REST degradation. *Nature*, **452**, 370–374.

Wu, J. and Xie, X. (2006). Comparative sequence analysis reveals an intricate network among REST, CREB and miRNA in mediating neuronal gene expression. *Genome Biology*, **7**, R85.

Yamada, Y., Aoki, H., Kunisada, T., and Hara, A. (2010). Rest promotes the early differentiation of mouse ESCs but is not required for their maintenance. *Cell Stem Cell*, **6**, 10–15.

Zhao, X., Ueba, T., Christie, B. R., *et al.* (2003). Mice lacking methyl-CpG binding protein 1 have deficits in adult neurogenesis and hippocampal function. *Proceedings of the National Academy of Sciences USA*, **100**, 6777–6782.

Zheng, D., Zhao, K., and Mehler, M. F. (2009). Profiling RE1/REST-mediated histone modifications in the human genome. *Genome Biology*, **10**, R9.

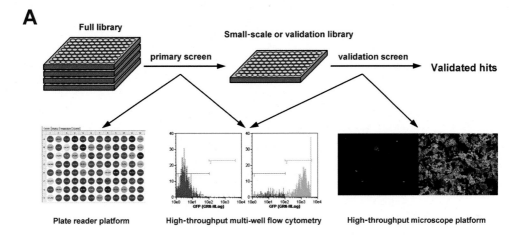

A

Full library

primary screen

Small-scale or validation library

validation screen

Validated hits

Plate reader platform

High-throughput multi-well flow cytometry

High-throughput microscope platform

GFP (GRN-HLog)

GFP (GRN-HLog)

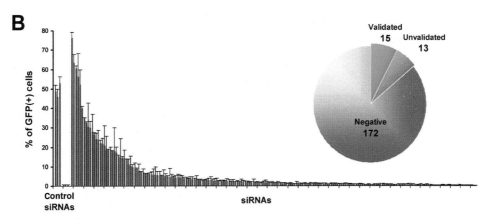

B

% of GFP(+) cells

Control
siRNAs

siRNAs

Validated
15 Unvalidated
13

Negative
172

Figure 3.2 SiRNA-based high-throughput functional screen for detection of epigenetic silencing factors. (A) A schematic representation of the siRNA screening procedure using the GFP-silent reporter cell population, showing results with different GFP detection methods. General representations of readout results with plate reader format, multi-well flow cytometry, and high-content microscopy are indicated. Flow cytometry profiles and microscope images represent samples with GAPDH and HDAC1 siRNA, as negative and positive controls, respectively. (B) Representative results of primary siRNA screen (Poleshko *et al.*, 2010). Screening results are scored as percentage of GFP-positive cells measured by 96-well flow cytometry. Triplicate siRNAs for HDAC1 and GAPDH served as positive and negative control, respectively. Error bars indicated range ($n = 2$) determined with duplicate plates. Pie chart illustrates the siRNA screen hit rate.

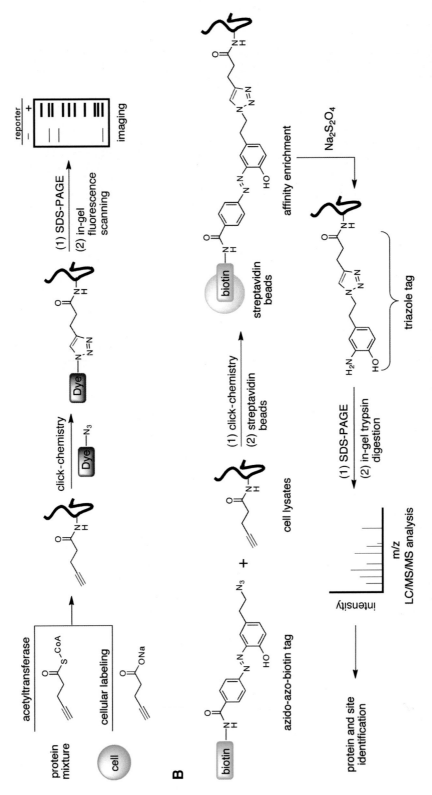

Figure 5.2 General strategy for the analysis of protein acetylation with 4-pentynoyl groups in vitro or in cells can be reacted with azido-functionalized dyes via click-chemistry to enable the in-gel fluorescence imaging of labeled proteins. (B) 4-Pentynoylated proteins in cell lysates can be reacted with a cleavable azido-azo-functionalized biotin tag and retrieved from protein mixtures on streptavidin beads. Bound proteins can be selectively eluted from the beads by selective reduction of the azo-bond using 25 mM $Na_2S_2O_4$, separated on SDS-PAGE and then trypsin-digested for mass spectrometric analysis to enable protein and site identification.

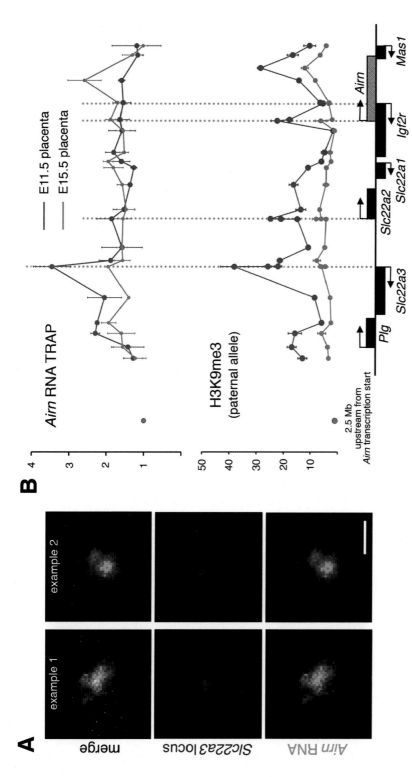

Figure 6.2 *Airn* accumulates specifically at the promoter of the *Slc22a3* gene in dynamic correlation with allele-specific gene silencing and repressive histone modification. (A) Simultaneous detection of *Airn* RNA FISH signals (bottom panels) and *Slc22a3* DNA FISH signals (middle panels). *Slc22a3* loci densely covered by *Airn* RNA (left column) are dominant in E11.5 placenta where paternal *Slc22a3* gene is silenced in *Airn*-dependent manner, whereas *Slc22a3* loci with little overlap with *Airn* RNA (right column) are observed more often in E15.5 placenta where paternal *Slc22a3* gene is relieved from silencing. Scale bar = 1 μm. (B) Association of *Airn* RNA with chromatin analyzed by RNA TRAP (tagging and recovery of associated proteins)(upper panel) and distribution of trimethylation of H3K9 on the paternal allele (lower panel), each normalized to 2.5 Mb upstream from *Airn* transcription start site, in E11.5 (pink lines) and E15.5 (blue lines) placentas.

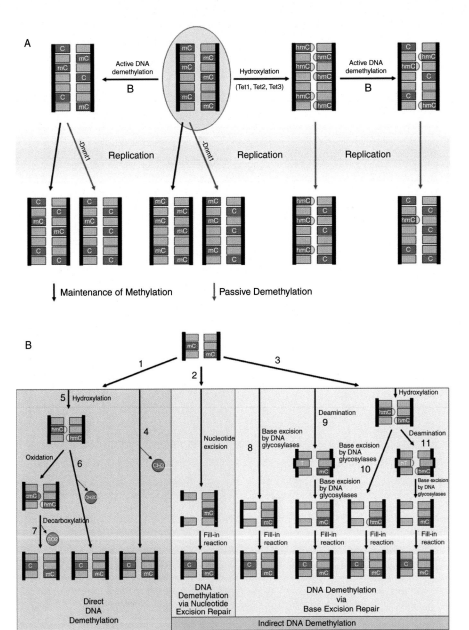

Figure 7.2 DNA demethylation pathways. (A) Demethylation of a methylated DNA molecule could include different pathways. It can be either an active mechanism (arrow left and right and 1B) or passive (red arrows) in the absence of Dnmt1 or passive due to a modification of the methylated cytosine on the parental strand (e.g., hydroxylation), which Dnmt1 cannot read as hemi-methylated sites. DNA demethylation might also include passive and active processes together. (B) Active demethylation can be performed directly (1) or indirectly, by nucleotide (2) or base (3) excision repair. Direct demethylation includes removal of the methyl group (3), or hydroxylation of the methyl group (4) followed by removal of the hydroxymethyl group (5) or further oxidation resulting in a final decarboxylation of the hydroxymethylgroup (6). The base excision repair (BER) pathway requires the recognition of the methylated cytosine by DNA glycosylases. Either DNA glycosylases directly recognize methylcytosine (7), or DNA deaminases could deaminate 5mC to T (8) and then the T–G mismatch is recognized and repaired. An alternative BER-mediated demethylation pathway includes the hydroxylation of 5mC, and a recognition of 5hmC by a DNA glycosylase (9) or a deamination (10) followed by BER.

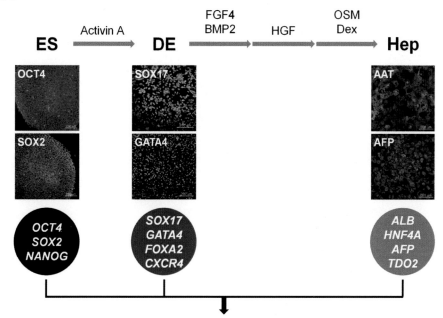

Analyses of histone modifications in the lineage-specific genes

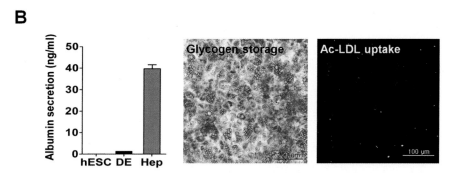

Figure 8.1 (A) Schematic representation of differentiation of hESCs into hepatocytes and analysis of histone modifications at lineage-specific genes in hESCs (ES), DE cells (DE), and hepatocytes (Hep). (B) Immunofluorescence detection of ES (OCT4 and SOX2), DE (SOX17 and GATA4), and hepatocyte (AAT and AFP) markers was performed in hESCs, DE cells, and hepatocytes, respectively.

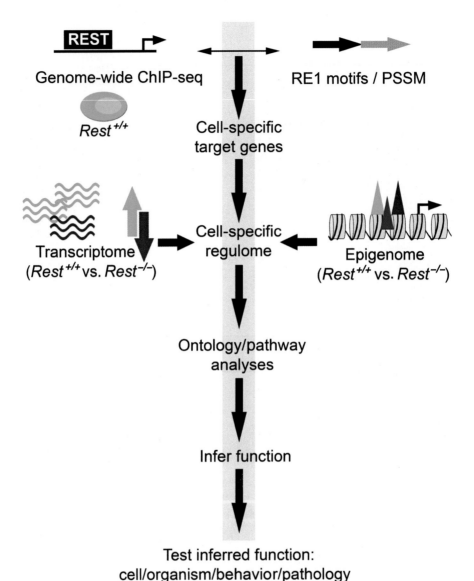

Figure 11.4 A REST 'binding-to-function' pipeline. The flow diagram depicts an integrative approach to investigate the role of REST in ESCs and NSCs. Identification of genome-wide REST binding by genome-wide ChIP sequencing (ChIPseq) and identification of RE1 motifs using a position-weighted matrix (PSSM) generates a set of list target genes. This is then combined with transcriptome data (RNA sequencing [RNA-seq] or microarray) and epigenome data (ChIPseq chromatin state maps) from $Rest^{+/+}$ and $Rest^{-/-}$ to determine ESC- and NSC-specific regulomes. These data are then interrogated with ontological and pathway analysis tools to infer REST function and generate testable hypotheses at the level of the cell, the organism, behavior or pathological state. Green and red arrows depict up- and downregulated genes. Green/blue/red triangles represent different chromatin marks.

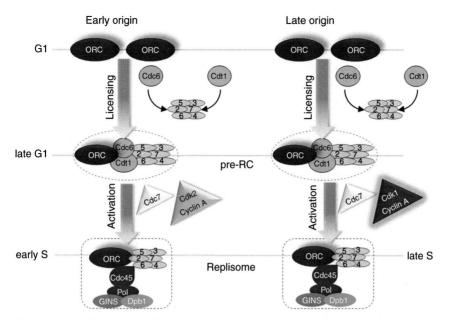

Figure 13.1 Replication model for eukaryotic DNA. During both early and late S phase, origins are recognized by ORC. The Cdc6 and Cdt1 proteins load the MCM helicases to form licensed pre-RC complexes during late G1 phase. The activation of pre-RCs might differ for early and late replication. Cdc7 and Cdk2 activate MCM helicases and load replisome complexes during early S phase. Cdc7 and Cdk1 activate MCM helicases and load replisome complexes during late S phase.

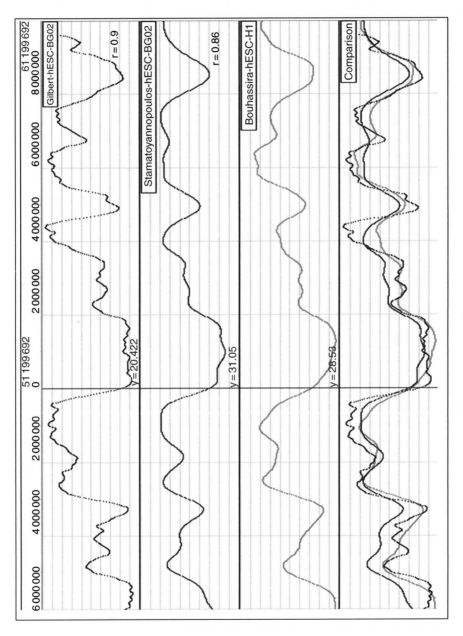

Figure 13.2 Comparison of replication timing in human embryonic stem cells using different methods. The three tracks represent replication timing measured with different methods from three different laboratories: Gilbert, Stamatoyannopoulos, and Bouhassira. The fourth track shows an overlay of all the results. The correlations between the results from different laboratories are high ($r = 0.9$ and $r = 0.86$), indicating the robustness of the methods.

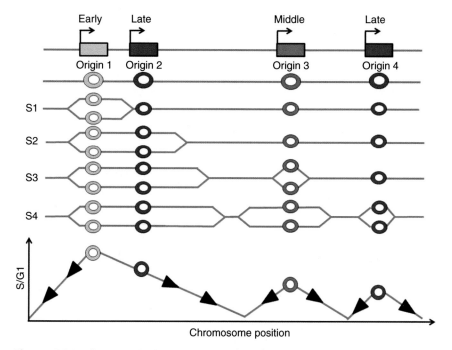

Figure 13.3 Replicon organization. Desprat *et al.* (2009) suggested on the basis of the analysis of their TimEX results that there might be an origin close to every gene and that the timing of the firing of the origin is proportional to the level of expression. The lighter-color genes represent highly expressed genes associated with early replicating origins. The intermediate and dark colors respectively represent genes with intermediate and low levels of expression which are associated with origin of replication programmed to fire in middle or late S phase. Poorly expressed genes located near active genes do not contribute to the timing of regulation since they are replicated early by forks emanating from the early origin associated with the neighboring highly expressed gene. In this model, forks proceed at a constant rate unidirectionally until they encounter another fork progressing in the opposite direction.

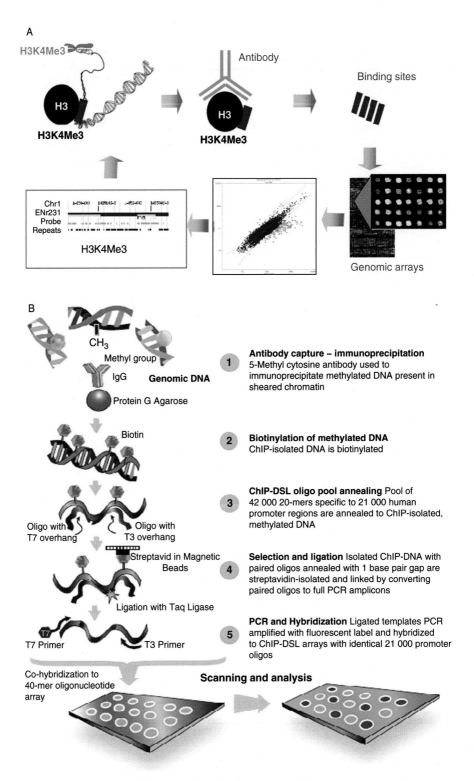

Figure 15.1 Overview of ChIP-chip and ChIP-DSL technologies. (A) Basic components of ChIP-chip technology. (B) Overview of key steps in the ChIP-DSL technology.

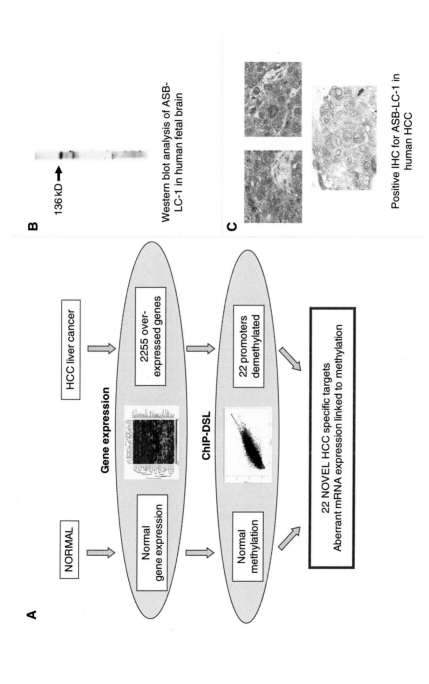

Figure 15.2 ChIP-DSL identification of cancer biomarkers in hepatocellular carcinoma (HCC) cells. (A) Identification of cancer biomarkers in HCCs using ChIP-DSL and gene-expression profiling to identify promoters where aberrant demethylation increases gene expression. (B) Western blot analysis of HCC ChIP-DSL identified candidate ASB LC-1 in human fetal brain cells. (C) Immunohistochemistry confirming expression of ASB LC-1 in normal brain cells and HCC liver cells.

Sequence to analyze:

C/TGTTTTGC/TGTTTC/TGAC/TGTTC/TG AGGTTTTC/TGC/TGGTGC/TG ATC/TGTT

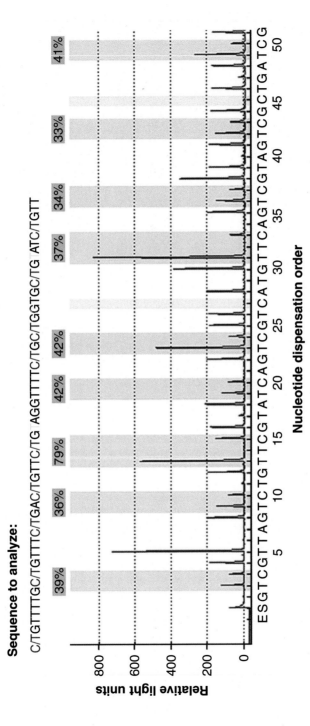

Figure 16.3 Analysis of multiple contiguous CpG sites. Methylation at nine independent CpG sites (highlighted in gray) is quantified in a single pyrosequencing run. Position-specific information in the context of an analyzed sequence presents broad sequence methylation patterns. Note the built-in quality control sites (highlighted in yellow) consisting of cytosines converted to thymines, demonstrating full bisulfite conversion of the treated DNA.

Single-molecule PCR-based sequencing

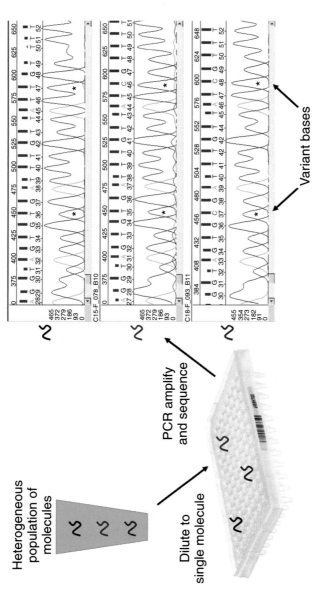

Figure 18.1 Schematic outline of smPCR bisulfite sequencing workflow wherein a gDNA has been bisulfite-converted to a heterogeneous population of molecules (black, blue, and red squiggles; only one of many molecules of each allele is shown) which are serially diluted such that aliquots transferred to a microtiter plate allow some wells to have a single molecule for PCR amplification and then fluorescence-based CE Sanger-sequencing to identify variant bases (asterisks) that are either T or C and thus indicate C or 5mC, respectively, at CpG positions in gDNA.

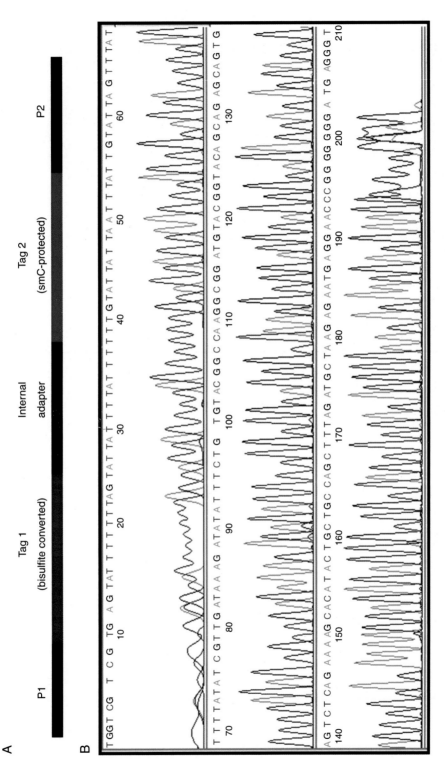

Figure 18.2 (A) Schematic representation of a SOLiD™ mate-pair library fragment comprised of adapters (black; P1, internal, and P2) and mate-pair sequences (tags) that undergo bisulfite conversion (red; Tag 1) or are 5mC-protected (blue; Tag 2). (B) Exemplary smPCR/CE-sequencing traces of a bisulfite-converted SOLiD™ mate-pair library fragment using a shortened P1 adapter sequence as the sequencing primer. Bases 0–12 are the P1 adapter extended from the shortened primer; bases 13–95 are bisulfite-converted Tag 1; bases 96–131 are the internal adapter (also protected with 5mC during nick translation); bases 132–179 are 5mC-protected Tag 2; bases 180–end are the reverse complement of the P2 adapter.

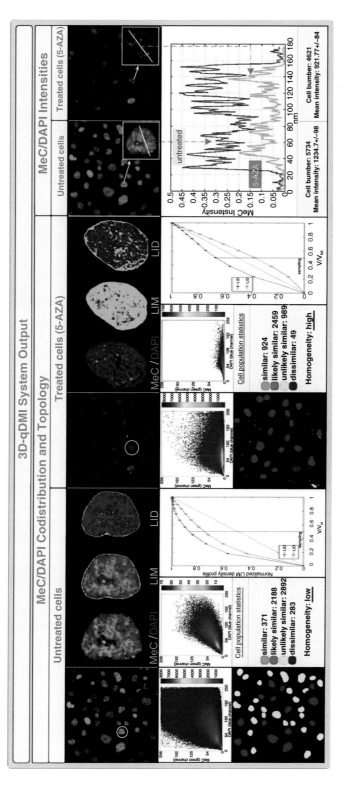

Figure 20.3 The 3D-qDMI system currently offers three features for the in situ assessment of drug-induced changes in global DNA methylation and relevant chromatin organization in cells (DU145 prostate cancer cells in here): (i) changes in MeC (green) and global gDNA (blue) displayed as their respective nuclear mean intensities as well as an intensity profile across different selectable cross-sections of individual nuclei (for simplicity only MeC signals shown in here); (ii) changes in the spatial distribution of MeC and global DNA as an indicator of chromatin reorganization in cells; and (iii) alterations in the topology of MeC and DAPI intensities displayed as graphs and maps, also indicating $LIM_{0.5}$ and $LID_{0.5}$. The graphical user interface (GUI) allows for each feature to be displayed for interactively selected nuclei (marked by a white ring) via clicking on the labeled ROI in the K–L map. A fourth module has been implemented for statistical (homogeneity) assessment of the population based on 3D chromatin texture and MeC load.

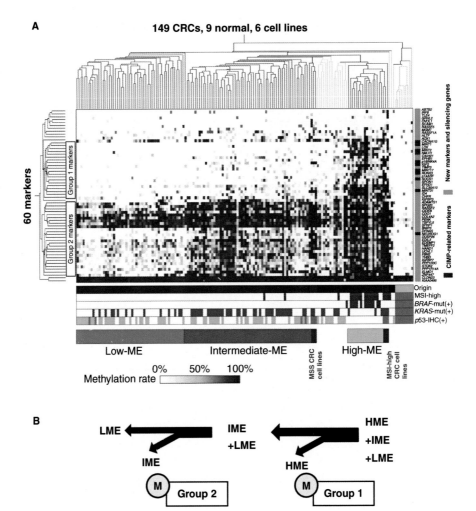

Figure 21.4 Epigenotyping of colorectal cancer (CRC). (A) Unsupervised two-way hierarchical clustering classified CRC cases into three epigenotypes: high-ME (HME) were clustered with MSI-high cell lines, and correlated strongly with MSI-high and *BRAF*-mutation(+). Intermediate-ME (IME) were clustered with microsatellite-stable (MSS) cell lines, and correlated with *KRAS*-mutation(+). Methylation markers were clustered into two major groups: Group 1 showing high methylation in HME, and Group 2 showing high methylation in HME and LME. In heat map: 100% methylation, red; 0% methylation, white; no data, gray. In Origin column at the bottom: red, CRC; dark blue, CRC cell lines; light blue, normal colon mucosa. (B) At the first step, colorectal cancer with Group 1 marker methylation is extracted as HME. At the second step, the remaining samples are divided into IME if Group 2 markers are methylated, and into LME if Group 2 markers are not methylated.

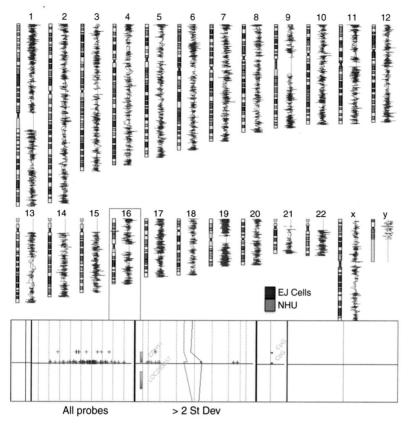

Figure 23.2 Global analysis of DNA methylation using the MeDIP technique reveals differences between non-transformed normal (NHU) and malignant urothelial cells (EJ). The highlighted region reveals probe enrichment for DNA hypermethylation in the cancer cell line at the CpG island adjacent to CDH11, when compared to NHU. This gene is known to be epigenetically silenced in other cancers (Sandgren *et al.*, 2010).

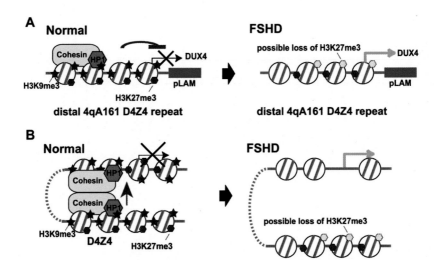

Figure 25.5 Putative models for the role of epigenetic change at D4Z4 in FSHD pathogenesis. (A) The effect on *DUX4* expression. Heterochromatin in normal D4Z4 inhibits the transcription of *DUX4*. In FSHD, accompanied by the disruption of this silencing structure, the transcription of *DUX4* ORF in D4Z4 repeats is activated. The loss of H3K9me3 and possibly H3K27me3 also favors the utilization of the poly(A) signal in the pLAM sequence of the 4qA161 variant through a splicing mechanism, which helps stabilize the *DUX4* mRNA transcribed from the distal D4Z4 repeat. The DUX4 protein may function in FSHD pathogenesis. (B) Looping model. HP1γ and cohesin may function in the physical interactions of the heterochromatic D4Z4 region with other genomic regions, which results in the spreading of the repressive effect to target genes in normal cells. In FSHD, the loss of H3K9me3 (and perhaps H3K27me3), HP1γ, and cohesin on D4Z4 leads to the disruption of these physical interactions and abnormal activation of these genes.

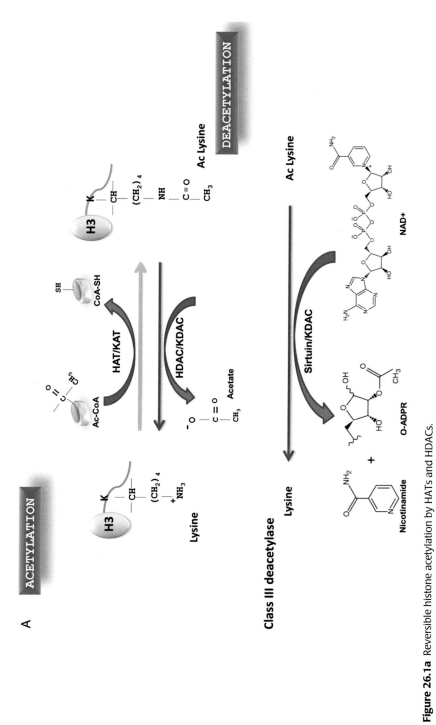

Figure 26.1a Reversible histone acetylation by HATs and HDACs.

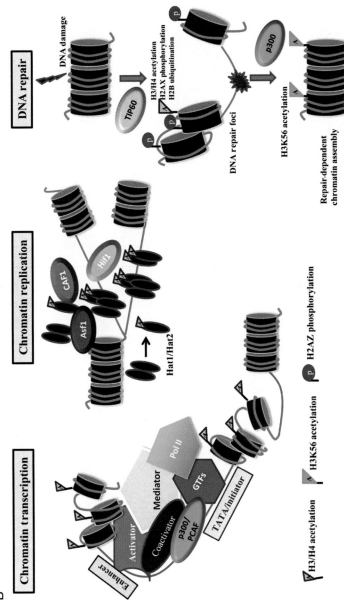

Figure 26.1b Role of histone acetylation in different chromatin-templated phenomena like transcription, replication, and DNA damage repair. Acetylation results in maintenance of an open form of chromatin that is essential for chromatin remodeling and transcription, replication, and DNA damage repair. It also helps to recruit specific factors depending on the context. Furthermore, the HATs/KATs acetylate several factors (for example GTFs, transcription factors, repair associated proteins, etc.) and modulate their activities.

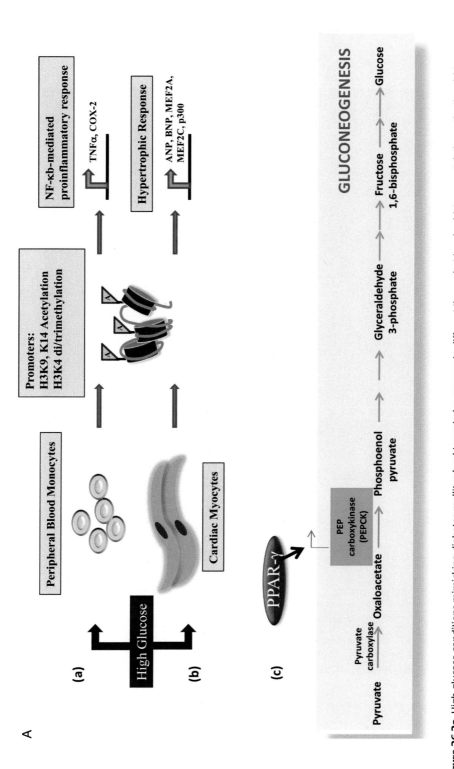

Figure 26.2a High-glucose conditions mimicking diabetes mellitus lead to varied responses in different tissues but involve histone acetylation activation. (a) In peripheral blood monocytes high-glucose treatment leads to increased recruitment of p300/CBP to NF-κB-dependent genes leading to promoter histone H3/H4 acetylation and hyperexpression of TNFα and COX-2. (b) In cardiac myocytes high-glucose treatment increases expression of p300 and pro-hypertrophic genes associated with increase in histone acetylation. (c) The gluconeogenesis pathway representing the different enzymes and the products. The coactivator PPARγ regulates the expression of PEPCK.

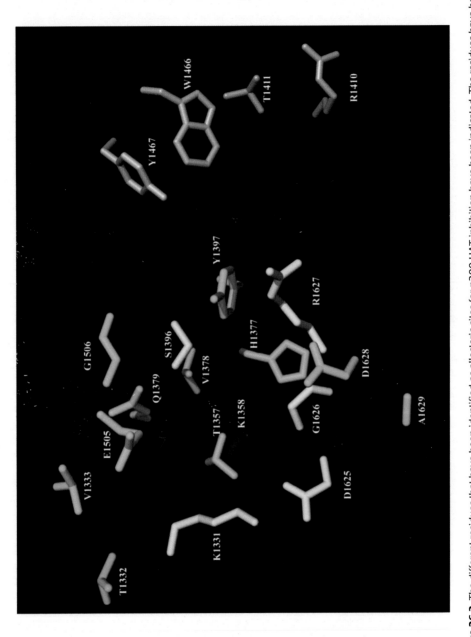

Figure 26.3 The different residues that have been identified as allosteric sites for p300 HAT inhibition have been indicated. The residues have been color-coded based on the reports as follows. White, residues identified in the C646 study; yellow, common residues for garcinol, isogarcinol, and LTK-14; cyan, common residues for isogarcinol and LTK-14; green, residues binding exclusive to garcinol; red, plumbagin.

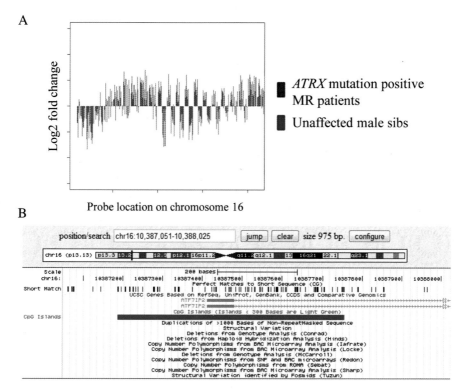

Figure 30.2 The DMR encompassing the probe CHR16FS010387819 identified by genome-wide promoter DNA methylation analysis in *ATRX*-mutation-positive patients and subsequent statistical analyses. (A) The output generated by statistical analysis also showing the hypermethylation of this 16p region in the ATR-X patients (red) as compared to the unaffected sibs (blue). (B) A UCSC exsert of the hypermethylated *ATF7IP2* locus in the ATR-X patients; this hypermethylation occurs in the CpG island upstream of *ATF7IP2*. No segmental duplications or CNVs are present in this region.

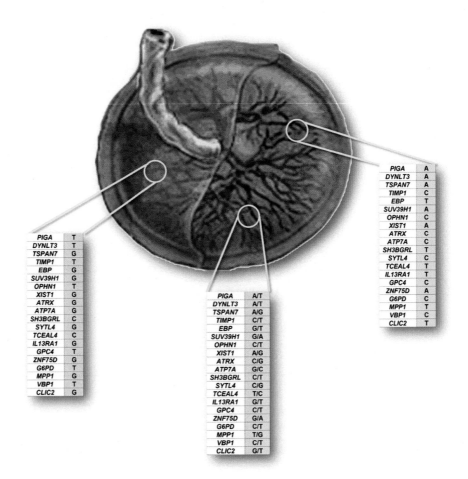

Gene	
PIGA	A
DYNLT3	A
TSPAN7	A
TIMP1	C
EBP	T
SUV39H1	A
OPHN1	C
XIST1	A
ATRX	C
ATP7A	C
SH3BGRL	T
SYTL4	C
TCEAL4	T
IL13RA1	T
GPC4	C
ZNF75D	A
G6PD	C
MPP1	T
VBP1	C
CLIC2	T

Gene	
PIGA	T
DYNLT3	T
TSPAN7	G
TIMP1	T
EBP	G
SUV39H1	G
OPHN1	T
XIST1	G
ATRX	G
ATP7A	G
SH3BGRL	C
SYTL4	G
TCEAL4	C
IL13RA1	G
GPC4	T
ZNF75D	G
G6PD	T
MPP1	G
VBP1	T
CLIC2	G

Gene	
PIGA	A/T
DYNLT3	A/T
TSPAN7	A/G
TIMP1	C/T
EBP	G/T
SUV39H1	G/A
OPHN1	C/T
XIST1	A/G
ATRX	C/G
ATP7A	G/C
SH3BGRL	C/T
SYTL4	C/G
TCEAL4	T/C
IL13RA1	G/T
GPC4	C/T
ZNF75D	G/A
G6PD	C/T
MPP1	T/G
VBP1	C/T
CLIC2	G/T

Figure 33.2 X-chromosome inactivation (XCI) in human term placenta. Pink and blue patches represent regions where the maternal and the paternal X is the active chromosome, respectively. Depending on the region analyzed, monoallelic or biallelic expression of the listed X-linked genes can be detected, suggesting imprinted or random XCI, respectively.

12 MicroRNAs in embryonic stem cells

Shari Orlanski and Yehudit Bergman*

12.1   Introduction

Embryonic stem (ES) cells are cells which are isolated from the inner cell mass of embryos at the blastocyst stage of development (Evans and Kaufman, 1981; Martin, 1981; Brook and Gardner, 1997). These cells are characterized by their unique capabilities including self-renewal and the ability to maintain themselves as undifferentiated cells, as well as their pluripotent capacity to develop into any cell type. It is these two unique characteristics that make embryonic stem cells so interesting, as they present potential to be important scientific models as well as important therapeutic tools. Much research has been done toward understanding the delicate balance that ES cells possess, allowing them to maintain themselves as both undifferentiated and pluripotent. Though much still remains to be clarified, it is clear that many epigenetic elements such as DNA methylation, histone modifications, and microRNAs play crucial roles in this regulation (Boyer *et al.*, 2005; Bernstein *et al.*, 2006; Spivakov and Fisher, 2007; Benetti *et al.*, 2008; Bibikova *et al.*, 2008; Sinkkonen *et al.*, 2008; Tiscornia and Izpisua Belmonte, 2010).

MicroRNAs (miRNAs) are short sequences of RNA, 21–24 nucleotides (nt) in length, which play an integral role in the regulation of protein expression. MicroRNAs were first described with the discovery of lin-4 in *Caenorhabditis elegans* (Lee *et al.*, 1993). However, the importance of this discovery was not fully appreciated until some years later, a decade ago, when what would be the second miRNA discovered, let-7, was described (Reinhart *et al.*, 2000). Only then was it understood that the phenomenon was not an isolated occurrence, and that these short RNAs could actually play very important roles in gene regulation. Today it is known that there are more than 700 miRNAs in the human genome, and it has been predicted that likely over 60% of coding genes are targets of

---

* Author to whom correspondence should be addressed.

Epigenomics: From Chromatin Biology to Therapeutics, ed. K. Appasani. Published by Cambridge University Press. © Cambridge University Press 2012.

miRNAs. Bioinformatic predictions as well as experimental approaches indicate that one miRNA can have many, perhaps even hundreds of targets (He and Hannon, 2004; Friedman *et al.*, 2009).

MicroRNAs are widely conserved between species, and are crucial for proper development in many organisms. This characteristic attests to their importance in various biological processes. MicroRNAs are often expressed in a tissue-dependent manner, and can even be used as a sort of "fingerprint" as to the identity of a particular cell. MicroRNAs are directed by tissue-specific transcription factors which have been nicknamed, by some, the "miR-ome" of each particular cell type (Selbach *et al.*, 2008). Embryonic stem cells are no exception and possess their own unique miRNA expression pattern (Houbaviy *et al.*, 2003; Suh *et al.*, 2004). Stem cells require a delicate balance of factors which allow for maintenance of stemness as well as pluripotency, and the precise regulation methods of miRNAs are ideally suited to this.

MicroRNAs are first transcribed by RNA polymerase II as long transcripts known as pri-miRNAs (Du and Zamore, 2005). These transcripts are both capped at the 5′ end as well as polyadenylated at their 3′ tail. The pri-miRNAs form hairpin loops which are cleaved in the nucleus by the microprocessor complex containing Drosha, an RNase III enzyme, and DGCR8, a double-stranded RNA binding protein (also known as Pasha). This cleavage occurs specifically at the base of the stem structure, and leaves the transcript as a hairpin loop approximately 60–80 nt in length now known as a pre-miRNA. These pre-miRNAs are then exported into the cytoplasm. In the cytoplasm, pre-miRNAs are recognized and processed by another member of the RNase III enzyme family, Dicer, and the double-stranded RNA binding protein TRBP. (It has been shown that phosphorylation of TRBP by the MAPK/ERK pathway promotes maturation of miRNAs by stabilizing Dicer [Paroo *et al.*, 2009].) Dicer cleaves off the hairpin loop of the transcript, and gives rise to single-stranded mature miRNAs. One of these strands associates with one of the members of the Argonaute (AGO) family of proteins in the RNA-induced silencing complex (RISC). Once formed, this complex binds directly to the 3′ untranslated region (UTR) of target mRNAs via complementary base-pairing between the mRNA sequence and the seed sequence (2–7 nt in length) of the miRNA. Upon binding, the complex can induce silencing of the mRNA either by cleavage, deadenylation, and degradation or physical translational inhibition (Kim, 2005; Mathonnet *et al.*, 2007; Wang *et al.*, 2009; Wu *et al.*, 2010).

Recently, there has been some evidence for miRNA regulation independent of seed sequence complementarity. For example, miR-328, a miRNA whose absence is known to be involved in chronic myelogenous leukemia, can simultaneously interact with the hnRNP E2 protein via its C-rich region, and with the mRNA encoding the survival factor PIM1 by complementary binding of its seed sequence to the 3′ UTR of the mRNA. Therefore, miR-328's interaction with hnRNP E2 is independent of the miRNA's seed sequence (Eiring *et al.*, 2010). This chapter will focus on those miRNA clusters which are known to be crucial for the regulation of ES cell maintenance and pluripotency.

12.2   Significance of miRNA in ES cells

Pivotal to understanding the significance of miRNAs in ES cells were studies done on ES cells null for Dicer and DGCR8, which as mentioned earlier, are two important players in the biogenesis of miRNAs. Dicer is required for the processing of both miRNAs as well as other small RNAs in the cell, whereas DGCR8 plays a role only in the maturation of miRNAs (Wang *et al.*, 2009).

Dicer knockout ES cells are viable and express the pluripotency marker *Oct4* at high levels, though they do not produce miRNAs or other endogenous interference RNA. These cells have severe defects in their ability to differentiate both in vitro and in vivo. In addition, loss of Dicer leads to a complete proliferation block, which can be rescued via continued cell culture which probably results in expression of compensatory factors. This Dicer knockout proliferation phenotype resembles the one resulting from deleting the Argonaute (Ago) family members (Su *et al.*, 2009). This phenotype can be rescued either upon introduction of Dicer (Kanellopoulou *et al.*, 2005; Murchison *et al.*, 2005), or through introduction of any of the Ago isoforms (Su *et al.*, 2009) into the respective ES cells.

In order to elucidate further the specific importance of miRNAs in ES cells, the role of DGCR8 was investigated (Wang *et al.*, 2007). In general, ES cells in which DGCR8 is knocked out have a similar phenotype to those which are deficient in Dicer, though to a much milder extent. While these cells were incapable of fully downregulating pluripotency genes such as *Oct* and *Nanog*, they did succeed in upregulating certain markers of differentiation, which the Dicer knockout cells could not (Kanellopoulou *et al.*, 2005). The differences between the ability of the two cell types to differentiate most probably suggest that other small RNAs may also play a role in ES cells maintenance and development.

In vivo knockout of Dicer leads to severe growth retardation and embryonic death by 7.5 dpc (Bernstein *et al.*, 2003). Insight into the involvement of miRNAs in early developmental processes is presented in a recent study. Using Dicer mutant embryos, this study shows that contrary to what occurs in the epiblast where miRNAs modulate cell survival, and in ES cells where they are required for differentiation, in the trophectoderm, primitive endoderm, and the stem cells derived from these tissues, miRNAs act to maintain self-renewal and multipotency. Taken together, these findings stress the importance of miRNAs as regulators of pluripotency, proliferation, and differentiation during early embryogenesis and in ES cells (Spruce *et al.*, 2010).

12.3   MicroRNAs highly expressed in ES cells

The first major profile of miRNA expression specific to ES cells showed different expression profiles of miRNAs in undifferentiated and differentiated mouse ES (mES) cells. It showed that miRNAs were likely important in the maintenance of ES cells and pluripotency as there were some miRNAs which were specific to ES cells and their expression was lost upon differentiation. Among others, they described the miR-290 cluster as ES-cell-specific; the members of the cluster shared related sequences and were transcribed in a cluster. Two interesting categories were observed: one including those miRNAs which are ES-cell-specific and important for ES cell pluripotency and embryonic development, and another containing miRNAs which are found in both ES cells as well as differentiated ES cells and adult tissues, such as miR-15a and miR-16 (Houbaviy et al., 2003). This work clearly proved the importance of miRNAs in ES cells, but the targets on which they acted still remained to be discovered.

A similar work in human ES (hES) cells showed that of the miRNAs they discovered in hES cells, three were conserved between the novel miRNAs discovered in mES and hES, miR-296, miR-301, and miR-302. They also reported 17 novel miRNAs in addition, including 14 ES-cell-specific: miR-302b*, miR-302b, miR-302c*, miR-302c, miR-302a*, miR-302d, miR-367, miR-200c, miR-368, miR-154*, miR-371, miR-372, miR-373*, and miR-373. They divided the miRNAs they discovered into four general categories; the first being miRNAs which are expressed in ES cells as well as in teratocarcinoma (EC) cells, including the miR-302 cluster; the second containing miRNAs which are expressed specifically in ES cells but not in other cells, even EC cells, including miRs-371, 372, 373*, and 373; thirdly, miRNAs which are involved in the various stages of development and therefore rare in ES cells but abundant in differentiated cells, including let-7a and miR-21; and lastly, those which are expressed in most cells and likely to be involved in basic cellular functions, such as miR-16, miR-17–5p, and miR-222. Interestingly, it was noted that many of the ES-specific miRNA genes found in mES and hES cells are highly related to each other. Members of the miR-302 cluster are homologous between the species, and miRs-371, 372, 373*, and 373 in hES are homologs of the mouse miR-290 cluster. They concluded that the conservation of the miRNAs showed that these miRNAs likely play important roles in regulating ES cells (Suh et al., 2004).

The miR-290 cluster accounts for the majority of all miRNAs in undifferentiated ES cells and is downregulated upon differentiation (Houbaviy et al., 2003; Chen et al., 2007). This cluster consists of family members miRs-290, 291a, 291b, 292, 293, 294, and 295. Some of these miRNAs share a common seed region, even with their human homologs. The miR-290 cluster shares a promoter and as such, is transcribed as a polycistronic RNA molecule (Suh et al., 2004). This promoter is transcriptionally regulated by Oct4, Nanog, and Sox2 (Card et al., 2008; Marson et al., 2008; Barroso-del Jesus et al., 2009) (Figure 12.1). The miR-290 cluster is indispensable in embryonic development, as mice that are null for miR-290 are embryonic lethal (Ambros and Chen, 2007). Since the miR-290 cluster is also expressed in germ cells (Hayashi et al., 2008), it could be interesting to study the effects of miR-290 cluster conditional deletion upon the germ-line.

12.4   MicroRNAs and pluripotency factors creating a regulatory network of ES cells

The role of pluripotency factors such as Oct4, Nanog, and Sox2 has been well elucidated with regard to regulation of pluripotency genes in ES cell regulation

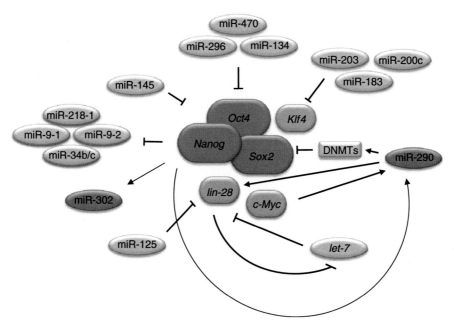

Figure 12.1 MicroRNAs and pluripotency factors create a regulatory network of ES cells. Pluripotency factors are both crucial regulators of miRNAs as well as targets of them. *Oct4*, *Sox2*, and *Nanog* upregulate miRNAs important for self-renewal and maintenance of pluripotency, the miR-290 and miR-302 clusters, while they downregulate those which are cell-specific. Likewise, miRNAs that support differentiation suppress the expression of the pluripotency factors. *Let-7* both suppresses *lin-28* and is suppressed by it; miR-290 is both activated by the pluripotency factors as well as necessary for their downregulation via DNMTs upon differentiation.

(Boyer *et al.*, 2005; Loh *et al.*, 2006). In addition to these factors, it was found that miRNAs are also crucial components of the pluripotency regulatory network of ES cells, as these core pluripotency factors both directly regulate the transcription of miRNAs and are targets of differentiation inducing miRNAs (Marson *et al.*, 2008; Tay *et al.*, 2008; Xu *et al.*, 2009; Tarantino *et al.*, 2010) (Figure 12.1).

In order to characterize the way in which these factors can regulate miRNAs directly, genome-wide analyses were performed to identify the transcriptional start sites of miRNAs in both mouse and human (Marson *et al.*, 2008). Interestingly, it was found that these factors associate with two sets of miRNAs, one that is actively expressed in wild-type ES cells and promotes pluripotency, including the miR-290 and miR-302 clusters, which are crucial for maintenance of stemness (Card *et al.*, 2008; Marson *et al.*, 2008; Barroso-del Jesus *et al.*, 2009), and the other that is repressed in ES cells, including tissue-specific expressed miRNAs, such as miR-34b/34c, miR-9–1, miR-9–2, and miR-218–1 (Marson *et al.*, 2008) (Figure 12.1). The presence of these factors at tissue-specific promoters may indicate a poising of these genes for expression later in development in specific lineages. Previously published studies have shown that the core pluripotent transcription factors occupy, together with polycomb group proteins (PcG), a set of transcriptionally repressed genes that are poised for expression upon differentiation (Bernstein *et al.*, 2006). Similarly, it was found that the PcG repressor complex co-occupies miRNAs that are silenced in ES cells. In fact, approximately one-quarter of the core transcription-factor-occupied miRNAs belong to the repressed set of miRNA genes bound by PcG in murine ES cells (Marson *et al.*, 2008).

MicroRNAs often function in a complex, multilayered manner, and sometimes form a feedback loop with the transcription factors which regulate them. One example is that of the feedback loop among miR-302, *Oct4*, and NR2F2 (also known as COUP-TFII) that is necessary for ES cell pluripotency and differentiation. In undifferentiated ES cells, *Oct4* positively regulates the miR-302 family, and negatively regulates NR2F2, both at the transcriptional level. NR2F2 is also posttranscriptionally controlled by miR-302, conferring a buffering system against perturbation of uncontrolled NR2F2 expression in ES cells. Conversely, NR2F2 directly inhibits *Oct4* upon differentiation (Ben-Shushan *et al.*, 1995; Schoorlemmer *et al.*, 1995), triggering a positive feedback loop for its own expression. The decrease of miR-302 during differentiation is necessary in order for NR2F2 to function and activate early neural genes such as *Pax6*, which is an early-expressed gene involved in development of the forebrain, as well as other early genes such as *Six3* and *Lhx2* which bind *Pax6* (Rosa and Brivanlou, 2010) (Figure 12.2).

Another important factor in the regulation of miRNAs in ES cells is *c-Myc*. The Myc proteins were previously known to be important in directing cell proliferation (Adhikary and Eilers, 2005; Cole and Nikiforov, 2006; Kleine-Kohlbrecher *et al.*, 2006). *C-Myc* has also been demonstrated to activate the expression of

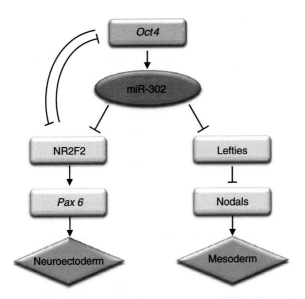

Figure 12.2 miR-302 cluster is crucial for cell lineage specification. The miR-302 cluster has been shown to play a role in cellular differentiation to the mesodermal lineage versus the neuroectodermal lineage. In the ES cells state, *Oct4* inhibits expression of NR2F2 and also activates miR-302, which inhibits NR2F2, but upon differentiation, NR2F2 inhibits *Oct4* in return, and activates neuroectoderm specific genes. Additionally, miR-302 inhibits Lefties, allowing Nodals to induce mesoderm formation.

several miRNAs in ES cells (Card *et al.*, 2008; Lin *et al.*, 2009; Smith and Dalton, 2010). For example, it has been shown that c-Myc binds at the promoter region of the mir-290 cluster (Figure 12.1) (Hanina *et al.*, 2010). In turn, mir-294 (a member of this cluster) can indirectly activate the expression of *c-Myc* (Melton *et al.*, 2010). This could be mediated by the ability of the miR-290 cluster to target the *Dkk-1* gene (Zovoilis *et al.*, 2009), a Wnt pathway inhibitor, since *c-Myc* is a downstream target of this pathway. *C-Myc* likely affects the ES cells phenotype by inhibiting genes necessary for differentiation. In fact, a number of genes that are important in differentiation have been shown to be targets of *c-Myc* and the miRNAs it regulates. For example, *Cdh11*, an adhesion molecule important in various developmental processes (Boscher and Mege, 2008), is downregulated by both *c-Myc* and miR-141 (Lin *et al.*, 2009).

Most recently, the importance of this cluster of miRNAs was demonstrated by introducing a specific member of the cluster, miR-294, into Dicer−/− cells. Via transcriptome profiling and isolation of mRNA targets associated with Ago2, followed by RNA sequencing, direct miR-294 targets were identified (Hanina *et al.*, 2010). Gene ontology (GO) analysis of the genes regulated by miR-294 showed enrichment for genes involved in development, as well as genes involved in the regulation of the G1–S transition of the cell cycle, for which the miR-290 cluster is known to be important as discussed below (Hanina *et al.*, 2010) (Figure 12.3). No substantial overlap between *Oct4-*, *Nanog-*, and *Sox2*-regulated

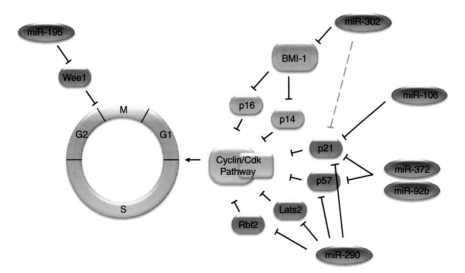

Figure 12.3 MicroRNAs as regulators of the cell cycle. The miR-290 cluster along with miRs-106, 372, and -92b inhibit inhibitors of the G1–S transition, allowing cells to proliferate rapidly and maintain the unique ES cell cycle. In addition, miR-195 inhibits inhibitors of the G2–M transition, also allowing the cell cycle to progress quickly. The miR-302 cluster plays a role in both activation and inhibition of the ES cells cycle. There is evidence that in mice, it may help p21 to promote the G1–S transition. This cluster is also important in allowing ES cells to exit this rapid cell cycle and enter a differentiated cell cycle. It does so by inhibiting BMI-1, which in turn inhibits inhibitors of the G1–S transition, which allows the cells to arrest in G1.

genes and miR-294-regulated transcripts was found. In contrast, it was shown that miR-294 and *c-Myc* share an overlapping set of target genes, which is consistent with their comparable roles during somatic cell reprogramming (see below). Additionally, miR-294 may act to repress a subset of genes induced by *c-Myc*. This phenomenon, in which *c-Myc* both activates a target, and inhibits it via miR-294, is known as "incoherent feedforward" regulation (Alon, 2007). The functional significance of such regulation is not fully understood, but it could assist in buffering against untimely stimuli that perturbs a critical network. Interestingly, miR-294 also promotes pluripotency in ES cells by upregulating pluripotency-associated genes such as *Lin-28*, an mRNA binding protein which promotes pluripotency (Hanina *et al.*, 2010) (Figure 12.1), which is also regulated directly by transcriptional activation by *c-Myc* (Melton *et al.*, 2010).

12.5   MicroRNAs as regulators of the cell cycle

Embryonic stem cells possess a unique cell cycle, crucial for their distinctive characteristics. In somatic cells, proteins regulating the cell cycle tend to vary, with activators and inhibitors fluctuating at different phases. In contrast, ES cells express activating factors of the cell cycle, such as Cdk2–cyclin E/A and Cdk6–cyclin D3 more steadily, and show a marked absence of inhibiting factors, such as p21cip, p27Kip1, and p16INK4a (Faast *et al.*, 2004). Moreover, in ES cells, the G1–S restriction is largely absent which enables cells to rapidly move through the G1 phase and enter S phase. This somewhat accelerated cell cycle allows for the rapid growth required in early embryogenesis (Mac Auley *et al.*, 1993; Stead *et al.*, 2002; Wang *et al.*, 2008; Wang and Blelloch, 2009). There is some evidence which supports the theory that ES cells can commence differentiation during the G1 phase if the proper external or internal signals are received, which is one explanation as to why regulation of the length of this phase is crucial for ES cell pluripotency and differentiation (Savatier *et al.*, 1996; Jirmanova *et al.*, 2002).

Further inquiry to better understand the DGCR8 null phenotype, described above, revealed that the cells were defective in the G1–S transition. A library of miRNAs were individually reintroduced into these cells, and it was discovered that there were 14 that could rescue the phenotype, including different clusters, such as miR-290, miR-302, and the miR-17–92 clusters. These 14 miRNAs are expressed in ES cells and downregulated upon differentiation. These miRNAs were referred to as ESC-specific cell-cycle-regulating (ESCC) miRNAs. These miRNAs share the same or similar seed sequences as a result of which they most likely share common targets. Targets of these miRNAs include inhibitors of the cyclin/Cdk pathways, which are responsible for regulating the G1–S transition. These ESCC miRNAs inhibit these inhibitors of the cell cycle, and in turn, increase cyclin/Cdk activity, which allows the cell cycle to proceed more rapidly (Wang *et al.*, 2008; Wang and Blelloch, 2009) (Figure 12.3).

The miR-290 cluster directly targets and suppresses multiple inhibitors of the G1–S transition, such as p21, Rbl2 (Retinoblastoma-like 2 protein), and Lats2

(Wang *et al.*, 2008), which allows G1 to remain relatively short in ES cells. As mentioned previously, the expression of the miR-290 cluster decreases upon differentiation, which correlates well with the fact that as the cells differentiate they need to achieve the cycle of a somatic cell. And in fact, when cells differentiate and the miR-290 cluster is no longer expressed, the G1 phase elongates (Wang *et al.*, 2008; Wang and Blelloch, 2009).

The miR-302–367 cluster is composed of nine miRNAs: miR-302a, miR-302a*, miR-302b, miR-302b*, miR-302c, miR-302c*, miR-302d, miR-367, and miR-67*. As with the miR-290 cluster, the miR-302 cluster is expressed at high levels in ES cells but is downregulated upon differentiation. The promoter of the (cluster that is transcriptionally regulated by *Oct4*, *Nanog*, and *Sox2*), as well as its structure have recently been described and characterized in human ES cells. This cluster is responsible in part for regulating the cell cycle in ES cells, allowing them to self-renew and maintain pluripotency (Barroso-del Jesus *et al.*, 2009). It has been shown that inhibition of members of the miR-302 cluster in hES cells promotes G1 arrest, inducing the ES cells cycle to become more similar to that of differentiated cells. Conversely, overexpression of miR-302 in differentiated cells induces exit from G1. In addition, it was shown that the miR-302 posttranscriptionally silences BMI-1, a marker of cancer stem cells which increases the function of tumor suppressors p16Ink4a and p14/p19Arf. They in turn inhibit the cyclin D1/Cdk4 and cyclin E/Cdk2 pathways thereby blocking the G1–S transition (Card *et al.*, 2008; Lin *et al.*, 2010b). Though miR-302 is conserved between humans and mice, and in both is crucial for inducing G1 phase arrest, the direct targets of the cluster can vary between species. In mice, miR-302 may also silence G1 checkpoint regulator p21cip1, while in humans it does not (Lin *et al.*, 2010b) (Figure 12.3).

While the miR-290 and 302 clusters play critical roles in regulating the ES cell cycle, other miRNAs are also involved in this process. Another member of the ESCC group, miR-106, promotes the cell cycle by downregulating p21cip to inhibit G1–S transition (Ivanovska *et al.*, 2008). Similarly, miR-372 and miR-92b which are both highly expressed in hES cells regulate p21cip and p57, both inhibitors of the G1–S transition, while miR-195 suppresses multiple inhibitors of the G2–M transition such as Wee1 (Qi *et al.*, 2009; Sengupta *et al.*, 2009). Taken together, several apparently redundant miRNAs are crucial regulators of the distinctive cell cycle profile of ES cells which is critical for fast growth during embryogenesis (Figure 12.3).

12.6   The involvement of miRNAs in differentiation of ES cells

Upon differentiation ES cells need to shut down the pluripotency network and switch to molecular programs allow differentiation to highly specialized lineages to occur. Using both Dicer- and DGCR8-knockout ES cells helped in deciphering miRNAs that play a key role in this process.

Investigation into the cause of the Dicer-null phenotype revealed that the inability of Dicer-null ES cells to silence the self-renewal program upon differentiation was due to the absence of the miR-290 cluster. The miR-290 cluster is required for expression of *de novo* DNA methyltransferases (DNMTs) and in its absence, as with Dicer-null cells, the *Oct4* promoter was unable to be properly silenced upon induction of differentiation due to improper methylation in the region (Benetti *et al.*, 2008; Sinkkonen *et al.*, 2008). Previous studies have clearly shown that these DNMTs are responsibly for the irreversible silencing of the *Oct4* promoter during ES cells differentiation (Feldman *et al.*, 2006; Epsztejn-Litman *et al.*, 2008). In fact, the protein Rbl2, a member of the Rb family of genes that inactivates DNMTs, was found to be a specific and direct target of the miR-290 cluster, and thus the cause of the reduced methylation in Dicer$^{-/-}$ cells. Reintroduction of the miR-290 cluster into the cells resulted in decreased levels of Rbl2, and thereby increased levels of the DNMTs and appropriate methylation and silencing of the *Oct4* promoter (Benetti *et al.*, 2008; Sinkkonen *et al.*, 2008) (Figure 12.1).

An additional miRNA that plays a role in ES cell differentiation is *let-7*, which upon introduction into DGCR8-knockout ES cells rescues the differentiation defect, allowing the cells to shut down the self-renewal program (Melton *et al.*, 2010). The *let-7* family is made up of nine miRNAs. It was first described in *C. elegans* as a heterochronic gene, one of the genes that regulate relative developmental timing and thus mutations in such genes perturb developmental stage-specific events. In the absence of *let-7* there was gene activity which induced the appearance of larval cell fates in the adult worm, and overexpression of *let-7* caused expression of adult cell lineages in larvae. It was shown that *let-7* encodes a 21-nt-long RNA molecule which was complementary to the 3′ UTR of other important heterochronic genes crucial for development such as *lin-14*, *lin-28*, *lin-41*, *lin-42*, and *daf-12*, which suggested the possibility that these genes might be regulated by this short molecule of RNA (Reinhart *et al.*, 2000).

The *let-7* family of miRNAs is a good example of the tight regulation of miRNA in ES cells as they are regulated at both the transcriptional and posttranscriptional levels. It has been shown that *let-7* functions in a feedback loop with *lin-28*. Lin-28 binds the pre-*let-7* RNA molecule and prevents its further processing by Dicer into a mature and functional miRNA. In contrast, *lin-28* is downregulated by mir-125 (*lin-4* homolog) and *let-7*, which allows *let-7* maturation to occur unperturbed (Figure 12.1). When *let-7* is depleted, *lin-28* levels increase and pre-*let-7* molecules do not mature. Mature *let-7* is not detectable in early embryogenesis but is strongly upregulated during organogenesis and tissue differentiation (Rybak *et al.*, 2008). This feedback loop is crucial for embryonic stem cell pluripotency and differentiation as *lin-28* is required in ES cells, and *let-7* facilitates differentiation.

In contrast to the ability of *let-7* to rescue the DGCR8-targeted phenotype, *let-7* had no effect on expression of the pluripotency genes in wild-type ES cells, leading to a model in which ESCC miRNAs oppose the function of *let-7* family miRNAs by repressing expression of many of the same target genes that *let-7* indirectly activate, resulting in self-renewal in ES cells. It is likely that *c-Myc* plays a role in this miRNA-mediated cell fate switch, since unlike wild-type ES cells, *c-Myc*-depleted ES cells respond to insertion of *let-7* miRNA and downregulate the pluripotency gene expression to a certain extent. It has been shown that *c-Myc* transcriptionally represses *let-7* expression and in certain cellular contexts, members of the *let-7* family directly target and downmodulate *c-Myc* (Kumar *et al.*, 2007) (Figure 12.3). In ES cells *let-7* strongly represses *n-Myc* and as a consequence downregulates a set of genes that are positively controlled by Myc.

Other miRNAs have been reported to target the ES cells transcriptional network. Upon differentiation with retinoic acid the expression of a number of miRNAs, including miR-296, miR-470, and miR-134, increases. They inhibit the expression of the pluripotency factors *Oct4*, *Nanog*, and *Sox2* (Tay *et al.*, 2008). Similarly in human ES cells, miR-145 has been shown to inhibit pluripotency by repressing the pluripotency factors *Sox2*, *Oct4*, and *Klf4* (Xu *et al.*, 2009). In addition, miR-200c, miR-203, and miR-183 repress *Sox2* and *Klf4* (Wellner *et al.*, 2009) (Figure 12.1). Similarly, miR-34a, miR-137, and miR-100 are required for proper differentiation of mouse ES cells upon withdrawal of LIF, and function in part by targeting genes involved in maintaining the undifferentiated state such as Sirt1, a deacetylase, Jarid1b, a histone demethylase, and Smarca5, a member of the SWI/SNF chromatin remodeling proteins (Tarantino *et al.*, 2010). These and other studies have proved the important role miRNAs play both in maintaining pluripotency as well as inhibiting differentiation in ES cells.

The miR-302 cluster is well conserved between species, but can display species-specific functions. It seems that the general role it serves is largely evolutionarily preserved with the variations observed among the particular targets within each species. In *Xenopus*, the miR-302 homolog, miR-427, targets *Xnr5* and *Xnr6b*, two of the Nodal ligands, and both of the Lefties, while in hES cells, miR-302 targets the Lefties, but not Nodals at all. Lefties are closely related to transforming growth factor beta (TGF-β) proteins and are responsible for left–right axis formation during development and for inhibiting Nodals; Nodals are members of the TGF-β family and are necessary for mesoderm formation. Reduction of miR-427 in *Xenopus* embryos hampers the organizer formation and leads to severe dorsal mesodermal patterning defects. Likewise, knockdown and overexpression of miR-302 in hES cells show that miR-302 promotes the mesendodermal lineage while suppressing neuroectoderm formation. These results are in line with the above described data showing that miR-302 directly represses NR2F2 which is necessary for proper activation of early neural genes (Rosa *et al.*, 2009). Altogether, these findings suggest a crucial role for the miR-427/302 in germ layer specification (Figure 12.2).

12.7 MicroRNAs in reprogramming to ES cell state

Reprogramming of cells to induced pluripotent stem cells (iPS) is achieved by introduction of key pluripotency genes into somatic cells. Recently miRNAs were found to play a role in reprogramming to iPS as well. The miR-290 cluster, which regulates the cell cycle of ES cells, can affect the efficiency of reprogramming of somatic cells into iPS cells. Mouse embryonic fibroblasts were induced to express *Oct4*, *Sox2*, and *Klf4*, and then molecules mimicking the cell cycle regulating miRNAs were introduced to the cells. It was observed that the miRNAs introduced increased the efficiency of the generation of iPS cells induced by the combination of *Sox2*, *Oct4*, and *Klf4*, with miR-294 exerting the greatest effect. However, when miR-294 was introduced with *Sox2*, *Oct4*, *Klf4*, and *c-Myc*, it had no effect on reprogramming. Therefore, cell-cycle-regulating miRNAs appear to promote *Sox2*, *Oct4*, and *Klf4* mediated reprogramming by substituting for *c-Myc* (Judson et al., 2009).

As *let-7* promotes differentiation, the effect of its inhibition on the reprogramming of somatic cells into iPS cells was investigated. *Let-7* antisense was introduced into mouse embryonic fibroblast cells, and reprogramming was tested in conjunction with *Sox2*, *Oct4*, and *Klf4*, in the presence or absence of *c-Myc*. Inhibition of *let-7* increased the efficiency of reprogramming via *Sox2*, *Oct4*, and *Klf4*. In contrast, the efficiency of *Sox2*, *Oct4*, *Klf4*, and *c-Myc* together yielded only a slight increase. This supports the theory that inhibition of *let-7* increases efficiency of reprogramming by targeting genes such as *c-Myc* and *lin-28* (Melton et al., 2010) (Figure 12.1).

By comparing genetically identical mouse ES cells and iPS cells it was shown that their transcriptosomes are extremely similar, with the exception of a paternally imprinted locus *Dlk1-Dio3* (Stadtfeld et al., 2010). Interestingly, the iPS cells that failed to appropriately activate the maternal locus failed to give rise to live mice by tetraploid complementation. This locus encodes about 50 miRNAs; 18 of them were expressed in ES cells but not in iPS, suggesting a role in this process (Mallanna and Rizzino, 2010).

Human skin cancer cells were successfully reprogrammed into a pluripotent ES-cell-like state with ectopic expression of the miR-302 cluster (Chang et al., 2008; Lin et al., 2008). More recently, it was shown that this miRNA led to reprogramming of human hair follicle cells to iPS cells (Lin et al., 2011). In these cells miR-302 was shown to suppress four epigenetic regulators; two histone demethylases; LSD1 (also known as AOF2 or KDM1) and AOF1, and methyl CpG-binding proteins; MECP1-p66 and MECP2, all enhancing global DNA demethylation, which apparently induce hES-specific gene expression essential for somatic cell reprogramming (Lin et al., 2011). Altogether, it seems that manipulation of miRNA levels in somatic cells might increase the efficiency of generating iPS cells.

12.8   Concluding remarks

Since the initial discovery of miRNAs a decade ago, the field has advanced by leaps and bounds towards understanding the mechanisms by which miRNAs function and are regulated. The discoveries with specific regard to their function in ES cell self-renewal and pluripotency have shown them to be crucial players in the gene network involved in ES cell maintenance, proliferation, and differentiation.

MicroRNAs have been shown both in pluripotent and differentiated ES cells as playing important roles in regulating expression of transcription factors, cell cycle proteins, histone methyltransferase, DNA methyltrasferases, and a myriad of other regulatory proteins all necessary for ES cell stemness or differentiation. Though the field has indeed come so far in this relatively short time, much work still remains to be done and to be uncovered about the function of miRNAs in regulating self-renewal and differentiation in stem cell biology. Moreover, it will be very informative to pursue the molecular mechanisms and factors controlling the expression of miRNAs at the transcriptional and posttranscriptional levels, as the regulatory power of these small molecules will likely have great impact on the use of ES cells both as scientific and therapeutic tools.

REFERENCES

Adhikary, S. and Eilers, M. (2005). Transcriptional regulation and transformation by Myc proteins. *Nature Reviews Molecular Cell Biology*, **6**, 635–645.

Alon, U. (2007). Network motifs: theory and experimental approaches. *Nature Reviews Genetics*, **8**, 450–461.

Ambros, V. and Chen, X. (2007). The regulation of genes and genomes by small RNAs. *Development*, **134**, 1635–1641.

Barroso-del Jesus, A., Lucena-Aguilar, G., and Menendez, P. (2009). The miR-302–367 cluster as a potential stemness regulator in ESCs. *Cell Cycle*, **8**, 394–398.

Ben-Shushan, E., Sharir, H., Pikarsky, E., and Bergman, Y. (1995). A dynamic balance between ARP-1/COUP-TFII, EAR-3/COUP-TFI, and retinoic acid receptor : retinoid X receptor heterodimers regulates *Oct-3/4* expression in embryonal carcinoma cells. *Molecular Cell Biology*, **15**, 1034–1048.

Benetti, R., Gonzalo, S., Jaco, I., *et al.* (2008). A mammalian microRNA cluster controls DNA methylation and telomere recombination via Rbl2-dependent regulation of DNA methyltransferases. *Nature Structural and Molecular Biology*, **15**, 998.

Bernstein, E., Kim, S. Y., Carmell, M. A., *et al.* (2003). Dicer is essential for mouse development. *Nature Genetics*, **35**, 215–217.

Bernstein, B. E., Mikkelsen, T. S., Xie, X., *et al.* (2006). A bivalent chromatin structure marks key developmental genes in embryonic stem cells. *Cell*, **125**, 315–326.

Bibikova, M., Laurent, L. C., Ren, B., Loring, J. F., and Fan, J. B. (2008). Unraveling epigenetic regulation in embryonic stem cells. *Cell Stem Cell*, **2**, 123–134.

Boscher, C. and Mege, R. M. (2008). Cadherin-11 interacts with the FGF receptor and induces neurite outgrowth through associated downstream signalling. *Cell Signal*, **20**, 1061–1072.

Boyer, L. A., Lee, T. I., Cole, M. F., *et al.* (2005). Core transcriptional regulatory circuitry in human embryonic stem cells. *Cell*, **122**, 947–956.

Brook, F. A., and Gardner, R. L. (1997). The origin and efficient derivation of embryonic stem cells in the mouse. *Proceedings of the National Academy of Sciences USA*, **94**, 5709–5712.

Card, D. A., Hebbar, P. B., Li, L., *et al.* (2008). Oct4/*Sox2*-regulated miR-302 targets cyclin D1 in human embryonic stem cells. *Molecular Cell Biology*, **28**, 6426–6438.

Chang, S. S., Jiang, W. W., Smith, I., *et al.* (2008). MicroRNA alterations in head and neck squamous cell carcinoma. *International Journal of Cancer*, **123**, 2791–2797.

Chen, C., Ridzon, D., Lee, C. T., *et al.* (2007). Defining embryonic stem cell identity using differentiation-related microRNAs and their potential targets. *Mammalian Genome*, **18**, 316–327.

Cole, M. D. and Nikiforov, M. A. (2006). Transcriptional activation by the Myc oncoprotein. *Current Topics in Microbiology and Immunology*, **302**, 33–50.

Du, T. and Zamore, P. D. (2005). microPrimer: the biogenesis and function of microRNA. *Development*, **132**, 4645–4652.

Eiring, A. M., Harb, J. G., Neviani, P., *et al.* (2010). miR-328 functions as an RNA decoy to modulate hnRNP E2 regulation of mRNA translation in leukemic blasts. *Cell*, **140**, 652–665.

Epsztejn-Litman, S., Feldman, N., Abu-Remaileh, M., *et al.* (2008). De novo DNA methylation promoted by G9a prevents reprogramming of embryonically silenced genes. *Nature Structural and Molecular Biology*, **15**, 1176–1183.

Evans, M. J. and Kaufman, M. H. (1981). Establishment in culture of pluripotential cells from mouse embryos. *Nature*, **292**, 154–156.

Faast, R., White, J., Cartwright, P., *et al.* (2004). Cdk6-cyclin D3 activity in murine ES cells is resistant to inhibition by p16(INK4a). *Oncogene*, **23**, 491–502.

Feldman, N., Gerson, A., Fang, J., *et al.* (2006). G9a-mediated irreversible epigenetic inactivation of *Oct-3/4* during early embryogenesis. *Nature Cell Biology*, **8**, 188–194.

Friedman, R. C., Farh, K. K., Burge, C. B., and Bartel, D. P. (2009). Most mammalian mRNAs are conserved targets of microRNAs. *Genome Research*, **19**, 92–105.

Hanina, S. A., Mifsud, W., Down, T. A., *et al.* (2010). Genome-wide identification of targets and function of individual microRNAs in mouse embryonic stem cells. *PLoS Genetics*, **6**, e1001163.

Hayashi, K., Chuva de Sousa Lopes, S. M., Kaneda, M., *et al.* (2008). MicroRNA biogenesis is required for mouse primordial germ cell development and spermatogenesis. *PLoS One*, **3**, e1738.

He, L. and Hannon, G. J. (2004). MicroRNAs: small RNAs with a big role in gene regulation. *Nature Reviews Genetics*, **5**, 522–531.

Houbaviy, H. B., Murray, M. F., and Sharp, P. A. (2003). Embryonic stem cell-specific microRNAs. *Developmental Cell*, **5**, 351–358.

Ivanovska, I., Ball, A. S., Diaz, R. L., *et al.* (2008). MicroRNAs in the miR-106b family regulate p21/CDKN1A and promote cell cycle progression. *Molecular Cell Biology*, **28**, 2167–2174.

Jirmanova, L., Afanassieff, M., Gobert-Gosse, S., Markossian, S., and Savatier, P. (2002). Differential contributions of ERK and PI3-kinase to the regulation of cyclin D1 expression and to the control of the G1/S transition in mouse embryonic stem cells. *Oncogene*, **21**, 5515–5528.

Judson, R. L., Babiarz, J. E., Venere, M., and Blelloch, R. (2009). Embryonic stem cell-specific microRNAs promote induced pluripotency. *Nature Biotechnology*, **27**, 459–461.

Kanellopoulou, C., Muljo, S. A., Kung, A. L., *et al.* (2005). Dicer-deficient mouse embryonic stem cells are defective in differentiation and centromeric silencing. *Genes and Development*, **19**, 489–501.

Kim, V. N. (2005). MicroRNA biogenesis: coordinated cropping and dicing. *Nature Reviews Molecular Cell Biology*, **6**, 376–385.

Kleine-Kohlbrecher, D., Adhikary, S., and Eilers, M. (2006). Mechanisms of transcriptional repression by Myc. *Current Topics in Microbiology and Immunology*, **302**, 51–62.

Kumar, M. S., Lu, J., Mercer, K. L., Golub, T. R., and Jacks, T. (2007). Impaired microRNA processing enhances cellular transformation and tumorigenesis. *Nature Genetics*, **39**, 673–677.

Lee, R. C., Feinbaum, R. L., and Ambros, V. (1993). The *C. elegans* heterochronic gene *lin-4* encodes small RNAs with antisense complementarity to *lin-14*. *Cell*, **75**, 843–854.

Lin, C. H., Jackson, A. L., Guo, J., Linsley, P. S., and Eisenman, R. N. (2009). Myc-regulated microRNAs attenuate embryonic stem cell differentiation. *EMBO Journal*, **28**, 3157–3170.

Lin, S. L., Chang, D. C., Chang-Lin, S., *et al.* (2008). Mir-302 reprograms human skin cancer cells into a pluripotent ES-cell-like state. *RNA*, **14**, 2115–2124.

Lin, S. L., Chang, D. C., Ying, S. Y., Leu, D., and Wu, D. T. (2010). MicroRNA miR-302 inhibits the tumorigenecity of human pluripotent stem cells by coordinate suppression of the CDK2 and CDK4/6 cell cycle pathways. *Cancer Research*, **70**, 9473–9482.

Lin, S. L., Chang, D. C., Lin, C. H., *et al.* (2011). Regulation of somatic cell reprogramming through inducible mir-302 expression. *Nucleic Acids Research*, **39**, 1054–1065.

Loh, Y. H., Wu, Q., Chew, J. L., *et al.* (2006). The *Oct4* and *Nanog* transcription network regulates pluripotency in mouse embryonic stem cells. *Nature Genetics*, **38**, 431–440.

Mac Auley, A., Werb, Z., and Mirkes, P. E. (1993). Characterization of the unusually rapid cell cycles during rat gastrulation. *Development*, **117**, 873–883.

Mallanna, S. K. and Rizzino, A. (2010). Emerging roles of microRNAs in the control of embryonic stem cells and the generation of induced pluripotent stem cells. *Developmental Biology*, **344**, 16–25.

Marson, A., Levine, S. S., Cole, M. F., *et al.* (2008). Connecting microRNA genes to the core transcriptional regulatory circuitry of embryonic stem cells. *Cell*, **134**, 521–533.

Martin, G. R. (1981). Isolation of a pluripotent cell line from early mouse embryos cultured in medium conditioned by teratocarcinoma stem cells. *Proceedings of the National Academy of Sciences USA*, **78**, 7634–7638.

Mathonnet, G., Fabian, M. R., Svitkin, Y. V., *et al.* (2007). MicroRNA inhibition of translation initiation in vitro by targeting the cap-binding complex eIF4F. *Science*, **317**, 1764–1767.

Melton, C., Judson, R. L., and Blelloch, R. (2010). Opposing microRNA families regulate self-renewal in mouse embryonic stem cells. *Nature*, **463**, 621–626.

Murchison, E. P., Partridge, J. F., Tam, O. H., Cheloufi, S., and Hannon, G. J. (2005). Characterization of Dicer-deficient murine embryonic stem cells. *Proceedings of the National Academy of Sciences USA* **102**, 12135–12140.

Paroo, Z., Ye, X., Chen, S., and Liu, Q. (2009). Phosphorylation of the human microRNA-generating complex mediates MAPK/Erk signaling. *Cell*, **139**, 112–122.

Qi, J., Yu, J. Y., Shcherbata, H. R., *et al.* (2009). microRNAs regulate human embryonic stem cell division. *Cell Cycle*, **8**, 3729–3741.

Reinhart, B. J., Slack, F. J., Basson, M., *et al.* (2000). The 21-nucleotide *let-7* RNA regulates developmental timing in *Caenorhabditis elegans*. *Nature*, **403**, 901–906.

Rosa, A. and Brivanlou, A. H. (2010). A regulatory circuitry comprised of miR-302 and the transcription factors OCT4 and NR2F2 regulates human embryonic stem cell differentiation. *EMBO Journal*, **30**, 237–248.

Rosa, A., Spagnoli, F. M., and Brivanlou, A. H. (2009). The miR-430/427/302 family controls mesendodermal fate specification via species-specific target selection. *Developmental Cell*, **16**, 517–527.

Rybak, A., Fuchs, H., Smirnova, L., *et al.* (2008). A feedback loop comprising *lin-28* and *let-7* controls pre-*let-7* maturation during neural stem-cell commitment. *Nature Cell Biology*, **10**, 987–993.

Savatier, P., Lapillonne, H., van Grunsven, L. A., Rudkin, B. B., and Samarut, J. (1996). Withdrawal of differentiation inhibitory activity/leukemia inhibitory factor up-regulates D-type cyclins and cyclin-dependent kinase inhibitors in mouse embryonic stem cells. *Oncogene*, **12**, 309–322.

Schoorlemmer, J., Jonk, L., Sanbing, S., *et al.* (1995). Regulation of *Oct-4* gene expression during differentiation of EC cells. *Molecular Biology Reports*, **21**, 129–140.

Selbach, M., Schwanhausser, B., Thierfelder, N., *et al.* (2008). Widespread changes in protein synthesis induced by microRNAs. *Nature*, **455**, 58–63.

Sengupta, S., Nie, J., Wagner, R. J., *et al.* (2009). MicroRNA 92b controls the G1/S checkpoint gene p57 in human embryonic stem cells. *Stem Cells*, **27**, 1524–1528.

Sinkkonen, L., Hugenschmidt, T., Berninger, P., *et al.* (2008). MicroRNAs control de novo DNA methylation through regulation of transcriptional repressors in mouse embryonic stem cells. *Nature Structural and Molecular Biology*, **15**, 259–267.

Smith, K. and Dalton, S. (2010). Myc transcription factors: key regulators behind establishment and maintenance of pluripotency. *Regenerative Medicine*, **5**, 947–959.

Spivakov, M. and Fisher, A. G. (2007). Epigenetic signatures of stem-cell identity. *Nature Reviews Genetics*, **8**, 263–271.

Spruce, T., Pernaute, B., Di-Gregorio, A., *et al.* (2010). An early developmental role for miRNAs in the maintenance of extraembryonic stem cells in the mouse embryo. *Developmental Cell*, **19**, 207–219.

Stadtfeld, M., Apostolou, E., Akutsu, H., *et al.* (2010). Aberrant silencing of imprinted genes on chromosome 12qF1 in mouse induced pluripotent stem cells. *Nature*, **465**, 175–181.

Stead, E., White, J., Faast, R., *et al.* (2002). Pluripotent cell division cycles are driven by ectopic Cdk2, cyclin A/E and E2F activities. *Oncogene*, **21**, 8320–8333.

Su, H., Trombly, M. I., Chen, J., and Wang, X. (2009). Essential and overlapping functions for mammalian Argonautes in microRNA silencing. *Genes and Development*, **23**, 304–317.

Suh, M. R., Lee, Y., Kim, J. Y., *et al.* (2004). Human embryonic stem cells express a unique set of microRNAs. *Developmental Biology*, **270**, 488–498.

Tarantino, C., Paolella, G., Cozzuto, L., *et al.* (2010). miRNA 34a, 100, and 137 modulate differentiation of mouse embryonic stem cells. *Federation of the American Society for Experimental Biologists Journal*, **24**, 3255–3263.

Tay, Y., Zhang, J., Thomson, A. M., Lim, B., and Rigoutsos, I. (2008). MicroRNAs to *Nanog, Oct4* and *Sox2* coding regions modulate embryonic stem cell differentiation. *Nature*, **455**, 1124–1128.

Tiscornia, G. and Izpisua Belmonte, J. C. (2010). MicroRNAs in embryonic stem cell function and fate. *Genes and Development*, **24**, 2732–2741.

Wang, Y. and Blelloch, R. (2009). Cell cycle regulation by microRNAs in embryonic stem cells. *Cancer Research*, **69**, 4093–4096.

Wang, Y., Baskerville, S., Shenoy, A., *et al.* (2008). Embryonic stem cell-specific microRNAs regulate the G1–S transition and promote rapid proliferation. *Nature Genetics*, **40**, 1478–1483.

Wang, Y., Keys, D. N., Au-Young, J. K., and Chen, C. (2009). MicroRNAs in embryonic stem cells. *Journal of Cellular Physiology*, **218**, 251–255.

Wang, Y., Medvid, R., Melton, C., Jaenisch, R., and Blelloch, R. (2007). DGCR8 is essential for microRNA biogenesis and silencing of embryonic stem cell self-renewal. *Nature Genetics*, **39**, 380–385.

Wellner, U., Schubert, J., Burk, U. C., *et al.* (2009). The EMT-activator ZEB1 promotes tumorigenicity by repressing stemness-inhibiting microRNAs. *Nature Cell Biology*, **11**, 1487–1495.

Wu, E., Thivierge, C., Flamand, M., *et al.* (2010). Pervasive and cooperative deadenylation of 3′ UTRs by embryonic microRNA families. *Molecular Cell*, **40**, 558–570.

Xu, N., Papagiannakopoulos, T., Pan, G., Thomson, J. A., and Kosik, K. S. (2009). MicroRNA-145 regulates *OCT4, SOX2*, and *KLF4* and represses pluripotency in human embryonic stem cells. *Cell*, **137**, 647–658.

Zovoilis, A., Smorag, L., Pantazi, A., and Engel, W. (2009). Members of the miR-290 cluster modulate in vitro differentiation of mouse embryonic stem cells. *Differentiation*, **78**, 69–78.

13 Regulation of timing of replication

Rituparna Mukhopadhyay and Eric E. Bouhassira*

13.1 Introduction to DNA replication

DNA replication is a process fundamental to cell proliferation. Eukaryotic cells must replicate their DNA once and only once per cell cycle. In addition, after each DNA replication, the epigenetic information, such as DNA methylation and histone modifications, that constitute the memory and the identity of the cells must also be replicated. These functional constraints might have caused eukaryotes to evolve complex tissue specific replication programs, perhaps because the timing during S phase at which each DNA segment replicates is critical for the transmission of epigenetic memory.

DNA replication begins by the recruitment of the replication machinery at specific sites in the genome called "origins." DNA replicated from a single origin is referred to as a replicon. DNA synthesis in each replicon is initiated at an origin and proceeds bidirectionally. These replicons eventually fuse resulting in complete genomic duplication. Initiation of DNA replication involves the following steps.

13.1.1 Origin recognition

The origin recognition complex (ORC) was originally identified in budding yeast and is evolutionarily conserved across species (Bell and Stillman, 1992). It is composed of six subunits (ORC1 to ORC6) that act as initiators and bind to the origin sites in eukaryotic DNA. Mammalian ORC consists of a stable core complex of subunits ORC 2, 3, 4, and 5 which interact feebly with ORC1 and ORC6.

ORC1 is tightly bound to DNA in G1 phase and is selectively degraded by ubiquitin dependent proteolysis during S phase. Binding of ORC1 to DNA origin sites is cell-cycle regulated in some species but not all. The sequences required for ORCs binding vary significantly among species. In budding yeast, ORCs recognize

---

* Author to whom correspondence should be addressed.

Epigenomics: From Chromatin Biology to Therapeutics, ed. K. Appasani. Published by Cambridge University Press. © Cambridge University Press 2012.

and bind to the 11-bp to 17-bp AT-rich consensus sequence. In humans and frogs, ORCs prefer AT-rich sequences – but there is no consensus. Local chromatin structure seems to be the major determinant of origin location in mammalian cells.

Preference for AT-rich regions may be explained in part by the AT-hook motif of the ORC4 subunit. In fission yeast, this subunit recognizes and preferentially binds to AT stretches of DNA. The AT content may also be important for unwinding as the AT base pair has two hydrogen bonds which separate easily. Other factors might also play a role in origin recognition and binding of ORC. These include transcription factor binding sites, dinucleotide repeats, matrix attachment sites, and asymmetrical purine–pyrimidine sequences (DePamphilis, 2005; Aladjem, 2007; Sclafani and Holzen, 2007; Masai *et al.*, 2010).

13.1.2   Pre-replication complex assembly and licensing

Pre-replication complexes (pre-RCs) assemble at replication origin sites throughout the genome during the G1 phase of the cell cycle. A subset of chromatin-bound ORC recruits Cdc6 and Cdt1 at potential replication origin sites. This subset of origins is further converted into pre-RCs while the other chromatin-bound ORCs are eventually degraded. The Cdc6 and Cdt1 proteins are conserved during evolution and are regulated during the cell cycle. They are abundant during G1 when pre-RCs are formed and are reduced or degraded during S, G2, and M phases. In late G1 phase, the minichromosome maintenance (MCM) 2–7 helicases are loaded onto the replication origin site by Cdc6 and Cdt1 to complete the formation of pre-RCs in a process referred to as DNA replication licensing. The MCM helicases were first identified in budding yeast and are conserved in eukaryotes. The MCM2 to MCM7 proteins are closely related and compose a hexameric ring shaped complex responsible for the unwinding of the DNA strand. All members of the pre-RCs including MCM2 to MCM7 proteins contain ATPase domains. The MCM2 to MCM7 proteins are required for both the initiation and elongation steps in DNA replication. After the licensing step, ORC and Cdc6 can be removed from the chromatin without affecting the subsequent DNA replication steps (DePamphilis, 2005; Machida *et al.*, 2005; Aladjem, 2007; Sclafani and Holzen, 2007; Masai *et al.*, 2010).

13.1.3   Activation of helicases and unwinding of DNA

The next step involves activation of the pre-RCs and unwinding of DNA. The pre-RCs are assembled in G1 phase but are activated only during S phase. Cyclin-dependent kinases (Cdk) and Cdc 7/Dbf 4 (DDK) are conserved protein kinases that play crucial roles in helicase activation and DNA unwinding. Cdk and Cdc7-Dbf contribute to MCM helicases activation by facilitating the binding of Cdc45 protein to the MCM complex. Cdc45 is then involved in the conversion of the pre-RCs into a large multiprotein complex called the pre-initiation complex (pre-IC). Pre-ICs contain Cdc45, MCM10, Sld2–3, Dbp11, and the GINS complex. Pre-ICs exist for a very short time and replication begins as soon as pre-ICs are formed. Studies in *Xenopus* have shown that MCM10 along with Cdc45 is important for unwinding of DNA (Bell and Dutta, 2002; Aladjem, 2007; Sclafani and Holzen, 2007; Masai *et al.*, 2010).

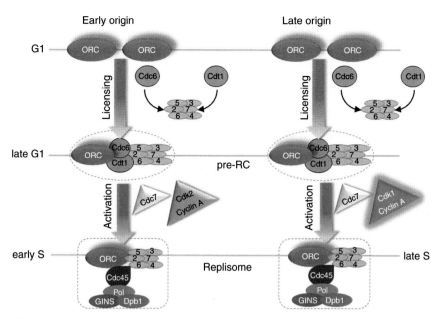

Figure 13.1 Replication model for eukaryotic DNA. During both early and late S phase, origins are recognized by ORC. The Cdc6 and Cdt1 proteins load the MCM helicases to form licensed pre-RC complexes during late G1 phase. The activation of pre-RCs might differ for early and late replication. Cdc7 and Cdk2 activate MCM helicases and load replisome complexes during early S phase. Cdc7 and Cdk1 activate MCM helicases and load replisome complexes during late S phase. See plate section for color version.

Importantly, a recent report by Katsuno *et al.* suggests that the regulation of the timing of replication during S phase might be controlled at this step. These authors have shown that cyclin A2–Cdk1 activity is low in early S phase and higher in mid and late S phase, and that ectopic expression of cyclin A2–Cdk1 leads to early firing of late origins. These results suggest that cyclin A2–Cdk1 activity is a regulator of the S phase program (Katsuno *et al.*, 2009) (Figure 13.1).

13.1.4 Elongation assembly (replisome loading)

The pre-IC becomes the replisome through a cascade of events including loading of the polymerase holoenzyme, the SSB protein, and the GINS complex. The GINS complex is composed of the Sld5, Psf1, Psf2, and Psf3 proteins. Along with Cdc45, MCM2–7, and two protein kinases (Cdk and DDK), the GINS complex is important for replisome loading at the replication fork. Loaded replisome complexes are called replisome progression complexes (RPCs) and are composed of more than 20 proteins. The main components of RPCs, in addition to the Cdc45, the MCM2–7 complex, and the GINS complex, are Mrc1, Tof1, and Csm3 (required for fork stabilization), Ctf4 (required for sister chromatid cohesion), and Spt16 and Pob3 (components of the histone chaperone FACT). DNA polymerase epsilon (ε) and DNA polymerase alpha (α)-primase complex are sequentially loaded onto origin sites by Cdc45–Sld3–Dbp11 complexes. DNA polymerase delta (δ) is also loaded around the same time. The replisome progression complex migrates with the replication fork. Previously mentioned Cdc45 also moves with the replication fork and has been found to play an important role for both initiation and fork elongation (Bell and Dutta, 2002; Sclafani and Holzen, 2007; Masai *et al.*, 2010).

13.2   Replication of the genome once per cycle

Replication of a complete genome must be restricted to once per cell cycle. This seems to be controlled by the segregation of the assembly of pre-RCs in the G1 phase and of the activation of pre-RCs in S phase. Inappropriate assembly of pre-RCs outside of G1 is regulated by Geminin, a protein that accumulates during S, G2, and M phases of the cell cycle. Geminin blocks pre-RCs formation by binding to Cdt1 and preventing loading of MCM2–7 onto chromatin hence blocking re-replication or over-replication of the genome. During G1, Geminin is destabilized by ubiquitination by the anaphase-promoting complex (APC) and degraded by proteosomes, thereby allowing formation of pre-replication complexes. In addition, the catalytic action of cyclin-dependent kinases (Cdks) degrades many components of pre-replication complexes. For instance, Cdc6 and Cdt1 are phosphorylated by Cdk and degraded by proteosomes. Levels of Cdk are lower in G1 phase but increase in S, G2, and M phases. In addition to avoiding over-replication, the replication machinery must ensure complete synthesis of the entire genome within the time-frame of a single S phase. The mechanisms that prevent under-replication remain poorly understood at the molecular level (Bell and Dutta, 2002; Machida *et al.*, 2005).

13.3   Replication and checkpoint control

As discussed above, the two S-phase-promoting kinases (SPKs), Cdk2 and Cdc7, play an important role in origin firing and initiation of DNA replication. Activation of the ATM- and ATR-dependent pathways by DNA damage during S phase prevents origin firing probably because activation of ATM and ATR inhibits Cdk2 and Cdc7 and therefore inhibits pre-RC activation. Recent studies have shown that caffeine and neutralizing antibodies that inhibit ATM and ATR result in increased Cdk2 activity and origin firing in *Xenopus* egg extract, suggesting that ATM- and ATR-mediated checkpoint pathways also regulate origin firing in unperturbed cells (Marheineke and Hyrien, 2004; Shechter *et al.*, 2004). These and other observations indicate that Cdk2 and Cdc7 exist in equilibrium between active and inactive states and that their activity is regulated by ATM and ATR during unperturbed cell cycles as well as after DNA damage (Shechter *et al.*, 2004).

13.4   Timing of DNA replication and gene expression

The complexity of the timing of the replication program was first glimpsed through the study of a few model gene systems. These early studies demonstrated that tissue-specific genes often replicated earlier in the cells in which they were expressed than in the cells in which they were inactive. For example, the β-globin gene is late-replicating and carries features of silent chromatin in non-erythroid cells but it is early-replicating and carries features of open chromatin in erythroid cells (Hiratani and Gilbert, 2009).

The transition between early- and late-replicating regions has been studied in detail at a few loci (Strehl *et al.*, 1997; Ermakova *et al.*, 1999; Norio *et al.*, 2005; Guan *et al.*, 2009). Focusing on a region several megabases long around the immunoglobulin locus, the Schildkraut laboratory demonstrated that in most cells, the IgH region had a tripartite organization with an early domain and a late domain joined by a transition region, devoid of active origin of replication in which the fork progressed unidirectionally. The same authors also showed that during B-cell differentiation, this organization was remodeled with the entire locus replicating early because of the activation of tissue-specific origins in the transition region (Ermakova *et al.*, 1999; Norio *et al.*, 2005; Guan *et al.*, 2009).

13.5   Genome-wide studies

In the last few years, several methods have been developed to study timing of replication genome-wide (MacAlpine *et al.*, 2004; White *et al.*, 2004; Woodfine *et al.*, 2004; Jeon *et al.*, 2005; Karnani *et al.*, 2007; Hiratani *et al.*, 2008). Two general approaches have been used: detection of newly replicated DNA and direct measurement of copy number variation during S phase.

The first approach involves pulse-labeling of newly synthesized DNA using BrdU, separation of the labeled cells in two or more fractions according to their position in the cell cycle, and immunoprecipitation of the labeled DNA followed by hybridization of the fraction to tiling microarrays or sequencing of the fragment. The advantage of this approach is that it can be very precise if multiple S phase fractions are analyzed but it is technically complex and requires an amplification step to detect the immunoprecipitated DNA. As an all PCR-based approach, this amplification step might affect the results. This approach was first reported by White *et al.*, using a microarray platform (White *et al.*, 2004). More recently, a similar approach was developed but detection of the newly synthesized DNA was performed by massively parallel sequencing (Hansen *et al.*, 2010).

The second approach relies on the detection of variation in copy number during S phase as an indicator for the timing of DNA replication. It therefore has the advantage of simplicity and requires minimal cell manipulation but it involves the detection of very small differences in copy number which can be difficult to quantify precisely, particularly in a high-throughput format. Woodfine *et al.* (2004) have pioneered the use of tiling arrays to measure copy number difference for measuring replication timing using an array containing about 3500 bacterial artificial chromosomes (BACs). These authors provided a genome-wide map of the timing of replication at a 1-Mb resolution in a transformed human cell line and detected a positive correlation between replication timing and a range of genomic parameters including GC content, gene density, and transcriptional activity. The Bouhassira group then developed TimEX (Timing Express), a method that is similar in principle to the BAC array method of Woodfine *et al.*, but which

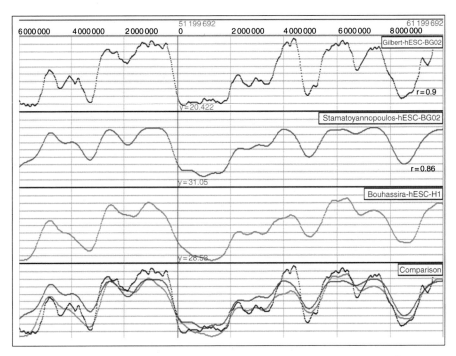

Figure 13.2 Comparison of replication timing in human embryonic stem cells using different methods. The three tracks represent replication timing measured with different methods from three different laboratories: Gilbert, Stamatoyannopoulos, and Bouhassira. The fourth track shows an overlay of all the results. The correlations between the results from different laboratories are high ($r=0.9$ and $r=0.86$), indicating the robustness of the methods. See plate section for color version.

provides higher-resolution timing maps with fine spatial resolution because it relies on precise estimation of copy number variation based on Gaussian convolution to integrate highly redundant massive numbers of individual measurements obtained either by hybridization to high-density arrays containing 400 000 tiled oligonucleotides or by massively parallel next-generation sequencing of genomic DNA libraries (Desprat *et al.*, 2009).

Correlations between the results obtained with the various methods are relatively high (Figure 13.2), suggesting that the methods are all quite robust. At the moment the micro-array based method might be somewhat cheaper but the sequencing approaches have the potential advantage of allowing the detection of allele-specific timing of replication and of being generally more precise and directly applicable to all species with a sequenced genome without having to develop new tiling arrays. Below we summarize the major findings that came out of the genome-wide timing studies.

13.5.1   Replication domains

Replication is regulated at the level of large chromosomal blocks known as replication domains. In studies of mice (Farkash-Amar *et al.*, 2008; Hiratani *et al.*, 2008, 2010), human (White *et al.*, 2004; Woodfine *et al.*, 2004; Jeon *et al.*, 2005; Desprat *et al.*, 2009; Hansen *et al.*, 2010) and flies (Schubeler *et al.*, 2002; MacAlpine *et al.*,

2004), it has been estimated that these replication domains can range from 0.5 to 2 Mb in size. These domains are replicated by synchronous firing of several origins which lie in close proximity to each other. Several studies have provided evidence that originless transition regions which have been observed using low-throughput regions might be a general feature of the genome, although this remains to be demonstrated at the molecular level (Ermakova *et al.*, 1999; Norio *et al.*, 2005; Guan *et al.*, 2009; Hiratani *et al.*, 2009).

A recent study in mouse embryonic stem cells (ESCs) suggested that domains larger than 2.5 Mb either replicate very early or very late, suggesting that domains of a size above certain threshold tend to replicate at extreme ends of the S phase. They also showed that molecular boundaries of replication domains are conserved between different ESCs and induced pluripotent stem cells (iPSCs) (Hiratani *et al.*, 2008).

13.5.2   Correlation of replication timing and genomic structure

Several studies have correlated replication timing with a range of chromosomal features. Microscopic studies on metaphase chromosome show that gene-rich R band replicates early and gene-poor G band replicates late (Hand, 1978). Using microarrays for genome-wide and chromosome-wide studies on human cell lines, different groups have reported a positive correlation between replication timing and GC content. Chromosomal regions with high GC content, rich in gene density, high CpG density, and Alu sequences, tend to replicate early whereas AT-rich, gene-poor regions with high number of LINE repeats tend to replicate late (White *et al.*, 2004; Woodfine *et al.*, 2004; Jeon *et al.*, 2005; Hansen *et al.*, 2010). This observation is consistent with the earlier microscopic analysis (Hand, 1978). However, there are some exceptions. For example, the GC content and replication timing become less correlated at the distal end of chromosome 22q13 since this region is GC-rich but replicates late (Woodfine *et al.*, 2004).

13.5.3   Replication timing and transcription

Microarray analysis in budding yeast showed no correlation between replication timing and transcriptional activity (Raghuraman *et al.*, 2001) but almost all studies in higher organisms have detected a genome-wide link between transcription and replication. An early study in *Drosophila* (Schubeler *et al.*, 2002; MacAlpine *et al.*, 2004) revealed a positive correlation between replication timing and gene expression using high-density tiling microarrays covering the left arm of *Drosophila* chromosome 2. This study showed a correlation between transcription and replication timing among large chromosomal domains but not at the level of individual genes. Genome-wide data in human and mouse revealed clear but relatively weak correlation between timing and transcription (Pearson's $r = 0.50$) (White *et al.*, 2004; Woodfine *et al.*, 2004; Jeon *et al.*, 2005; Desprat *et al.*, 2009; Hansen *et al.*, 2010).

Importantly, Desprat *et al.* (2009) have shown that the correlation between timing and replication is much stronger if the linear arrangement of the genes on the chromosome is taken into consideration because the correlation between active transcription and early replication is in part masked by the presence of silent genes near active genes. Desprat *et al.* (2009) proposed a model that takes into consideration the distance between neighboring genes in addition to their levels of expression and demonstrated that the relative position of active and inactive genes plays a major role in determining their influence on timing of replication. To quantify these hypotheses, these authors developed a mechanistic model and an algorithm to predict timing of replication genome-wide. The model postulates: (1) that replication origins are unevenly distributed in the genome and can fire at any time in S phase according to a predefined tissue-specific program; (2) that once an origin has fired, the polymerase continues at a constant rate until it meets a fork going in the opposite direction; (3) that there is an origin of replication close to all genes in the genome; and (4) that timing of the firing of the origin of replication is proportional to the level of transcription (Figure 13.3).

Although this model is an oversimplification of replication it is supported by a detailed analysis of replication patterns performed by Farkash-Amar *et al.* (2008) and by the fact that predicted timing of replication profiles based on the model

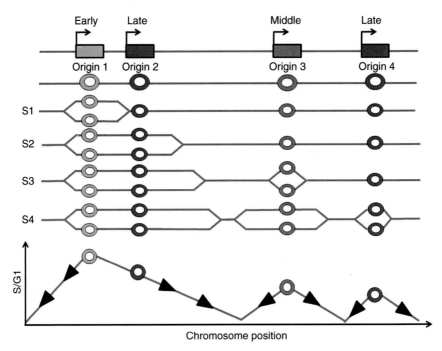

Figure 13.3 Replicon organization. Desprat *et al.* (2009) suggested on the basis of the analysis of their TimEX results that there might be an origin close to every gene and that the timing of the firing of the origin is proportional to the level of expression. The lighter-color genes represent highly expressed genes associated with early replicating origins. The intermediate and dark colors respectively represent genes with intermediate and low level of expression which are associated with origin of replication programmed to fire in middle or late S phase. Poorly expressed genes located near active genes do not contribute to the timing of regulation since they are replicated early by forks emanating from the early origin associated with the neighboring highly expressed gene. In this model, forks proceed at a constant rate unidirectionally until they encounter another fork progressing in the opposite direction. See plate section for color version.

correlates with experimental profiles with a coefficient of correlation of 0.75. This suggests that the link between transcriptional activity and timing of replication is stronger than generally reported and provides a molecular explanation for the observation that timing seems to correlate with expression only when large domains are examined (see above). The model also suggests that the timing of replication is proportional to the level of transcriptional activity.

Importantly, the model implies that regions that are gene-rich are likely to be replicating early in all tissues because there is high probability that at least some of the gene will be expressed in any given tissue. Conversely, gene-poor regions are likely to be replicating late in most cell types except when the rare genes present are expressed (Desprat *et al.*, 2009).

A study in *Drosophila* has shown that ORCs localize to defined chromosome sites, many of which coincide with early origins of replication. ORC binding sites were shown to be AT-rich and coincide with a subset of RNA polymerase II binding sites strengthening the link between replication and transcription (MacAlpine *et al.*, 2004). A study of human chromosome 22 based on a tiling array analysis of total RNA also demonstrated a correlation between early replication and transcription (including non-coding RNA) (White *et al.*, 2004). Interestingly these authors detected several genomic regions that appeared heavily transcribed but that replicated late, in apparent contradiction with the model of Desprat *et al.* (2009).

13.5.4 Replication timing and asynchronous replication

Asynchronous replication can theoretically be caused by unregulated firing of origins throughout S phase but can also be associated with monoallelic expression, a highly regulated epigenetic process. We will first discuss the evidence that timing can be unregulated and then the highly regulated asynchronous replication patterns associated with monoallelic expression.

Studies of 1% of the genome of transformed cells (HeLa and breast epithelial cell line MCF10A) suggested that up to 20% of the genome replicates at random time in S phase in an asynchronous manner (Karnani *et al.*, 2007). In addition, a genome-wide study in mouse L1210 cells (Farkash-Amar *et al.*, 2008) suggested that 9% of the genome replicates asynchronously but often two peaks of replication were detected, suggesting either that one allele replicates early and the other one late, or that both alleles replicate asynchronously in all cells. In both studies, fluorescence in situ hybridization (FISH) analysis suggested that most of the cases of asynchronous replication were due to divergence between the two alleles (Karnani *et al.*, 2007; Farkash-Amar *et al.*, 2008). Desprat *et al.* have used a refinement of the TimEX methods in which the S phase was sorted into three subfractions to assess the extent of regulation of the timing of replication and concluded that in these cells the large majority of the genome replicated in a highly synchronous, regulated manner and that origins were programmed to fire in very narrow windows of time (Desprat *et al.*, 2009).

In conclusion, these studies strongly suggest that the timing of replication is highly regulated and that the instances of asynchronous replication are associated with monoallelic expression.

Asynchronous replication timing was first detected in low-throughput studies of the X chromosome (Takagi *et al.*, 1982) and of a few imprinted genes which demonstrated that expressed alleles replicate in early S phase whereas silent alleles often replicate later (Kitsberg *et al.*, 1993; Simon *et al.*, 1999).

Three types of monoallelic expression have been described. The first type is X-chromosome inactivation in which monoallelic expression is established by random silencing of one allele. The second type is parental imprinting where one allele is silenced in a parent-of-origin-specific manner. The third type includes autosomal genes with random monoallelic expression. Examples of these genes include T-cell receptor, odorant receptor, immunoglobulins, interleukins, natural killer cell receptor, and pheromone receptors (Dutta *et al.*, 2009).

Asynchronous replication of both X-linked genes and monoallelic autosomal genes has been reported in two undifferentiated human ES cell types (Dutta *et al.*, 2009) using a FISH-based technique. Hansen *et al.* (2010), using a method (BrdU coupled to sequencing) that is particularly well adapted to the detection of asynchronous replication, reported that in hES cells and a female lymphoblastoid cell line, about 0.3% of the autosomal genome exhibited biphasic regions that are characteristic of asynchronous replication. These asynchronously replicating regions include some of imprinted domains in hES cells and a large part of the X chromosome in the female lymphoblastoid cell line. However, the correlation between biphasically replicating regions and reported loci with non-parentally imprinted monoallelic expression was very poor with only three of 371 genes known to exhibit random monoallelic expression located within a biphasic replication zone. The authors concluded that random monoallelic expression does not necessarily perturb replication (Hansen *et al.*, 2010). This conclusion is in agreement with the model of Desprat *et al.* (2009) which suggests that active and inactive genes in gene-rich regions have little influence on timing of replication because they are often near other active genes which have a dominant effect on timing.

13.5.5   Correlation of replication timing with epigenetic marks

It has long been suspected that there is strong correlation between replication timing and some chromatin modifications. This is because euchromatin carries active marks such as H3K4me3 and generally replicates early within the interior compartment of the nucleus while heterochromatin carries repressive marks and generally replicates late near the nuclear periphery. A recent genome-wide study showed that early replication timing correlates with active histone marks (H3K4me3 and H3K36me3) but late replication timing does not correlate with repressive marks (H3K27me3, H3K9me3, and H4K20me3) in both ESCs and neural precursor cells (NPCs). In that study, the strong positive correlation of replication timing with active marks decreases with differentiation. Moreover, subsets of promoters in ESCs have bivalent marks (i.e. they are co-occupied by both active and repressive marks). Upon differentiation, either the active or the repressive mark remains (Hiratani *et al.*, 2008). Interestingly, in most of these regions the timing of replication was found to be early, suggesting that active marks might be dominant over the presence of inactive marks.

Similar study with lymphoblastoid cells showed correlation between early domains and active marks (H3K4me1, 2, 3; H3K9Ac; H3K27Ac; H3K36me3; H3K20me1) but no correlation between late domains and repressive marks (H3K9me2, 3; H3K27me3). Boundaries of early-replicating domains are also shown to be enriched with active marks (Ryba *et al.*, 2010). Another study showed that the late replicating nuclear periphery compartment is enriched in repressive H3K9me2 mark (Yokochi *et al.*, 2009). The insulator protein CTCF, present in the chromatin boundary, was not enriched at the boundaries of early-replicating domains (Ryba *et al.*, 2010). Similarly, in *Drosophila*, the acetylated H4K16 mark is associated with early replication on the male X chromosome as well as other initiation sites on autosomes (Schwaiger *et al.*, 2009). These studies indicate a link between replication timing and active epigenetic marks associated with transcriptional activity. However, it still remains unclear whether initial assembly of heterochromatic marks is a cause or consequence of replication timing change. It is not clear at this point if the association between histone modifications and replication can be attributed entirely to transcriptional activity. Cedar *et al.* proposed that different types of chromatin might be assembled at different times in S phase explaining in part the epigenetic inheritance of chromatin modifications (Zhang *et al.*, 2002). However, this attractive hypothesis needs to be tested experimentally.

13.5.6 Replication timing and developmental regulation

Replication timing correlates with gene expression and hence it is subjected to developmental control. Not surprisingly, almost all of the high-throughput approaches confirmed that the timing of replication is tissue-specific. Dynamic changes in replication timing were observed for several gene loci during differentiation of ESCs to NPCs. Genes residing in AT-rich isochors and LINE sequences showed changes in replication timing after differentiation whereas genes within GC-rich isochors remained unchanged (Hiratani *et al.*, 2004). Further, a comprehensive genome-wide study of replication timing during differentiation of mouse ESCs to NPCs was done using three independent cell lines. Differentiated cells show greater positive correlation between replication timing and transcription at the level of large chromosomal domains. Finally, replication timing changes are correlated to rearrangement in subnuclear compartments where early-replicating genes move towards the interior of the nucleus and late-replicating genes move towards the nuclear periphery (Hiratani *et al.*, 2008). In a similar study, 21% of the *Drosophila* genome showed changes in replication timing between embryonic versus wing disk cell lines (Schwaiger *et al.*, 2009). A study using the TimEX method showed that undifferentiated hESC, mesenchymal stem cells, and erythroid cells have distinct replication timing profiles and that about 20% of the genome exhibited differential timing between these three cell types (Desprat *et al.*, 2009). A genome-wide replication profile was done to model early mouse development using 22 independent cell lines. A cell-type-specific replication profile was generated where 45% of mouse genome reveals changes at some point during development. Replication timing changes are coordinated with changes in transcription (Hiratani *et al.*, 2010). A recent study compared mouse and human replication timing profiles and showed that a developmentally regulated replication timing program is evolutionary conserved among mammals (Ryba *et al.*, 2010).

One microarray-based comparison of human chromosome 22 between fibroblast and lymphoblastoid cells revealed only 1% of chromosomal fragments that differed in replication timing (White *et al.*, 2004). Since chromosome 22 is a particularly gene-rich chromosome, these results are therefore in complete agreement with the model of Desprat *et al.* (2009), which suggest that gene-rich regions should have an invariant timing.

13.5.7   Deregulation of replication timing and diseases

Well-regulated DNA replication is essential for normal development of an organism because the entire genome must faithfully be duplicated during a single cell cycle. Failure to do so may result in over-replication or under-replication of genome giving rise to various abnormalities. Studies indicate that the ORC mutants in *Drosophila* and *Xenopus* model systems have altered replication timing and die at late larval stages. In a normal cell, euchromatic DNA replicates early and heterochromatic DNA replicates late. However, ORC mutants show abnormal chromosomal replication patterns and some of the euchromatic regions in these mutants replicate later than heterochromatic regions. Late replication of euchromatin gives rise to a defect in chromosome condensation. This results in defective metaphase and uneven segregation of chromosomes. Therefore, defect in replication timing is associated with defects in mitotic condensation and sister chromatid cohesion, suggesting that replication timing plays an important role in chromatin assembly (Loupart *et al.*, 2000; Pflumm and Botchan, 2001).

Chromosomal rearrangements like translocations and deletions occur in a variety of human cancers. It has been shown that translocation of the *BCL2* gene from chromosome 18 to chromosome 14 (t(14;18) *BCL2*-IGH translocation) changes the replication timing of the translocated gene. The *BCL2* gene plays an important role in the apoptotic pathway. The wild-type *BCL2* allele usually replicates in mid S phase but the translocated allele replicates earlier at the G1–S boundary. This difference in the replication timing is associated with allele-specific changes in B2 expression. Thus, a shift in the replication timing is associated with an important step in carcinogenesis (Sun *et al.*, 2001). Whether the switch in timing is a consequence or a cause of the increased expression of BCL2 after translocation has not been determined. In other instances, several studies have shown that chromosome translocations found in different tumor types result in delay in replication timing (DRT). This subsequently resulted in delayed mitotic chromosome condensation (DMC). This process might trigger additional translocations at higher frequencies resulting in chromosomal instability (Smith *et al.*, 2001; Breger *et al.*, 2005; Chang *et al.*, 2007).

Synchronous replication is characteristic of biallelically expressed genes whereas asynchronous replication is associated with monoallelic expression of genes. In many malignant cancers and inherited human diseases, asynchronous replication has been documented for many important biallelically expressed genes. For example patients with essential thrombocytosis and polycythemia vera show a high rate of asynchronous replication. In these patients the *TP53*, *C-MYC*, and *HER-2/neu* loci replicate synchronously in normal diploid cells but in malignant cells show asynchronous pattern of replication. The loss of synchrony in allelic replication in many cases is accompanied by aneuploidy and genomic instability (Amiel *et al.*, 2001, 2002; Korenstein-Ilan *et al.*, 2002; Reish *et al.*, 2002). Chromosome-wide studies have linked the replication transition regions to disease-related genes. Almost 15 disease-related genes mapped on chromosome 21q and 11q are located in or near the vicinity of the early/late replication transition regions. Eight of the 15 genes are oncogenes/tumor suppressors. Another 21 genes associated with well-characterized diseases such as Alzheimer's disease also belong to the replication transition region. These transition regions have higher probability of DNA damage and hence can generate genomic instability. The model of Desprat *et al.* (2009) suggests that in gene-poor regions, change of expression of a single genes can affect the timing of replication over megabase regions. If the timing of replication affects gene expression or the probability of gene activation or inactivation after drug treatment, this property of timing of replication might contribute to regional deregulation in cancer. Genome-wide studies of the replication transition regions might become useful for the development of medical diagnostics (Watanabe *et al.*, 2002, 2004).

REFERENCES

Aladjem, M. I. (2007). Replication in context: dynamic regulation of DNA replication patterns in metazoans. *Nature Reviews Genetics*, **8**, 588–600.

Amiel, A., Elis, A., Sherker, S., *et al.* (2001). The influence of cytogenetic aberrations on gene replication in chronic lymphocytic leukemia patients. *Cancer Genetics and Cytogenetics*, **125**, 81–86.

Amiel, A., Elis, A., Maimon, O., *et al.* (2002). Replication status in leukocytes of treated and untreated patients with polycythemia vera and essential thrombocytosis. *Cancer Genetics and Cytogenetics*, **133**, 34–38.

Bell, S. P. and Dutta, A. (2002). DNA replication in eukaryotic cells. *Annual Review of Biochemistry*, **71**, 333–374.

Bell, S. P. and Stillman, B. (1992). ATP-dependent recognition of eukaryotic origins of DNA replication by a multiprotein complex. *Nature*, **357**, 128–134.

Breger, K. S., Smith, L., and Thayer, M. J. (2005). Engineering translocations with delayed replication: evidence for *cis* control of chromosome replication timing. *Human Molecular Genetics*, **14**, 2813–2827.

Chang, B. H., Smith, L., Huang, J., and Thayer, M. J. (2007). Chromosomes with delayed replication timing lead to checkpoint activation, delayed recruitment of Aurora B and chromosome instability. *Oncogene*, **26**, 1852–1861.

DePamphilis, M. L. (2005). Cell cycle dependent regulation of the origin recognition complex. *Cell Cycle*, **4**, 70–79.

Desprat, R., Thierry-Mieg, D., Lailler, N., *et al.* (2009). Predictable dynamic program of timing of DNA replication in human cells. *Genome Research*, **19**, 2288–2299.

Dutta, D., Ensminger, A. W., Zucker, J. P., and Chess, A. (2009). Asynchronous replication and autosome-pair non-equivalence in human embryonic stem cells. *PLoS One*, **4**, e4970.

Ermakova, O. V., Nguyen, L. H., Little, R. D., *et al.* (1999). Evidence that a single replication fork proceeds from early to late replicating domains in the IgH locus in a non-B cell line. *Molecular Cell*, **3**, 321–330.

Farkash-Amar, S., Lipson, D., Polten, A., *et al.* (2008). Global organization of replication time zones of the mouse genome. *Genome Research*, **18**, 1562–1570.

Guan, Z., Hughes, C. M., Kosiyatrakul, S., *et al.* (2009). Decreased replication origin activity in temporal transition regions. *Journal of Cell Biology*, **187**, 623–635.

Hand, R. (1978). Eucaryotic DNA: organization of the genome for replication. *Cell*, **15**, 317–325.

Hansen, R. S., Thomas, S., Sandstrom, R., *et al.* (2010). Sequencing newly replicated DNA reveals widespread plasticity in human replication timing. *Proceedings of the National Academy of Sciences USA*, **107**, 139–144.

Hiratani, I. and Gilbert, D. M. (2009). Replication timing as an epigenetic mark. *Epigenetics*, **4**, 93–97.

Hiratani, I., Leskovar, A., and Gilbert, D. M. (2004). Differentiation-induced replication-timing changes are restricted to AT-rich/long interspersed nuclear element (LINE)-rich isochores. *Proceedings of the National Academy of Sciences USA*, **101**, 16 861–16 866.

Hiratani, I., Ryba, T., Itoh, M., *et al.* (2008). Global reorganization of replication domains during embryonic stem cell differentiation. *PLoS Biology*, **6**, e245.

Hiratani, I., Takebayashi, S., Lu, J., and Gilbert, D. M. (2009). Replication timing and transcriptional control: beyond cause and effect. Part II. *Current Opinion in Genetics and Development*, **19**, 142–149.

Hiratani, I., Ryba, T., Itoh, M., *et al.* (2010). Genome-wide dynamics of replication timing revealed by in vitro models of mouse embryogenesis. *Genome Research*, **20**, 155–169.

Jeon, Y., Bekiranov, S., Karnani, N., *et al.* (2005). Temporal profile of replication of human chromosomes. *Proceedings of the National Academy of Sciences USA*, **102**, 6419–6424.

Karnani, N., Taylor, C., Malhotra, A., and Dutta, A. (2007). Pan-S replication patterns and chromosomal domains defined by genome-tiling arrays of ENCODE genomic areas. *Genome Research*, **17**, 865–876.

Katsuno, Y., Suzuki, A., Sugimura, K., *et al.* (2009). Cyclin A-Cdk1 regulates the origin firing program in mammalian cells. *Proceedings of the National Academy of Sciences USA*, **106**, 3184–3189.

Kitsberg, D., Selig, S., Brandeis, I., *et al.* (1993). Allele-specific replication timing of imprinted gene regions. *Nature*, **364**, 459–463.

Korenstein-Ilan, A., Amiel, A., Lalezari, S., *et al.* (2002). Allele-specific replication associated with aneuploidy in blood cells of patients with hematologic malignancies. *Cancer Genetics and Cytogenetics*, **139**, 97–103.

Loupart, M. L., Krause, S. A., and Heck, M. S. (2000). Aberrant replication timing induces defective chromosome condensation in *Drosophila* ORC2 mutants. *Current Biology*, **10**, 1547–1556.

MacAlpine, D. M., Rodriguez, H. K., and Bell, S. P. (2004). Coordination of replication and transcription along a *Drosophila* chromosome. *Genes and Development*, **18**, 3094–3105.

Machida, Y. J., Hamlin, J. L., and Dutta, A. (2005). Right place, right time, and only once: replication initiation in metazoans. *Cell*, **123**, 13–24.

Marheineke, K. and Hyrien, O. (2004). Control of replication origin density and firing time in *Xenopus* egg extracts: role of a caffeine-sensitive, ATR-dependent checkpoint. *Journal of Biological Chemistry*, **279**, 28 071–28 081.

Masai, H., Matsumoto, S., You, Z., Yoshizawa-Sugata, N., and Oda, M. (2010). Eukaryotic chromosome DNA replication: where, when, and how? *Annual Review of Biochemistry*, **79**, 89–130.

Norio, P., Kosiyatrakul, S., Yang, Q., *et al.* (2005). Progressive activation of DNA replication initiation in large domains of the immunoglobulin heavy chain locus during B cell development. *Molecular Cell*, **20**, 575–587.

Pflumm, M. F. and Botchan, M. R. (2001). ORC mutants arrest in metaphase with abnormally condensed chromosomes. *Development*, **128**, 1697–1707.

Raghuraman, M. K., Winzeler, E. A., Collingwood, D., *et al.* (2001). Replication dynamics of the yeast genome. *Science*, **294**, 115–121.

Reish, O., Gal, R., Gaber, E., *et al.* (2002). Asynchronous replication of biallelically expressed loci: a new phenomenon in Turner syndrome. *Genetics in Medicine*, **4**, 439–443.

Ryba, T., Hiratani, I., Lu, J., *et al.* (2010). Evolutionarily conserved replication timing profiles predict long-range chromatin interactions and distinguish closely related cell types. *Genome Research*, **20**, 761–770.

Schubeler, D., Scalzo, D., Kooperberg, C., *et al.* (2002). Genome-wide DNA replication profile for *Drosophila melanogaster*: a link between transcription and replication timing. *Nature Genetics*, **32**, 438–442.

Schwaiger, M., Stadler, M. B., Bell, O., *et al.* (2009). Chromatin state marks cell-type- and gender-specific replication of the *Drosophila* genome. *Genes and Development*, **23**, 589–601.

Sclafani, R. A. and Holzen, T. M. (2007). Cell cycle regulation of DNA replication. *Annual Review of Genetics*, **41**, 237–280.

Shechter, D., Costanzo, V., and Gautier, J. (2004). ATR and ATM regulate the timing of DNA replication origin firing. *Nature Cell Biology*, **6**, 648–655.

Simon, I., Tenzen, T., Reubinoff, B. E., *et al.* (1999). Asynchronous replication of imprinted genes is established in the gametes and maintained during development. *Nature*, **401**, 929–932.

Smith, L., Plug, A., and Thayer, M. (2001). Delayed replication timing leads to delayed mitotic chromosome condensation and chromosomal instability of chromosome translocations. *Proceedings of the National Academy of Sciences USA*, **98**, 13 300–13 305.

Strehl, S., LaSalle, J. M., and Lalande, M. (1997). High-resolution analysis of DNA replication domain organization across an R/G-band boundary. *Molecular and Cellular Biology*, **17**, 6157–6166.

Sun, Y., Wyatt, R. T., Bigley, A., and Krontiris, T. G. (2001). Expression and replication timing patterns of wildtype and translocated *BCL2* genes. *Genomics*, **73**, 161–170.

Takagi, N., Sugawara, O., and Sasaki, M. (1982). Regional and temporal changes in the pattern of X-chromosome replication during the early post-implantation development of the female mouse. *Chromosoma*, **85**, 275–286.

Watanabe, Y., Fujiyama, A., Ichiba, Y., *et al.* (2002). Chromosome-wide assessment of replication timing for human chromosomes 11q and 21q: disease-related genes in timing-switch regions. *Human Molecular Genetics*, **11**, 13–21.

Watanabe, Y., Ikemura, T., and Sugimura, H. (2004). Amplicons on human chromosome 11q are located in the early/late-switch regions of replication timing. *Genomics*, **84**, 796–805.

White, E. J., Emanuelsson, O., Scalzo, D., *et al.* (2004). DNA replication-timing analysis of human chromosome 22 at high resolution and different developmental states. *Proceedings of the National Academy of Sciences USA*, **101**, 17 771–17 776.

Woodfine, K., Fiegler, H., Beare, D. M., *et al.* (2004). Replication timing of the human genome. *Human Molecular Genetics*, **13**, 191–202.

Yokochi, T., Poduch, K., Ryba, T., *et al.* (2009). G9a selectively represses a class of late-replicating genes at the nuclear periphery. *Proceedings of the National Academy of Sciences USA*, **106**, 19 363–19 368.

Zhang, J., Xu, F., Hashimshony, T., *et al.* (2002). Establishment of transcriptional competence in early and late S phase. *Nature*, **420**, 198–202.

Part III

Epigenomic assays and sequencing technology

14 Detection of CpG methylation patterns by affinity capture methods

Luis G. Acevedo*, Ana Sanz, Dylan Maixner, Kornel Schuebel,
Mary A. Jelinek, David Goldman, and Joseph M. Fernandez

14.1   Introduction

The 29 million CpG dinucleotides in the mammalian genome have long piqued biologists' interest because they encode a blueprint both of the cell's past as well as its future. Cytosine nucleotides within these CpG dinucleotide pairs exist in one of two distinct states whose conversion is accomplished via a highly evolved enzymatic machinery. The default state is unmodified cytosine, while modification with a methyl group on the 5th carbon of the pyrimidine ring unleashes a variety of biological effects. Since no change in the primary genetic structure of DNA is observed during this transition, the conversion from unmethylated to methylated state has been coined an epigenetic or "above genetic" transition.

Broadly speaking there are two categories of CpG dinucleotides – clustered and unclustered – with the clustered CpG dinucleotides found primarily within and near gene loci. The methylation state of CpG dinucleotides in clustered regions of the mammalian genome – termed CpG islands (CGI) – mediates a variety of biological processes such as gene expression, X-chromosome inactivation, imprinting, cellular differentiation, aging, and chromatin structure (Ferguson-Smith and Surani, 2001; Issa, 2003; Lee, 2003; Robertson, 2005). Cellular roles for aberrant DNA methylation are now well established in disease states such as in the initiation and progression of cancer (Jones and Baylin, 2007) and clues are beginning to emerge for a potential role of DNA methylation in neuropsychiatric disorders (Plazas-Mayorca and Vrana, 2011). Central to these studies has been the emergence of methodologies that permit accurate discrimination between the unmethylated and methylated states. Thus, these methods may have useful applications during both discovery and validation phases of an experimental study. Here we describe one way to monitor these states that may promise to

---

* Author to whom correspondence should be addressed.

Epigenomics: From Chromatin Biology to Therapeutics, ed. K. Appasani. Published by Cambridge University Press. © Cambridge University Press 2012.

render deep mechanistic insight into the cellular transcriptome and lead to new and important discoveries in human diseases.

14.2 Protein-mediated capture of methylated DNA

A variety of methods for detection of methylated DNA have been developed and we present a brief summary of these in Table 14.1. They fall into two broad categories differing by their ability to measure cytosine methylation directly,

Table 14.1 Comparison of the features and potential sources of bias for different DNA methylation assays

TECHNOLOGY	Features								Potential sources of bias								
	Unambiguous identification of CpG measured	In *cis* co-methylation information	Non-CpG methylation information	Allele-specific measurement capability	Covers regions with low CpG density	Compatible with low amounts of input DNA	Full repeat-masked genome coverage	Discrimination between 5-hmC and 5-mC	Copy-number variation bias	Fragment size bias	Denaturation of DNA bias	Incomplete bisulfite conversion bias	Bisulfite PCR bias	Cross-hybridization bias	DNA methylation status bias	GC content bias	CpG density bias
Infinium	○					○						○	○	○			
Enzyme–chip	○	○			○						●			●		●	
MeDIP–chip									●	●	●			●		●	●
MIRA–chip						●			●	●				●		●	●
CHARM									●	●	●			●		●	●
MBD-affinity					○	○	○	●	○							●	●
BC–seq	●	●	●	●							●	●			●		
RRBS	●	●	●	●		●					●	●					
Enzyme–seq	●	●		●	○	●						●					
MeDIP–seq						●			●	●	●				●	●	●
MIRA–seq						●			●	●					●	●	●
WGSBS	●	●	●	●	●	●	●					●	●				

● indicates that the method has this feature or bias.
○ indicates that the method has this feature or bias to a limited extent.
MeDIP, methyl DNA immunoprecipitation; MIRA, methylated CpG island recovery assay; CHARM, comprehensive high-throughput arrays for relative methylation; MBD-affinity, methyl binding domain enrichment assays (MethylCollector, MethylCollector Ultra, MethylMiner, MethylCap); BC-seq, bisulfite conversion followed by capture and sequencing; RRBS, reduced representation bisulfite sequencing; -chip, followed by DNA microarray; -seq, followed by sequencing; WGSBS, whole-genome shotgun bisulfite sequencing.
Source: Adapted and modified from Laird (2010).

either through chemical or enzymatic detection, or indirectly, through enrichment of methylated DNA via antibody- or protein-mediated "capture." One of the more recent advances in capturing methylated DNA takes advantage of the high affinity of methyl-CpG binding domain containing proteins (MBDs) for methylated DNA. In their natural cellular role these proteins translate the epigenetic code embedded in methylated CpG dinucleotides (Bird, 1986) into direct gene expression silencing (Ballestar and Wolffe, 2001). In the mammalian genome, several proteins encoded at unique genetic loci including MeCP2, MBD1, MBD2, and MBD4 (Hendrich and Bird, 1998) have this domain and of these, an isoform of MBD2, MBD2b, appears to have the highest affinity for methylated DNA and the greatest capacity to discriminate methylated from unmethylated DNA (Fraga *et al.*, 2003) making it the currently preferred protein to capture methylated DNA.

A proof-of-principle experiment utilizing MBD proteins to isolate methylated DNA was first presented by Adrian Bird and colleagues (Cross *et al.*, 1994). This approach centered around using the MBD domain from the related methyl-binding protein MeCP2 linked to sepharose beads and immobilized on columns to demonstrate a significant and efficient recovery of CGI-enriched genomic compartments. Subsequent studies by Rauch and Pfeiffer (2005) and Ting and colleagues (Serre *et al.*, 2010) have expanded on the original experiments by utilizing MBD domains from MBD2, and/or complexes of MBD2 and MBD3, to increase the affinity and specificity of methylcytosine capture. Of the MBD-based methods, the methylated-CG island recovery assay (MIRA) (Rauch *et al.*, 2006), is based on a MBD2b/MBD3-Like-1 (MBD3L1) heterodimer. MBD3L1 is a protein with substantial homology to MBD2 and MBD3, but it lacks the methyl-CG binding domain which has been shown to form heterodimers with MBD2b resulting in a protein complex with increased affinity for methylated DNA (Jiang *et al.*, 2004). A significant advantage of the MBD-based capture approach is that it can be experimentally designed to exclusively query CGI methylation, thereby significantly enriching these genomic regions and streamlining downstream molecular detection (Nair *et al.*, 2011). Importantly methylated DNA fragments captured in this away may be labeled and hybridized with oligonucleotide arrays or sequenced directly to generate methyl CpG frequency "maps" of the genome.

14.3 Protein-mediated capture of unmethylated DNA

A dedicated enzymatic machinery for recognizing unmethylated cytosines is present in organisms that methylate their DNA and consists of three distinct multi-functional proteins: DNA methyltransferase 3a (DNMT3a), DNMT3b, and DNMT1 (Mortusewicz *et al.*, 2005; Chen and Li, 2006). These proteins can bind unmethylated DNA as well as double-stranded DNA that is methylated on only one strand (Ooi and Bestor, 2008). DNMT3a and DNMT3b feature a cysteine-rich zinc-binding domain comprising six CXXC motifs (C is cysteine and X is any amino acid) (Cheng and Blumenthal, 2008) (reviewed in Acevedo *et al.*, 2011) with similarity to MBD1a, a major mouse MBD1 isoform detected by virtue of its third CXXC domain (CXXC-3) (Cross *et al.*, 1997; Hendrich and Bird, 1998) which is able, by transfection studies, to target to nonmethylated CpG sites in vivo (Jorgensen *et al.*, 2004).

Methods to detect unmethylated DNA (Illingworth *et al.*, 2008) have recently been developed. One approach to enrich for nonmethylated CG-rich DNA – the CXXC affinity purification (CAP) method – uses the cysteine-rich CXXC domain from MBD1 linked to a sepharose matrix to fractionate fragments according to their CpG density and methylation status using salt gradients. The development of CXXC-based methods for the analysis of DNA hypomethylation has provided an important control for identifying false-negative results obtained with MBD enrichment techniques. While the absence of a target DNA sequence in the MBD enrichment pool may be due its lack of mCpG content, inefficient capture, or underrepresentation of the target sequence in the sample DNA following fragmentation, will yield negative results in MBD-based methods. The ability to simultaneously enrich for both methylated and unmethylated DNA fragments allows for significantly more accurate CpG methylation assessment. This principle forms the basis of the methylation ratio assessment (MRA) assay (Figure 14.1A) which consists of parallel pull-down reactions performed to separately capture either methylated DNA or hypomethylated DNA. MethylCollector Ultra™, a commercialized adaptation of the MIRA assay developed by Rauch and colleagues (Rauch *et al.*, 2006), is used to capture hypermethylated DNA using the MBD2b/MBD3L1 heterodimer coupled to magnetic beads leaving hypomethylated DNA fragments in the supernatant or unbound fraction, while hypomethylated DNA is similarly captured using Unmethylcollector™, a commercialized version of CAP (Illingworth *et al.*, 2008), leaving hypermethylated DNA in the supernatant. The relative distribution of hypermethylated DNA and hypomethylated DNA in the enriched (bead-bound or captured) and unbound fractions is quantitated

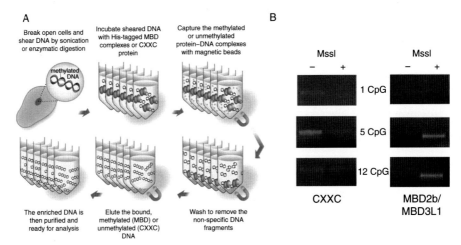

Figure 14.1 Capture and DNA-binding specificity of CXXC and MBD2b/MBD3L1 proteins. (A) Schematic representation of the MRA assays. In parallel reactions, either CXXC or MBD2b/MBD3L1 complexes were used to capture DNA. DNA–protein complexes are isolated using nickel-coated magnetic beads. (B) Capture of 14–3–3 DNA fragments containing 1, 5, or 12 CpG sites. 14–3–3 DNA fragments were generated by PCR amplification. Half of the amplified templates were set aside as unmethylated (–) and the other half were methylated in vitro with the restriction endonuclease MssI (+). In separate reactions, DNA templates were allowed to bind to either CXXC or MBD2b/MBD3L1 using nickel-coated magnetic beads. Bound DNA was eluted and evaluated by endpoint PCR.

using real time (RT)-PCR and DNA based on a standard curve using the input sample DNA.

DNA captured by either method is expressed as a percentage of the initial input DNA using formula 1:

$$\text{DNA}_{\text{captured}}/(\text{DNA}_{\text{input}}) \times 100\%$$

DNA remaining in unbound by either method, in the supernatant fraction is expressed as a percentage of the initial input using formula 2:

$$\text{DNA}_{\text{unbound}}/(\text{DNA}_{\text{input}}) \times 100\%$$

The MRA for each locus consists of a ratio between the DNA captured by MBD2b/MBD3L1 (by MethylCollector Ultra™) and the DNA captured by CXXC protein domain (by UnmethylCollector™) expressed in percentages as it is shown in formula 3 below:

$$\text{MRA} = \% \text{ DNA}_{\text{captured by MBD2b/MBD3L1}}/\%\text{DNA}_{\text{captured by CXXC}}$$

14.4   Validation of binding specificities with synthetic templates

The binding specificity of the His-tagged recombinant heterodimer MBD2b/MBD3L1 and CXXC proteins for methylated or non-methylated DNA was established using PCR templates containing a defined number of CpG dinucleotides in a series of pull-down experiments (Figure 14.1B). 14–3–3 promoter DNA templates containing either 1, 5, or 12 CG were generated by PCR. Half of the amplified templates were methylated in vitro with the restriction endonuclease MssI, while the other half remained unmethylated. Methylated and unmethylated PCR templates were incubated in optimized binding buffers (containing 200 mM NaCl for CXXC and 400 mM NaCl for MBD2b/MBD3L1) with either 2 μg MBD2b/MBD3L1 or 6 μg CXXC in the presence of 5 μg of polydAdT, added as DNA carrier. DNA–protein complexes were collected with nickel-coated magnetic beads and washed to remove unbound material. The captured DNA was eluted and subjected to PCR analysis.

CXXC was able to capture DNA fragments containing as little as one unmethylated CpG, whereas the ability of MBD2b/MBD3L1 to capture methylated DNA was limited to templates containing 5 and 12 mCpG and not the template containing 1 mCpG (Figure 14.1B). None of the unmethylated templates were captured by the MBD2b/MBD3L1 beads, demonstrating the specificity of this heterodimer for methylated DNA. By contrast, some MssI-treated templates containing 5 and 12 CpG were captured in the CXXC-containing pull-downs. Use of a binding buffer containing 0.3 M NaCl resulted in increased specificity of CXXC for non-methylated PCR-generated templates containing three or more CpG sites, but as a consequence of the increased stringency, some of the unmethylated DNA remains in the unbound, supernatant fraction (data not shown). For following experiments a more stringent binding buffer (300 mM NaCl) was chosen.

14.5   Assessing CpG methylation status by measuring the MRA ratio

CXXC and MBD2b/MBD3L1 capture experiments were next performed with fragmented genomic DNA isolated from the human breast cancer cell line MCF7. Genomic DNA was extracted from log-phase MCF7 cells according to the established procedures described in the MethylCollector Ultra™ Kit manual, then exhaustively digested with the MseI restriction endonuclease which cleaves in the sequence 5′-TTAA-3′ to generate fragments with an average size of 500 bp. Then 200 ng of the fragmented genomic DNA was incubated with the MBD2b/MBD3L1 heterodimer (1 µg of each protein) in a binding reaction containing 400 mM NaCl for 1 hour at 4 °C in a 100-µl volume in the presence of nickel-charged magnetic beads. In a separate but parallel reaction, an equivalent amount of genomic DNA was incubated with 6 µg CXXC in a 300 mM NaCl binding buffer containing nickel-charged magnetic beads for 30 minutes at room temperature. Protein–DNA complexes were then captured by a magnet placed at the side of the tube. Supernatant fractions containing unbound DNA were removed and retained. The beads were then subsequently washed four times with the respective binding buffers, and bound DNA was extracted from the beads using proteinase K. Capture experiments were performed in replicates of four to eight. Isolated DNA was pooled then purified using a Qiaquick PCR purification kit and an aliquot of the DNA was analyzed by RT-PCR in triplicate. One representative experiment in which 20 loci were analyzed in consecutive PCR reactions is depicted in Figure 14.2. PCR primer pairs were selected for CpG islands whose methylation status in MCF7 cells has been independently reported in the literature (Paz *et al.*, 2003; Rivenbark *et al.*, 2006; Chung *et al.*, 2008; Hodges *et al.*, 2009). For both enrichment reactions (MBD2b/MBD3L1 [Figure 14.2A] and CXXC [Figure 14.2B]) both unbound and captured DNA was quantified by RT-PCR using genomic input DNA as a reference standard curve.

Enrichments were calculated as the percentage of DNA captured by either method relative to the amount of input DNA (as per formula 3) and are represented in Table 14.2 as MRA ratios. Genomic regions that contained a high level of methylated CpGs and a low level of unmethylated CpGs produced a high value for the MRA ratio (value >2.5), while regions that contained a low level of methylated CpGs and a high level of unmethylated CpGs produced a low value for the MRA ratio (value <0.1). The low and high cut-off values were set empirically.

For 11 of the 20 loci examined, the calculated MRA values were consistent with CpG methylation status reported in the literature (last column of Table 14.2). Moreover, promoters with MRA ratios between 0.1 and 2.5 (classified as unclear) led to the determination that those DNA regions contained either partial methylation, were imprinted genes, or change-point regions. Importantly, this type of information could not have been obtained if DNA had been captured by either of the methods alone. The majority of these transition methylation states were also confirmed within the scientific literature (Table 14.2).

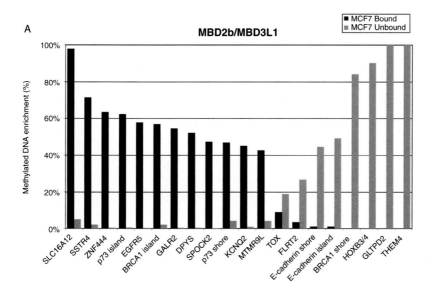

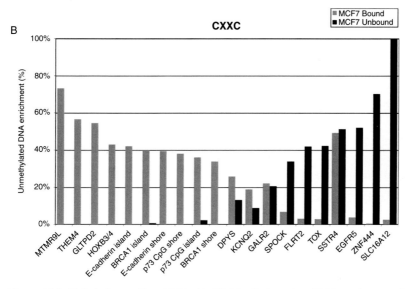

Figure 14.2 Methylation profiles of 20 loci in MCF7 cells. Samples of 200 ng of enzymatically digested MCF7 genomic DNA were incubated with 1 µg MBD2b/MBD3L1 protein complex (A) to isolate methylated DNA or 6 µg CXXC (B) to isolate unmethylated DNA. DNA that was methylated (black bars) was collected in the bound fraction of MBD2/MBD3L1 and in the unbound fraction of CXXC. DNA that was unmethylated (gray bars) remained in the unbound fraction of MBD2/MBD3L1 and was collected in the bound fraction of CXXC. All collected fractions were purified and subsequently analyzed by RT-PCR and plotted as a percentage of input.

The MRA assay also enables determination of the percentage of DNA that is not present in the captured or unbound fractions. This represents the DNA lost during the washes or that was initially bound to the beads but with low affinity. Since this assay is quantitative and generally the amount of input DNA is known before performing the assay, additional parameters can be used to aid in the data analysis

Table 14.2 Comparison of methylation ratio of MCF7 DNA loci and reported methylation status. Shaded rows denote agreement of the methylation status determined by the MRA assay with that reported in the literature

Target	MethylCollector Ultra™		UnmethylCollector™		MRA	MRA methylation status	Reported methylation status[a]
	Bound (%)	Unbound (%)	Bound (%)	Unbound (%)			
SLC16A12	97.8	5.1	2.6	117.8	37.9	++	++
GALR2	54.5	0.4	22.1	20.6	2.5	++	++
TOX	9	18.8	2.9	42.2	3.1	++	++
SPOCK	47.4	0.2	6.8	33.8	7.0	++	++
EGFR5	57.8	0.3	3.8	52.1	15.3	++	++
DPYS	52.1	0.5	25.8	13.1	2.0	−/+	++
NBR2 (BRCA1-CpG)	56.9	2.1	39.9	0.7	1.4	−/+	++
BRCA1 shore	0.1	84.1	33.8	0.1	0.0	−	
E-CAD island	1.1	49.2	42	0	0.0	−	−
E-CAD shore	1.1	44.6	39.4	0.1	0.0	−	
p73 island	62.3	0.5	36	2.2	1.7	−/+	−
p73 shore	46.8	4.2	38.1	0.1	1.2	−/+	
THEM4	0.1	102.4	56.6	0.1	0.0	−	−
ZNF444	6.5	0.4	0.5	70.2	119.7	++	++
FLRT2	3.5	26.7	3.1	41.9	1.2	−/+	−
KCNQ2	45.1	0.9	18.9	8.8	2.4	−/+	pm
HOXB3/4	0	90.2	43	0.1	0.0	−	pm
SSTR4	71.3	2.1	49.3	51.2	1.4	−/+	cp
MTMR9L	42.6	4.2	73.2	0	0.6	−/+	cp
GLTPD2	0	107.2	54.6	0	0.0	−	cp

++ denotes hypermethylated; −/+ unclear; − denotes hypomethylated; pm denotes partial methylation; cp denotes change point.
[a] Based on methylation status reported in Paz *et al.* (2003), Rivenbark *et al.* (2006), Chung *et al.* (2008), and Hodges *et al.* (2009).

which will help to better understand the transitional methylation degree. We defined "Non-specific DNA" as the DNA lost during the washes; this type of DNA binds initially to the beads but due to its low affinity to the beads is removed from the beads during the washing steps. The formula used to define "Non-specific DNA" is described as:

Non-specific DNA = (input DNA) − (bound fraction) − (unbound fraction)

Another parameter used was "Delta" which is simply the absolute difference between bound and unbound fractions. This variable helps to classify the true methylated and unmethylated genes from the "unclear" category. For example, a true methylated gene will have a Delta greater than 40 absolute percentage points, whereas for an unclear gene the Delta value will be lower:

$$\text{Delta} = (\text{bound fraction}) - (\text{unbound fraction})$$

14.6 Generating genome-wide CpG methylation maps using MethylCollector Ultra™ and the ABI SOLiD sequencing platform

Previous sections have discussed methodologies for capturing fragments of genomic DNA based upon their CpG methylation status and reducing background capture due to non-specific binding. Once captured, CGIs and other genomic compartments can then be further assessed for methylation at individual CpG dinucleotides by traditional PCR-based, enzymatic, or chemical approaches. These later methodologies are well established in the literature, having been developed specifically to query the methylation status of individual loci, and provide a wealth of information about changes in DNA methylation during normal to disease state progression and during organismal development.

In this section we describe how the affinity capture approach can be used to generate genome-wide maps of CpG methylation "peaks" corresponding with areas of high-frequency DNA methylation and "valleys" having little or no methylation (Bock *et al.*, 2010). We follow the original nomenclature for this approach – MiGS (MBD isolated genome sequencing) (Serre *et al.*, 2010) – to describe coupling of MBD affinity approaches with high-throughput sequencing (Figure 14.3).

All of the sequencing shown here was performed using ABI's SOLiD instrument; however, the basic principles and workflow are applicable to other platforms. MiGS consists of four parts – first, construction of an unbiased library of nucleic acids that represents all of the captured DNA fragments (Figure 14.3A); second, a series of ligations and clonal amplifications of DNA fragments linked to beads and immobilization of the beads on slides (Figure 14.3A); third, the physical sequencing of nucleic acid fragments with the ABI SOLiD instrument utilizing sequential ligation based sequencing with dye-labeled oligonucleotides (Figure 14.3A); and finally, data processing (Figure 14.3B).

The first step requires preparation and shearing of DNA into fragments suitable for capture. We prefer to isolate DNA using techniques that do not employ phenol which may cause oxidative damage to nucleic acids. Nucleic acid shearing with a target size of 150 to 250 base pairs is performed with the Covaris S2 system which uses focused acoustics technology. This process works by sending acoustic energy wave packets from a dish-shaped transducer that converges and focuses to a small localized area and allows mechanical energy to be applied to a sample without directly contacting the sample. Importantly, since the process occurs in a closed vessel, there is no cross-contamination between samples.

Capture of methylated DNA fragments with MBD2b/MBD3L1 heterodimer represents a critical juncture in experimental design. Under high-affinity binding conditions and with low ionic conditions of washing and elution it is possible to capture large portions of the genome including methylated non-genic repetitive DNA, regions of low-density CpG methylation, and the desired targets such as genic CpG

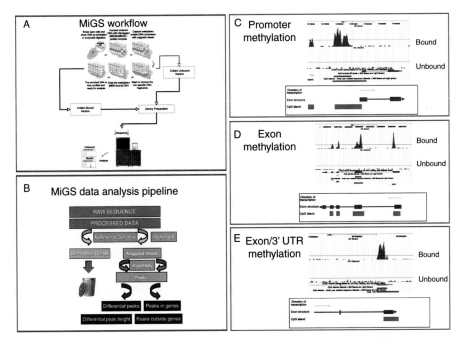

Figure 14.3 MBD-isolated genomic sequencing (MiGS) workflow and DNA methylation mapping results. (A) General workflow of MiGS (MBD-isolated genome sequencing). DNA samples are sheared and incubated with immobilized MBD2b/MBD3L1. The protein complexes are captured with beads and unbound and bound fractions are collected. Following MBD-mediated capture, libraries are constructed of the bound and unbound fragments and sequenced on an ABI SOLiD Next Generation Sequencer. (B) Schematic of the MiGS data analysis pipeline. Raw sequence is generated from the SOLiD Next Generation Sequencer and aligned to a known reference genome. Depending on user-selected criteria, mapped reads are assembled into peaks. Further approaches are utilized to determine location, height, and other information on those peaks. (C–E) Sequencing data obtained from the Active Motif MethylCollector Ultra Kit. The uppermost two tracts represent the bound and unbound fractions of adult rat hippocampal stem cells. In the case of promoter methylation (C), three distinct peaks are observed. The lower tract shows the unbound fraction with minimal peaks and thus low background. These peaks overlap with know CpG islands in two different CpG tracts (light-colored rectangle found in lower panel). The CpG islands shown in the top tract have a content of 50% or greater GC, length greater than 200 bp, and a ratio greater than 0.6 for observed over expected number of CG dinucleotides (Gardiner-Garden and Frommer, 1987). The second CpG island tract shows CpG islands identified by the modified Andy Law algorithms. (D) DNA methylation peaks overlapping known CpG islands located in the exonic regions or (E) Exon/3′ untranslated region (UTR) of an individual gene.

dinucleotides and CGIs (Robinson *et al.*, 2010). With increasing stringency of washes the repetitive DNA is excluded in a CpG-density-dependent manner (Maunakea *et al.*, 2010; Nair *et al.*, 2011) leaving predominantly CpG-rich regions and CGIs. Importantly, we also collect DNA that is not captured – the unbound fraction – for further processing. Pools of DNA fragments either captured and enriched for methylated DNA or not methylated and therefore comprising the unbound fraction, are then ready for library preparation and sequencing according to established protocols for SOLiD fragment library construction (Bormann Chung *et al.*, 2010).

Sequencing runs are structured to generate information for 50 million individual DNA molecules ("tags") providing appropriate representation of captured

CpG methylated DNA or unmethylated DNA (the "unbound tags"). After processing and mapping to a reference genome approximately 20 to 30 million of each of these tags are used for downstream analysis. A critical step for accurate CpG methylation identification involves appropriate filtering of mapped tags during data analysis to generate normalized peak heights. Normalization is an essential process for allowing direct quantitative comparisons of methylated regions within a genome and across genomes derived from different samples.

One powerful option for data analysis is provided by the CLC Bio Genomic Workbench (http://wwwclcbio.com/index.php) which allows direct alignment with reference genomes, peak calling, and analysis software for desktop computers. We have found these Genomic Workbench settings to produce excellent results:

(1) Normalize bound "tags" with unbound "tags"
(2) Set threshold settings for peak heights greater than 17 tags
(3) Set the FDR (false discovery rate) at 2%
(4) A correlation coefficient of greater than 0.95 significantly reduces background.

After mapping, peaks are assigned to various genomic compartments. Incorporating these stringent parameters allows for identification of several thousand peaks of CpG enriched methylated genomic fragments.

Analysis of peak location confirms the efficacy of methylated DNA capture with MBD2b/MBD3L1 and deep sequencing approach (Figure 14.3C, D, and E). Abundant peaks are found in promoter regions, introns and 3′ untranslated regions consistent with other molecular approaches (Ball *et al.*, 2009) and recently also identified by capture sequencing approaches using the Illumina G2aX high-throughput sequencer (Harris *et al.*, 2010; Serre *et al.*, 2010). Intragenic regions are also represented as these may contain the so-called "orphan" CGIs recently identified by Bird and colleagues (Illingworth *et al.*, 2010).

14.7 Conclusion

The apparent ease with which protein-based affinity capture methods are applied to the various analytic platforms used in analysis of DNA methylation is suggestive that their use will continue to grow. Perhaps their greatest potential lies in their application in the clinical setting where their use to detect the methylation status of clinically significant targets will aid in the development of diagnostic or therapeutic assays.

REFERENCES

Acevedo, L. G., Sanz, A., and Jelinek, M. A. (2011). Novel DNA binding domain-based assays for the detection of methylated and non-methylated DNA. *Epigenomics*, **3**, 93–101.
Ball, M. P., Li, J. B., Gao, Y., *et al.* (2009). Targeted and genome-scale strategies reveal gene-body methylation signatures in human cells. *Nature Biotechnology*, **27**, 361–368.

Ballestar, E. and Wolffe, A. P. (2001). Methyl-CpG-binding proteins: targeting specific gene repression. *European Journal of Biochemistry*, **268**, 1–6.

Bird, A. P. (1986). CpG-rich islands and the function of DNA methylation. *Nature*, **321**, 209–213.

Bock, C., Tomazou, E. M., Brinkman, A. B., *et al.* (2010). Quantitative comparison of genome-wide DNA methylation mapping technologies. *Nature Biotechnology*, **28**, 1106–1114.

Bormann Chung, C. A., Boyd, V. L., Mckernan, K. J., *et al.* (2010). Whole methylome analysis by ultra-deep sequencing using two-base encoding. *PLoS One*, **5**, e9320.

Chen, T. and Li, E. (2006). Establishment and maintenance of DNA methylation patterns in mammals. *Current Topics in Microbiology and Immunology*, **301**, 179–201.

Cheng, X. and Blumenthal, R. M. (2008). Mammalian DNA methyltransferases: a structural perspective. *Structure*, **16**, 341–350.

Chung, W., Kwabi-Addo, B., Ittmann, M., *et al.* (2008). Identification of novel tumor markers in prostate, colon and breast cancer by unbiased methylation profiling. *PLoS One*, **3**, e2079.

Cross, S. H., Charlton, J. A., Nan, X., and Bird, A. P. (1994). Purification of CpG islands using a methylated DNA binding column. *Nature Genetics*, **6**, 236–244.

Cross, S. H., Meehan, R. R., Nan, X., and Bird, A. (1997). A component of the transcriptional repressor MeCP1 shares a motif with DNA methyltransferase and HRX proteins. *Nature Genetics*, **16**, 256–259.

Ferguson-Smith, A. C. and Surani, M. A. (2001). Imprinting and the epigenetic asymmetry between parental genomes. *Science*, **293**, 1086–1089.

Fraga, M. F., Ballestar, E., Montoya, G., *et al.* (2003). The affinity of different MBD proteins for a specific methylated locus depends on their intrinsic binding properties. *Nucleic Acids Research*, **31**, 1765–1774.

Gardiner-Garden, M. and Frommer, M. (1987). CpG islands in vertebrate genomes. *Journal of Molecular Biology*, **196**, 261–282.

Harris, R. A., Wang, T., Coarfa, C., *et al.* (2010). Comparison of sequencing-based methods to profile DNA methylation and identification of monoallelic epigenetic modifications. *Nature Biotechnology*, **28**, 1097–1105.

Hendrich, B. and Bird, A. (1998). Identification and characterization of a family of mammalian methyl-CpG binding proteins. *Molecular and Cellular Biology*, **18**, 6538–6547.

Hodges, E., Smith, A. D., Kendall, J., *et al.* (2009). High-definition profiling of mammalian DNA methylation by array capture and single molecule bisulfite sequencing. *Genome Research*, **19**, 1593–1605.

Illingworth, R., Kerr, A., Desousa, D., *et al.* (2008). A novel CpG island set identifies tissue-specific methylation at developmental gene loci. *PLoS Biology*, **6**, e22.

Illingworth, R. S., Gruenewald-Schneider, U., Webb, S., *et al.* (2010). Orphan CpG islands identify numerous conserved promoters in the mammalian genome. *PLoS Genetics*, **6**, e1001134.

Issa, J. P. (2003). Age-related epigenetic changes and the immune system. *Clinical Immunology*, **109**, 103–108.

Jiang, C. L., Jin, S. G., and Pfeifer, G. P. (2004). MBD3L1 is a transcriptional repressor that interacts with methyl-CpG-binding protein 2 (MBD2) and components of the NuRD complex. *Journal of Biological Chemistry*, **279**, 52 456–52 464.

Jones, P. A. and Baylin, S. B. (2007). The epigenomics of cancer. *Cell*, **128**, 683–692.

Jorgensen, H. F., Ben-Porath, I., and Bird, A. P. (2004). Mbd1 is recruited to both methylated and nonmethylated CpGs via distinct DNA binding domains. *Molecular and Cellular Biology*, **24**, 3387–3395.

Laird, P. W. (2010). Principles and challenges of genome-wide DNA methylation analysis. *Nature Reviews Genetics*, **11**, 191–203.

Lee, J. T. (2003). Molecular links between X-inactivation and autosomal imprinting: X-inactivation as a driving force for the evolution of imprinting? *Current Biology*, **13**, R242–254.

Maunakea, A. K., Nagarajan, R. P., Bilenky, M., *et al.* (2010). Conserved role of intragenic DNA methylation in regulating alternative promoters. *Nature,* **466**, 253–257.

Mortusewicz, O., Schermelleh, L., Walter, J., Cardoso, M. C., and Leonhardt, H. (2005). Recruitment of DNA methyltransferase I to DNA repair sites. *Proceedings of the National Academy of Sciences USA,* **102**, 8905–8909.

Nair, S. S., Coolen, M. W., Stirzaker, C., *et al.* (2011). Comparison of methyl-DNA immuno-precipitation (MeDIP) and methyl-CpG binding domain (MBD) protein capture for genome-wide DNA methylation analysis reveals CpG sequence coverage bias. *Epigenetics,* **6**, 34–44.

Ooi, S. K. and Bestor, T. H. (2008). Cytosine methylation: remaining faithful. *Current Biology,* **18**, R174–176.

Paz, M. F., Fraga, M. F., Avila, S., *et al.* (2003). A systematic profile of DNA methylation in human cancer cell lines. *Cancer Research,* **63**, 1114–1121.

Plazas-Mayorca, M. D. and Vrana, K. E. (2011). Proteomic investigation of epigenetics in neuropsychiatric disorders: a missing link between genetics and behavior? *Journal of Proteome Research,* **10**, 58–65.

Rauch, T. and Pfeifer, G. P. (2005). Methylated-CpG island recovery assay: a new technique for the rapid detection of methylated-CpG islands in cancer. *Laboratory Investigations,* **85**, 1172–1180.

Rauch, T., Li, H., Wu, X., and Pfeifer, G. P. (2006). MIRA-assisted microarray analysis, a new technology for the determination of DNA methylation patterns, identifies frequent methylation of homeodomain-containing genes in lung cancer cells. *Cancer Research,* **66**, 7939–7947.

Rivenbark, A. G., Jones, W. D., Risher, J. D., and Coleman, W. B. (2006). DNA methylation-dependent epigenetic regulation of gene expression in MCF-7 breast cancer cells. *Epigenetics,* **1**, 32–44.

Robertson, K. D. (2005). DNA methylation and human disease. *Nature Review Genetics,* **6**, 597–610.

Robinson, M. D., Stirzaker, C., Statham, A. L., *et al.* (2010). Evaluation of affinity-based genome-wide DNA methylation data: effects of CpG density, amplification bias, and copy number variation. *Genome Research,* **20**, 1719–1729.

Serre, D., Lee, B. H. and Ting, A. H. (2010). MBD-isolated genome sequencing provides a high-throughput and comprehensive survey of DNA methylation in the human genome. *Nucleic Acids Research,* **38**, 391–399.

15 Genome-wide ChIP-DSL profiling of promoter methylation patterns associated with cancer and stem cell differentiation

Jeffrey D. Falk

15.1 Introduction

The effect of epigenetic modifications on the regulation of gene expression and their relationship to human diseases has become increasingly apparent and a major focal area of biological studies in recent years. Past studies have intimated that there is a direct correlation between epigenetic modifications, such as DNA methylation and histone methylation and acetylation, and aberrant gene expression in cells associated with diseases such as cancer (Weber *et al.*, 2005). However, in past years the appropriate tools and methodologies did not exist for accurately identifying and assessing the biological effects of these epigenetic modifications, which significantly impeded advances in the epigenetics field. A number of technologies have evolved in recent years that have enhanced our abilities to identify and characterize epigenetic modifications on a genome-wide scale, which have facilitated comparative studies to elucidate the functional relationship between epigenetic modifications and gene expression. During the past decade the development of chromatin immunoprecipitation (ChIP)-based technologies has emerged as one of the predominant methodologies for high-throughput genome-wide mapping of epigenetic modifications such as histone modifications or DNA methylation. These technologies utilize antibodies to capture specific epigenetic modifications which are subsequently mapped using genomics-based detection technologies such as DNA microarrays (ChIP-chip), qPCR (ChIP-qPCR), and more recently next-generation sequencing (ChIP-Seq). ChIP-chip has emerged as one of the predominant technologies available for comparative genome-wide epigenetic mapping. Traditional ChIP-chip experiments (Figure 15.1A) utilize standard chromatin immunoprecipitation with an antibody to a specific transcription factor or epigenetic modification to capture specific regions within the chromatin where the modification occurs or the factor is bound. These captured binding sites are then labeled and used to probe

Epigenomics: From Chromatin Biology to Therapeutics, ed. K. Appasani. Published by Cambridge University Press. © Cambridge University Press 2012.

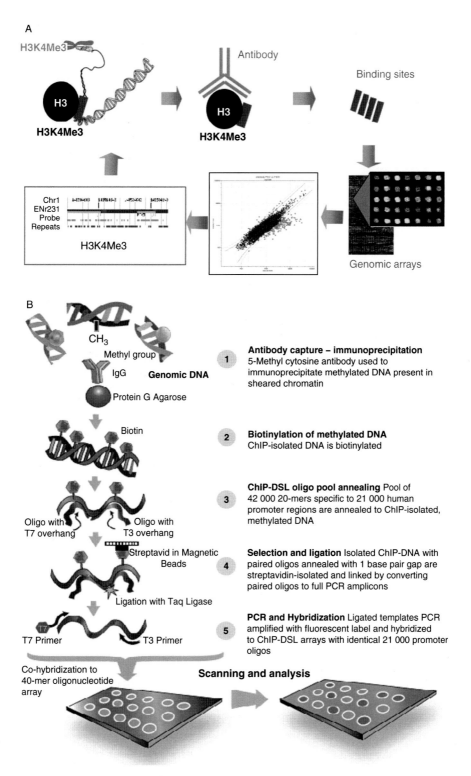

Figure 15.1 Overview of ChIP-chip and ChIP-DSL technologies. (A) Basic components of ChIP-chip technology. (B) Overview of key steps in the ChIP-DSL technology. See plate section for color version.

microarrays containing specific portions of the genome in order to visualize and quantitate precisely where these epigenetic modifications are occurring or factors are binding throughout the genome. Typically, ChIP-chip studies can be conducted with a set of microarrays that span the entire genome, followed by subsequent tiling arrays that perform high-density mapping focusing in on very specific binding-site regions. Through the use of ChIP-chip studies researchers have been able to correlate changes in epigenetic modifications with physiological changes in disease-relevant cells in order to ascertain the nature of the differences that may be contributing to the development or progression of the diseases, and to identify specific biomarkers that can be used as diagnostic predictors of disease development.

Early ChIP-Chip studies focused on using antibodies to key transcription factors (TF) in order to identify promoters that interact with these TFs that are important in modulating cancer-related activities. This is exemplified by the first ChIP-chip studies that were conducted by Ren *et al.* (2000), which identified genome-wide *myc* oncogene binding site interactions in Burkitt lymphoma cells and exemplified how ChIP-chip could be effectively used to map promoters that are important in modulating *myc*-dependant oncogenic transformation. The success of these transcription factor-based ChIP-chip studies rapidly led researchers to start applying ChIP-chip to the characterization of global epigenetic modifications through the use of antibodies to specific histone methylation sites such as H3K4Me3, H3K9me3, and H3K27me3, as well as methylated DNA which is commonly found in promoter region CpG islands that are associated with chromatin remodeling and modulating gene expression (Barski *et al.*, 2007; Koch *et al.*, 2007). These epigenetic ChIP-chip studies resulted in key discoveries such as the correlation of H3KMe4 histone modifications with promoter upregulation (Koch *et al.*, 2007), and the direct correlation of promoter DNA hypermethylation in cancer cells (Esteller, 2007).

Despite the obvious utility of ChIP-chip technology there are several key obstacles and roadblocks which are inherent in the traditional ChIP-chip methodology that hindered the widespread use of the technology. For example, ChIP-chip studies typically required 10^8 cells of starting material which made ChIP-chip studies impractical for studying systems such as stem cells or cancer issue samples where the amount of starting material was limited. These traditional studies also require large sets of 10–30 microarrays in order to provide complete genome coverage. This chapter summarizes the development and application of a novel ChIP-chip technology, ChIP-DSL (Chromatin immunoprecipitation – DNA selection and ligation) which was developed to overcome many of the deficiencies that are inherent in traditional ChIP-chip methodology (Kwon *et al.*, 2007). The primary advantage of the ChIP-DSL technology is that it utilizes a single microarray, and demonstrates significantly increased sensitivity that is about three orders of magnitude greater than traditional ChIP-chip methods while at the same time it has dramatically decreased the starting material requirements ($\sim 10^6$ cells) so that ChIP-DSL can and has been effectively applied to both clinical and diagnostic sample analysis.

ChIP-DSL has been effectively applied to global profiling of both epigenetic modification and transcription factor interactions. This includes published studies utilizing ChIP-DSL to characterize estrogen receptor (Kwon *et al.*, 2007), androgen receptor (Li *et al.*, 2008), and p15ink4b tumor suppressor (Thillainadesan *et al.*, 2008), and LSD1 demethylase interactions (Nunez *et al.*, 2008), as well as the effects of histone methylation/demethylation on estrogen receptor expression (Garcia-Bassets *et al.*, 2007), and DNA methylation profiling in leukemia (Ying *et al.*, 2007; Ganetzgy *et al.*, 2008). ChIP-DSL has also has been particularly useful in deciphering global DNA methylation patterns. This chapter describes several of these global DNA methylation studies that have successfully identified diagnostic and prognostic biomarkers in cancer cells, as well as characterized the methylome of embryonic stem cells and hematopoetic stem cells associated with acute myelogenous leukemia (AML) patients.

15.2   Methodology

15.2.1   ChIP-DSL technology

The primary impetus for the development of the ChIP-DSL technology was to modify existing ChIP-chip methodology to facilitate single microarray genome-wide detection of epigenetic modifications and factor interactions with significantly increased specificity and sensitivity, while concomitantly decreasing the starting material that is necessary for each experiment. To accomplish this several key changes were incorporated into the ChIP-DSL methodology.

(a) Assays were designed to specifically monitor epigenetic interactions within promoter regions localized throughout the genome thereby decreasing the complexity of the sample to be assayed and dramatically increasing the signal-to-noise ratio of the reactions, as compared to other methods that scan the entire genome.

(b) Incorporation of a biotin bead-based DNA selection step dramatically increased the fidelity of the assay by removing extraneous background material.

(c) PCR amplifications of captured sequences containing epigenetic modifications were conducted using target-specific PCR amplicons derived from a DNA selection and ligation step which significantly increased the fidelity and specificity of the assay.

(d) Detection of captured epigenetic modifications was enhanced by deconvoluting the assay system by arraying the same promoter target sequences used for PCR amplicon amplification on the ChIP-DSL microarrays.

The basic steps involved in the ChIP-DSL assay (Kwon *et al.*, 2007) are shown in Figure 15.1B and summarized below. The initial stages of the experiment involve traditional ChIP-based immunoprecipitation of cross-linked chromatin that has been fractionated by either sonication (factor-based ChIP-DSL) or restriction-digested DNA fragments (epigenetic modifications). DNA fragments containing the appropriate epigenetic modification or bound factor are then captured using an antibody to 5-methylcytosine to capture methylated DNA, or to a specific histone modification for capturing other epigenetic modification, or a specific transcription factor in order to capture the factor DNA binding-site complex. The captured modification/binding-site-specific fragments are then isolated and used for subsequent DSL analysis and detection.

The key steps in the following DSL analysis and detection process revolve around the use of a set of ChIP-DSL 40-mer promoter sequences that represent the promoter regions (localized within the −300 to −500 bp promoter region) within 21 000 different genes throughout the human genome. In the next step of the ChIP-DSL procedure, the captured modification/binding-site fragments are annealed with a pool of 42 000 ChIP-DSL 20-mers where each oligo represent one half of a specific 40-mer promoter sequence, and each promoter is represented by a pair of consecutive adjoining half site 20-mers with a 1 bp gap. After annealing binding/modification amplicons templates are created by a subsequent Taq ligase DNA ligation step that will only result in viable templates for PCR amplification when the two ChIP-DSL promoter 20-mer oligos have annealed properly with a 1 bp gap that is ligated to form a consecutive 40 bp promoter amplicon. After ligation the resulting biotinylated templates are isolated with streptavidin beads and PCR amplified incorporating Cy5 and Alexa dyes. This ChIP-DSL selection and ligation process dramatically increases the sensitivity and fidelity of the reaction.

The amplified, labeled binding fragments are then used as probes in a final detection step that adds an additional level of sensitivity to the ChIP-DSL technology whereby the amplified, labeled probes are hybridized to ChIP-DSL microarrays containing the same 21 000 Chip-DSL 40-mers that were used for the selection and ligation process. This results in a coordinated amplification detection system where gene-specific amplification only occurs with binding fragments that are localized in the 21 000 promoter regions and these same regions are also contained on the ChIP-DSL detection array. As a consequence the complexity of the amplification process is reduced significantly, and the specificity of the amplification and subsequent detection are dramatically increased in comparison to traditional ChIP-chip that uses blunt-end primer ligation resulting in amplification of the entire myriad of sequences that are present after chromatin immunoprecipitation. The resulting ChIP-DSL binding profile is then used to identify and quantitate the promoters within the genome where the specific epigenetic modifications are occurring and exerting their regulatory effects. Comparative studies can then be used to identify significant changes in factor binding or epigenetic modifications at specific sites throughout the genome in order to highlight specific target genes of interest as biomarkers or predictors of specific cellular events. Once these target candidates have been identified, the observed change in binding profile can then be confirmed using ChIP-qPCR analysis (Haring *et al.*, 2007).

As noted above, the ChIP-DSL method has been effectively applied to studies looking at a wide range of changes in chromatin morphology including transcription factor binding, and epigenetic modifications such as DNA methylation, and histone modification (methylation, acetylation). The method has proven to be particularly effective in characterizing genome-wide profiles of changes in the global DNA methylation patterns in relevant cell types in order to define the methylome and highlight key targets that may be implicated in important processes such as stem cell differentiation, and disease resistance to therapeutic treatments. The following sections illustrate several examples where ChIP-DSL analysis of DNA methylation sites has been effectively used to identify cancer related biomarkers, as well as target genes that are implicated in stem cell differentiation and involvement in AML.

15.3   Results and discussion

15.3.1   ChIP-DSL identification of cancer biomarkers

The correlation between changes in DNA methylation in cancer cells is a well-characterized phenomenon that has led to increasing interest in determining changes in the global methylation patterns that occur in the transition from normal to neoplastic cells in order to gain insight into the biological mechanism involved in the oncological transition. A number of technologies have been developed for looking at changes in DNA methylation at a selected set of specific targets such as bisulfite sequencing (Fraga *et al.*, 2002), me-PCR (Herman *et al.*, 1996), and mass spectrometry (Erlich *et al.*, 2005). However very few technologies are capable of characterizing genome-wide changes in methylation. ChIP-DSL has been very effective in characterizing these genome-wide changes in DNA methylation, particularly in cancer-related cells. This has been especially useful in using comparative DNA methylation profiling of normal and cancer-related cells in order to characterize significant changes that are occurring in the methylome which have enabled us to identify potential biomarkers for a number of common cancer subtypes.

Traditionally, cancer methylation studies have focused on identifying important tumor suppressor targets by identifying instances where promoter hypermethylation in cancer cells has silenced the tumor suppressor thereby facilitating the transition to cancer formation. We have taken a different approach to identifying cancer biomarkers whereby our studies have focused on correlating changes in methylation patterns with changes in gene expression in order to identify instances where promoter hypomethylation has resulted in increased gene expression which is specifically associated with the cancerous cells. The following comparative study of ChIP-DSL DNA methylation patterns observed in normal and cancerous liver cells exemplifies these type of experiments where changes in the methylome were used in order to identify these biomarkers (Figure 15.2). In this particular study both normal liver and hepatocellular carcinoma (HCC) cells were compared at two levels. The first level involved is the standard microarray analysis to characterize genome-wide gene-expression patterns in order to identify a pool of target genes that exhibited elevated levels of gene expression in the HCC cells. In the second level of investigation ChIP-DSL

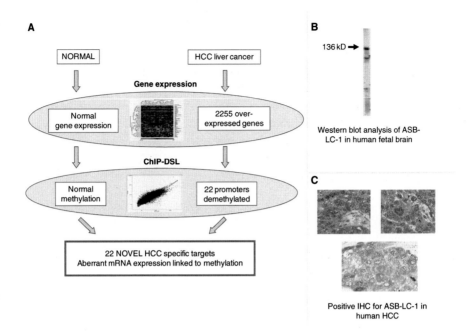

Figure 15.2 ChIP-DSL identification of cancer biomarkers in hepatocellular carcinoma (HCC) cells. (A) Identification of cancer biomarkers in HCCs using ChIP-DSL and gene-expression profiling to identify promoters where aberrant demethylation increases gene expression. (B) Western blot analysis of HCC ChIP-DSL identified candidate ASB LC-1 in human fetal brain cells. (C) Immunohistochemistry confirming expression of ASB LC-1 in normal brain cells and HCC liver cells. See plate section for color version.

DNA methylation analysis was conducted using standard ASB human 21K ChIP-DSL arrays containing 21 000 human promoters in order to identify which of these promoters exhibited a significant decrease in the DNA methylation in the cancerous cells. Once these data were merged we were able to identify 22 "candidate genes" that fit our experimental criteria and demonstrated increased levels of gene expression as a result of aberrant demethylation of their promoters. Subsequent ChIP-DSL studies were then employed to further characterize these candidates to determine the nature of their biological activities and their potential significance in the cancer cells. An example of this is shown in Figure 15.2B where liver cancer biomarker candidate ASB LC-1 was further characterized and shown to be a transcription factor with unknown function that interacts at 227 sites within the genome and is expressed in high levels in the brain but virtually non-existent in liver in normal cells, but is expressed at high levels in the liver in HCC-derived cells (Figure 15.2C). These characteristics clearly indicate that ASB LC-1 is a potential biomarker worthy of further characterization that may have prognostic, diagnostic, and therapeutic value.

Similar studies have been conducted employing ChIP-DSL DNA methylation analysis in conjunction with gene-expression profiling to identify other oncology-based biomarkers in other tissues such as lung, prostate, and brain. Comparison of promoter methylation in glioblastoma and normal brain cells identified five potential brain cancer biomarkers including a phosphatase, protease inhibitor, neural migration protein, and a metalloproteinase. The protease behaved in a similar manner as the previously described liver cancer biomarker in that it is normally expressed in salivary tissue but aberrant demethylation in glioblastoma cells results in overexpression of this gene in cancerous brain cells.

15.3.2 Stem cell promoter DNA methylation profiling

Stem cells have recently come to the forefront of biological studies because of their ability to be manipulated and differentiated into many types of different cells and tissue that are medically relevant. A thorough understanding of the biological processes is essential to understanding the key components involved in these transformations and how they bring about these changes at the cellular level. ChIP-DSL's ability to provide a comprehensive map of global promoter DNA methylation patterns has proven to be valuable in analyzing stem cell methylation and identifying potentially important biomarkers and targets in these cells. An example of these studies is shown below where ChIP-DSL was used to study both embryonic stem (ES) cells and hematopoietic stem cells.

15.3.2.1 DNA methylation profiling of embryonic stem cells

The transition of native embryonic stem cells into differentiated tissue is a complex process with enormous therapeutic potential. To gain a better understanding of the differentiation process, genome-wide ChIP-DSL promoter DNA methylation profiling studies were conducted to characterize the methylome of human ES cells at three stages of differentiation: native ES cells, cells at 36 hours into differentiation, and cells after 94 hours of differentiation to ectoderm. The results from these studies are summarized in Table 15.1 and clearly demonstrated that ChIP-DSL was able to identify unique sets of methylated promoters that associated with different stages of ES cell differentiation, as well as those uniquely associated with CpG islands. Approximately 400–600 methylated promoters were identified in the three differentiation stages that were tested. Of these 183 promoters were uniquely methylated only in the native ES cells, whereas 85 promoters were uniquely methylated only in the 94-hour ectoderm-differentiated stem cells. In each case the methylated promoters are likely to reflect promoters whose target gene expression has been silenced by the methylation process, and may correspond to key components of pathways associated with the differentiation process. Further analysis of these methylated promoters showed that in each of the differentiation cell types approximately 75% of the methylated promoters correlated to known CpG islands which are known to be "hotspots" for epigenetic modulation of gene expression. Once again distinct sets of these methylated promoters localized within CpG islands were found to be uniquely methylated in the specific differentiation cell type studied, with 116 methylated CpG island promoters being identified in native ES cells and 62 promoters occurring in the differentiated stem cells.

Since transcription factors are key modulators of transcriptional activation, an analysis was also conducted to see if there was any correlation between TF

Table 15.1 ChIP-DSL characterization of stem cell DNA methylation patterns: comparative ChIP-DSL mapping of promoter DNA methylation in native, 36-hour differentiated, and 92-hour differentiated embryonic stem cells

Class of promoter		Embryonic stem cells			Commonly methylated genes	Housekeeping genes
	Shared	Non-differentiated	36-hour differentiated	92-hour differentiated		
Methylated promoters identified						
Total	282	470	654	407		
Unique		183	182	85		
Methylated promoters in CpG islands						
Total	211	326	503	301		
Unique		116	151	62		
Methylated transcription factor promoters						
Differentiation and development		9	–	–		
Zinc finger proteins		8	–	3		
Homeobox		4	–	1		
Nuclear receptor		3	–	1		
Gene silenceing		3	–	–		
Transcription		3	–	2		
Activator		1	–	–		
Cell cycle		1	–	–		
Oncogene		1	–	–		
ES cells and	4			407	157	
commonly genes	9			407		319

promoter DNA methylation patterns and ES cell differentiation (Table 15.1). It was interesting to note that many TFs associated with the differentiation process such as differentiation and development, zinc finger protein, and Homeobox genes displayed preferential promoter methylation in the native state, and were demethylated upon transition to the differentiated state. This as would be expected since it is indicative of these "differentiation associated" TFs being silenced in the native ES cells and actively expressed once the cells had differentiated.

Through the course of comprehensive profiling of many different tissues we have been able to compile a set of 157 promoters that have been observed to be highly methylated in highly differentiated tissues. A comparison of these commonly methylated promoters with the methylated promoters in native ES cells showed they had very little in common with only four methylated ES cell promoters belonging to this set of commonly methylated promoters. An additional comparison of promoters methylated in ES cells to 319 typical housekeeping genes also showed very little overlap, with only nine methylated promoters in common. These results suggest that the methylated promoters associated with ES cells are in fact generally functionally important genes that are involved in the differentiation process.

15.3.2.2 DNA methylation of hematopoietic stem cells associated with leukemia

The importance of methylation in regulating gene expression and its correlation in cancer cells has been well established (Esteller, 2007). In fact, there are several cancers where therapeutic demethylating agents have become standard components of the treatment regime. However, resistance to the effects of these demethylating agents is a distinct problem that often arises in patients being treated with these demethylating drugs (Stressman *et al.*, 2008). In order to demonstrate the clinical applications of the ChIP-DSL technology, studies were conducted where ChIP-DSL DNA methylation profiling was used to define the methylome of normal and malignant hematopoetic stem cells associated with AML, and to assess the effects of treatment with the classical therapeutic demethylating agent decitabine (Kantarjian *et al.*, 2003). In order to accomplish this global ChIP-DSL methylation profiles were obtained from normal and AML clinical patients, as well as a variety of different AML cell lines, as well as patients and cell lines that had been treated versus untreated with the decitabine demethylating agent. As shown in Table 15.2, ChIP-DSL was very effective in defining the general

Table 15.2 ChIP-DSL profiling of DNA methylation patterns associated with acute myelogenous leukemia (AML): comparison of DNA methylation patterns associated with AML in normal CD34$^+$ and AML patients, AML cell lines, and cells treated with the demethylation drug decitabine

Sample type	Sample	Methylation status	Methylated promoters	Targets affected
CD34$^+$ Blastocysts	CD34$^+$ methylome	Hypermethylated	534	Transcriptional regulation, nucleotide metabolism, cytoskeleton remodeling
	AML	Hypermethylated	335	mRNA transcription, leukocyte regulation, *Wnt* signaling pathway
CD34$^+$ AML patients	AML vs. normal	Hypermethylated	89/117	Transcription factors
	AML vs. normal	Hypomethylated	84	Transcription factors
AML cell lines		Hypermethylated in any cell line	552	Transcription factors
		Hypermethylated in all seven cell lines tesed	115	Transcription factors
Decitabine-treated cell lines	Normal CD34$^+$	Hypomethylation of previously hypermethylated	195	DNA repair/replication, metabolic processes
	Kasumi AML	Hypomethylation of previously hypermethylated	75	Differentiation, transcription, cell proliferation

methylome in both normal CD34$^+$ cells as well as CD34$^+$ cells isolated from AML cancer patients, with 534 and 335 promoters identified respectively within these cells that exhibited hypermethylation (>2.6× increase in methylation). Hypermethylation in normal CD34$^+$ cells was most prevalent in genes involved in transcription regulation, nucleotide catabolism, and cytoskeleton remodeling, whereas the hypermethylated AML cells showed a preference for genes involved in the *Wnt* signaling pathway, leukocyte regulation, and mRNA transcription. A comparison of the observed promoter methylomes in AML patients as compared to normal patients revealed that 89 promoters were hypermethylated in AML patient, and 84 promoters exhibited hypomethylation, with transcription factors being overrepresented in both cases. These promoters are clearly exhibiting a significant shift in their methylation status in AML patients and are indicative of genes where promoter hypermethylation is most likely inhibiting expression and hypomethylation is inducing expression of these targets. As a consequence, these methylation-modulated targets represent excellent candidates for potential biomarkers, as well as critical regulators of key pathways that may be involved in the AML phenotype and that could serve as therapeutic targets.

ChIP-DSL DNA methylation profiles were also used to characterize promoter methylation in several additional AML cell types. One set of studies focused on characterizing promoter methylation patterns in a variety of well-established AML cell lines including Kasumi, KG1, and TF1 cell lines, as well as the more differentiated AML cell lines K562, HL60, and U937. These studies identified a wide range of 552 hypermethylated genes present in at least one cell line within the collective group of cell lines. However, only a distinctive set of 115 promoters demonstrated concordant hypermethylation in all the AML cell lines tested. Studies were also conducted to asses the genome-wide effects of promoter methylation by the hypomethylating anti-cancer drug decitabine, which is used therapeutically to treat leukemias, but patients oftentimes develop resistance to this therapy. These studies were conducted in both normal CD34$^+$ cells as well Kasumi AML cell lines that had been treated with decitabine. In both cases distinct sets of genes were identified where decitabine had enacted a shift from hypermethylation to hypomethylation in the treated cells. These genes represented DNA repair and metabolic processes in the normal cells and genes involved in differentiation, transcription, and cell proliferation in the AML cell lines. Further analysis of these decitabine-affected genes may be helpful in determining how this drug exerts its anti-cancer effects and how to avert the drug resistance that is often encountered therapeutically.

15.4   Conclusion

The ChIP-DSL technology has been developed to facilitate rapid global profiling of epigenetic modifications and factor binding sites with significantly enhanced sensitivity and decreased starting material requirements with respect to conventional ChIP-chip technologies. This review has focused on the utility of ChIP-DSL profiling of DNA methylation sites in order to identify key pathways and biomarkers that may be involved in disease progression, or key cellular pathways such as differentiation. Application of ChIP-DSL DNA methylation profiling to cancer biomarker discovery studies was described using hepatocellular carcinoma as an example. However, to date ChIP-DSL methylation profiling has been applied to a wide variety of cancers resulting in the identification of biomarkers for prostate, lung, brain, and breast cancers, as well 300 tissue-specific marker genes. Additional applications of ChIP-DSL DNA methylation mapping of key components and pathways involved in embryonic stem cell differentiation and hematopoietic stem cells associated with AML demonstrate the utility of such global methylation studies in characterization of important biological systems such as stem cells. With the recent advances in the development of next-generation sequencing technologies, epigenetic profiling using ChIP-Seq, and Next Gen Sequencing is developing at a rapid pace. However, these efforts are still in their infancy and are not yet well developed for global profiling, and consequently ChIP-DSL has maintained its place as an excellent technology for rapid, economical global profiling of epigenetic modifications.

REFERENCES

Barski, A., Cuddapah, S., Cui, K., *et al.* (2007). High-resolution profiling of histone methylations in the human genome. *Cell*, **129**, 823–837.

Ehrich, M., Nelson, M. R., Stanssens, P., *et al.* (2005). Quantitative high-throughput analysis of DNA methylation patterns by base-specific cleavage and mass spectrometry. *Proceedings of the National Academy of Sciences USA*, **102**, 15 785–15 790.

Esteller, M. (2007). Cancer epigenomics: DNA methylomes and histone-modification maps. *Nature Reviews Genetics*, **8**, 286–297.

Fraga, M. F. and Esteller, M. (2002). DNA methylation: a profile of methods and applications. *BioTechniques*, **33**, 632–636.

Ganetzky, R. D., Ying, J., Falk, J. D., *et al.* (2008). Differences between normal and leukemic stem cell-specific methylome indicates aberrantly silenced genes involved in the pathogenesis of malignant evolution. *Blood*, **112**, 599.

Garcia-Bassets, I., Kwon, Y. S., Telese, F., *et al.* (2007). Histone methylation and de-methylation establish ligand-dependant gene activation by Estrogen Receptor α (ERα). *Cell*, **128**, 505–518.

Haring, M., Offermann, S., Danker, T., *et al.* (2007). Chromatin immunoprecipitation: optimization, quantitative analysis and data normalization. *Plant Methods*, **3**, 11.

Herman, J. G., Graff, J. R., Myöhänen, S., Nelkin, B. D., and Baylin, S. B. (1996). Methylation-specific PCR: a novel PCR assay for methylation status of CpG islands. *Proceedings of the National Academy of Sciences USA*, **93**, 9821–9826.

Kantarjian, H. M., O'Brien, S., Cortes, J., *et al.* (2003). Results of decitabine (5-aza-2′deoxycytidine) therapy in 130 patients with chronic myelogenous leukemia. *Cancer*, **98**, 522–528.

Koch, C. M., Andrews, R. M., Flicek, P., *et al.* (2007). The landscape of histone modifications across 1% of the human genome in five human cell lines. *Genome Research*, **17**, 691–707.

Kwon, Y. S., Garcia-Bassets, I., Hutt, K. R., *et al.* (2007). Sensitive ChIP-DSL technology reveals an extensive estrogen receptor alpha-binding program on human gene promoters. *Proceedings of the National Academy of Sciences USA*, **104**: 4852–4857.

Li, H., Lovci, M. T., Kwon, Y. S., *et al.* (2008). Determination of tag density required for digital transcriptome analysis: application to an androgen-sensitive prostate cancer model. *Proceedings of the National Academy of Sciences USA*, **105**, 20 179–20 184.

Nunez, E., Kwon, Y. S., Hutt, K. R., *et al.* (2008). Nuclear receptor-enhanced transcription requires motor- and LSD1-dependent gene networking in interchromatin granules. *Cell*, **132**, 996–1010.

Park, P. J. (2009). ChIP-Seq: advantages and challenges of a maturing technology. *Nature Review Genetics*, **10**, 669–680.

Ren, B., Robert, F., Wyrick, J. J., *et al.* (2000). Genome-wide location and function of DNA binding proteins. *Science*, **290**, 2306–2309.

Stresemann, C. and Lyko, F. (2008). Modes of action of the DNA methyltransferase inhibitors azacytidine and decitabine. *International Journal of Cancer*, **123**, 8–13.

Thillainadesan, G., Isovic, M., Loney, E., *et al.* (2008). Genome analysis identifies the p15ink4b tumor suppressor as a direct target of the ZNF217/CoREST complex. *Molecular and Cellular Biology*, **19**, 6066–6077.

Weber, M., Davies, J. J., Wittig, D., *et al.* (2005). Chromosome-wide and promoter-specific analyses identify sites of differential DNA methylation in normal and transformed human cells. *Nature Genetics*, **37**, 853–862.

Ying, J., Falk, J. D., Jin, M., Liu, D., and Maciejewski, J. P. (2007). Global methylome of normal and malignant hematopoietic stem cells. *Blood*, **110**, 2118.

16 Quantitative, high-resolution CpG methylation assays on the pyrosequencing platform

Dirk Löffert, Ralf Peist, Thea Rütjes, Norbert Hochstein, Dorothee Honsel, Frank Narz, Ioanna Andreou, Richard Kroon, Andreas Missel, Andrea Linnemann-Florl, Lennart Suckau, and Gerald Schock*

16.1   Introduction

CpG methylation is a heritable epigenetic mechanism in eukaryotes that affects gene regulation during the development of an organism in many ways: through transcriptional regulation, CpG methylation is involved in differentiation and proliferation of cells, imprinting of genomic regions and X-chromosome inactivation, silencing of repetitive elements, and in the maintenance of genomic stability (Reik and Lewis, 2005; Kacem and Feil, 2009; Straussman et al., 2009). However, it must be noted that considerable differences in methylation patterns exist among eukaryotes. Plant genomes can reach a methylation level as high as 50%, and methylation may be targeted to transposable elements (non-CpG methylation); they may display a mosaic DNA methylation type and may also show high levels of CpG methylation in transcribed regions of genes (Suzuki and Bird, 2008). Although methylation in mammalian genomes has been recently discovered to additionally target gene bodies, DNA methylation in mammals apparently plays an important role for gene regulation targeted to CpG islands, genomic regions comprising a high frequency of CpG sites and stretching from a few hundred to up to 2000–3000 bp in size, often located within or close to the promoter region of a gene (Ehrlich et al., 1982; Gardiner-Garden and Frommer, 1987; Saxonov et al., 2006). Aberrant CpG methylation correlates in mammals with many diseases like cancer, e.g., by inactivation of tumor suppressor genes or activation of proto-oncogenes, or in developmental disorders such as Prader–Willi syndrome caused by microdeletions in an imprinting center located on the paternal allele of human chromosome 15q11–q13 and the accompanied silencing of paternally expressed genes by DNA methlyation (Glenn et al., 1993; Reis et al., 1994; Sutcliffe et al., 1994; Buiting et al., 1995; Falls et al., 1999; Feinberg and

---

* Author to whom correspondence should be addressed.

Epigenomics: From Chromatin Biology to Therapeutics, ed. K. Appasani. Published by Cambridge University Press. © Cambridge University Press 2012.

Tycko, 2004; Esteller, 2005; Fraga and Esteller, 2007). More recently, DNA methylation has also been shown to affect regulation of miRNA expression, another key molecule in gene regulation (Lujambio *et al.*, 2008). Determining methylation of specific genomic sequences may also be useful as a diagnostic marker: One example is the MGMT promoter methylation that may serve as a prognostic as well as predictive marker for treatment of glioblastoma patients with alkylating agents for chemotherapy (Weller *et al.*, 2010).

Thus, the apparent importance of CpG methylation in the complex process of gene regulation and as a potential diagnostic marker in routine testing demands methods that can accurately quantify DNA methylation levels as well as provide high-resolution sequence information on the level of individual CpG positions. Sensitive methods are required because biological samples often represent a rather heterogeneous mixture of cells with varying degrees of CpG methylation at specific CpG islands. As CpG methylation is a dynamic process, methylation of CpG sites across a CpG island may also vary significantly. This makes sequencing of several CpG sites the preferred method over others such as real-time PCR or methylation-sensitive PCR that only cover a few CpG dinucleotides in a particular assay.

Sanger sequencing has been a valuable technology to elucidate DNA methylation patterns. However, since direct Sanger sequencing poorly discriminates subtle differences in DNA methylation levels and has significant limitations to accurately measure lower CpG methylation levels, cloning of PCR-amplified CpG stretches followed by sequencing of a statistically sufficient number of individual clones has been necessary, and this renders this method cumbersome and expensive (Reed *et al.*, 2010).

With the invention of pyrosequencing by Nyren (Ronaghi *et al.*, 1998), a technology has become available that overcomes many of the limitations inherent to Sanger sequencing providing a cost-effective and sensitive real-time sequencing method that can accurately quantify minor differences of modified CpG dinucleotides even at levels below 5–10%. Not surprisingly, pyrosequencing has been quickly adopted for DNA methylation analysis and has become the gold standard technology for the analysis of this type of epigenetic modification. Due to its unprecedented capability to quantify subtle differences in the DNA code, pyrosequencing is also frequently employed for numerous other molecular biology applications that aim to analyze DNA variations such as detecting and quantifying point mutations or insertions and deletions, allele quantification, single nucleotide polymorphism (SNP) typing, gene copy number variation, microbial identification, and drug-resistance typing, to name only a few (Marsh, 2007).

16.2   Methodology

16.2.1   Pyrosequencing: an introduction to the sequencing technology

Pyrosequencing is a simple and robust technology that makes use of unlabeled nucleotides, thereby providing a cost-efficient sequencing technique. It builds on the so-called "sequencing-by-synthesis" method, following the activity of a DNA polymerase in real time while it extends a primer bound to the sequencing template DNA. This is achieved by measuring the release of pyrophosphate once a complementary matching base is incorporated into the growing DNA strand of the primer extension product. In contrast to a Sanger sequencing reaction which contains all four fluorescent-labeled dideoxyribonucleotides, pyrosequencing makes use of an ordered dispensation of the four unlabeled deoxyribonucleotides on the PyroMark instrument. The assay starts with the amplification of a DNA segment in a regular PCR reaction in which one of the PCR primers is biotinylated. The biotinylated DNA strand that serves as the pyrosequencing template is quickly separated from the other DNA strand by alkaline denaturation to generate a single-stranded DNA sequencing template. During this fast single-strand DNA separation process, the biotinylated DNA strand is bound to streptavidin-coated sepharose beads which are carried through a denaturation and washing step. This procedure is altogether finished in about 10 minutes for up to 96 samples. In the next step, the sequencing primer is hybridized to the sequencing template and subsequently incubated with the enzymes DNA polymerase, ATP sulfurylase, luciferase, and apyrase as well as the substrates adenosine 5′-phosphosulfate (APS) and luciferin. Upon the dispensation of a first deoxyribonucleotide triphosphate (dNTP), the DNA polymerase catalyzes the addition of the nucleotide to the sequencing primer, if it is complementary to the base in the template strand. Each incorporation event is accompanied by the release of pyrophosphate (PPi) in a quantity equimolar to the amount of incorporated nucleotide. In the next step of the enzyme cascade, ATP sulfurylase converts PPi to ATP in the presence of APS. This ATP drives the luciferase-mediated conversion of luciferin to oxyluciferin which generates visible light in amounts that are proportional to the amount of ATP. The light produced in the luciferase-catalyzed reaction is detected by CCD sensors on the PyroMark platform and seen as a peak in the raw data output (Pyrogram®). The height of each such peak (light signal) is again proportional to the number of nucleotides incorporated. During the last step of a single dNTP dispensation cycle, apyrase continuously degrades unincorporated nucleotides and ATP. When degradation is complete, another nucleotide is added to the reaction. Addition of dNTPs is performed sequentially taking about 1 minute per bp to complete a single dispensation cycle. As the process continues, the complementary DNA strand is elongated and the nucleotide sequence is determined from the signal peaks in the Pyrogram trace. All incubation steps and data acquisition are done with the PyroMark instruments, providing a throughput of either 24 or 96 simultaneous sequencing reactions. The instrument provides all means for conducting the sequencing reactions. The sequencing primers are mixed with the purified sequencing template in a special microtiter plate. Enzymes, substrates, and dNTPs are loaded into a cartridge that is placed into the instrument and the dispensation order of nucleotides is easily programmed according to the sequence to be tested. Quantification of genetic variation at a given site (e.g., DNA methylation) is achieved by the addition of all nucleotides that can occur in the variation. For *de novo* sequencing, a cyclic dispensation order of all four nucleotides is used. In general, each individual well of the microtiter plate can hold a different sequencing reaction, giving utmost flexibility and throughput. When the sequencing reaction is finished, an application-specific software module automatically calls the sequence information and provides also a quality assessment of the analyzed sequence, indicating quantitative measurements such as the CpG methylation levels at various positions in the template sequence.

16.2.2   DNA isolation

A broad variety of biological samples are analyzed for DNA methylation levels, including samples as different as cell culture cells, white blood cells, tumor tissue from different sources such as breast, prostate, or brain, and processed biological samples like formalin-fixed paraffin-embedded (FFPE) tissue sections. To provide highest conversion efficiency in the subsequent bisulfite treatment reaction, highly purified DNA must be used since incomplete denaturation of DNA, potentially due to protein impurities or contaminating salts, will result in lowered conversion efficiency and consequently inaccurate measurements of methylation levels. Today, a broad range of DNA preparation technologies and commercial kits, such as the QIA amp DNA Mini Kit or the QIAamp DNA FFPE Tissue Kit (QIAGEN, Hilden, Germany), are available and provide the appropriate method for high-quality DNA isolation. Those technologies are usually available for manual extraction and often also for automated isolation on instruments providing a standardized mean for DNA preparation such as the QIAcube or QIAsymphony instruments (both QIAGEN).

16.2.3   Bisulfite conversion of DNA

To characterize the methylation status of a DNA sequence via pyrosequencing or other methods, the DNA is first incubated with a high concentration of sodium bisulfite. As a result, unmethylated cytosine residues are converted into uracils while methylated cytosines are protected from the conversion process and remain unchanged. Thus, the bisulfite conversion reaction eventually gives rise to two DNA strands that are at least partially no longer complementary and whose sequence differences can be distinguished by sequence analysis. The most critical step for correct determination of a methylation pattern is the complete conversion of unmethylated cytosines. This is achieved by incubating the DNA with bisulfite in high concentrations, at high temperature and low pH. These harsh conditions usually lead to a high degree of DNA fragmentation and subsequent loss of DNA during purification. Purification is necessary to remove bisulfite salts and chemicals used in the conversion process that can inhibit PCR amplification prior to sequencing procedures. Common bisulfite procedures usually require high amounts of input DNA. However, due to DNA degradation during conversion and DNA loss during purification, such procedures often lead to low DNA yield, highly fragmented DNA, and irreproducible conversion rates. A novel technical solution, the EpiTect Bisulfite Kit (QIAGEN) technology, results in lower DNA degradation through the use of a novel radical scavenger contained in the mixture which minimizes DNA strand breaks. Yield of purified DNA is increased with a special low-elution volume compatible silica membrane spin column technology, which can also be automated on QIAGEN's QIAcube platform. Dedicated solutions are also available to directly process FFPE material without an intermediate DNA purification procedure (EpiTect Plus Bisulfite FFPE Bisulfite Kit, QIAGEN). The overall bisulfite conversion efficiency has been accurately measured by sequencing non-CpG nucleotides, which should all be unmethylated and thus be converted into uracil. Conversion efficiencies were determined to be greater than 99% (Meissner *et al.*, 2008).

16.2.4 CpG methylation assay design for pyrosequencing

Assay development for various applications is supported by the PyroMark Assay Design Software 2.0, which also contains a module for CpG methylation assay design. The software predicts the bisulfite converted sequence and highlights CpG sites and non-CpG cytosines that can serve as controls. The program automatically generates primer sets that include both PCR and sequencing primers. Each primer set is given a quality score based on several parameters that are specific for pyrosequencing analysis. In general, one of the PCR primers must be biotin-labeled at its 5' end to enable immobilization of the PCR product to streptavidin-coated beads during the preparation of the single-stranded pyrosequencing template. High-performance liquid chromatography (HPLC) purification or an equivalent procedure to purify the biotinylated primer is recommended. The orientation of the assay can either be forward or reverse. PCR primers should be 18 to 24 bases in length, with annealing temperatures that are similar and typically in the range of 50 °C to 68 °C (nearest-neighbor method). The optimal PCR amplicon length for pyrosequencing is between 80 and 200 bp, although products up to 500 bp might work well for pyrosequencing assays on genomic DNA. When not using the PyroMark Assay Design software, primers need to be carefully checked to ensure that they do not form strong hairpin loops or dimers with themselves or with the other primers. The biotinylated primer should not form any hairpin loops and duplexes with the sequencing primer, as excess biotinylated primer might cause background in pyrosequencing assays. If possible, placing primers over polymorphic positions should be avoided. The sequencing primer should be about 20 bases long and have an annealing temperature in the range of 50 °C to 55 °C. Ideally, the sequencing primer should differ from the PCR primer by at least one additional, specific base at the 3' end. In contrast to Sanger sequencing, pyrosequencing can determine the immediate next base following the 3' end of the sequencing primer.

Specific considerations should be taken for CpG methylation assays due to the reduced DNA code complexity in bisulfite-treated DNA. In CpG assays, the optimal length of the primers is slightly increased to function on bisulfite-treated DNA, which has a high proportion of A and T. PCR primers should usually be 22–30 bases and sequencing primers should be 18–23 bases in length. When using PyroMark Assay Design Software 2.0 for primer design, an annealing temperature of 56 °C for bisulfite-converted DNA gives good results in most cases. The amplicons for CpG assays should ideally be shorter than 200 bp. When working with DNA isolated from FFPE tissue, even shorter PCR products may be beneficial, since DNA from FFPE tissue is usually already highly degraded prior to the bisulfite treatment. PCR amplicon sizes should usually not exceed 150 bp in this case. For CpG methylation analysis, it is extremely important to place primers not across CpG sites to avoid amplification bias towards a particular DNA strand, thus compromising the quantification of the CpG methylation level. If the region of interest is densely packed with CpG sites, a PCR primer may be also placed over a CpG site; however, no more than one CpG site should be involved and this CpG site must be located in the 5′ part of the primer.

16.2.5   PCR amplification of bisulfite-treated DNA

Bisulfite treatment of genomic DNA converts non-methylated cytosines to uracils leading to DNA mainly consisting of three bases. This less-complex DNA is a challenging PCR template and the PCR products obtained are likely to have low yields. Non-specific products might be generated due to increased mispriming probability caused by the reduced complexity of the template DNA. Hot-start PCR methods can overcome some of these challenges. However, careful design of PCR and sequencing primers is a prerequisite as excess of biotinylated primers and misprimed artifacts can interfere with the pyrosequencing reaction. A PCR master mix specifically developed for pyrosequencing, the PyroMark PCR Kit (QIAGEN), prevents accumulation of an excess of biotinylated primer and minimizes generation of artifacts through a built-in PCR hot start and a balanced cation combination in the buffer that increases the stringency of the PCR reaction. The high yields of specific PCR products obtained ensure reliable pyrosequencing results. The PCR reaction contains typically between 10 and 20 ng of bisulfite-converted DNA and 0.2 μM of each PCR primer. PCR is performed on a standard PCR cycler with a typical PCR cycling profile as follows: 15 minutes reactivation of HotStarTaq DNA polymerase (PyroMark PCR Kit, QIAGEN) at 95 °C followed by 45 cycles of denaturing the template DNA at 94 °C for 30 seconds, PCR primer annealing at 56 °C for 30 seconds, and primer extension at 72 °C for 30 seconds. The CoralLoad Concentrate contained in the reagent kit to facilitate agarose-gel loading and contains gel tracking dyes should also be included in the PCR reaction even if the PCR product is not checked by agarose gel electrophoresis after the PCR. The additive can greatly improve PCR product yield and thereby reduce the amount of non-incorporated biotinylated primer in the course of the PCR reaction. Additionally, it serves as a visual control during the immobilization step as the dye easily indicates which wells have already received an aliquot of a PCR reaction. Adjustment of Mg^{2+} concentration is usually not required when using the PyroMark PCR Kit (QIAGEN). For optimal results, specificity and yield of the PCR product can be checked by agarose-gel analysis. In most cases, 5–10 μl of the PCR product gives satisfactory Pyrosequencing results when using the PyroMark MD instrument, 10–20 μl when using the PyroMark Q24 instrument, and 20 μl when using the PyroMark ID instrument. Conventional PCR carry-over decontamination by dUTP and use of UNG in the PCR reaction is not applicable to the analysis of bisulfite-converted DNA as unmethylated cytosines become converted to uracil in the template strand which would lead to destruction of the PCR template prior to the amplification reaction.

16.2.6   Preparation of single-stranded template DNA for pyrosequencing

PCR products are made single-stranded and subsequently purified from residual contaminants such as PCR reaction buffer, enzyme and PCR primers, and dNTPs on the PyroMark Vacuum Workstation (QIAGEN). Before starting the purification process, a master mix is prepared consisting of 2 μl of streptavidin-coated sepharose beads, 40 μl binding buffer, and 18–33 μl high-purity water per sample. Depending on the sample volume, 60–75 μl master mix is pipetted into each necessary well of a PCR plate to give a total volume of 80 μl per well. In the next step, 5–20 μl PCR product is added to each well. The wells are sealed and the PCR plate is agitated at 1400 rpm for 5–10 minutes at room temperature (15–25 °C) using an orbital shaker. Before starting the single-strand DNA preparation, sequencing primers are diluted to 0.3 μM with PyroMark Annealing Buffer, 25 μl is pipetted into each necessary well of a PyroMark Sequencing Plate (PyroMark Q24), and the plate is positioned on the workstation which is filled with 70% ethanol, denaturation solution, wash buffer, and high-purity water. A vacuum tool is used to capture the sepharose beads from the PCR plate and transfer the beads through the buffer troughs to obtain single-stranded PCR product. At the end of the quick procedure, the vacuum tool is lowered into the wells of the PyroMark plate and the beads are released by gentle shaking from side to side.

In the next step, the sequencing primers are annealed to the sequencing template. The PyroMark Sequencing Plate is placed onto a pre-warmed PyroMark Plate Holder and heated on a heating block at 80 °C for 2 minutes. The plate is removed from the holder and the samples are allowed to cool to room temperature (15–25 °C) for at least 10 minutes. The cooled plate is then directly processed on the PyroMark instrument.

16.2.7   Pyrosequencing reaction

PyroMark Gold Reagents (QIAGEN) are a set of reagents that are optimized for pyrosequencing technology. The reagents are designed to generate Pyrogram® traces with sharp and distinct peaks and low background. PyroMark Gold Reagents especially improve assays with longer sequencing read-lengths, such as with CpG methylation analysis, as well as providing optimal conditions for mutation and SNP analyses. A set of PyroMark Gold reagents contains all components that are needed in the pyrosequencing cascade: an enzyme mix, a substrate mix, and nucleotides. The enzyme mix includes the DNA polymerase for incorporation of nucleotides, ATP sulfurylase for conversion of pyrophosphate to ATP, luciferase for generation of the light signal, and apyrase to degrade ATP and unincorporated nucleotides, which switches off the light signal and regenerates the reaction solution. In addition, single-stranded binding protein (SSB) has been added to prevent secondary structures in the template and primers. The substrate mix consists of APS needed for generation of ATP, and luciferin to serve as a substrate for luciferase. This ATP drives the luciferase-mediated conversion of luciferin to oxyluciferin, which generates visible light in amounts that are proportional to the amount of ATP. Nucleotides included in the PyroMark Gold Reagents are dissolved in a balanced buffer to prevent degradation of the nucleotides. It should be noted that deoxyadenosine alpha-thio triphosphate (dATPαS) is used as a substitute for the natural deoxyadensine triphosphate (dATP) since it is efficiently used by the DNA polymerase but not recognized by luciferase. The reagents are loaded in the appropriate compartments of the PyroMark Cartridge. The sequencing run is programmed, the sequencing plate and PyroMark Cartridge are placed into the instrument, and the sequencing run is started.

16.3   Quantitative CpG methylation assays with built-in bisulfite conversion quality control

Because it allows unbiased quantitative analysis of DNA methylation, pyro-sequencing has become the gold standard technology to determine the global methylation content of a sample or for gene-specific CpG methylation analysis (Figures 16.1 and 16.2).

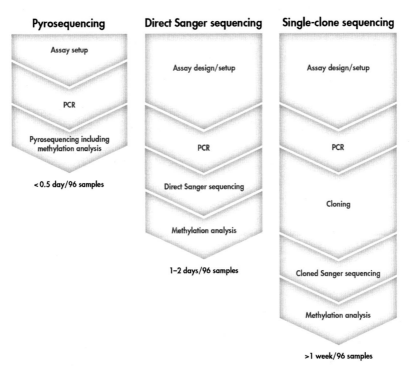

Figure 16.1 Comparison of pyrosequencing versus direct Sanger sequencing and single-clone sequencing. Typically, Sanger sequencing of bisulfite-converted DNA is used to analyze methylation of CpG dinucleotides. Direct Sanger sequencing of PCR products can provide sequence information but fails to provide reliable and sensitive quantitative methylation results. Furthermore, the analysis of the chromatograms obtained is time-consuming and error-prone. Therefore, cloning of DNA after bisulfite conversion and sequencing of single clones is often used in order to get more reliable methylation results. However, due to various steps necessary for cloning, this procedure is extremely time-consuming and expensive, since the sensitivity of the results depends on the number of clones sequenced. In contrast, pyrosequencing provides a fast and easy way to determine DNA methylation at levels below 5–10%. After PCR, up to 96 samples can be analyzed in parallel in as little as 1 hour including data analysis. Pyrosequencing assay design can be further streamlined by the use of predesigned PyroMark CpG Assays (QIAGEN), available for virtually any CpG island of the human genome.

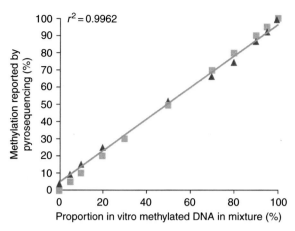

Figure 16.2 Linearity of methylation quantification by pyrosequencing. PCR products from varying mixtures of unmethylated genomic DNA and methylated DNA (EpiTect Control DNAs, QIAGEN) were analyzed by pyrosequencing. A tight correlation between the known percentage of methylated DNA in the mixtures (■) and the methylation percentage reported by pyrosequencing (▲) was observed ($r^2 = 0.9962$). The graph represents the quantification of methylation at a single CpG site in the *p16* gene.

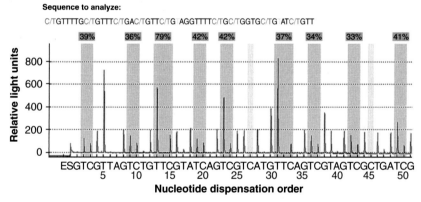

Figure 16.3 Analysis of multiple contiguous CpG sites. Methylation at nine independent CpG sites (highlighted in gray) is quantified in a single pyrosequencing run. Position-specific information in the context of an analyzed sequence presents broad sequence methylation patterns. Note the built-in quality control sites (highlighted in light gray) consisting of cytosines converted to thymines, demonstrating full bisulfite conversion of the treated DNA. See plate section for color version.

Unlike ambiguous Sanger sequencing, the peak heights in the resulting Pyrogram report the ratio of cytosine to thymine at each analyzed CpG site even at low methylation levels, which reflects the proportion of methylated DNA. Assay design is flexible – it can be performed in forward or reverse orientations, on either the top or the bottom strand. In addition, contiguous CpG sites are analyzed independently and within the same run, which enables assessment of sequence-wide methylation patterns while retaining details of position-specific methylation (Figure 16.3).

One important feature of CpG methylation analysis with pyrosequencing is the simultaneous analysis of the overall bisulfite conversion efficiency of the employed template DNA. As all cytosines in the mammalian genome that are not followed by a guanine base are supposed to be unmethylated, one expects that cytosines at those positions are completely converted to uracils, i.e., being displayed as a thymine base if the genomic DNA has been converted with high efficiency. Detectable methylation levels at these positions indicate incomplete bisulfite-mediated conversion and failure of reliable CpG analysis. Further analysis of DNA methylation patterns is supported with a software package that includes methylation frequency analysis, quality assessment, mean methylation values per well and replicates, statistical methylation patterns of multiple consecutive sites, and deviation from expected methylation patterns. To further aid DNA methylation analysis, genome-wide, predesigned pyrosequencing methylation assays can be used to assess the methylation status of any human CpG island (PyroMark CpG Assays, QIAGEN).

Global DNA methylation can be determined by analyzing the methylation level of the LINE-1 retrotransposable element. LINE-1 retrotransposable elements make up about 20% of the human genome (Furano et al., 2004). DNA methylation within the promoter region of human LINE-1 elements is important for maintaining transcriptional inactivation and for inhibiting transposition. Genome-wide losses of DNA methylation within the promoter region of human LINE-1 elements have been regarded as a common epigenetic event in malignancies and may play crucial roles in carcinogenesis (Ogino et al., 2008). Thus, a methylation assay that amplifies a region of the LINE-1 element can serve as a marker for global DNA methylation of the human genome that can be analyzed by pyrosequencing.

16.4   Concluding remarks

Studying epigenetic modifications and specifically DNA methylation has become an extensive field of research as this heritable though dynamic DNA modification relates to the organism's response to environmental conditions, plays an important role in the organism's development and aging, and can be an underlying factor or indicator of serious diseases such as cancer. As DNA methylation is a dynamic process and DNA methylation levels can significantly vary at individual CpG sites within a target sequence or differ for specific sites within a cell population, reliable techniques that allow unequivocal analysis and quantitation of DNA methylation are required. Sequencing is the method of choice to determine consecutive methylation status of CpG sites in a target sequence, providing more reliable methylation levels than techniques such as TaqMan probe-based, real-time PCR which determines the methylation level of only one or few CpG sites in a target sequence. Pyrosequencing has been proven to be a highly reliable and accurate technology that is also superior to Sanger sequencing as the method is fast, highly accurate, and much less labor-intensive and more cost-effective than bisulfite sequencing of individual cloned PCR products (Reed et al., 2010). Furthermore, built-in quality control capability and published performance data indicate that the technology can be a valuable diagnostic tool not only in methylation analysis but also for other applications such as mutation analysis in cancer research and diagnosis (Tsiatis *et al.*, 2010) or resistance testing in microbial detection (Jureen *et al.*, 2006).

REFERENCES

Buiting, K., Saitoh, S., Gross, S., *et al.* (1995). Inherited microdeletions in the Angelman and Prader–Willi syndromes define an imprinting centre on human chromosome 15. *Nature Genetics*, **9**, 395–400.

Ehrlich, M., Gama-Sosa, M. A., and Huang, L. H. (1982). Amount and distribution of 5-methylcytosine in human DNA from different types of tissues of cells. *Nucleic Acids Research*, **10**, 2709–2721.

Esteller, M. (2005). Aberrant DNA methylation as a cancer-inducing mechanism. *Annual Review of Pharmacology and Toxicology*, **45**, 629–656.

Falls, J. G., Pulford, D. J., Wylie, A. A., and Jirtle, R. L. (1999). Genomic imprinting: implications for human disease. *American Journal of Pathology*, **154**, 635–647.

Feinberg, A. P. and Tycko, B. (2004). The history of cancer epigenetics. *Nature Reviews Cancer*, **4**, 143–153.

Fraga, M. F. and Esteller, M. (2007). Epigenetics and aging: the targets and the marks. *Trends in Genetics*, **23**, 413–418.

Furano, A. V., Duvernell D. D., and Boissinot S. (2004). L1 (LINE-1) retrotransposon diversity differs dramatically between mammals and fish. *Trends in Genetics*, **20**, 9–14.

Gardiner-Garden, M. and Frommer, M. (1987). CpG islands in vertebrate genomes. *Journal of Molecular Biology*, **196**, 261–282.

Glenn C. C., Nicholls R. D., Robinson W. P., *et al.* (1993). Modification of 15q11–q13 DNA methylation imprints in unique Angelman and Prader–Willi patients. *Human Molecular Genetics*, **2**, 1377–1382.

Jureen, P., Engstrand, L., Eriksson, S., *et al.* (2006). Rapid detection of rifampin resistance in *Mycobacterium tuberculosis* by Pyrosequencing technology. *Journal of Clinical Microbiology*, **44**, 1925–1929.

Kacem, S. and Feil, R. (2009). Chromatin mechanisms in genomic imprinting. *Mammalian Genome*, **20**, 544–556.

Lujambio, A., Calinc, G. A., Villanuevad, A., *et al.* (2008). A microRNA DNA methylation signature for human cancer metastasis. *Proceedings of the National Academy of Sciences USA*, **105**, 13 556–13 561.

Marsh, S. (ed.) (2007). *Pyrosequencing Protocols*. Totowa, NJ: Humana Press.

Meissner, A., Mikkelsen, T. S., Gu, H., *et al.* (2008). Genome-scale DNA methylation maps of pluripotent and differentiated cells. *Nature*, **454**, 766–770.

Ogino, S., Kawasaki, T., Nosho, K., Ohnishi, M., and Suemoto, Y. (2008). LINE-1 hypomethylation is inversely associated with microsatellite instability and CpG island methylator phenotype in colorectal cancer. *International Journal of Cancer*, **122**, 2767–2773.

Reed, K., Poulin, M. L., Yan, L., and Parissenti, A. M. (2010). Comparison of bisulfite sequencing PCR with pyrosequencing for measuring differences in DNA methylation. *Analytical Biochemistry*, **397**, 96–106.

Reik, W. and Lewis, A. (2005). Co-evolution of X-chromosome inactivation and imprinting in mammals. *Nature Reviews Genetics*, **6**, 403–410.

Reis, A., Dittrich, B., Greger, V., *et al.* (1994). Imprinting mutations suggested by abnormal DNA methylation patterns in familial Angelman and Prader–Willi syndromes. *American Journal of Human Genetics*, **54**, 741–747.

Ronaghi, M., Uhlen, M., and Nyren, P. (1998). A sequencing method based on real-time pyrophosphate detection. *Science*, **281**, 363–365.

Saxonov, S., Berg, P., and Brutlag, D. L. (2006). A genome-wide analysis of CpG dinucleotides in the human genome distinguishes two distinct classes of promoters. *Proceedings of the National Academy of Sciences USA*, **103**, 1412–1417.

Straussman, R., Nejman, D., Roberts, D., *et al.* (2009). Developmental programming of CpG island methylation profiles in the human genome. *Nature Structural and Molecular Biology*, **16**, 564–571.

Sutcliffe, J. S., Nakao, M., Christian, S., *et al.* (1994). Deletions of a differentially methylated CpG island at the *SNRPN* gene define a putative imprinting control region. *Nature Genetics*, **8**, 52–58

Suzuki, M. M. and Bird, A. (2008). DNA methylation landscapes: provocative insights from epigenomics. *Nature Reviews Genetics*, **9**, 465–476.

Tsiatis, A. C., Norris-Kirby, A., Rich, R. G., *et al.* (2010). Comparison of Sanger sequencing, Pyrosequencing, and melting curve analysis for the detection of *KRAS* mutations. *Journal of Molecular Diagnostics*, **12**, 425–432.

Weller, M., Stupp, R., Reifenberger, G., *et al.* (2010). *MGMT* promoter methylation in malignant gliomas: ready for personalized medicine? *Nature Reviews Neurology*, **6**, 39–51.

17 DNA methylation profiling using Illumina BeadArray platform

Marina Bibikova,* Jian-Bing Fan, and Kevin L. Gunderson

17.1 Introduction

DNA methylation in various regions of the genome is a characteristic of gene activity, tissue type, disease state, and underlying epigenetic regulation of the genome. Knowledge of DNA methylation patterns can greatly advance our ability to elucidate and diagnose the molecular basis of human diseases. A joint effort by an international collaboration, The Human Epigenome Project (www.epigenome. org/), set out to identify, catalog, and interpret genome-wide DNA methylation patterns of all human genes in all major tissues (Jones *et al.*, 2008).

Interest in monitoring methylation states of cytosines in CpG dinucleotides has led to the development of various techniques for DNA methylation profiling. These include methylation-specific restriction enzyme digestion (Singer-Sam *et al.*, 1990), bisulfite DNA sequencing (Frommer *et al.*, 1992; Clark *et al.*, 1994), methylation-specific PCR (MSP) (Herman *et al.*, 1996), MethyLight (Eads *et al.*, 2000; Cottrell *et al.*, 2004), restriction landmark genomic scanning (RLGS) (Kawai *et al.*, 1993; Okazaki *et al.*, 1995; Akama *et al.*, 1997), pyrosequencing (Colella *et al.*, 2003; Dupont *et al.*, 2004) and MALDI mass spectrometry (Tost *et al.*, 2003; Ehrich *et al.*, 2005, 2008), to name a few. In the recent years, many microarray and next-generation sequencing based technologies have emerged, which attempted to provide high-throughput access to specific sequences in the genome and enable analysis of methylation profiles at high resolution in large sample sets. The microarray-based approaches are largely based on three major techniques: restriction enzyme digest, sodium bisulfite conversion of genomic DNA, and affinity-based assays. The well-known methods based on restriction enzyme digest include differential methylation hybridization (DMH) with various modifications (Huang *et al.*, 1999; Gitan *et al.*, 2002; Schumacher *et al.*, 2006), methylated CpG

---

* Author to whom correspondence should be addressed.

Epigenomics: From Chromatin Biology to Therapeutics, ed. K. Appasani. Published by Cambridge University Press. © Cambridge University Press 2012.

island amplification (MCA) (Kuang *et al.*, 2008; Omura *et al.*, 2008), and methylation-specific fractionation after McrBC digestion coupled with tiling arrays (Ordway *et al.*, 2006, 2007). To overcome the limitation of enzyme-recognition sites, affinity-based technologies have been developed to enrich the methylated fraction of the genome; these include methylated DNA immunoprecipitation (MeDIP) (Weber *et al.*, 2005) or affinity chromatography over a methyl-binding domain (MBD) (Yamada *et al.*, 2004; Ibrahim *et al.*, 2006). The affinity-based enrichment methods can also be combined with next-generation sequencing readout (Down *et al.*, 2008; Ruike *et al.*, 2010). However, none of these methods combine custom access to specific sequences in the genome with high throughput and low cost, which is needed for analyzing genome-wide methylation profiles across large sample sets. Several recent reviews compared these approaches, and discussed the strengths and weaknesses associated with microarray and next-generation sequencing for DNA methylation profiling (Beck, 2010; Huang *et al.*, 2010; Laird, 2010).

Bisulfite sequencing of genomic DNA (gDNA) is considered to be a gold standard for analyzing methylation state of CpG sites within the genome (Eckhardt *et al.*, 2006). Bisulfite treatment of gDNA converts all cytosine bases to uracils except those protected by methylation (Clark *et al.*, 1994). Methylation in the human genome is generally limited to 5-methyl cytosine in the context of CpG sites, although recent studies uncovered non-CpG methylation, primarily in pluripotent cells (Lister *et al.*, 2009; Laurent *et al.*, 2010). While whole-genome bisulfite sequencing (WGBS) can access every CpG site in the genome, the method is still quite expensive and is limited to a low number of samples. A variation of this method, called reduced-representation bisulfite sequencing (RRBS), uses restriction enzyme digestion to fractionate the genome and targets regions of high CpG density (Meissner *et al.*, 2005).

Methylation analysis on DNA arrays can be accomplished by measuring the DNA methylation level, i.e., the fraction of methylated vs. unmethylated cytosines, using quantitative genotyping of the "pseudo-SNP" created by bisulfite conversion of gDNA. We have modified well-established Illumina genotyping assays to analyze the methylation state of CpG sites in bisulfite converted genomic DNA. We utilized the GoldenGate® genotyping assay and universal arrays to assess the methylation level of over 1500 CpG sites in cancer-related genes (Bibikova *et al.*, 2006b; Bibikova and Fan, 2009). Similarly, we used the Infinium® whole-genome genotyping assay in combination with bisulfite DNA conversion to enable quantitative methylation analysis on a genome-wide scale (Bibikova *et al.*, 2009, 2011).

In this chapter, we discuss our approaches to genome-wide DNA methylation array design and some techniques to overcome array design limitations.

17.2   Methodology

While each method described above can provide information about methylation states across the genome, they all have limitations. Methods that use methylation-sensitive restriction enzyme digest have to rely on the presence of recognition sites within the region of interest. Affinity-based methods cannot determine DNA methylation level at a unique CpG site, and have bias towards CpG-dense regions. Most next-generation sequencing platforms require large amounts of sample input, intensive labor, and complex bioinformatic analyses; in addition, most of these technologies can only analyze a few samples at a time. The above limitations make them difficult to use in large-scale studies where sample materials may be limited.

Illumina has developed two methylation profiling platforms that provide high sample throughput with single-CpG resolution (Bibikova *et al.*, 2006b, 2009). These assays are based upon the adaptation of the Infinium and GoldenGate SNP genotyping platforms, using bisulfite-converted DNA.

17.2.1 Illumina BeadArray™ technology

The concept behind Illumina's BeadArrays is straightforward in principle: arrays made of oligos linked to silica beads randomly assemble into wells (Fan *et al.*, 2006). Among the advantages of this bead-based approach are that oligo probes are synthesized in solution, using standard phosphoamidite chemistry. Only full-length oligos are attached to beads and only beads that can hybridize to a target can be decoded (determining which bead is in which well). The necessity of decoding the randomly distributed beads serves as an automatic quality control for the array. The decoding process involves iterative hybridizations with fluorescent oligo pools, and has been described in detail by Gunderson *et al.* (2004). The assignment of fluorescent labels in the oligo pool for each hybridization step allows the unique identification of each bead by its particular sequence of fluors across a series of hybridization steps. Depending on the array density, there are about 15–30 beads of each type (having the same oligo sequence) present in each array, offering redundancy in analytical measurements.

Illumina currently manufactures two types of bead arrays: a universal iCode array consisting of short address sequence oligos, and locus-specific arrays consisting of long 50-mer locus-specific sequences concatenated to a decoding sequence. The GoldenGate assay employs iCode arrays and the iCode address sequence serves as both decoding and hybridization capture probe. In iCode arrays, loci of interest can be revised by changing the chimeric assay oligos, rather than changing array content, which provides assay flexibility and enables easy custom design (Shen *et al.*, 2005; Fan *et al.*, 2006). In contrast, beads with long target-specific probes are primarily used to build arrays with a high density of features of fixed content, such as Infinium genome-wide methylation arrays (Steemers and Gunderson, 2007). The high density of the Infinium BeadChips is achieved with a Micro-Electro-Mechanical Systems (MEMS)-patterned slide substrate which consists of multiple sample sections into which beads are assembled from a pool containing hundreds of thousands of different bead types. Each section receives a same bead pool, thus allowing the analysis of many different samples in parallel. Beads provide a better substrate than slides for bulk surface modifications and immobilization of assay oligonucleotides. These bulk processes, in which beads of a particular type are created in one immobilization event, greatly improve array-to-array feature consistency. This is particularly important for Infinium Methylation assay with Infinium I probes that employ a ratiometric comparison between two bead types (see below), with this ratio relatively invariant from one array to another. Bead arrays are also scalable in density. A reduction in bead size and spacing can greatly increase the number of assays per bead chip. The choice of slide substrate has also contributed to the success of the array-based enzymatic Infinium assay.

17.2.2   DNA methylation assays

There are several challenges to implementing analysis of bisulfite-converted DNA on a microarray genotyping platform. First of all, most of the cytosines in the genome are converted to uracils, and the uniqueness of any given sequence within the bisulfite-converted genome decreases dramatically. A second issue arises when designing probes to the methylated and unmethylated "alleles" of a particular CpG site in which flanking CpG sites will affect the hybridization and extension of the probes. Furthermore, after bisulfite conversion opposite strands are no longer complementary and the whole-genome amplification (WGA) step required for Infinium assay effectively creates four different strands reducing the effective concentration and uniqueness of any given locus.

17.2.2.1   GoldenGate assay

We adapted a high-throughput single nucleotide polymorphism (SNP) GoldenGate genotyping system (Fan *et al.*, 2003) to DNA methylation detection, based on "genotyping" of bisulfite-converted genomic DNA. Specific CpG dinucleotides with oligos linked to unique address sequences are designed to hybridize to their complementary strand on the universal arrays (Figure 17.1). The assay requires a relatively short target sequence of about 50 nucleotides for query oligonucleotide annealing and gap filling. The assay uses one set of universal PCR primers to amplify all of the targets and generates amplicons of ~120 bp. This uniformity results in a relatively unbiased amplification of the "methylated" and "unmethylated" PCR template populations. The assay multiplexes from 96 to over 1500 CpG sites.

The assay procedure is similar to that described previously for standard SNP genotyping (Fan *et al.*, 2003), except that four oligos, two allele-specific oligos (ASOs) and two locus-specific oligos (LSOs), are required for each assay site rather than three (Figure 17.1). Briefly, bisulfite-treated, biotinylated gDNA is immobilized on paramagnetic beads. Pooled query oligos are annealed to the immobilized gDNA under stringent hybridization conditions, and then washed to remove excess or mishybridized oligos. Adjacent hybridized oligos are then extended and ligated to generate amplifiable templates. The requirement that two fragments need to join to create a PCR template provides an additional level of locus specificity. It is unlikely that any incorrectly hybridized ASOs and LSOs will be adjacent, and therefore should not be able to ligate after ASO extension. The PCR reaction is performed with fluorescently labeled universal PCR primers.

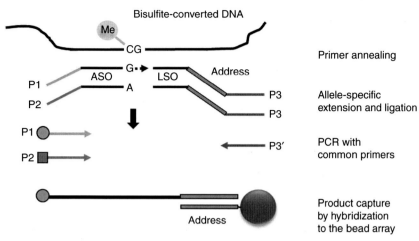

Figure 17.1 GoldenGate assay scheme. For each CpG site, two pairs of probes are designed: an allele-specific oligo (ASO) and locus-specific oligo (LSO) probe pair for the methylated state of the CpG site and a corresponding ASO–LSO pair for the unmethylated state. Each ASO consists of a 3' part that hybridizes to the bisulfite-converted genomic DNA, with the 3' base complementary to either the "C" or "T" allele of the targeted CpG site, and a 5' part that incorporates a universal PCR primer sequence P1 or P2. The LSOs consist of three parts: a CpG locus-specific sequence is at the 5' end, an address sequence is in the middle, and a universal PCR priming site (P3) is at the 3' end. The assay oligos are annealed to bisulfite-converted genomic DNA, and then an allele-specific primer extension step is carried out, followed by ligation of the extended ASOs to their corresponding LSOs, to create PCR templates. The ligated products are amplified by PCR using common primers P1, P2, and P3', and hybridized to a microarray bearing the complementary address sequences. P1 and P2 were fluorescently labeled, each with a different dye, and associated with the "T" (unmethylated) allele or the "C" (methylated) allele, respectively.

Methylation status of an interrogated CpG site is determined by calculating the "beta-value," which is defined as the ratio of the fluorescent signal from the methylated allele to the sum of the fluorescent signals of both methylated and unmethylated alleles. The beta-value provides a continuous measure of levels of DNA methylation in samples, ranging from 0 in the case of completely unmethylated sites, to 1 in completely methylated sites.

High reproducibility of beta-values is observed in the methylation assay ($R^2 > 0.98$), and differences in methylation status are readily detectable and effective for classifying experimental samples in different biological states.

For cancer research, we selected over 370 genes known to be differentially methylated in various cancer types and/or normal tissues during development. Over 1500 CpG sites from the 370 genes were measured simultaneously, which was a significant step forward compared to single-gene analysis or approaches. The GoldenGate platform also enabled high-throughput analysis of multiple samples in parallel. For more information, the details of the GoldenGate DNA methylation profiling protocol can be found in Bibikova *et al.* (2006b).

17.2.2.2 Infinium assay
The Infinium assay greatly increased the number of sites that can be analyzed in parallel. The assay complexity is limited only by the number of beads which are

assembled on the slide section. This assay uses beads with long target-specific probes designed corresponding to individual CpG sites. Similar to the idea of using GoldenGate genotyping assay for DNA methylation analysis, we adapted the Illumina Infinium whole-genome genotyping (WGG) assay for measuring CpG methylation using quantitative "genotyping" of bisulfite-converted genomic DNA. The (C/T) polymorphism in the bisulfite-converted DNA resulting from different sensitivity of methylated and unmethylated cytosines to the sodium bisulfite treatment can be queried either using standard Infinium I methylation-specific assay design consisting of two probes per CpG locus: one "unmethylated" and one "methylated" query probe (Figure 17.2A), or Infinium II design with one probe per locus, where the underlying CpG sites are represented by a "degenerate" R-base, allowing multiple combinations of oligos attached to the bead (Figure 17.2B).

The Infinium assay comprises the following major steps: first, bisulfite-converted DNA is whole-genome amplified. This amplified DNA is enzymatically

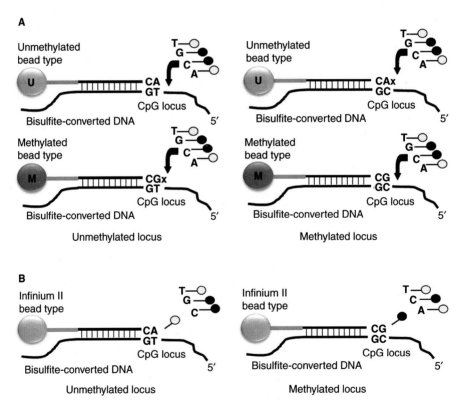

Figure 17.2 Infinium assay scheme. (A) Infinium I assay. Two bead types correspond to each CpG locus: one bead type to methylated (C), another bead type to unmethylated (T) state of the CpG site. Probe design assumes same methylation status for adjacent CpG sites. Both bead types for the same CpG locus will incorporate the same type of labeled nucleotide, determined by the base preceding the interrogated "C" in the CpG locus, and therefore will be detected in the same color channel. (B) Infinium II assay. One bead type corresponds to each CpG locus. Probe can contain up to three underlying CpG sites, with degenerate R-base corresponding to C in the CpG position. Methylation state is detected by single-base extension. Each locus will be detected in two colors.

fragmented and hybridized to a CpG-specific bead array. After hybridization, the array is processed through a primer extension and an immunohistochemistry staining protocol to allow detection of a single-base extension reaction (Gunderson *et al.*, 2006; Steemers and Gunderson, 2007; Gunderson, 2009). For the Infinium I assays, allele-specific single-base extension is used to determine methylation state of the query locus, requiring two beads for each CpG site – one corresponding to the unmethylated, and the other to the methylated state, with underlying CpG sites "in phase" with the queried site. The perfectly matched bead type preferentially extends over the mismatched bead type. For the Infinium II assay, one bead type is used, and the metylation state is determined by single-base extension (SBE) using hapten-labeled terminators.

17.2.3   Development of methylation reference samples

We developed a set of reference standards to assess the quality of Infinium methylation probes (Bibikova *et al.*, 2009). Unmethylated (U), hemi-methylated (H), and methylated (M) genomic reference samples were created from genomic DNAs received from Coriell DNA repository. Reference samples were created starting with 50 ng of gDNA that was amplified with the REPLI-g Mini Kit (QIAGEN, Hilden, Germany) following manufacturer's recommendations. Limiting the WGA reaction to a 100-fold amplification minimized representation bias. The WGA-amplified DNA was subjected to mung bean nuclease treatment to remove single-stranded DNA. The resultant unmethylated DNA was treated with SssI methylase, which globally methylates all double-stranded CpG sites, to create a nearly completely methylated reference standard. The hemi-methylated reference was created by mixing U and M samples in a 1 : 1 stoichiometric ratio. The standards were used for the development of both HumanMethylation27 and HumanMethylation450 arrays. The methylation reference samples were also used to calibrate the GoldenGate methylation assay.

17.2.4   Challenges and limitations with probe design and content selection

Illumina DNA methylation arrays are based on "genotyping" of bisulfite-converted genomic DNA. Treating a DNA sample with bisulfite results in a change in sequence from C to U for unmethylated cystosines, while methylated cytosines remain unchanged. Therefore, the methylation status of a given cytosine base in a CpG site can be interrogated using a genotyping assay for a C/T polymorphism after bisulfite treatment. We have developed algorithms for designing assay oligos that minimize the impact of the greatly diminished sequence complexity caused by the loss of most cytosines in the genome. Assay probes can be designed for a large fraction of the CpG sites in the genome, and to either strand.

One of the challenges with designing probes for bisulfite-converted DNA is dealing with underlying CpG sites. Degenerate bases can be employed, but concentration of the perfectly matched oligos on the beads would be diluted depending on the number of the underlying CpG sites. For the probe design with multiple underlying CpG sites, we assumed methylation is regionally correlated and resolved underlying CpG sites to be in phase with either the "methylated" (C) or "unmethylated" (U) query site. The co-methylation assumption is based on a previous study by Eckhardt *et al.* in which they bisulfite sequenced chromosomes 6, 20, and 22 (Eckhardt *et al.*, 2006). Our probes have a span of 50 bases; within this distance, methylation should be highly correlated. But there are likely to be exceptions to this regional methylation rule. There are over 28 million CpG sites in the human genome. Using a set of empirical rules, we were able to bioinformatically design Infinium methylation probes for over 16 million of these CpG loci. For the first Infinium methylation array, HumanMethylation27, we designed two "allele-specific" probes for each CpG site. The 3′ terminus of the probe was designed to match either the protected cytosine (methylated design) or the uracil base resulting from bisulfite conversion (unmethylated design). Underlying CpG sites were resolved to their respective "methylated" (C) or "unmethylated" (U) phase in the probe design. There are likely exceptions to this local co-methylation rule; nonetheless, this approach allowed us to design probes in CpG-dense regions and inside of CG islands (CGIs).

We selected a set of 27 578 CpG sites located within the proximal promoter regions (1 kb upstream and 500 bases downstream of transcription start sites) of 14 475 consensus-coding DNA sequence (CCDS) genes and well-known cancer genes. In addition, we included 254 assays covering 110 miRNA promoters. On average, we designed two assays per CCDS promoter; for a subset of 180 cancer-related genes, we assayed from three to up to 20 CpG sites per promoter region. Assays were preferentially designed to sites within CpG islands whenever possible. We employed a NCBI "relaxed" definition in which CGIs are identified bioinformatically as DNA sequences (200 base window) with a GC base composition greater than 50% and a CpG ratio of more than 0.6 over expected (Takai and Jones, 2002). Using this relaxed definition, 60% of CCDS genes contain one or more CGIs, and 40% contained no CGI.

For the next generation HumanMethylation450 array, we decided to use a degenerate base (R = A or G) in positions complementary to the "C" base in the underlying CpG sites. We first designed probes for over 1000 CpG loci to compare Infinium I (two probes per locus) and Infinium II (one probe per locus with degenerate base) designs. We found that probes with up to three underlying CpG sites can be designed as Infinium II probes, which allowed us to significantly increase the number of covered CpG sites with the same number of beads in the pool.

The HumanMethylation450 array covers not only majority of RefSeq genes, but also multiple regions outside of known coding areas. Over 650 000 bead types are designed to interrogate more than 480 000 query sites, including a small subset of non-CpG loci reported to be differentially methylated in pluripotent cell lines (Lister *et al.*, 2009; Laurent *et al.*, 2010).

17.2.5 Data analysis

Each methylation data point is represented by fluorescent signals from the M (methylated) and U (unmethylated) alleles of each targeted site. The beta-value (β), the ratio of intensities between M and U alleles, is calculated as

$$\beta = Max(M,0)/Max(U,0)+Max(M,0)+100$$

and used to estimate the methylation level of the CpG sites. Background intensity computed from a set of negative controls is subtracted from each analytical data point. An absolute value is used in the denominator of the formula, as a compensation for any "negative signals" which may arise from over-subtraction (a constant bias of 100 was added to regularize β; this is particularly useful when both U and M values are small) (Bibikova *et al.*, 2006b). In Infinium I assays, signal U and signal M are derived from two different bead types and measured with the same color. In Infinium II assays, U corresponds to the signal in the Red channel and M corresponds to the signal in the Green channel. The beta-value calculation results in a number between 0 and 1. In ideal situations, a value of 0 indicates that all copies of the CpG site in the sample are completely unmethylated, and a value of 1 indicates that every copy of the site is methylated.

17.3   Results and discussion

17.3.1   Testing of the Infinium methylation assay performance with reference DNA samples

Unlike genotyping, methylation data are not based on binary discrimination between unmethylated, hemi-methylated, and methylated loci, but rather represented by continuously varying beta-values. Since it is not possible to find a collection of biological samples where every single CpG site will be present in three different methylation states, we had to rely on artificially generated genomic DNA reference standards. We created three methylation state controls: unmethylated (U), 50% hemi-methylated (H), and a 100% methylated (M) control. These three reference standards were created from standard Coriell gDNA (see Section 17.2.3 for details). The reference standards were run on the Infinium methylation arrays, and the corresponding methylation beta-values were extracted. As shown in Figure 17.3, the distribution of beta-values measured on

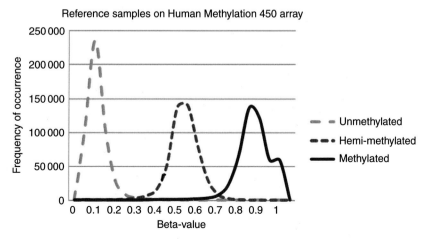

Figure 17.3 Methylation reference samples. The graph shows distribution of un-normalized beta-values for 482 421 CpG sites.

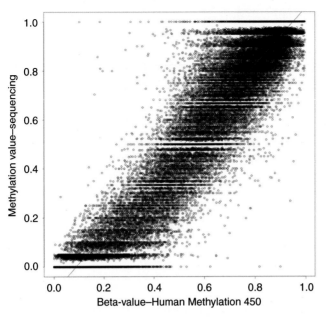

Figure 17.4 Correlation between methylation states (beta-values) for 51 843 CpG loci in human normal lung sample measured on HumanMethylation450 array and by whole-genome bisulfite sequencing data for the same sample collected on Illumina GAIIx. Loci with 20–90 × coverage in the sequencing data set and loci with detection *p*-value <0.01 were selected for comparison. Correlation *R* = 0.96.

HumanMethylation450 arrays are consistent with expected results in which the unmethylated (U) standard shows low beta-values, the hemi-methylated (H) standard intermediate beta-values, and the methylated (M) standard high beta-values.

17.3.2 Correlation between the array measurements and whole-genome bisulfite sequencing

High-throughput sequencing technologies enabled generation of single-base-resolution DNA methylation maps for various organisms (Lister and Ecker, 2009). The whole-genome bisulfite sequencing (WGBS) method provides the most comprehensive measurement of the DNA methylation state at every CpG site, but it is still expensive and allows analysis of only a few samples at a time. We compared DNA methylation profiles generated by WGBS and by Infinium assay using HumanMethylation450 bead chip for the same normal lung tissue sample. We observed high correlation between methylation measurements from the two independent platforms (Figure 17.4).

17.3.3 GoldenGate assay for methylation applications

To prove the assay utility, we designed probes to analyze methylation profiles of 1536 CpG sites from 371 genes and demonstrated distinct differential methylation profiles in cancer cell lines, lung cancers, and normal tissues, and identified a panel of lung adenocarcinoma-specific methylation markers (Bibikova *et al.*, 2006b). This method has also been used to identify a unique epigenetic signature for human embryonic stem cells (Bibikova *et al.*, 2006a). Since the assay became commercially available, many research groups, including the NIH Cancer Genome Atlas (TCGA) initiative, used Illumina GoldenGate methylation technology for high-throughput methylation profiling (TCGA, 2008). It opened up new avenues to large-scale discovery, validation, and clinical application of methylation markers in many human diseases (Byun *et al.*, 2009; Hinoue *et al.*, 2009; Killian *et al.*, 2009; Wolff *et al.*, 2010).

Currently, the robust GoldenGate methylation assay is deployed on Illumina VeraCode technology (Lin *et al.*, 2009) and offers an economical and high-throughput solution for 96–384 multiplexing level and presents an ideal platform for focused studies or biomarker validation.

17.3.4   Major advantage of Infinium methylation assay

The Infinium methylation assay offers a quick and cheap solution for methylation profiling of large numbers of CpG sites in the genome. It provides a true genome-wide scan, at single-CpG-site resolution. Data generated from the Infinium methylation arrays can be easily compared to other data, such as those generated with genomic bisulfite sequencing. The array probe design is very flexible; it is possible to target almost any specific region of interest, such as CpG islands and genomic regions with low CpG density. For example, on the HumanMethylation450 array, we have designed probes to interrogate methylation status of 99% RefSeq genes, 96% of all CpG islands, and 95% of CpG island shores, and have provided extensive coverage of other genomic regions (Table 17.1). In addition, methylation status at non-CpG sites can be measured without any change to the experimental workflow.

Table 17.1 HumanMethylation450 array content

Feature type	Included on array
Total number of sites	485 577
RefSeq genes	99%
CpG islands	96%
CpG island shores (0–2 kb from CGI)	92%
CpG island shelves (2–4 kb from CGI)	86%
Hidden Marker model (HMM) islands[a]	62 600
FANTOM 4 promoters (high CpG content)[a]	9 426
FANTOM 4 promoters (low CpG content)[a]	2 328
Differentially methylated regions (DMRs)[a]	16 232
Informatically predicted enhancers[a]	80 538
DNAse hypersensitive sites	59 916
Ensemble regulatory features[a]	47 257
Loci in major histocompatibility complex (MHC) region	12 334
Non-CpG loci	3 009

[a] Features may contain multiple assay probes. One probe may belong to several content categories

17.3.5   Infinium methylation assay design limitations

There are several challenges in designing Infinium assay probes to query the state of a CpG site in bisulfite-converted DNA. Since most of the cytosines in the genome are converted to uracils, the uniqueness of any given sequence within the bisulfite-converted genome decreases dramatically. This can potentially affect assay specificity, and the fact that opposite strands are no longer complementary reduces the effective concentration of any given locus by a factor of 2. Nonetheless, we found that, in general, the specificity and sensitivity of the Infinium assay was sufficient to read out the requisite loci.

Underlying SNPs can also affect the assay by reducing hybridization efficiency of the probe containing a mismatch nucleotide. By avoiding SNPs in the first 10 bases adjacent to the query site, this effect can be significantly reduced.

17.4   Concluding remarks

We have developed a low-cost, high-throughput, and true genome-wide DNA methylation profiling technology using Illumina BeadArray platform. With as little as 500 ng to 1 µg of gDNA, the GoldenGate and Infinium methylation assays can measure from 1536 to over 480 000 CpG sites, and are highly expandable and customizable. Both of the methylation assays provide single-CpG-site resolution data, which can be easily compared to other data such as genomic bisulfite sequencing data. The assay can be automated, which greatly increases assay robustness and throughput, thus enabling parallel processing of large numbers of samples.

Unlike the whole-genome bisulfite sequencing approach, which has access to every cytosine in the genome, array-based DNA methylation analysis is limited to the targeted CpG loci represented by either assay oligos in the case of GoldenGate assay, or by assay probes immobilized on beads in the Infinium arrays. Nevertheless, we have shown that arrays can access individual CpG sites across both CpG islands and genomic regions with low CpG density, and therefore give a good overview of epigenetic profiles on a genome-wide scale.

REFERENCES

Akama, T. O., Okazaki, Y., Ito, M., *et al.* (1997). Restriction landmark genomic scanning (RLGS-M)-based genome-wide scanning of mouse liver tumors for alterations in DNA methylation status. *Cancer Research,* **57**, 3294–3299.

Beck, S. (2010). Taking the measure of the methylome. *Nature Biotechnology,* **28**, 1026–1028.

Bibikova, M. and Fan, J. B. (2009). GoldenGate assay for DNA methylation profiling. *Methods in Molecular Biology,* **507**, 149–163.

Bibikova, M., Chudin, E., Wu, B., *et al.* (2006a). Human embryonic stem cells have a unique epigenetic signature. *Genome Research,* **16**, 1075–1083.

Bibikova, M., Lin, Z., Zhou, L., *et al.* (2006b). High-throughput DNA methylation profiling using universal bead arrays. *Genome Research,* **16**, 383–393.

Bibikova, M., Le, J., Barnes, B., *et al.* (2009). Genome-wide DNA methylation profiling using Infinium® assay. *Epigenomics,* **1**, 177–200.

Bibikova, M., Barnes, B., Tsan, C., *et al.* (2011). High density DNA methylation array with single CpG site resolution. *Genomics,* **98**, 288–295.

Byun, H. M., Siegmund, K. D., Pan, F., *et al.* (2009). Epigenetic profiling of somatic tissues from human autopsy specimens identifies tissue- and individual-specific DNA methylation patterns. *Human Molecular Genetics,* **18**, 4808–4817.

Clark, S. J., Harrison, J., Paul, C. L., *et al.* (1994). High-sensitivity mapping of methylated cytosines. *Nucleic Acids Research*, **22**, 2990–2997.

Colella, S., Shen, L., Baggerly, K. A., *et al.* (2003). Sensitive and quantitative universal Pyrosequencing methylation analysis of CpG sites. *BioTechniques*, **35**, 146–150.

Cottrell, S. E., Distler, J., Goodman, N. S., *et al.* (2004). A real-time PCR assay for DNA-methylation using methylation-specific blockers. *Nucleic Acids Research*, **32**, e10.

Down, T. A., Rakyan, V. K., Turner, D. J., *et al.* (2008). A Bayesian deconvolution strategy for immunoprecipitation-based DNA methylome analysis. *Nature Biotechnology*, **26**, 779–785.

Dupont, J. M., Tost, J., Jammes, H., *et al.* (2004). De novo quantitative bisulfite sequencing using the Pyrosequencing technology. *Analytical Biochemistry*, **333**, 119–127.

Eads, C. A., Danenberg, K. D., Kawakami, K., *et al.* (2000). MethyLight: a high-throughput assay to measure DNA methylation. *Nucleic Acids Research*, **28**, e32.

Eckhardt, F., Lewin, J., Cortese, R., *et al.* (2006). DNA methylation profiling of human chromosomes 6, 20 and 22. *Nature Genetics*, **38**, 1378–1385.

Ehrich, M., Nelson, M. R., Stanssens, P., *et al.* (2005). Quantitative high-throughput analysis of DNA methylation patterns by base-specific cleavage and mass spectrometry. *Proceedings of the National Academy of Sciences USA*, **102**, 15 785–15 790.

Ehrich, M., Turner, J., Gibbs, P., *et al.* (2008). Cytosine methylation profiling of cancer cell lines. *Proceedings of the National Academy of Sciences USA*, **105**, 4844–4849.

Fan, J. B., Oliphant, A., Shen, R., *et al.* (2003). Highly parallel SNP genotyping. *Cold Spring Harbor Symposia on Quantitative Biology*, **68**, 69–78.

Fan, J. B., Gunderson, K. L., Bibikova, M., *et al.* (2006). Illumina universal bead arrays. *Methods in Enzymology*, **410**, 57–73.

Frommer, M., McDonald, L. E., Millar, D. S., *et al.* (1992). A genomic sequencing protocol that yields a positive display of 5-methylcytosine residues in individual DNA strands. *Proceedings of the National Academy of Sciences USA*, **89**, 1827–1831.

Gitan, R. S., Shi, H., Chen, C. M., *et al.* (2002). Methylation-specific oligonucleotide microarray: a new potential for high-throughput methylation analysis. *Genome Research*, **12**, 158–164.

Gunderson, K. L. (2009). Whole-genome genotyping on bead arrays. *Methods in Molecular Biology*, **529**, 197–213.

Gunderson, K. L., Kruglyak, S., Graige, M. S., *et al.* (2004). Decoding randomly ordered DNA arrays. *Genome Research*, **14**, 870–877.

Gunderson, K. L., Steemers, F. J., Ren, H., *et al.* (2006). Whole-genome genotyping. *Methods in Enzymology*, **410**, 359–376.

Herman, J. G., Graff, J. R., Myohanen, S., *et al.* (1996). Methylation-specific PCR: a novel PCR assay for methylation status of CpG islands. *Proceedings of the National Academy of Sciences USA*, **93**, 9821–9826.

Hinoue, T., Weisenberger, D. J., Pan, F., *et al.* (2009). Analysis of the association between CIMP and BRAF in colorectal cancer by DNA methylation profiling. *PLoS One*, **4**, e8357.

Huang, T. H., Perry, M. R., and Laux, D. E. (1999). Methylation profiling of CpG islands in human breast cancer cells. *Human Molecular Genetics*, **8**, 459–470.

Huang, Y. W., Huang, T. H., and Wang, L. S. (2010). Profiling DNA methylomes from microarray to genome-scale sequencing. *Technology in Cancer Research and Treatment*, **9**, 139–147.

Ibrahim, A. E., Thorne, N. P., Baird, K., *et al.* (2006). MMASS: an optimized array-based method for assessing CpG island methylation. *Nucleic Acids Research*, **34**, e136.

Jones, P. A., Archer, T. K., Baylin, S. B., *et al.* (2008). Moving AHEAD with an international human epigenome project. *Nature*, **454**, 711–715.

Kawai, J., Hirotsune, S., Hirose, K., *et al.* (1993). Methylation profiles of genomic DNA of mouse developmental brain detected by restriction landmark genomic scanning (RLGS) method. *Nucleic Acids Research*, **21**, 5604–5608.

Killian, J. K., Bilke, S., Davis, S., *et al.* (2009). Large-scale profiling of archival lymph nodes reveals pervasive remodeling of the follicular lymphoma methylome. *Cancer Research*, **69**, 758–764.

Kuang, S. Q., Tong, W. G., Yang, H., *et al.* (2008). Genome-wide identification of aberrantly methylated promoter associated CpG islands in acute lymphocytic leukemia. *Leukemia*, **22**, 1529–1538.

Laird, P. W. (2010). Principles and challenges of genome-wide DNA methylation analysis. *Nature Reviews Genetics*, **11**, 191–203.

Laurent, L., Wong, E., Li, G., *et al.* (2010). Dynamic changes in the human methylome during differentiation. *Genome Research*, **20**, 320–331.

Lin, C. H., Yeakley, J. M., McDaniel, T. K., *et al.* (2009). Medium- to high-throughput SNP genotyping using VeraCode microbeads. *Methods in Molecular Biology*, **496**, 129–142.

Lister, R. and Ecker, J. R. (2009). Finding the fifth base: genome-wide sequencing of cytosine methylation. *Genome Research*, **19**, 959–966.

Lister, R., Pelizzola, M., Dowen, R. H., *et al.* (2009). Human DNA methylomes at base resolution show widespread epigenomic differences. *Nature*, **462**, 315–322.

Meissner, A., Gnirke, A., Bell, G. W., *et al.* (2005). Reduced representation bisulfite sequencing for comparative high-resolution DNA methylation analysis. *Nucleic Acids Research*, **33**, 5868–5877.

Okazaki, Y., Hirose, K., Hirotsune, S., *et al.* (1995). Direct detection and isolation of restriction landmark genomic scanning (RLGS) spot DNA markers tightly linked to a specific trait by using the RLGS spot-bombing method. *Proceedings of the National Academy of Sciences USA*, **92**, 5610–5614.

Omura, N., Li, C. P., Li, A., *et al.* (2008). Genome-wide profiling of methylated promoters in pancreatic adenocarcinoma. *Cancer Biology and Therapy*, **7**, 1146–1156.

Ordway, J. M., Bedell, J. A., Citek, R. W., *et al.* (2006). Comprehensive DNA methylation profiling in a human cancer genome identifies novel epigenetic targets. *Carcinogenesis*, **27**, 2409–2423.

Ordway, J. M., Budiman, M. A., Korshunova, Y., *et al.* (2007). Identification of novel high-frequency DNA methylation changes in breast cancer. *PLoS One*, **2**, e1314.

Ruike, Y., Imanaka, Y., Sato, F., *et al.* (2010). Genome-wide analysis of aberrant methylation in human breast cancer cells using methyl-DNA immunoprecipitation combined with high-throughput sequencing. *BMC Genomics*, **11**, 137.

Schumacher, A., Kapranov, P., Kaminsky, Z., *et al.* (2006). Microarray-based DNA methylation profiling: technology and applications. *Nucleic Acids Research*, **34**, 528–542.

Shen, R., Fan, J. B., Campbell, D., *et al.* (2005). High-throughput SNP genotyping on universal bead arrays. *Mutation Research*, **573**, 70–82.

Singer-Sam, J., LeBon, J. M., Tanguay, R. L., and Riggs, R. L., (1990). A quantitative HpaII-PCR assay to measure methylation of DNA from a small number of cells. *Nucleic Acids Research*, **18**, 687.

Steemers, F. J. and Gunderson, K. L. (2007). Whole-genome genotyping technologies on the BeadArray platform. *Biotechnology Journal*, **2**, 41–49.

Takai, D. and Jones, P. A. (2002). Comprehensive analysis of CpG islands in human chromosomes 21 and 22. *Proceedings of the National Academy of Sciences USA*, **99**, 3740–3745.

TCGA (2008). Comprehensive genomic characterization defines human glioblastoma genes and core pathways. *Nature*, **455**, 1061–1068.

Tost, J., Schatz, P., Schuster, M., *et al.* (2003). Analysis and accurate quantification of CpG methylation by MALDI mass spectrometry. *Nucleic Acids Research*, **31**, e50.

Weber, M., Davies, J. J., Wittig, D., *et al.* (2005). Chromosome-wide and promoter-specific analyses identify sites of differential DNA methylation in normal and transformed human cells. *Nature Genetics*, **37**, 853–862.

Wolff, E. M., Chihara, Y., Pan, F., *et al.* (2010). Unique DNA methylation patterns distinguish noninvasive and invasive urothelial cancers and establish an epigenetic field defect in premalignant tissue. *Cancer Research*, **70**, 8169–8178.

Yamada, Y., Watanabe, H., Miura, F., *et al.* (2004). A comprehensive analysis of allelic methylation status of CpG islands on human chromosome 21q. *Genome Research*, **14**, 247–266.

18 Advances in capillary electrophoresis-based methods for DNA methylation analysis

Benjamin G. Schroeder, Victoria L. Boyd, and Gerald Zon*

18.1 Introduction

Capillary electrophoresis (CE) can be used with various methods for DNA methylation analysis. Methods for quantification of global DNA methylation by CE following enzymatic digestion of genomic DNA (gDNA) without prior bisulfite conversion (Berdasco *et al.*, 2009), and targeted analysis of bisulfite-converted/PCR-amplified DNA by single-strand conformation polymorphism (bisulfite/PCR-SSCP) using CE (Xu *et al.*, 2008) have been recently reported. Bisulfite conversion followed by PCR amplification, bacterial cloning, and sequencing continues to be employed for the targeted analysis of methylated genomic regions. For example, a recent large study used this technique to report the methylation status of 580 427 CG sites from 28 626 subclones of 190 gene promoter regions on chromosome 21 for five human cell types (Zhang *et al.*, 2009).

The present account focuses on advances in various methods that employ bisulfite conversion and PCR in conjunction with fluorescence-based CE and either Sanger sequencing or fragment-length analysis. Notable among these are relatively new single-molecule PCR (smPCR) techniques as advantageous replacements for traditional cloning. The studies featured herein are largely those carried out by the present co-authors and their colleagues, with selected references to related publications due to page limitations.

18.2 Single-molecule PCR

While both PCR amplification and bacterial cloning are fraught with potential artifacts, it turns out that PCR amplification starting from single DNA molecules largely eliminates these artifacts (Kraytsberg and Khrapko 2005). Bisulfite

---

* Author to whom correspondence should be addressed

Epigenomics: From Chromatin Biology to Therapeutics, ed. K. Appasani. Published by Cambridge University Press. © Cambridge University Press 2012.

sequencing (Frommer *et al.*, 1992) is one application that can clearly benefit from single-molecule PCR. Traditionally, after bisulfite conversion and PCR amplification, the pool of amplified fragments is analyzed by bacterial cloning and sequencing. PCR amplification of a mixed population of bisulfite-converted molecules can often result in biased amplification (Warnecke *et al.*, 1997; Wojdacz *et al.*, 2008), due to some sequences in the mixture amplifying more efficiently than others. In addition, cloning efficiency of the AT-rich amplicons from bisulfite-converted templates can be problematic, leading to a second source of bias (Warnecke *et al.*, 2002).

The cost, time, and bias associated with bacterial cloning can be eliminated by the use of techniques that directly analyze the amplified pool of molecules; however, such techniques still suffer from the possibility of PCR amplification artifacts, and have the additional drawback that they cannot provide information about the methylation haplotype, that is, the methylation status of each CpG within an individual DNA molecule. In contrast, (smPCR) bisulfite sequencing (Chhibber *et al.*, 2008) (Figure 18.1) eliminates bacterial cloning as well as PCR bias, since there is no competition between template molecules for amplification. This technique provides a detailed, unbiased, molecule-by-molecule survey of the methylation status of every CpG within the amplified region.

Capillary electrophoresis sequencing has been used to reveal the sequence changes responsible for differences in melting temperatures observed for different

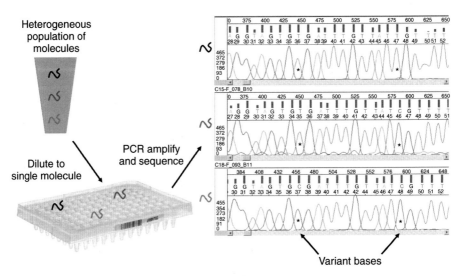

Figure 18.1 Schematic outline of smPCR bisulfite sequencing workflow wherein a gDNA has been bisulfite-converted to a heterogeneous population of molecules (black, blue, and red squiggles; only one of many molecules of each allele is shown) which are serially diluted such that aliquots transferred to a microtiter plate allow some wells to have a single molecule for PCR amplification and then fluorescence-based CE Sanger-sequencing to identify variant bases (asterisks) that are either T or C and thus indicate C or 5mC, respectively, at CpG positions in gDNA. See plate section for color version.

molecules analyzed with digital methylation-sensitive high-resolution melting (MS-HRM) (Snell *et al.*, 2008). The Laird group also reported successful use of the single-molecule bisulfite PCR sequencing method and described it as cost- and time-effective (Weisenberger *et al.*, 2008).

18.3 Whole-genome analysis

Massively parallelized sequencing technologies (Mardis, 2008) require construction of genomic DNA (gDNA) libraries starting with fragmented DNA. Determining whole-genome methylation (i.e., methylome: Beck *et al.*, 2008) status further requires bisulfite conversion of a library. SOLiD™ sequencing of a bisulfite-converted genome is less error-prone than alternative massively parallelized sequencing technologies due to sequence-matching in color-space (Bormann Chung *et al.*, 2010). The largely three-base (i.e., T, A, and G) genome following bisulfite conversion is represented as four colors in color-space, thus enhancing matching accuracy to the color-space representation of the bisulfite-converted reference sequence. During development of methods for preparation of SOLiD fragment libraries, smPCR/CE-sequencing was used to determine the quality of SOLiD fragment library construction, and the extent of bisulfite conversion, prior to carrying out steps of clonal amplification and sequencing (Ranade *et al.*, 2009).

More recently, smPCR/CE-sequencing was used to confirm the quality of a novel (McKernan *et al.*, 2010) SOLiD™ mate-pair library construction method that allows unambiguous methylation analysis of complex genomes, such as those for humans. This library construction closely follows the SOLiD 2.0 mate-pair (2 × 50) library preparation (Applied Biosystems). However, one of the pairs (tags) of the mate-pair is native DNA, and the other tag is protected from bisulfite conversion during a nick-translation step by substituting 5mC-dNTP for dCTP. After capture of the biotinylated mate-pair library fragments onto streptavidin-polystyrene (in place of MyOne™) beads, the non-biotinylated strand is released into solution, bisulfite converted, and recovered using a Microcon 10 spin-filtration cartridge to ensure retention of library fragments of ~200 nucleotides or greater (Ranade *et al.*, 2009; Zon *et al.*, 2009). An aliquot of the amplified library is then diluted for smPCR/CE-sequencing, as exemplified by the results shown in Figure 18.2. A significant advantage of this sequencing method is that it permits non-redundant reference-sequence matching of the bisulfite tag by using the non-bisulfite converted tag (5mC-protected) as a locator in the genome. The application of SOLiD™ and other commercially available next-generation sequencing technologies for DNA methylation analysis has been reviewed very recently (Zhang *et al.*, 2010).

18.4 Enrichment of differentially methylated genomic regions

5-Methylcytosine (5mC) content in mammalian genomes is a small percentage; for example, in the human genome, only ~1% is 5mC. In many cases, such as application of high-density deep-coverage sequencing, it is neither cost-effective nor

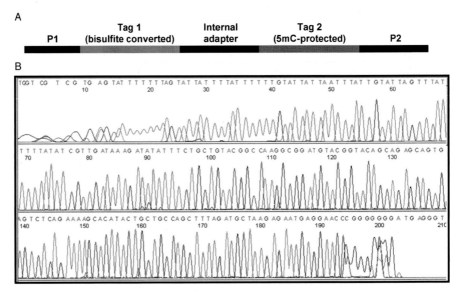

Figure 18.2 (A) Schematic representation of a SOLiD™ mate-pair library fragment comprised of adapters (black; P1, internal, and P2) and mate-pair sequences (tags) that undergo bisulfite conversion (red; Tag 1) or are 5mC-protected (blue; Tag 2). (B) Exemplary smPCR/CE-sequencing traces of a bisulfite-converted SOLiD™ mate-pair library fragment using a shortened P1 adapter sequence as the sequencing primer. Bases 0–12 are the P1 adapter extended from the shortened primer; bases 13–95 are bisulfite-converted Tag 1; bases 96–131 are the internal adapter (also protected with 5mC during nick translation); bases 132–179 are 5mC-protected Tag 2; bases 180–end are the reverse complement of the P2 adapter. See plate section for color version.

necessary to analyze the entire genome and then "extract" information concerning 5mC patterns. Consequently, methods for selective enrichment of 5mC-containing gDNA fragments can be advantageously used prior to bisulfite conversion and sequencing (or other types of analysis). Detailed protocols for antibody-based methylated DNA immunoprecipitation (MeDIP) enrichment have been recently reported (Mohn *et al.*, 2009); however, the present account will focus on MethylMiner™ (Invitrogen) which employs the 5mCpG binding domain (MBD) of human MBD2 protein coupled to paramagnetic Dynabeads® M-280 streptavidin via a biotin linker. In contrast to MeDIP which requires use of single-stranded DNA, MethylMiner is performed directly on double-stranded DNA. A detailed application note for MethylMiner (www.invitrogen.com) describes a genome-scale study on patterns of DNA methylation in the MCF-7 breast cancer cell line using MethylMiner enrichment in conjunction with the SOLiD sequencing system (a related study has been very recently published: Serre *et al.*, 2010). MethylMiner enrichment/SOLiD sequencing revealed that 25% of unenriched sequenced gDNA fragments had no CpGs. In contrast, only 0.65% of the 500 mM NaCl-enriched sequences and 0.11% of the 1 M NaCl-enriched sequences contain no CpGs. Although these MethylMiner enrichment/SOLiD sequencing findings clearly demonstrate that 5mC content increases in gDNA fractions eluted with higher concentration of NaCl, a single-molecule PCR/CE-sequencing method (Ranade *et al.*, 2009) has been used to examine CpG motifs in randomly selected (albeit not bisulfite-converted) single molecules in such fractions. Twenty single-molecule amplicons thus derived from elution with 400 mM NaCl, and an equal number of single-molecule amplicons from elution with 2 M NaCl, that all gave >90-nucleotide CE sequences were inspected for various CpG-related motifs (V. L. Boyd and G. Zon, unpublished results). While the CG dinucleotide motif was qualitatively less dense in the low-salt fraction, as expected from MethylMiner enrichment/SOLiD sequencing findings, the surprising observation was that CGCG, CGNCG, CGNNCG, and CGNNNCG motifs were found almost exclusively in the high-salt fraction. Further analysis of MethylMiner fractions using SOLiD sequence data may confirm and extend these initial observations that are presumably related to dynamics of MBD binding (Ohki *et al.*, 2001).

18.5   Targeted bisulfite/smPCR sequencing of FFPE clinical samples

Among factors that play a role in successful bisulfite conversion, incomplete denaturation of gDNA is widely recognized to be a likely cause of incomplete conversion of C to U. A commonly used protocol (Clark *et al.*, 2006) employs NaOH to denature gDNA followed by quickly cooling on ice and then addition of acidic, pH 5 bisulfite solution. A novel denaturing process in which pre-denaturing with NaOH is replaced entirely by using only formamide has been recently reported (Zon *et al.*, 2009) to offer the following advantages:

(1) Use of formamide eliminates heating with a caustic reagent.
(2) It avoids variable neutralization of hydroxide ion upon addition of pH 5 bisulfite reagent, thus ensuring maintenance of the acidity that is required for optimal conversion kinetics.
(3) The presence of formamide before addition of pH 5 bisulfite reagent also mitigates partial reannealing of DNA, thus lessening time constraints and/ or need for ice-cooling during sample handling.
(4) Finally, the presence of formamide during sulfonation maintains and/or promotes formation of single-stranded DNA, thus leading to more complete sulfonation.

Application of formamide-mediated bisulfite conversion has been investigated (Zon *et al.*, 2009) in conjunction with the use of RecoverAll™ (Invitrogen) for gDNA extraction of clinically relevant formalin-fixed/paraffin-embedded (FFPE) prostate cancer samples for CE-sequencing methylation analysis, following PCR amplification of targeted regions of interest. Primers for PCR were designed to amplify sequences as close as possible to regions in genes previously found to exhibit correlations between hypermethylation and prostate cancer, namely, glutathione S-transferase-pi (GSTPi) and retinoic acid receptor-beta2 (RARβ2). These primers were designed to (1) anneal to CpG-free regions of template to ensure that amplicons are produced regardless of methylation status, and (2) exclude homopolymer regions exceeding nine nucleotides, which are prone to "slippage" of DNA polymerase, as previously reported (Boyd *et al.*, 2006) in another context. Several of such candidate primer-pairs were screened for relative performance in order to select those that amplify "cleanly", i.e., no significant primer-dimer or mispriming amplicons, which would interfere with sequence analysis by CE. In contrast to sequencing either conventional, clonal PCR amplicons or single-molecule PCR amplicons (see above) that provide 100% C or T base-calling at CG loci of interest for enumerating percent-methylation, direct bisulfite/PCR-sequencing provides ensemble-average sequence data for each CG locus of interest. The latter data derived from overlapping, mixed C and T peaks, provide C/T peak-height ratios that are approximate values of percent-methylation, and were semi-arbitrarily grouped as 0, 25, 50, 75 or 100% with an estimated error equal to ±10%. Nevertheless, unmethylated GSTPi and RARβ2 were detected in 22 of 22 (100%) cases of prostate benign hyperplasia or inflammatory prostate disease. The frequency of these cases was significantly associated with unmethylated genes in comparison to primary prostate cancer (Fisher's exact $p < 0.001$): prostate cancer at the primary site exhibited methylated GSPTi in 37 of 43 (86%) and methylated RARβ2 in 39 of 43 (91%) cases. A highly significant association was observed between methylated GSTPi or methylated RARβ2 and primary prostate cancer as compared to the benign prostate (Fisher's exact $p < 0.001$). However, methylated GSTPi or methylated RARβ2 was not detected in 14% or 9%, respectively, cases of prostate cancer. The sensitivity and specificity of detecting methylated GSTPi or methylated RARβ2 were 86% and 100% or 91% and 100%, respectively, at 95% confidence interval. Moreover, finding either methylated GSTPi or methylated RARβ2 resulted in the detection of prostate cancer with sensitivity of 100% and specificity of 100%. It was concluded (Zon *et al.*, 2009) that these promising initial findings warrant additional studies on a larger cohort of patients to confirm the results and thus provide a basis for prostate cancer diagnostics using bisulfite/PCR- sequencing. Expert commentary on prospects for epigenomics-based diagnostics has been recently published (Diamandis *et al.*, 2010).

Several of the clinical samples used in the targeted sequencing described above were also investigated by smPCR. The FFPE-extracted and bisulfite-converted DNA was diluted to less than a single copy per well. Although the DNA concentration (pre-bisulfite) was determined by UV, a single copy of DNA of the region of interest required empirical determination due to unknown fragmentation and DNA damage caused by the FFPE preservation and bisulfite treatment. The smPCR and sequencing results for the two targeted regions, GSTPi and RARβ2, for a representative non-malignant (E) and a prostate cancer sample (D) are shown in Figure 18.3. The non-malignant sample for both gene targets was nearly devoid of methylated CpGs. By contrast, the cancerous sample was a mixture of DNA molecules with CpGs that were either fully methylated in the amplified region, or fully unmethylated. The bimodal distribution of DNA molecules into mostly methylated or non-methylated regions is consistent with previous literature reports (Rakyan *et al.*, 2004). The ratio of methylated and non-methylated CpGs is also consistent with the "bulk" (ensemble average between 25% and 75% methylated) sequencing analysis results generally observed for the prostate cancer samples.

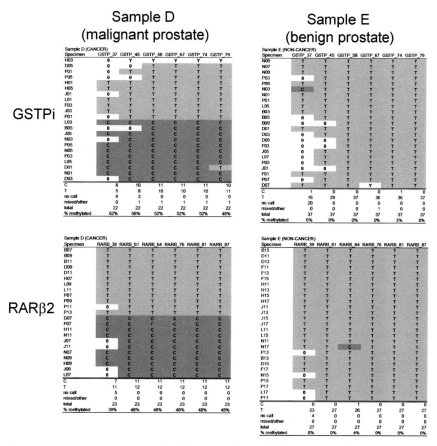

Figure 18.3 Tabulated results for bisulfite conversion/smPCR/sequencing of two targeted regions, GSTPi and RARβ2, in malignant prostate (D) clinical specimens and benign prostate (E) clinical specimens. Numbered column headings indicate six CpG loci classified as C, T, no call (0), or mixed/other (Y) to thus calculate values of percent methylated (C = 5mCpG and T = CpG). Note that for GSTPi in E representative data for 23 of 37 total specimens are shown.

18.6   Targeted analysis of single-CG methylation status

Bisulfite conversion of a CG locus in gDNA followed by PCR amplification produces CG or TG depending on whether C at this locus was methylated or not, respectively. In CE-based analysis of this (C/T)G single-nucleotide polymorphism (SNP) in bisulfite-converted gDNA, a primer is hybridized immediately adjacent to the C/T position of interest to allow polymerase-mediated incorporation of 2′,3′-dideoxy-G/A 5′-triphosphate terminators having different fluorescent labels (Uhlmann et al., 2002). This straightforward adaptation of the SNaPshot™ (Applied Biosystems, Foster City, CA, USA) method, which was originally described for quantitative SNP allele frequency measurement in gDNA pools (Norton et al., 2002), advantageously avoids use of radioactive ^{32}P-labeled terminators (Gonzalgo et al., 1997). Capillary electrophoresis analysis of resultant single-base (G or A) primer-extension products provides peak areas for calculating 0–100%

methylation (Uhlmann *et al.*, 2002). Percent methylation thus determined for 32 tumor samples when compared to results independently obtained by pyrosequencing (Elahi and Ronaghi, 2004) gave datasets that were largely in agreement, although some systematic differences between these two sequencing methods were noted (Uhlmann *et al.*, 2002). Additional examples of methylation analysis using this basic SNaPshot method have been recently described (Xu *et al.*, 2008).

In the above methodology, hybridization of SNaPshot primers is restricted to non-polymorphic regions of template. Overcoming this limitation was investigated (Kaminsky *et al.*, 2005) using synthetic templates having one to five polymorphic positions in the SNaPshot primer hybridization region. For example, shown in Figure 18.4A is a 10-mer segment of an arbitrary template having four polymorphic C/T positions, upstream of the c/t query position, with separation by either one or two nucleotides (N). Also shown are corresponding mixed-base G/A-containing 9-mer primers, which are simultaneously synthesized as a mixture of 16 possible sequences using mixed-base G/A couplings. While this approach can be used successfully, some instances were reported (Kaminsky *et al.*, 2005) to be non-trivially complicated due to differences in primer composition that can bias hybridization efficiency and/or cause CE mobility shifts. As exemplified below, these problems have been addressed (Boyd *et al.*, 2008) by using novel SNaPshot primers modified with N^6-methoxy-2,6-diaminopurine (K), which is a unique G/A-analog that in principle can participate in three H-bonds with either C or T (Figure 18.4B), in contrast to G or A which form three or two H-bonds with C or T, respectively. Consequently, K-primers can mitigate bias in hybridization, as well

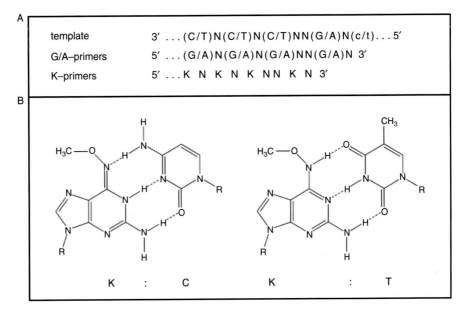

Figure 18.4 (A) 10-Mer segment of a template having four polymorphic C/T positions, upstream of a c/t query position, with separation by either one or two nucleotides (N). Also shown is a G/A-containing 9-mer representing a mixture of 16 possible primer sequences, and a single K-containing 9-mer primer sequence. (B) H-Bonding structures for K–C base pairing and K–T base pairing.

as eliminate stoichiometric imbalance in mixed-base G/A-primers that may arise due to G vs. A coupling differences and/or inadvertent sequence selection during primer purification. In addition, SNaPshot using a single-sequence K-primer precludes problematic CE mobility shifts that have been occasionally observed (Kaminsky *et al.*, 2005; Boyd and Zon, 2008) for mixtures of primers.

18.7   Targeted methylation analysis by methylation-dependent fragment separation

A novel method called methylation-dependent fragment separation (MDFS) (Boyd *et al.*, 2006) involves bisulfite conversion of gDNA followed by PCR amplification using a primer-pair designed to hybridize to non-CpG-containing sequences flanking a CpG-rich region of interest. The forward primer has a fluorescein (FAM) label, and CE is used to separate the faster-migrating C-rich strand from the slower-migrating T-rich strand derived from methylated and unmethylated gDNA, respectively. Fragment mobilities are precisely determined by use of ROX 500 size-standard, a 36-cm capillary array, POP 4 polymer at 60 °C, and GeneMapper software for data collection (all from Applied Biosystems).

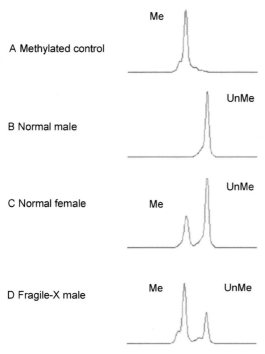

Figure 18.5 Capillary-electrophoresis-based MDFS analysis of methylation status of the FMR1 gene in four individual gDNA samples. (A) methylated (Me) control gDNA showing only the amplicon from methylated gDNA; (B) normal male gDNA wherein only the amplicon from unmethylated (UnMe) gDNA is seen; (C) normal female gDNA that has amplicons for both methylated and unmethylated gDNA due to X-chromosome silencing; (D) fragile-X male gDNA that has amplicons for both methylated and unmethylated gDNA, in contrast to normal male gDNA wherein only unmethylated gDNA is detected.

Although MDFS alone does not provide information on possible heterogeneity of methylation, such data can be easily obtained by carrying out bisulfite/CE sequencing using MDFS amplicons either as a mixed population of molecules or as single molecules (see Section 18.2.)

The MDFS technique is predicated on the cumulative effect of lower mass/charge ratio for C vs. T in CpG-rich sequences, which can lead to near baseline resolution during high-resolution CE under appropriate conditions. A series of synthetic oligonucleotides was used to derive a mathematical algorithm (Boyd *et al.*, 2006) which predicts the extent of such resolution, and can thus be applied to pre-select MDFS amplicons as candidates for this method of methylation analysis. In this manner, excellent agreement was obtained between calculated and observed lengths for a test panel of 18 gene regions of interest in DNA methylation analysis, including the fragile-X gene (FMR1) for which MDFS results are shown in Figure 18.5 (Boyd *et al.*, 2006).

REFERENCES

Beck, S. and Rakyan, V. K. (2008). The methylome: approaches for global DNA methylation profiling. *Trends in Genetics*, **24**, 231–237.

Berdasco, M., Fraga, M. F., and Esteller, M. (2009). Quantification of global DNA methylation by capillary electrophoresis and mass spectrometry. *Methods in Molecular Biology*, **507**, 23–34.

Bormann Chung, C. A., Boyd, V. L., Mckernan, K. J., *et al.* (2010). Whole-methylome analysis by ultra-deep sequencing using two-base encoding. *PLoS One*, **5**, e9320.

Boyd, V. L. and Zon, G. (2008). Capillary electrophoretic analysis of methylation status in CpG-rich regions by single-base extension of primers modified with N6-methoxy-2,6-diaminopurine. *Analytical Biochemistry*, **380**, 13–20.

Boyd, V. L., Moody, K. I., Karger, A. E., *et al.* (2006). Methylation-dependent fragment separation: direct detection of DNA methylation by capillary electrophoresis of PCR products from bisulfite-converted genomic DNA. *Analytical Biochemistry*, **354**, 266–273.

Chhibber, A. and Schroeder, B. G. (2008). Single-molecule polymerase chain reaction reduces bias: application to DNA methylation analysis by bisulfite sequencing. *Analytical Biochemistry*, **377**, 46–54.

Clark, S. J., Statham, A., Stirzaker, C., Molloy, P. L., and Frommer, M. (2006). DNA methylation: bisulphite modification and analysis. *Nature Protocols*, **1**, 2353–2364.

Diamandis, E. P., Sidransky, D., Laird, P. W., Cairns, P., and Bapat, B. (2010). Epigenomics-based diagnostics. *Clinical Chemistry*, **56**, 1216–1219.

Elahi, E. and Ronaghi, M. (2004). Pyrosequencing: a tool for DNA sequencing analysis. *Methods in Molecular Biology*, **255**, 211–219.

Frommer, M., Mcdonald, L. E., Millar, D. S., *et al.* (1992). A genomic sequencing protocol that yields a positive display of 5-methylcytosine residues in individual DNA strands. *Proceedings of the National Academy of Sciences USA*, **89**, 1827–1831.

Gonzalgo, M. L. and Jones, P. A. (1997). Rapid quantitation of methylation differences at specific sites using methylation-sensitive single nucleotide primer extension (Ms-SNuPE). *Nucleic Acids Research*, **25**, 2529–2531.

Kaminsky, Z. A., Assadzadeh, A., Flanagan, J., and Petronis, A. (2005). Single nucleotide extension technology for quantitative site-specific evaluation of metC/C in GC-rich regions. *Nucleic Acids Research*, **33**, e95.

Kraytsberg, Y. and Khrapko, K. (2005). Single-molecule PCR: an artifact-free PCR approach for the analysis of somatic mutations. *Expert Review of Molecular Diagnostics*, **5**, 809–815.

Mardis, E. R. (2008). Next-generation DNA sequencing methods. *Annual Review of Genomics and Human Genetics*, **9**, 387–402.

Mckernan, K. J., Schroeder, B. G., and Boyd, V. L. (2010). Methylation analysis of mate pairs. US Patent Application WO/2010/003153 A2.

Mohn, F., Weber, M., Schubeler, D., and Roloff, T. C. (2009). Methylated DNA immunoprecipitation (MeDIP). *Methods in Molecular Biology*, **507**, 55–64.

Norton, N., Williams, N. M., Williams, H. J., *et al.* (2002). Universal, robust, highly quantitative SNP allele frequency measurement in DNA pools. *Human Genetics*, **110**, 471–478.

Ohki, I., Shimotake, N., Fujita, N., *et al.* (2001). Solution structure of the methyl-CpG binding domain of human MBD1 in complex with methylated DNA. *Cell*, **105**, 487–497.

Rakyan, V. K., Hildmann, T., Novik, K. L., *et al.* (2004). DNA methylation profiling of the human major histocompatibility complex: a pilot study for the human epigenome project. *PLoS Biology*, **2**, e405.

Ranade, S. S., Chung, C. B., Zon, G., and Boyd, V. L. (2009). Preparation of genome-wide DNA fragment libraries using bisulfite in polyacrylamide gel electrophoresis slices with formamide denaturation and quality control for massively parallel sequencing by oligonucleotide ligation and detection. *Analytical Biochemistry*, **390**, 126–135.

Serre, D., Lee, B. H., and Ting, A. H. (2010). MBD-isolated genome sequencing provides a high-throughput and comprehensive survey of DNA methylation in the human genome. *Nucleic Acids Research*, **38**, 391–399.

Snell, C., Krypuy, M., Wong, E. M., Loughrey, M. B., and Dobrovic, A. (2008). BRCA1 promoter methylation in peripheral blood DNA of mutation negative familial breast cancer patients with a BRCA1 tumour phenotype. *Breast Cancer Research*, **10**, R12.

Uhlmann, K., Brinckmann, A., Toliat, M. R., Ritter, H., and Nurnberg, P. (2002). Evaluation of a potential epigenetic biomarker by quantitative methyl-single nucleotide polymorphism analysis. *Electrophoresis*, **23**, 4072–4079.

Warnecke, P. M., Stirzaker, C., Melki, J. R., *et al.* (1997). Detection and measurement of PCR bias in quantitative methylation analysis of bisulphite-treated DNA. *Nucleic Acids Research*, **25**, 4422–4426.

Warnecke, P. M., Stirzaker, C., Song, J., *et al.* (2002). Identification and resolution of artifacts in bisulfite sequencing. *Methods*, **27**, 101–107.

Weisenberger, D. J., Trinh, B. N., Campan, M., *et al.* (2008). DNA methylation analysis by digital bisulfite genomic sequencing and digital MethyLight. *Nucleic Acids Research*, **36**, 4689–4698.

Wojdacz, T. K., Hansen, L. L., and Dobrovic, A. (2008). A new approach to primer design for the control of PCR bias in methylation studies. *BMC Research Notes*, **1**, 54.

Xu, G., Shi, X., Zhao, C., *et al.* (2008). Capillary electrophoresis of gene mutation. *Methods in Molecular Biology*, **384**, 441–455.

Zhang, Y. and Jeltsch, A. (2010). The application of next generation sequencing in DNA methylation analysis. *Genes*, **1**, 85–101.

Zhang, Y., Rohde, C., Tierling, S., *et al.* (2009). DNA methylation analysis of chromosome 21 gene promoters at single base pair and single allele resolution. *PLoS Genetics*, **5**, e1000438.

Zon, G., Barker, M. A., Kaur, P., *et al.* (2009). Formamide as a denaturant for bisulfite conversion of genomic DNA: bisulfite sequencing of the GSTPi and RARbeta2 genes of 43 formalin-fixed paraffin-embedded prostate cancer specimens. *Analytical Biochemistry*, **392**, 117–125.

19 Genome-wide methylome analysis based on new high-throughput sequencing technology

Mingzhi Ye,* Fei Gao, Xu Han, Guanyu Ji, Zhixiang Yan, and Honglong Wu

19.1 Introduction

DNA methylation is an epigenetic mark found in organisms across all domains of life. Eukaryotes mainly contain 5-methylcytosine (5mC) in genomes, and the spectrum of methylation levels is very broad (Bird, 2002). For example, vertebrate genomes often present highly globally methylated profiles in the vast majority of cell types, while invertebrate genomes show diversity of DNA methylation from undetectable levels to highly methylated levels, which show mosaic DNA methylation patterns (Tweedie *et al.*, 1997). Although the biological function of DNA methylation has been mainly established as a stable epigenetic mark that regulates chromatin structure and gene expression in many eukaryotes, the variety of animal DNA methylation patterns highlights the possibility that different distributions reflect different functions for the DNA methylation system (Colot and Rossignol, 1999).

Methods for DNA methylation analysis are mostly based on bisulfite conversion, methylation-sensitive restriction enzymes digestion, and affinity purification of methylated DNA. Bisulfite conversion of 5mC is the most informative way to analyze DNA methylation patterns. Formerly, converted PCR products were analyzed by Sanger sequencing (Eckhardt *et al.*, 2006), pyrosequencing (Tost and Gut, 2007), or mass spectrometry (Ehrlich *et al.*, 2006) to quantify the extent of methylation at each cytosine. The most common pitfall of this method is incomplete C–U conversion; therefore it is essential to ensure the complete conversion of the bisulfite-treated samples. Methylation-sensitive restriction endonucleases are classic tools of DNA methylation analysis (Bird and Southern, 1978). Most of these are inhibited by their recognition site. Many variations of restriction-enzyme-based methods have been used in conjunction with genomic analysis

---

* Author to whom correspondence should be addressed.

Epigenomics: From Chromatin Biology to Therapeutics, ed. K. Appasani. Published by Cambridge University Press. © Cambridge University Press 2012.

(Khulan *et al.*, 2006; Rollins *et al.*, 2006; Schumacher *et al.*, 2006). Affinity purification is the simplest way to enrich methylated DNA, which takes advantage of the methyl-binding domain (MBD) (Zhang *et al.*, 2006); alternatively, commercially available monoclonal antibody that specifically recognizes methylated cytosine can be used (Keshet *et al.*, 2006; Zilberman *et al.*, 2007). However, the density of methylation in a given region will influence the efficiency of purification, which means that regions with sparse methylation target sites will be difficult to purify (Weber *et al.*, 2007). Based on these three techniques, microarray technology and high-throughput sequencing technology have been applied broadly to analyze the methylation state of an organism. Further, methylated DNA immunoprecipitation (MeDIP) (Weber *et al.*, 2005) is used with tiling microarrays for evaluating specific genetic loci and disease states (Cheng *et al.*, 2008). Many microarray variants have also been used to analyze the methylation status of specific loci of interest (Hoque *et al.*, 2008) or whole genomes (Hayashi *et al.*, 2007; Penterman *et al.*, 2007). However, drawbacks of tiling array design and production have confounded its application on account of inaccurate hybridization signals and the impossibility of analyzing methylated regions of repeat DNA. Thus, microarray technology is being replaced by high-throughput next-generation sequencing as the method of choice.

The next-generation sequencing technology has been developed since 2005, represented by 454–FLX (Roche), SOLiD (Applied Biosystems), and Genome Analyzer (Illumina). To date, 454–FLX (Margulies *et al.*, 2005; Schuster, 2008) which utilizes pyrosequencing for readout has raised the mark to more than 400 bp recently, approaching the Sanger sequencing read-length. The specificity and reproducibility of SOLiD has been improved by water-in-oil emulsion PCR, and it has been applied in many genomic studies (Diehl *et al.*, 2006; Williams *et al.*, 2006; Chen *et al.*, 2010). Solexa was brought into production by Illumina scientists, and upgraded to Genome Analyzer from 2006. In 2009, Illumina replaced the Genome Analyzer system with a new sequencer, Hiseq, with higher sequencing speed and lower cost (Quail *et al.*, 2008).

The Beijing Genomics Institute, a sequencing center in China, has established its own technical platforms based on large-scale genome sequencing and efficient bioinformatics analyses (Li *et al.*, 2010). We obtained the DNA methylome of human peripheral blood mononuclear cells by performing whole-genome bisulfite sequencing, and have developed other high-throughput sequencing techniques including MBD sequencing (MBD-seq), MeDIP sequencing (MeDIP-seq), reduced representation-bisulfite sequencing (RRBS), and ChIP-bisulfite sequencing (ChIP-BS). In this chapter, we introduce these techniques and evaluate all the data generated by these techniques from the same genome.

19.2   Materials and methodology

19.2.1   Samples

We isolated genomic DNA from the same peripheral blood mononuclear cells (PBMCs) named YH cells from a consenting donor whose genome was deciphered in the YH project (Wang *et al.*, 2008) and applied them in constructing sequencing libraries for BS-seq, MBD-seq, MeDIP-seq, and RRBS, while for ChIP-seq and ChIP-BS library construction, we cultured the cells with RPMI1640 (C22400500BT, Invitrogen, Carlstad, CA, USA) and collected them when their number reached 1×10^7.

19.2.2   Construction of BS-seq library

At least 100 ng starting DNA was applied following the Illumina (San Diego, CA, USA) paired-end library construction. Simply, adaptor-ligated DNA was treated with bisulfite sodium and PCR products were sequenced with the Illumina high-throughput sequencer (Yingrui *et al.*, 2010).

19.2.3   MBD-seq library construction

A sample of 10 µg DNA was applied following Illumina single-end or paired-end sequencing library preparation protocols. Simply, adaptor-ligated DNA was incubated with MBD proteins–Dynabeads® and eluted in a stepwise gradient of increasing NaCl concentration (Li *et al.*, 2010). The recovered DNA was largely double-stranded and ready for Illumina GAII sequencing.

19.2.4   MeDIP-seq library construction

A sample of 5 µg DNA was used for library construction. After paired-end adaptor ligation, immunoprecipitation was performed by Ning Li *et al.*'s method (Li *et al.*, 2010). Amplification products were sequenced by Illumina GA paired-end method.

19.2.5   RRBS library construction

PBMCs from a consenting donor whose genome had been deciphered in the YH project (Wang *et al.*, 2008) were separated by Ficoll–Paque (GE Healthcare) gradient centrifugation; total DNA was prepared by proteinase K/phenol extraction, followed by 100 U MspI (NEB R0106L) digestion before library construction. The sequencing libraries from the generated DNA fragments were constructed following Li Wang and colleagues' method (Wang *et al.*, 2012). Finally, amplified libraries were purified, recovered and sequenced according to Illumina guidelines.

19.2.6   ChIP-BS library construction

We start with 5×10^7 cultured cells cross-linked with 1% formaldehyde at 37 °C. Then the cells were rinsed with 1× cold phosphate buffered saline (PBS) plus bovine serum albumin (BSA) (5 mg/ml) and added lysis buffer for chromosome fragmentation through sonication. Then 10% of the fragmentated products were taken out as input DNA and reversed cross-linking before running on agarose gel to evaluate DNA size and quantity. The rest was immunoprecipitated overnight with antibodies which were prebound to A/G Dynal-magnetic-beads (Invitrogen). Enriched DNA was collected and rinsed six times with RIPA buffer, followed by 2× Tris-EDTA (TE) wash and reversed cross-linking at 65 °C overnight. Reversed DNA was purified and prepared according to the Illumina paired-end sequencing library protocol, then treated with bisulfite sodium; PCR products were purified with QIAquick (QIAGEN, Hilden, Germany) purification kit.

19.3 Results and discussion

We generated data on the DNA methylation pattern of the YH genome by the techniques of bisulfite sequencing, MBD-seq, MeDIP-seq, RRBS, and ChIP-BS based on the Illumina GAII and Hiseq sequencers. In earlier studies, Li *et al.* (2010) evaluated the first three methods; here we have added MeDIP-BS and ChIP-BS, and have evaluated these techniques systematically as alternative techniques to map DNA methylation that may possess different strengths and weaknesses. Our results may help researchers to select the most appropriate technology for any given project. Figure 19.1 gives a summary of these methylation mapping methods.

19.3.1 Sequencing data generation and quality evaluation

The methylome data (YH DNA methylome project, *PLoS* online) were generated from PBMCs of the same donor whose genome was published in the YH project. In total 103.5 Gbp of paired-end data were generated by using whole-genome bisulfite sequencing. Of the 18 962 679 CpGs present in the YH reference genome sequence, approximately 99.86% were covered by at least one unambiguously mapped read and 92.62% were unambiguously covered on both strands. The

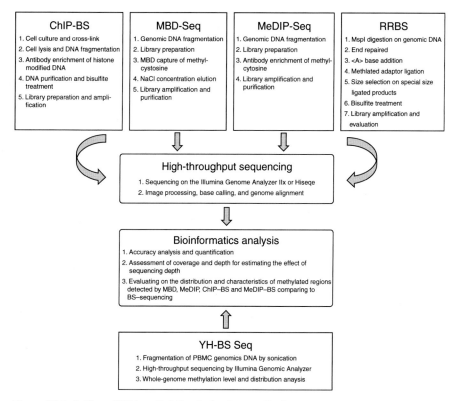

Figure 19.1 Outline of DNA methylation technology methods.

bisulfite conversion rate was determined to be 99.8%, ensuring reliable ascertainment of CpG methylcytosines with a false-positive rate of <0.5%. The analysis in the YH DNA methylome excluded non-CpG, and this single-base-resolution methylome is taken as the standard reference to estimate other methylated methods.

The MBD protocol generated ~750 Mbp 35-bp single-end reads from each salt concentration library, and MACS identified peaks associated with enrichment (Weber and Schubeler, 2007). Finally, 700 mM had been shown to be the most useful salt concentration as it struck a good balance between CpG density and DNA methylation level (Li *et al.*, 2010). In this study, we took 136 324 734 reads for complementary evaluation about this method (Table 19.1), and reached up to 91.20% reads that had been mapped back to the reference as well as a 59.48% unique map rate, showing the high sequencing quality and good enrichment of methylated DNA tags.

The MeDIP-seq protocol generated in total 7.9 G 45-bp paired-end data by Illumina Genome Analyzer II. The protocol recommended that about 2–3 Gbp MeDIP data is the most reasonable point considering the expensive sequencing cost in our reported studies. Here, we brought 127 877 818 reads for further detection and comparison with MBD (Table 19.1). Of these 110 548 559 reads were mapped with a map rate of 86.45%, and 61.64% of them were uniquely mapped, which also demonstrated good sequencing quality and enrichment efficiency.

The RRBS library generated 139 950 000 paired-end reads, and 75.22% of them were successfully aligned to the methylome reference (Table 19.1). Of the generated reads, 97 844 444 were unique reads with a mapped rate of 69.91%, as well as a high conversion rate which showed good library and sequencing quality.

The ChIP-BS library generated 142 493 860 paired-end reads. Of these 76 039 736 were mapped to the reference and 26.04% of them were uniquely mapped (Table 19.1). The unique map rate of 26.04% and the fact that all of the peaks covered 0.05% of the total region of the genome seemed a little lower than other methods; we are still seeking reasons for this.

19.3.2   Sequencing data assessment

Our reported data (Li *et al.*, 2010) show that ~4.4 Gbp MBD and ~4 G bp MeDIP data could produce peaks of 447 598 and 447 598, respectively. Of these peaks, there were 4 787 790 CpG numbers in MBD-seq data and 4 787 790 in MeDIP-seq data, 73.24% and 42.70% of which showed a methylation rate above 60%. For further validation, we took out 136 324 734 MBD reads and 127 877 818 MeDIP reads for deeper peak information assessment. Peaks of 279 229 and 500 554 had been called for gene ontology (GO) enrichment analysis, and these peaks covered 6.8% and 7.32% of genome regions, showing reasonable enrichment and coverage of gene units.

Reads from each of the biological replicates were aligned to genome reference as well as the size-selected MspI fragments generated by of RRBS. The alignments were carried out with BGI SOAPaligner v. 2.01, allowing up to two mismatches for successful mapping. We assessed the coverage, mean coverage depth, genomic

Table 19.1 Data alignment and assessment of bisulfite sequencing, MBD-seq, MeDIP-seq, MeDIP-BS, and ChIP-BS

Library	Total reads	Map reads	Map rate(%)	Unique reads	Unique map rate (%)	Peak counts	Average peak length	Total region length	Total region percentage in genome(%)	Methylation rate (%)	Conversion rate (%)
YH-methylome	1 587 460 142	1 174 566 564	73.99	1 035 028 056	65.20	–	–	–	–	66.87	99.80
MBD	136 324 734	124 325 155	91.20	81 087 446	59.48	279 229	730.74	204 043 516	6.80	–	–
MeDIP	127 877 818	110 548 559	86.45	78 821 377	61.64	500 554	438.82	219 654 160	7.32	–	–
RRBS	139 950 000	105 270 039	75.22	97 844 444	69.91						
ChIP-BS	142 493 860	76 039 736	53.36	37 110 148	26.04	4 639	274.7	1 274 421	0.05	39.11	98.13

distribution of the detected CpG sites and reproducibility of RRBS with an increasing quantity of the data set. We also carried out comparisons between RRBS and BS-seq data considering their concordance of DNA methylation level. Our annotation of coverage and coverage depth of RRBS to PBMC of YH genome with an increasing data set size (0.5–5 Gbp) indicated that 3 Gbp could sufficiently guarantee good results. Given a 3 Gb data quantity, RRBS could cover around 70% of the theoretical maximum of the 6.8 million CpG sites with at least one read where 82.2% and 61.4% of these covered CpG sites enjoy the coverage depth ≥4× or ≥10×, respectively (Wang *et al.*, 2012).

On the other hand, we obtained a 39.11% methylation rate and 98.13% conversion of the YH ChIP-BS sample, which revealed effective bisulfite treatment. While coverage and depth evaluation of reads were distributed on gene units, it appeared that the ChIP-BS reads could cover most gene units with a reasonable distribution under lower depth. And the annotated ChIP-BS data showed that 0.5× depth could cover more than 30% Up5K, intron, down5k unit regions and at least 20% 5'-UTR, 3'-UTR and CDSs (Figure 19.2A and B). Concordance was also evaluated between ChIP-BS and ChIP-seq data. To test this objective, we made a quantitative assessment on the coding and intergenic regions as well as repeat regions. Figure 19.2C and D showed that it was well confirmed for the concordance in gene units, intergenic regions, and repeatitve sequences, suggesting that we enriched the same DNA fragments with the two techniques, and that the bisulfite treatment exerted insignificant effect on the captured reads.

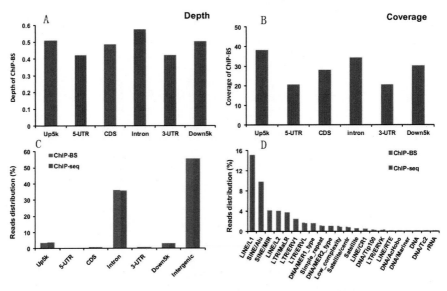

Figure 19.2 Analysis of coverage, depth, and reads of RRBS and ChIP-BS. (A) RRBS coverage with 3.5 Gbp YH meythylome data. (B) MeDIP-BS depth comparison with 3.5 Gbp YH meythylome. (C) Coverage of repeated units of RRBS. (D) Evaluation of ChIP-BS and ChIP-seq reads.

19.3.3   Evaluation of distribution and characteristics of methylated regions detected by MBD-seq, MeDIP-seq, RRBS, and ChIP-BS compared to bisulfite sequencing

19.3.3.1   Characteristics of the YH methylome

Applying the bisulfite sequencing method, we generated an essentially complete (92.62%) version of the PBMC methylome and carried out a global analysis, finding the overall CpG methylation level to be 68.8%. We found the majority of CpG sites (68.4% at a false-positive rate of 0.46%) and only <0.2% of non-CpG sites to be methylated, demonstrating that non-CpG cytosine methylation is negligible in human PBMCs. Analysis of the PBMC methylome revealed a rich landscape of epigenomic data for 20 distinct features including regulatory, protein-coding, RNA-gene-coding, non-coding, and repeat sequences. Using a conservative cut-off ($p < 0.001$), we identified 599 haploid differentially methylated regions (hDMRs) covering 287 genes; these data suggest that imprinting may be more common than previously thought. Our study not only provides a comprehensive resource for future epigenomic research but also demonstrates a paradigm for large-scale epigenomics studies through new sequencing technology.

19.3.3.2   Evaluation of MBD-seq and MeDIP-seq

To evaluate MBD-seq and MeDIP-seq, we referred to the PMBC methylome and screened the CpG profiles in regulatory regions using eligible windows, including 2 kbp upstream of genes, 5′ untranslated region (UTR), and CDS, on the basis of which we compared these three methods (Li *et al.*, 2010). In the same regions, the bisulfite sequencing profile presented relatively lower CpG counts but a higher methylation level compared to MeDIP and MBDs. Similarly, MeDIP and MBD profiles enriched more CpG-rich fragments that had been highly methylated. The relative sensitivities of MeDIP and MBD enrichment profiles were evaluated. MBD was inclined to CpG-rich regions, while MeDIP covered the intermediate CpG density regions which were highly methylated. Followed with cellular component, molecular function and biological process GO analysis (data not shown), both results showed that biological process units occupied nearly half of the enriched sequences, which validated the most accepted conclusion that most regulatory gene units were modified by methylation. Interestingly, genes enriched by these two methods were similarly distributed, which hinted to us that they both possess affinity for highly methylated regions. To conclude, MBD was more sensitive than MeDIP for high CpG densities, and was not related to the DNA methylation level; conversely, MeDIP was generally more efficient than MBD in enriching for regions of medium CpG density but only when they were highly methylated.

19.3.3.3   Evaluation of MeDIP-BS and ChIP-BS

As ChIP-BS and RRBS were both newly developed technologies, we tested the concordance of ChIP-seq and ChIP-BS data, which have both been applied to the detection of DNA fragments that interact with histone modified by methylation, and between RRBS and bisulfite-sequencing data, which both focused on genome-wide DNA methylation status. To evaluate the efficiency of MeDIP-BS, we compared it with the MeDIP-seq data and MeDIP-BS data with non-removal of repeat sequences. These data will be presented in future studies as time permits. As for ChIP-BS, recent studies indicate that the proteins directly involved in DNA methylation and H3K4/K9/K27 methylation present molecular interactions, which suggests that both modifications could be the basis for recruiting the other, thus they cooperate to maintain the repressive state of genes (Ikegami *et al.*, 2009). To date, ChIP-seq is the best technology to study histone–DNA interaction; however, it only enriches DNA fragments interacting with histone, without permitting the detection of their methylation status. Therefore we combined ChIP-seq with bisulfite treatment to achieve this objective. Thus we validated the methylation level of peak regions for evaluating data requirement and GO analysis by comparing ChIP-seq and bisulfite sequencing data. Results showed that they enriched nearly the same gene region, which might suggest to us that such 3.5 Gbp ChIP-BS data could basically satisfy this kind of bioinformatics analysis (data not shown).

While our reported RRBS studies revealed that a 3 Gbp data set could cover 40% of the CpG islands, 27% of the 1000 bp upstream and 45% of the 5′-UTR with at least 4 reads and 32%, 20%, and 34% of them with at least 10 reads. Also, a 5 Gbp data set could improve the coverage by 1–2%. There was a relatively lower coverage of the flanking regions of the CpG islands and the other gene regions such as CDS, introns, 3′-UTR, and the downstream as well as the repetitive sequences. For all the regions covered by both RRBS and BS-seq within the YH genome, RRBS showed far greater coverage depth (>33×) compared with BS-seq (>6×). It was known that the BS-seq (YH methylome) achieved an average 6× coverage depth over these regions at 103.5 Gbp whereas RRBS needed 33× which was 6 Gbp. However, this was offset when we considered the genome-wide coverage of the cytosines by RRBS compared with the complete coverage of BS-seq.

19.3.4   Comparison of bisulfite-based methods

A major advantage of such bisulfite-based methods is that they allow quantitative comparisons of methylation levels at single-base resolution. We calculated the methylation level for all data generated from three bisulfite-based methods by defining the methylation level of a specific cytosine as the proportion of reads covering each mC to the total reads covering the site. As Table 19.2 and Figure 19.3 show, results from BS-seq indicate that 2.63% of all the cytosines in the whole genome are methylated, among which the average methylation level of

Table 19.2 Methylation patterns of bisulfite sequencing, MeDIP-BS, and ChIP-BS

Pattern	C	CG	CHG	CHH
Whole-genome methylation	2.63	66.87	0.22	0.21
MeDIP-BS	18.21	81.16	2.03	7.24
ChIP-BS	5.15	39.11	1.65	2.4

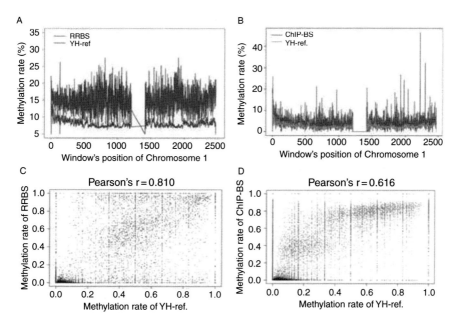

Figure 19.3 Methylation level distribution and Spearman correlation analysis. (A) Methylation level distribution of RRBS and YH methylome at a 200-kbp window. (B) Chromosome 1 methylation level distribution of ChIP-BS and YH methylome. (C and D) Spearman correlation analysis of MeDIP-BS, ChIP-BS, and YH methylome.

all CpG dinucleotides is 66.87%; in contrast the methylation levels for CHG and CHH are minimal, only about 0.22% and 0.21%, repectively. Comparatively, enzyme-digested fragments presented higher methylation levels of 3.6% among all cytocines and 12.17% among all CpG dinucleotides; especially in non-CpG sites, the methylation levels are 10 times higher than the average level of the whole genome, indicating that RRBS can enrich general 5mC especially the CpG islands and promoter regions in a genome-wide scale. The methylation rates of the aligned data on the same enriched gene region by RRBS and BS-seq were compared over a 500-Kbp sequence window (Figure 19.3A). A similar profile of DNA methylation throughout the genome was observed between them, though the methylation rate of regions captured by RRBS was obviously higher than the counterpart data from BS-seq. And the curve changed on lower methylation regions (1200–1500 Kbp) which might due to the MspI's affinity for some characteristic bases.

ChIP-enriched fragments presented a relatively higher methylation level among all enriched cytosines (5.15%) than the average methylation level in the whole genome (2.63%), while the methylation level in all CpG dinucleotides was relatively lower (39.11%) than the average level in the genome (Table 19.2), though non-CpG sites also presented a higher methylation level than average, indicating a special pattern of DNA methylation in these histone–DNA fragments. To further study methylation status on ChIPed DNA, which represents the interaction sites for histone–DNA, we tested the methylation level on the same enriched gene region in chromosome 1 by bisulfite sequencing and ChIP-BS. As shown in Figure 19.3B, bisulfite sequencing detected a much higher density of CpG sites than ChIP-BS, while ChIP-BS obtained relatively fewer CpG sites but with higher methylation levels, suggesting deeper sequencing and more detailed information. More importantly, the same tendency for both bisulfite sequencing and ChIP-BS profile was observed, indicated good sensitivity of the ChIP-BS method.

For a more quantitative assessment of the accuracy of measurement, we isolated the methylation level information of certain DNA fragments from the YH methylome corresponding to the fragments enriched by the RRBS and ChIP-BS techniques, and performed Pearson correlation analysis on them. According to the results, a Pearson's correlation of 0.95 was observed between RRBS, and BS-seq, indicating a high degree of concordance (Wang *et al.*, 2012). However, when the DNA methylation level of a particular cytosine was considered, concordance between RRBS and BS was relatively lower (Figure 19.3C). One possible reason could be that the coverage depth of BS-seq for a particular cytosine was not sufficient for accurate estimation of the DNA methylation state. However, by taking the DNA methylation level of a region into account, most of the background noise could be reduced. Similarly, for ChIP-BS, a relatively low correlation score of 0.616 was obtained (Figure 19.3D), which is possibly acceptable, since between the blood genome and the genome from cultured cell state, the DNA–histone interaction and the DNA methylation profile might be accordingly changed. However, it would be a good choice for studies on histone and DNA interaction, as reads captured and treated by ChIP-BS are without evident bias.

19.4   Concluding remarks

In this chapter, we have evaluated five high-throughput sequencing methods for evaluating DNA methylation on chromatin level, including bisulfite sequencing, MeDIP-seq, MBD-seq, RRBS, and ChIP-BS. For MeDIP-seq and MBD-seq analysis, we found that MeDIP was generally more efficient than MBD in enriching regions of medium CpG density but only when they were highly methylated. While RRBS can be the method of choice for investigating DNA methylation if the studies are intended to involve multiple samples and if CpG islands or promoter regions are the regions of interest, we now expect to apply RRBS in DNA methylation studies on a genome-wide and large-sample scale, since it has been suggested that variation in DNA methylation levels of CpG islands relates to the process of cellular development and cyclical fluctuation of DNA methylation level in the promoter regions. As for ChIP-BS, data concordance with ChIP-seq told us that information has not been lost via bisulfite treatment and that genes enriched by ChIP do not have bias on methylation levels on peak regions. Therefore the ChIP-BS method could be efficiently used to evaluate the methylation status of the DNA fragments interacting with methylation-modified histone.

Overall, scientists could evaluate their studied targets from different modification levels using these methods, depending on time and cost. On the other hand, single-molecular DNA sequencing technologies such as Helicos and SMRT have been invented to accelerate development in genetics (Clarke *et al.*, 2009; Hart *et al.*, 2010). Studies taking advantage of single-molecular DNA sequencing on DNA methylation sequencing have already been published, which give an indication for a new direction for development (Harris *et al.*, 2008; Efcavitch and Thompson, 2010; Flusberg *et al.*, 2010).

REFERENCES

Bird, A. (2002). DNA methylation patterns and epigenetic memory. *Genes and Development*, **16**, 6–21.

Bird, A. P. and Southern, E. M. (1978). Use of restriction enzymes to study eukaryotic DNA methylation. I. The methylation pattern in ribosomal DNA from *Xenopus laevis*. *Moleculor Biology*, **118**, 27–47.

Chen, W., Ullmann, R., Langnick, C., *et al.* (2010). Breakpoint analysis of balanced chromosome rearrangements by next-generation paired-end sequencing. *European Journal of Human Genetics*, **18**, 539–543.

Cheng, A. S., Culhane, A. C., Chan, M. W., *et al.* (2008). Epithelial progeny of estrogen-exposed breast progenitor cells display a cancer-like methylome. *Cancer Research*, **68**, 1786–1796.

Clarke, J., Wu, H. C., Jayasinghe, L., *et al.* (2009). Continuous base identification for single-molecule nanopore DNA sequencing. *Nature Nanotechnology*, **4**, 265–270.

Colot, V. and Rossignol, J. L. (1999). Eukaryotic DNA methylation as an evolutionary device. *BioEssays*, **21**, 402–411.

Diehl, F., Li, M., He, Y., *et al.* (2006). BEAMing: single-molecule PCR on microparticles in water-in-oil emulsions. *Nature Methods*, **3**, 551–559.

Eckhardt, F., Lewin, J., Cortese, R., *et al.* (2006). DNA methylation profiling of human chromosomes 6, 20 and 22. *Nature Genetics*, **38**, 1378–1385.

Efcavitch J. W. and Thompson, J. F. (2010). Single-molecule DNA analysis. *Annual Review of Analytical Chemistry*, **3**, 109–128.

Ehrlich, M., Field, J. K., Liloglou, T., *et al.* (2006). Cytosine methylation profiles as a molecular marker in non-small cell lung cancer. *Cancer Research*, **66**, 10911–10918.

Flusberg, A., Webster, D. R., Lee, J. H., *et al.* (2010). Direct detection of DNA methylation during single-molecule, real-time sequencing. *Nature Methods*, **7**, 461–465.

Harris, T. D., Buzby, P. R., Babcock, H., *et al.* (2008). Single-molecule DNA sequencing of a viral genome. *Science*, **320**, 106–119.

Hart, C., Lipson, D., Ozsolak, F., *et al.* (2010). Single-molecule sequencing: sequence methods to enable accurate quantitation. *Methods in Enzymology*, **472**, 407–430.

Hayashi, H., Nagae, G., Tsutsumi, S., *et al.* (2007). High-resolution mapping of DNA methylation in human genome using oligonucleotide tiling array. *Human Genetics*, **120**, 701–711.

Hoque, M. O., Kim, M. S., Ostrow, K. L., *et al.* (2008). Genome-wide promoter analysis uncovers portions of the cancer methylome. *Cancer Research*, **68**, 2661–2670.

Ikegami, K., Ohgane, J., Tanaka, S., *et al.* (2009). Interplay between DNA methylation, histone modification and chromatin remodeling in stem cells and during development. *International Journal of Developmental Biology*, **53**, 203–214.

Keshet, I., Schlesinger, Y., Farkash, S., *et al.* (2006). Evidence for an instructive mechanism of de novo methylation in cancer cells. *Nature Genetics*, **38**, 149–153.

Khulan, B., Thompson, R. F., Ye, K., *et al.* (2006). Comparative isoschizomer profiling of cytosine methylation: the HELP assay. *Genome Research*, **16**, 1046–1055.

Li, N., Ye, M., Li, Y., *et al.* (2010). Whole genome DNA methylation analysis based on high throughput sequencing technology. *Methods*, **52**, 203–212

Margulies, M., Egholm, M., Altman, W. E., *et al.* (2005). Genome sequencing in microfabricated high-density picolitre reactors. *Nature*, **437**, 376–380.

Penterman, J., Zilberman, D., Huh, J. H., *et al.* (2007). DNA demethylation in the *Arabidopsis* genome. *Proceedings of the National Academy of Sciences USA*, **104**, 6752–6757.

Quail, M. A., Kozarewa, I., Smith, F., *et al.* (2008). A large genome center's improvements to the Illumina sequencing system. *Nature Methods*, **5**, 1005–1010.

Rollins, R. A., Haghighi, F., Edwards, J. R., *et al.* (2006). Large-scale structure of genomic methylation patterns. *Genome Research*, **16**, 157–163.

Schumacher, A., Kapranov, P., Kaminsky, Z., *et al.* (2006). Microarray-based DNA methylation profiling: technology and applications. *Nucleic Acids Research*, **34**, 528–542.

Schuster, S. C. (2008). Next-generation sequencing transforms today's biology. *Nature Methods*, **5**, 16–28.

Tost, J. and Gut, I. G. (2007). Analysis of gene-specific DNA methylation patterns by pyrosequencing technology. *Methods in Molecular Biology*, **373**, 89–102.

Tweedie, S., Charlton, J., Clark, V., *et al.* (1997). Methylation of genomes and genes at the invertebrate-vertebrate boundary. *Molecular Cell Biology*, **17**, 1469–1475.

Wang, J., Wang, W., Li, R., *et al.* (2008). The diploid genome sequence of an Asian individual. *Nature*, **456**, 60–65.

Wang, L., Sun, J., Wu, H., *et al.* (2012). Systematic assessment of reduced representation bisulfite sequencing to human blood samples: a promising method for large-sample-scale epigenomic studies. *Biotechnology*, **157**, 1–6.

Weber, M. and Schubeler, D. (2007). Genomic patterns of DNA methylation: targets and function of an epigenetic mark. *Current Opinion in Cell Biology*, **19**, 273–280.

Weber, M., Davies, J. J., Wittig, D., *et al.* (2005). Chromosome-wide and promoter-specific analyses identify sites of differential DNA methylation in normal and transformed human cells. *Nature Genetics*, **37**, 853–862.

Weber, M., Hellmann, I., Stadler, M. B., *et al.* (2007). Distribution, silencing potential and evolutionary impact of promoter DNA methylation in the human genome. *Nature Genetics*, **39**, 457–466.

Williams, R., Peisajovich, S. G., Miller, O. J., *et al.* (2006). Amplification of complex gene libraries by emulsion PCR. *Nature Methods*, **3**, 545–550.

Yingrui L., Jingde, Z., Geng T., *et al.* (2010). The DNA methylome of human peripheral blood mononuclear cells. *PLoS Biology*, **8**, e1000533

Zhang, X., Yazaki, J., Sundaresan, A., *et al.* (2006). Genome-wide high-resolution mapping and functional analysis of DNA methylation in arabidopsis. *Cell*, **126**, 1189–1201.

Zilberman, D., Gehring, M., Tran, R. K., *et al.* (2007). Genome-wide analysis of *Arabidopsis thaliana* DNA methylation uncovers an interdependence between methylation and transcription. *Nature Genetics*, **39**, 61–69.

20 Three-dimensional quantitative DNA methylation imaging for chromatin texture analysis in pharmacoepigenomics and toxicoepigenomics

Jian Tajbakhsh* and Arkadiusz Gertych

20.1   Introduction

Since epigenetic changes are thought to underlie a wide range of complex diseases, the scope of epigenetic drug therapy is likely to expand, as epigenetic phenomena, in contrast to pure genetic mechanisms, are reversible. This nature of epigenetic imbalances in various types of cancers constitutes an attractive therapeutic target. The disruption of epigenetic misregulation via enzyme inhibition has considerable pharmacological impact that fuels the development of epigenetic therapies with DNA methyltransferase inhibitors (DMNTi) and histone deacetylase inhibitors (HDACi) (Yoo *et al.*, 2006; Szyf, 2009; Herceg, 2010). In addition to their physiologic role these drugs may also affect chromatin architecture and related gene-expression programs in mammalian cells (Haaf, 1995; de Capoa *et al.*, 1996; Stresemann *et al.*, 2008; Jefferson *et al.*, 2010; Tajbakhsh *et al.*, 2010; Tong *et al.*, 2010). These effects can be targeted but can also occur as unwanted adverse side effects that might put cells, organs, and eventually the whole organism at risk (Csoka and Szyf, 2009). The boost in epigenetic drug development calls for high-throughput cell-based screening assays to be implemented in the early phases of drug discovery for assessing the agents' genotoxic side effects on chromatin architecture and the correlated risk of genome instability in targeted cells and tissues in order to improve the agents' clinical utility. These new approaches, which have been termed *pharmacoepigenomics* or *toxicoepigenomics* (Szyf, 2004; Peedicayil, 2008; Ingelman-Sundberg and Gomez, 2010) in analogy to pharmaco-/toxicogenomics, if introduced into the drug development program, might also significantly reduce attrition in the early phases of drug development.

DNA methylation is considered the most robust among the different types of chromatin modifications (Best *et al.*, 2010) and is therefore an attractive

---

* Author to whom correspondence should be addressed.

Epigenomics: From Chromatin Biology to Therapeutics, ed. K. Appasani. Published by Cambridge University Press. © Cambridge University Press 2012.

pharmacological target in the reactivation of ectopically methylated genes (Issa, 2007; Toyota *et al.*, 2009; Yang *et al.*, 2010). A number of DNMTi of different categories have been designed, with 5-azacytidine (5-AZA) and its analog 5-aza-2-deoxycytidine (decitabine) being approved for treatment of patients with myelo-dysplatic syndrome (MDS) (Luebbert, 2000; Momparler, 2005). Recently a new type of cytosine analog has been introduced: 1-β-D-ribofuranosyl-2(1H)-pyrimidone, also known as zebularine, with less toxicity to cultured cells than the other two azanucleosides (Marquez *et al.*, 2005), and significantly stronger anti-proliferative effect on cancer cells than on normal cells (Balch *et al.*, 2005). Nevertheless it is important to recognize that hypermethylation of single-gene promoters occurs against a strong background of general DNA hypomethylation including a decrease in the methylation load of heterochromatic regions of the genome (55% of the human genome). The loss of methyl groups is achieved mainly by hypomethylation of heterochromatin-residing repetitive DNA sequences, including transposable elements such as long interspersed nuclear elements (LINE), especially LINE-1 as part of the facultative heterochromatin, and satellite 2 (Sat2) DNA as part of the constitutive heterochromatin (Ehrlich, 2005). The extent of genome-wide hypomethylation correlates closely with the degree of malignancy, in a tumor-type-dependent manner (Costello and Plass, 2001). In normal cells, the activity and interaction of these classes of DNA with neighboring chromatin regions (in the nuclear space) are strongly suppressed by methylation and compaction through histone modifications. Therefore, there are multiple risks associated with an increase in global hypomethylation (Carr *et al.*, 1984; Chen *et al.*, 1998; Cui *et al.*, 2002; Gaudet *et al.*, 2003; Holm *et al.*, 2005), especially demethylation of repetitive elements, which can lead to their decondensation and related consequences – with side effects, such as transcriptional activation of oncogenes, activation of latent retrotransposons, chromosomal instability, and telomere elongation of chromosomes (Holliday and Pugh, 1975; Feinberg and Vogelstein, 1983; Bestor and Tycko, 1996; Ehrlich, 2000; Yamada *et al.*, 2005; Vera *et al.*, 2008). This is in fact a critical issue, as malignant cells already contain 20–60% less genomic methylcytosine than their normal counterparts. Inhibitors of DNMT and HDAC are far from perfect, as they show little consistent efficacy against solid tumors with their extraordinary degree of complexity in chromatin modifications and signaling, which makes the identification of suitable compounds challenging.

These types of cancer show a significant disruption in chromatin phenotype (texture) demonstrated in many morphologic studies on cancer by the application of quantitative digital texture analysis – supporting the role of chromatin texture as a biomarker for diagnosis and prognosis (Huisman *et al.*, 2007; Orr and Hamilton, 2007). Given the prevalence and load of DNA methylation imbalances, especially hypomethylation of repetitive elements, cellular imaging of global nuclear DNA patterns may become extremely supportive in texture analysis, as the underlying molecular processes involve large-scale chromatin reorganization, visible by light microscopy (Espada and Esteller, 2007). Indeed high-resolution optical imaging has evolved into an essential tool for moving new chemical entities through the pharmaceutical discovery pipeline utilizing cell-based assays. Cellular imaging, when combined with systems biology, could promote a more biology-driven environment for compound progression, therefore compelling us to think in terms of cell genotype–phenotype relationships. Since heterochromatin decondensation bears a serious risk in epigenetic therapy, it is important to consider the testing of new compounds and drugs regarding their effects on the higher-order structure of chromatin by cell-based assays in conjunction with imaging in the preclinical phases of drug discovery. The authors of this chapter introduce a novel cytometric approach termed 3D quantitative DNA methylation imaging (3D-qDMI), which addresses this matter (Tajbakhsh *et al.*, 2008, 2010; Gertych *et al.* 2009, 2010). The technology applies image-analysis algorithms for extraction of fluorescence signals from 3D images of chromatin texture to visualize and measure drug-induced changes in global DNA methylation and related chromatin reorganization in nuclei of thousands of cells in parallel. The 3D-qDMI system connects fluorescence techniques including indirect immunofluorescence (IF) and fluorescence in situ hybridization (FISH) with computational techniques for image-data analysis and interpretation. The method is amenable to scale and can be flexibly implemented for drug screening in research and in the pharmaceutical industry.

20.2   Methodology

The workflow of 3D-qDMI comprises three major steps: (1) a cell-based fluorescence assay to delineate nuclear chromatin and associated targets such as methylated cytosine and global DNA (gDNA), (2) high-resolution 3D imaging of labeled cells, and (3) image analysis including extraction of features such as MeC/gDNA, as outlined in Figure 20.1.

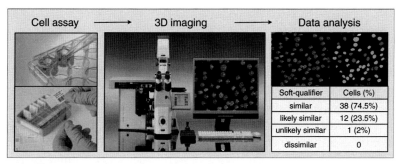

Figure 20.1 The workflow of 3D quantitative DNA methylation imaging (3D-qDMI) in cell-based assays comprises three steps: (1) cytological specimen preparation: retrieval of cultured cells and immunofluorescence assay, (2) 3D imaging of specimens with microscopes or high-throughput scanners, and (3) image/data analysis for specimen characterization. The workflow can be integrated and automated for industrial scale.

20.2.1   Cell-based fluorescence assay

In mammalian cells, the fluorescence-based covisualization of methylated CpG dinucleotides (MeC) and global DNA (Habib *et al.*, 1999; Mayer *et al.*, 2000; Barton *et al.*, 2001; Santos *et al.*, 2002; Tajbakhsh *et al.*, 2008) should include the following steps: (i) fixation of cells, (ii) permeabilization of the cell and nuclear membrane, (iii) antigen retrieval of MeC sites by HCl treatment, (iv) RNase digestion, (v) MeC-specific immunostaining, and (vi) DNA counter-staining, for subsequent confocal microscopy and 3D imaging. The assay protocol supports three essential points: (1) the preservation of structural integrity of the cells, (2) the accessibility promotion of targeted epitopes, and (3) the insurance of detection specificity. Fixation of the cells with cross-linking agents such as 4% paraformaldehyde for 10–30 minutes at room temperature has been proven to preserve the 3D structure of cells, making it suitable for in situ fluorescence assays, 3D visualization, and topological analysis of fixed cells (Kurz *et al.*, 1996; Tajbakhsh *et al.*, 2000; Scheuermann *et al.*, 2004). We recommend avoiding the use of dehydrating agents such as organic solvents (alcohols, acetone) as they can cause nucleic acid precipitation, and thus perturb the higher-order spatial chromatin organization and nuclear architecture. Visualization of MeC patterns requires the accessibility of the antibody to its epitope on the single-stranded DNA. This access is achieved by denaturing double-stranded DNA (dsDNA), for which different techniques have been applied: UV irradiation (Miller *et al.*, 1974), enzymatic DNA digestion with different endonucleases (Bensaada *et al.*, 1998), and nucleic acid hydrolysis through HCl treatment (Barbin *et al.*, 1994; Miniou *et al.*, 1994; Montpellier *et al.*, 1994; de Capoa *et al.*, 1996; Rougier *et al.*, 1998; Soares *et al.*, 1999; Piyathilake *et al.*, 2001). Treatment with HCl at room temperature is very effective and simplest to automate as it does not require additional technical components such as a UV-lamp or sophisticated temperature controlling elements. The use of HCl requires adjustment as there is a trade-off between stoichiometric antigen (MeC) labeling, DNA decomposition, and loss of structural integrity visualized through interference with binding of DNA intercalating dyes such as 4′,6-diamidino-2-phenylindole (DAPI) and propidium iodide, as previously reported (Sasaki *et al.*, 1998; Kennedy *et al.*, 2000). We have experienced that 2N HCl used at exposure times between 35 and 45 minutes can yield satisfying in situ detection of nuclear methylcytosine residues. These conditions can be used with a variety of different cell types. Significant divergence from this combination may result in either weaker MeC signals or obscuring of the DAPI pattern and/or the pattern of covisualized cellular proteins (data not shown). Antibody specificity for methylcytosine can be assessed with small DNA test arrays that contain immobilized oligonucleotides with a defined number of methylcytosine residues and a negative control representing the unmethylated copy sequence. Figure 20.2 displays the microarray we designed and utilized in the specificity assessment of the anti-MeC antibodies that we used in immuno-fluorescence cell-based assays.

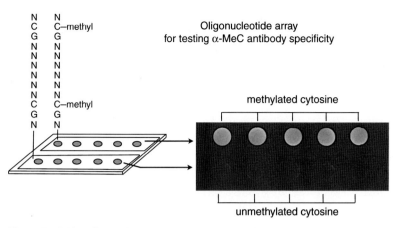

Figure 20.2 The oligonucleotide microarray designed for testing of α-methylcytosine antibodies helps in evaluating the specificity of the antibodies by a rapid Southwestern hybridization. The schematic (not in scale) on the left side of this figure illustrates the array – manufactured through standard surface chemistry – which consists of two types of synthesized 24-mer oligonucleotides that are deposited on a microscopic glass slide. The two single strands have the same nucleotide sequence, including two CpGs which only differ in their methylation statuses: both methylated and unmethylated cytosine analogs were incorporated into the synthesized strands by phosphoramadite chemistry. The fluorescence image taken with a CCD-camera confirms the specificity of the applied anti-MeC antibody for methylated cytosine: arbitrary spot intensity was on average 20-fold over the background signal seen for the unmethylated spots. The signal dynamic range can be increased by utilizing oligonucleotides with a higher number of MeCs.

20.2.2 Immunofluorescence protocol

Cells were plated at 1×10^5 cells on round cover slips (18 mm, 1 ounce) in a 12-well microplate and allowed to attach for 24 hours. Then cells were treated with demethylating agents such as 5-azacytidine, decitabine, and zebularine (Sigma-Aldrich, St. Louis, MO, USA) for 48 hours or more – an untreated control population was kept in pure medium, in parallel – before being washed with phosphate buffered saline (PBS) prior to fixation. This protocol has been optimized for applications with 12-well microplates ($\varnothing = 3.5$ cm). We recommend adjusting volumes and reagent concentrations when using other sizes of dishes. In order to preserve the cells' 3D structure, cells cultured on coverslips were fixed with 4% paraformaldehyde/PBS. Permeabilization to facilitate probe penetration into cells and nuclei was achieved by incubation with a mixture of 0.5% saponin/0.5% triton X-100/PBS. The use of the detergent mixture at this concentration has been demonstrated to have no adverse effect on the cellular distribution of nucleic acids and proteins (Tajbakhsh *et al.*, 2000; Scheuermann *et al.*, 2004). Even though hydrolysis is known to also disintegrate cellular RNA species, we introduced a subsequent RNase A treatment step to avoid any residual transfer RNAs (tRNA) that have escaped complete digestion and removal from the cells, as tRNA can contain methylated cytosine. Although tRNA is not expected within the mammalian nucleus, this step should be used for eliminating cytosolic tRNA species in favor of a better demarcation and contrasting of the nucleus against the cytosol (in the MeC image channel). Furthermore, in cases such as HIV-infected cells, it has been shown that defective tRNA is shuttled into the nucleus (Zaitseva *et al.*, 2006). Cells were subsequently treated with HCl, then washed with 0.1% bovine serum albumin fraction V (BSA)/PBS. The cells were blocked with 3% BSA/PBS before incubation with an unconjugated monoclonal mouse anti-5-methylcytosine antibody (GeneTex, Irvine, CA, USA), at 1–2 µg/ml for 1–2 hours at 37 °C (or overnight at 4 °C), followed by a secondary Alexa488-linked goat (or donkey) anti-mouse polyclonal IgG (Invitrogen, Carlsbad, CA, USA), at the concentration of 5–10 µg/ml for 1–2 hours at 37 °C. Antibodies were diluted in blocking solution. Intermediate stringency washes after antibody incubation were performed with 0.1% BSA/0.1% Tween 20/PBS. For all other washing steps 0.1% BSA/0.1% Tween 20/PBS was used. The specimens were counterstained for 15 minutes at room temperature with a 1.43 µM DAPI solution (FluoroPure grade, Invitrogen), and finally embedded in ample mounting solution (ProLong Gold, Invitrogen) on glass slides. Specimens should be kept at room temperature (protected from light) for a minimum of 6 hours to ensure hardening of embedding resin before fluorescence microscopy.

20.2.3   High-resolution three-dimensional imaging

Specimens were analyzed by confocal laser scanning microscopy using two different systems: (1) a TCS SP2 (Leica Microsystems, Mannheim, Germany) equipped with a multi-line argon laser (465 nm, 488 nm, 514 nm), three HeNe lasers (540 nm, 590 nm, 633 nm) for Alexa488 (MeC) and Alexa568 (ACTH); and (2) a TCS SP5 X Supercontinuum microscope (Leica Microsystems), equipped with a white laser. The latter system provides full freedom and flexibility in excitation and emission, within the continuous range of 470 to 670 nm in 1-nm increments. Both microscopes are coupled with a 405-nm diode laser line for excitation of DAPI fluorescence. Serial optical sections were collected at increments of 200–300 nm with a Plan-Apo 63X 1.4 oil-immersion lens (for TCS SP2) and a Plan-Apo 63X 1.3 glycerol-immersion lens (for TCS SP5 X). The pinhole size was consistently 1.0 airy unit. To avoid bleed-through, the imaging of each of the two channels (MeC and DAPI) was acquired sequentially. The typical image size was either 1024 × 1024 (for TCS SP2) or 2048 × 2048 (for TCS SP5 X) with a respective voxel size of 116 nm × 116 nm × 230.5 nm (*x*, *y*, and *z* axes) or 120 nm × 120 nm × 250 nm, and a dynamic intensity range of 8–12 bits per pixel in both channels. The MeC and DAPI signals from optical sections were recorded into separate 3D channels. All images were acquired under nearly identical conditions and modality settings. The drift of the settings during acquisition was considered minimal and therefore neglected.

20.2.4   Image analysis and feature extraction

Image files of cells originally saved in Leica format (*.lif) were converted to a series of TIFFs using the open source ImageJ™ package. Output files were sequentially analyzed with a dedicated software we developed that contains two modules: (I) preprocessing and (II) in-depth analysis. Preprocessing entails nuclear segmentation by adaptive seeded watershed resulting in the delineation of a 3D region of interest (ROI) for each individual nucleus. In-depth analysis focuses on the extraction of MeC and DAPI features within each ROI. The 3D-qDMI system currently offers three features for the in situ assessment of drug-induced changes in global DNA methylation and relevant chromatin organization in cells: (1) the MeC and gDNA global intensities displayed as the nuclear mean intensities of MeC and DAPI, as well as an intensity profile across different selectable cross-sections of individual nuclei, (2) the intensity codistribution of MeC and DAPI, and (3) the topology of MeC and DAPI intensities. The latter two can serve as an indicator of chromatin reorganization in cells. Additionally 3D-qDMI is equipped with a fourth module, namely the statistical (homogeneity) assessment of the population based on MeC and gDNA codistributions.

20.2.4.1 Nuclear segmentation

The main goal of nuclear segmentation is to separate nuclear ROIs connected or located in close proximity. This procedure delineates individual nuclear ROIs in DAPI images using adaptive seeded watershed segmentation. First the DAPI image cube is thresholded and two binary image cubes containing small and large binary objects are distinguished. Each image cube is filtered separately by a Gaussian filter to suppress non-uniformities in each class of objects. To obtain optimal effects of smoothing, the mask of the Gaussian filter is adaptively adjusted in 3D according to the mean volume of objects in each class. Next, the filtered cubes are thresholded again to yield quasi-ellipsoidal seeds for small and large objects, respectively. The large and small seeds are combined into one binary cube and serve as an input for seeded watershed-based segmentation, which yields spatially separated ROIs. The last step of preprocessing is the labeling of nuclei with information for further identification and reference in the in-depth image analysis.

20.2.4.2 Global intensity

Overall intensities of MeC and DAPI fluorescence signals also referred to as MeC and DAPI load represent nuclear contents of the two classes of DNA. Their quantification can provide an estimate of a drug's effect on nuclear DNA based on changes in response to drug application. Global intensities provide a more crude way of measuring epigenetic effects such as demethylation. In contrast, the measurement of the distribution of signal fluctuations within each nucleus provides a means towards a more differential assessment of signal alterations. We have implemented a semi-quantitative feature that displays intensity profiles along selectable ROI cross-sections. A more quantitative form of intensity analysis studies the respective densities within a nucleus through signal codistribution assessment or signal topology.

20.2.4.3 Codistribution of MeC and DAPI

A codistribution of two signals can be useful in the analysis of cellular patterns, and can essentially be displayed as a scatter plot. Scatter plots can depict mixture models of simple relationships between variables. These relationships can reflect cellular patterns as specific signatures, in which the variables can be nuclear structures as shown in the case of more specific methylcytosine patterns versus globally DAPI-stained DNA. These nuclear entities are not static and can reorganize during the application of demethylating agents. We have shown that such reorganizations can be dynamically monitored by scatter plotting the two classes of DNA, with their differential distribution becoming visible as changes in the plotted patterns (Tajbakhsh *et al.*, 2008; Gertych *et al.*, 2009, 2010).

20.2.4.4 Nuclear topology

To address 3D topology of DNA methylation signals a specific feature was implemented, which allows for the mapping of MeC and DAPI densities within the entire nucleus. By selecting specific thresholds different subclasses of target intensities can be localized within an ROI. The thresholding is followed by a complete shell-by-shell binary erosion of each nucleus to define all contained shells and record target signals in each shell. The erosion is performed from the nuclear boundary towards its center, and allows for quantification of thresholded MeC- and DAPI-related sites (as targets) in predefined nuclear compartments to identify target zones. In our application, the algorithm is tuned to localize a specific subclass of signals we termed low-intensity MeC (LIM) sites and low-intensity DAPI (LID) sites within the nuclei, which can depict results of cell treatment with demethylating agents. To obtain the LIM and LID signal components, the respective sites were first defined as voxels with signal amplitudes between two adaptively adjusted thresholds. This procedure yielded LIM/LID density profiles for each nucleus. Each profile sampled at half of the nuclear volume (border between the peripheral and the interior part of the nucleus) yielded two parameters, $LIM_{0.5}$ and $LID_{0.5}$, which are related to demethylation and changes in the higher-order nuclear organization of hypomethylated global DNA, specifically of DAPI-intensive heterochromatic sites. Therefore, these parameters are valuable measures in the detection of unwanted side effects of epigenetic drugs.

20.2.4.5 Homogeneity assessment

Cell population homogeneity can be defined as the degree of similarity of cells in a population regarding one or multiple structural and/or functional properties. Isolation of cells which display a common (similar) pattern under certain criteria, or those that can be flagged as outliers, is usually the primary purpose of cell screening. Therefore an automated homogeneity assessment is highly valuable in high-content screening, in which a rapid evaluation of a large number of cells is required (Gertych and Tajbakhsh, 2010). The Kullback–Leibler's (K–L) divergence (Kullback, 1997) can be used in this case to measure the distance between various kinds of distributions or in our case codistributions. This measure can be applied to any multicolor cellular assay that utilizes intensity-based information to assess cellular response. Our comparison of the K–L metric with other metrics most frequently used for similarity measurements – such as Mahalanobis and Bhattacharyya distances, and Dice's index – demonstrated that the K–L method produces the highest certainty (least uncertainty) for the nuclear MeC/DAPI pattern analysis within the imaged cell populations. Moreover, the Pearson's correlation coefficient between two distributions (such as MeC and DAPI) can be directly calculated from the K–L divergence, if the distributions are normal. We observed the robustness of the K–L divergence against potential intra-experimental data variability introduced through the biochemical processing of specimens and/or the modality settings in between imaging sessions, which may compound alterations of the intensity levels within the MeC and the DAPI channels. Additionally, the K–L divergence measurement has the advantage of being independent from image rotation and the inherent anisotropy of confocal microscopy images, as well as being flexible in dealing with a high dynamic range in sample size. This characteristic is highly beneficial in connection with the current limited capabilities of imaging systems that are restricted in the field of view size when acquiring highest-resolution images. Thus, it is necessary to collect and tile multiple image stacks in order to obtain a complete picture of the entire sample. The robustness of the K–L measurement allows it to be applied across the complete assembled image. Such an approach could be helpful in the assessment of relationships between single cells or cell cohorts and their neigh-borhoods: i.e., intra- and inter-population functional relationships through epigenetic effects such as DNA methylation via tissue diagnostics in disease pathology and cell-based assays for compound screening in drug development. In our approach we calculate a K–L divergence value using the normalized scatter plots of an individual nucleus and a reference scatter plot, constructed from all individual plots in the analyzed cells of a population. To make the K–L values more descriptive, and to allow intra-population assessment of cells we introduced four soft-qualifiers defining similarity degrees of cells in the entire population. These degrees are associated with particular ranges of K–L divergen-ces: *similar* for $0 \leq KL < 0.5$, *likely similar* for $0.5 \leq KL < 2$, *unlikely similar* for $2 \leq KL < 4.5$, and *dissimilar* for $KL \geq 4.5$. Figures 20.1 and 20.3 show sample results of cell population homogeneity assessments based on nuclear MeC/DAPI signal co-distributions. The outcome characterizes each individual cell, and provides statistical information about the number of cells that fall into these four cate-gories. The homogeneity is currently evaluated using MeC and DAPI codistribu-tions, only. However, it can be flexibly applied in conjunction with a more complex set of nuclear markers.

20.2.4.6   System output

The program's graphical user interface (GUI) allows for the parallel display of all results from the different analysis features and the export of the graphical data for documentation purposes, as shown in Figure 20.3. Additionally, tabular data are automatically generated and output as txt-files that can be easily imported into a spreadsheet. More details about the system components can be found in our recent publications (Gertych *et al.*, 2009, 2010; Tajbakhsh *et al.*, 2010).

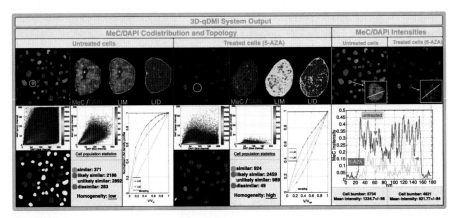

Figure 20.3 The 3D-qDMI system currently offers three features for the in situ assessment of drug-induced changes in global DNA methylation and relevant chromatin organization in cells (DU145 prostate cancer cells in here): (i) changes in MeC (green) and global gDNA (blue) displayed as their respective nuclear mean intensities as well as an intensity profile across different selectable cross-sections of individual nuclei (for simplicity only MeC signals shown in here); (ii) changes in the spatial distribution of MeC and global DNA as an indicator of chromatin reorganization in cells; and (iii) alterations in the topology of MeC and DAPI intensities displayed as graphs and maps, also indicating $LIM_{0.5}$ and $LID_{0.5}$. The graphical user interface (GUI) allows for each feature to be displayed for interactively selected nuclei (marked by a white ring) via clicking on the labeled ROI in the K−L map. A fourth module has been implemented for statistical (homogeneity) assessment of the population based on 3D chromatin texture and MeC load. See plate section for color version.

20.3   Results and discussion

20.3.1   Measuring spatial codistribution of methylcytosine and global DNA in mouse cell nuclei

Our initial experience with mouse pituitary tumor cells confirms that demethylating agents can exert the two known effects: (i) a decrease in the number of MeCs in global DNA, and (ii) the subsequent reorganization of highly compact heterochromatic regions of the genome, that affect nuclear architecture (Tajbakhsh *et al.*, 2008; Gertych *et al.*, 2009) (Figure 20.4). These results are in agreement with previous observations made through genetic manipulation of CpG methylation in mouse embryonic stem cells that lack *Dnmt1* (Ma *et al.*, 2005), a maintenance DNA methyltransferase that converts hemi-methylated CpGs on *de novo* synthesized DNA strands to symmetrically methylated dinucleotides during genome replication in cells. The image analysis solution we developed detects and quantifies these coexisting phenomena to measure and display the relevant changes in intensity distribution of the two types of signals that reflect the two phenomena: (a) MeC signals created through immunofluorescence targeting of methylated cytosine and (b) DAPI signals generated by subsequent counterstaining of the same cells, as DAPI intercalates into AT-rich DNA, the main component of highly repetitive and compact heterochromatic sequences. In particular, heterochromatin decondensation, as a secondary effect of global demethylation, results in the relocation of heterochromatic sites within the nucleus (which is associated with genome destabilization). As a consequence of these conformational and

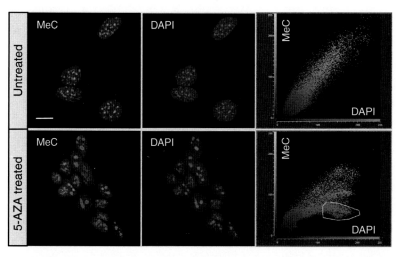

Figure 20.4 Effect of drug-induced demethylation on the chromatin of AtT20 pituitary tumor cells in culture. 5-AZA (at 1 μM concentration) causes both (i) demethylation of global DNA including DAPI-intensive heterochromatic regions and (ii) significant reorganization of the chromatin in delineated cell nuclei. Upper panel: untreated cells show 20–25 small MeC foci (diameter ~1 μm), which almost fully overlap with the nuclear DAPI signals, except for some heterochromatic regions at the nuclear border (scale bar is 10 μm). Lower panel: in contrast, cells treated for 48 hours with 5-AZA display a change in chromatin phenotype; with some nuclei that have reached the endpoint and only show a few (two to five) large MeC-foci (diameter ~3–5 μm), having a unique MeC distribution: the heterochromatic core is hypomethylated and surrounded by a hypermethylated ring; and other nuclei that seem to be in transition. This phenomenon is represented by changes in the distribution of the MeC and DAPI signals displayed as a 2D scatter plot. The outlined area represents signals of hypomethylated heterochromatin that occur in response to drug treatment.

organizational changes of the DAPI-intensive nuclear sites, the same DAPI signal intensity is spread out over a higher number of voxels. Thus, both MeC and DAPI have dynamic patterns in the cell nucleus that become more discernible in a joint 2D plot than in a 1D MeC plot, or even when the two signals are separately displayed in one dimension. Our approach directly illustrates the distribution of voxel intensities. The changes of these distributions are derived from the under-lying changes in the topology (spatial patterns) of global DNA in response to drug treatment. Consequently, we were able to demonstrate that when the topological patterns of MeC and gDNA (DAPI) are converted into 2D histograms, they can be utilized as differential biosignatures in the evaluation of cellular response to treatment with demethylating anti-cancer agents. This characteristic is in line with the larger purpose of our approach, namely to create a rapid image analysis method that is of low complexity and therefore computationally inexpensive for high-throughput cell screening tasks. In previous studies, we observed that untreated and drug-treated cells of the same kind display different sets of MeC/ DAPI codistributions (Gertych *et al.*, 2009). This led to the development of an algorithm which utilizes the resulting scatter plots in a statistical fashion to assess drug response indicated by the structural variance of chromatin within a cell population. Our notion was to develop software that can be meaningfully and robustly applied to the evaluation of demethylating agents. However, the developed algorithm can be flexibly utilized for a variety of cell biological applications with different targets. Especially, the K–L divergence measure is a powerful statistical feature when the targets do not have a consistent location within the considered ROI, such as the nucleus.

20.3.2 Topological quantification of MeC/DAPI sites in human cell nuclei

In comparison to mouse tumor cells, human cancer cells show a much finer granulation of MeC and gDNA textures. Therefore, we upgraded our imaging-based cytometrical approach with additional features for the analysis of human cells by 3D-qDMI. The upgraded methodology combines the aforementioned image-processing routines with the newly added topological integration of low-intensity MeC and DAPI signals (a consequence of drug-induced demethylation) in consecutive nuclear sections obtained by 3D erosion. The additional features provide two new parameters in the evaluation of demethylating drug effects: (1) region-specific changes in MeC load and (2) alterations in density distributions of gDNA. Both parameters yielded highly differential values in feasibility studies between naive control cell populations and derived populations that were treated with the well-characterized demethylating agents 5-azacytidine and zebularine; the latter is known for being milder and significantly less cytotoxic and genotoxic to cultured cells (Cheng *et al.*, 2003; Balch *et al.*, 2005; Marquez *et al.*, 2005). Thereby, two interesting observations were made: (1) in treated as well as untreated cancer cells the highest LIM density was always found at the nuclear border, and (2) the degree of demethylation was concordant with an increase in LIM density beyond the nuclear border into the interior of the nucleus, meaning that the stronger the demethylating effect of the drug was the more LIM sites could be registered within the inner shells of the nucleus. Azacytidine nuclei display significantly higher LIM densities even in the areas deep inside the nuclei compared to Zebularine cells. As the interior of the mammalian nucleus harbors a large portion of the highly compact constitutive heterochromatin, it is assumed that these areas of the genome have been largely demethylated by azacytidine but not as much by zebularine. However, both drugs seem to also affect global DNA organization. Therefore, the results for MeC/DAPI codistributions could be strengthened with the topological findings obtained through the newly added feature. These results provide evidence that mapping of low-intensity MeC and DAPI sites in cancer cells can serve as a potent indicator in the quantitative assessment of demethylating drug actions that include targeted effects (DNA demethylation) and accompanying side effects (chromatin reorganization) of such therapeutic manipulations. Both effects are considered in therapy as perturbation of the higher-order chromatin organization and need to be tested for eventual risks of causal genome instability in relevant cell models (Ma *et al.*, 2005; Yamada *et al.*, 2005; Csoka and Szyf, 2009).

20.3.3 3D-qDMI versus other DNA methylation analysis methods

All analytical modules were initially developed as a tool to be used in basic epigenetics-related research with the notion of subsequent translation into clinical diagnostics and therapy, as 3D-qDMI is scalable, and it can be used for high-throughput cell-based assays as well as pathological diagnostics. Furthermore the approach described can be supportive to molecular methods by adding more descriptive information to the (epi)genotype of cells, towards the convergence of imaging and molecular data. Global DNA methylation could be first analyzed by 3D-qDMI, which provides a holistic estimate of DNA methylation changes in a cell-by-cell mode. Then individual target cells or groups of cells that share a phenotype or represent outliers can be selected and captured for high-resolution methylation-specific genotyping: with a plethora of existing molecular methods including PCR-based approaches, whole-genomic tiling arrays, and massively parallel sequencing (Ammerpohl *et al.*, 2009; Lister and Ecker, 2009; Smith *et al.*, 2009; Laird *et al.*, 2010), as illustrated in Figure 20.5. The cell-similarity assessment feature – implemented in 3D-qDMI – is therefore extremely valuable in two ways: (1) drug efficiency can be estimated from the degree of cellular response based on the homogeneity of the cell population in spatial MeC/DAPI codistribution, and (2) the selection of phenotypically similar cells leverages the generation of molecular data with higher confidence. The accuracy and performance of this technique is only constrained by parameters of imaging modalities such as the spatial resolution and point spread function, and the speed of computational unit. However, through automation of imaging, mostly practiced in the pharmaceutical industry, imaging modalities can be kept much more consistently than in

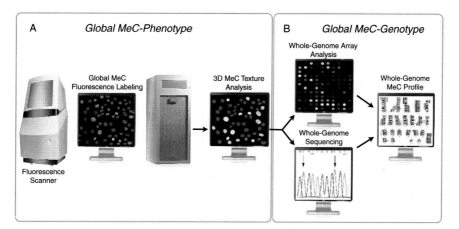

Figure 20.5 Cost-effective convergence of imaging and molecular data through phenotype-to-genotype analysis of cells. (A) First, drug-treated and control cells can be rapidly analyzed for differential chromatin texture by 3D-qDMI in a cell-by-cell mode. (B) Then, individual or groups of cells with specific MeC-DAPI features can be selected for deeper analysis of MeC profiles with whole-genomic technologies such as DNA microarrays and massively parallel sequencing. The technologies exist for identifying and capturing 3D-analyzed cells in situ for the subsequent DNA retrieval and molecular analysis.

research environments, and computational capacity can be ramped up, internally or by crowd computing to compensate for high-volume parallel throughputs.

20.4   Concluding remarks

In summary, 3D-qDMI has proven itself as a method that could reconcile the reported effects of the drugs obtained by more elaborate molecular analyses with a rapid imaging-based approach at single-cell resolution. Our recent investigations summarized in this chapter strongly support the fact that spatial nuclear DNA methylation patterns, utilizing 3D-qDMI, could serve as a potential surrogate pharmacodynamic biomarker of demethylating drug action, making structure–genotoxicity relationships an additional valuable means for drug and compound characterization in the preclinical phases of therapeutical strategies. The implementation of an algorithm that localizes low-intensity methylcytosine sites delivers an actual (phenotype) map of differential demethylation in the nucleus alongside with accompanying changes in the organization of global DNA. Specifically, this method can detect the more subtle changes of global DNA methylation features caused by milder drugs such as zebularine. The information of LIM and LID topology can therefore be applied in the assessment of risks associated with genome-wide demethylation, specifically leading to hypomethylation of repetitive elements causing an adverse reorganization of the genome. In combination with the visualized cell population homogeneity this cytometrical approach can provide information towards drug efficacy. In its current version 3D-qDMI is able to deliver a quick shot of structure–function relations for researchers in the field of epigenetics. Moreover, the development of 3D-qDMI as an automated high-throughput screening method in the preclinical profiling of second-generation epigenetic drugs in development – that act more DNA sequence-specifically – can benefit the pharmaceutical industry by significantly reducing attrition in early development phases. This type of *toxicoepigenomic* assessment could be practically expanded to any drug beyond anti-cancer drugs, for which we have no current knowledge of epigenetic side effects.

REFERENCES

Ammerpohl, O., Martin-Subero, J. I., Richter, J., Vater, I., and Siebert, R. (2009). Hunting for the 5th base: techniques for analyzing DNA methylation. *Biochimica et Biophysica Acta*, **1790**, 847–862.

Balch, C., Yan, P., Craft, T., *et al.* (2005). Antimitogenic and chemosensitizing effects of the methylation inhibitor zebularine in ovarian cancer. *Molecular Cancer Therapeutics*, **4**, 1505–1514.

Barbin, A., Montpellier, C., Kokalj-Vokac, N., *et al.* (1994). New sites of methylcytosine-rich DNA detected on metaphase chromosomes. *Human Genetics*, **94**, 684–692.

Barton, S. C., Arney, K. L., Shi, W., *et al.* (2001). Genome-wide methylation patterns in normal and uniparental early mouse embryos. *Human Molecular Genetics*, **10**, 2983–2987.

Bensaada, M., Kiefer, H., Tachdjian, G., *et al.* (1998). Altered patterns of DNA methylation on chromosomes from leukemia cell lines: identification of 5-methylcytosines by indirect immunodetection. *Cancer Genetics and Cytogenetics*, **103**, 101–109.

Best, J. D. and Carey, N. (2010). Epigenetic opportunities and challenges in cancer. *Drug Discovery Today*, **15**, 65–70.

Bestor, T. H. and Tycko, B. (1996). Creation of genomic methylation patterns. *Nature Genetics*, **12**, 363–367.

Carr, B. I., Reilly, J. G., Smith, S. S., Winberg, C., and Riggs, A. (1984). The tumorigenicity of 5-azacytidine in the male Fischer rat. *Carcinogenesis*, **5**, 1583–1590.

Chen, R. Z., Pettersson, U., Beard, C., Jackson-Grusby, L., and Jaenisch, R. (1998). DNA hypomethylation leads to elevated mutation rates. *Nature*, **395**, 89–93.

Cheng, J. C., Yoo, C. B., Weisenberger, D. J., *et al.* (2003). Preferential response of cancer cells to zebularine. *Cancer Cell*, **6**, 151–158.

Costello, J. F. and Plass, C. (2001). Methylation matters. *Journal of Medical Genetics*, **38**, 285–303.

Csoka, A. B. and Szyf, M. (2009). Epigenetic side-effects of common pharmaceuticals: a potential new field in medicine and pharmacology. *Medical Hypotheses*, **73**, 770–780.

Cui, H., Onyango, P., Brandenburg, S., *et al.* (2002). Loss of imprinting in colorectal cancer linked to hypomethylation of H19 and IGF2. *Cancer Research*, **62**, 6442–6446.

de Capoa, A., Menendez, F., Poggesi, I., *et al.* (1996). Cytological evidence for 5-azacytidine-induced demethylation of the heterochromatic regions of human chromosomes. *Chromosome Research*, **4**, 271–276.

Ehrlich, M. (2000). DNA methylation: normal development, inherited diseases, and cancer. *Journal of Clinical Ligand Assay*, **23**, 144–146.

Ehrlich, M. (2005). DNA methylation and cancer-associated genetic instability. In: *Genome Instability in Cancer Development*, eds. Back, N., Cohen, I. R., Kritchevsky, D. Heidelberg: Springer.

Espada, J. and Esteller, M. (2007). Epigenetic control of nuclear architecture. *Cellular and Molecular Life Sciences*, **64**, 449–457.

Feinberg, A. P. and Vogelstein, B. (1983). Hypomethylation distinguishes genes of some human cancers from their normal counterparts. *Nature*, **301**, 89–92.

Gaudet, F., Hodgson, J. G., Eden, A., *et al.* (2003). Induction of tumors in mice by genomic hypomethylation. *Science*, **300**, 489–492.

Gertych, A. and Tajbakhsh, J. (2010). Homogeneity assessment of cell populations for high-content screening platforms. In *Information Technology in Biomedicine*, vol. **2**, eds. Pietka, E. and Jacek, K., pp. 309–319. Heidelberg: Springer.

Gertych, A., Wawrowsky, K. A., Lindsley, E., *et al.* (2009). Automated quantification of DNA demethylation effects in cells via 3D mapping of nuclear signatures and population homogeneity assessment. *Cytometry A*, **75**, 569–583.

Gertych, A., Farkas, D. L., and Tajbakhsh, J. (2010). Measuring topology of low-intensity DNA methylation sites for high throughput assessment of epigenetic drug-induced effects in cancer cells. *Experimental Cell Research*, **316**, 3150–3160.

Habib, M., Fares, F., Bourgeois, C. A., *et al.* (1999). DNA global hypomethylation in EBV-transformed interphase nuclei. *Experimental Cell Research*, **249**, 46–53.

Haaf, T. (1995). The effects of 5-azacytidine and 5-azadeoxycytidine on chromosome structure and function: implications for methylation-associated cellular processes. *Pharmacology and Therapeutics*, **65**, 19–46.

Herceg, Z. (2010). Epigenetic drugs on the rise: new promises for cancer therapy and prevention. *International Drug Discovery*, June/July, 24–29.

Holliday, R. and Pugh, J. E. (1975). DNA modification mechanisms and gene activity during development. *Science*, **187**, 226–232.

Holm, T. M., Jackson-Grusby, L., Brambrink, T., *et al.* (2005). Global loss of imprinting leads to widespread tumorigenesis in adult mice. *Cancer Cell*, **8**, 275–285.

Huisman, A., Ploeger, L. S., Dullens, H. F., *et al.* (2007). Discrimination between benign and malignant prostate tissue using chromatin texture analysis in 3-D by confocal laser scanning microscopy. *Prostate*, **15**, 248–254.

Ingelman-Sundberg, M. and Gomez, A. (2010). The past, present and future of pharmacoepigenomics. *Pharmacogenomics*, **11**, 625–627.

Issa, J. P. (2007). DNA methylation as a therapeutic target in cancer. *Clinical Cancer Research*, **13**, 1634–1637.

Jefferson, A., Colella, S., Moralli, D., *et al*. (2010). Altered intra-nuclear organisation of heterochromatin and genes in ICF syndrome. *PLoS One*, **5**, e11364.

Kennedy, B. K., Barbie, D. A., Classon, M., Dyson, N., and Harlow, E. (2000). Nuclear organization of DNA replication in primary mammalian cells. *Genes and Development*, **14**, 2855–2868.

Kullback, S. (1997). *Information Theory and Statistics*. New York: Dover.

Kurz, A., Lampel, S., Nickolenko, J. E., *et al*. (1996). Active and inactive genes localize preferentially in the periphery of chromosome territories. *Journal of Cell Biology*, **135**, 1195–1205.

Laird, P. W. (2010). Principles and challenges of genome-wide DNA methylation analysis. *Nature Reviews Genetics*, **11**, 191–203.

Lister, R. and Ecker, J. R. (2009). Finding the fifth base: genome-wide sequencing of cytosine methylation, *Genome Research*, **19**, 959–966.

Luebbert, M. (2000). DNA methylation inhibitors in the treatment of leukemias, myelodysplastic syndromes and hemoglobinopathies: clinical results and possible mechanismas in action. *Current Topics in Microbiology and Immunology*, **249**, 135–164.

Ma, Y., Jacobs, S. B., Jackson-Grusby, L., *et al*. (2005). DNA CpG hypomethylation induces heterochromatin reorganization involving the histone variant macroH2A. *Journal of Cell Science*, **118**, 1607–1616.

Marquez, V. E., Kelley, J. A., Agbaria, R., *et al*. (2005). Zebularine: a unique molecule for an epigenetically based strategy in cancer chemotherapy. *Annals of the New York Academy of Sciences*, **1058**, 246–254.

Mathiassen, J. R., Skavhaug, A., and Bø, K. (2002). Texture similarity measure using Kullback–Leibler divergence between gamma distributions. *Lecture Notes in Computer Science*, **2352**, 133–147.

Mayer, W., Niveleau, A., Walter, J., Fundele, R., and Haaf, T. (2000). Demethylation of the zygotic paternal genome. *Nature*, **403**, 501–502.

Miller, O. J., Schnedl, W., Allen, J., and Erlanger, B. F. (1974). 5-Methylcytosine localised in mammalian constitutive heterochromatin. *Nature*, **251**, 636–637.

Miniou, P., Jeanpierre, M., Blanquet, V., *et al*. (1994). Abnormal methylation pattern in constitutive and facultative (X inactive chromosome) heterochromatin of ICF patients. *Human Molecular Genetics*, **3**, 2093–2102.

Momparler, R. L. (2005). Epigenetic therapy of cancer with 5-aza-2′-deoxycytidine (decitabine). *Seminars in Oncology*, **32**, 443–451.

Montpellier, C., Burgeois, C. A., Kokalj-Vokac, N., *et al*. (1994). Detection of methylcytosine-rich heterochromatin on banded chromosomes: application to cells with various status of DNA methylation. *Cancer Genetics and Cytogenetics*, **78**, 87–93.

Orr, J. A. and Hamilton, P. W. (2007). Histone acetylation and chromatin pattern in cancer: a review. *Analytical and Quantitative Cytology and Histology*, **29**, 17–31.

Peedicayil, J. (2008). Pharmacoepigenetics and pharmacoepigenomics. *Pharmacogenomics*, **9**, 1785–1786.

Piyathilake, C. J., Frost, A. R., Bell, W. C., *et al*. (2001). Altered global methylation of DNA: an epigenetic difference in susceptibility for lung cancer is associated with its progression. *Human Pathology*, **32**, 856–862.

Rougier, N., Bourc'his, D., Gomes, D. M., *et al*. (1998). Chromosome methylation patterns during mammalian preimplantation development. *Genes and Development*, **12**, 2108–2113.

Santos, F., Hendrich, B., Reik, W., and Dean, W. (2002). Dynamic reprogramming of DNA methylation in the early mouse embryo. *Developmental Biology*, **241**, 172–182.

Sasaki, K., Adachi, S., Yamamoto, T., *et al*. (1998). Effects of denaturation with HCl on the immunological staining of bromodeoxyuridine incorporated into DNA. *Cytometry*, **9**, 93–96.

Scheuermann, M. O., Tajbakhsh, J., Kurz, A., *et al*. (2004). Topology of genes and nontranscribed sequences in human interphase nuclei. *Experimental Cell Research*, **301**, 266–279.

Smith, Z. D., Gu, H., Bock, C., Gnirke, A., and Meissner, A. (2009). High-throughput bisulfite sequencing in mammalian genomes. *Methods*, **48**, 226–232.

Soares, J., Pinto, A. E., Cunha, C. V., *et al.* (1999). Global DNA hypomethylation in breast carcinoma: correlation with prognostic factors and tumor progression. *Cancer*, **85**, 112–118.

Stresemann, C., Bokelmann, I., Mahlknecht, U., and Lyko, F. (2008). Azacytidine causes complex DNA methylation responses in myeloid leukemia. *Molecular Cancer Therapy*, **7**, 2998–3005.

Szyf, M. (2004). Toward a discipline of pharmacoepigenomics. *Current Pharmacogenomics*, **2**, 357–377.

Szyf, M. (2009). Epigenetics, DNA methylation, and chromatin modifying drugs. *Annual Review of Pharmacology and Toxicology*, **49**, 243–263.

Tajbakhsh, J., Luz, H., Bornfleth, H., *et al.* (2000). Spatial distribution of GC- and AT-rich DNA sequences within human chromosome territories. *Experimental Cell Research*, **255**, 229–237.

Tajbakhsh., J., Wawrowsky, K. A., Gertych, A., *et al.* (2008). Characterization of tumor cells and stem cells by differential nuclear methylation imaging. *Proceedings of the SPIE*, **6859**, 6859F1–6859F10.

Tajbakhsh, J., Gertych, A., and Farkas, D. L. (2010). Utilising 3D nuclear DNA methylation patterns in cell-based assays for epigenetic drug screening. *Drug Discovery World*, Spring 2010, 27–35.

Tong, W. G., Wierda, W. G., Lin, E., *et al.* (2010). Genome-wide DNA methylation profiling of chronic lymphocytic leukemia allows identification of epigenetically repressed molecular pathways with clinical impact. *Epigenetics*, **5**, 499–508.

Toyota, M., Suzuki, H., Yamashita, T., *et al.* (2009). Cancer epigenomics: implications of DNA methylation in personalized cancer therapy. *Cancer Science*, **100**, 787–791.

Vera, E., Canela, A., Fraga, M. F., Esteller, M., and Blasco M. A. (2008). Epigenetic regulation of telomeres in human cancer. *Oncogene*, **27**, 6817–6833.

Yang, X., Lay, F., Han, H., and Jones, P. A. (2010). Targeting DNA methylation for epigenetic therapy. *Trends in Pharmacological Science*, **31**, 536–546.

Yamada, Y., Jackson-Grusby, L., Linhart, H., *et al.* (2005). Opposing effects of DNA hypo-methylation on intestinal and liver carcinogenesis. *Proceedings of the National Academy of Sciences USA*, **102**, 13 580–13 585.

Yoo, C. B. and Jones, P. A. (2006). Epigenetic therapy of cancer: past, present and future. *Nature Reviews Drug Discovery*, **5**, 37–50.

Zaitseva, L., Myers, R., and Fassati, A. (2006). tRNAs promote nuclear import of HIV-1 intracellular reverse transcription complexes. *PLoS Biology*, **4**, e332.

Part IV

Epigenomics in disease biology

21 Cancer classification by genome-wide and quantitative DNA methylation analyses

Atsushi Kaneda

21.1   Introduction

Cancer is known to arise through the accumulation of epigenetic alterations and genetic alterations (Feinberg *et al.*, 2006; Jones and Baylin, 2007). Gene silencing is a major epigenetic gene-inactivation mechanism by DNA methylation of its promoter region, and is involved in the initiation and progression of cancer (Feinberg and Tycko, 2004; Jones and Baylin, 2007). Genome-wide approaches to search for aberrantly methylated regions in cancer have been developed since 1993, mainly since 1997 (Hayashizaki *et al.*, 1993; Ushijima, 2005). Identification of novel inactivated genes in cancer using aberrant methylation as markers is useful to identify novel tumor-suppressor genes and methylation markers (Yoshikawa *et al.*, 2001; Kaneda *et al.*, 2004; Suzuki *et al.*, 2004; Yu *et al.*, 2005). Most of these methods utilized methylation-sensitive restriction enzymes, however, and thus only limited regions of genome could be analyzed (Kaneda *et al.*, 2003; Ushijima, 2005), which means that detected methylation markers may not be comprehensive enough.

For classification of cancer cases using DNA methylation data, a subset of colorectal cancer was found to show accumulation of CpG island methylation, so-called "CpG island methylator phenotype (CIMP)", in 1999 by Toyota *et al.* (1999). CIMP-low was proposed by Ogino *et al.* (2006) as a subgroup with less extensive methylation of CIMP-related markers, which was associated with *KRAS* mutation compared with CIMP-high and CIMP(–) subgroups. Shen *et al.* (2007) reported through genetic and epigenetic analysis that colon cancer was classified into three subsets, CIMP-1, CIMP-2, and CIMP-negative. In Shen's report, however, methylation levels of 27 regions were analyzed using four different methylation detection methods: 13 regions by pyrosequencing, seven by combined bisulfite restriction analysis (COBRA), six by methylated CpG island amplification, and one by methylation-specific PCR. Moreover,

Epigenomics: From Chromatin Biology to Therapeutics, ed. K. Appasani. Published by Cambridge University Press.© Cambridge University Press 2012.

previously reported and frequently analyzed markers such as *p16*, *hMLH1*, and MINT (methylated-in-tumor) loci had been analyzed in most studies so far, and Ogino and Goel (2008) suggested that CIMP markers were considered to be specific for CIMP-high and not ideal for the identification of CIMP-low. New specific markers for CIMP-low and/or any other methylation phenotypes therefore needed to be identified if they really existed. To classify cancers using a more comprehensive approach, e.g., hierarchical clustering using highly quantitative methylation data by a single detection method using genome-widely selected novel regions, was required to be performed to classify cancers comprehensively.

To enable high-resolution mapping of DNA methylation, methylated DNA immunoprecipitation (MeDIP)-on-array (chip) analysis using unbiased amplification of MeDIP products by in vitro transcription on short oligonucleotide tiling arrays was developed (Hayashi *et al.*, 2007). The method was applied to promoter tiling arrays for genome-wide identification of genes methylated in cancer cell lines and clinical cancer samples. For analysis of colorectal cancer, a microsatellite instability (MSI)-high cell line HCT116 and a microsatellite-stable cell line SW480 were analyzed by MeDIP-chip, and gene re-expression after treatment with 5-aza-2′-deoxycytidine (5-AZA) and/or Trichostatin A (TSA) was also analyzed using expression microarray. Candidate genes silenced by promoter methylation were selected using these methylation mapping and re-expression data, and methylation of these marker genes was analyzed using MassARRAY (Sequenom, San Diego, CA, USA). MassARRAY utilizes MALDI-TOF mass spectrometry in combination with RNA base-specific cleavage, and has a potential for high-throughput analysis of hundreds of samples and genes, and allows us to classify cancer cases by hierarchical clustering (Yagi *et al.*, 2010). If cancer cell content is high enough, MeDIP-chip analysis using clinical cancer samples is also possible. Clinical hepatocellular carcinoma was analyzed using MeDIP-chip, for example, and candidate methylation markers were analyzed quantitatively by MassARRAY (Deng *et al.*, 2010).

21.2   Methodology

21.2.1   Sample preparation

Clinical cancer samples were obtained from patients who underwent surgical operation, with written informed consent, and kept frozen until use. Cancer samples were microscopically examined for determination of cancer cell content by two independent pathologists, and were dissected to enrich cancer cells when necessary. This step is critical for the study, and cancer cell content should be at least 40% for quantitative DNA methylation analysis. For use of clinical cancer samples in MeDIP-chip analysis, cancer cell content should be more than 90%, and hepatocellular carcinoma was chosen for the analysis. DNA of clinical samples and cell lines was extracted using a QIAamp DNA Micro Kit (QIAGEN, Hilden, Germany). RNA was extracted using TRIzol (Invitrogen, Carlsbad, CA, USA).

21.2.2 MeDIP-chip analysis

To identify candidate genes with promoter methylation, we performed MeDIP-chip using Human Promoter 1.0R tiling array (Affymetrix, Santa Clara, CA, USA), which covered 199 543 165 bp and 4 071 296 CpG sites around over 25 500 promoter regions (Figure 21.1). Genomic DNA of cancer cell lines or clinical samples was fragmented by sonication, immunoprecipitated by anti 5-methylcytocine polyclonal antibody (Megabase Research Products, Lincoln, NE, USA) or anti 5-methylcytocine monoclonal antibody (kindly supplied by Dr. K. Watanabe, Toray Research Center, Inc.). MeDIPed sample and Input sample underwent unbiased amplification by in vitro transcription. Amplified cRNA was converted

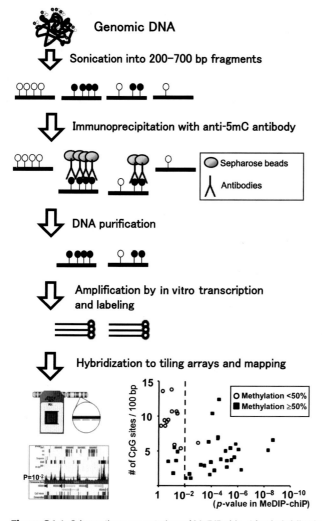

Figure 21.1 Schematic representation of MeDIP-chip. After hybridization to Promoter tiling array, enrichment of MeDIPed product compared to input sample was calculated using the Wilcoxon rank sum test, and regions with $p < 10^{-2}$ were considered to be methylated. The figure at the bottom right showed that among 27 regions with $p < 10^{-2}$, 26 were methylated (solid square, methylation rate $\geq 50\%$ by bisulfite sequencing). Among regions with $p > 10^{-2}$ in MeDIP-chip, high-CpG regions were not methylated (open circle, methylation rate $< 50\%$ by bisulfite sequencing), but low-CpG regions were mostly methylated. MeDIP-chip is therefore not suitable to map methylated low-CpG regions, but quite accurate for identifying candidate methylated high-CpG regions such as promoter CpG islands.

into cDNA, which was then hybridized to Promoter tiling array for both MeDIPed sample and Input sample. Hybridization was done twice. Within a window of 550 bp, the duplicated data from MeDIP DNA were compared with duplicated data from input DNA using the Wilcoxon rank sum test to calculate p-values to detect candidate methylation sites (CMS). Sites with $p < 10^{-2}$ were considered to be CMS.

In the previous analysis within ENCODE regions (Hayashi et $al.$, 2007), false-positive CMS was rarely found and 26 out of 27 regions with $p < 10^{-2}$ were confirmed to be of methylation rate $\geq 50\%$ (Figure 21.1). Methylated low-CpG regions, however, showed false-negative results often in MeDIP, i.e., $p > 10^{-2}$, and were not regarded as CMS. Therefore MeDIP-chip is not considered to be suitable to map methylated low-CpG regions, but to be a powerful tool for identifying candidate methylated high-CpG regions such as promoter CpG islands (Figure 21.1).

For analysis of colorectal cancer, HCT116 and SW480 cell lines were analyzed. For hepatocellular carcinoma, clinical cancer samples were used in MeDIP-chip analysis.

21.2.3   Treatment with 5-AZA and TSA
HCT116 and SW480 were seeded at a density of 3×10^5 cells/10-cm dish on day 0, and exposed to 3 μM 5-AZA on days 1, 2, and 3. Treatment with TSA was performed on day 3 at a dose of 300 nM. Medium was changed every 24 hours, and cells were harvested on day 4.

21.2.4   Expression microarray analysis
Expression of mRNA in HCT116 and SW480 with or without 5-AZA/TSA treatment and normal adult and fetal colon samples was analyzed on GeneChip Human Genome U133 plus 2.0 oligonucleotide arrays (Affymetrix). For global normalization, the average signal in an array was made equal to 100.

21.2.5   Bisulfite treatment
Bisulfite conversion of DNA was performed as previously described (Kaneda et $al.$, 2002). After ultrasonic fragmentation of DNA in 30 seconds, 1 μg of DNA was denatured in 0.3 N NaOH then subjected to 15 cycles of 30 seconds at 95 °C and 15 minutes incubation in 3.6 M sodium bisulfite and 0.6 mM hydroquinone at 50 °C. The samples were desalted with the Wizard DNA Clean-Up system (Promega, Madison, WI, USA), desulfonated in 0.3 N NaOH at room temperature for 5 minutes, then purified by ethanol precipitation. Finally, bisulfite-treated DNA was dissolved in 80 μl of distilled water.

21.2.6   Quantitative methylation analysis using MassARRAY
Methylation status of clinical samples, cell lines, and control samples was analyzed quantitatively using MassARRAY (Ehrich et $al.$, 2005). Bisulfite-treated DNA was amplified by PCR, and the PCR product was transcribed by in vitro transcription and the RNA was cleaved by RNaseA. Unmethylated cytosine (C) was converted to uracil (U) by bisulfite treatment, i.e., thymine (T) in PCR product, and finally adenine (A) in the in vitro transcription product. Methylated cytosine (mC) was not converted, i.e., cytosine (C) in PCR product, and finally guanine (G) in the in vitro transcription product. RNaseA cleaves RNA at the 3′ site of both U and C.

T (U) specific cleavage was possible when using dC instead of C during in vitro transcription. Methylation status was determined by mass difference between A and G in a cleaved RNA product. Methylation rate was calculated quantitatively for each cleaved product, and this analytic unit containing several CpG sites in a cleaved product was called "CpG unit."

Primers were designed to include no CpG site or only one CpG site in the 5′ region (Yagi et al., 2010). Control samples were prepared as follows: first, human peripheral lymphocyte DNA (Coriell Cell Repositories, Camden, NJ, USA) was used as diploid human DNA, and amplified by GenomiPhi v2 DNA amplification kit (GE Healthcare Lifescience, Piscataway, NJ, USA). The amplified DNA was not methylated at all in any CpG sites, and used as unmethylated (0%) control. The amplified DNA was also methylated by SssI methylase (New England Biolabs, Beverly, MA, USA) and used as fully methylated (100%) control. By mixing the 0% and 100% control samples, partially methylated control samples (25%, 50%, and 75%) were generated. The quantitation of primer pairs was validated by analyzing these methylation control samples (Figure 21.2).

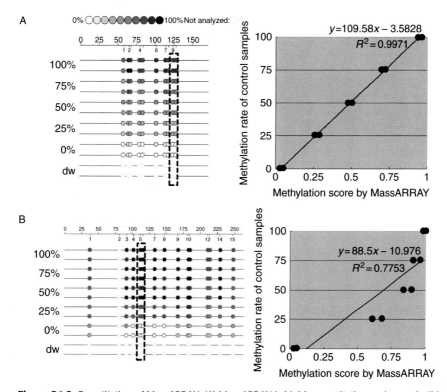

Figure 21.2 Quantitation of MassARRAY. (A) MassARRAY is highly quantitative and reproducible. Duplicated methylation control samples (methylation 0%, 25%, 50%, 75%, 100%) showed a linear standard curve with a high correlation coefficient (R^2). (B) Because of the sequence difference between methylated and unmethylated DNA after bisulfite treatment, their amplification efficiencies can be different and PCR bias is often observed. Methylated allele tends to be more efficiently amplified, as shown. When R^2 was low (≤ 0.9), the primer pairs were excluded or redesigned.

Because of the sequence difference between methylated DNA and unmethylated DNA after bisulfite treatment, amplification efficiencies of the two can be different and PCR bias is often observed. The linear standard curve was drawn and the correlation coefficient (R^2) was calculated at each CpG unit. When there was no PCR bias, methylated and unmethylated alleles were amplified equally, and R^2 would be high (>0.9) (Figure 21.2A). When there was PCR bias, R^2 would be low (≤0.9), and the methylated allele tends to be more efficiently amplified (Figure 21.2B). CpG units with $R^2 ≤ 0.9$ were excluded, and primer pairs whose amplicon contained fewer than three CpG units with $R^2 > 0.9$ were excluded or redesigned. When three or more CpG units with $R^2 > 0.9$ existed, the primer pairs were considered to be quantitative enough and kept to be used for further analyses.

21.2.7   Statistical analysis

Unsupervised two-way hierarchical clustering was performed based on standard correlation and average linkage clustering algorithm in sample direction, and Euclid distance and complete linkage clustering algorithm in marker direction using GeneSpring 7.3.1 software (Agilent Technology, Santa Clara, CA, USA). Correlations between epigenotypes and clinicopathological factors were analyzed by Fisher's exact test. Kaplan–Meier survival analysis was performed by JMP 7 software (www.jmp.com/) and the p-value was calculated by log-rank test.

21.3   Results and discussion

21.3.1   Generation of methylation markers for colorectal cancer

To select candidate silencing genes, we set up four criteria (Figure 21.3A): (i) genes which possess candidate methylation sites within 1 kb from the transcription start site (TSS), (ii) GeneChip score of normal adult or fetal colon >50, (iii) GeneChip score of the methylated cancer cell line <50, and (iv) upregulation >1.5-fold after 5-AZA or 5-AZA and TSA treatment. These criteria are important to select marker regions not only methylated but also playing a silencing role in cancer by the methylation. Among candidate silencing genes fulfilling the four criteria in HCT116 only, in SW480 only, and in both cell lines, we selected 21, 10, and 24 genes randomly as candidate new markers respectively (Figure 21.3B), for which primer pairs were designed. In addition, we also selected 13 previously reported CIMP markers and six previously reported silenced genes for primer design.

21.3.2   Validation of primer pairs

We first validated quantitativity of 74 primer pairs by analyzing control samples (0%, 25%, 50%, 75%, and 100% methylation) using MassARRAY to select primer pairs that guarantee highly quantitative analysis without PCR bias. Finally, primers for 60 regions, including 44 candidate new markers (Figure 21.3B), 11 CIMP markers, and five reported silencing genes, were validated and used in further analysis. The exclusion of primer pairs that are less quantitative due to PCR bias is important in quantitative DNA methylation analysis, as well as keeping the

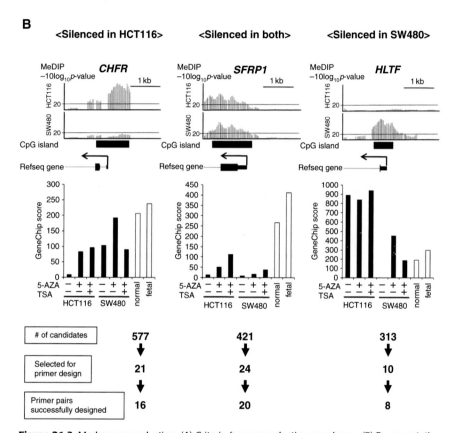

A   **Selection criteria**

(i)   Genes methylated around TSS (by MeDIP-chip data)
(i)   Genes expressed in normal adult colon or fetal colon
(iii) Genes silenced in the methylated cell line(s)
(vi) Genes re-expressed after 5-AZA or 5-AZA/TSA treatment

Figure 21.3 Marker gene selection. (A) Criteria for gene selection are shown. (B) Representative genes, *CHFR* selected from HCT116, *HLTF* from SW480, and *SFRP1* from both cell lines, are shown. In each category, number of candidate genes, number of selected genes for primer design, and number of genes whose primer pairs were validated and used for further analysis.

cancer cell content >40% in the clinical samples in determination of methylation status by quantitative analysis.

21.3.3 Epigenotyping of colorectal cancer

In order to epigenotype colorectal cancer by DNA methylation, 149 colorectal cancer samples along with nine normal colon mucosa samples and six colorectal cancer cell lines were classified by unsupervised two-way hierarchical clustering using methylation rates of the above 60 markers (Figure 21.4A). There are three methylation clusters identified: high-methylation epigenotype (HME), intermediate-methylation epigenotype (IME), and low-methylation epigenotype (LME). Nine normal samples were clustered outside these three epigenotypes of cancer. Three MSI-high cell lines, HCT116, DLD1, and LoVo, were clustered with HME. Three microsatellite-stable cell lines, SW480, T84, and Caco-2, were clustered with IME (Figure 21.4A). The Bayesian information criterion curve constructed by model-based clustering revealed that the optimal number of clusters for the analyzed 149 cancer samples was three, supporting the above classification into three epigenotypes.

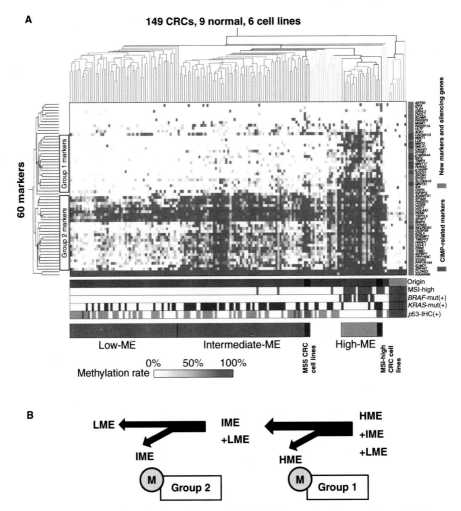

Figure 21.4 Epigenotyping of colorectal cancer (CRC). (A) Unsupervised two-way hierarchical clustering classified CRC cases into three epigenotypes: high-ME (HME) were clustered with MSI-high cell lines, and correlated strongly with MSI-high and *BRAF*-mutation(+). Intermediate-ME (IME) were clustered with microsatellite-stable (MSS) cell lines, and correlated with *KRAS*-mutation(+). Methylation markers were clustered into two major groups: Group 1 showing high methylation in HME, and Group 2 showing high methylation in HME and LME. In heat map: 100% methylation, red; 0% methylation, white; no data, gray. In Origin column at the bottom: red, CRC; dark blue, CRC cell lines; light blue, normal colon mucosa. (B) At the first step, colorectal cancer with Group 1 marker methylation is extracted as HME. At the second step, the remaining samples are divided into IME if Group 2 markers are methylated, and into LME if Group 2 markers are not methylated. See plate section for color version.

We found that HME is strongly associated with MSI-H, *BRAF*-mutation(+), and absence of p53-IHC. Microsatellite-stable (MSS) CRC is classified into two epigeno-types, IME and LME. Of these, IME is strongly associated with *KRAS*-mutation(+). These strong associations suggest that there are three distinct molecular geneses of colorectal cancer.

21.3.4 Two groups of methylation markers

The two-way hierarchical clustering also classified methylation markers into two major groups, Group 1 and Group 2 (Figure 21.4A). Group 1 markers are charac-terized as methylated in HME only. Group 2 markers are highly methylated in both HME and IME, but not or less in LME. Although MSS and MSI-H cell lines were used to select new methylation markers, there was neither an MSS-specific (i.e., methylated in both IME and LME) nor an IME-specific methylation marker identified. No LME-specific methylation marker was identified, either. Therefore, two-step classification was required to classify three epigenotypes: the first step to extract HME using Group 1 markers, and the second step to classify IME and LME using Group 2 markers (Figure 21.4B).

In the previous study, we examined all random combinations of 60 markers by Weka software, and developed two marker panels to decide epigenotypes with 95% accuracy: the first panel consisting of three Group 1 markers (*CACNA1G*, *LOX*, *SLC30A10*) to extract HME, and the second panel consisting of four Group 2 markers (*ELMO1*, *FBN2*, *THBD*, *HAND1*) and *SLC30A10* again to divide the remains into IME and LME (Yagi *et al.*, 2010).

Previous CIMP-related markers were all categorized into Group 1 markers except *NEUROG1*, indicating that CIMP-related markers were suitable to distinguish HME from others, but hardly classifier markers for IME or *KRAS*-mutation(+) colorectal cancers. It had been reported that *BRAF* mutation showed a non-random pattern of CpG island methylation while *KRAS* mutation showed random pattern, by analyz-ing 16 regions (Nosho *et al.*, 2008). In these 16 regions, 13 regions were analyzed in our study, and 12 out of the 13 were Group 1 markers. We agree that while *BRAF*-mutation(+) CRC should show non-random pattern of methylation of Group 1 markers, *KRAS*-mutation(+) CRC should show only rare and random methylation of Group 1 markers and would show a non-random pattern of methylation of Group 2 markers if analyzed. Selection of novel markers on genome-wide scale is thus important to avoid any bias in classification. In CRC analysis, one-step classification using Group 1 markers should not give accurate results, but two-step classification using Group 1 and Group 2 markers is recommended for accurate classification.

Shen *et al.* (2007) reported that colon cancer was classified into three groups using methylation data of 27 previously reported regions. Correlation of our epigenotypes with MSI status, *BRAF* and *KRAS* mutation was similar to Shen's CIMPs. But our study showed no correlation of LME with p53 status, so our IME and their CIMP-2 might be different clusters. In fact, they used several MINT loci as classifiers of CIMP-2, by analyzing methylation using a competitive PCR method, methylated CpG island amplification. MINT loci showed no or very small differences in methylation rate between IME and LME and belonged to

Group 1 markers, so were considered not to classify IME and LME by quantitative methylation data. They also used *NEUROG1* in their study, and this single Group 2 marker might have worked in classifying properly. For exact classification using a single, quantitative methylation detection method, multiple classifier markers are recommended for use. Selection of markers on genome-wide scale is again important.

21.3.5   Epigenotypes and oncogene mutation status

It is interesting that *BRAF*-mutation(+) and *KRAS*-mutation(+) CRCs clearly showed different epigenotypes; *BRAF*-mutation(+) strongly correlated to HME, and *KRAS*-mutation(+) strongly correlated to IME (Figure 21.4). Activation of oncogenes may perhaps induce specific epigenetic alterations. In fact, silencing of genes occurred in *Kras*-transformed NIH3T3 but not in untransformed NIH3T3, and essential effectors for the epigenetic gene silencing in *Kras*-transformed NIH3T3 were reported (Gazin *et al.*, 2007). Another possibility is that oncogene activation itself may not cause DNA methylation, but gene inactivation by methylation might be required to escape from oncogene-induced senescence to cause cancer, as disruption of p16 and/or p53 leads to escape from oncogene-induced senescence (Sharpless and DePinho, 2005). For example, *IGFBP3*, reported to be induced in replicatively senescent human umbilical vein endothelial cells (Kim *et al.*, 2007), showed high methylation rate in IME and HME colorectal cancer (Figure 21.4), though colon epithelium is different from endothelial cell.

21.3.6   Poorer survival of *KRAS*-mutation(+) IME

Previous meta-analysis showed that MSS CRC had a worse prognosis than MSI-high cases (Popat *et al.*, 2005) which were correlated to HME. Our interest was whether epigenotypes could predict prognosis among the relatively unfavorable MSS CRC, and *KRAS*-mutation(+) IME colorectal cancer significantly correlated to poorer survival ($p = 0.027$, log-rank test). When survival analysis by Cox proportional hazard model was also performed, *KRAS*-mutation(+) IME showed a significantly worse prognosis ($p = 0.027$, unadjusted analysis). When analyzing Stage III and IV cases only, the significant difference was reproduced. There was no significant difference in survival rate between *KRAS*-mutation(+) and *KRAS*-mutation(–) in MSS CRC. Although the prognostic significance of IME needs to be investigated by additional studies including a larger cohort of patients, epigenotype may be useful in predicting more unfavorable groups in MSS cases.

This poor prognosis was observed without anti-EGFR (epidermal growth factor receptor) or anti-VEGF (vascular endothelial growth factor) therapy. Anti-EGFR antibody Cetuximab (Erbitux®) was reported to be effective in *KRAS*-mutation(–) cases, but not for *KRAS*-mutation(+) cases (Karapetis *et al.*, 2008). It means that the relatively better prognosis of *KRAS*-mutation(–) cases could be even improved by use of antibody therapy. No improvement is expected for the most unfavorable *KRAS*-mutation(+) IME group, therefore a novel target therapy based on *KRAS* activation and/or the specific epigenetic alteration of IME is expected to be developed.

21.3.7 MeDIP-chip using clinical hepatocellular carcinoma samples

If cancer cell content is high enough (>90%), clinical cancer samples can be analyzed by MeDIP-chip. When clinical hepatocellular carcinoma samples were analyzed, MeDIP-chip data distinguished hepatitis C virus (HCV)-related cases from hepatitis B virus (HBV)-related cases, suggesting methylation information can divide the two and HCV-related cases might have aberrant methylation more frequently (Figure 21.5A). Analysis with MeDIP-chip just gives p-values for enrichment of the immunoprecipitated products in the analyzed regions, so that it shows the probability of methylation, but not the quantitative methylation status. For

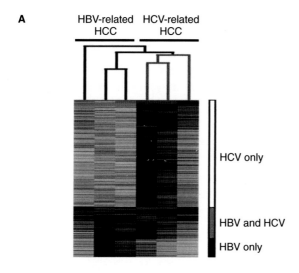

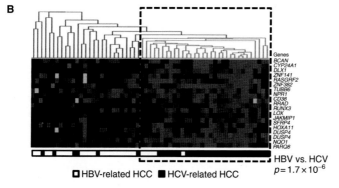

Figure 21.5 Analysis of hepatocellular carcinoma (HCC). (A) MeDIP-chip of three normal liver tissues, six hepatocellular carcinoma tissues, and their six matched surrounding liver tissues was performed. Heat map of genes showing methylation ($p < 10^{-2}$) in any of six cancers, but no methylation ($p > 10^{-2}$) in all the six surrounding liver tissues and three normal liver tissues, was shown. Hepatitis C virus (HCV)-related samples were distinguished from hepatitis B virus (HBV)-related samples, and showed more frequent methylation than HBV-related samples. (B) Candidate marker genes showing methylation in all the three HCV-related samples but not in HBV-related samples were selected and analyzed quantitatively by MassARRAY. Hierarchical clustering using the quantitative methylation data classified a methylated group (dotted square) where HCV-related samples were significantly more enriched than HBV ($p = 1.7 \times 10^{-6}$, Fisher's exact test).

validation, it is necessary to select candidate marker genes showing methylation in all the three HCV-related samples but in no HBV-related samples in MeDIP-chip, and analyze these markers quantitatively by MassARRAY. Hierarchical clustering using quantitative methylation data classified a highly methylated group, where HCV-related cases were strongly enriched (Figure 21.5B). It was indicated that there is a group of genes showing methylation preferentially in HCV-related cases rather than HBV-related cases, and hepatocellular carcinoma showed different methylation phenotypes by related viruses (Deng *et al.*, 2010).

21.4   Concluding remarks

Selection of novel markers on a genome-wide scale allows us to epigenotype cancer cases comprehensively, avoiding a bias of marker selection in previous studies. While MeDIP-chip may give false-negative data for low CpG regions, it is still a powerful genome-wide tool for identifying methylated promoter CpG islands in cancer cell lines and clinical cancer samples. High-throughput and highly quantitative methylation assay, e.g., MassARRAY, is useful for unsupervised hierarchical clustering, a comprehensive classification method. Successful classification needs careful preparation, such as microscopic examination of cancer cell content in clinical cancer samples and their dissection when necessary, and careful validation of quantitativity of MassARRAY primer pairs.

ACKNOWLEDGMENTS

This work was supported by Grants-in-Aid for Scientific Research from the Ministry of Education, Culture, Sports, Science and Technology of Japan, and by the JST PRESTO program.

REFERENCES

Deng, Y. B., Nagae, G., Midorikawa, Y., *et al.* (2010). Identification of genes preferentially methylated in hepatitis C virus-related hepatocellular carcinoma. *Cancer Science*, **101**, 1501–1510.

Ehrich, M., Nelson, M. R., Stanssens, P., *et al.* (2005). Quantitative high-throughput analysis of DNA methylation patterns by base-specific cleavage and mass spectrometry. *Proceedings of the National Academy of Sciences USA*, **102**, 15 785–15 790.

Feinberg, A. P. and Tycko, B. (2004). The history of cancer epigenetics. *Nature Reviews Cancer*, **4**, 143–153.

Feinberg, A. P., Ohlsson, R., and Henikoff, S. (2006). The epigenetic progenitor origin of human cancer. *Nature Reviews Genetics*, **7**, 21–33.

Gazin, C., Wajapeyee, N., Gobeil, S., Virbasius, C. M., and Green, M. R. (2007). An elaborate pathway required for Ras-mediated epigenetic silencing. *Nature*, **449**, 1073–1077.

Hayashi, H., Nagae, G., Tsutsumi, S., *et al.* (2007). High-resolution mapping of DNA methylation in human genome using oligonucleotide tiling array. *Human Genetics*, **120**, 701–711.

Hayashizaki, Y., Hirotsune, S., Okazaki, Y., *et al.* (1993). Restriction landmark genomic scanning method and its various applications. *Electrophoresis*, **14**, 251–258.

Jones, P. A. and Baylin, S. B. (2007). The epigenomics of cancer. *Cell*, **128**, 683–692.

Kaneda, A., Kaminishi, M., Yanagihara, K., Sugimura, T., and Ushijima, T. (2002). Identification of silencing of nine genes in human gastric cancers. *Cancer Research*, **62**, 6645–6650.

Kaneda, A., Takai, D., Kaminishi, M., Okochi, E., and Ushijima, T. (2003). Methylation-sensitive representational difference analysis and its application to cancer research. *Annals of the New York Academy of Sciences*, **983**, 131–141.

Kaneda, A., Wakazono, K., Tsukamoto, T., *et al.* (2004). Lysyl oxidase is a tumor suppressor gene inactivated by methylation and loss of heterozygosity in human gastric cancers. *Cancer Research*, **64**, 6410–6415.

Karapetis, C. S., Khambata-Ford, S., Jonker, D. J., *et al.* (2008). K-ras mutations and benefit from cetuximab in advanced colorectal cancer. *New England Journal of Medicine*, **359**, 1757–1765.

Kim, K. S., Kim, M. S., Seu, Y. B., *et al.* (2007). Regulation of replicative senescence by insulin-like growth factor-binding protein 3 in human umbilical vein endothelial cells. *Aging Cell*, **6**, 535–545.

Nosho, K., Irahara, N., Shima, K., *et al.* (2008). Comprehensive biostatistical analysis of CpG island methylator phenotype in colorectal cancer using a large population-based sample. *PLoS One*, **3**, e3698.

Ogino, S. and Goel, A. (2008). Molecular classification and correlates in colorectal cancer. *Journal of Molecular Diagnostics*, **10**, 13–27.

Ogino, S., Kawasaki, T., Kirkner, G. J., Loda, M., and Fuchs, C. S. (2006). CpG island methylator phenotype-low (CIMP-low) in colorectal cancer: possible associations with male sex and KRAS mutations. *Journal of Molecular Diagnostics*, **8**, 582–588.

Popat, S., Hubner, R., and Houlston, R. S. (2005). Systematic review of microsatellite instability and colorectal cancer prognosis. *Journal of Clinical Oncology*, **23**, 609–618.

Sharpless, N. E. and DePinho, R. A. (2005). Cancer: crime and punishment. *Nature*, **436**, 636–637.

Shen, L., Toyota, M., Kondo, Y., *et al.* (2007). Integrated genetic and epigenetic analysis identifies three different subclasses of colon cancer. *Proceedings of the National Academy of Sciences USA*, **104**, 18 654–18 659.

Suzuki, H., Watkins, D. N., Jair, K. W., *et al.* (2004). Epigenetic inactivation of SFRP genes allows constitutive WNT signaling in colorectal cancer. *Nature Genetics*, **36**, 417–422.

Toyota, M., Ahuja, N., Ohe-Toyota, M., *et al.* (1999). CpG island methylator phenotype in colorectal cancer. *Proceedings of the National Academy of Sciences USA*, **96**, 8681–8686.

Ushijima, T. (2005). Detection and interpretation of altered methylation patterns in cancer cells. *Nature Reviews Cancer*, **5**, 223–231.

Yagi, K., Akagi, K., Hayashi, H., *et al.* (2010). Three DNA methylation epigenotypes in human colorectal cancer. *Clinical Cancer Research*, **16**, 21–33.

Yoshikawa, H., Matsubara, K., Qian, G. S., *et al.* (2001). SOCS-1, a negative regulator of the JAK/STAT pathway, is silenced by methylation in human hepatocellular carcinoma and shows growth-suppression activity. *Nature Genetics*, **28**, 29–35.

Yu, L., Liu, C., Vandeusen, J., *et al.* (2005). Global assessment of promoter methylation in a mouse model of cancer identifies ID4 as a putative tumor-suppressor gene in human leukemia. *Nature Genetics*, **37**, 265–274.

22 Promoter CpG island methylation in colorectal cancer: biology and clinical applications

Sarah Derks and Manon van Engeland*

22.1   Introduction

Colorectal cancer (CRC) is the third leading cause of cancer death in both the USA and Europe and incidence and mortality rates are still on the rise (Parkin *et al.*, 2005).

   Colorectal cancers are thought to arise from pluripotent stem cells located in intestinal crypts which can develop into aberrant crypt foci (ACF) (Humphries and Wright, 2008) and premalignant adenomas of which about 5–6% will develop into a carcinoma with invasive and metastatic potential (Hermsen *et al.*, 2002). Steps that transform normal epithelium into adenomas and carcinomas were first described in the model of Fearon and Vogelstein in 1988 (Fearon and Vogelstein, 1990). In this model CRC was considered mainly a genetic disease characterized by accumulating abnormalities such as *TP53* (Baker *et al.*, 1989), *KRAS* (Bos *et al.*, 1987), and *APC* (Nakamura *et al.*, 1991) mutations and allelic deletions of chromosomes 5, 17, and 18 (Fearon and Vogelstein, 1990). In the following years it became apparent that CRC is a heterogeneous disease. About 85% of CRCs can be characterized by a condition of aneuploidy and an increased rate of loss of heterozygosity (LOH) which occurs in a non-random pattern of associated genomic changes (Hermsen *et al.*, 2002; Douglas *et al.*, 2004). The remaining 15% of CRCs are characterized by a defective mismatch repair (MMR) which is commonly achieved by DNA hypermethylation of the DNA mismatch repair gene *MHL1* in sporadic CRCs (Cunningham *et al.*, 1998; Herman *et al.*, 1998). Defective mismatch repair leads to accumulating frameshift mutations at simple repeated nucleotide sequences affecting genes such as *TGFβRII*, *IGFIIR*, and *BAX*, thereby leading to a mutator phenotype (Parsons *et al.*, 1995; Malkhosyan *et al.*, 1996; Souza *et al.*, 1996; Rampino *et al.*, 1997).

* Author to whom correspondence should be addressed.

Epigenomics: From Chromatin Biology to Therapeutics, ed. K. Appasani. Published by Cambridge University Press. © Cambridge University Press 2012.

In the last decades a revolution in epigenetic cancer research has added epigenetic abnormalities to the model. Promoter CpG island hypermethylation is the most extensively studied epigenetic modification in CRC which plays a pivotal role in carcinogenesis by transcriptional silencing of DNA-repair and tumor-suppressor genes (Herman and Baylin, 2003). DNA hypermethylation is already observed in ACF (Chan *et al.*, 2002) and adenomas (Derks *et al.*, 2006) and is thereby one of the earliest alterations in colorectal carcinogenesis. It has become clear that DNA methylation occurs in a network of other epigenetic alterations such as histone tail modifications, nucleosomal occupancy, chromosome looping, and small non-coding RNAs which interact and influence each other in a specific way (van Engeland *et al.*, 2011).

Nowadays CRC is one of the best-studied malignancies and provides an excellent model for studying the complexity of epigenetics and its driving role in colorectal carcinogenesis.

This chapter will outline the current knowledge of promoter CpG island hypermethylation in CRC and its promising role as biomarker for early disease detection, and the prediction of prognosis and response to therapy.

22.2 Promoter CpG hypermethylation and transcriptional silencing

DNA hypermethylation is the postreplicative addition of a methyl group to the 5-carbon position of the cytosine ring which occurs almost exclusively in the context of cytosine-phosphate-guanine (CpG) dinucleotides (Gardiner-Garden and Frommer, 1987). CpG dinucleotides cluster in regions called CpG islands, defined as regions of more than 200 bp, a G+C content of at least 50%, and a ratio of observed : expected CpGs of at least 0.6 (Gardiner-Garden and Frommer, 1987). CpG islands are associated with promoter or gene-regulatory regions of approximately 60% of all genes (Issa, 2000).

In normal cells promoter CpG island regions are protected from methylation, with the exception of tissue-specific genes which become hypermethylated early in development and differentiation (Straussman *et al.*, 2009). In cancer cells, specific promoter CpG islands become hypermethylated which results in transcriptional silencing of tumor-suppressor genes.

DNA hypermethylation occurs also in regions with less dense CpG dinucleotides situated in the proximity (~2 kb) of CpG islands, called CpG shores (Irizarry *et al.*, 2009). CpG shores are highly conserved regions associated with tissue differentiation. Hypermethylation at these sites is associated with gene silencing which is a predominant mechanism in normal differentiation as well as carcinogenesis (Irizarry *et al.*, 2009).

CpG hypermethylation is mediated by a family of DNA methyltransferases (DNMTs) which catalyze the transfer of a methyl group of S-adenosyl methionine to a cytosine. DNMT3a and 3b regulate *de novo* methylation and are responsible for the establishment of methylation patterns while DNMT1 is active in S phase and responsible for the maintenance of methylation.

Cytosine methylation influences gene transcription by the recruitment of methyl-CpG-binding domain (MBD) proteins which inhibit the binding of other DNA-binding proteins such as CTCF (Hark *et al.*, 2000) and transcription factors to their target sequences (Figure 22.1). MBD proteins such as MBD1, MBD2, and MeCP2 recruit chromatin remodeling complexes. This will lead to the generation of a repressive chromatin state characterized by trimethylation of histone 3 lysine 9 (H3K9me3) and lysine 27 (H3K27me3) and deacetylation of H3 and H4 (Ohm *et al.*, 2007). Due to higher-order chromatin structures and subsequent chromatin looping multiple hypermethylated CpG islands can physically interact, cluster, and cooperate in attracting repressor proteins and ensure a transcriptional silenced state. This can occur within genes as described for a specific gene (*GATA4*) (Tiwari *et al.*, 2008) but also for an entire chromosomal region (Easwaran *et al.*, 2010).

Long-term transcriptional repression of genes is under control of the Polycomb group of proteins (PRC) which associate within two multimeric complexes PRC1 and PRC2. PRC2 is involved in the initiation of silencing and contains EZH2 (among SUZ12 and EED), which functions as a histone methyltransferase that methylates H3K27, and to a lesser extent H3K9. This mark can attract PRC1 containing BMI1, which is involved in recognition of the H3K27me3. PRC1 has a role in subsequent maintenance of Polycomb-group-mediated long-term

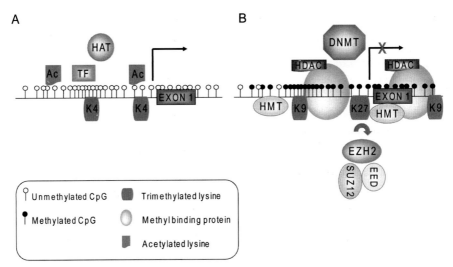

Figure 22.1 CpG island promoter methylation and transcriptional silencing. Shown are schematic gene promoters. Filled circles represent methylated CpG dinucleotides, and open circles represent unmethylated CpG dinucleotides. Under normal conditions (A) CpG dinucleotides are unmethylated and packaged with active modified histone proteins containing trimethyl histone 3 lysine 4 (H3K4me3) as well as histone acetylation accomplished by histone acetyltransferases (HATs). These epigenetic modifications constitute an "open" chromatin structure and allow transcription. In cancer cells (B) CpG dinucleotides can become hypermethylated which is catalyzed by DNMTs and bound by methyl-binding proteins (MBPs). Some genes are targeted by Polycomb proteins such as EZH2 which catalyzes H3K27 methylation and bound by histone deacetylase (HDAC) proteins to remove histone acetylation and histone methyltransferase to methylate H3K9 as well. These epigenetic modifications constitute a "closed" chromatin structure and a transcriptional silenced state.

silencing and includes besides BMI1 the CBX and HPH family of proteins, YY1, RING1/1a, and RING2/1b. Enrichment of EZH2 and the H3K27 mark has been shown to be a characteristic of DNA hypermethylated promoters and silenced genes (Squazzo *et al.*, 2006; Ohm *et al.*, 2007).

22.3 Promoter CpG island methylation is a tissue- and gene-specific process

22.3.1 What determines gene specificity?

The occurrence of DNA hypermethylation in the genome is a non-random, tissue-specific process (Bird, 2002). It is estimated that within individual cancers only 1–5% of all CpG islands are aberrantly methylated and that different cancer types greatly differ in methylation profile (Weber *et al.*, 2005). It is unknown what determines this gene specificity but several explanations have been proposed.

One proposed mechanism involves local DNA sequence features contributing to the susceptibility of CpG islands to become methylated (Feltus *et al.*, 2003). In a genome-wide analysis in which genes became hypermethylated after DNMT1 overexpression, five DNA sequence patterns were identified based on their association with hypermethylated gene promoters. Together with eight other motifs methylation-prone CpG islands can be differentiated from methylation-resistant CpG islands with 82% accuracy (Feltus *et al.*, 2006). Others have also identified overrepresented sequence patterns to discriminate cancer-specific from tissue-specific hypermethylated genes and have used bioinformatic tools to identify novel epigenetically silenced genes in cancer (Hoque *et al.*, 2008). Whether these sequences indicate the involvement of sequence-specific binding factors or a common secondary DNA structure remains to be elucidated (Feltus *et al.*, 2006).

A recent study demonstrates that genes with a methylation-prone sequence motif are strongly related to genes that are marked by PcG proteins (Lee *et al.*, 2006) and bivalent chromatin marks (H3K4me3 and H3K27me3) in embryonic stem cells (McCabe *et al.*, 2009). These PcG target genes are important in stem cells and development (Ohm *et al.*, 2007; Schlesinger *et al.*, 2007) and have been shown to be vulnerable for aberrant *de novo* methylation later in life. Furthermore, PcG proteins are able to directly interact with DNMT (Vire *et al.*, 2006), but it is not clear whether PcG binding initiates DNA methylation. Although elevated DNMT expression levels trigger DNA hypermethylation, DNMTs themselves have little sequence specificity and therefore require sequence-specific binding proteins, such as transcription factors or gene repressors, to target specific promoters (Yoder *et al.*, 1997). This mechanism has been shown for the oncogenic transcription factor PML-RAR (Di Croce *et al.*, 2002) and has been proposed for the proto-oncogenes *c-MYC* and *KRAS* as well (Brenner *et al.*, 2005; Gazin *et al.*, 2007). Whether these oncogenes induce widespread DNA methylation is unknown.

An additional possibility is the spreading of methylation from one hyper-methylated gene to surrounding chromosomal regions containing multiple genes, a mechanism termed long-range epigenetic silencing (LRES) (Frigola *et al.*, 2006). This mechanism has been shown for genes located on a 1-Mb domain on chromosomal region 2q14 which are coordinately epigenetically suppressed by H3K9 methylation independently of promoter CpG island methylation. The same mechanism was observed for a 4-Mb chromosomal region on 3p22 which flanks the *MLH1* gene (Hitchins *et al.*, 2007) and chromosomal region 5q32.2 (Rodriguez *et al.*, 2008). The mechanism underlying LRES is not known. It is not clear whether LRES is initiated by a critical target gene and that a silenced hetero-chromatin state is spread to innocent bystanders or whether neighboring genes are coordinately suppressed due to a specific spatial organization in the nucleus (Easwaran *et al.*, 2010). Whether LRES is a general mechanism involving more chromosomes needs to be unraveled.

22.3.2   Not the entire but a specific region within the promoter CpG island determines transcriptional silencing

Within the hypermethylated promoter region the location influencing gene silencing is highly specific. For *P16INK4A* (Gonzalgo *et al.*, 1998) and *MLH1* (Deng *et al.*, 2002) it has been shown that not the entire CpG island, but hyper-methylation of a small region in exon 1 and a region proximal of the transcription start site, designated as core regions, determines transcriptional activity. Also for *hTERT* (Zinn *et al.*, 2007), *NDRG4* (Melotte *et al.*, 2009), and *WIF1* (Licchesi *et al.*, 2010) among others, core regions have been identified which were located upstream and downstream of the transcription start site (TSS) and can theoret-ically even be present at distant loci.

Methylation outside these core regions is frequently observed in CRC as well and has been described to be associated with aging and inflammation but does not affect gene transcription (Issa, 1999; Christensen *et al.*, 2009). This non-functional methylation can spread to neighboring regions and to core regions which might explain the linkage between aging, inflammation, and cancer.

22.4   The colorectal cancer hypermethylome

22.4.1   Genes affected by CpG island methylation in colorectal cancer

High-throughput genome-wide analyses have shown that in CRC about 400–600 genes are hypermethylated (Schuebel *et al.*, 2007). Promoter methylation of some genes is associated with aging but hypermethylation of cancer associated (CAN) genes are highly prevalent as well and affect the regulation of proliferation, migration, differentiation, adhesion, angiogenesis, apoptosis, DNA stability, and repair (Weber *et al.*, 2005; Schuebel *et al.*, 2007). Disrupted key pathways driving colorectal carcinogenesis are WNT signaling, transforming growth factor β (TGFβ) signaling, epidermal growth factor (EGFR) signaling, RAS/RAF/MAPK signaling, and phosphatidylinositol 3-kinase (PI3K) but also other pathways affecting DNA repair cell cycle control, p53 network, and apoptosis are affected. In these path-ways multiple genes are hypermethylated, among which are *APC, MLH1, MGMT, CHFR, RASSF1A, P14ARF*, and *P16INK4A*, among many others (Table 22.1).

Table 22.1 Genes and microRNAs hypermethylated in colorectal cancer

Signaling pathways	Hypermethylated genes / miRNAs
WNT	*APC, SFRP1, SFRP2, SFRP4, SFRP5, SOX17, WNT5a, DKK1, DKK3, WIF1, AXIN2*
RAS/RAF/MAPK	*RASSF1A, RASSF2A, EPHB2, RAB32, NORE1*
TP53	*HIC1, miRNA34a, miRNA34b, miRNA34c*
PI3K	*PIK3CG*
Cell cycle regulation	*CDKN2A, KLF4*
Transcription regulation	*GATA4, GATA5, RUNX3, CDX1, HLTF, FOXL2, ALX4*
DNA repair/stability	*MLH1, MGMT, WRN, CHFR*
Apoptosis	*BNIP3, IRF8, DAPK1, HRK*

Note: This table is not exhaustive and represents genes that function in key signaling pathways and are transcriptionally silenced by CpG island promoter methylation in CRC.

Besides CpG-island-containing promoters and CpG island shores microRNAs can also be transcriptionally silenced by DNA hypermethylation. MiRNA-34a promotes apoptosis by targeting SIRT, an enzyme responsible for the inactivation of p53 and is frequently hypermethylated in CRC (Yamakuchi *et al.*, 2008). The same is true for miRNA-34b, miRNA-34c (Toyota *et al.*, 2008), miRNA-148a, and miRNA-9, of which hypermethylation has been associated with metastastic CRC (Lujambio *et al.*, 2008).

22.4.2 The CpG island methylator phenotype

Besides CpG hypermethylation of individual genes, a distinct subset of CRCs display an exceptionally high frequency of concordantly hypermethylated genes, also designated as the "CpG island methylation phenotype" (CIMP). CIMP is strongly associated with microsatellite instability (MSI) by CpG island hypermethylation of mismatch repair gene *MLH1* and characteristics including older age, female sex, proximal tumor location, poor differentiation, *BRAF* mutation, and wild-type *TP53* (Hawkins *et al.*, 2002; van Rijnsoever *et al.*, 2002; Samowitz *et al.*, 2005; Weisenberger *et al.*, 2006; Ogino *et al.*, 2007a; Samowitz *et al.*, 2007). Since the introduction of CIMP by Toyota in 1999 (Toyota *et al.*, 1999), however, the concept is under debate.

The use of different CIMP markers throughout the years has caused confusion by reporting conflicting associations. Different attempts have been made to standardize the gene panel; among these is a study by Weisenberger (Weisenberger *et al.*, 2006) who systematically screened 195 CpG islands in 295 CRCs, for a high specificity to a subclass with a high number of methylated genes. Using different algorithms and bimodality assessment five gene promoters (*CACNA1G, IGF2, NEUROG1, RUNX3,* and *SOCS1*) were selected as markers for CIMP (Weisenberger *et al.*, 2006). Weisenberger's marker panel has been validated in different large series and has outperformed other definitions of CIMP in respect of an association with specific phenotypic characteristics (Ogino *et al.*, 2007a; Lee *et al.*, 2008; Nosho *et al.*, 2008). Later studies, however, have shown that Weisenberger's bimodal distribution of methylation is only present in MSI CRC and not in microsatellite-stable (MSS)/CIMP CRCs. To overcome this problem a new panel of markers containing *CACNA1G, IGF2, RUNX3, HTR6, RIZ1, MINT31,* and *MAP1B* was identified which allowed a more distinct classification of CIMP tumors in MSS CRC (Ferracin *et al.*, 2008). Although these studies assisted the understanding of the concept of CIMP, the absence of a consensus panel defining CIMP makes it difficult to compare literature.

Besides CIMP, a CIMP-low phenotype has been identified. CIMP-low describes a subgroup of CRCs with a lower number of hypermethylated genes (Iacopetta *et al.*, 2006; Derks *et al.*, 2008) but frequent *MGMT* CpG island promoter methylation, mutant *KRAS* (Ogino *et al.*, 2006, 2007b), and a poor prognosis (Barault *et al.*, 2008).

Why some genes are methylated within the concept of CIMP(-high) and others in CIMP-low is not known. Different experimental and correlative studies have suggested a role for DNMT3b (Ibrahim *et al.*, 2010) and mutated *BRAF* in the establishment of CIMP(-high) while *KRAS* mutation is associated with CIMP-low (Tanaka *et al.*, 2010). In colorectal carcinogenesis *KRAS*– and *BRAF* mutation occur in a mutually exclusive fashion (Rajagopalan *et al.*, 2002). *KRAS*– and *BRAF* mutation both occur early in carcinogenesis and *KRAS* mutation has been linked to the induction of DNA hypermethylation (Gazin *et al.*, 2007). A clear mechanistic basis for these oncogenes in locus specific CpG island methylation, however, is still under investigation.

22.5 The promise of methylation biomarkers

In the last five years it has become increasingly clear that promoter CpG island methylation markers are promising biomarkers for early detection and risk stratification.

Clinicopathologic staging within the tumor–node–metastasis (TNM) classification system is the cornerstone for prognosis and treatment selection by measuring the depth of invasion, counting the number of lymph nodes involved, and searching for distant metastasis. When CRC is detected at an early stage and confined to the bowel wall (stage I) the five-year survival is over 90%. In case of lymph node involvement (stage III) or metastatic disease (stage IV), however, five-year survival decreases to around 68% and 10% respectively (Levin *et al.*, 2008). At early-stage disease surgical resection is sufficient for curative treatment while later-stage disease requires adjuvant chemotherapy. Therefore it is expected that a great reduction in CRC incidence and mortality can be achieved by early detection and removal of premalignant lesions.

Furthermore, patients within the same TNM stage can demonstrate considerable variation in survival. It has been clearly shown that not all patients benefit from chemotherapeutic treatment, and that they may experience unneeded toxicity. Biomarkers to assist in the selection of optimal treatment regimes are needed and methylation markers are promising candidates.

22.5.1 Diagnostic markers

Methods used nowadays to detect CRC at an early stage include the (immuno-logical) fecal occult blood test (FOBT), sigmoidoscopy, and (virtual) colonoscopy.

Unfortunately, (i)FOBT lacks sufficient sensitivity and specificity (Imperiale *et al.*, 2004) and although colonoscopy is considered the gold standard, screening compliance is low because of the invasive character of the procedure (Gatto *et al.*, 2003).

Since CpG island promoter methylation is prevalent and occurs early in CRC methylation biomarkers are attractive candidates for early detecion and risk stratification. Several studies have shown proof-of-principle for the detection of methylation biomarkers in colorectal adenoma- or carcinoma-derived DNA in stool (Belshaw *et al.*, 2004; Leung *et al.*, 2004; Muller *et al.*, 2004; Chen *et al.*, 2005; Petko *et al.*, 2005) or blood (Ebert *et al.*, 2006; Frattini *et al.*, 2006).

The detection of biomarkers in stool is based on cellular shedding from colorectal neoplasms. Although these tests suffer from degeneration of DNA, different methylation biomarkers have been described as having high sensitivity and specificity.

One of the first diagnostic methylation biomarkers in stool was *SFRP2* CpG island promoter methylation. *SFRP2* promoter methylation was tested in cancer patients and healthy controls and detected CRC with a sensitivity of 77–90% and a specificity of 77% (Muller *et al.*, 2004). Later hypermethylation of exon 1 of Vimentin (VIM) which occurs in 53–84% of CRCs was added to the list and could be detected in stool with a sensitivity of 83% and specificity of 82% and is commercially available in the USA (Itzkowitz *et al.*, 2007, 2008; Li *et al.*, 2009). Further, CpG island promoter methylation of other genes such as *MGMT*, *P16INK4A* (Petko *et al.*, 2005), *GATA4*, *GATA5*[5], *NDRG4*[6], *OSMR* (Kim *et al.*, 2009), and *TFPI* (Glockner *et al.*, 2009) among others has been shown to give sensitive biomarkers of which the latter is able to detect CRCs in stages I–III with a sensitivity and specificity of respectively 76% to 89% and 79% to 93% (Glockner *et al.*, 2009).

The same accounts for a diversity of blood markers, single or in panel, which can detect CRCs with promising sensitivity and specificity (Table 22.2). Circulating cancer cells in peripheral blood may predict the presence of a malignancy and might be a marker of metastasis and recurrence. Cancer embryonic antigen (CEA) is a conventional serum marker which is widely used to monitor disease progression and response to therapy but lacks sensitivity and specificity in early-stage disease (Duffy, 2001). PCR-based methods are highly sensitive and able to detect small amounts of circulating cancer DNA in blood. Different methylation blood markers have shown proof of principle, among which is *SEPT9* promoter methylation which has been shown to detect CRCs with a sensitivity and specificity of 69% and 86% respectively (Lofton-Day *et al.*, 2008).

Although these studies have provided impressive preliminary data about promising epigenetic biomarkers in stool and blood, before adoption into routine clinical practice markers need to be validated in independent large population-based studies using standardized protocols. At the moment, the available results are difficult to compare due to differences in study designs, patient population, adjustment for tumor stage, and treatment. The ultimate goal is to design an ideal screening tool which is non-invasive, sensitive, specific, cost-effective, and easy to

Table 22.2 Overview of promising epigenetic biomarkers in colorectal cancer

Diagnostic	Study design, sensitivity/specificity
Stool	
SFRP2	CRC vs. controls, 77–90%/77% (Muller et al., 2004)
SFRP1	CRC and CRA vs. controls, 42%-77% and 31–48%/73%-100% (Zhang et al., 2007)
VIMENTIN	CRC vs. controls, 83%/82% (Li et al., 2009)
GATA4	CRC vs. controls, 51%–71%/84–93% (Hellebrekers et al., 2009)
NDRG4	CRC vs. controls, 53%/100% (Melotte et al., 2009)
OSMR	CRC vs. controls, 38%/95% (Kim et al., 2009)
TFPI	Stage I–III CRC vs. controls, 77–90%/77% (Glockner et al., 2009)
Blood	
ALX4	CRC vs. controls, 83%/70% (Ebert et al., 2006)
SEPT9	CRC and precursor lesions vs. controls, 69–72%/86–93% (Lofton-Day et al., 2008; deVos et al., 2009)
MLH1	Positive blood test vs. primary cancer in MSI CRCs, 33%/100% (Grady et al., 2001)
DAPK4	Positive blood test vs. primary cancer, 21%/100% (Yamaguchi et al., 2003)
APC, MLH1 and HLTF	CRC vs. controls, >1/3 markers positive, 57%/90% (Leung et al., 2005)
APC, MGMT, RASSF2A and WIF1	CRC vs. controls, 86%/93% (Lee et al., 2009)
ALX4 and SEPT9	Precursor lesions vs. controls, 71%/95% (Tanzer et al., 2010)
ANXA3, CLEC4D, LMNB1, PRRG4, TNFAIP6, VNN1 and IL2RB	CRC vs. controls, 82%/64% (Marshall et al., 2010)

implement across large populations, and which detects all different CRC epi-genotypes without false-positive or false-negative results.

22.5.2   Prognostic and predictive markers

In addition to markers for risk assessment and secondary prevention, it is imperative to identify prognostic and predictive biomarkers to improve personalized cancer treatment.

Different anti-cancer agents are used in metastatic CRC. These include the intravenous and oral fluoropyrimidines, 5-fluorouracil (5-FU) and capecitabine, the platinum derivative oxaliplatin, the camptothecin derivative and topoisomerase inhibitor irinotecan, and the monoclonal antibodies directed against epidermal growth factor receptor (EGFR) (cetuximab and panitumumab) and vascular endothelial growth factor (VEGF) (bevacizumab).

Recently it has become clear that response to treatment with chemotherapeutic agents cetuximab and panitumumab is confined to patients with wild-type *KRAS* (Karapetis et al., 2008; McNeil, 2008) and *BRAF* (Di Nicolantonio et al., 2008; Wong and Cunningham, 2008). Activating mutation of *KRAS* would replace the dependency of the tumor cells on increased signaling from upstream EGFR which explains why only patients with wild-type *KRAS* benefit from EGFR inhibitors. Whether other downstream EGFR signaling genes such as *RASSF1A* and *RASSF2A*, both affected by CpG island promoter methylation in CRC, influence treatment response remains to be elucidated.

Furthermore, it has been shown that the presence of CIMP as well as CpG island promoter methylation (Dahlin *et al.*, 2010) of individual genes is associated with a poor prognosis (Table 22.3). On the other hand CIMP has been reported to be an indicator of good response to 5-FU (Iacopetta *et al.*, 2008) (Table 22.4). Other studies, however, have reported conflicting results (Ogino *et al.*, 2007d; Shen *et al.*, 2007) which might be explained by the use of different marker sets in defining CIMP, differences in patient populations but also confounding effects of other molecular events. The prognostic effect of CIMP, for instance, is dependent upon the presence of MSI. In contrast to MSS/CIMP CRCs, MSI/CIMP CRCs show a more favorable outcome (Popat *et al.*, 2005) but a reduced sensitivity to 5-FU (Warusavitarne and Schnitzler, 2007).

The sensitivity to the topoisomerase inhibitor irinotecan turns out to be dependent upon CpG island methylation of Werner syndrome gene (*WRN*). Silencing of *WRN* leads to increased chromosomal instability and apoptosis as well as sensitivity to irinotecan (Agrelo *et al.*, 2006). The same is true of the *UGT1A1* gene which influences cellular inactivation of the active metabolite of irinotecan. CpG island promoter hypermethylation of *UGT1A1* leads to decreased expression and an increased response to irinotecan (Gagnon *et al.*, 2006).

Furthermore CpG island promoter methylation of the DNA repair gene *MGMT* is associated with resistance to alkylating agents (Weller *et al.*, 2010). Functional

Table 22.3 Epigenetic markers for prognosis of colorectal cancer

Prognostic	Outcome
CIMP	Good prognosis (Ogino *et al.*, 2007c) Poor prognosis in MSS CRC (Dahlin *et al.*, 2010)
P16INK4A	Poor prognosis (Esteller *et al.*, 2001; Maeda *et al.*, 2003) in CIMP CRCs (Shima *et al.*, 2011)
LINE-1 hypomethylation	Poor prognosis stage II–III MSS/CIMP CRC (Kawakami *et al.*, 2011)
IGF2 DMR0 hypomethylation	Poor prognosis (Baba *et al.*, 2010)

Table 22.4 Prediction of outcomes of treatment in colorectal cancer

Predictive	Outcome
CIMP	Good response to treatment with oral 5-FU (Warusavitarne and Schnitzler, 2007) Resistance to treatment with oral 5-FU (Ogino *et al.*, 2007d; Shen *et al.*, 2007)
MLH1	Resistence to treatment with oral 5-FU (Arnold *et al.*, 2003)
WRN	Sensitivity to treatment with irinotecan (Agrelo *et al.*, 2006)
UGT1A1	Sensitivity to treatment with irinotecan (Gagnon *et al.*, 2006)
MGMT	Sensitivity to treatment with oxaliplatin (Park *et al.*, 2010)
LINE-1 hypomethylation	Good response to treatment with 5-FU in MSS/CIMP CRC (Kawakami *et al.*, 2011)

MGMT reverses cross-linking of DNA induced by alkylating agents and thereby causes resistance to these drugs. Transcriptional silencing of *MGMT* by CpG island promoter methylation has been clearly shown to be a predictor of response to alkylating agents in gliomas (Esteller *et al.*, 2000) and also in CRC decreased *MGMT* expression has been proposed as marker of sensitivity for oxaliplatin (Park *et al.*, 2010).

Although further prospective trials are warranted before bringing them into use in clinical practice, these potential epigenetic biomarkers are promising for selecting patients who might benefit from treatment and thereby avoid the unnecessary toxicity of adjuvant treatment. Predictive markers can help in tailoring therapy by selecting effective agents and make personalized medicine a clinical reality.

22.6 Conclusion

This chapter has described the current state of knowledge of the complexity and driving role of epigenetics in CRC as well as its potential clinical applications. Extensive research in the last decades has shown that promoter CpG island methylation occurs early in colorectal carcinogenesis and involves a great number of genes in a tissue- and gene-specific fashion. CpG island promoter methylation functions in a complex network of DNA- and chromatin-modifying processes which together lead to a highly disrupted cancer epigenome. Due to their early and prevalent occurrence hypermethylated genes are promising biomarkers for disease detection, prognosis, and response to therapy. Although prospective clinical trials are warranted before introducing them in clinical practice, epigenetic biomarkers are promising to bring us closer towards the ultimate goal of decreasing CRC mortality.

REFERENCES

Agrelo, R., Cheng, W. H., Setien, F., *et al.* (2006). Epigenetic inactivation of the premature aging Werner syndrome gene in human cancer. *Proceedings of the National Academy of Sciences USA*, **103**, 8822–8827.

Arnold, C. N., Goel, A., and Boland, C. R., (2003). Role of hMLH1 promoter hypermethylation in drug resistance to 5-fluorouracil in colorectal cancer cell lines. *International Journal of Cancer*, **106**, 66–73.

Baba, Y., Nosho, K., Shima, K., *et al.* (2010). Hypomethylation of the IGF2 DMR in colorectal tumors, detected by bisulfite pyrosequencing, is associated with poor prognosis. *Gastroenterology*, **139**, 1855–1864.

Baker, S. J., Fearon, E. R., Nigro, J. M., *et al.* (1989). Chromosome 17 deletions and p53 gene mutations in colorectal carcinomas. *Science*, **244**, 217–221.

Barault, L., Charon-Barra, C., Jooste, V., *et al.* (2008). Hypermethylator phenotype in sporadic colon cancer: study on a population-based series of 582 cases. *Cancer Research*, **68**, 8541–8546.

Belshaw, N. J., Elliott, G. O., Williams, E. A., *et al.* (2004). Use of DNA from human stools to detect aberrant CpG island methylation of genes implicated in colorectal cancer. *Cancer Epidemiology and Biomarkers Prevention*, **13**, 1495–1501.

Bird, A. (2002). DNA methylation patterns and epigenetic memory. *Genes and Development*, **16**, 6–21.

Bos, J. L., Fearon, E. R., Hamilton, S. R., *et al.* (1987). Prevalence of *ras* gene mutations in human colorectal cancers. *Nature*, **327**, 293–297.

Brenner, C., Deplus, R., Didelot, C., *et al.* (2005). Myc represses transcription through recruitment of DNA methyltransferase corepressor. *EMBO Journal*, **24**, 336–346.

Chan, A. O., Broaddus, R. R., Houlihan, P. S., *et al.* (2002). CpG island methylation in aberrant crypt foci of the colorectum. *American Journal of Pathology*, **160**, 1823–1830.

Chen, W. D., Han, Z. J., Skoletsky, J., *et al.* (2005). Detection in fecal DNA of colon cancer-specific methylation of the nonexpressed vimentin gene. *Journal of the National Cancer Institute*, **97**, 1124–1132.

Christensen, B. C., Houseman, E. A., Marsit, C. J., *et al.* (2009). Aging and environmental exposures alter tissue-specific DNA methylation dependent upon CpG island context. *PLoS Genetics*, **5**, e1000602.

Cunningham, J. M., Christensen, E. R., Tester, D. J., *et al.* (1998). Hypermethylation of the hMLH1 promoter in colon cancer with microsatellite instability. *Cancer Research*, **58**, 3455–3460.

Dahlin, A. M., Palmqvist, R., Henriksson, M. L., *et al.* (2010). The role of the CpG island methylator phenotype in colorectal cancer prognosis depends on microsatellite instability screening status. *Clinical Cancer Research*, **16**, 1845–1855.

Deng, G., Peng, E., Gum, J., *et al.* (2002). Methylation of hMLH1 promoter correlates with the gene silencing with a region-specific manner in colorectal cancer. *British Journal of Cancer*, **86**, 574–579.

Derks, S., Postma, C., Moerkerk, P. T., *et al.* (2006). Promoter methylation precedes chromosomal alterations in colorectal cancer development. *Cell Oncology*, **28**, 247–257.

Derks, S., Postma, C., Carvalho, B., *et al.* (2008). Integrated analysis of chromosomal, microsatellite and epigenetic instability in colorectal cancer identifies specific associations between promoter methylation of pivotal tumour suppressor and DNA repair genes and specific chromosomal alterations. *Carcinogenesis*, **29**, 434–439.

Devos, T., Tetzner, R., Model, F., *et al.* (2009). Circulating methylated SEPT9 DNA in plasma is a biomarker for colorectal cancer. *Clinical Chemistry*, **55**, 1337–1346.

Di Croce, L., Raker, V. A., Corsaro, M., *et al.* (2002). Methyltransferase recruitment and DNA hypermethylation of target promoters by an oncogenic transcription factor. *Science*, **295**, 1079–1082.

Di Nicolantonio, F., Martini, M., Molinari, F., *et al.* (2008). Wild-type *BRAF* is required for response to panitumumab or cetuximab in metastatic colorectal cancer. *Journal of Clinical Oncology*, **26**, 5705–5712.

Douglas, E. J., Fiegler, H., Rowan, A., *et al.* (2004). Array comparative genomic hybridization analysis of colorectal cancer cell lines and primary carcinomas. *Cancer Research*, **64**, 4817–4825.

Duffy, M. J. (2001). Carcinoembryonic antigen as a marker for colorectal cancer: is it clinically useful? *Clinical Chemistry*, **47**, 624–630.

Easwaran, H. P., Van Neste, L., Cope, L., *et al.* (2010). Aberrant silencing of cancer-related genes by CpG hypermethylation occurs independently of their spatial organization in the nucleus. *Cancer Research*, **70**, 8015–8024.

Ebert, M. P., Model, F., Mooney, S., *et al.* (2006). Aristaless-like homeobox-4 gene methylation is a potential marker for colorectal adenocarcinomas. *Gastroenterology*, **131**, 1418–1430.

Esteller, M., Garcia-Foncillas, J., Andion, E., *et al.* (2000). Inactivation of the DNA-repair gene *MGMT* and the clinical response of gliomas to alkylating agents. *New England Journal of Medicine*, **343**, 1350–1354.

Esteller, M., Gonzalez, S., Risques, R. A., *et al.* (2001). K-ras and p16 aberrations confer poor prognosis in human colorectal cancer. *Journal of Clinical Oncology*, **19**, 299–304.

Fearon, E. R. and Vogelstein, B. (1990). A genetic model for colorectal tumorigenesis. *Cell*, **61**, 759–767.

Feltus, F. A., Lee, E. K., Costello, J. F., *et al.* (2003). Predicting aberrant CpG island methylation. *Proceedings of the National Academy of Sciences USA*, **100**, 12 253–12 258.

Feltus, F. A., Lee, E. K., Costello, J. F., *et al.* (2006). DNA motifs associated with aberrant CpG island methylation. *Genomics*, **87**, 572–579.

Ferracin, M., Gafa, R., Miotto, E., *et al.* (2008). The methylator phenotype in microsatellite stable colorectal cancers is characterized by a distinct gene expression profile. *Journal of Pathology*, **214**, 594–602.

Fraga, M. F., Ballestar, E., Montoya, G., *et al.* (2003). The affinity of different MBD proteins for a specific methylated locus depends on their intrinsic binding properties. *Nucleic Acids Research*, **31**, 1765–1774.

Frattini, M., Gallino, G., Signoroni, S., *et al.* (2006). Quantitative analysis of plasma DNA in colorectal cancer patients: a novel prognostic tool. *Annals of the New York Academy of Sciences*, **1075**, 185–190.

Frigola, J., Song, J., Stirzaker, C., *et al.* (2006). Epigenetic remodeling in colorectal cancer results in coordinate gene suppression across an entire chromosome band. *Nature Genetics*, **38**, 540–549.

Gagnon, J. F., Bernard, O., Villeneuve, L., *et al.* (2006). Irinotecan inactivation is modulated by epigenetic silencing of UGT1A1 in colon cancer. *Clinical Cancer Research*, **12**, 1850–1858.

Gardiner-Garden, M. and Frommer, M. (1987). CpG islands in vertebrate genomes. *Journal of Molecular Biology*, **196**, 261–282.

Gatto, N. M., Frucht, H., Sundararajan, V., *et al.* (2003). Risk of perforation after colonoscopy and sigmoidoscopy: a population-based study. *Journal of the National Cancer Institute*, **95**, 230–236.

Gazin, C., Wajapeyee, N., Gobeil, S., *et al.* (2007). An elaborate pathway required for Ras-mediated epigenetic silencing. *Nature*, **449**, 1073–1077.

Glockner, S. C., Dhir, M., Yi, J. M., *et al.* (2009). Methylation of TFPI2 in stool DNA: a potential novel biomarker for the detection of colorectal cancer. *Cancer Research*, **69**, 4691–4699.

Gonzalgo, M. L., Hayashida, T., Bender, C. M., *et al.* (1998). The role of DNA methylation in expression of the p19/p16 locus in human bladder cancer cell lines. *Cancer Research*, **58**, 1245–1252.

Grady, W. M., Rajput, A., Lutterbaugh, J. D., *et al.* (2001). Detection of aberrantly methylated hMLH1 promoter DNA in the serum of patients with microsatellite unstable colon cancer. *Cancer Research*, **61**, 900–902.

Hark, A. T., Schoenherr, C. J., Katz, D. J., *et al.* (2000). CTCF mediates methylation-sensitive enhancer-blocking activity at the H19/Igf2 locus. *Nature*, **405**, 486–489.

Hawkins, N., Norrie, M., Cheong, K., *et al.* (2002). CpG island methylation in sporadic colorectal cancers and its relationship to microsatellite instability. *Gastroenterology*, **122**, 1376–1387.

Hellebrekers, D. M., Lentjes, M. H., van den Bosch, S. M., *et al.* (2009). GATA4 and GATA5 are potential tumor suppressors and biomarkers in colorectal cancer. *Clinical Cancer Research*, **15**, 3990–3997.

Herman, J. G. and Baylin, S. B., (2003). Gene silencing in cancer in association with promoter hypermethylation. *New England Journal of Medicine*, **349**, 2042–2054.

Herman, J. G., Umar, A., Polyak, K., *et al.* (1998). Incidence and functional consequences of hMLH1 promoter hypermethylation in colorectal carcinoma. *Proceedings of the National Academy of Sciences USA*, **95**, 6870–6875.

Hermsen, M., Postma, C., Baak, J., *et al.* (2002). Colorectal adenoma to carcinoma progression follows multiple pathways of chromosomal instability. *Gastroenterology*, **123**, 1109–1019.

Hitchins, M. P., Lin, V. A., Buckle, A., *et al.* (2007). Epigenetic inactivation of a cluster of genes flanking *MLH1* in microsatellite-unstable colorectal cancer. *Cancer Research*, **67**, 9107–9116.

Hoque, M. O., Kim, M. S., Ostrow, K. L., *et al.* (2008). Genome-wide promoter analysis uncovers portions of the cancer methylome. *Cancer Research*, **68**, 2661–2670.

Humphries, A. and Wright, N. A. (2008). Colonic crypt organization and tumorigenesis. *Nature Reviews in Cancer*, **8**, 415–424.

Iacopetta, B., Grieu, F., Li, W., *et al.* (2006). APC gene methylation is inversely correlated with features of the CpG island methylator phenotype in colorectal cancer. *International Journal of Cancer*, **119**, 2272–2278.

Iacopetta, B., Kawakami, K., and Watanabe, T. (2008). Predicting clinical outcome of 5-fluorouracil-based chemotherapy for colon cancer patients: is the CpG island methylator phenotype the 5-fluorouracil responsive subgroup? *International Journal of Clinical Oncology*, **13**, 498–503.

Ibrahim, A. E., Arends, M. J., Silva, A. L., *et al.* (2010). Sequential DNA methylation changes are associated with DNMT3B overexpression in colorectal neoplastic progression. *Gut*, **60**, 499–508.

Imperiale, T. F., Ransohoff, D. F., Itzkowitz, S. H., *et al.* (2004). Fecal DNA versus fecal occult blood for colorectal-cancer screening in an average-risk population. *New England Journal of Medicine*, **351**, 2704–2714.

Irizarry, R. A., Ladd-Acosta, C., Wen, B., *et al.* (2009). The human colon cancer methylome shows similar hypo- and hypermethylation at conserved tissue-specific CpG island shores. *Nature Genetics*, **41**, 178–186.

Issa, J. P. (1999). Aging, DNA methylation and cancer. *Critical Reviews in Oncology and Hematology*, **32**, 31–43.

Issa, J. P. (2000). The epigenetics of colorectal cancer. *Annals of the New York Academy of Sciences*, **910**, 140–153; discussion 153–155.

Itzkowitz, S. H., Jandorf, L., Brand, R., *et al.* (2007). Improved fecal DNA test for colorectal cancer screening. *Clinical Gastroenterology and Hepatology*, **5**, 111–117.

Itzkowitz, S., Brand, R., Jandorf, L., *et al.* (2008). A simplified, noninvasive stool DNA test for colorectal cancer detection. *American Journal of Gastroenterology*, **103**, 2862–2870.

Karapetis, C. S., Khambata-Ford, S., Jonker, D. J., *et al.* (2008). *K-ras* mutations and benefit from cetuximab in advanced colorectal cancer. *New England Journal of Medicine*, **359**, 1757–1765.

Kawakami, K., Matsunoki, A., Kaneko, M., *et al.* (2011). Long interspersed nuclear element-1 hypomethylation is a potential biomarker for the prediction of response to oral fluoropyrimidines in microsatellite stable and CpG island methylator phenotype-negative colorectal cancer. *Cancer Science*, **102**, 166–174.

Kim, M. S., Louwagie, J., Carvalho, B., *et al.* (2009). Promoter DNA methylation of oncostatin m receptor-beta as a novel diagnostic and therapeutic marker in colon cancer. *PLoS One*, **4**, e6555.

Lee, B. B., Lee, E. J., Jung, E. H., *et al.* (2009). Aberrant methylation of *APC, MGMT, RASSF2A*, and *Wif-1* genes in plasma as a biomarker for early detection of colorectal cancer. *Clinical Cancer Research*, **15**, 6185–6191.

Lee, S., Cho, N. Y., Yoo, E. J., *et al.* (2008). CpG island methylator phenotype in colorectal cancers: comparison of the new and classic CpG island methylator phenotype marker panels. *Archives of Pathology and Laboratory Medicine*, **132**, 1657–1665.

Lee, T. I., Jenner, R. G., Boyer, L. A., *et al.* (2006). Control of developmental regulators by Polycomb in human embryonic stem cells. *Cell*, **125**, 301–313.

Leung, W. K., To, K. F., Man, E. P., *et al.* (2004). Detection of epigenetic changes in fecal DNA as a molecular screening test for colorectal cancer: a feasibility study. *Clinical Chemistry*, **50**, 2179–2182.

Leung, W. K., To, K. F., Man, E. P., *et al.* (2005). Quantitative detection of promoter hypermethylation in multiple genes in the serum of patients with colorectal cancer. *American Journal of Gastroenterology*, **100**, 2274–2279.

Levin, B., Lieberman, D. A., McFarland, B., *et al.* (2008). *Screening and Surveillance for the Early Detection of Colorectal Cancer and Adenomatous Polyps, 2008*: A joint guideline from the American Cancer Society, the US Multi-Society Task Force on Colorectal Cancer, and the American College of Radiology. *CA Cancer Journal Clinic*, **58**, 130–160.

Li, M., Chen, W. D., Papadopoulos, N., *et al.* (2009). Sensitive digital quantification of DNA methylation in clinical samples. *Nature Biotechnology*, **27**, 858–863.

Licchesi, J. D., Van Neste, L., Tiwari, V. K., *et al.* (2010). Transcriptional regulation of Wnt inhibitory factor-1 by Miz-1/c-Myc. *Oncogene*, **29**, 5923–5934.

Lofton-Day, C., Model, F., Devos, T., *et al.* (2008). DNA methylation biomarkers for blood-based colorectal cancer screening. *Clinical Chemistry*, **54**, 414–423.

Lujambio, A., Calin, G. A., Villanueva, A., *et al.* (2008). A microRNA DNA methylation signature for human cancer metastasis. *Proceedings of the National Academy of Sciences USA*, **105**, 13 556–13 561.

Maeda, K., Kawakami, K., Ishida, Y., *et al.* (2003). Hypermethylation of the *CDKN2A* gene in colorectal cancer is associated with shorter survival. *Oncology Reports*, **10**, 935–938.

Malkhosyan, S., Rampino, N., Yamamoto, H., *et al.* (1996). Frameshift mutator mutations. *Nature*, **382**, 499–500.

Marshall, K. W., Mohr, S., Khettabi, F. E., *et al.* (2010). A blood-based biomarker panel for stratifying current risk for colorectal cancer. *International Journal of Cancer*, **126**, 1177–1186.

McCabe, M. T., Lee, E. K., and Vertino, P. M. (2009). A multifactorial signature of DNA sequence and polycomb binding predicts aberrant CpG island methylation. *Cancer Research*, **69**, 282–291.

McNeil, C. (2008). *K-Ras* mutations are changing practice in advanced colorectal cancer. *Journal of the National Cancer Institute*, **100**, 1667–1669.

Melotte, V., Lentjes, M. H., van den Bosch, S. M., *et al.* (2009). N-Myc downstream-regulated gene 4 (*NDRG4*): a candidate tumor suppressor gene and potential biomarker for colorectal cancer. *Journal of the National Cancer Institute*, **101**, 916–927.

Muller, H. M., Oberwalder, M., Fiegl, H., *et al.* (2004). Methylation changes in faecal DNA: a marker for colorectal cancer screening? *Lancet*, **363**, 1283–1285.

Nakamura, Y., Nishisho, I., Kinzler, K. W., *et al.* (1991). Mutations of the adenomatous polyposis coli gene in familial polyposis coli patients and sporadic colorectal tumors. *Princess Takamatsu Symposium*, **22**, 285–292.

Nosho, K., Irahara, N., Shima, K., *et al.* (2008). Comprehensive biostatistical analysis of CpG island methylator phenotype in colorectal cancer using a large population-based sample. *PLoS One*, **3**, e3698.

Ogino, S., Kawasaki, T., Kirkner, G. J., *et al.* (2006). CpG island methylator phenotype-low (CIMP-low) in colorectal cancer: possible associations with male sex and *KRAS* mutations. *Journal of Molecular Diagnostics*, **8**, 582–588.

Ogino, S., Kawasaki, T., Kirkner, G. J., *et al.* (2007a). Evaluation of markers for CpG island methylator phenotype (CIMP) in colorectal cancer by a large population-based sample. *Journal of Molecular Diagnostics*, **9**, 305–314.

Ogino, S., Kawasaki, T., Kirkner, G. J., *et al.* (2007b). Molecular correlates with MGMT promoter methylation and silencing support CpG island methylator phenotype-low (CIMP-low) in colorectal cancer. *Gut*, **56**, 1564–1571.

Ogino, S., Meyerhardt, J. A., Kawasaki, T., *et al.* (2007c). CpG island methylation, response to combination chemotherapy, and patient survival in advanced microsatellite stable colorectal carcinoma. *Virchows Archives*, **450**, 529–537.

Ohm, J. E., McGarvey, K. M., Yu, X., *et al.* (2007). A stem cell-like chromatin pattern may predispose tumor suppressor genes to DNA hypermethylation and heritable silencing. *Nature Genetics*, **39**, 237–242.

Park, J. H., Kim, N. S., Park, J. Y., *et al.* (2010). MGMT-535G>T polymorphism is associated with prognosis for patients with metastatic colorectal cancer treated with oxaliplatin-based chemotherapy. *Journal of Cancer Research and Clinical Oncology*, **136**, 1135–1142.

Parkin, D. M., Bray, F., Ferlay, J., *et al.* (2005). Global cancer statistics, 2002. *CA Cancer Journal for Clinicians*, **55**, 74–108.

Parsons, R., Myeroff, L. L., Liu, B., *et al.* (1995). Microsatellite instability and mutations of the transforming growth factor beta type II receptor gene in colorectal cancer. *Cancer Research*, **55**, 5548–5550.

Petko, Z., Ghiassi, M., Shuber, A., *et al.* (2005). Aberrantly methylated CDKN2A, MGMT, and MLH1 in colon polyps and in fecal DNA from patients with colorectal polyps. *Clinical Cancer Research*, **11**, 1203–1209.

Popat, S., Hubner, R., and Houlston, R. S. (2005). Systematic review of microsatellite instability and colorectal cancer prognosis. *Journal of Clinical Oncology*, **23**, 609–618.

Rajagopalan, H., Bardelli, A., Lengauer, C., *et al.* (2002). Tumorigenesis: RAF/RAS oncogenes and mismatch-repair status. *Nature*, **418**, 934.

Rampino, N., Yamamoto, H., Ionov, Y., *et al.* (1997). Somatic frameshift mutations in the *BAX* gene in colon cancers of the microsatellite mutator phenotype. *Science*, **275**, 967–969.

Rodriguez, J., Munoz, M., Vives, L., *et al.* (2008). Bivalent domains enforce transcriptional memory of DNA methylated genes in cancer cells. *Proceedings of the National Academy of Sciences USA*, **105**, 19 809–19 814.

Samowitz, W. S., Albertsen, H., Herrick, J., *et al.* (2005). Evaluation of a large, population-based sample supports a CpG island methylator phenotype in colon cancer. *Gastroenterology*, **129**, 837–845.

Samowitz, W. S., Slattery, M. L., Sweeney, C., *et al.* (2007). *APC* mutations and other genetic and epigenetic changes in colon cancer. *Molecular Cancer Reserch*, **5**, 165–170.

Schlesinger, Y., Straussman, R., Keshet, I., *et al.* (2007). Polycomb-mediated methylation on Lys27 of histone H3 pre-marks genes for de novo methylation in cancer. *Nature Genetics*, **39**, 232–236.

Schuebel, K. E., Chen, W., Cope, L., *et al.* (2007). Comparing the DNA hypermethylome with gene mutations in human colorectal cancer. *PLoS Genetics*, **3**, 1709–1723.

Shen, L., Catalano, P. J., Benson, A. B., III *et al.* (2007). Association between DNA methylation and shortened survival in patients with advanced colorectal cancer treated with 5-fluorouracil based chemotherapy. *Clinical Cancer Research*, **13**, 6093–6098.

Shima, K., Nosho, K., Baba, Y., *et al.* (2011). Prognostic significance of CDKN2A (p16) promoter methylation and loss of expression in 902 colorectal cancers: cohort study and literature review. *International Journal of Cancer*, **128**, 1080–1094.

Souza, R. F., Appel, R., Yin, J., *et al.* (1996). Microsatellite instability in the insulin-like growth factor II receptor gene in gastrointestinal tumours. *Nature Genetics*, **14**, 255–257.

Squazzo, S. L., O'Geen, H., Komashko, V. M., *et al.* (2006). Suz12 binds to silenced regions of the genome in a cell-type-specific manner. *Genome Research*, **16**, 890–900.

Straussman, R., Nejman, D., Roberts, D., *et al.* (2009). Developmental programming of CpG island methylation profiles in the human genome. *Nature Structural and Molecular Biology*, **16**, 564–571.

Tanaka, N., Huttenhower, C., Nosho, K., *et al.* (2010). Novel application of structural equation modeling to correlation structure analysis of CpG island methylation in colorectal cancer. *American Journal of Pathology*, **177**, 2731–2740.

Tanzer, M., Balluff, B., Distler, J., *et al.* (2010). Performance of epigenetic markers SEPT9 and ALX4 in plasma for detection of colorectal precancerous lesions. *PLoS One*, **5**, e9061.

Tiwari, V. K., McGarvey, K. M., Licchesi, J. D., *et al.* (2008). PcG proteins, DNA methylation, and gene repression by chromatin looping. *PLoS Biology*, **6**, 2911–2927.

Toyota, M., Ahuja, N., Ohe-Toyota, M., *et al.* (1999). CpG island methylator phenotype in colorectal cancer. *Proceedings of the National Academy of Sciences USA*, **96**, 8681–8686.

Toyota, M., Suzuki, H., Sasaki, Y., *et al.* (2008). Epigenetic silencing of microRNA-34b/c and B-cell translocation gene 4 is associated with CpG island methylation in colorectal cancer. *Cancer Research*, **68**, 4123–4132.

van Engeland, M., Derks, S., Smits, K. M., *et al.* (2011). Colorectal cancer epigenetics: complex simplicity. *Journal of Clinical Oncology*, **29**, 1382–1391.

van Rijnsoever, M., Grieu, F., Elsaleh, H., *et al.* (2002). Characterization of colorectal cancers showing hypermethylation at multiple CpG islands. *Gut*, **51**, 797–802.

Vire, E., Brenner, C., Deplus, R., *et al.* (2006). The Polycomb group protein EZH2 directly controls DNA methylation. *Nature*, **439**, 871–874.

Warusavitarne, J. and Schnitzler, M. (2007). The role of chemotherapy in microsatellite unstable (MSI-H) colorectal cancer. *International Journal of Colorectal Diseases*, **22**, 739–748.

Weber, M., Davies, J.J., Wittig, D., *et al.* (2005). Chromosome-wide and promoter-specific analyses identify sites of differential DNA methylation in normal and transformed human cells. *Nature Genetics*, **37**, 853–862.

Weisenberger, D.J., Siegmund, K.D., Campan, M., *et al.* (2006). CpG island methylator phenotype underlies sporadic microsatellite instability and is tightly associated with *BRAF* mutation in colorectal cancer. *Nature Genetics*, **38**, 787–793.

Weller, M., Stupp, R., Reifenberger, G., *et al.* (2010). MGMT promoter methylation in malignant gliomas: ready for personalized medicine? *Nature Reviews in Neurology*, **6**, 39–51.

Wong, R. and Cunningham, D. (2008). Using predictive biomarkers to select patients with advanced colorectal cancer for treatment with epidermal growth factor receptor antibodies. *Journal of Clinical Oncology*, **26**, 5668–5670.

Yamaguchi, S., Asao, T., Nakamura, J., *et al.* (2003). High frequency of DAP-kinase gene promoter methylation in colorectal cancer specimens and its identification in serum. *Cancer Letters*, **194**, 99–105.

Yamakuchi, M., Ferlito, M., and Lowenstein, C.J. (2008). miR-34a repression of SIRT1 regulates apoptosis. *Proceedings of the National Academy of Sciences USA*, **105**, 13 421–13 426.

Yoder, J.A., Soman, N.S., Verdine, G.L., *et al.* (1997). DNA (cytosine-5)-methyltransferases in mouse cells and tissues: studies with a mechanism-based probe. *Journal of Molecular Biology*, **270**, 385–395.

Zhang, W., Bauer, M., Croner, R.S., *et al.* (2007). DNA stool test for colorectal cancer: hypermethylation of the secreted frizzled-related protein-1 gene. *Diseases of Colon and Rectum*, **50**, 1618–1626; discussion 1626–1627.

Zinn, R.L., Pruitt, K., Eguchi, S., *et al.* (2007). hTERT is expressed in cancer cell lines despite promoter DNA methylation by preservation of unmethylated DNA and active chromatin around the transcription start site. *Cancer Research*, **67**, 194–201.

23 The epigenetic profile of bladder cancer

Ewa Dudziec and James W. F. Catto*

23.1 Introduction

Urothelial carcinoma of the bladder (UCC) is the fourth commonest male malignancy in the United States with 70 530 new cases and 14 680 deaths in 2010 (Jemal *et al.*, 2010). Due to its natural history, UCC is the third most prevalent cancer and one of the most expensive to treat (Avritscher *et al.*, 2006). Estimates suggest that the disease costs around $3.7 billion per annum in the USA. The majority of this cost is spent on diagnosis and surveillance (Sangar *et al.*, 2005). The etiology of the disease is acquired carcinogen exposure with little or no familial component. Known urothelial carcinogens are found in cigarette smoke (e.g., 2-naphthylamine) and in occupational exposure to aromatic amines (e.g., benzedine). Whilst many high-risk carcinogens have been identified, and their use replaced or restricted, around 20% of contemporary UCC still arise from occupational exposure (Reulen *et al.*, 2008). This suggests ongoing occupational carcinogen exposure and these may be unknown carcinogens or the uncontrolled use of known carcinogens.

Clinicopathological observations show that bladder cancer arises by at least two different molecular pathways, with divergent clinical behaviors. The disease is best separated into low- and high-grade tumors to reflect these pathways. Low-grade tumors are commonest (70%) and recur frequently within the bladder. They rarely progress to become invasive tumors and are safely managed by endoscopic resection and intravesical chemotherapy. In contrast, high-grade UCC have a poor prognosis and can present with or after the onset of muscle invasion. Non-invasive high-grade tumors frequently become invasive (50%) and after the development of muscle invasion the overall survival at five years is less than 50% (Knowles, 2006; Catto *et al.*, 2007).

---

* Author to whom correspondence should be addressed.

Epigenomics: From Chromatin Biology to Therapeutics, ed. K. Appasani. Published by Cambridge University Press. © Cambridge University Press 2012.

Low- and high-grade UCC are distinct at the genetic and epigenetic level (Dyrskjot *et al.*, 2004; Dhawan *et al.*, 2006; Goebell and Knowles, 2010) (Figure 23.1). Genetically low-grade tumours are characterized by FGFR3 mutation and loss of regions from chromosome 9 (Van Oers *et al.*, 2009). Infrequently they develop HRAS mutations, in a mutually exclusive manner to FGFR3 mutation (Jebar *et al.*, 2005). High-grade tumours are characterized by multiple genetic events including chromosomal instability and loss of p53/Rb pathway. These alterations occur early in the disease pathway and can be detected in carcinoma in situ, the non-invasive precursor of invasive tumours (Spruck *et al.*, 1994).

Epigenetic gene regulation occurs at three levels including biochemical modification of DNA, histone modifications, and non-coding RNA (ncRNA) expression. These are interrelated and can act synchronously to regulate gene expression. Numerous epigenetic alterations have been observed in UCC, and these represent potential biomarkers or therapeutic targets due to their reversible nature. In general, low-grade tumours are characterized by infrequent aberrant

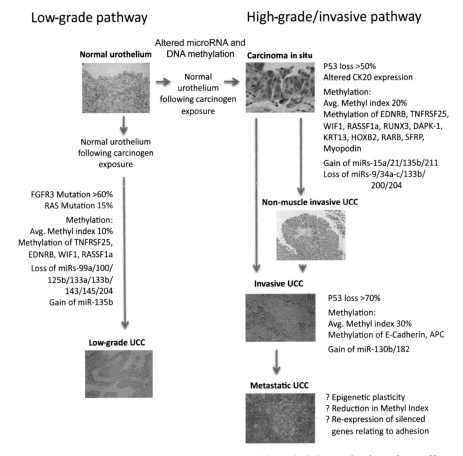

Figure 23.1 High- and low-grade bladder cancers arise through distinct molecular pathways. Here we have adapted a genetic map to show epigenetic changes in the disease. (Adapted from Kim *et al.*, 2005; Christoph *et al.*, 2006; Catto *et al.*, 2009; Goebell and Knowles, 2010; Marsit *et al.*, 2010; Dudziec *et al.*, 2011.)

hypermethylation, generalized DNA hypomethylation, and the downregulation of many microRNAs. High-grade tumours have extensive promoter hypermethylation and upregulation of many microRNAs (Catto *et al.*, 2009; Dyrskjot *et al.*, 2009; Wolff *et al.*, 2010).

23.2   DNA methylation

DNA methylation is the best-studied epigenetic modification in cancer. It consists of the covalent addition of methyl groups to cytosine residues within CpG dinucleotides. In malignant cells the epigenetic landscape changes to produce hypermethylation of previously unmethylated CpG islands and an overall decrease in cellular methyl content as many solitary CpG dinucleotides become demethylated. Hypermethylation of CpG islands represses gene transcription directly by preventing binding of transcription factors (Watt and Molloy, 1988) or indirectly by recruiting methyl-CpG binding proteins (Nan *et al.*, 1997). Hypomethylation of CpG-poor regions may activate oncogenes or repetitive elements in the genome and lead to chromosome instability.

Around 40% of genes (both protein coding and non-coding) are located within 10 kb of a CpG island. This makes them potentially susceptible to epigenetic regulation. Numerous reports have examined the role of DNA methylation in urothelial carcinogenesis. Both candidate gene and global profiling experiments have been performed and the list of aberrantly methylated genes is extensive (examples in Table 23.1). Susceptible genes involve most cellular pathways but also have shared ontologies. To better understand the global effect of candidate gene silencing we interrogated this list using an integrative gene ontology database (Huang da *et al.*, 2009) (Table 23.2). Of note, these candidate genes are biased towards cancer-related pathways as they were originally identified in malignant cells or from familial cancer syndromes. We found significant gene enrichment for many common carcinogenic pathways (such as apoptosis avoidance, signal transduction, proliferation, and cell cycle regulation), suggesting a cooperative effect from DNA hypermethylation. Evidence to implicate the epigenetic silencing of these genes in UCC is mainly derived from the close association between gene hypermethylation and bladder cancer phenotype, and functional in vitro assays. The strong association between CpG hypermethylation and UCC phenotype enables this epigenetic trait to function as a predictive biomarker (Yates *et al.*, 2007).

Many epigenetic alterations occur early in the disease's biology. To examine the timing of alterations in UCC, Dhawan *et al.* (2006) measured aberrant DNA methylation in patients with urothelial carcinoma in situ and matching subsequent invasive cancers. When compared with the normal urothelium from control patients, it was seen that many alterations of DNA methylation occurred in histologically normal urothelium (in bladders with cancers). The extent and frequency of methylation increased proportionally with tumour stage. Wolff *et al.* (2010) have recently investigated the methylation profiles of 1370 CpG loci in UCC. This represents one of the first published epigenetic profiling studies

Table 23.1 Examples of genes susceptible to hypermethylation in bladder cancer and their ontological pathways

Function	Symbol	Gene	Percent methylation	References
Cell cycle control	CDKN2A	Cyclin-dependent kinase inhibitor 2A, isoform 4	0–56	Dominguez *et al.*, 2002; Serizawa *et al.*, 2011
	CDKN2B	Cyclin-dependent kinase inhibitor 2B	13	Chan *et al.*, 2002
	CDKN2A	Cyclin-dependent kinase inhibitor 2A	1–73	Hoque *et al.*, 2006; Serizawa *et al.*, 2011
	CCND2	Cyclin D2	22	Yates *et al.*, 2007
	SOCS1	Suppressor of cytokine signaling 1	33	Friedrich *et al.*, 2005
	STAT1	Signal transducer and activator of transcription 1	11	Friedrich *et al.*, 2005
DNA repair	COX2	Cyclooxygenase 2	24–75	Friedrich *et al.*, 2005; Ellinger *et al.*, 2008
	GSTP1	Glutathione S-transferase pi1	11–59	Maruyama *et al.*, 2001; Ellinger *et al.*, 2008
	MGMT	O-6-methylguanine-DNA methyltransferase	2–69	Maruyama *et al.*, 2001; Abbosh *et al.*, 2008
	XPC	Xeroderma pigmentosum	32	Yang *et al.*, 2010
Differentiation	RARB	Retinoic acid receptor beta	2–88	Chan *et al.*, 2002; Yates *et al.*, 2007
	TAL1	T-cell acute lymphocytic leukemia 1	N/A	Wolff *et al.*, 2010
	PAX6	Paired box 6	23	Hellwinkel *et al.*, 2008
	RARRES1	Retinoic acid receptor responder 1	32	Ellinger *et al.*, 2008
Apoptosis	DAPK1	Death-associated protein kinase	2–74	Christoph *et al.*, 2006; Ellinger *et al.*, 2008
	BCL2	B-cell CLL/lymphoma 2	45–65	Friedrich *et al.*, 2004; Yates *et al.*, 2007
	EDNRB	Endothelin receptor type B	57–69	Friedrich *et al.*, 2004; Friedrich *et al.*, 2005
	TNFRSF21	Tumor necrosis factor receptor superfamily, member 21	6–7	Friedrich *et al.*, 2004, 2005
	TNFRSF25	Tumor necrosis factor receptor superfamily, member 25	75	Yates *et al.*, 2007
	PYCARD	Apoptosis-associated speck-like protein (ASC)	0–4	Friedrich *et al.*, 2005; Marsit *et al.*, 2007
	CASP8	Caspase 8	4	Christoph *et al.*, 2006
	APAF1	Apoptotic protein-activating factor	100	Christoph *et al.*, 2006
	TERT	Human telomerase reverse transcriptase	21	Yates *et al.*, 2007
Cell invasion	CDH1	E-cadherin	4–87	Hoque *et al.*, 2006; Marsit *et al.*, 2007
	CDH4	Cadherin 4	15	Yates *et al.*, 2007
	CDH13	Cadherin 13	18–29	Maruyama *et al.*, 2001; Marsit *et al.*, 2007
	LAMA3	Laminin alpha 3	45	Sathyanarayana *et al.*, 2004
	LAMB3	Laminin belta 3	21	Sathyanarayana *et al.*, 2004
	LAMC2	Laminin gamma 2	23	Sathyanarayana *et al.*, 2004

Table 23.1 (cont.)

Function	Symbol	Gene	Percent methylation	References
Tumour suppressors	RASS1F1A	Ras association domain family member 1	33–67	Hoque et al., 2006; Marsit et al., 2007
	RUNX3	Runt-related transcription factor 3	73	Kim et al., 2005
	SOX9	Sex-determining region Y box 9	56	Aleman et al., 2008
	MLH1	MutL homolog 1	3–25	Friedrich et al., 2005; Marsit et al., 2007
	VHL	Hippel–Lindau tumor suppressor	4	Tada et al., 2002
	FHIT	Fragile histidine triad gene	16	Maruyama et al., 2001
	HIC1	Hypermethylated in cancer 1	22	Yates et al., 2007
	DBC1	Deleted in bladder cancer 1	52	Habuchi et al., 2001
	PRSS3	Protease, serine, 3	34	Marsit et al., 2007
	GALR1	Galanin receptor 1	N/A	Wolff et al., 2010
	TPEF	Transmembrane protein EF	33	Hellwinkel et al., 2008
	IGFBP3	Insulin-like growth factor binding protein 3	43–66	Christoph et al., 2006; Yates et al., 2007
Wnt signaling	APC	Adenomatous polyposis coli	31–73	Hoque et al., 2006; Yates et al., 2007
	WIF1	WNT inhibitory factor 1	1–57	Yates et al., 2007; Serizawa et al., 2011
	SFRP1	Secreted frizzled-related protein 1	22	Khin et al., 2009
	FRZB (SFRP3)	Secreted frizzled-related protein 3	N/A	Marsit et al., 2010
	SFRP2	Secreted frizzled-related protein 2	32	Khin et al., 2009
	SFRP4	Secreted frizzled-related protein 4	12	Khin et al., 2009
	SFRP5	Secreted frizzled-related protein 5	50	Khin et al., 2009
Cell adhesion	ICAM1	Intercellular adhesion molecule 1	90	Friedrich et al., 2005
	TIMP3	TIMP metallopeptidase inhibitor 3	23–93	Hoque et al., 2006; Yates et al., 2007
Others	HOXB2	Homeobox 2	N/A	Marsit et al., 2010
	KRT13	Keratin 13	N/A	Marsit et al., 2010
	BAMBI	Activin membrane-bound inhibitor	31	Khin et al., 2009
	IPF1	Insulin promoter factor 1	N/A	Wolff et al., 2010
	ZO2 (TJP2)	Tight junction protein 2	N/A	Wolff et al., 2010
	PENK	Preproenkephalin A	N/A	Wolff et al., 2010

within the disease. The authors observed that around 12% of loci were hypermethylated in the normal urothelium from patients with cancer when compared to controls. Profiles of multiple cancers within the same patient that develop years after the primary lesion reveal epigenetic and genetic stability over time (Catto et al., 2006). Few authors have examined epigenetic events in metastatic UCC. In other cancers, epigenetic silencing changes with the onset of metastases (so-called epigenetic plasticity). For example, in breast cancer epigenetic silencing of *E-Cadherin* produces a loss of cellular adhesion and the ability to migrate in the primary focus. This enables a tumour to become invasive and metastatic. Once

Table 23.2 Ontological analysis using genes susceptible to hypermethylation in bladder cancer (n = 56 from Table 23.1) reveals enrichment for various malignant pathways (Huang da *et al.*, 2009)

Cluster	Process / pathway	Count	*p* value
Cluster 1	Regulation of apoptosis	23	4.00E-14
Cluster 2	Wnt signaling pathway	8	3.10E-09
Cluster 3	Regulation of epithelial cell proliferation	6	7.40E-06
Cluster 4	Regulation of cell adhesion	7	1.30E-05
Cluster 5	Epidermis development	10	2.90E-08
Cluster 6	Small-cell lung cancer	8	1.80E-06
Cluster 7	Regulation of cell cycle	8	8.20E-08
Cluster 8	Anterior/posterior pattern formation	5	2.00E-03
Cluster 9	Response to organic cyclic substance	5	1.20E-03
Cluster 10	Mesenchymal cell differentiation	3	1.60E-02
Cluster 11	Urogenital system development	4	8.70E-03
Cluster 12	Reproductive structure development	5	1.40E-03
Cluster 13	Cytokine-mediated signaling pathway	4	2.40E-03
Cluster 14	Cadherin prodomain like	3	2.00E-04
Cluster 15	Epithelium development	5	1.10E-02
Cluster 16	Adherens junction	5	1.80E-03
Cluster 17	Regulation of binding	4	2.10E-02
Cluster 18	DNA binding	13	3.50E-03
Cluster 19	DNA metabolic process	7	1.30E-02
Cluster 20	Ubl conjugation	6	2.10E-02
Cluster 21	Regulation of protein stability	3	1.10E-02
Cluster 22	ANK repeat	3	6.90E-02
Cluster 23	Protein homodimerization activity	4	1.00E-01
Cluster 24	Positive regulation of macromolecule metabolic process	10	4.90E-03
Cluster 25	Protein kinase cascade	4	1.70E-01
Cluster 26	Endometrial cancer	3	5.40E-02
Cluster 27	Visual perception	3	2.00E-01
Cluster 28	Cell fraction	5	5.00E-01
Cluster 29	Calcium ion binding	6	2.00E-01
Cluster 30	Organelle membrane	3	6.20E-01
Cluster 31	Topological domain: Extracellular	9	4.60E-01
Cluster 32	ATP binding	5	7.50E-01
Cluster 33	Metal binding	6	9.20E-01

cancerous cells enter the circulation they re-express *E-Cadherin* to promote fixation to the target tissue (Graff *et al.*, 2000).

When low- and high-grade UCC are compared they appear distinct at the global epigenetic level. This difference complements a genetic model for urothelial carcinogenesis (Figure 23.1), although changes in DNA methylation are not as mutually exclusive. It appears that most cancer-specific CpG regions are susceptible to differential methylation in all UCC, regardless of tumour phenotype. For example, nearly 89% of CpG loci examined by Wolff *et al.* (2010) shared methyl patterns between superficial and invasive UCC. The remaining 11% appeared specific to the invasive UCC pathway. Thus, low- and high-grade UCC mostly differ in the extent and quantity of aberrant hypermethylation rather than in specific targets. When compared epigenetic silencing typically affects 10% of loci in low-grade UCC and 20–30% for high-grade non-invasive and invasive UCC (Catto *et al.*, 2005; Dhawan *et al.*, 2006; Wolff *et al.*, 2010). There do appear to be some genes with a relative phenotype-specific pattern of hypermethylation (including *FRZB, KRT13, HOXB2,* and *Myopodin*) (Cebrian *et al.*, 2008; Marsit *et al.*, 2010). Interestingly, low-grade tumours are characterized by distinct patterns of hypomethylation when compared to high-grade tumours and control samples which may explain inability of these tumours to progress further to muscles (Wolff *et al.*, 2010).

Global epigenetic profiling may be accomplished by various methods. One method is to use CpG island microarrays to interrogate immunoprecipitated DNA. Depending upon the antibody for immunoprecipitation this may reveal methylated patterns of histone alteration or DNA methylation. We have immunoprecipitated sheared DNA with an antibody to 5-methylcytosine (methylated DNA immunoprecipitation [MeDIP] technique) (Weber *et al.*, 2005) and hybridized this to a 28 000 CpG island microarray. In Figure 23.2 we compared a normal non-transformed urothelial cell line (NHU) with a bladder cancer cell line (EJ). As can be seen the cancer cell line has higher levels of hypermethylation throughout the genome and at specific regions.

DNA hypomethylation has also been detected in UCC. As discussed, Wolff *et al.* (2010) observed hypomethylation outside of CpG islands in low-grade UCC, and

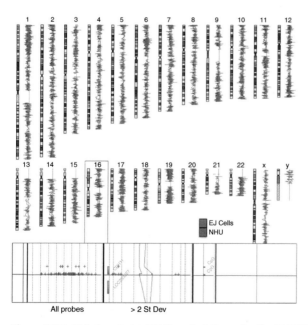

Figure 23.2 Global analysis of DNA methylation using the MeDIP technique reveals differences between non-transformed normal (NHU) and malignant urothelial cells (EJ). The highlighted region reveals probe enrichment for DNA hypermethylation in the cancer cell line at the CpG island adjacent to CDH11, when compared to NHU. This gene is known to be epigenetically silenced in other cancers (Sandgren *et al.*, 2010). See plate section for color version.

previously Byun *et al.* (2007) detected biallelic expression of imprinted genes *IGF2* and *H19* in around one-fifth of cancers. Analysis of repetitive DNA elements revealed that hypomethylation of LINE-1, Alu Yb8, Sat-alpha, and NBL-2 is common in bladder cancer (Choi *et al.*, 2009). Hypomethylation of LINE-1 leads to the expression of alternate transcript of the *MET* oncogene in bladder tumors and was found across the entire urothelium from patients with cancer (Wolff *et al.*, 2010). Wilhelm *et al.* (2010) suggested this epigenetic event could be a potential biomarker for diagnosis and treatment. Of interest, Moore *et al.* (2008) reported global DNA hypomethylation in leukocytes which was associated with the risk of developing bladder cancer.

23.3 Modifications of histone proteins

The fundamental unit of the chromosome is the nucleosome, consisting of DNA wrapped around an octamer of histone proteins. Histone protein tails can be covalently modified (e.g., with methylation, acetylation, phosphorylation) to facilitate or inhibit gene expression. Alterations in the enzymes responsible for establishing and maintaining these histone modifications are common in cancer and lead to changes in chromatin structure. The most investigated repressive histone marks are histone 3 lysine 9 (H3K9) and histone 3 lysine 27 (H3K27) trimethylation. These have been analyzed in bladder cancer, with respect to candidate tumor-suppressor genes. For example, the histones adjacent to aberrantly silenced p14ARF/p16INK4a exhibit repressive marks such as H3K9 hypermethylation and histone 3 lysine 4 (H3K4) demethylation (Nguyen *et al.*, 2002). These epigenetic events could be reversed when treated with a DNA methyltransferase inhibitor (5-aza-2-deoxycytidine). H3K27 trimethylation is deposited by EZH2, a component of Polycomb repressive complex 2 (PCR2). EZH2 appears to have oncogenic properties and is frequently overexpressed in many cancers (Varambally *et al.*, 2002; Weikert *et al.*, 2005). In bladder cancer, EZH2 can induce transcriptional silencing of E-Cadherin through H3K27 hypermethylation, independently of DNA methylation (Cao *et al.*, 2008).

The reversible nature of histone modifications makes them appealing therapeutic targets. For instance, lysine-specific demethylase 1 (LSD1) catalyzes demethylation of mono- and dimethyl H3K4, a chromatin mark associated with transcriptional activation. In colon cancer, treatment with an LSD1 inhibitor increases H3K4 methylation and decreases repressive H3K9 dimethylation (Huang *et al.*, 2009). LSD1 inhibition in combination with DNA methyltransferase (DNMT) inhibitors can restore expression of SFRP2 and reduce the tumour growth. Promising preclinical studies suggest that treatment with an S-adenosyl homocysteine hydrolase inhibitor (3-deazaneplanocin A (DZNep)) depletes components of PCR2 complex (EZH2, EED, and SUZ12) and reduces H3K27 trimethylation (Tan *et al.*, 2007). This effect was more prominent in cancerous cells, when compared to normal cells. In bladder cancer, treatment with a histone deacetylase inhibitor (HDACi) (e.g., Trichostatin A) has been shown to increase H3K9 acetylation, reduce global DNA hypomethylation and restore the expression of genes associated with DNA hypermethylation (e.g., *E-Cadherin*) (Ou *et al.*, 2007). Sachs *et al.* (2004) found upregulation of Coxsackie and adenovirus receptor (CAR), potential gene delivery vectors, when treated with HDACi.

23.4 MicroRNAs expression

MicroRNAs (miRNAs) are short ncRNA molecules that posttranscriptionally modulate protein expression. Over one-third of human genes are conserved as their targets and consequently miRs represent one of the most abundant classes of regulatory genes in humans (Lewis *et al.*, 2005). Altered miR expression is implicated in most human cancers and may arise from either genetic or epigenetic events (Croce, 2009).

The first study to evaluate miRNA in bladder cancer was performed by Gottardo *et al.* (2007) and revealed upregulation of many microRNAs. To date, several large miR profiling experiments have been reported. Dyrskjot *et al.* (2009) analyzed 117 samples and detailed changes in the expression of many miRs (examples in Table 23.3). These including downregulation of miR-145 and upregulation of

Table 23.3 MicroRNAs with altered expression in bladder cancer and their defined mRNA targets

Upregulated microRNAs	Downregulated microRNAs	Identified target	Phenotype/cells	References
	miR-1/133a/218	LASP1	Not specified/ all grades	Chiyomaru *et al.*, 2011
	miR-19a	PTEN a	T24 cells	Cao *et al.*, 2010
	miRs-30a-3p/133a/ 199a	KRT7	Not specified/all grades	Ichimi *et al.*, 2009
	miR-34a	CDK6 a	Not specified	Lodygin *et al.*, 2008
	miR-99a/100	FGFR3	Low-grade	Catto *et al.*, 2009
	miR-101	EZH2	Not specified/all grades	Friedman *et al.*, 2009
	miR-125b	E2F3/Cyclin A2 pathway	All grades	Huang *et al.*, 2010
miR-129		SOX4, GALNT1	High-grade	Dyrskjot *et al.*, 2009
	miR-145/133a	FSCN1	All grades	Chiyomaru *et al.*, 2010
	miR-145	CBFB, PPP3CA, CLINT1	All grades	Ostenfeld *et al.*, 2010
	miR-200s/205	ZEB1 and ZEB2 a	High-grade	Wiklund *et al.*, 2011
	miR-200 family	ERRFI-1 / EMT process b	Invasive	Adam *et al.*, 2009
miR-221		TRAIL pathway	Invasive	Lu *et al.*, 2010

[a] Target supported by observational data or from other cancers.
[b] EMT, epithelial-to-mesenchymal transition.

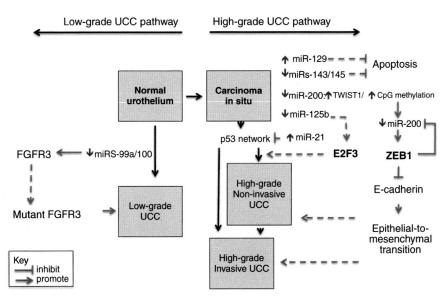

Figure 23.3 Phenotypic expression of miRNAs in UCC reveals molecular events within the disease. (Adapted from Catto *et al.*, 2009, 2011; Dyrskjot *et al.*, 2009.)

miR-21, but miR-129 appeared most interesting. The authors demonstrated miR-129 targeted the SOX4 transcription factor and GALNT1 to aid apoptosis avoidance. More recently Catto *et al.* (2009) analyzed 75 samples and detailed phenotype-specific alterations in miRNA expression that reflected genetic events within the disease. As with DNA methylation, the authors found almost half of the carcinogenic miRNA alterations were already present in the normal urothelium from bladder cancer patients, when compared with benign controls. Specifically, low-grade tumours were characterized by downregulation of many miRs, including miR-99a/100 which directly target the 3′ untranslated region (UTR) of FGFR3 to produce FGFR3 upregulation prior to acquisition of mutation. High-grade tumors were characterized by upregulation of many miRs, including miR-21 (known to suppress p53) and miR-135b. Interestingly, many of these alterations occurred in high-grade UCC before the onset of muscle invasion and were associated with a worse outcome (Catto *et al.*, 2009).

This phenotype-specific expression of miRNAs can illustrate the biology of UCC (Figure 23.3). Invasive tumours are characterized by epithelial-to-mesenchymal transition (EMT). To explore this Adam *et al.* (2009) evaluated the miR-200 family. The expression of MiR-200 is reduced in high-grade UCC, and the authors demonstrated that this targets ERRF-1, a regulator of EGFR-independent growth and mediator of EMT. Wiklund *et al.* (2011) found that miR-200 was associated with a pro-mesenchymal transcription factor, TWIST1, and silenced with CpG hypermethylation. Kenney *et al.* (2010) revealed ZEB1 expression in one in five of UCCs and that this was probably regulated by miR-200b/c expression. ZEB1 is a mediator of EMT through E-cadherin repression. This appears to be a feedback loop with ZEB1 also negatively regulating miR-141/200 (Burk *et al.*, 2008).

23.5 The interaction between miRNA expression and DNA methylation

As stated in our introduction, the three tiers of epigenetic regulation can interact to modulate gene expression. We have seen that DNA methylation and histone alterations often match each other. This is also true for microRNA expression and DNA methylation. Around one-fifth and one-third of mapped miRNA genes are located within 3 kb and 10 kb of a CpG island. Saito *et al.* (2006) investigated epigenetic regulation of microRNA genes in UCC using DNMT and HDAC inhibitors. The authors found that 5% of miRNAs were upregulated, including miR-127 that targets Bcl2. Consequently, miR-127 silencing is a mechanism for Bcl2 oncogene activation. In vitro transfection of miR-127 into deficient cells restored its expression and reduced Bcl2, suggesting a potential treatment strategy. A similar pattern of silencing was demonstrated for miR-126 and its oncogenic target, EGFL7 (Saito *et al.*, 2009). As discussed, Wiklund *et al.* (2011) have recently reported DNA hypermethylation and histone H3K27 trimethylation for members of the miR-200 family and miR-205 in muscle invasive bladder tumors.

MiRNAs can also regulate the DNA and histone methylation machinery. In lung cancer the miR-29 family directly target *de novo* DNA methyltransferases (DNMT3a and 3b) to increase promoter DNA hypermethylation and silence susceptible genes (Fabbri *et al.*, 2007). MiR-29 can also indirectly target the maintenance DNA methyltransferase DNMT1 via downregulation of *Sp1*, a known transactivating factor of the *DNMT1* gene (Garzon *et al.*, 2009). In UCC, downregulation of miR-101 leads to the overexpression of EZH2 and epigenetic silencing through trimethylation of H3K27–me3 (Friedman *et al.*, 2009). This enhances tumorogenesis.

23.6 Conclusions

Bladder cancer is a common disease characterized by two disease phenotypes. These have distinct molecular pathways that produce different clinical outcomes. Epigenetic and genetic events characterize these molecular pathways. Whilst genetic events are often discrete, many epigenetic changes are shared between tumor phenotypes. Epigenetic alterations occur in the normal urothelium from patients with bladder cancer and this supports the field effect suggested within the disease. The three tiers of epigenetic regulation interact in UCC and thus are amenable to therapeutic targeting or their use as translational biomarkers.

REFERENCES

Abbosh, P. H., Wang, M., Eble, J. N., *et al.* (2008). Hypermethylation of tumor-suppressor gene CpG islands in small-cell carcinoma of the urinary bladder. *Modern Pathology*, **21**, 355–362.

Adam, L., Zhong, M., Choi, W., *et al.* (2009). miR-200 expression regulates epithelial-to-mesenchymal transition in bladder cancer cells and reverses resistance to epidermal growth factor receptor therapy. *Clinical Cancer Research*, **15**, 5060–5072.

Aleman, A., Adrien, L., Lopez-Serra, L., *et al.* (2008). Identification of DNA hypermethylation of *SOX9* in association with bladder cancer progression using CpG microarrays. *British Journal of Cancer*, **98**, 466–473.

Avritscher, E. B., Cooksley, C. D., Grossman, H. B., *et al.* (2006). Clinical model of lifetime cost of treating bladder cancer and associated complications. *Urology*, **68**, 549–553.

Burk, U., Schubert, J., Wellner, U., *et al.* (2008). A reciprocal repression between ZEB1 and members of the miR-200 family promotes EMT and invasion in cancer cells. *EMBO Reports*, **9**, 582–589.

Byun, H. M., Wong, H. L., Birnstein, E. A., *et al.* (2007). Examination of IGF2 and H19 loss of imprinting in bladder cancer. *Cancer Research*, **67**, 10 753–10 758.

Cao, Q., Yu, J., Dhanasekaran, S. M., *et al.* (2008). Repression of E-cadherin by the polycomb group protein EZH2 in cancer. *Oncogene*, **27**, 7274–7284.

Cao, Y., Yu, S. L., Wang, Y., *et al.* (2010). MicroRNA-dependent regulation of PTEN after arsenic trioxide treatment in bladder cancer cell line T24. *Tumour Biology*, **32**, 179–188.

Catto, J. W., Azzouzi, A. R., Rehman, I., *et al.* (2005). Promoter hypermethylation is associated with tumor location, stage, and subsequent progression in transitional cell carcinoma. *Journal of Clinical Oncology*, **23**, 2903–2910.

Catto, J. W., Hartmann, A., Stoehr, R., *et al.* (2006). Multifocal urothelial cancers with the mutator phenotype are of monoclonal origin and require panurothelial treatment for tumor clearance. *Journal of Urology*, **175**, 2323–2330.

Catto, J. W., Yates, D. R., Rehman, I., *et al.* (2007). Behavior of urothelial carcinoma with respect to anatomical location. *Journal of Urology*, **177**, 1715–1720.

Catto, J. W., Miah, S., Owen, H. C., *et al.* (2009). Distinct microRNA alterations characterize high- and low-grade bladder cancer. *Cancer Research*, **69**, 8472–8481.

Catto, J. W. F., Alcaraz, A., Bjartell, A., *et al.* (2011). MicroRNA in prostate, bladder and kidney cancer: a systematic review. *European Urology*, **59**, 671–681.

Cebrian, V., Alvarez, M., Aleman, A., *et al.* (2008). Discovery of myopodin methylation in bladder cancer. *Journal of Pathology*, **216**, 111–119.

Chan, M. W., Chan, L. W., Tang, N. L., *et al.* (2002). Hypermethylation of multiple genes in tumor tissues and voided urine in urinary bladder cancer patients. *Clinical Cancer Research*, **8**, 464–470.

Chiyomaru, T., Enokida, H., Tatarano, S., *et al.* (2010). miR-145 and miR-133a function as tumour suppressors and directly regulate FSCN1 expression in bladder cancer. *British Journal of Cancer*, **102**, 883–891.

Chiyomaru, T., Enokida, H., Kawakami, K., *et al.* (2011). Functional role of LASP1 in cell viability and its regulation by microRNAs in bladder cancer. *Urologic Oncology* (epub ahead of publication).

Choi, S. H., Worswick, S., Byun, H. M., *et al.* (2009). Changes in DNA methylation of tandem DNA repeats are different from interspersed repeats in cancer. *International Journal of Cancer*, **125**, 723–729.

Christoph, F., Kempkensteffen, C., Weikert, S., *et al.* (2006). Methylation of tumour suppressor genes *APAF-1* and *DAPK-1* and in vitro effects of demethylating agents in bladder and kidney cancer. *British Journal of Cancer*, **95**, 1701–1707.

Croce, C. M. (2009). Causes and consequences of microRNA dysregulation in cancer. *Nature Reviews Genetics*, **10**, 704–714.

Dhawan, D., Hamdy, F. C., Rehman, I., *et al.* (2006). Evidence for the early onset of aberrant promoter methylation in urothelial carcinoma. *Journal of Pathology*, **209**, 336–343.

Dominguez, G., Carballido, J., Silva, J., *et al.* (2002). p14ARF promoter hypermethylation in plasma DNA as an indicator of disease recurrence in bladder cancer patients. *Clinical Cancer Research*, **8**, 980–985.

Dudziec, E., Goepel, J. R., and Catto, J. W. F. (2011). Global epigenetic profiling in bladder cancer. *Epigenomics*, **3**, 35–45.

Dyrskjot, L., Kruhoffer, M., Thykjaer, T., *et al.* (2004). Gene expression in the urinary bladder: a common carcinoma in situ gene expression signature exists disregarding histopathological classification. *Cancer Research*, **64**, 4040–4048.

Dyrskjot, L., Ostenfeld, M. S., Bramsen, J. B., *et al.* (2009). Genomic profiling of microRNAs in bladder cancer: miR-129 is associated with poor outcome and promotes cell death in vitro. *Cancer Research*, **69**, 4851–4860.

Ellinger, J., El Kassem, N., Heukamp, L. C., *et al.* (2008). Hypermethylation of cell-free serum DNA indicates worse outcome in patients with bladder cancer. *Journal of Urology*, **179**, 346–352.

Fabbri, M., Garzon, R., Cimmino, A., *et al.* (2007). MicroRNA-29 family reverts aberrant methylation in lung cancer by targeting DNA methyltransferases 3A and 3B. *Proceedings of the National Academy of Sciences USA*, **104**, 15 805–15 810.

Friedman, J. M., Liang, G., Liu, C. C., *et al.* (2009). The putative tumor suppressor microRNA-101 modulates the cancer epigenome by repressing the polycomb group protein EZH2. *Cancer Research*, **69**, 2623–2629.

Friedrich, M. G., Weisenberger, D. J., Cheng, J. C., *et al.* (2004). Detection of methylated apoptosis-associated genes in urine sediments of bladder cancer patients. *Clinical Cancer Research*, **10**, 7457–7465.

Friedrich, M. G., Chandrasoma, S., Siegmund, K. D., *et al.* (2005). Prognostic relevance of methylation markers in patients with non-muscle invasive bladder carcinoma. *European Journal of Cancer*, **41**, 2769–2778.

Garzon, R., Liu, S., Fabbri, M., *et al.* (2009). MicroRNA-29b induces global DNA hypomethylation and tumor suppressor gene reexpression in acute myeloid leukemia by targeting directly DNMT3A and 3B and indirectly DNMT1. *Blood*, **113**, 6411–6418.

Goebell, P. J. and Knowles, M. A. (2010). Bladder cancer or bladder cancers? Genetically distinct malignant conditions of the urothelium. *Urologic Oncology*, **28**, 409–428.

Gottardo, F., Liu, C. G., Ferracin, M., *et al.* (2007). Micro-RNA profiling in kidney and bladder cancers. *Urologic Oncology*, **25**, 387–392.

Graff, J. R., Gabrielson, E., Fujii, H., *et al.* (2000). Methylation patterns of the E-cadherin 5′ CpG island are unstable and reflect the dynamic, heterogeneous loss of E-cadherin expression during metastatic progression. *Journal of Biological Chemistry*, **275**, 27272732.

Habuchi, T., Takahashi, T., Kakinuma, H., *et al.* (2001). Hypermethylation at 9q32–33 tumour suppressor region is age-related in normal urothelium and an early and frequent alteration in bladder cancer. *Oncogene*, **20**, 531–537.

Hellwinkel, O. J., Kedia, M., Isbarn, H., *et al.* (2008). Methylation of the TPEF- and PAX6-promoters is increased in early bladder cancer and in normal mucosa adjacent to pTa tumours. *British Journal of Urology International*, **101**, 753–757.

Hoque, M. O., Begum, S., Topaloglu, O., *et al.* (2006). Quantitation of promoter methylation of multiple genes in urine DNA and bladder cancer detection. *Journal of the National Cancer Institute*, **98**, 996–1004.

Huang da, W., Sherman, B. T., and Lempicki, R. A. (2009). Systematic and integrative analysis of large gene lists using DAVID bioinformatics resources. *Nature Protocols*, **4**, 44–57.

Huang, L., Luo, J., Cai, Q., *et al.* (2010). MicroRNA-125b suppresses the development of bladder cancer by targeting E2F3. *International Journal of Cancer*, **128**, 1758–1769.

Huang, Y., Stewart, T. M., Wu, Y., *et al.* (2009). Novel oligoamine analogues inhibit lysine-specific demethylase 1 and induce reexpression of epigenetically silenced genes. *Clinical Cancer Research*, **15**, 7217–7228.

Ichimi, T., Enokida, H., Okuno, Y., *et al.* (2009). Identification of novel microRNA targets based on microRNA signatures in bladder cancer. *International Journal of Cancer*, **125**, 345–352.

Jebar, A. H., Hurst, C. D., Tomlinson, D. C., *et al.* (2005). *FGFR3* and *Ras* gene mutations are mutually exclusive genetic events in urothelial cell carcinoma. *Oncogene*, **24**, 5218–5225.

Jemal, A., Siegel, R., Xu, J., *et al.* (2010). Cancer statistics, 2010. *Cancer Journal for Clinicians*, **60**, 277–300.

Kenney, P. A., Wszolek, M. F., Rieger-Christ, K. M., *et al.* (2010). Novel ZEB1 expression in bladder tumorigenesis. *British Journal of Urology International*, **107**, 656–663.

Khin, S. S., Kitazawa, R., Win, N., *et al.* (2009). *BAMBI* gene is epigenetically silenced in subset of high-grade bladder cancer. *International Journal of Cancer*, **125**, 328–338.

Kim, W. J., Kim, E. J., Jeong, P., *et al.* (2005). RUNX3 inactivation by point mutations and aberrant DNA methylation in bladder tumors. *Cancer Research*, **65**, 9347–9354.

Knowles, M. A. (2006). Molecular subtypes of bladder cancer: Jekyll and Hyde or chalk and cheese? *Carcinogenesis*, **27**, 361–373.

Lewis, B. P., Burge, C. B., and Bartel, D. P. (2005). Conserved seed pairing, often flanked by adenosines, indicates that thousands of human genes are microRNA targets. *Cell*, **120**, 15–20.

Lodygin, D., Tarasov, V., Epanchintsev, A., *et al.* (2008). Inactivation of miR-34a by aberrant CpG methylation in multiple types of cancer. *Cell Cycle*, **7**, 2591–2600.

Lu, Q., Lu, C., Zhou, G. P., *et al.* (2010). MicroRNA-221 silencing predisposed human bladder cancer cells to undergo apoptosis induced by TRAIL. *Urologic Oncology*, **28**, 635–641.

Marsit, C. J., Houseman, E. A., Schned, A. R., *et al.* (2007). Promoter hypermethylation is associated with current smoking, age, gender and survival in bladder cancer. *Carcinogenesis*, **28**, 1745–1751.

Marsit, C. J., Houseman, E. A., Christensen, B. C., *et al.* (2010). Identification of methylated genes associated with aggressive bladder cancer. *PLoS One*, **5**, e12334.

Maruyama, R., Toyooka, S., Toyooka, K. O., *et al.* (2001). Aberrant promoter methylation profile of bladder cancer and its relationship to clinicopathological features. *Cancer Research*, **61**, 8659–8663.

Moore, L. E., Pfeiffer, R. M., Poscablo, C., *et al.* (2008). Genomic DNA hypomethylation as a biomarker for bladder cancer susceptibility in the Spanish Bladder Cancer Study: a case-control study. *Lancet Oncology*, **9**, 359–366.

Nan, X., Campoy, F. J., and Bird, A. (1997). MeCP2 is a transcriptional repressor with abundant binding sites in genomic chromatin. *Cell*, **88**, 471–481.

Nguyen, C. T., Weisenberger, D. J., Velicescu, M., *et al.* (2002). Histone H3-lysine 9 methylation is associated with aberrant gene silencing in cancer cells and is rapidly reversed by 5-aza-2'-deoxycytidine. *Cancer Research*, **62**, 6456–6461.

Ostenfeld, M. S., Bramsen, J. B., Lamy, P., *et al.* (2010). miR-145 induces caspase-dependent and -independent cell death in urothelial cancer cell lines with targeting of an expression signature present in Ta bladder tumors. *Oncogene*, **29**, 1073–1084.

Ou, J. N., Torrisani, J., Unterberger, A., *et al.* (2007). Histone deacetylase inhibitor Trichostatin A induces global and gene-specific DNA demethylation in human cancer cell lines. *Biochemical Pharmacology*, **73**, 1297–1307.

Reulen, R. C., Kellen, E., Buntinx, F., *et al.* (2008). A meta-analysis on the association between bladder cancer and occupation. *Scandinavian Journal of Urology and Nephrology*, **218** (Suppl.), 64–78.

Sachs, M. D., Ramamurthy, M., Poel, H., *et al.* (2004). Histone deacetylase inhibitors upregulate expression of the coxsackie adenovirus receptor (CAR) preferentially in bladder cancer cells. *Cancer Gene Therapy*, **11**, 477–486.

Saito, Y., Liang, G., Egger, G., *et al.* (2006). Specific activation of microRNA-127 with downregulation of the proto-oncogene *BCL6* by chromatin-modifying drugs in human cancer cells. *Cancer Cell*, **9**, 435–443.

Saito, Y., Friedman, J. M., Chihara, Y., *et al.* (2009). Epigenetic therapy upregulates the tumor suppressor microRNA-126 and its host gene EGFL7 in human cancer cells. *Biochemical and Biophysical Research Communications*, **379**, 726–731.

Sandgren, J., Andersson, R., Rada-Iglesias, A., *et al.* (2010). Integrative epigenomic and genomic analysis of malignant pheochromocytoma. *Experimental Molecular Medicine*, **7**, 484–502.

Sangar, V. K., Ragavan, N., Matanhelia, S. S., *et al.* (2005). The economic consequences of prostate and bladder cancer in the UK. *British Journal of Urology International*, **95**, 59–63.

Sathyanarayana, U. G., Maruyama, R., Padar, A., *et al.* (2004). Molecular detection of non-invasive and invasive bladder tumor tissues and exfoliated cells by aberrant promoter methylation of laminin-5 encoding genes. *Cancer Research*, **64**, 1425–1430.

Serizawa, R. R., Ralfkiaer, U., Steven, K., *et al.* (2011). Integrated genetic and epigenetic analysis of bladder cancer reveals an additive diagnostic value of *FGFR3* mutations and hypermethylation events. *International Journal of Cancer*, **129**, 78–87.

Spruck, C. H., III, Ohneseit, P. F., Gonzalez-Zulueta, M., *et al.* (1994). Two molecular pathways to transitional cell carcinoma of the bladder. *Cancer Research*, **54**, 784–788.

Tada, Y., Wada, M., Taguchi, K., *et al.* (2002). The association of death-associated protein kinase hypermethylation with early recurrence in superficial bladder cancers. *Cancer Research*, **62**, 4048–4053.

Tan, J., Yang, X., Zhuang, L., *et al.* (2007). Pharmacologic disruption of Polycomb-repressive complex 2-mediated gene repression selectively induces apoptosis in cancer cells. *Genes and Development*, **21**, 1050–1063.

Van Oers, J. M., Zwarthoff, E. C., Rehman, I., *et al.* (2009). FGFR3 mutations indicate better survival in invasive upper urinary tract and bladder tumours. *European Urology*, **55**, 650–657.

Varambally, S., Dhanasekaran, S. M., Zhou, M., *et al.* (2002). The polycomb group protein EZH2 is involved in progression of prostate cancer. *Nature*, **419**, 624–629.

Watt, F. and Molloy, P. L. (1988). Cytosine methylation prevents binding to DNA of a HeLa cell transcription factor required for optimal expression of the adenovirus major late promoter. *Genes and Development*, **2**, 1136–1143.

Weber, M., Davies, J. J., Wittig, D., *et al.* (2005). Chromosome-wide and promoter-specific analyses identify sites of differential DNA methylation in normal and transformed human cells. *Nature Genetics*, **37**, 853–862.

Weikert, S., Christoph, F., Kollermann, J., *et al.* (2005). Expression levels of the EZH2 polycomb transcriptional repressor correlate with aggressiveness and invasive potential of bladder carcinomas. *International Journal of Molecular Medicine*, **16**, 349–353.

Wiklund, E. D., Bramsen, J. B., Hulf, T., *et al.* (2011). Coordinated epigenetic repression of the miR-200 family and miR-205 in invasive bladder cancer. *International Journal of Cancer*, **128**, 1327–1334.

Wilhelm, C. S., Kelsey, K. T., Butler, R., *et al.* (2010). Implications of LINE-1 methylation for bladder cancer risk in women. *Clinical Cancer Research*, **16**, 1682–1689.

Wolff, E. M., Chihara, Y., Pan, F., *et al.* (2010). Unique DNA methylation patterns distinguish noninvasive and invasive urothelial cancers and establish an epigenetic field defect in premalignant tissue. *Cancer Research*, **70**, 8169–8178.

Yang, J., Xu, Z., Li, J., *et al.* (2010). XPC epigenetic silence coupled with p53 alteration has a significant impact on bladder cancer outcome. *Journal of Urology*, **184**, 336–343.

Yates, D. R., Rehman, I., Abbod, M. F., *et al.* (2007). Promoter hypermethylation identifies progression risk in bladder cancer. *Clinical Cancer Research*, **13**, 2046–2053.

24 Genome-scale DNA methylation analyses of cancer in children

Nicholas C. Wong* and David M. Ashley

24.1 Introduction

Although rare, cancer is a leading cause of disease mortality in children after accidents, homicides, and suicides (Ries *et al.*, 1999). The biology of cancer in children is markedly different to cancer in adults. This is highlighted by the disparate incidence and etiology of common cancers found between children and adults (AIHW and AACR, 2004). In children, leukemia and brain tumors predominate in newly diagnosed cases, while breast, colorectal, melanoma, and lung cancer are common cancers in adults. While there seems to be a genetic basis for adult cancers, where familial cases of breast (reviewed in Narod and Foulkes, 2004) and colorectal (reviewed in Jasperson *et al.*, 2010) cancer have identified mutations of candidate risk genes, the same can not be said for childhood cancer.

The precise causes of childhood cancer still remain unknown. With the scarce number of cases of familial or inherited cancers in children, it would seem that genetics plays a minor role in childhood cancer. Tumours acquire somatic mutation during their progression to cancer and it has long been hypothesized that selected environmental exposures can accelerate mutation rate in tumours that have given rise to their presentation as cancer in children (Knudson, 1976). There is a proposed list of high-risk exposures including radiation, infection, and pesticides associated with childhood cancer (reviewed in Anderson, 2006); however, the small study numbers used to identify these factors have led to small effect sizes and have been difficult to reproduce. Large, multi-center prospective cohort studies have begun to address this issue (Brown *et al.*, 2007).

A number of animal studies linking environmental exposures to changes in epigenetic modifications such as DNA methylation suggest that changes in

---

* Author to whom correspondence should be addressed.

Epigenomics: From Chromatin Biology to Therapeutics, ed. K. Appasani. Published by Cambridge University Press. © Cambridge University Press 2012.

phenotype can arise from the environment via changes in the epigenome (reviewed in Sutherland and Costa, 2003). The link between environment and epigenetics is becoming clearer in cancer (reviewed in Ulrich and Grady, 2010). There are an increasing number of studies characterizing DNA methylation changes in cancer with early studies focusing on the promoters of known tumour suppressor and oncogenes. With the advent of second-generation sequencing technology, methods are rapidly being developed for genome-scale characterization of the epigenome (Feinberg, 2010). Genome-scale analysis offers an unbiased approach in cataloging DNA methylation changes associated with disease and could be readily applied here for childhood cancer.

24.2   Methods for DNA methylation analysis

A variety of protocols are available to analyze DNA methylation (summarized in Figure 24.1). Here we will focus on only those used to analyze DNA methylation in childhood cancer, while the remaining protocols are reviewed in Fraga and Esteller (2002). Methylation-specific PCR (MSP) (Herman *et al.*, 1996) is the most widely used loci-specific DNA methylation assay to look at the promoters of known tumour suppressors and oncogenes. A range of methods to look at genome-scale DNA methylation in childhood cancer have been used and these include DNA methylation enrichment, such as methylated DNA immunoprecipitation (MeDIP) and methylated CpG island recovery assay (MIRA), methylation-sensitive restriction enzyme mediated techniques such as *Hpa*II enrichment by ligation-mediated PCR (HELP) and differential methylation hybridization (DMH), and bisulfite-mediated methods such as GoldenGate bead array (reviewed in Laird, 2010).

24.3   Childhood cancer and DNA methylation

Research in childhood cancer has always lagged behind adult cancer. One of the reasons for this is the rarity of cancer in children compared to adults, with an incidence rate of 17.3 cases per 100 000 compared to a rate of 1021.7 per 100 000 (AIHW and AACR, 2004). With such a low incidence in children, sourcing cases and samples for analysis is more challenging for childhood cancers than for adult cancers. As a result, there are only a handful of studies looking at DNA methylation at a genome scale.

Acute leukemia and central nervous system cancers represent the bulk number of tumours in children accounting for 34.1% and 30.2% of diagnosed cases respectively (Kaatsch, 2010). The greatest number of genome-scale DNA methylation analyses and loci-specific DNA methylation studies have been performed for childhood acute lymphoblastic leukemia (Table 24.1). As well as being the most common cancer site in children, bone marrow is easier to access for analysis and manipulation than is the case for brain biopsies.

Acute lymphoblastic leukemia (ALL) describes a disease of the hematopoietic system and involves a block in differentiation of the lymphoid lineage. It is quite heterogeneous with a range of subtypes categorized by common chromosome

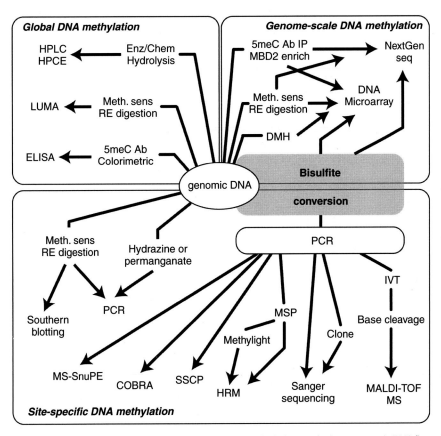

Figure 24.1 Overview of methods available for DNA methylation analysis on genomic DNA (based on Fraga and Esteller, 2002). Analysis methods can be divided into three categories: global, genome-scale, and loci-specific DNA methylation methods. Global methods measure DNA methylation of the genome as a whole while genome-scale approaches attempt to analyze most CpG sites across the genome. Loci-specific methods look at DNA methylation at targeted regions/genes within the genome. All genome-scale methods have been employed to investigate DNA methylation in childhood cancer while the most popular loci-specific method used is MSP. HPLC, high-performance liquid chromatography; HPCE, high-performance capillary electrophoresis; LUMA, luminometric methylation assay; ELISA, enzyme-linked immunosorbent assay; DMH, differential methylation hybridization; RE, restriction endonuclease; PCR, polymerase chain reaction; MSP, methylation-specific PCR; MS-SNuPE, methylation-specific single-nucleotide primer extension; COBRA, combined bisulfite and restriction analysis; SSCP, single-stranded conformation polymorphism; HRM, high-resolution melt; IVT, in-vitro transcription; MALDI-TOF MS, matrix-assisted laser desorption ionization time of flight mass spectrometry.

aberrations present in the tumour and white cell count on presentation. The difference in distribution of ALL subtypes and relative incidence between adults and children implies that these are two separate diseases with distinct mechanisms of action (Pui *et al.*, 2004). Genomic studies looking for recurrent gene mutations in childhood ALL have accounted for approximately 40% of cases (Mullighan, 2009) suggesting other changes such as DNA methylation could be at play. Differential DNA methylation hybridization (DMH) was one of the first methods used to profile DNA methylation at genome scale (Stumpel *et al.*, 2009)

Table 24.1 Summary of genome-scale DNA methylation profiling in childhood cancer

Cancer	Proportion of all cancers (Kaatsch, 2010)	Subtype	DNA methylation protocol[a]	Genome scale	Genes identified	Reference
Leukemias	34.1%	Acute lymphoblastic leukemia	MeDIP	Yes	Over 90 genes	Davidsson et al., 2009
			MIRA	Yes	(reviewed in Chatterton et al., 2010)	Dunwell et al., 2010
			GoldenGate	Yes		Milani et al., 2010
			HELP	Yes		Schafer et al., 2010
			DMH	Yes		Stumpel et al., 2009
			DMH	Yes		Dunwell et al., 2009
			Various single genes	No		
Central nervous system cancers	30.2%	Glioma/astrocytoma	MSP	No	MGMT	Buttarelli et al., 2010
			MSP	No	RASSF1A	Michalowski et al., 2006
		Medulloblastoma	DAMD	Yes	PRDM8, AXIN2, HIC1 and PTCH1	Diede et al., 2010
			aPRIMES	Yes	ZIC1	Pfister et al., 2007
			MSP	No	SFRP family	Kongkham et al., 2010
			MSP	No	Caspase-8,	Ebinger et al., 2004
Lymphomas	11.5%	Hodgkin's lymphoma	MSP	No	RASSF1A	Murray et al., 2004
			MSP	No	DLC1	Ying et al., 2007
		Non-Hodgkin's lymphoma	MSP	No	DAPK1	Katzenellenbogen et al., 1999
Renal tumors	5%	Wilm's tumour	MSP	No	IGF2/H19	Hubertus et al., 2010
			MeDIP	Yes	PDCH	Dallosso et al., 2009
Bone tumors	4.6%	Osteosarcoma	MeDIP/ChiP of cell lines	Yes	WTI, PCDHB8, LHX9, KCND3	Sadikovic et al., 2008
		Ewing's sarcoma	MSP	No	RASSF1A	Avigad et al., 2009
Hepatic tumors	1.1%	Hepatoblastoma	MSP	No	MT1G, APC, CDH1, MT1G, RASSF1A, and SOCS1	Sakamoto et al., 2010
			MSP	No	SOCS1	Sugawara et al., 2007
			MSP	No	RASSFIA, SOCS1 and FGFR1	Nagai et al., 2003
Others	14.5%	Rhabdomyosarcoma	Bisulfite sequencing	No	PAX3	Goldstein et al., 2007
			MSP	No	MGMT	Kurmasheva et al., 2005
			MSP	No		Yeager et al., 2003

[a] MeDIP, methylated DNA immunoprecipitation; MIRA, methylated CpG island recovery assay; HELP, HpaII enrichment by ligation-mediated PCR; DMH, differential DNA methylation hybridization; MSP, methylation-specific PCR; DAMD, denaturation analysis of methylation differences; aPRIMES, array-based profiling of reference-independent methylation status; ChIP, chromatin immunoprecipitation.

and at chromosome 3 (Dunwell *et al.*, 2009). Since then HELP (Schafer *et al.*, 2010), MIRA (Dunwell *et al.*, 2010), and MeDIP (Davidsson *et al.*, 2009) have been used and each has given its own unique list of candidate genes of interest. Numerous other studies in DNA methylation in childhood ALL involve loci-specific approaches such as MSP and are reviewed in more detail in Chatterton *et al.* (2010). The diversity of genes affected by DNA methylation in childhood ALL is likely due to the range of methods used to profile DNA methylation in childhood leukemia. Furthermore, each study to date has focused on a unique ALL subtype and compared different groups to identify DNA methylation changes of interest. Therefore, no consensus has been reached between studies (Chatterton *et al.*, 2010).

The central nervous system is the next most common site of cancer in children accounting for 30.2% of newly diagnosed cases (Kaatsch, 2010) (Table 24.1). However, there are only a few studies investigating genome-scale DNA methylation (Table 24.1). Denaturation analysis of methylation differences (DAMD) based on the higher melting temperatures of methylated DNA was used to identify common DNA methylation markers associated with medulloblastoma (Diede *et al.*, 2010). The promoters of PRDM8, AXIN2, and HIC1 negative regulators of the Notch, Sonic hedgehog, and Wingless (Wnt) pathways were methylated in medulloblastoma. This was expected because of their known roles in other cancers. The same study identified DNA methylation of a previously uncharacterized upstream promoter of PTCH1 by this method (Diede *et al.*, 2010) and demonstrates the power of genome-scale DNA methylation analysis in identifying novel candidate regions. Another study using genome-scale restriction-enzyme-based microarray analysis in medulloblastoma described differential DNA methylation of the *ZIC2* gene promoter (Pfister *et al.*, 2007). Loci-specific studies describe DNA methylation at SFRP (Kongkham *et al.*, 2010) and Caspase-8 (Ebinger *et al.*, 2004) gene promoters in medulloblastoma while P16INK4A, MGMT, TIMP-3, and E-cadherin were unmethylated (Ebinger *et al.*, 2004). The remaining studies in childhood cancer of the central nervous system involve DNA methylation of known tumour suppressors and oncogenes such as MGMT (Buttarelli *et al.*, 2010) and RASSFA1 (Michalowski *et al.*, 2006) in childhood glioma. Both studies utilized MSP, and it was found that only a small proportion of cases demonstrated DNA methylation at MGMT (Buttarelli *et al.*, 2010) while up to 60% of cases analyzed were methylated at RASSF1A (Michalowski *et al.*, 2006).

Lymphomas are the next most prevalent form of childhood cancer and to date genome-scale analysis have not been performed on childhood cases while more extensive studies have been performed on adult cases. Gene-specific analyses have described DNA methylation of RASSF1A (Murray *et al.*, 2004) and DLC1 (Ying *et al.*, 2007) in a significant proportion of Hodgkin's lymphoma cases using MSP. Methylation of DAPK1 in non-Hodgkin's lymphoma has also been described with the majority of cases analyzed being methylated at the DAPK1 gene promoter using MSP analysis (Katzenellenbogen *et al.*, 1999).

Renal tumours account for 5% of newly diagnosed cases of childhood cancer with Wilm's tumour (WT) being the major type in children (Kaatsch, 2010).

Ongoing intensive studies in WT have identified genetic mutations in the WT gene where mutations account for up to 30% of cases (Ruteshouser *et al.*, 2008). This tumor is associated with a number of over-growth and under-growth syndromes in which loss of imprinting at the H19 and IGF2 locus is observed (reviewed in Nakamura and Ritchey, 2010). Given the strong linkage between these aberrations, the current focus has been observing DNA methylation changes at imprinted loci with the thinking that loss of imprinting in WT could be a general phenomenon across the genome of WT cases. Further, MeDIP and DNA microarray analysis has been performed on WT cases and a chromosomal region of DNA methylation was identified at 5q31 containing the protocadherin gene cluster (Dallosso *et al.*, 2009). Later work on imprinted loci across the genome confirmed DNA methylation changes at H19 but also revealed by gene expression analyses an associated loss of imprinting at IGF2, NNAT, and MEST (Hubertus *et al.*, 2010).

Bone tumors including osteosarcoma and Ewing's sarcoma account for 4.6% of childhood cancer cases (Kaatsch, 2010). Tumours are characterized by chromosomal abberations within the Ewing's sarcoma gene (EWS) generating fusion proteins with the most common fusion being EWS-FLI1 (reviewed in Riggi and Stamenkovic, 2007). EWS-FLI1 is thought to behave as an aberrant transcription factor dysregulating downstream gene targets. Limited DNA methylation analysis has been performed in this tumour type with only one description of MeDIP microarray analysis on osteosarcoma model cell lines identifying WTI, PCDHB8, LHX9, and KCND3 genes methylated in these cell lines when compared to normal controls (Sadikovic *et al.*, 2008). An MSP analysis of the RASSF1A gene promoter in Ewing's sarcoma found methylation of RASSF1A was associated with the tumour and a poor prognosis (Avigad *et al.*, 2009).

Hepatoblastoma is the common form of hepatic cancers in children, accounting for 1.1% of cases in children (Kaatsch, 2010). Given their rarity, very few studies in DNA methylation have been performed. All have utilized MSP on candidate gene promoter regions of genes including APC, CDH1, MT1G, RASSF1A, and SOCS1 whose disruption in expression have been implicated in other cancers. In particular, MT1G was found to be associated with poor prognosis in this form of childhood cancer (Sakamoto *et al.*, 2010). Previous studies have also looked at RASSF1A (Sugawara *et al.*, 2007) and SOCS1 (Nagai *et al.*, 2003) in childhood hepatoblastoma with similar findings.

The remaining 14.5% of childhood cancer cases include a wide range of types that individually are very rare in occurrence (Kaatsch, 2010). However, a number of DNA methylation studies have been performed on rhabdomyosarcoma at the gene promoters of FGFR1 (Goldstein *et al.*, 2007), PAX3 (Kurmasheva *et al.*, 2005), and MGMT (Yeager *et al.*, 2003). Bisulfite sequencing was used to describe hypomethylation of the FGFR1 gene promoter and concomitant upregulation of this gene in tumours (Goldstein *et al.*, 2007). In a similar study, hypomethylation of MGMT promoter was found in the same tumor type using MSP (Yeager *et al.*, 2003) while PAX3 was found to be hypermethylated (Kurmasheva *et al.*, 2005).

24.4   Conclusions and perspectives

The number and scale of studies investigating DNA methylation analysis in childhood cancer seems to decrease with increasing rarity of the cancer type. Cancer is already a rare disease in children and access to high-quality, clinically annotated samples for proper genome-scale analyses will be challenging for the rarest forms of childhood cancer. Given the rarity, one study looked at caspase-8 and caspase-10 gene promoter methylation in a wide range of childhood cancers and found this gene to be highly methylated in various proportions of cases with a particular cancer type (Harada *et al.*, 2002b) while another investigated RASSF1A promoter methylation with similar outcomes (Harada *et al.*, 2002a). With second-generation sequencing technologies maturing, analysis methods exploiting this technology are requiring less sample input for analysis (Feinberg, 2010). There is a great opportunity to study genome-scale DNA methylation in childhood cancer with a properly curated bank of samples. The Cancer Genome Atlas Research Network (2008) is investigating DNA mutations in a range of cancers including glioblastoma multiforme (http://cancergenome.nih.gov/). Moreover, epigenomic studies have recently been completed for this tumor type (Noushmehr *et al.*, 2010). However, childhood cancers have not been included in the Cancer Genome Atlas research program. The International Human Epigenome Consortium (IHEC: www.ihec-epigenomes.org/) and the NIH Epigenome RoadMap (Bernstein *et al.*, 2010) have set out ambitious goals to characterize entire epigenomes in normal and diseased cells; however, there are currently no efforts that are focusing on childhood cancer. The clear differences between adult and childhood cancers preclude direct extrapolation of findings in adult studies to their childhood counterparts. Therefore there is a need to investigate the epigenomes of childhood cancer in addition to current efforts.

REFERENCES

AIHW and AACR (2004). *Cancer in Australia 2001*, CAN 23 edn. Canberra: Australian Institute of Health and Welfare & Australasian Association of Cancer Registries.

Anderson, L. M. (2006). Environmental genotoxicants/carcinogens and childhood cancer: bridgeable gaps in scientific knowledge. *Mutation Research*, **608**, 136–156.

Avigad, S., Shukla, S., Naumov, I., *et al.* (2009). Aberrant methylation and reduced expression of RASSF1A in Ewing sarcoma. *Pediatric Blood and Cancer*, **53**, 1023–1028.

Bernstein, B. E., Stamatoyannopoulos, J. A., Costello, J. F., *et al.* (2010). The NIH Roadmap Epigenomics Mapping Consortium. *Nature Biotechnology*, **28**, 1045–1048.

Brown, R. C., Dwyer, T., Kasten, C., *et al.* (2007). Cohort profile: the International Childhood Cancer Cohort Consortium (I4C). *International Journal of Epidemiology*, **36**, 724–730.

Buttarelli, F. R., Massimino, M., Antonelli, M., *et al.* (2010). Evaluation status and prognostic significance of O6-methylguanine-DNA methyltransferase (MGMT) promoter methylation in pediatric high grade gliomas. *Child's Nervous System*, **26**, 1051–1056.

Cancer Genome Atlas Research Network (2008). Comprehensive genomic characterization defines human glioblastoma genes and core pathways. *Nature*, **455**, 1061–1068.

Chatterton, Z., Morenos, L., Saffery, R., *et al.* (2010). DNA methylation and miRNA expression profiling in childhood B-cell acute lymphoblastic leukemia. *Epigenomics*, **2**, 697–708.

Dallosso, A. R., Hancock, A. L., Szemes, M., *et al.* (2009). Frequent long-range epigenetic silencing of protocadherin gene clusters on chromosome 5q31 in Wilms' tumor. *PLoS Genetics*, **5**, e1000745.

Davidsson, J., Lilljebjorn, H., Andersson, A., *et al.* (2009). The DNA methylome of pediatric acute lymphoblastic leukemia. *Human Molecular Genetics*, **18**, 4054–4065.

Diede, S. J., Guenthoer, J., Geng, L. N., *et al.* (2010). DNA methylation of developmental genes in pediatric medulloblastomas identified by denaturation analysis of methylation differences. *Proceedings of the National Academy of Sciences USA*, **107**, 234–239.

Dunwell, T. L., Hesson, L. B., Pavlova, T., *et al.* (2009). Epigenetic analysis of childhood acute lymphoblastic leukemia. *Epigenetics*, **4**, 185–193.

Dunwell, T., Hesson, L., Rauch, T. A., *et al.* (2010). A genome-wide screen identifies frequently methylated genes in haematological and epithelial cancers. *Molecular Cancer*, **9**, 44.

Ebinger, M., Senf, L., Wachowski, O., and Scheurlen, W. (2004). Promoter methylation pattern of caspase-8, P16INK4A, MGMT, TIMP-3, and E-cadherin in medulloblastoma. *Pathology and Oncology Research*, **10**, 17–21.

Feinberg, A. P. (2010). Genome-scale approaches to the epigenetics of common human disease. *Virchows Archive*, **456**, 13–21.

Fraga, M. F. and Esteller, M. (2002). DNA methylation: a profile of methods and applications. *Biotechniques*, **33**, 632, 634, 636–649.

Goldstein, M., Meller, I., and Orr-Urtreger, A. (2007). FGFR1 over-expression in primary rhabdomyosarcoma tumors is associated with hypomethylation of a 5' CpG island and abnormal expression of the AKT1, NOG, and BMP4 genes. *Genes, Chromosomes and Cancer*, **46**, 1028–1038.

Harada, K., Toyooka, S., Maitra, A., *et al.* (2002a). Aberrant promoter methylation and silencing of the RASSF1A gene in pediatric tumors and cell lines. *Oncogene*, **21**, 4345–4349.

Harada, K., Toyooka, S., Shivapurkar, N., *et al.* (2002b). Deregulation of caspase 8 and 10 expression in pediatric tumors and cell lines. *Cancer Research*, **62**, 5897–5901.

Herman, J. G., Graff, J. R., Myohanen, S., Nelkin, B. D., and Baylin, S. B. (1996). Methylation-specific PCR: a novel PCR assay for methylation status of CpG islands. *Proceedings of the National Academy of Sciences USA*, **93**, 9821–9826.

Hubertus, J., Lacher, M., Rottenkolber, M., *et al.* (2010). Altered expression of imprinted genes in Wilms tumors. *Oncology Reports*, **25**, 817–823.

Jasperson, K. W., Tuohy, T. M., Neklason, D. W., and Burt, R. W. (2010). Hereditary and familial colon cancer. *Gastroenterology*, **138**, 2044–2058.

Kaatsch, P. (2010). Epidemiology of childhood cancer. *Cancer Treatment Reviews*, **36**, 277–285.

Katzenellenbogen, R. A., Baylin, S. B., and Herman, J. G. (1999). Hypermethylation of the DAP-kinase CpG island is a common alteration in B-cell malignancies. *Blood*, **93**, 4347–4353.

Knudson, A. G., Jr. (1976). Genetics and the etiology of childhood cancer. *Pediatric Research*, **10**, 513–517.

Kongkham, P. N., Northcott, P. A., Croul, S. E., *et al.* (2010). The SFRP family of WNT inhibitors function as novel tumor suppressor genes epigenetically silenced in medulloblastoma. *Oncogene*, **29**, 3017–3024.

Kurmasheva, R. T., Peterson, C. A., Parham, D. M., *et al.* (2005). Upstream CpG island methylation of the PAX3 gene in human rhabdomyosarcomas. *Pediatric Blood and Cancer*, **44**, 328–337.

Laird, P. W. (2010). Principles and challenges of genome-wide DNA methylation analysis. *Nature Reviews Genetics*, **11**, 191–203.

Michalowski, M. B., de Fraipont, F., Michelland, S., *et al.* (2006). Methylation of RASSF1A and TRAIL pathway-related genes is frequent in childhood intracranial ependymomas and benign choroid plexus papilloma. *Cancer Genetics and Cytogenetics*, **166**, 74–81.

Milani, L., Lundmark, A., Kiialainen, A., *et al.* (2010). DNA methylation for subtype classification and prediction of treatment outcome in patients with childhood acute lymphoblastic leukemia. *Blood*, **115**, 1214–1225.

Mullighan, C. G. (2009). Genomic analysis of acute leukemia. *International Journal of Laboratory Hematology*, **31**, 384–397.

Murray, P. G., Qiu, G. H., Fu, L., *et al*. (2004). Frequent epigenetic inactivation of the RASSF1A tumor suppressor gene in Hodgkin's lymphoma. *Oncogene*, **23**, 1326–1331.

Nagai, H., Naka, T., Terada, Y., *et al*. (2003). Hypermethylation associated with inactivation of the SOCS-1 gene, a JAK/STAT inhibitor, in human hepatoblastomas. *Journal of Human Genetics*, **48**, 65–69.

Nakamura, L. and Ritchey, M. (2010). Current management of Wilms' tumor. *Current Urology Reports*, **11**, 58–65.

Narod, S. A. and Foulkes, W. D. (2004). BRCA1 and BRCA2: 1994 and beyond. *Nature Reviews Cancer*, **4**, 665–676.

Noushmehr, H., Weisenberger, D. J., Diefes, K., *et al*. (2010). Identification of a CpG island methylator phenotype that defines a distinct subgroup of glioma. *Cancer Cell*, **17**, 510–522.

Pfister, S., Schlaeger, C., Mendrzyk, F., *et al*. (2007). Array-based profiling of reference-independent methylation status (aPRIMES) identifies frequent promoter methylation and consecutive downregulation of ZIC2 in pediatric medulloblastoma. *Nucleic Acids Research*, **35**, e51.

Pui, C. H., Relling, M. V., and Downing, J. R. (2004). Acute lymphoblastic leukemia. *New England Journal of Medicine*, **350**, 1535–1548.

Ries, L., Smith, M., Gurney, J., *et al*. (eds.) (1999). *Cancer Incidence and Survival among Children and Adolescents: United States SEER Program 1975–1995*. Bethesda, MD: National Cancer Institute, Surveillance Research Program.

Riggi, N. and Stamenkovic, I. (2007). The biology of Ewing sarcoma. *Cancer Letters*, **254**, 1–10.

Ruteshouser, E. C., Robinson, S. M., and Huff, V. (2008). Wilms' tumor genetics: mutations in WT1, WTX, and CTNNB1 account for only about one-third of tumors. *Genes, Chromosomes and Cancer*, **47**, 461–470.

Sadikovic, B., Yoshimoto, M., Al-Romaih, K., *et al*. (2008). In vitro analysis of integrated global high-resolution DNA methylation profiling with genomic imbalance and gene expression in osteosarcoma. *PLoS One*, **3**, e2834.

Sakamoto, L. H., de Camargo, B., Cajaiba, M., Soares, F. A., and Vettore, A. L. (2010). MT1G hypermethylation: a potential prognostic marker for hepatoblastoma. *Pediatric Research*, **67**, 387–393.

Schafer, E., Irizarry, R., Negi, S., *et al*. (2010). Promoter hypermethylation in MLL-r infant acute lymphoblastic leukemia: biology and therapeutic targeting. *Blood*, **115**, 4798–4809.

Stumpel, D. J., Schneider, P., van Roon, E. H., *et al*. (2009). Specific promoter methylation identifies different subgroups of MLL-rearranged infant acute lymphoblastic leukemia, influences clinical outcome, and provides therapeutic options. *Blood*, **114**, 5490–5498.

Sugawara, W., Haruta, M., Sasaki, F., *et al*. (2007). Promoter hypermethylation of the RASSF1A gene predicts the poor outcome of patients with hepatoblastoma. *Pediatric Blood and Cancer*, **49**, 240–249.

Sutherland, J. E. and Costa, M. (2003). Epigenetics and the environment. *Annals of the New York Academy of Sciences*, **983**, 151–160.

Ulrich, C. M. and Grady, W. M. (2010). Linking epidemiology to epigenomics: where are we today? *Cancer Prevention Research*, **3**, 1505–1508.

Yeager, N. D., Dolan, M. E., Gastier, J. M., *et al*. (2003). O6-methylguanine-DNA methyltransferase activity and promoter methylation status in pediatric rhabdomyosarcoma. *Journal of Pediatric Hematology/Oncology*, **25**, 941–947.

Ying, J., Li, H., Murray, P., *et al*. (2007). Tumor-specific methylation of the 8p22 tumor suppressor gene DLC1 is an epigenetic biomarker for Hodgkin, nasal NK/T-cell and other types of lymphomas. *Epigenetics*, **2**, 15–21.

25 The epigenetics of facioscapulohumeral muscular dystrophy

Weihua Zeng, Alexander R. Ball, Jr., and Kyoko Yokomori*

25.1 Introduction

Muscular dystrophies are a group of hereditary muscle diseases marked by muscle weakness and muscle tissue loss. The causative gene mutations for certain muscular dystrophies have been identified, including in the emerin gene for X-linked Emery–Dreifuss muscular dystrophy (EDMD), in the dystrophin gene for Duchenne/Becker muscular dystrophy (DMD/BMD), and in the lamin A/C gene for limb-girdle muscular dystrophy type 1B (LGMD1B)) (Rocha and Hoffman, 2010). Facioscapulohumeral muscular dystrophy (FSHD) is one of the most common muscular dystrophies, but identifying its pathogenic genes has been challenging. We recently discovered specific epigenetic changes uniquely associated with FSHD that may be critical for disease development, providing strong evidence that FSHD is an "epigenetic abnormality" disease. Here we review the epigenetic changes found in FSHD, and how they relate to the disease's pathogenesis.

25.2 D4Z4 repeat sequences on chromosome 4q are critically involved in FSHD pathogenesis

25.2.1 Contraction of 4q D4Z4 repeats is genetically linked to the majority of FSHD cases

FSHD is an autosomal dominant myopathy and is the third most common muscular dystrophy with a prevalence of at least 1 in 20 000 people. The clinical symptoms usually commence with weakness of facial and shoulder girdle muscles. This muscular wasting, usually asymmetric, often progresses in a descending manner to the upper extremities, trunk, and the lower limbs. The

---

* Author to whom correspondence should be addressed.

Epigenomics: From Chromatin Biology to Therapeutics, ed. K. Appasani. Published by Cambridge University Press. © Cambridge University Press 2012.

clinical severity varies dramatically between individual cases, ranging from asymptomatic carriers to wheelchair-dependent patients (Tawil, 2008). The age of clinical onset is also broad (i.e., from early infancy to late fifties), though most patients become symptomatic in their teens (van der Maarel *et al.*, 2007). About 75% of patients suffer from some high-frequency hearing loss, and asymptomatic retinal telangiectasias also affect 60% of patients (Fitzsimons *et al.*, 1987; Padberg *et al.*, 1995).

Genome-wide microsatellite scanning revealed that FSHD is linked to the subtelomeric region of the long arm of chromosome 4, specifically the 4q35 region (Wijmenga *et al.*, 1990). The genetic defect in this region was detected as a size reduction in an EcoRI fragment that consists of a series of 3.3 kb macrosatellite repeat units termed "D4Z4" (Figure 25.1A). While the D4Z4 repeat number in unaffected individuals is usually between 11 and 150, more than 95% of FSHD patients have at least one 4q35 allele containing fewer than 10 D4Z4 repeats (van Deutekom *et al.*, 1993). There are D4Z4-like sequences in several other chromosomes (Tam *et al.*, 2004). The D4Z4 repeat at 10q26 is highly homologous to 4q35 D4Z4, but contraction of 10q D4Z4 does not lead to FSHD (van der Maarel *et al.*, 2007). Thus, the majority of FSHD appears to be caused by contraction of 4q D4Z4 (Figure 25.1B). This type of FSHD is called "4q-linked" FSHD, FSHD1, or FSHD1A.

The onset age and the severity of FSHD symptoms are inversely correlated with the number of remaining D4Z4 repeats. However, patients with the same number of remaining D4Z4 repeats within a family can have different disease presentations (Tawil, 2008). This implies the involvement of other factors in

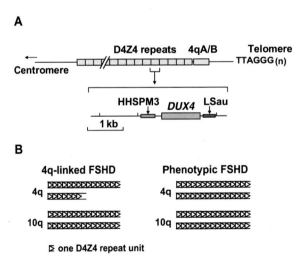

Figure 25.1 D4Z4 repeats and FSHD. (A) Schematic diagram of 4qter D4Z4 repeats (Zeng *et al.*, 2009). The location of D4Z4 in the subtelomeric region of chromosome 4 long arm is indicated, together with the relative position of 4qA/B allelic variants. The detailed structure of a single D4Z4 repeat unit is shown in the lower panel. (B) The status of D4Z4 repeats in 4q-linked and phenotypic FSHD. One D4Z4 repeat unit is represented by one triangle. The deletion of monoallelic 4q D4Z4 in 4q-linked FSHD patients is shown in the diagram, as well as the undeleted 4q and 10q D4Z4 alleles in phenotypic FSHD patients.

determining symptom severity, which have not been determined. No FSHD patients have been found to completely lack D4Z4 repeats in one 4q35 allele, suggesting that the remaining repeat(s) is important for FSHD pathogenesis (Tupler *et al.*, 1996).

Interestingly, in less than 5% of FSHD patients, there is no detectable D4Z4 deletion (Figure 25.1B). However, the clinical features are largely identical to those observed in the 4q-linked FSHD patients (de Greef *et al.*, 2010). Therefore, this form of the disease is termed "phenotypic" FSHD, FSHD2, or FSHD1B, and the etiology remains unclear.

25.2.2   The role of D4Z4 and neighboring regions in FSHD pathogenesis

25.2.2.1   Contraction of D4Z4 upregulates the expression of neighboring 4q35 genes

How does the contraction of D4Z4 repeats (only on chromosome 4q) cause FSHD? D4Z4 is able to recruit a repressor complex consisting of YY1, HMG2B, and nucleolin (Gabellini *et al.*, 2002). It was observed that in comparison to unaffected individuals, FSHD patients show higher expression of genes located proximal to D4Z4 at the 4q35 region, including FSHD related gene-1 (*FRG1*, a gene involved in RNA processing), *FRG2*, and adenine nucleotide translocator-1 (*ANT1*) (Gabellini *et al.*, 2002). *FRG2*, *FRG1*, and *ANT1* are 37 kb, 120 kb, and 3.5 Mb upstream of the 4q D4Z4 repeats, respectively (Figure 25.2A). Depletion of the repressor complex by siRNA upregulates *FRG2* expression (Gabellini *et al.*, 2002). It was therefore proposed that in FSHD patients, due to D4Z4 deletion and decreased recruitment of the repressor complex, silenced D4Z4-flanking genes are abnormally activated in a manner reminiscent of the loss of position-effect variegation (PEV) in *Drosophila* (Gabellini *et al.*, 2002) (Figure 25.2A). While overexpression of either *FRG2* or *ANT1* in transgenic mice does not result in any muscle atrophy, *FRG1* transgenic mice indeed developed muscular dystrophy, and abnormal splicing of muscle-specific mRNA was detected in the skeletal muscle of these mice as in FSHD patient myoblasts (Gabellini *et al.*, 2006). In *Xenopus laevis*, FRG1 knock-down impaired myotome organization, while overexpression of FRG1 resulted in abnormal epaxial and hypaxial muscle development (Hanel *et al.*, 2009). In addition, FRG1 is important for vascular development in *Xenopus,* and over-expression of FRG1 in tadpoles causes blood vessel branching, dilation, and edema, reminiscent of vascular findings in the muscles and retinas of some FSHD patients (Wuebbles *et al.*, 2009). Despite these intriguing observations that highlight the functional relevance of FRG1, other studies failed to consistently observe any significant overexpression of 4q35 genes, including *FRG1*, in FSHD myoblasts and muscle biopsies (Jiang *et al.*, 2003; Winokur *et al.*, 2003; Osborne *et al.*, 2007; Klooster *et al.*, 2009; Masny *et al.*, 2010). In addition, it is difficult to apply the repressor model to the FSHD2 patients with no D4Z4 repeat deletion, and the repressor complex components were ruled out as pathogenic genes in FSHD2 (Bastress *et al.*, 2005). Thus, to what extent *FRG1* is involved in FSHD remains unclear.

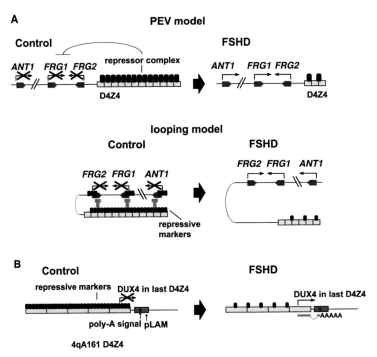

Figure 25.2 Hypothetical models explaining the role of D4Z4 in FSHD pathogenesis. (A) The downregulation of 4q D4Z4 proximal genes is disrupted in FSHD. The involvement of D4Z4 in this mechanism is proposed in two models. In the upper panel is the position effect variegation (PEV) model. In this model, normal D4Z4 repeats recruit repressor complexes and the inhibitory effect spreads to the flanking genes and downregulates their expression, while in FSHD, with the decreased recruitment of repressor complexes, the flanking genes are abnormally activated. The looping model is shown in the lower panel, in which the repressive effect from D4Z4 affects proximal genes through long-distance looping, not by direct spreading. (B) The distal D4Z4 repeat in the 4qA161 variant expresses stable *DUX4* mRNA in FSHD. In the 4qA161 variant, the pLAM sequence just downstream of the 4q D4Z4 repeats contains a functional poly(A) signal. In FSHD, with the loss of repressive markers on D4Z4, the *DUX4* open reading frame (ORF) in the last repeat unit is activated and it utilizes the poly(A) signal to produce stable *DUX4* mRNA, which may act as a pathogenic gene.

25.2.2.2 The *DUX4* gene encoded within the D4Z4 repeat is not silenced in FSHD

The D4Z4 repeat itself contains a double homeodomain gene called *DUX4* with a putative promoter (Figure 25.1A). The first homeodomain of DUX4 is highly homologous to those of Pax3 and Pax7, key transcriptional regulators that control myogenic lineage commitment. Overexpression of recombinant *DUX4* led to apoptotic cell death and alteration of emerin distribution in the nuclear envelope, which may explain the clinical similarity between FSHD and EDMD (Kowaljow *et al.*, 2007). *Pitx1*, a putative pathogenic gene upregulated in FSHD, contains a DUX4 binding site in its promoter and is indeed upregulated upon *DUX4* overexpression in the mouse C2C12 myoblast cell line, further supporting the notion that *DUX4* expression contributes to FSHD pathogenesis (Dixit *et al.*, 2007).

There are no introns and no polyadenylation (poly(A)) signal in the *DUX4* gene in each D4Z4 repeat, and it was unclear whether the endogenous *DUX4* gene is actually transcribed (Gabriels *et al.*, 1999; Winokur *et al.*, 2003; Osborne *et al.*, 2007). In the 4q subtelomeric region distal to the D4Z4 repeats, there are two prominent allelic variants termed 4qA and 4qB. These two variants have equal frequency, with the major difference being the presence of pLAM and 6 kb of beta-satellite repeats distal to D4Z4 in 4qA, but not in 4qB (van Geel *et al.*, 2002). Curiously, it was found that FSHD is associated only with D4Z4 deletion in the 4qA allele, while shortened D4Z4 repeats in the 4qB allele do not appear to result in the disease (Lemmers *et al.*, 2002). Further investigation of haplotype-specific single-nucleotide polymorphisms (SNPs) at this 4q distal locus identified at least 17 variants. Interestingly, only D4Z4 contraction in 4qA161 (and in two other rare variants: 4qA159 and 4qA168) is associated with FSHD (Figure 25.2B). Thus, these alleles are "permissive" 4q variants (Lemmers *et al.*, 2010). Significantly, this study showed that these variants specifically contain a poly(A) signal after the last copy of D4Z4, allowing generation of functional *DUX4* mRNA in C2C12 cells. A *DUX4* transcript from the most distal D4Z4 repeat was indeed found in differentiating FSHD myoblasts (Dixit *et al.*, 2007). Thus, the expression of *DUX4* appears to correlate specifically with the development of the disease (Figure 25.2B).

Further analysis of the *DUX4* transcripts, however, revealed that the correlation between *DUX4* expression and FSHD is complex. At least three different splicing variants are produced from the *DUX4* gene region in the last D4Z4 copy, some of which are expressed even in normal cells (Snider *et al.*, 2009, 2010). There is a tendency for the full-length transcript to be expressed in FSHD both in vivo in muscle biopsies and in vitro in differentiating myoblasts (Snider *et al.*, 2010). However, not all the FSHD samples show this transcript. Analysis of its protein product indicated that only ~0.1% of FSHD muscle cells express a relatively high level of the full-length DUX4 protein, and that these cells appear to undergo apoptosis (Snider *et al.*, 2010). While this may explain the inconsistent detection of DUX4 in the patient samples, the reason for this low incidence and how it contributes to FSHD pathogenesis remains an open question.

It is currently unclear to what extent the abnormal expression of *FRG1* and/or *DUX4* contributes to FSHD, and whether an additional factor(s) is involved. Nevertheless, it is interesting to note that, in both cases, it is the failure to appropriately silence these genes that appears to be linked to the disease.

25.3   Epigenetic changes of D4Z4 repeats in FSHD

25.3.1   DNA hypomethylation of D4Z4 is associated with, but is not sufficient for, FSHD development

Based on the fact that many repeat sequences are heterochromatic, and that the D4Z4 repeat contains the repressor complex binding site (Figure.25.2A), it was postulated early on that D4Z4 repeats may have transcriptionally silenced heterochromatic structure in normal cells. Indeed, DNA hypomethylation was found at the first proximal D4Z4 repeat of the contracted 4q allele in 4q-linked FSHD patients (van Overveld *et al.*, 2003). This DNA hypomethylation can be observed in both muscle and peripheral blood cells, which does not change during aging, indicating that the DNA methylation pattern at D4Z4 is established at an early developmental stage and is stably maintained. Importantly, D4Z4 DNA hypomethylation was also found in phenotypic FSHD, but not in other muscular dystrophies (DMD/BMD, myotonic dystrophy [DM], oculopharyngeal muscular dystrophy [OPMD], and LGMD2) (van Overveld *et al.*, 2003). The degree of DNA hypomethylation in the proximal D4Z4 repeat is inversely correlated with the number of remaining D4Z4 repeats, revealing the close relationship between the level of D4Z4 DNA methylation and FSHD symptom severity (van Overveld *et al.*, 2005).

DNA hypomethylation was also confirmed in the internal D4Z4 repeats of the contracted 4q allele in 4q-linked FSHD and in both 4q and 10q alleles in phenotypic FSHD patients (de Greef *et al.*, 2009). Surprisingly, however, the unaffected individuals with either contracted 10q D4Z4 or contracted non-permissive 4q haplotypes also had DNA hypomethylation in the internal D4Z4 repeats (de Greef *et al.*, 2009). Furthermore, severe DNA hypomethylation at 4q D4Z4 was observed in the clinically unrelated "immunodeficiency, centromere instability, and facial anomaly" (ICF) syndrome, which is caused by a mutation of DNA methyltransferase 3B (DNMT3B) (van Overveld *et al.*, 2003). No mutation of DNMTs has been identified in phenotypic FSHD (de Greef *et al.*, 2007). Thus, these findings indicate that DNA hypomethylation is not sufficient for the development of the disease.

25.3.2   There are no significant epigenetic changes in D4Z4 proximal genes

Chromatin immunoprecipitation (ChIP) assays on the genomic region proximal to D4Z4 showed that the *FRG1* and *ANT1* genes are hyperacetylated on histone H4 in both control and FSHD cells (Jiang *et al.*, 2003). Cytological analyses by immunofluorescent staining and fluorescence in situ hybridization (FISH) (immuno-FISH) using probes covering the region proximal to D4Z4 revealed no significant association of the 4q35 region with histone modification marks characteristic of either heterochromatin or transcriptionally active euchromatin (H3K9me3 and its binding protein HP1α, or H4K8Ac and H3K4me, respectively) in both control and FSHD myoblasts (Yang *et al.*, 2004). Taken together, there appears to be no significant epigenetic alteration of the neighboring genes in FSHD, which argues against the PEV model (Figure 25.2A).

25.3.3   The loss of H3K9me3 at D4Z4 is a marker for FSHD

25.3.3.1   PCR primer specificity for ChIP-PCR

Analysis of the chromatin structure of 4q D4Z4 itself was technically difficult due to the presence of similar sequences on other chromosomes (Tam *et al.*, 2004). We recently developed a ChIP-PCR assay approach that does allow analysis of D4Z4 chromatin specifically at chromosome 4q as well as at 10q (Zeng *et al.*, 2009). We found that it is essential not only to demonstrate that the PCR primers can amplify 4q D4Z4, but also to confirm that they do not amplify products from any of the other chromosomes. We used a DNA mapping panel to test the specificity of a collection of PCR primers against all 24 chromosomes individually. Interestingly, while primers against other parts of the D4Z4 repeat amplify many other sequences, presumably D4Z4 repeat-like, on other chromosomes, primers that bind to the 5'-end of the *DUX4* gene region are highly specific for 4q and 10q D4Z4. These primers were used in our subsequent studies (Zeng *et al.*, 2009) (Figure 25.3A).

25.3.3.2 4q and 10q D4Z4 repeats contain similar histone modification marks representing both heterochromatic and euchromatic domains

Unlike what was predicted from both the repressor model and DNA methylation, we found that normal D4Z4 chromatin contains not only transcriptionally repressive histone modifications (H3K9me3 and H3K27me3) but also permissive marks (H3K4me2 and H3Ac). Using sequential ChIP analyses, we confirmed that heterochromatin and euchromatin in D4Z4 are present in two distinct domains (Zeng *et al.*, 2009). Using PCR primers specific for the most proximal D4Z4 repeat, we found that this first repeat lacks heterochromatic marks, suggesting that the euchromatic domain may lie proximal to the heterochromatic domain, which is consistent with the chromatin landscape found in the neighboring region (Jiang *et al.*, 2003; Yang *et al.*, 2004). Although it is currently unclear how these two chromatin domains are divided and arranged within the D4Z4 repeat segment, the results indicate that D4Z4 repeats are not uniformly heterochromatic.

Although our PCR primers amplified both 4q and 10q D4Z4 repeat regions due to high sequence similarity, the corresponding PCR products included SNPs that can distinguish 4q- and 10q-derived D4Z4 (Figure 25.3B). Cloning and sequencing of the ChIP-PCR products revealed that both 4q and 10q SNPs can be found, indicating that chromatin landscapes within 4q and 10q D4Z4 repeats are similar. Interestingly, the ratio of the appearance of 4q and 10q SNPs roughly correlates

Figure 25.3 ChIP PCR primers and nucleotide polymorphisms used to distinguish between 4q and 10q D4Z4 and D4Z4 homologs on other chromosomes (Zeng *et al.*, 2009). (A) PCR analysis on a DNA mapping panel consisting of each individual human chromosome's genomic DNA. 4qHox and QPCR primer pairs show PCR signal specifically from chromosomes 4 and 10 D4Z4. The non-specific primer pair binds to a different subdomain within D4Z4. In addition to chromosomes 4 and 10, this pair also amplified D4Z4-like repeat sequences on several other chromosomes. Therefore, this primer set was not used for the experiments. (B) Sequence polymorphisms between 4q and 10q D4Z4. These polymorphisms are present in the PCR products with the QPCR primer pairs. Sequencing of the products reveals from which chromosome they are derived. The positions (nt) of the polymorphisms are designated according to sequence AF117653 in the GenBank/EMBL Nucleotide Sequence Database.

with the ratio of the D4Z4 copy number in each chromosome, further supporting the notion that 4q and 10q D4Z4 are epigenetically very similar (Zeng *et al.*, 2009).

25.3.3.3   H3K9me3 is specifically lost in both 4q-linked and phenotypic FSHD patient cells

Significantly, we found that H3K9me3 is specifically lost from D4Z4 in both 4q-linked and phenotypic FSHD patient cells (Figure 25.4A). There is not a generalized reduction of H3K9me3 in the cell, and H3K9me3 is intact at other repeat sequences. The loss of H3K9me3 can be detected even in FSHD patient lymphoblasts, indicating that this is a preexisting feature in FSHD patients and not an epiphenomenon associated with the dystrophic phenotype of muscles. This change is unique to FSHD, but not other muscular dystrophies, similar to DNA hypomethylation (see Section 25.3.1). Unlike DNA hypomethylation, however, ICF patient cells (with severe DNA hypomethylation at D4Z4) still retain H3K9me3. Thus, the loss of H3K9me3 in FSHD is not a mere consequence of DNA hypomethylation. Taken together, the loss of H3K9me3 at D4Z4 may potentially serve as a useful diagnostic marker for FSHD.

25.3.3.4   H3K9me3 is diminished at both 4q and 10q D4Z4

Based on the SNP analysis, we found that the loss of H3K9me3 occurs on both alleles of 4q D4Z4, as well as on the 10q D4Z4 alleles, in FSHD patient cells (Figure 25.4A). This provided the first evidence that the epigenetic abnormality associated with the contraction of D4Z4 in one allele spreads to other D4Z4 clusters. The mechanism of this process is currently unclear. Chromosome pairing between 4q and 10q subtelomeric regions in FSHD was reported previously (Stout *et al.*, 1999; Pirozhkova *et al.*, 2008). It is possible, therefore, that somatic pairing between the two alleles of 4q D4Z4, as well as 4q and 10q D4Z4, may regulate H3K9me3 dynamics at D4Z4, reminiscent of transvection in *Drosophila* (Duncan, 2002). Interestingly, while DNA hypomethylation was observed at both 4q and 10q D4Z4, similar to the loss of H3K9me3, in phenotypic FSHD, it was restricted to the contracted 4q allele in 4q-linked FSHD (de Greef *et al.*, 2009). Thus, the mechanism of loss of H3K9me3 and its relationship to DNA methylation may be distinct in 4q-linked and phenotypic FSHD. We found that SUV39H1 is the HMTase responsible for D4Z4 H3K9me3. No mutation of SUV39H1, however, has been identified in FSHD (de Greef *et al.*, 2007).

25.3.3.5   HP1γ and cohesin bound at D4Z4 are the cell-type-specific downstream effectors of H3K9me3, and are also lost in FSHD

The methylated histone H3K9 residue in heterochromatin directly recruits HP1, which plays an important role in transcriptional silencing. Swi6, an HP1 homolog in *Schizosaccharomyces pombe*, was reported to recruit the sister chromatid cohesion complex "cohesin" to the pericentromeric heterochromatin where it mediates centromeric sister chromatid cohesion (for review, see Zeng *et al.*, 2010). While cohesin was originally discovered to be essential for proper chromosome segregation in mitosis, recent studies have demonstrated that cohesin has

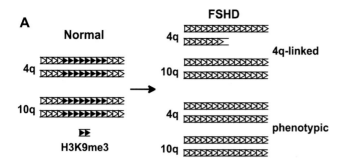

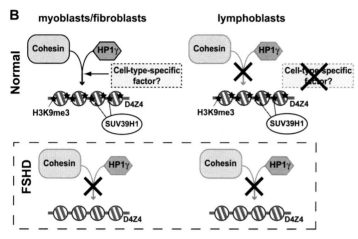

Figure 25.4 Chromatin organization of the normal 4q/10q D4Z4 regions and their changes in FSHD (Zeng *et al.*, 2009). (A) Coordinated loss of H3K9me3 on 4q and 10q D4Z4 in 4q-linked and phenotypic FSHD. H3K9me3, represented by black solid triangles, is clustered in the D4Z4 subdomains in normal cells, but is lost in both 4q-linked and phenotypic FSHD. (B) Cell-type-specific D4Z4 chromatin organization and its loss in FSHD. H3K9me3 on normal D4Z4 is established by SUV39H1. HP1γ and cohesin, in an interdependent way, are recruited to D4Z4 H3K9me3 in certain cell types, including myoblasts and fibroblasts. In normal lymphoblasts, in spite of the presence of H3K9me3, HP1γ and cohesin do not bind to D4Z4, implying a putative cell-type-specific factor(s) necessary for their recruitment. The loss of D4Z4 H3K9me3 in FSHD causes the abolishment of HP1γ/cohesin binding in myoblasts, which may be responsible for the tissue-specific manifestations of FSHD.

additional roles in long-distance chromatin–chromatin interactions important for developmental gene regulation (Nasmyth and Haering, 2009). We found that HP1γ, one of the three HP1 variants in mammalian cells, and cohesin are recruited to D4Z4 in an H3K9me3-dependent manner in control myoblasts and fibroblasts, while their binding is lost in the corresponding FSHD cells (Zeng *et al.*, 2009). Unlike the pericentromeric heterochromatin in *S. pombe*, HP1γ and cohesin binding to D4Z4 is interdependent and cell type-specific. Despite the presence of H3K9me3 at D4Z4, HP1γ and cohesin are absent at D4Z4 in normal lymphoblasts. This suggests that a cell-type-specific factor(s) or modification is required for HP1γ/cohesin loading to D4Z4, which may be important for cell-type-specific chromatin organization and explain the tissue-specific manifestations of the disease (Figure 25.4B).

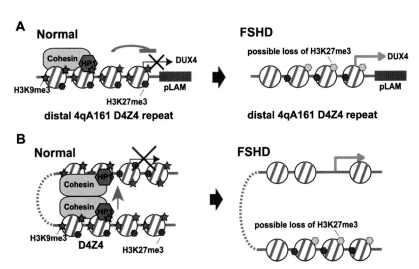

Figure 25.5 Putative models for the role of epigenetic change at D4Z4 in FSHD pathogenesis. (A) The effect on *DUX4* expression. Heterochromatin in normal D4Z4 inhibits the transcription of *DUX4*. In FSHD, accompanied by the disruption of this silencing structure, the transcription of *DUX4* ORF in D4Z4 repeats is activated. The loss of H3K9me3 and possibly H3K27me3 also favors the utilization of the poly(A) signal in the pLAM sequence of the 4qA161 variant through a splicing mechanism, which helps stabilize the *DUX4* mRNA transcribed from the distal D4Z4 repeat. The DUX4 protein may function in FSHD pathogenesis. (B) Looping model. HP1γ and cohesin may function in the physical interactions of the heterochromatic D4Z4 region with other genomic regions, which results in the spreading of the repressive effect to target genes in normal cells. In FSHD, the loss of H3K9me3 (and perhaps H3K27me3), HP1γ, and cohesin on D4Z4 leads to the disruption of these physical interactions and abnormal activation of these genes. See plate section for color version.

25.3.3.6   The loss of H3K9me3–HP1γ–cohesin and development of disease

We demonstrated that the loss of H3K9me3 and HP1γ/cohesin binding to D4Z4 are specific epigenetic changes marking both 4q-linked and phenotypic FSHD. However, exactly how these epigenetic changes relate to disease development remains unclear. Two possible models, which are not mutually exclusive, are (1) the loss of the transcriptionally repressive mark may lead to the abnormal expression of *DUX4* at D4Z4 (Figure 25.5A), and (2) loss of cohesin and HP1γ may impact long-distance chromatin:chromatin interactions involving D4Z4 that alter distant target gene expression (Figure 25.5B). In the first scenario, despite the significant loss of H3K9me3, the number of FSHD muscle cells derived from patient biopsies that actually express *DUX4* is no more than 0.1% (Snider *et al.*, 2010). Furthermore, there is no significant reduction of H3K27me3, which coincides with H3K9me3, at D4Z4, and thus the loss of H3K9me3 does not necessarily mean a total loss of heterochromatin. It is possible that stochastic and simultaneous loss of both marks may occur at a low frequency, leading to *DUX4* over-expression in a small number of cells (Figure 25.5A).

The second scenario is supported by recent findings that cohesin participates in cell-type- and differentiation-stage-specific long-distance chromatin interactions to regulate developmental gene expression (Hadjur *et al.*, 2009; Kagey *et al.*, 2010). It is possible that heterochromatic D4Z4 spreads its silencing effect to distant target genes through long-distance interactions, and the loss of H3K9me3, cohesin, and HP1γ may result in abnormal expression of these genes resulting in the dystrophic phenotype (Figure 25.5B). However, evidence for such long-distance interactions involving D4Z4 is still very limited (see below).

Interestingly, genes involved in normal myogenesis are critically regulated by changes of epigenetic marks similar to what was observed at D4Z4. In satellite cells, the promoters of muscle-specific genes (e.g., *MyoD*) are silenced by associating with transcriptional repressors, such as YY1, polycomb repressive complex 2 (PRC2), and SUV39H1, which together confer a repressive chromatin environment (i.e., H3K27 and K9 methylation, respectively) (Perdiguero *et al.*, 2009). Thus, although further study is needed to understand the cause and effect of D4Z4 chromatin alterations, these changes at D4Z4 may nonetheless impact epigenetic regulation of myogenesis and muscle regeneration.

25.3.4 Alteration of long-distance chromatin interactions surrounding D4Z4

Several recent studies investigated higher-order chromatin organization surrounding the 4q D4Z4 region, which may affect expression of proximal genes. The CCCTC-binding factor (CTCF) and lamin A/C were shown to bind to D4Z4 to mediate its function as an insulator (Ottaviani *et al.*, 2009). Since this activity is lost upon multimerization of D4Z4, it was proposed that the contracted D4Z4 may gain insulator function in 4q-linked FSHD. However, since there is no apparent spreading of heterochromatin from D4Z4 to the neighboring genes, it is currently unclear how this induced CTCF binding affects gene expression.

A scaffold/nuclear matrix attachment region (S/MAR) was identified proximal to D4Z4 (Petrov *et al.*, 2006). This S/MAR, which normally separates *FRG1/FRG2* and D4Z4 into two DNA loops, detaches from the nuclear matrix and keeps *FRG1* and *FRG2* in the same loop with D4Z4 in the contracted allele in 4q-linked FSHD (Petrov *et al.*, 2006). It was found that the region immediately distal to D4Z4 contains a strong enhancer activity that appears to cause overexpression of the *FRG1/FRG2* genes (Petrov *et al.*, 2008; Pirozhkova *et al.*, 2008). Interestingly, however, direct interaction between D4Z4 repeats and the *FRG1* promoter was also reported, which appears to be important for proper silencing of *FRG1* in myoblasts (Bodega *et al.*, 2009). While the interaction is weakened during normal myotube differentiation correlating with the loss of repressive histone modification and upregulation of *FRG1*, the interaction apparently is weaker in FSHD myoblasts resulting in the premature expression of *FRG1*.

It remains to be determined how these different results can be integrated into one picture and how they relate to the loss of H3K9me3/HP1γ/cohesin within the D4Z4 region in FSHD. Furthermore, the significance of these findings in those FSHD cases with no apparent *FRG1* overexpression is unclear.

25.4 Conclusion and future directions

Epigenetic chromatin regulation is crucial for myogenesis and muscle regeneration. In addition to FSHD described above, specific epigenetic changes were also shown to be associated with other muscular dystrophies, including myotonic dystrophy type 1 (DM1) and DMD (Cho *et al.*, 2005; Colussi *et al.*, 2009). Thus, development of therapies that may control these epigenetic chromatin altera-tions, both to stimulate muscle regeneration and to alleviate the pathological changes, is an important direction of research. Exposure of undifferentiated myoblasts to the histone deacetylase (HDAC) inhibitors trichostatin A (TSA) or valproic acid (VPA) significantly enhanced the formation of myotubes, in part by increasing histone acetylation in the enhancers of *MyoD* and other muscle-related genes (Iezzi *et al.*, 2002). In DMD and α-sarcoglycan muscular dystrophy (α-SGD) mouse models, TSA treatment increased the myofiber size by inducing the expres-sion of follistatin, an antagonist of the dystrophy marker myostatin, in satellite cells (Minetti *et al.*, 2006). Thus, therapeutic epigenetic manipulation may be of clinical value in the treatment of muscular dystrophies.

Although the genetic change (D4Z4 contraction) in FSHD has been known for two decades, the disease's pathogenesis remains enigmatic and there have not been any mechanism-based therapeutic strategies tailored specifically for FSHD patients. Understanding the specific epigenetic program that is compromised in FSHD, therefore, is essential for the design of future therapeutic approaches.

ACKNOWLEDGMENTS

The work in the Yokomori laboratory is supported in part by National Institutes of Health AR058548, Muscular Dystrophy Association (MDA4026), and the David and Helen Younger Research Fellowship from the FSH Society (FSHS-DHY-001) to KY, and the FSH Society Helen Younger and David Younger Fellowship Research Grant FSHS-DHY-002 to WZ. We apologize to the researchers whose work we could not cite in this chapter due to space limitations.

REFERENCES

Bastress, K. L., Stajich, J. M., Speer, M. C., and Gilbert, J. R. (2005). The genes encoding for D4Z4 binding proteins HMGB2, YY1, NCL, and MYOD1 are excluded as candidate genes for FSHD1B. *Neuromuscular Disorders*, **15**, 316–320.

Bodega, B., Ramirez, G. D., Grasser, F., *et al.* (2009). Remodeling of the chromatin structure of the facioscapulohumeral muscular dystrophy (FSHD) locus and upregulation of FSHD-related gene 1 (FRG1) expression during human myogenic differentiation. *BMC Biology*, **7**, 41.

Cho, D. H., Thienes, C. P., Mahoney, S. E., *et al.* (2005). Antisense transcription and hetero-chromatin at the DM1 CTG repeats are constrained by CTCF. *Molecular Cell*, **20**, 483–489.

Colussi, C., Gurtner, A., Rosati, J., *et al.* (2009). Nitric oxide deficiency determines global chromatin changes in Duchenne muscular dystrophy. *Federation of the American Society for Experimental Biology Journal*, **23**, 2131–2141.

de Greef, J. C., Wohlgemuth, M., Chan, O. A., *et al.*, (2007). Hypomethylation is restricted to the D4Z4 repeat array in phenotypic FSHD. *Neurology*, **69**, 1018–1026.

de Greef, J. C., Lemmers, R. J., van Engelen, B. G., *et al.* (2009). Common epigenetic changes of D4Z4 in contraction-dependent and contraction-independent FSHD. *Human Mutation*, **30**, 1449–1459.

de Greef, J. C., Lemmers, R. J., Camano, P., *et al.* (2010). Clinical features of facioscapulo-humeral muscular dystrophy 2. *Neurology*, **75**, 1548–1554.

Dixit, M., Ansseau, E., Tassin, A., *et al.* (2007). *DUX4*, a candidate gene of facioscapulohum-eral muscular dystrophy, encodes a transcriptional activator of PITX1. *Proceedings of the National Academy of Sciences USA*, **104**, 18 157–18 162.

Duncan, I. W. (2002). Transvection effects in *Drosophila*. *Annual Reviews of Genetics*, **36**, 521–556.

Fitzsimons, R. B., Gurwin, E. B., and Bird, A. C. (1987). Retinal vascular abnormalities in facioscapulohumeral muscular dystrophy: a general association with genetic and therapeutic implications. *Brain*, **110**, 631–648.

Gabellini, D., Green, M. R., and Tupler, R. (2002). Inappropriate gene activation in FSHD: a repressor complex binds a chromosomal repeat deleted in dystrophic muscle. *Cell*, **110**, 339–348.

Gabellini, D., D'Antona, G., Moggio, M., *et al.* (2006). Facioscapulohumeral muscular dys-trophy in mice overexpressing *FRG1*. *Nature*, **439**, 973–977.

Gabriels, J., Beckers, M. C., Ding, H., *et al.* (1999). Nucleotide sequence of the partially deleted D4Z4 locus in a patient with FSHD identifies a putative gene within each 3.3 kb element. *Gene*, **236**, 25–32.

Hadjur, S., Williams, L. M., Ryan, N. K., *et al.* (2009). Cohesins form chromosomal *cis*-interactions at the developmentally regulated *IFNG* locus. *Nature*, **460**, 410–413.

Hanel, M. L., Wuebbles, R. D., and Jones, P. L. (2009). Muscular dystrophy candidate gene *FRG1* is critical for muscle development. *Developmental Dynamics*, **238**, 1502–1512.

Iezzi, S., Cossu, G., Nervi, C., Sartorelli, V., and Puri, P. L. (2002). Stage-specific modulation of skeletal myogenesis by inhibitors of nuclear deacetylases. *Proceedings of the National Academy of Sciences USA*, **99**, 7757–7762.

Jiang, G., Yang, F., van Overveld, P. G., *et al.* (2003). Testing the position-effect variegation hypothesis for facioscapulohumeral muscular dystrophy by analysis of histone mod-ification and gene expression in subtelomeric 4q. *Human Molecular Genetics*, **12**, 2909–2921.

Kagey, M. H., Newman, J. J., Bilodeau, S., *et al.* (2010). Mediator and cohesin connect gene expression and chromatin architecture. *Nature*, **467**, 430–435.

Klooster, R., Straasheijm, K., Shah, B., *et al.* (2009). Comprehensive expression analysis of FSHD candidate genes at the mRNA and protein level. *European Journal of Human Genetics*, **17**, 1615–1624.

Kowaljow, V., Marcowycz, A., Ansseau, E., *et al.* (2007). The *DUX4* gene at the FSHD1A locus encodes a pro-apoptotic protein. *Neuromuscular Disorders*, **17**, 611–623.

Lemmers, R. J., de Kievit, P., Sandkuijl, L., *et al.* (2002). Facioscapulohumeral muscular dystrophy is uniquely associated with one of the two variants of the 4q subtelomere. *Nature Genetics*, **32**, 235–236.

Lemmers, R. J., van der Vliet, P. J., Klooster, R., *et al.* (2010). A unifying genetic model for facioscapulohumeral muscular dystrophy. *Science*, **329**, 1650–1653.

Masny, P. S., Chan, O. Y., de Greef, J. C., *et al.* (2010). Analysis of allele-specific RNA tran-scription in FSHD by RNA–DNA FISH in single myonuclei. *European Journal of Human Genetics*, **18**, 448–456.

Minetti, G. C., Colussi, C., Adami, R., *et al.* (2006). Functional and morphological recovery of dystrophic muscles in mice treated with deacetylase inhibitors. *Nature Medicine*, **12**, 1147–1150.

Nasmyth, K. and Haering, C. H. (2009). Cohesin: its roles and mechanisms. *Annual Reviews of Genetics*, **43**, 525–558.

Osborne, R. J., Welle, S., Venance, S. L., Thornton, C. A., and Tawil, R. (2007). Expression profile of FSHD supports a link between retinal vasculopathy and muscular dystrophy. *Neurology*, **68**, 569–577.

Ottaviani, A., Rival-Gervier, S., Boussouar, A., *et al.* (2009). The D4Z4 macrosatellite repeat acts as a CTCF and A-type lamins-dependent insulator in facio-scapulo-humeral dys-trophy. *PLoS Genetics*, **5**, e1000394.

Padberg, G. W., Brouwer, O. F., de Keizer, R. J., *et al.* (1995). On the significance of retinal vascular disease and hearing loss in facioscapulohumeral muscular dystrophy. *Muscle Nerve*, **2**, S73–S80.

Perdiguero, E., Sousa-Victor, P., Ballestar, E., and Munoz-Canoves, P. (2009). Epigenetic regulation of myogenesis. *Epigenetics*, **4**, 541–550.

Petrov, A., Pirozhkova, I., Carnac, G., *et al.* (2006). Chromatin loop domain organization within the 4q35 locus in facioscapulohumeral dystrophy patients versus normal human myoblasts. *Proceedings of the National Academy of Sciences USA*, **103**, 6982–6987.

Petrov, A., Allinne, J., Pirozhkova, I., *et al.* (2008). A nuclear matrix attachment site in the 4q35 locus has an enhancer-blocking activity in vivo: implications for the facio-scapulohumeral dystrophy. *Genome Research*, **18**, 39–45.

Pirozhkova, I., Petrov, A., Dmitriev, P., *et al.* (2008). A functional role for 4qA/B in the structural rearrangement of the 4q35 region and in the regulation of *FRG1* and *ANT1* in facioscapulohumeral dystrophy. *PLoS One*, **3**, e3389.

Rocha, C. T. and Hoffman, E. P. (2010). Limb-girdle and congenital muscular dystrophies: current diagnostics, management, and emerging technologies. *Current Neurology and Neuroscience Reports*, **10**, 267–276.

Snider, L., Asawachaicharn, A., Tyler, A. E., *et al.* (2009). RNA transcripts, miRNA-sized fragments and proteins produced from D4Z4 units: new candidates for the pathophysiology of facioscapulohumeral dystrophy. *Human Molecular Genetics*, **18**, 2414–2430.

Snider, L., Geng, L. N., Lemmers, R. J., *et al.* (2010). Facioscapulohumeral dystrophy: incomplete suppression of a retrotransposed gene. *PLoS Genetics*, **6**, e1001181.

Stout, K., van der Maarel, S., Frants, R. R., *et al.* (1999). Somatic pairing between subtelomeric chromosome regions: implications for human genetic disease? *Chromosome Research*, **7**, 323–329.

Tam, R., Smith, K. P., and Lawrence, J. B. (2004). The 4q subtelomere harboring the FSHD locus is specifically anchored with peripheral heterochromatin unlike most human telomeres. *Journal of Cell Biology*, **167**, 269–279.

Tawil, R. (2008). Facioscapulohumeral muscular dystrophy. *Neurotherapeutics*, **5**, 601–606.

Tupler, R., Berardinelli, A., Barbierato, L., *et al.* (1996). Monosomy of distal 4q does not cause facioscapulohumeral muscular dystrophy. *Journal of Medical Genetics*, **33**, 366–370.

van der Maarel, S. M., Frants, R. R., and Padberg, G. W. (2007). Facioscapulohumeral muscular dystrophy. *Biochimica et Biophysica Acta*, **1772**, 186–194.

van Deutekom, J. C., Wijmenga, C., van Tienhoven, E. A., *et al.* (1993). FSHD associated DNA rearrangements are due to deletions of integral copies of a 3.2 kb tandemly repeated unit. *Human Molecular Genetics*, **2**, 2037–2042.

van Geel, M., Dickson, M. C., Beck, A. F., *et al.* (2002). Genomic analysis of human chromosome 10q and 4q telomeres suggests a common origin. *Genomics*, **79**, 210–217.

van Overveld, P. G., Lemmers, R. J., Sandkuijl, L. A., *et al.* (2003). Hypomethylation of D4Z4 in 4q-linked and non-4q-linked facioscapulohumeral muscular dystrophy. *Nature Genetics*, **35**, 315–317.

van Overveld, P. G., Enthoven, L., Ricci, E., *et al.* (2005). Variable hypomethylation of D4Z4 in facioscapulohumeral muscular dystrophy. *Annals of Neurology*, **58**, 569–576.

Wijmenga, C., Frants, R. R., Brouwer, O. F., *et al.*, (1990). Location of facioscapulohumeral muscular dystrophy gene on chromosome 4. *Lancet*, **336**, 651–653.

Winokur, S. T., Chen, Y. W., Masny, P. S., *et al.* (2003). Expression profiling of FSHD muscle supports a defect in specific stages of myogenic differentiation. *Human Molecular Genetics*, **12**, 2895–2907.

Wuebbles, R. D., Hanel, M. L., and Jones, P. L. (2009). FSHD region gene 1 (*FRG1*) is crucial for angiogenesis linking FRG1 to facioscapulohumeral muscular dystrophy-associated vasculopathy. *Disorders and Model Mechanisms*, **2**, 267–274.

Yang, F., Shao, C., Vedanarayanan, V., and Ehrlich, M. (2004). Cytogenetic and immuno-FISH analysis of the 4q subtelomeric region, which is associated with facioscapulo-humeral muscular dystrophy. *Chromosoma*, **112**, 350–359.

Zeng, W., de Greef, J.C., Chen, Y.Y., *et al.* (2009). Specific loss of histone H3 lysine 9 trimethylation and HP1gamma/cohesin binding at D4Z4 repeats is associated with facioscapulohumeral dystrophy (FSHD). *PLoS Genetics*, **5**, e1000559.

Zeng, W., Ball, A.R., Jr., and Yokomori, K. (2010). HP1: heterochromatin binding proteins working the genome. *Epigenetics*, **5**, 287–292.

26 Modulating histone acetylation with inhibitors and activators

B. Ruthrotha Selvi, D. V. Mohankrishna, and Tapas K. Kundu*

26.1   Introduction

Protein lysine acetylation has emerged as an important regulator of gene expression. The acetylation of both histones and non-histone proteins modulate essential cellular processes ranging from transcription, DNA repair, replication, the cell cycle, cytoskeletal organization, inflammatory responses, as well as metabolism. Recent evidence implicates this modification in several pathophysiological conditions including cancer, inflammatory diseases, cardiovascular disorders, neurodegenerative states, and infectious diseases. Modulation of the levels of acetylation in these disease states is being explored as an alternative to conventional therapeutic strategies. The identification of several histone acetylation inhibitors as well as few activators in the past decade has indeed strengthened this option. Various preclinical investigations using these modulators have provided an impetus to the field of histone acetylation modulation as a possible therapeutic strategy. The role of histone acetylation in physiology and a few pathophysiological states will be described, thus setting the stage for the modulators of histone acetylation both as biological probes and therapeutic tools.

26.2   Histone acetylation

Protein lysine acetylation is one of the important covalent modifications, which was, incidentally, first observed on the histones and subsequently after about three decades also identified on non-histone proteins. Histone acetylation and its correlation with transcriptional activation were reported by Vincent Allfrey in 1964 (Allfrey et al., 1964). However, the resurrection of the field had to wait for

---

* Author to whom correspondence should be addressed.

Epigenomics: From Chromatin Biology to Therapeutics, ed. K. Appasani. Published by Cambridge University Press. © Cambridge University Press 2012.

30 years until the first histone acetyltransferase (HAT) GCN5 was identified from a *Tetrahymena* by David Allis's group in the mid-1990s (Brownell and Allis, 1995). This was followed by the identification of several acetyltransferases most of which are positive regulators of transcription. All these enzymes are also capable of acetylating different non-histone substrates with physiological consequences. The process of histone acetylation is reversible, in which the forward reaction is catalyzed by the HATs/KATs and the reverse reaction by the histone deacetylases (HDACs/KDACs). Nucleosomal histone acetylation has been considered as an indispensable component of transcriptionally active chromatin. Subsequently, specific acetylation marks have been correlated to other physiological events such as repair and replication (Shahbazian and Grunstein, 2007). Non-histone protein acetylation opened up an entirely new concept regarding the role of this modification in non-nuclear processes. The status of histone modifications is often altered in the pathophysiological conditions, and hence these marks are also being considered as diagnostic markers for several diseases including cancer (Seligson *et al.*, 2009). Thus modulators of histone acetylation are the most sought-after resources to decipher, alter, and decode histone acetylation in physiology and pathophysiology. This chapter highlights the role of these modulators as biological tools and therapeutic measures too.

26.2.1 The machineries and mechanism of acetylation

The HATs and HDACs are found embedded in large multiprotein complexes near euchromatic regions of the chromatin. The HATs/KATs are divided into two types on the basis of their cellular localization and are numbered from KAT1 to KAT13D (Allis *et al.*, 2007). The most abundantly present and the most studied HAT group is the nuclear localized type A HAT/KAT, which is further subclassified into five main classes based on their functional characteristics. These are: (1) the first-identified HAT GCN5 family which is represented by GCN5 and PCAF; (2) the master regulators p300/CBP family consisting of p300, CBP, and Rtt109; (3) the developmental regulators the MYST family comprising MOZ, MOF, Sas1, and TIP60; (4) transcription factors possessing HAT activity such as ATF2, TAF1, TAFII250, and TFIIIC; and finally (5) the receptor coactivators such as SRC and ACTR which also exhibit acetyltransferase ability. These enzymes exist in distinct complexes with specific functional attributes, listed in Table 26.1. The other type of HATs is cytoplasmic and is referred to as the type B HATs/KATs. The function of type B HATs such as Hat1 is restricted to the modification of newly synthesized histones as well as other cytosolic proteins (Qin and Parthun, 2002). The presence of HAT enzymes in complexes such as the Esa1-containing NuA4 and the Gcn5-containing SAGA complexes allows the histone acetyltransferase activity to be targeted to promoters of active genes as well as for assigning substrate specificity. It is important to note that all these modifications are reversibly mediated by the HDACs/KDACs. There are four classes of HDACs: class I (HDAC1–3, 8), II (HDAC4–7, 9–10), and IV (HDAC 11) which possess an active-site metal-dependent catalytic mechanism (Gregoretti *et al.*, 2004). Class III HDACs (or sirtuins) utilize a distinct nicotinamide adenine dinucleotide (NAD+) dependent

Table 26.1 Different HAT complexes

Species	HAT complex	Catalytic subunit	Histone substrate
Saccharomyces cerevisiae	SAGA	Gen5	H2B,H3/ H4
	SLIK	Gen5	H2B, H3/ H4
	ADA	Gen5	H3
	HAT-A2	Gen5	H3
	NuA4	Esa1	H2A, H4
	Pic.NuA4	Esa1	H2A,H4
	NuA3	Sas3	H3
	SAS	Sas2	H4
	HATB	Hat1	H2A,H4
	Elongator	Elp3	H3
	Hpa2	Hpa2	H3, H4
	TFIID	TAFII145	H3, H4
	Hat1	Hat1	Free H4
Drosophila melanogaster	SAGA	GCN5	H3
	ATAC	GCN5	H3, H4
	TIP60	TIP60	H2A, H4
	MSL	MOF	H4
	TFIID	dTFII230	H3, H4
Homo sapiens	PCAF	PCAF	H3, H4
	STAGA	GCN5L	H3, H4
	TFTC	GCN5L	H3, H4
	HBO1	TIP60	H2A, H4
	MOZ/ MORF	MOZ/ MORF	H3
	TFIID	TAFII250	H3, H4
	TFIIIC	p220, p110, p90	H2A, H3, H4
	Hat1	Hat1	Free H4

catalytic mechanism and are well conserved across species, with seven human homologs (Sirt1–7) (Frye, 2000).

The histone acetylation reaction is a bisubstrate reaction that involves the transfer of an acetyl group from the pseudo substrate, acetyl-coenzyme A (CoA) onto the ε-amino group of the lysine residue on protein substrates as depicted in Figure 26.1A. Presumably, all the members of a HAT family follow a similar kinetic mechanism which is an ordered sequential bi–bi kinetic mechanism wherein the subtle differences between families may affect substrate specificity but not the overall mechanism of catalysis (Smith and Denu, 2009). During the process of acetylation, the HATs catalyze the transfer of the acetyl group onto the substrate releasing the CoA with a free thiol followed by the release of the acetylated product. The deacetylation reaction involves the removal of the acetyl group from the modified substrates. The class III deacetylases, sirtuins, cleave the nicotinamide ribosyl bond of NAD^+ and transfer the acetyl group from proteins to CoA. The sirtuin deacetylation reaction generates nicotinamide, deacetylated protein, and a mixture of 2′ and 3′-O-acetyl-ADP-ribose (OAADPR) (Liou et al., 2005). Overall, the HAT–HDAC activities establish a balance, owing to their involvement in the

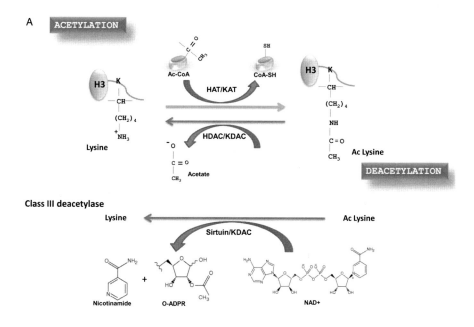

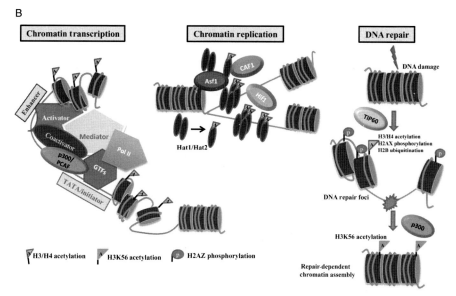

Figure 26.1 (A) Reversible histone acetylation by HATs and HDACs. (B) Role of histone acetylation in different chromatin-templated phenomena like transcription, replication, and DNA damage repair. Acetylation results in maintenance of an open form of chromatin that is essential for chromatin remodeling and transcription, replication, and DNA damage repair. It also helps to recruit specific factors depending on the context. Furthermore, the HATs/KATs acetylate several factors (for example GTFs, transcription factors, repair associated proteins, etc.) and modulate their activities. (C) Crosstalk of histone modifications: crosstalk involving acetylation, methylation, and phosphorylation sites within H3 and H4 in *cis*, and between H3 and H4 in *trans*. A, acetylation; me1/2/3, mono/ di/ tri methylation; P, phosphorylation ⟶ activation, inhibition. See plate section for color version.

C

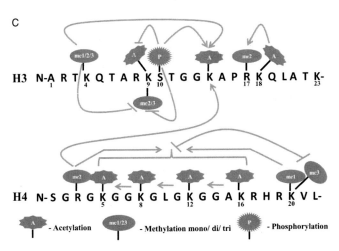

H3 N-A R T K Q T A R K S T G G K A P R K Q L A T K-

H4 N-S G R G K G G K G L G K G G A K R H R K V L-

- Acetylation   - Methylation mono/ di/ tri   - Phosphorylation

Figure 26.1 (cont.)

turnover of histone and transcription machinery regulation, and thereby act as one of the essential regulatory switches of gene expression. In addition to transcriptional regulation, HAT–HDAC balance is also postulated to modulate other chromatin-associated processes like replication, site-specific recombination, and DNA repair, thereby playing a major role in modulating overall cellular homeostasis.

Several theories exist with respect to the exact role of acetylation on chromatin dynamics and transcription which include both biophysical and biochemical evidence. Acetylation of histones results in deposition of a partial negative charge on the lysine residue which acts as a neutralizer of the strong positive charge of lysine to which DNA (with a negatively charged phosphate backbone) is electrostatically bound and thus loosens the histone–DNA contacts (Bode *et al.*, 1980; Bertrand *et al.*, 1984; Lee *et al.*, 1993). Acetylation of histones also leads to the recruitment of bromodomain-containing ATP-dependent chromatin remodeling complexes that read the acetylation mark and act as effectors by loosening the compaction. Additionally, the acetylation of transcription factors such as RelA, E2F, p53, and GATA1 further enhances their DNA-binding ability thereby facilitating efficient transcription. Contrarily, HDACs attenuate transcription process at a particular site by deacetylating both histones and the non-histone proteins.

26.2.2 Role of histone acetylation in cellular function

Acetylation has emerged as an important regulator of gene expression which can manifest either by promoter histone acetylation (which modulates the local chromatin environment) or by non-histone protein acetylation (facilitating functional efficiency in terms of signaling, protein–protein, and protein–DNA interactions) (Yang and Seto, 2008). Histone acetylation occurs at distinct lysine residues on all the core histones, H3, H4, H2A, and H2B. The promoter histone acetylation has been shown to be a mark of active transcription. P300-mediated acetylation is a prerequisite for the pre-initiation complex (PIC) formation. The role of acetylation in the process of transcription has been observed to be at multiple levels. The HAT complexes facilitate the activator recruitment in an ordered manner which is followed by the action of the remodeling complexes depending upon the cellular signals and the gene under transcription. These events lead to the loosening of the chromatin favoring histone replacement/displacement, coordinated with the action of histone chaperones. Thus, the HAT complexes mediated histone and non-histone protein acetylation act as signals, facilitating the recruitment and subsequent functioning of the transcription machinery. Although it was considered that histone acetylation had a predominant role in transcription initiation, it has now been shown to possess modulatory effects on transcription elongation (Li *et al.*, 2007) as well as in the histone deposition process, especially after the polymerase machineries pass through the chromatin template. Additionally, acetylation also regulates RNA Pol I transcription wherein p300/CREB-binding protein-associated factor (PCAF) acetyltransferase interacts with the transcription termination factor TTF1 and also acetylates TAFI68, the second largest subunit of the TATA box binding protein (TBP)-containing factor TIF-IB/SL1. This acetylation event has been shown to enhance TAFI68 binding to the rDNA promoter. Furthermore, PCAF-mediated acetylation also regulates the RNA Pol I transcription in an in vitro reconstituted system (Muth *et al.*, 2001).

Apart from transcription, histone acetylation also plays a crucial role in other processes such as replication and repair (Figure 26.1B). Chromatin in general forms a barrier to the passage of replication machinery through DNA. Hence chromatin modifications not only offer regulatory points for the process of replication but also act as marks with some functional significance. In fact, pre-replication complexes are marked by H4 acetylation. This event is majorly regulated by the deposition of histones by histone chaperones CAF-1, Rtt106, and Asf1 onto newly replicated DNA. Most often, the histone modification that acts as the recognition mark for this event is H3K56 acetylation (Fillingham and Greenblatt, 2008). The type B HAT is one of the major acetyltransferases that is functional during the process of replication. A recent study has very elegantly characterized the acetylated histone fraction from the cytoplasm and tracked the events beginning from the acetylation of histones by the Hat1 holoenzyme, followed by its transfer mediated by the histone chaperones, finally resulting in their karyopherin-mediated nuclear import (Campos *et al.*, 2010).

DNA repair requires the access of repair machineries to the damaged DNA foci. The earliest known histone modification in the process of repair is phosphorylation of the variant H2AX at the repair foci. However, recently the role of histone acetylation in DNA repair has also been recognized. The acetylation of H4K16 by TIP60 is a mark associated with DNA repair (Miyamoto *et al.*, 2008). H4K12 acetylation by Esa1 (Chang and Pillus, 2009) is implicated in DNA repair while p300-mediated H3K56 acetylation is a mark involved in DNA damage facilitating chromatin reassembly. It has also been shown to be required for resumption of cell proliferation after DNA repair (Tjeertes *et al.*, 2009). A few residues of histone H2A and H2B also get acetylated, mostly by p300. For example, H2BK15 acetylation, which is a mark of non-apoptotic cells, is decreased in apoptosis (Ajiro *et al.*, 2010). Several non-histone proteins including c-myc, p53, and cohesin, involved in the cell cycle and DNA damage response, are also regulated by acetylation.

Apart from acetylation other modifications such as phosphorylation, methylation, ubiquitination, etc., also aid in these events. The different marks together are postulated to form a 'histone code'. There are arguments both in favor as well as against the histone code, but it is so far clear that the overall effect on gene expression or physiology is not the result of a single modification but rather is the outcome of a set of marks which finally lead to the functional output (reviewed in Selvi *et al.*, 2010a).

26.2.3   Crosstalk of histone acetylation with other histone modifications

In general, in the context of transcription, promoter H3 and H4 tail acetylation leads to increased transcription. It has been observed that upon growth factor stimulation, there is H3S10 phosphorylation in the promoters of immediate early genes like *c-fos*. This is followed by removal of the H3K9 dimethylation mark. H3K9 dimethylation and H3K9 acetylation are mutually exclusive and result in functionally opposing outcomes. While H3K9 acetylation assists in a permissive chromatin for transcription, H3K9 dimethylation is a hallmark of transcriptionally silent/repressive chromatin. It has been observed that the presence of H3S10 phosphorylation makes it convenient for docking enzymes and acetylation of H3K14. In fact, the phosphorylated form of H3 is a better substrate for H3K14 acetylation by most HATs like PCAF and p300. Thus, a combination of H3S10 phosphorylation and H3K14 acetylation brings about activation of gene transcription. However, crosstalk between H3K9 acetylation and H3S10 phosphorylation is interesting. Though these marks coexist in factor-induced and activated gene promoters, largely, H3S10 phosphorylation antagonizes H3K9 acetylation by changing the structure of the substrate H3, thereby decreasing acetylation (reviewed in Selvi *et al.*, 2010a). The exact contexts during which these crosstalks function are as yet unclear.

Trimethylation at the H3K4 site is consistently associated with active gene expression. This modification is enriched at the 5′ ends of the coding region and as expected is antagonistic to H3K9 trimethylation which is a mark of pericentric heterochromatin. However, H3K4 trimethylation is also associated with hyperacetylated histones. When histone deacetylase inhibitors are added to cells H3K4 trimethylation always follows hyperacetylation (Nightingale *et al.*, 2007). It has further been noted that acetylated histone H3 is a better substrate for MLL-4 methyltransferase in vitro (also true for phosphoacetylated H3) and this could be the reason for the crosstalk. However, trimethylated H3K4 is also observed to be a better substrate for acetylation than non-methylated or trimethylated H3K9, for HATs like p300. H3K4 methylation by MLL is also a signal for MOF-mediated acetylation of H4K16 (Dou *et al.*, 2005). Crosstalk is also known between arginine methylation at H3R17 and acetylation at H3K18. Prior acetylation of H3K18 upon estrogen stimulation is essential for the recruitment of CARM1 and methylation of H3R17, with resulting expression of estrogen-responsive genes.

Cells that are deficient in CREB (CAMP response element binding) binding protein (CBP) acetyltransferase cannot recruit CARM1, mainly due to absence of H3K18 acetylation, thus inhibiting estrogen-dependent gene expression. Acetylation of histone H4 has also been implicated in gene expression. The N-terminal tail of H4 gets acetylated at H4K16, H4K12, H4K8, and H4K5 sites in a sequential order thus establishing another crosstalk in *cis*. H4K16 acetylation also affects higher-order chromatin structure, resulting in an open form of chromatin that can further facilitate recruitment of various factors and positively modulate transcription. Coactivator PRMT1 methylates histone H4R3, and this methylation enhances acetylation of H4 by p300 in human cells while the acetylation of H4 inhibits methylation of H4R3 both in vitro and in vivo (reviewed in Selvi *et al.*, 2010a). Thus gene expression is regulated by an orchestra of events that is invariably associated with histone acetylation (Figure 26.1C). Methyltransferase disruptor of telomeric silencing-1 (Dot1) is responsible for methylating histone H3 at K79. This methylation is dependent on the ability of Dot1 to interact with a short basic region on the N-terminal tail of histone H4. Silent information regulator (Sir3) is known to associate with unmodified H3 and H4 tails, and competes with Dot1 by interacting with the same basic region on H4 and by binding H3 adjacent to K79. GCN5-mediated H4K16 acetylation displaces Sir3 on H4, thereby allowing Dot1 to interact with H4 and subsequently methylate H3K79 (Altaf *et al.*, 2007). H3K79 methylation in turn further blocks Sir3–H3 interactions. This series of events serves to define a heterochromatin boundary and allows for transcriptional elongation.

The knowledge of histone posttranslational modification (PTM) crosstalks and their complete elucidation is vital for a thorough understanding of gene expression and thereby various pathophysiological states. With increasing evidence of their involvement in various disorders, it is only prudent to think that the same crosstalks could act as targets of new generation therapeutics. Histone deacetylase inhibitors and DNA methylation inhibitors are already in clinical trials. However, modulation of HATs has also received considerable attention. HAT inhibitors decrease global acetylation levels and also affect other histone PTMs that are essential for transcriptional activation, very much contradictory to the actions of histone deacetylase inhibitors and thus are attractive molecules for chromatin based therapeutics (D. V. Mohankrishna and T. K. Kundu, unpublished data).

26.3   Histone acetylation in pathophysiology

The process of histone acetylation as described above is one of the important modifications that regulates gene expression at various stages both in a signal-dependent as well as independent manner. Hence, it is obvious that histone acetylation would also be a determinant in pathophysiological states. Recently the acetylation field has been flooded with reports implicating acetylation in diseases such as cancer, inflammatory disorders, retroviral pathogenesis, cardiovascular defects, and, most importantly, in neurodegenerative states as well. The following paragraphs highlight the role of histone acetylation in different disease conditions.

26.3.1 Acetylation and cancer

Histone acetylation as indicated in the above section regulates the replication process as well as the DNA damage response. Among the six hallmarks of cancerous cells, one of the important aspects is the loss of cell cycle control and increased proliferation (Hanahan and Weinberg, 2000). Another hallmark is the ability to avoid apoptosis that results in accumulated DNA damage. Acetylation is known to regulate these processes and hence the alteration in acetylation levels could significantly contribute towards cancer. For a long time, it was considered that cancer is characterized by hypoacetylation at the promoters of the tumor-suppressor genes and hence increasing this acetylation was considered as a therapeutic strategy. Subsequently, an interesting study reported the histone hypoacetylation at H4K16 in prostate cancer (Seligson *et al.*, 2005) which correlated with the defective DNA damage response in cancerous cells. Recently, increased acetylation of H3K56, a target of p300/CBP, was also observed in multiple cancers (Das *et al.*, 2009). Although the exact mechanisms have not been identified in these reports, there have been a few studies which have addressed the molecular details involving histone acetylation and cancer. It has been found that adenoviral protein small early region 1 (e1a) causes cellular transformation that is associated with epigenetic reprogramming which is dependent on histone acetylation. Adenoviral e1a causes global relocalization and restriction of the Rb (retinoblastoma) proteins (RB, p130, and p107) and p300/CBP histone acetyltransferases to specific promoters, thereby restricting the acetylation of H3K18ac to a limited set of genes that are responsible for cell cycle promotion and growth. Hyperacetylation of H3K18 is a result of e1a-dependent enrichment of p300 and PCAF on the promoters which is coupled to a decrease in Rb thus paving way for a genome-wide reprogramming of epigenetic states towards cellular transformation (Horwitz *et al.*, 2008). It has to be noted that histone acetylation plays a vital role in this case.

Incidentally, histone hyperacetylation has been identified in many other cancerous cell lines and cancer forms like the hepatocellular carcinoma cell lines and oral cancer cell lines/patient samples. It has been shown that the hyperacetylation of histones is brought about by hyperacetylated p300 through interferon gamma (IFN-γ)-dependent, nitric-oxide mediated signaling (Arif *et al.*, 2010), wherein acetylated histone chaperone NPM1 by regulating genes such as tumor necrosis factor alpha (TNFα) helps in oral cancer manifestation (Shandilya *et al.*, 2009). In oral squamous cell carcinoma, the acetyltransferase machinery is in a hyperactive form; p300 HAT is overexpressed and is enzymatically hyperactive due to increased autoacetylation capability by virtue of GAPDH and NPM1. Further, p300 hyperacetylates NPM1, which is a known activator of acetylation-dependent transcription due to its function as a histone chaperone, and NPM1 itself is overexpressed in oral cancer, thus establishing a vicious positive-feedback loop that could result in cancer progression. This mechanism of cancer manifestation based on histone hyperacetylation is a very attractive therapeutic target and the phenomenon has been successfully exploited to counter oral cancer progression using an efficient inhibitor of histone acetyltransferase p300, CTK7A (see Section 26.4.3).

26.3.2   Acetylation and diabetes

Histone acetylation has been very closely linked to the inflammatory gene expression. It is reported that hyperacetylation of H4 plays an important role in the manifestation of inflammation (Ito, 2007). In a human epithelial cell line, it has been observed that activation of nuclear factor kappa B (NF-κB) by exposing the cell to pro-inflammatory signals like interleukin (IL)-1β, TNF-α, or endotoxins resulted in acetylation of specific lysine residues on histone H4. This hyperacetylation also correlated with an increased expression of inflammatory genes, such as granulocyte macrophage colony stimulating factor (GM-CSF). These results established a strong link between epigenetic alterations especially acetylation and inflammatory disorders such as diabetes mellitus (Goh and Sum, 2010). It has been identified that histones are hyperacetylated in NF-κB responsive pro-inflammatory genes like TNF-α. Indeed, H4K8 hyperacetylation was observed at the TNF-α promoter in response to high glucose in monocytes. Increase in histone H4 acetylation has also been observed in circulating monocytes of Type I diabetes patients (Chen *et al.*, 2009). It is proposed that the activities of HATs/HDACs affect the regulation of transcription of genes critical for beta-cell function and metabolic homeostasis, and therefore may play a crucial role in the pathogenesis and/or management of diabetes (Gray and De Meyts, 2005). In a mutant mouse model of a HAT, mice heterozygous for the CBP demonstrated an increased insulin sensitivity and glucose tolerance even while demonstrating a marked lipodystrophy of white adipose tissue (Yamauchi *et al.*, 2002). More recently, it was also demonstrated that the regulation of expression of insulin by glucose is under the control of histone hyperacetylation, suggesting an important role for HATs and HDACs in the regulation of these critical genes.

26.3.3   Acetylation and disorders of the central nervous system

The role of histone acetylation in the nervous system is a relatively new field as compared to the other processes. In fact, the understanding with respect to the functioning of the brain and the nervous system is very limited, and there is compelling evidence that links histone acetylation to these events. The brain functions in a stimulation-dependent way, constantly establishing newer circuits and removing older circuits and synapses that depend majorly on environmental stimulation or lack of it. These events finally lead to the outcomes such as senses, perception, behavior, learning, memory, and even forgetfulness. Any disturbance in this process leads to neurological and psychiatric disorders. One characteristic feature of the brain is to learn and store the information in the form of memory. These events require the transcription factors, CREB, and coactivators like CBP along with their HAT activity (reviewed in Levenson and Sweatt, 2005). CBP is a very close homolog of e1a binding protein and histone acetyltransferase, p300/KAT3B, with which it shares conserved functions. In a learning model called contextual fear conditioning (a hippocampus-dependent learning model), in which an animal learns to associate a novel context with an aversive stimulus, specific acetylation of histone H3 is significantly increased after an animal undergoes contextual fear conditioning

(Levenson *et al.*, 2004). Also, formation of long-term contextual fear memories requires *N*-methyl-D-aspartate (NMDA)-receptor-dependent synaptic transmission and the MEK–ERK/MAPK signaling cascade (where MEK refers to MAPK/ERK kinase) in the hippocampus (Malenka and Bear, 2004). Inhibition of either of these processes blocks the increase in acetylation of H3. Two different studies have generated CBP-deficient mice that lack the severe developmental problems of the CBP dominant-negative animals. The first study linked the dominant-negative allele of CBP to an inducible promoter (Korzus *et al.*, 2004). Activation of the dominant-negative allele after animals had developed normally led to impaired learning of the spatial water maze task and novel object recognition. In the second study, mice that lacked one allele of CBP (CBP+/–) had impairments in contextual and cued fear memory, and novel object recognition (Alarcón *et al.*, 2004). In both studies, administration of an HDAC inhibitor, which blocks deacetylase activity and increases histone acetylation, restored normal long-term memory formation. Also, the induction of late-phase long-term potentiation, which requires transcription, was significantly impaired in CBP+/– animals. These studies indicate that the epigenetic state of the genome, more importantly chromatin acetylation, affects the induction of long-term forms of mammalian synaptic plasticity and thus influences memory.

Thus, epigenetic modifications of both DNA and histone proteins have emerged as fundamental mechanisms by which neurons direct their transcriptional response to developmental and environmental cues. While histone acetylation has been implicated in neurodevelopment and physiology, as expected, altered histone acetylation status has been implicated in neuropathological and psychiatric disorders. Many studies strongly support the maintenance of precise balance between HATs and HDACs as a prerequisite of neuronal survival in normal conditions (reviewed in Saha and Pahan, 2006). However, decrease in histone acetylation, which is postulated to result in neurodegenerative disorders, seems to be due to loss of function of a single acetyltransferase, CREB binding protein (CBP/KAT3A). CBP mutations render the HAT enzymatically inactive which is probably the root cause of Rubenstein–Taybi syndrome. This syndrome is a developmental disorder associated with mental retardation and childhood cancers of neural crest origin (Petrij *et al.*, 1995). Various mechanisms are now known to reduce CBP HAT availability in several models of neuronal insult. Deacetylation of histones is observed in apoptotic conditions of the neurons where this loss is an early event during apoptosis (Rouaux *et al.*, 2003) and is specific for affected neurons only. This decrease in acetylation is correlated with loss of CBP in the neurons by caspase-mediated degeneration. Nuclear translocation of expanded polyglutamine-containing neurotoxins (like mutated huntingtin protein), which are implicated in neurodegenerative diseases, selectively enhance ubiquitination and degradation of CBP by a proteosomal pathway (Jiang *et al.*, 2003). Also, redistribution of CBP from normal nuclear location into huntingtin protein aggregates compromises their availability for normal functions (Nucifora *et al.*, 2001). Decrease in expression of CBP is observed when neurons are stressed by oxidative conditions. Thus a decrease in expression of CBP, altered compartmentalization, or an increase in degradation of CBP has been implicated in the hypoacetylated states observed in neurodegeneration. This unique mechanism is an attractive target for utilizing enzyme-specific activators as a therapeutic option for neurodegenerative disorders.

26.4   Modulating histone acetylation

Due to this increasing evidence implicating histone acetylation in several disease conditions, modulating histone acetylation as a corrective measure has emerged as an attractive approach. Within the cellular system, there are several strategies to regulate the modification and thereby cellular homeostasis. One such method involves the reversibility of the reaction wherein the HAT–HDAC activities maintain a balance. Apart from this, the autoactivation ability of acetyltransferases, such as p300, PCAF, and TIP60, also fine-tunes their activities by autoacetylation. There also exist interacting proteins which also activate the enzyme activities (for example, NPM1 and GAPDH activate p300 autoacetylation). A protein complex like the INHAT (inhibitor of histone acetyltransferases) complex is also known to inhibit HAT activity by blocking the access of the HATs to substrates such as histones (Seo *et al.*, 2001). This complex has been characterized as being composed of Set/TAF1β, a myeloid-leukemia-associated oncoprotein, and pp32, a nuclear phosphoprotein. The INHAT complex by virtue of its inhibitory action on the HAT complexes also affects transcriptional activation, indicating the existence of several regulatory features within the cellular system itself. Recently, several small-molecule modulators of histone acetyltransferases have been discovered, which act as potential tools to probe HAT activity (Table 26.2). Physiologically acetyl CoA and NAD^+ levels are critical for acetylation per se. The recent addition to this group of modulators is microRNAs, which are 22-nt RNAs capable of binding to target mRNAs thereby regulating gene expression. Although no direct evidence linking microRNAs and acetylation has so far been identified, there are indications that they are intricately connected.

Apart from the physiological small-molecule modulators, several natural as well as synthetic inhibitors and activators have been identified and characterized as modulators of histone acetylation (Table 26.2). The subsequent paragraph details these discoveries along with the existing data about their therapeutic potential.

26.4.1   Histone acetylation inhibition

Diseases such as oral squamous cell cancer and inflammatory conditions including diabetes exhibit histone hyperacetylation. Hence, it has been proposed by several researchers that inhibiting histone acetylation in these cases might help a great deal in not just controlling the disease but also restricting the other complications. Approaches based on small interfering RNA (siRNA) have revealed a non-specific mechanism of action, since the absence of any acetyltransferase does not only interfere with the disease-associated abnormal gene expression but also with the normal transcriptional events brought about by the protein–protein interactions. This is because HATs/KATs mostly are transcriptional coactivators, that

Table 26.2 Small-molecule modulators of histone acetyltransferases and their targets

interact with several transcription factors and positively regulate their function which may or may not depend upon the intrinsic HAT activity of the coactivator. Hence, inhibitors targeting these acetyltransferases are being actively explored. As of now, there are about a dozen HAT inhibitors with characteristic chemical scaffolds which are found both in the natural form as well as synthesized. These inhibitors can be classified into two categories based on their spectrum of action which is also represented in Table 26.2.

26.3.1.1   Broad-spectrum HAT inhibitors

This class of inhibitors refers to the small molecules which have shown specificity towards more than one acetyltransferase. These include both synthetic as well as naturally occurring molecules as will be described below. Anacardic acid was the first naturally occurring HAT inhibitor (HATi), identified from the cashew nut (*Anacardium occidentale*) shell liquid with a broad-spectrum HATi potential (Balasubramanyam *et al.*, 2003). The first report showed its broad spectrum of activity against both p300 and PCAF acetyltransferases with a low micromolar half-maximal inhibitory concentration (IC_{50}). Subsequently, it has also been shown to inhibit activity of TIP60, a MYST family acetyltransferase, and sensitizes tumor cells to ionizing radiation (Sun *et al.*, 2006). It has been shown to possess a sensitizing effect along with genotoxic agents such as doxorubicin and cisplatin, thereby potentiating apoptosis. This activity of anacardic acid has been correlated with its ability to inhibit histone acetylation. It has also been shown to inhibit both inducible and constitutive NF-κB activation (Sung *et al.*, 2008).

Another naturally occurring broad-spectrum HATi is garcinol, isolated from the *Garcinia indica* fruit rind. Chemically, the molecule is a polyisoprenylated benzophenone with several reactive chiral centers (Balasubramanyam *et al.*, 2004a). Garcinol is found to be generally toxic to mammalian cells and it inhibits various histone acetyltransferases with a low micromolar IC_{50}. Subsequent derivatization of garcinol has led to the identification of a p300-specific inhibitor which is described in the next section. The natural green tea polyphenol EGCG is another HAT inhibitor which decreases histone acetylation by acting on more than one histone acetyltransferase. EGCG abrogates p300-induced p65 acetylation in vitro and in vivo, increases the level of cytosolic inhibitor kappa B kinase-α (IKKα), and suppresses TNF-α induced NF-κB activation and downstream gene expression (Choi *et al.*, 2009). One of the most recently identified HATis is a water-soluble curcumin derivative, hydrazinocurcumin (CTK7A), which inhibits both p300 and PCAF acetyltransferase activity with minimal effect on the other modifying enzymes (Arif *et al.*, 2010). It has been shown to induce senescence-like growth arrest in oral cancer cell lines thereby inducing polyploidy. Yet another class of HATi includes the synthetic bisubstrate analogs for TIP60 which have not been well characterized in vivo. But one promising candidate of this category of inhibitors is spermidinyl CoA which has been identified as a general inhibitor of histone acetylation which induces cell cycle arrest by sensitizing cells to radiation (Bandyopadhyay *et al.*, 2009).

26.4.1.2 Specific HAT inhibitors

The first HATi reported were synthetic and specific HATis; one of the earliest was lysyl CoA, a bisubstrate inhibitor synthesized based on rational design, which showed high selectivity to p300 HAT. Based on the same rationale, H3 CoA-20 was synthesized which was specific to PCAF HAT (Lau *et al.*, 2000). Its only shortcoming was cellular impermeability which was later overcome by including a "tat" peptide along with the molecule (Zheng *et al.*, 2005). Subsequently, a p300-specific HATi, curcumin, was identified from the natural source (Balasubramanyam *et al.*, 2004b). The molecular scaffold of curcumin is a diferuloylmethane moiety. Further derivatization of curcumin whereby it has been converted to hydrazinocurcumin (CTK7A) to improve its bioavailability changed its spectrum of HAT inhibition (Arif *et al.*, 2010). The broad-spectrum HATi garcinol has been chemically modified to obtain an intramolecular cyclized molecule, isogarcinol (IG). This has been further derivatized to monosusbtituted 14-isopropoxy IG (LTK-13) and 14-methoxy IG (LTK-14) as well as disubstituted IG, to give 13,14-disulfoxy IG (LTK-19), which are all p300-specific HATis. These derivatizations dramatically reduced the toxicity of garcinol. Subsequently, LTK-14 has been used to understand the role of p300-mediated acetylation of p53 acetylation upon DNA damage as well as in global gene expression (Mantelingu *et al.*, 2007b). Recently, two more p300-specific inhibitors have been identified and characterized from natural sources. These are the commercially available C646 (Bowers *et al.*, 2010) and grape-seed-extracted procyanidin B3 (Choi *et al.*, 2010). C646 consists of aromatic rings and a benzoic acid, with nanomolar p300 inhibitory potential. It has also been shown to possess anti-proliferative activity on cancerous cell lines. Procyanidin B3 inhibits p300 and has been shown to possess anti-cancerous activity on prostate cancer cell lines. Another HATi from a natural source is plumbagin, a naphthoquinone molecule, with preference towards p300 activity. Chemical substitutions of its single hydroxyl group have helped in identifying the functional group responsible for HAT inhibition (Ravindra *et al.*, 2009). However, due to its highly reactive nature, it is proposed to have other targets and different mechanisms of action in the in vivo context (D.V. Mohankrishna and T. K. Kundu, unpublished data).

Several attempts for synthesizing PCAF/GCN5-specific inhibitors have not been very successful to date. The γ-butyrolactones were synthesized based on the enzyme catalytic center of the GCN5 acetyltransferase (Biel *et al.*, 2004), which showed in vitro inhibition at a high IC_{50} concentration of 100 μM. However, this class of compound was not tested in the cellular system. Subsequently, cycloalkylidene-(4-phenylthiazol-2-yl) hydrazone derivatives were shown to inhibit GCN5 in yeast based on a chemical genetic screen (Chimenti *et al.*, 2009), which unfortunately exhibited inhibitory activity at millimolar concentrations (0.2–0.8 mM), thus limiting its possibility of use in physiological conditions. Another class of inhibitors claimed to have specificity against PCAF are the isothiazolones. These inhibitors could also inhibit p300, and no in vivo specificity was shown (Stimson *et al.*, 2005). Thus, apart from H3 CoA-20, specific, cell-permeable efficient inhibitors against PCAF/GCN5 acetyltransferase are yet to be identified.

Histone acetylation can also be inhibited by modulating the reversible process, i.e., by activating HDACs. There are few examples in the category and the best recognized is resveratrol, a sirtuin activator (Howitz *et al.*, 2003) with implications in calorie restriction and aging. Although theoretically inhibiting histone acetylation is possible by activating the deacetylation reaction, it is an indirect route and may not have the desired effect. However, another indirect method of modulating histone acetylation is the use of HDACi, which has several examples and its pros and cons will be discussed in the next section.

26.4.2   HAT inhibitors as biological probes

One of the main applications of enzyme inhibitors has been in deciphering the role of that enzyme activity in physiological processes. The inhibitors of HATs are also no exception. Several intriguing aspects of histone acetylation have been elucidated by using HATi. Lysyl CoA has been used to elucidate the role of p300/CBP HAT activity in muscle cell differentiation, p73 acetylation, *Caenorhabditis elegans* development, cyclooxygenase-2 gene regulation, melanocyte growth, HIV gene regulation, and nuclear hormone receptor induced gene expression (reviewed in Cole, 2008). The cell-permeable form of lysyl-CoA has also been used for identifying the role of p300 as a transcriptional repressor. A similar phenomenon was also identified in glucose metabolism involving TORC2 acetylation. However, the most compelling evidence for the role of p300-mediated acetylation in transcriptional repression on a global scale was provided by the p300-specific inhibitor, LTK-14. Interestingly, HAT inhibition by LTK-14 upregulated a few genes like *PLIN* and *GLDN* (Mantelingu *et al.*, 2007b). The relation between HAT inhibition and gene upregulation is yet to be understood. The PCAF-specific inhibitor H3 CoA-20 has helped in understanding the role of acetylation in Pol III transcription. The use of p300-specific inhibitors, curcumin and LTK-14, to repress HIV multiplication has indeed opened up a new view towards the role of p300 in retroviral pathogenesis. The small molecule plumbagin, by virtue of its single hydroxyl group, has led to the identification of the chemical moiety essential for HAT inhibition. The role of acetylation in various disease manifestations have also been very elegantly elucidated with the help of HAT inhibitors, however, since these aspects argue in favor of the therapeutic potential of HATi, it is discussed in the subsequent paragraph.

26.4.3 Therapeutic potential of HAT inhibitors

Histone acetylation is a major determinant of gene expression in physiological and pathophysiological states. Several disease states exhibit a marked increase in histone acetylation. Hence, several preclinical studies have been attempted and all the observations have provided an encouraging thrust to the therapeutic intervention using HATi. Some of these attempts will be highlighted in this section.

The most promising effect of HATi is its anti-tumor effect on several cell line models. Both broad-spectrum HATi and the specific HATi have shown antiproliferative activity on cancer cells. Anacardic acid has been shown to possess a sensitizing effect along with other genotoxic agents (Sun *et al.*, 2006). Among HAT inhibitors, curcumin has been shown to be effective against lung cancer (Moghaddam *et al.*, 2009), plumbagin against hepatocellular carcinoma (Shih *et al.*, 2009), CTK7A against oral cancer (Arif *et al.*, 2010), C646 against different cancer origins (Bowers *et al.*, 2010), and procyanidin B3 against prostate cancer (Choi *et al.*, 2010). Various HAT inhibitors have been shown to be effective against diverse cancerous conditions. S-substituted CoA (Spd-CoA) is an inhibitor of histone acetylation. Exposure of cancer cells to Spd-CoA causes inhibition of histone acetylation that correlates with a transient arrest of DNA synthesis, which is linked to a transient delay in S phase progression, and an inhibition of nucleotide excision repair and DNA double-strand break repair. Application of this molecule to cancer cells renders them more susceptible to DNA-targeted chemotherapeutic drugs like cisplatin and 5-fluorouracil, to the DNA-damaging drug camptothecin, and also to UV-C irradiation (Bandyopadhyay *et al.*, 2009). This sensitization effect of Spd-CoA has not been observed in normal cells due to a barrier to molecule uptake.

Most of the HATi have been shown to possess an anti-inflammatory effect mainly due to inhibition of NF-κB acetylation. These include curcumin, EGCG, plumbagin, and several more. Hence, it is possible to control the inflammatory responses and thereby diseases using HATi. Incidentally, one such area which has initial leads but is relatively less explored is the field of acetylation and diabetes.

Though it has been proven that NF-κB activity and CBP/p300 activity increase in high-glucose conditions (Miao *et al.*, 2004) in THP cells as well as cardiac myocytes (Figure 26.2A), and that these were effectively modulated using p300 siRNA and curcumin treatment (Chiu *et al.*, 2009), the therapeutic efficacy of HATi towards diabetes has been less exploited. The potential target locations for these HATi for studying the effect on diabetes include monocytes, liver cells, and the adipocytes. Preliminary data from our laboratory suggest of that HATi treatment to the HepG2 cells (liver origin), leads to an upregulation of gluconeogenic enzymes such as phosphoenolpyruvate carboxykinase (PEPCK) as well as coactivators such as peroxisome proliferator-activated receptor gamma (PPAR-γ) (Figure 26.2A(c) and 26.2B). Incidentally, PPAR-γ modulates the expression of PEPCK in the adipocytes, thereby modulating gluconeogenesis (Tontonoz *et al.*, 1995). The best effect was observed when toxicity was induced using palmitate followed by HATi (curcumin) treatment. This observation seems to fit well with the earlier data obtained with curcumin. However, the exact molecular mechanisms need further investigation. Curcumin and LTK-14 have been successfully used to repress syncytia formation in the HIV life cycle, thereby implicating their anti-retroviral activity (Balasubramanyam *et al.*, 2004b; Mantelingu *et al.*, 2007b).

Thus it is evident that the HAT inhibitors could be promising anti-cancer agents and anti-inflammatory agents, as well as potential anti-viral agents against HIV. The only limitation for these molecules as tested in clinical trials was their apparent cytotoxicity, low bioavailability, and non-specific targets. Several attempts have now been made to overcome these difficulties, as in the case of curcumin, for which a water-soluble derivative CTK7A has been made to combat its bioavailability problem. The p300-specific inhibitor curcumin was shown to possess anti-malarial activity when used along with artemisinin (Nandakumar *et al.*, 2006). Incidentally, anacardic acid, another HATi, has also been shown to inhibit the growth of the malaria-causing *Plasmodium falciparum* (Cui *et al.*, 2008), suggesting the use of other HATi as anti-malarial agents.

Yet another intriguing aspect of histone acetylation is the role of specific residue modification in determining important gene expression patterns. H4K5 and H4K8 acetylation are marks corresponding to the inflammatory response, H4K16 acetylation is correlated to DNA damage response, and H3K56 acetylation has unique functionalities with respect to DNA replication as well as repair. In an interesting study on the fungal pathogen *Candida albicans,* the targeted reduction of the H3K56 acetylation levels had a better anti-fungal effect with a drastic reduction in the virulence of the pathogen in a mouse model (Wurtele *et al.*, 2010). Thus modulation of specific residue acetylation might have better therapeutic effects. Although such specific inhibitors have not yet been characterized for acetylation, a recent study has very elegantly used histone acetylation mimics to target the inflammatory pathway specifically (Nicodeme *et al.*, 2010). In this report, the acetylation mimic compounds were used to compete with the acetylated lysine in mediating the remodeling processes specific to the inflammatory pathway and thereby achieving an anti-inflammatory effect. Most interestingly, in spite of this class of compound having a significant effect on gene expression, it could not

A

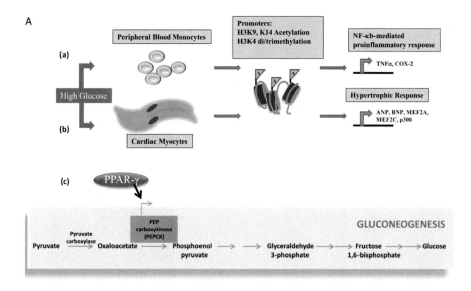

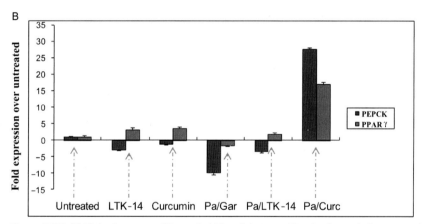

Figure 26.2 (A) High-glucose conditions mimicking diabetes mellitus lead to varied responses in different tissues but involve histone acetylation activation. (a) In peripheral blood monocytes high-glucose treatment leads to increased recruitment of p300/CBP to NF-κB-dependent genes leading to promoter histone H3/H4 acetylation and hyperexpression of TNFα and COX-2. (b) In cardiac myocytes high-glucose treatment increases expression of p300 and pro-hypertrophic genes associated with increase in histone acetylation. (c) The gluconeogenesis pathway representing the different enzymes and the products. The coactivator PPARγ regulates the expression of PEPCK. See plate section for color version. (B) HepG2 cells were treated with the compounds as indicated for 24 hours. The RNA was isolated and subjected to cDNA synthesis followed by real-time PCR analysis using primers for the represented genes (PEPCK and PPAR-γ), involved in gluconeogenesis.

inhibit the activity of the HATs. Thus inhibiting histone acetylation by different methodologies is gaining importance as an effective therapeutic strategy.

26.5   Activating histone acetylation

Histone acetylation activation is one phenomenon that has been attempted for a long time, and very encouraging results have been obtained with this approach. Perhaps the only shortcoming in this strategy was the mode of achieving histone hyperacetylation, by inhibiting the HDACs, an indirect approach, albeit with successful results. Researchers have been struggling to overcome this indirect mode by actively exploring the possibility of direct HAT activation. Within physiological systems, overexpression of the acetyltransferases has resulted in apoptosis indicating that cellular homeostasis is probably lost. Very few small-molecule modulators capable of activating histone acetylation have been identified, and only one of them has been properly characterized. The following section will highlight these discoveries in the light of their therapeutic potential.

26.5.1   HAT activation

Lysine acetyltransferase activity can be activated by intrinsic and extrinsic mechanisms. Intrinsic activation of the enzyme refers to intracellular signals, interacting proteins, and posttranslational modifications. Extrinsic mechanisms refer to overexpression strategies and the use of small-molecule activators.

26.5.1.1   Intrinsic mechanisms of HAT activation

Within the physiological system, the HATs and the HDACs regulate each others' activities, thereby maintaining homeostasis. Hence, the HATs are turned on (activated) on the basis of several regulatory events. The extracellular signals, such as heat shock or nitric oxide or cAMP signaling, leads to the activation of a signaling cascade, culminating in enhanced acetyltransferase activity. The acetyltransferases undergo several modifications such as acetylation, phosphorylation, ubiquitination, etc. Several acetyltransferases including p300, PCAF, and TIP60 undergo autoacetylation (reviewed in Selvi *et al.*, 2010b), an event in which the enzyme itself gets acetylated and becomes active towards other substrates. Phosphorylation has also been shown to play an important role in the acetyltransferase activity as in the case of CBP where the phosphorylated form represents an active acetyltransferase. Methylation is yet another modification which modulates the acetyltransferase activity. CBP methylation by the arginine methyltransferase CARM1 has been shown to have both activation and repression effects based on the residue being methylated. Thus posttranslational modifications play a significant role in modulating the acetyltransferase function. Yet another method of activating histone acetylation that exists in the cellular milieu is the role of regulatory proteins. The histone chaperone NPM1 (Arif *et al.*, 2010) and the metabolic enzyme GAPDH (Sen *et al.*, 2008) have been shown to be positive regulators of p300 autoacetylation. Several kinases and coactivators also act as regulatory proteins.

26.5.1.2 Extrinsic mechanisms of HAT activation

One of the areas where the role of loss of histone acetylation can be directly related to a disease state is the neurodegenerative condition wherein it was found that the CBP HAT was sequestered. To overcome this condition, CBP was overexpressed in these cells and surprisingly it was found that this led to an apoptotic event (Rouaux *et al.*, 2003). Subsequently, when HDACi was used in similar conditions, mild improvement in the response was observed. Hence, apart from overexpressing the acetyltransferase, the use of HDAC inhibitors have also proven to be efficient approaches for activating HAT. HDAC inhibitors have been extensively used from cellular to animal studies to repress excess deacetylase function, and to activate histone acetylation and thereby the downstream pathways. One major drawback in this strategy is that the class II HDACs function together as complexes and are generally redundant, whereas each acetyltransferase has a specific regulatory role at different stages of normal cell physiology. The functions of HDACs are generally redundant, and single knockouts with the class II members have been unsuccessful for gaining any knowledge since the other family members compensate for their absence. Hence the need for specific HAT activators has been very strongly felt.

26.5.2 Small-molecule activators of histone acetyltransferases

The first p300-specific HAT activator, CTPB (a flutamide), was synthesized from a HATi, anacardic acid (Balasubramanyam *et al.*, 2003). Although the molecule showed a preference towards p300/CBP acetyltransferase activity in vitro, further in vivo characterization was not possible since the molecule was highly impermeable to the cells. Several chemical substitutions and structural studies revealed the significance of the $-CF_3$ and $-Cl$ groups in the *para* position as important determinants for the activation potential (Mantelingu *et al.*, 2007a). The long hydrocarbon chain of CTPB when removed gave CTB which was also shown to be an activator of p300 acetyltransferase. The cell impermeability to the small molecule was overcome by employing nanotechnology. By conjugating CTPB with glucose-derived carbon nanospheres, CTPB could be successfully delivered into the nucleus of the cell so that histone hyperacetylation could be obtained. A similar approach was also used to induce hyperacetylation in mice. Since the carbon nanospheres could cross the blood–brain barrier, by conjugating the CSP to CTPB histone acetylation could be enhanced in the mouse brain (Selvi *et al.*, 2008). This observation is a very significant lead for the area of neurodegenerative disorders which will be discussed in the next section. Very recently, some derivatives of CTPB have been synthesized and cellular studies revealed their HAT activation potential, especially towards CBP and PCAF (Souto *et al.*, 2010). However, these derivatives need further characterization. A significant addition to the field of HAT activators includes the newly identified HAT activator nemorosone from natural sources (Dal Piaz *et al.*, 2010). The molecule has been convincingly shown to activate histone acetylation both in vitro and in vivo. The most important observation from this study was the surface plasmon resonance (SPR) analysis of the HAT activators CTPB and nemorosone with the p300 HAT domain, which showed that both these molecules bind to the same region on the p300 acetyltransferase.

26.5.3   Therapeutic potential of HAT activators

The small-molecule HAT activators (HATa) could be another set of important molecules to normalize the cellular balance of acetylation which is altered in pathophysiological conditions. As discussed in the previous section, activating histone acetylation is an important therapeutic strategy for the various central nervous system disorders which exhibit a decrease in histone acetylation. To date, there have been no studies with HAT activators on disease models. However, the data obtained with the nanospheres–CTPB conjugate which activates histone acetylation in mouse brain is indeed very encouraging. Subperitoneal injection of CSP-CTPB activated histone acetylation in the hippocampus of rat without showing any apparent toxicity to the animal (Selvi *et al.*, 2010b). Furthermore, though the well-established anti-epileptic drug valproic acid (HDACi) and the HAT CTPB could both activate bulk histone acetylation in the rat hippocampal tissue, p300-specific acetylation on histone H2B showed a drastic increase only in the case of the CTPB conjugate owing to specific activation. This result is a clear indication that the HATa and HDACi mediated histone acetylation are not the same and need to be considered as separate events. Hence, it is quite possible that the downstream gene expression pattern may also differ based on the mode of activating histone acetylation since specific residue modifications have distinct functional outcomes (see Section 26.2)

26.6   Future perspectives

The histone acetylation mark is not just a tag but rather acts as a signaling platform for the recruitment of various modifying and remodeling machineries. Alteration of the reversible acetylation balance may lead to abnormal gene expression and thereby several diseases. This "acetylation" is an integral component of a code or the "epigenetic language" rather than a mere "mark." The small-molecule modulators of acetyltransferases could be useful both as biological probes to elucidate the epigenetic language of cellular functions as well as therapeutic tools to target disease conditions.

Based on the present understanding a few aspects that could be considered for designing and synthesizing novel, specific, and more efficient HAT inhibitors include the following:

- *Competitive versus non-competitive or uncompetitive inhibitors* Several allosteric sites in HATs have been now identified (represented in Figure 26.3). This information could be used for the rational design of site blockers or interaction motifs so as to inhibit the enzyme activity.

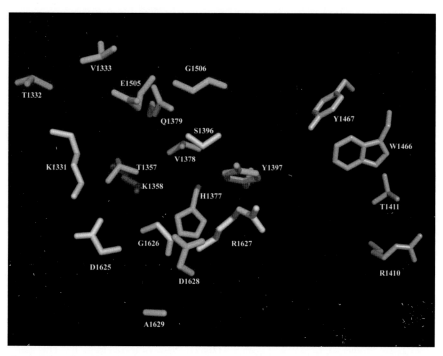

Figure 26.3 The different residues that have been identified as allosteric sites for p300 HAT inhibition have been indicated. The residues have been color-coded based on the reports as follows. White, residues identified in the C646 study; yellow, common residues for garcinol, isogarcinol, and LTK-14; cyan, common residues for isogarcinol and LTK-14; green, residues binding exclusive to garcinol; red, plumbagin. See plate section for color version.

- *Chemical functional groups* Reactive chiral centers increase the spectrum of HAT inhibition. Apparently, hydroxyl groups play an important role in determining the inhibitory potential.
- *Cellular uptake and effects* Salts of the compounds have better water solubility, and subsequent better bioavailability. Quinone moieties render the molecule highly cytotoxic.
- *Modification specific inhibition* Use of homology modeling studies to identify the interacting residues and subsequent synthesis of mimics or blockers could have very specific effects.
- *Global versus local effects* Histone acetylation is an important regulatory modification for gene expression. Hence, modulating the levels of histone acetylation on a global scale might have detrimental effects rather than being helpful. Hence, the inhibition needs to be more targeted which can partly be achieved by having enzyme-specific inhibitors/modification-specific inhibitors/specific mimics, etc.

With respect to HAT activation, very little has been done and lot more needs to be investigated. The chemical nature of HAT activators and the binding data available need to be exploited to synthesize better and more efficient HAT activators. Nanotechnology has been used for achieving histone hyperacetylation in cells as well as animal tissues. This can be further developed for bringing about targeted delivery of both activators and inhibitors. Thus histone acetylation can be turned on or off using the modulators for addressing epigenetics, epigenomics, and chromatin dynamics as well as therapeutics.

ACKNOWLEDGMENTS

Work done in the authors' laboratory is supported by Jawaharlal Nehru Centre for Advanced Scientific Research and Department of Biotechnology, Government of India. TKK is a Sir. J. C. Bose National Fellow (DST, Government of India).

REFERENCES

Ajiro, K., Scoltock, A. B., Smith, L. K., Ashasima, M., and Cidlowski, J. A. (2010). Reciprocal epigenetic modification of histone H2B occurs in chromatin during apoptosis in vitro and in vivo. *Cell Death and Differentiation*, **17**, 984–993.

Alarcón, J. M., Malleret, G., Touzani, K., *et al.* (2004). Chromatin acetylation, memory, and LTP are impaired in CBP+/− mice: a model for the cognitive deficit in Rubinstein–Taybi syndrome and its amelioration. *Neuron*, **42**, 947–959.

Allfrey, V. G., Faulkner, R., and Mirsky, A. E. (1964). Acetylation and methylation of histones and their possible role in the regulation of RNA synthesis. *Proceedings of the National Academy of Sciences USA*, **51**, 786–794.

Allis, C. D., Berger, S. L., Cote, J., *et al.* (2007). New nomenclature for chromatin-modifying enzymes. *Cell*, **131**, 633–636.

Altaf, M., Utley, R. T., Lacoste, N., *et al.* (2007). Interplay of chromatin modifiers on a short basic patch of histone H4 tail defines the boundary of telomeric heterochromatin. *Molecular Cell*, **28**, 1002–1014.

Arif, M., Vedamurthy, B. M., Choudhari, R., *et al.* (2010). Nitric oxide-mediated histone hyperacetylation in oral cancer: target for a water-soluble HAT inhibitor, CTK7A. *Chemistry and Biology*, **17**, 903–913.

Balasubramanyam, K., Swaminathan, V., Ranganathan, A., and Kundu, T. K. (2003). Small molecule modulators of histone acetyltransferase p300. *Journal of Biological Chemistry*, **278**, 19 134–19 140.

Balasubramanyam, K., Altaf, M., Varier, R. A., *et al.* (2004a). Polyisoprenylated benzophenone, garcinol, a natural histone acetyltransferase inhibitor, represses chromatin transcription and alters global gene expression. *Journal of Biological Chemistry*, **279**, 33 716–33 726.

Balasubramanyam, K., Varier, R. A., Altaf, M., *et al.* (2004b). Curcumin, a novel p300/CREB-binding protein-specific inhibitor of acetyltransferase, represses the acetylation of histone/nonhistone proteins and histone acetyltransferase-dependent chromatin transcription. *Journal of Biological Chemistry*, **279**, 51 163–51 171.

Bandyopadhyay, K., Banères, J. L., Martin, A., *et al.* (2009). Spermidinyl-CoA-based HAT inhibitors block DNA repair and provide cancer-specific chemo- and radiosensitization. *Cell Cycle*, **8**, 2779–2788.

Bertrand, E., Erard, M., Gómez-Lira, M., and Bode, J. (1984). Influence of histone hyperacetylation on nucleosomal particles as visualized by electron microscopy. *Archives of Biochemistry and Biophysics*, **229**, 395–398.

Biel, M., Kretsovali, A., Karatzali, E., Papamatheakis, J., and Giannis, A. (2004). Design, synthesis, and biological evaluation of a small-molecule inhibitor of the histone acetyltransferase Gcn5. *Angewandte Chemie (International Edition)*, **43**, 3974–3976.

Bode, J., Henco, K., and Wingender, E. (1980). Modulation of the nucleosome structure by histone acetylation. *European Journal of Biochemistry*, **110**, 143–152.

Bowers, E. M., Yan, G., Mukherjee, C., *et al.* (2010). Virtual ligand screening of the p300/CBP histone acetyltransferase: identification of a selective small molecule inhibitor. *Chemistry and Biology*, **17**, 471–482.

Brownell, J. E. and Allis, C. D. (1995). An activity gel assay detects a single, catalytically active histone acetyltransferase subunit in *Tetrahymena* macronuclei. *Proceedings of the National Academy of Sciences USA*, **92**, 6364–6368.

Campos, E. I., Fillingham, J., Li, G., *et al.* (2010). The program for processing newly synthesized histones H3.1 and H4. *Nature Structural and Molecular Biology*, **17**, 1343–1351.

Chang, C. S. and Pillus, L. (2009). Collaboration between the essential Esa1 acetyltransferase and the Rpd3 deacetylase is mediated by H4K12 histone acetylation in *Saccharomyces cerevisiae*. *Genetics*, **183**, 149–160.

Chen, S. S., Jenkins, A. J., and Majewski, H. (2009). Elevated plasma prostaglandins and acetylated histone in monocytes in Type 1 diabetes patients. *Diabetic Medicine*, **26**, 182–186.

Chimenti, F., Bizzarri, B., Maccioni, E., *et al.* (2009). A novel histone acetyltransferase inhibitor modulating Gcn5 network: cyclopentylidene-[4-(4′-chlorophenyl) thiazol-2-yl]hydrazone. *Journal of Medicinal Chemistry*, **52**, 530–536.

Chiu, J., Khan, Z. A., Farhangkhoee, H., and Chakrabarti, S. (2009). Curcumin prevents diabetes-associated abnormalities in the kidneys by inhibiting p300 and nuclear factor-kappaB. *Nutrition*, **25**, 964–972.

Choi, K. C., Jung, M. G., Lee, Y. H., *et al.* (2009). Epigallocatechin-3-gallate, a histone acetyltransferase inhibitor, inhibits EBV-induced B lymphocyte transformation via suppression of RelA acetylation. *Cancer Research*, **69**, 583–592.

Choi, K. C., Park, S., Lim, B. J., *et al.* (2010). Procyanidin B3, an inhibitor of histone acetyltransferase, enhances the action of antagonist for prostate cancer cells via inhibition of p300-dependent acetylation of androgen receptor. *Biochemical Journal*, **433**, 235–244.

Cole, P. A. (2008). Chemical probes for histone-modifying enzymes. *Nature Chemical Biology*, **4**, 590–597.

Cui, L., Miao, J., Furuya, T., *et al.* (2008). Histone acetyltransferase inhibitor anacardic acid causes changes in global gene expression during in vitro *Plasmodium falciparum* development. *Eukaryotic Cell*, **7**, 1200–1210.

Dal Piaz, F., Tosco, A., Eletto, D., *et al.* (2010). The identification of a novel natural activator of p300 histone acetyltranferase provides new insights into the modulation mechanism of this enzyme. *Chembiochem*, **11**, 818–827.

Das, C., Lucia, M. S., Hansen, K. C., and Tyler, J. K. (2009). CBP/p300-mediated acetylation of histone H3 on lysine 56. *Nature*, **459**, 113–117.

Dou, Y., Milne, T. A., Tackett, A. J., *et al.* (2005). Physical association and coordinate function of the H3 K4 methyltransferase MLL1 and the H4 K16 acetyltransferase MOF. *Cell*, **121**, 873–885.

Fillingham, J. and Greenblatt, J. F. (2008). A histone code for chromatin assembly. *Cell*, **134**, 206–208.

Frye, R. A. (2000). Phylogenetic classification of prokaryotic and eukaryotic Sir2-like proteins. *Biochemical and Biophysical Research Communications*, **273**, 793–798.

Goh, K. P. and Sum, C. F. (2010). Connecting the dots: molecular and epigenetic mechanisms in type 2 diabetes. *Current Diabetes Reviews*, **6**, 255–265.

Gray, S. G. and De Meyts, P. (2005). Role of histone and transcription factor acetylation in diabetes pathogenesis. *Diabetes/Metabolism Research and Reviews*, **21**, 416–433.

Gregoretti, I. V., Lee, Y. M., and Goodson, H. V. (2004). Molecular evolution of the histone deacetylase family: functional implications of phylogenetic analysis. *Journal of Molecular Biology*, **338**, 17–31.

Hanahan, D. and Weinberg, R. A. (2000). The hallmarks of cancer. *Cell*, **100**, 57–70.

Horwitz, G. A., Zhang, K., McBrian, M. A., *et al.* (2008). Adenovirus small e1a alters global patterns of histone modification. *Science*, **321**, 1084–1085.

Howitz, K. T., Bitterman, K. J., Cohen, H. Y., *et al.* (2003). Small molecule activators of sirtuins extend *Saccharomyces cerevisiae* lifespan. *Nature*, **425**, 191–196.

Ito, K. (2007). Impact of post-translational modifications of proteins on the inflammatory process. *Biochemical Society Transactions*, **35**, 281–283.

Jiang, H., Nucifora, F. C., Jr., Ross, C. A., and DeFranco, D. B. (2003). Cell death triggered by polyglutamine-expanded huntingtin in a neuronal cell line is associated with degradation of CREB-binding protein. *Human Molecular Genetics*, **12**, 1–12.

Korzus, E., Rosenfeld, M. G., and Mayford, M. (2004). CBP histone acetyltransferase activity is a critical component of memory consolidation. *Neuron*, **42**, 961–972.

Lau, O. D., Kundu, T. K., Soccio, R. E., *et al.* (2000). HATs off: selective synthetic inhibitors of the histone acetyltransferases p300 and PCAF. *Molecular Cell*, **5**, 589–595.

Lee, D. Y., Hayes, J. J., Pruss, D., and Wolffe, A. P. (1993). A positive role for histone acetylation in transcription factor access to nucleosomal DNA. *Cell*, **72**, 73–84.

Levenson, J. M. and Sweatt, J. D. (2005). Epigenetic mechanisms in memory formation. *Nature Reviews Neuroscience*, **6**, 108–118.

Levenson, J. M., O'Riordan, K. J., Brown, K. D., *et al.* (2004). Regulation of histone acetylation during memory formation in the hippocampus. *Journal of Biological Chemistry*, **279**, 40545–40559.

Li, B., Carey, M., and Workman, J. L. (2007). The role of chromatin during transcription. *Cell*, **128**, 707–719.

Liou, G.-G., Tanny, J. C., Kruger, R. G., Walz, T., and Moazed, D. (2005). Assembly of the SIR complex and its regulation by O-acetyl-ADP-ribose, a product of NAD dependent histone deacetylation. *Cell*, **121**, 515–527.

Malenka, R. C. and Bear, M. F. (2004). LTP and LTD: an embarrassment of riches. *Neuron*, **44**, 5–21.

Mantelingu, K., Kishore, A. H., Balasubramanyam, K., *et al.* (2007a). Activation of p300 histone acetyltransferase by small molecules altering enzyme structure: probed by surface enhanced raman spectroscopy. *Journal of Physical Chemistry B*, **111**, 4527–4534.

Mantelingu, K., Reddy, B. A., Swaminathan, V., *et al.* (2007b). Specific inhibition of p300-HAT alters global gene expression and represses HIV replication. *Chemistry and Biology*, **14**, 645–657.

Miao, F., Gonzalo, I. G., Lanting, L., and Natarajan, R. (2004). In vivo chromatin remodeling events leading to inflammatory gene transcription under diabetic conditions. *Journal of Biological Chemistry*, **279**, 18091–18097.

Miyamoto, N., Izumi, H., Noguch, T., *et al.* (2008). Tip60 is regulated by circadian transcription factor clock and is involved in cisplatin resistance. *Journal of Biological Chemistry*, **83**, 18218–18226.

Moghaddam, S. J., Barta, P., Mirabolfathinejad, S. G., *et al.* (2009). Curcumin inhibits COPD-like airway inflammation and lung cancer progression in mice. *Carcinogenesis*, **30**, 1949–1956.

Muth, V., Nadaud, S., Grummt, I., and Voit, R. (2001). Acetylation of TAFI68, a subunit of TIF-IB/SL1, activates RNA polymerase I transcription. *EMBO Journal*, **20**, 1353–1362

Nandakumar, D. N., Nagaraj, V. A., Vathsala, P. G., Rangarajan, P., and Padmanaban, G. (2006). Curcumin–artemisinin combination therapy for malaria. *Antimicrobial Agents and Chemotherapy*, **50**, 1859–1860.

Nicodeme, E., Jeffrey, K. L., Schaefer, U., *et al.* (2010). Suppression of inflammation by a synthetic histone mimic. *Nature*, **468**, 1119–1128.

Nightingale, K. P., Gendreizig, S., White, D. A., *et al.* (2007). Cross-talk between histone modifications in response to histone deacetylase inhibitors: MLL4 links histone H3 acetylation and histone H3K4 methylation. *Journal of Biological Chemistry*, **282**, 4408–4416.

Nucifora, F. C., Jr., Sasaki, M., Peters, M. F., *et al.* (2001). Interference by huntingtin and atrophin-1 with cbp mediated transcription leading to cellular toxicity. *Science*, **291**, 2423–2428.

Petrij, F., Giles, R. H., Dauwerse, H. G., *et al.* (1995). Rubinstein–Taybi syndrome caused by mutations in the transcriptional co-activator CBP. *Nature*, **376**, 348–351.

Qin, S. and Parthun, M. R. (2002) Histone H3 and the histone acetyltransferase Hat1p contribute to DNA double-strand break repair. *Molecular and Cellular Biology*, **22**, 8353–8365.

Ravindra, K. C., Selvi, B. R., Arif, M., *et al.* (2009). Inhibition of lysine acetyltransferase KAT3B/p300 activity by a naturally occurring hydroxynaphthoquinone, plumbagin. *Journal of Biological Chemistry*, **284**, 24 453–24 464.

Rouaux, C., Jokic, N., Mbebi, C., *et al.* (2003). Critical loss of CBP/p300 histone acetylase activity by caspase-6 during neurodegeneration. *EMBO Journal*, **22**, 6537–6549.

Saha, R. N. and Pahan, K. (2006). HATs and HDACs in neurodegeneration: a tale of disconcerted acetylation homeostasis. *Cell Death and Differentiation*, **13**, 539–550.

Seligson, D. B., Horvath, S., Shi, T., *et al.* (2005). Global histone modification patterns predict risk of prostate cancer recurrence. *Nature*, **435**, 1262–1266.

Seligson, D. B., Horvath, S., McBrian, M. A., *et al.* (2009). Global levels of histone modifications predict prognosis in different cancers. *American Journal of Pathology*, **174**, 1619–1628.

Selvi, B. R., Jagadeesan, D., Suma, B. S., *et al.* (2008). Intrinsically fluorescent carbon nanospheres as a nuclear targeting vector: delivery of membrane-impermeable molecule to modulate gene expression *in vivo*. *Nano Letters*, **8**, 3182–3188.

Selvi, B. R., Mohankrishna, D. V., Ostwal, Y. B., and Kundu, T. K. (2010a). Small molecule modulators of histone acetylation and methylation: A disease perspective. *Biochimica et Biophysica Acta*, **1799**, 810–828.

Selvi, B. R., Cassel, J. C., Kundu, T. K., and Boutillier, A. L. (2010b). Tuning acetylation levels with HAT activators: therapeutic strategy in neurodegenerative diseases. *Biochimica et Biophysica Acta*, **1799**, 840–853.

Sen, N., Hara, M. R., Kornberg, M. D., *et al.* (2008). Nitric oxide-induced nuclear GAPDH activates p300/CBP and mediates apoptosis. *Nature Cell Biology*, **10**, 866–873.

Seo, S. B., McNamara, P., Heo, S., *et al.* (2001). Regulation of histone acetylation and transcription by INHAT, a human cellular complex containing the set oncoprotein. *Cell*, **104**, 119–130.

Shahbazian, M. D. and Grunstein, M. (2007). Functions of site-specific histone acetylation and deacetylation. *Annual Reviews of Biochemistry*, **76**, 75–100.

Shandilya, J., Swaminathan, V., Gadad, S. S., *et al.* (2009). Acetylated NPM1 localizes in the nucleoplasm and regulates transcriptional activation of genes implicated in oral cancer manifestation. *Molecular and Cellular Biology*, **29**, 5115–5127.

Shih, Y. W., Lee, Y. C., Wu, P. F., Lee, Y. B., and Chiang, T. A. (2009). Plumbagin inhibits invasion and migration of liver cancer HepG2 cells by decreasing production of matrix metalloproteinase-2 and urokinase plasminogen activator. *Hepatology Research*, **39**, 998–1009.

Smith, B. C. and Denu, J. M. (2009). Chemical mechanisms of histone lysine and arginine modifications. *Biochimica et Biophysica Acta*, **1789**, 45–57.

Souto, J. A., Benedetti, R., Otto, K., *et al.* (2010). New anacardic acid-inspired benzamides: histone lysine acetyltransferase activators. *Chemical and Medicinal Chemistry*, **5**, 1530–1540.

Stimson, L., Rowlands, M. G., Newbatt, Y. M., *et al.* (2005). Isothiazolones as inhibitors of PCAF and p300 histone acetyltransferase activity. *Molecular Cancer Therapeutics*, **4**, 1521–1532.

Sun, Y., Jiang, X., Chen, S., and Price, B. D. (2006). Inhibition of histone acetyltransferase activity by anacardic acid sensitizes tumor cells to ionizing radiation. *FEBS Letters*, **580**, 4353–4356.

Sung, B., Pandey, M. K., Ahn, K. S., *et al.* (2008). Anacardic acid (6-nonadecyl salicylic acid), an inhibitor of histone acetyltransferase, suppresses expression of nuclear factor-kappaB-regulated gene products involved in cell survival, proliferation, invasion, and inflammation through inhibition of the inhibitory subunit of nuclear factor-kappaB alpha kinase, leading to potentiation of apoptosis. *Blood*, **111**, 4880–4891.

Tjeertes, J. V., Miller, K. M., and Jackson, S. P. (2009). Screen for DNA-damage-responsive histone modifications identifies H3K9Ac and H3K56Ac in human cells. *EMBO Journal*, **28**, 1878–1889.

Tontonoz, P., Hu, E., Devine, J., Beale, E. G., and Spiegelman, B. M. (1995). PPAR-gamma 2 regulates adipose expression of the phosphoenolpyruvate carboxykinase gene. *Molecular and Cellular Biology*, **15**, 351–357.

Wurtele, H., Tsao, S., Lépine, G., *et al.* (2010). Modulation of histone H3 lysine 56 acetylation as an antifungal therapeutic strategy. *Nature Medicine*, **16**, 774–780.

Yamauchi, T., Oike, Y., Kamon, J., *et al.* (2002). Increased insulin sensitivity despite lipodystrophy in Crebbp heterozygous mice. *Nature Genetics*, **30**, 221–226.

Yang, X. J. and Seto, E. (2008). Lysine acetylation: codified crosstalk with other posttranslational modifications. *Molecular Cell*, **31**, 449–461.

Zheng, Y., Balasubramanyam, K., Cebrat, M., *et al.* (2005). Synthesis and evaluation of a potent and selective cell-permeable p300 histone acetyltransferase inhibitor. *Journal of the American Chemical Society*, **127**, 17 182–17 188.

Part V

Epigenomics in neurodegenerative diseases

27 Study design considerations in epigenetic studies of neuropsychiatric disease

Fatemeh Haghighi,* Sephorah Zaman, and Yurong Xin

27.1   Introduction

In recent years, emerging evidence has revealed the importance of epigenetic processes in neurodevelopment and disease. Neurodevelopmental disorders such as autism and schizophrenia are complex disorders that likely arise from the interaction of alleles at multiple loci with environmental factors. However, additional information that affects phenotype is encoded in the distribution of epigenetic marks, including DNA methylation and histone modifications. Traditionally, the term "epigenetics" refers to the ensemble of such alterations that are heritable through both mitosis and meiosis. This definition is limiting for epigenetic studies of the brain. A prevailing idea is that functional states of neurons, which can be stable for many years, involve epigenetic phenomena (Hong *et al.*, 2005), but these states will not be transmitted to daughter cells because almost all neurons rarely divide. Such epigenetic phenomena have been reported in dynamic regulation of DNA methylation within differentiated neurons of human cerebral cortex throughout development, maturation, and aging. Striking DNA methylation changes have been observed in regulatory regions of a panel of 50 genes implicated in neurodevelopment (Siegmund *et al.*, 2007). Therefore, for neuroepigenetic studies, a comprehensive definition can be adopted that is in line with the contemporary use of the term epigenetics that encompasses both DNA methylation and histone modification marks. A unifying definition of epigenetic events that has been proposed by Bird (2007) states: "the structural adaptation of chromosomal regions so as to register, signal or perpetuate altered activity states." In the overview, we focus on DNA and histone methylation marks, because these marks are highly stable in postmortem brain. However, to simplify the subject matter in demonstrating relevant study design

* Author to whom correspondence should be addressed.

Epigenomics: From Chromatin Biology to Therapeutics, ed. K. Appasani. Published by Cambridge University Press. © Cambridge University Press 2012.

considerations in neuroepigenetic studies, we will in subsequent sections focus on DNA methylation.

Epigenetic marks can function in concert through multiple feedforward and feedback mechanisms, facilitating enhanced chromatin condensation for transcriptional silencing or chromatin opening for transcriptional activity (Li *et al.*, 2007; Ruthenburg *et al.*, 2007). The repertoire of DNA and histone modifications is established by such enzymes as DNA methyltransferases (DNMTs), histone acetyltransferases (HATs), histone deacetylases (HDACs), histone methyltransferases (HMTs), and histone demethylases (HDMs) (Shiio and Eisenman, 2003; Levenson and Sweatt, 2005; Shilatifard, 2006). These enzymes operate both together and independently to establish epigenetic marks that are highly dynamic and flexible in determining the pattern of gene expression (Jenuwein and Allis, 2001; Turner, 2002). As such, epigenetic mechanisms play a fundamental role in neuronal function and in the nervous system. They contribute to developmental and differentiation processes, and influence communication and signaling in neuronal networks (for review see Levenson *et al.*, 2006).

Epigenetic regulation of gene expression is thought to establish distinct domains of active and inactive chromatin structures, in part through posttranslational modifications of histone proteins. Histone modifications include acetylation, methylation, phosphorylation, and ubiquitylation (Jenuwein and Allis, 2001; Kouzarides, 2002). As compared with histone acetylation, which is generally associated with transcriptional activity, histone lysine methylation can serve as either repressive or active marks (Martin and Zhang, 2005). Specifically, methylation of lysine 4 (K4) and 27 (K27) on histone H3 has been shown to associate with active and repressed gene transcriptional states, respectively (Kirmizis *et al.*, 2004; Bracken *et al.*, 2006). These modifications are mediated through the Trithorax (TrxG) and Polycomb (PcG) protein complexes and play critical roles in lineage-specific developmental programs (Ringrose and Paro, 2004; Mohn *et al.*, 2008), thus making them ideal candidates for studies of neurodevelopmental disorders.

Accumulating evidence reveals that components of the histone methylation machinery are critical for normal brain function and development. Mutations in the AT-rich interactive domain 1C of Jarid1c, an X-linked gene *Jumonji* that encodes an H3K4me3-specific demethylase, results in mental retardation and autism (Iwase *et al.*, 2007; Adegbola *et al.*, 2008). Additionally, BRAF, a high mobility group (HMG)-containing protein, is thought to promote neuronal differentiation by recruiting an H3K4-specific methyltransferase, mixed-lineage leukemia (MLL1), to sites of neuronal gene promoters, thus activating neuronal gene expression (Wynder *et al.*, 2005). Mice lacking both Mll1 alleles die during early intrauterine development, whereas mice carrying only one intact allele show a defect in synaptic plasticity in the hippocampus (Kim *et al.*, 2007). Interestingly, there is evidence that the atypical anti-psychotic clozapine upregulates MLL1 occupancy and H3K4 methylation at selected gene promoters in vivo (Huang and Akbarian, 2007; Huang *et al.*, 2007).

DNA methylation is also an important epigenetic mark in studies of neuro-developmental disorders. In mammalian cells, DNA methylation occurs at the 5′ position of cytosine within CpG dinucleotides and is faithfully propagated on the daughter strand following DNA replication by the maintenance DNA methyltransferase 1 enzyme (DNMT1). This form of information is flexible enough to be adapted for different somatic cell types, yet is stable enough to be retained during mitosis and/or meiosis. DNA methylation is commonly associated with transcriptional silencing because it can directly inhibit the binding of transcription factors or regulators, or recruit methyl-CpG binding proteins (MBPs) with repressive chromatin-remodeling functions (Bird, 2002; Klose and Bird, 2006). DNA methylation plays a role in the protection against intragenomic parasites (Yoder et al., 1997), in genomic imprinting (Li et al., 1992), and in X-chromosome inactivation in females. Methylation of CpG dinucleotides is critical in genome defense and chromosomal structural integrity. Examination of distribution of CpG dinucleotides within the genome has revealed that most reside within parasitic DNA elements, such as endogenous retroviruses, L1 elements, and Alu elements (which are quite CpG-rich) (Yoder et al., 1997). These parasitic DNA elements constitute about 40% of the human genome, and so it has been posited that DNA methylation may have arisen to silence the activity and limit the proliferation of these elements (referred to as the "host-defense hypothesis") (Yoder et al., 1997; Bestor, 2003). Parasitic elements may pose a significant threat to genome stability and structure by mediating recombination between non-allelic repeats resulting in chromosome rearrangements or translocations (Yoder et al., 1997; Chen et al., 1998; Montagna et al., 1999). Methylation changes at repetitive elements can alter the expression of nearby genes resulting in phenotypic changes that can alter the risk of developing disease (Gaudet et al., 2004).

Recent studies have revealed involvement of DNA methylation in neural development and regulation of synaptic plasticity (Levenson et al., 2006). Conditional knockout of the *Dnmt1* gene in mouse forebrain disrupts sensory map formation and synaptic plasticity (Golshani et al., 2005). Synaptic plasticity is the ability of neuronal cells to strengthen or weaken their connections following neuronal activation. Inhibition of DNA methylation by DNMT inhibitors such as zebularine or 5-aza-2-deoxycytidine (5-aza) impairs induction of long-term potentiation (LTP) in hippocampus (Levenson et al., 2006). Deficiency of MBD1, a DNA methyl-binding protein and transcriptional repressor, also reduces LTP in hippocampus (Zhao et al., 2003). Further, administration of phorbol ester appears to induce activation of protein kinase C (PKC), which decreases DNA methylation, specifically at the promoter region of *reelin*, a gene implicated in induction of synaptic plasticity. Activation of PKC is further accompanied by increased transcriptional activation of *c-fos*, an immediate early gene involved in synaptic transmission, and of DNA methyltransferase DNMT3a, suggesting the involvement of DNA methylation in synaptic signaling cascades. Although substantial work still remains to be done, these findings are intriguing because they reveal that DNA methylation, a process initially thought to be static, is dynamically modulated during synaptic plasticity in mammalian brain. Therefore, the delicate balance of DNA and histone methylation marks in brain chromatin is essential for proper brain development and function.

Epigenetic alterations at selected genomic loci may affect social cognition (Isles et al., 2006), learning and memory (Miller and Sweatt, 2007), and stress-related behaviors (Weaver et al., 2004), and contribute to aberrant gene expression in a range of neurodevelopmental disorders, including autism, schizophrenia, depression, and Alzheimer's disease (Scarpa et al., 2003; Abdolmaleky et al., 2005; Grayson et al., 2005; Nagarajan et al., 2006; Polesskaya et al., 2006). In this chapter, we will outline key issues for consideration in design of epigenetic studies in the brain and brain-based disorders.

27.2   DNA methylation approaches and design considerations

In recent years, many new approaches have been developed to study genome-wide DNA methylation patterns, providing substantial insight into the role of cytosine methylation in genome organization and function. Some approaches depend on the use of restriction enzymes sensitive to methylated CG sites where the level of DNA methylation is quantified by hybridization to high-density oligonucleotide arrays or sequencing via next-generation sequencing platforms (Lippman et al., 2005; Khulan et al., 2006; Yuan et al., 2006; Irizarry et al., 2008; Kerkel et al., 2008; Edwards et al., 2010). Other approaches capture methylated genomic DNA using immunoprecipitation via an antibody that recognizes 5-methylcytosine, followed by array hybridization or sequencing (Wynder et al., 2005; Keshet et al., 2006; Weber et al., 2007; Down et al., 2008). These approaches can determine chromosomal methylation levels and patterns, but cannot detect DNA methylation at single-base resolution. Direct sequencing of bisulfite-treated DNA allows mapping of methylation of individual cytosines in a genome-wide fashion (Lister and Ecker, 2009). However, the depth of sequencing necessary for such DNA methylation mapping is still quite costly in applications of large numbers of primary samples. This approach was recently used for cross-species comparative DNA methylation analyses of eight divergent species, including green algae, plants, insects, and animals (Feng et al., 2010; Zemach et al., 2010). Due to limited depth of sequencing coverage for the larger genomes (such as the mouse), quantification of methylation levels could not be extended to individual cytosine bases, but were limited to broad genomic features, like genes and repetitive elements. Nevertheless, these studies revealed intriguing trends in both the conserved and divergent features of DNA methylation in eukaryotic evolution. Cytosine methylation was detected throughout the majority of repetitive sequences and transposons, but also within the body of protein-coding genes. Within gene bodies, CpG methylation appeared to be favored in exons over introns. Although the biological function of gene body methylation or mechanisms by which gene bodies are targeted by the methylation machinery are not well understood, the preferential methylation of exons in plant and animal species appears to be an evolutionarily conserved phenomenon (Feng et al., 2010). Therefore, depending on the specific biological question to be investigated, varying experimental approaches for DNA methylation analyses can be employed.

Findings from these whole-genome methylation-profiling assays require further experimental validation. Bisulfite sequencing and cloning and methylation pyrosequencing are two such approaches that allow direct interrogation of CpG methylation at single-base resolution within specific genomic regions of interest. These approaches are based on sodium bisulfite conversion of cytosine residues to uracil under conditions in which 5-methylcytosine remains unaltered. The bisulfite-treated DNA is then amplified, cloned, and sequenced (in bisulfite mapping) or amplified and directly sequenced (in methylation pyrosequencing). The cytosine residues in the sequence represent methylated cytosines in the genome that were protected from deamination by bisulfite (Clark *et al.*, 2006). Whole-genome DNA methylation studies, nonetheless, typically identify hundreds or thousands of CpG methylation changes that appear to be interesting biologically. Array-based technologies provide an alternative solution for probing samples for large panels of CpG sites across regions of interest. Illumina (San Diego, CA, USA) methylation chips are one such platform that can be adapted to include cytosine bases within specific regions of the sample and experiment at hand (Smith *et al.*, 2009). Design of such custom arrays is ideal for use in validation and subsequent replication studies. Moreover, Illumina offers methylation chips that provide limited coverage of CpG dinucleotides within gene features including promoters, exons, introns, and 5′ and 3′ untranslated regions. These include the Illumina Infinium HumanMethylation 27K BeadChip, and the recently released HumanMethylation 450K BeadChip with 27 578 and 485 553 CpG sites, respectively. Although the number of CpG sites captured within each of the gene features is very low, these chips provide highly reliable and reproducible data in a cost-effective fashion and importantly require only a small amount of starting DNA (~500 ng per sample). Therefore, they are an attractive option for studies that involve large numbers of clinical samples or studies that involve limited quantities of rare and valuable tissue. Thus, applications that involve human postmortem brain tissue specimens may be ideally suited. This is not only because human postmortem brain tissue is a very rare and valuable resource, but also because the dynamic range for CpG methylation changes in the brain appears to be very small.

A number of studies involving both human and animal models support this observation, revealing that the magnitude of DNA methylation changes associated with behavioral outcomes is small. In a large-scale study investigating the role of methylation in psychosis, Mill *et al.* analyzed methylation levels at approximately 7800 loci in the frontal cortex of postmortem brain tissue from individuals diagnosed with schizophrenia or bipolar disorder and control subjects. This study, which analyzed loci largely within CpG islands of promoter regions, found relatively small differences in absolute methylation levels among control and schizophrenia samples (Mill *et al.*, 2008). Further, in a rat model of posttraumatic stress disorder (PTSD), changes in average DNA methylation levels in exon IV of the brain-derived neurotrophic factor (*Bdnf*) gene assessed in dorsal hippocampus ranged from 5% to 20% (Roth *et al.*, 2011). A rat model of contextual fear memory also showed between 10% and 20% change in methylation levels at individual CpG sites near the transcription start site of *Bdnf* exon IV (Lubin *et al.*, 2008). DNA methylation levels of the pituitary adenylate cyclase-activating polypeptide (PACAP) receptor genes (encoded by ADCYAP1R1) analyzed in a cohort of more than 1200 highly traumatized subjects with and without PTSD showed that at the first site within the ADCYAP1R1 CpG island, methylation levels ranged from 2% to 10% with 0–40 increase in PTSD symptom scale (Ressler *et al.*, 2011). Interrogation of the methylation levels of the serotonin receptor 5HTR1A gene promoter region in schizophrenia and bipolar subjects compared to controls also exhibits small changes (Carrard *et al.*, 2011).

The extent of DNA methylation variability that may be detectable is highly dependent on genomic context. DNA methylation patterns are generally conserved at the transcription start site (TSS) of genes, yet significantly diverge proximal to the TSS. These regions of divergence are also referred to as "shores" in the context of CpG-rich CpG island promoters. It has been posited that such regions may contain regulatory sequences with tissue-specific methylation pattern (Irizarry *et al.*, 2009), and emerging data from our own work suggest that these regions may also show methylation patterns that are species-specific and/or disease-specific. Furthermore, such genomic regions with intermediate to low CpG density also exhibit regional and inter-individual variability, as demonstrated in whole-genome DNA methylation profiling of two human cortical regions, namely prefrontal and auditory cortices. These cortical specimens were obtained from the same individuals for the DNA methylation analyses.

The Methyl-MAPS method (methylation mapping analysis by paired-end sequencing) was used to profile DNA methylation patterns among these samples consisting of eight subjects with no history of neurological or neuropsychiatric disorders. Methyl-MAPS can delineate the methylation status of greater than 80% of CpG sites genome-wide (Rollins *et al.*, 2006; Edwards *et al.*, 2010). It is an enzymatic-based method that uses a battery of methylation-sensitive and -dependent endonucleases to profile DNA methylation patterns throughout the genome in an unbiased fashion. With the Methyl-MAPS method, we were able to map the DNA methylation state of up to 82% of CpG sites for 10 normal human adult cortical specimens (with $\geq 8\times$ sequence reads per CpG site). Using this coverage constraint, 36%, i.e., 10 262 160, CpG sites were mapped across all 10 specimens. These samples consist of six ventral prefrontal cortex (VPFC) and four auditory cortex (AC) specimens, where for four subjects we have tissue specimens for both dorsolateral PFC (VLPFC) and AC. Data from these analyses show that DNA methylation variability increases relative to CpG density (Figure 27.1). In comparing the variability among brain regions and individuals, the data show that regions with high CpG content, like CpG islands, show conserved methylation patterns (CpG density ≥ 4 CpGs/100 bp window). In contrast, CpG island shores with intermediate to low CpG density show greater DNA methylation variability (Figure 27.1). While in studies of brain and behavior DNA methylation changes of small magnitude may be intrinsic, they also may be attributed to

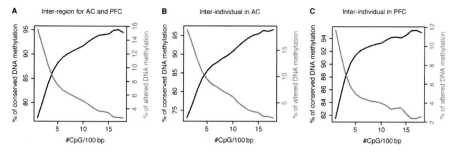

Figure 27.1 Inter-individual methylation differences show similar patterns to regional differences (across brain regions). DNA methylation conservation increases with increasing CpG density, whereas DNA methylation alteration increases with decreasing CpG density. Across individuals and brain regions examined, DNA methylation is considered as conserved when the difference is less than 0.2, and altered when the difference is greater than 0.25.

genomic context or specifically the CpG content within the regions investigated. Expanding the DNA methylation analyses to those regions of the genome with intermediate to low CpG content will likely allow detection of methylation changes of greater magnitude across samples. However, this gain has an associated cost, mainly increase in inter-individual methylation variability. This is demonstrated by the power analyses, where increase in DNA methylation variability ($\sigma = 0.2$) results in a dramatic reduction in power especially when differences in methylation changes among disease and control samples are small (i.e., methylation difference $= 0.05$) (Figure 27.2). The increase in methylation variability is not only due to the expanded genomic coverage, but also may be confounded by the experimental method used for whole-genome methylation profiling. Typically, DNA methylation analyses based on next-generation sequencing provide genome-wide coverage. Yet, due to variable depth of sequencing across samples, methylation estimates are highly variable (Robinson *et al.*, 2010).

27.3 Tissue and cellular specificity of epigenetic patterns

The transcriptional specificity observed in tissues and cells in the human body is determined by epigenetic factors. Differences in gene expression among different cells reflect differences in epigenetic patterns. We and others have shown that DNA methylation patterns vary among brain regions (Ladd-Acosta *et al.*, 2007; Xin *et al.*, 2010). DNA methylation patterns also vary by chronological age (Hernandez *et al.*, 2011) Further, within each brain region, there are a multitude of cell populations each also with varying epigenetic profiles. Epigenetic data from whole tissue provides a crude picture because brain tissue contains a complex array of cells with potentially varied DNA methylation signatures. This cellular heterogeneity impacts methylation estimates throughout the genome and thus makes comparison and interpretation of methylation data across samples difficult. Given the cellular complexity of the brain, investigations that focus on the study of specific cell populations are ideal. Established methods for isolation of neuronal cells (Spalding *et al.*, 2005; Jiang *et al.*, 2008) have been used to

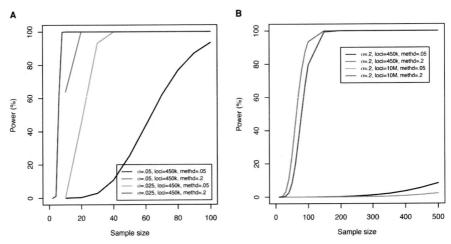

Figure 27.2 Power analyses using two-sample *t*-test. Sample size (*x*-axis) represents number of samples in each group, specifically case (n_1) vs. control (n_2) groups, assuming equal numbers of cases and controls ($n_1 = n_2$). The parameters including DNA methylation variation (σ), Type I error, and methylation difference (methd) between case and control groups were based on data from Illumina Infinium HumanMethylation Beadchip and Methyl-MAPS. The Type I error rates were adjusted using Bonferroni correction for multiple testing. (A) Parameters were estimated from data from Illumina methylation assay, where σ-values reflect methylation variation in tissue ($\sigma = 0.05$) and neurons ($\sigma = 0.025$). (B) Parameters were estimated from Methyl-MAPS genome-wide methylation data from brain tissue, where the methylation variation is larger than those observed in the Illumina methylation platform. Although Methyl-MAPS and other whole-genome methylation approaches can interrogate CpG methylation for greater than 90% of the human genome, for illustrative purposes we show power analyses for only 10 million CpG sites.

isolate neuronal nuclei by immunotagging and subsequent fluorescence-activated sorting. Laser capture microdissection (LCM) is another widely used approach for isolating neuronal populations or brain regions (Suarez-Quian *et al.*, 1999). These approaches are technically demanding, typically yielding limited amounts of DNA; hence to date their application for epigenetic analyses has been limited. Nevertheless, they are important in facilitating detailed epigenetic investigations of neuronal cell populations.

27.4   Strategies for designing epigenetic studies

Broadly there are two distinct design strategies that have been adopted in epigenetic studies to date. In these studies, there is a compromise between depth and breadth. In-depth investigations involve the use of well-characterized postmortem human brain tissue from regions implicated in disease neuropathology. In addition to examination of epigenetic patterns within specific brain regions, epigenetic patterns can also be studied in specific cell populations. Postmortem samples also require availability of detailed psychological and toxicological autopsy data, which are critical in postmortem studies of neuropsychiatric disorders. It is extremely difficult successfully to execute such in-depth studies, owing to the limited availability of well-characterized frozen human brain specimens and the technical difficulty in isolating sufficient amounts of starting DNA from specific cell populations of interest. Consequently, these studies are necessarily based on small sample sizes. The underlying assumption in these studies is that the epigenetic signal will be greater than the inter-individual variability. This is because the potential noise related to tissue cellular heterogeneity has been eliminated by focusing on cell-specific epigenetic signatures, where in principle the observed signal should be binary (i.e., 1 or 0 for methylated or unmethylated states, respectively). We demonstrate the power of such in-depth studies under various conditions. Based on our own DNA methylation data from tissue and neurons obtained from human cortex, we estimated DNA methylation variability for both the Illumina Infinium HumanMethylation BeadChip and whole-genome methylation profiling by Methyl-MAPS. These empirically derived variance parameters were then used for the power analyses in Figure 27.2. For in-depth postmortem studies with limited sample sizes, we observed $\geq$77% power for even modest sample sizes of 80 cases and 80 controls for $\sigma = 0.05$ and methylation difference $= 0.05$ for tissue-specific methylation patterns. The observed power is even greater for cell-specific DNA methylation patterns, because there is a twofold reduction in methylation variability in neuronal cells as compared to tissue (Figure 27.2A). In contrast, less-focused investigations that utilize large samples combining brain, blood, and saliva from samples of cases and controls will likely exhibit greater DNA methylation variability (Figure 27.2B). This is largely due to the tissue-specific nature of the epigenome. Still, given the limited availability of postmortem brain, these studies are far more tractable. These approaches can employ large samples, albeit at a cost of increase in signal-to-noise ratio, both with respect to sample and tissue heterogeneity. The underlying assumption is that the convergence of the signal from various tissue types will strengthen the findings and will identify potential biomarkers of disease susceptibility. Such studies may be of utility in identifying potential epigenetic alterations resulting from early embryogenesis or identifying inherited epigenetic variations (Pidsley and Mill, 2011). From a practical standpoint, peripheral tissue is far easier to obtain and would be of wide utility in prospective longitudinal epigenetic studies in monitoring changes associated with disease progression. To date, no study has systematically mapped epigenetic variations between specific brain regions and correlated these with DNA methylation in peripheral tissues from the same individual. Given the paucity of human postmortem tissues, such studies will be important in establishing the feasibility of future large-scale epigenetic studies of neuropsychiatric disorders.

27.5 Conclusion

The burgeoning field of neuroepigenetics holds great promise in revealing molecular abnormalities associated with mental illness. Together with genetic mutations, environmentally induced epigenetic mutations may result in neuropsychiatric disease. Epigenetics provides a mechanism for mediating the interaction of our genes with our environment. Although we have discussed the nature of epigenetic variation, we have not addressed the contribution of the environment in inducing epigenetic modifications. These are study-dependent and typically difficult to model in human-based research. However, studies involving animal models reveal that the epigenome is vulnerable to environmentally induced modifications, particularly during prenatal development because that is when elaborate DNA methylation patterning and chromatin structure remodeling required for normal tissue development are established (Dolinoy *et al.*, 2007). Therefore, in designs of neuroepigenetic studies, it is essential to adjust for environmental confounds when possible. Emerging evidence reveals that in depressed suicide victims, early childhood abuse and deprivation are associated with DNA methylation alterations (Mann and Haghighi, 2010). In addition, intrauterine exposure to infections, starvation, micronutrient deficiencies, and other factors are associated with an increased incidence of schizophrenia later in life – possibly through epigenetic modifications during prenatal brain development (Penner and Brown, 2007; Brown and Derkits, 2010; Brown and Patterson, 2011). The influence of the environment in inducing epigenetic alterations during sensitive periods of early development and throughout the lifespan could have an indelible effect on the mental health of an individual.

REFERENCES

Abdolmaleky, H. M., Cheng, K. H., Russo, A., *et al.* (2005). Hypermethylation of the reelin (RELN) promoter in the brain of schizophrenic patients: a preliminary report. *American Journal of Medical Genetics, Part B, Neuropsychiatric Genetics*, **134**, 60–66.

Adegbola, A., Gao, H., Sommer, S., and Browning, M. (2008). A novel mutation in JARID1C/SMCX in a patient with autism spectrum disorder (ASD). *American Journal of Medical Genetics, Part A*, **146**, 505–511.

Bestor, T. H. (2003). Cytosine methylation mediates sexual conflict. *Trends in Genetics*, **19**, 185–190.

Bird, A. (2002). DNA methylation patterns and epigenetic memory. *Genes and Development*, **16**, 6–21.

Bird, A. (2007). Perceptions of epigenetics. *Nature*, **447**, 396–398.

Bracken, A. P., Dietrich, N., Pasini, D., Hansen, K. H., and Helin, K. (2006). Genome-wide mapping of Polycomb target genes unravels their roles in cell fate transitions. *Genes and Development*, **20**, 1123–1136.

Brown, A. S. and Derkits, E. J. (2010). Prenatal infection and schizophrenia: a review of epidemiologic and translational studies. *American Journal of Psychiatry*, **167**, 261–280.

Brown, A. S. and Patterson, P. H. (2011). Maternal infection and schizophrenia: implications for prevention. *Schizophrenia Bulletin*, **37**, 284–290.

Carrard, A., Salzmann, A., Malafosse, A., and Karege, F. (2011). Increased DNA methylation status of the serotonin receptor 5HTR1A gene promoter in schizophrenia and bipolar disorder. *Journal of Affective Disorders*, **132**, 450–453.

Chen, R. Z., Pettersson, U., Beard, C., Jackson-Grusby, L., and Jaenisch, R. (1998). DNA hypomethylation leads to elevated mutation rates. *Nature*, **395**, 89–93.

Clark, S. J., Statham, A., Stirzaker, C., Molloy, P. L., and Frommer, M. (2006). DNA methylation: bisulphite modification and analysis. *Nature Protocols*, **1**, 2353–2364.

Dolinoy, D. C., Weidman, J. R., and Jirtle, R. L. (2007). Epigenetic gene regulation: linking early developmental environment to adult disease. *Reproductive Toxicology*, **23**, 297–307.

Down, T. A., Rakyan, V. K., Turner, D. J., *et al.* (2008). A Bayesian deconvolution strategy for immunoprecipitation-based DNA methylome analysis. *Nature Biotechnology*, **26**, 779–785.

Edwards, J. R., O'Donnell, A. H., Rollins, R. A., *et al.* (2010). Chromatin and sequence features that define the fine and gross structure of genomic methylation patterns. *Genome Research*, **20**, 972–980.

Feng, S., Cokus, S. J., Zhang, X., *et al.* (2010). Conservation and divergence of methylation patterning in plants and animals. *Proceedings of the National Academy of Sciences USA*, **107**, 8689–8694.

Gaudet, F., Rideout, W. M., III, Meissner, A., *et al.* (2004). *Dnmt1* expression in pre- and postimplantation embryogenesis and the maintenance of IAP silencing. *Molecular and Cellular Biology*, **24**, 1640–1648.

Golshani, P., Hutnick, L., Schweizer, F., and Fan, G. (2005). Conditional *Dnmt1* deletion in dorsal forebrain disrupts development of somatosensory barrel cortex and thalamo-cortical long-term potentiation. *Thalamus and Related Systems*, **3**, 227–233.

Grayson, D. R., Jia, X., Chen, Y., *et al.* (2005). *Reelin* promoter hypermethylation in schizophrenia. *Proceedings of the National Academy of Sciences USA*, **102**, 9341–9346.

Hernandez, D. G., Nalls, M. A., Gibbs, J. R., *et al.* (2011). Distinct DNA methylation changes highly correlated with chronological age in the human brain. *Human Molecular Genetics*, **20**, 1164–1172.

Hong, E. J., West, A. E., and Greenberg, M. E. (2005). Transcriptional control of cognitive development. *Current Opinion in Neurobiology*, **15**, 21–28.

Huang, H. S. and Akbarian, S. (2007). GAD1 mRNA expression and DNA methylation in prefrontal cortex of subjects with schizophrenia. *PLoS One*, **2**, e809.

Huang, H. S., Matevossian, A., Whittle, C., *et al.* (2007). Prefrontal dysfunction in schizophrenia involves mixed-lineage leukemia 1-regulated histone methylation at GABAergic gene promoters. *Journal of Neuroscience*, **27**, 11 254–11 262.

Irizarry, R. A., Ladd-Acosta, C., Carvalho, B., *et al.* (2008). Comprehensive high-throughput arrays for relative methylation (CHARM). *Genome Research*, **18**, 780–790.

Irizarry, R. A., Ladd-Acosta, C., Wen, B., *et al.* (2009). The human colon cancer methylome shows similar hypo- and hypermethylation at conserved tissue-specific CpG island shores. *Nature Genetics*, **41**, 178–186.

Isles, A. R., Davies, W., and Wilkinson, L. S. (2006). Genomic imprinting and the social brain. *Philosophical Transactions of the Royal Society, Series B*, **361**, 2229–2237.

Iwase, S., Lan, F., Bayliss, P., *et al.* (2007). The X-linked mental retardation gene *SMCX/ JARID1C* defines a family of histone H3 lysine 4 demethylases. *Cell*, **128**, 1077–1088.

Jenuwein, T. and Allis, C. D. (2001). Translating the histone code. *Science*, **293**, 1074–1080.

Jiang, Y., Matevossian, A., Huang, H. S., Straubhaar, J., and Akbarian, S. (2008). Isolation of neuronal chromatin from brain tissue. *BMC Neuroscience*, **9**, 42.

Kerkel, K., Spadola, A., Yuan, E., *et al.* (2008). Genomic surveys by methylation-sensitive SNP analysis identify sequence-dependent allele-specific DNA methylation. *Nature Genetics*, **40**, 904–908.

Keshet, I., Schlesinger, Y., Farkash, S., *et al.* (2006). Evidence for an instructive mechanism of de novo methylation in cancer cells. *Nature Genetics*, **38**, 149–153.

Khulan, B., Thompson, R. F., Ye, K., *et al.* (2006). Comparative isoschizomer profiling of cytosine methylation: the HELP assay. *Genome Research*, **16**, 1046–1055.

Kim, S. Y., Levenson, J. M., Korsmeyer, S., Sweatt, J. D., and Schumacher, A. (2007). Developmental regulation of Eed complex composition governs a switch in global histone modification in brain. *Journal of Biological Chemistry*, **282**, 9962–9972.

Kirmizis, A., Bartley, S. M., Kuzmichev, A., *et al.* (2004). Silencing of human polycomb target genes is associated with methylation of histone H3 Lys 27. *Genes and Development*, **18**, 1592–1605.

Klose, R. J. and Bird, A. P. (2006). Genomic DNA methylation: the mark and its mediators. *Trends in Biochemical Sciences*, **31**, 89–97.

Kouzarides, T. (2002). Histone methylation in transcriptional control. *Current Opinion in Genetics and Development*, **12**, 198–209.

Ladd-Acosta, C., Pevsner, J., Sabunciyan, S., *et al.* (2007). DNA methylation signatures within the human brain. *American Journal of Human Genetics*, **81**, 1304–1315.

Levenson, J. M. and Sweatt, J. D. (2005). Epigenetic mechanisms in memory formation. *Nature Reviews Neuroscience*, **6**, 108–118.

Levenson, J. M., Roth, T. L., Lubin, F. D., *et al.* (2006). Evidence that DNA (cytosine-5) methyltransferase regulates synaptic plasticity in the hippocampus. *Journal of Biological Chemistry*, **281**, 15 763–15 773.

Li, B., Carey, M., and Workman, J. L. (2007). The role of chromatin during transcription. *Cell*, **128**, 707–719.

Li, E., Bestor, T. H., and Jaenisch, R. (1992). Targeted mutation of the DNA methyltransferase gene results in embryonic lethality. *Cell*, **69**, 915–926.

Lippman, Z., Gendrel, A. V., Colot, V., and Martienssen, R. (2005). Profiling DNA methylation patterns using genomic tiling microarrays. *Nature Methods*, **2**, 219–224.

Lister, R. and Ecker, J. R. (2009). Finding the fifth base: genome-wide sequencing of cytosine methylation. (Review). *Genome Research*, **19**, 959–966.

Lubin, F. D., Roth, T. L., and Sweatt, J. D. (2008). Epigenetic regulation of BDNF gene transcription in the consolidation of fear memory. *Journal of Neuroscience*, **28**, 10 576–10 586.

Mann, J. J. and Haghighi, F. (2010). Genes and environment: multiple pathways to psychopathology. *Biological Psychiatry*, **68**, 403–404.

Martin, C. and Zhang, Y. (2005). The diverse functions of histone lysine methylation. *Nature Reviews Molecular Cell Biology*, **6**, 838–849.

Mill, J., Tang, T., Kaminsky, Z., *et al.* (2008). Epigenomic profiling reveals DNA-methylation changes associated with major psychosis. *American Journal of Human Genetics*, **82**, 696–711.

Miller, C. A. and Sweatt, J. D. (2007). Covalent modification of DNA regulates memory formation. *Neuron*, **53**, 857–869.

Mohn, F., Weber, M., Rebhan, M., *et al.* (2008). Lineage-specific polycomb targets and de novo DNA methylation define restriction and potential of neuronal progenitors. *Molecular Cell*, **30**, 755–766.

Montagna, M., Santacatterina, M., Torri, A., *et al.* (1999). Identification of a 3 kb Alu-mediated *BRCA1* gene rearrangement in two breast/ovarian cancer families. *Oncogene*, **18**, 4160–4165.

Nagarajan, R. P., Hogart, A. R., Gwye, Y., Martin, M. R., and LaSalle, J. M. (2006). Reduced MeCP2 expression is frequent in autism frontal cortex and correlates with aberrant MECP2 promoter methylation. *Epigenetics*, **1**, e1–11.

Penner, J. D. and Brown, A. S. (2007). Prenatal infectious and nutritional factors and risk of adult schizophrenia. *Expert Review of Neurotherapeutics*, **7**, 797–805.

Pidsley, R. and Mill, J. (2011). Epigenetic studies of psychosis: current findings, methodological approaches, and implications for postmortem research. *Biological Psychiatry*, **69**, 146–156.

Polesskaya, O. O., Aston, C., and Sokolov, B. P. (2006). Allele C-specific methylation of the 5-HT2A receptor gene: evidence for correlation with its expression and expression of DNA methylase DNMT1. *Journal of Neuroscience Research*, **83**, 362–373.

Ressler, K. J., Mercer, K. B., Bradley, B., *et al.* (2011). Post-traumatic stress disorder is associated with PACAP and the PAC1 receptor. *Nature*, **470**, 492–497.

Ringrose, L. and Paro, R. (2004). Epigenetic regulation of cellular memory by the Polycomb and Trithorax group proteins. *Annual Review of Genetics*, **38**, 413–443.

Robinson, M. D., Statham, A. L., Speed, T. P., and Clark, S. J. (2010). Protocol matters: which methylome are you actually studying? *Epigenomics*, **2**, 587–598.

Rollins, R. A., Haghighi, F., Edwards, J. R., *et al.* (2006). Large-scale structure of genomic methylation patterns. *Genome Research*, **16**, 157–163.

Roth, T. L., Zoladz, P. R., Sweatt, J. D., and Diamond, D. M. (2011). Epigenetic modification of hippocampal *Bdnf* DNA in adult rats in an animal model of post-traumatic stress disorder. *Journal of Psychiatric Research*, **45**, 919–926.

Ruthenburg, A. J., Allis, C. D., and Wysocka, J. (2007). Methylation of lysine 4 on histone H3: intricacy of writing and reading a single epigenetic mark. *Molecular Cell*, **25**, 15–30.

Scarpa, S., Fuso, A., D'Anselmi, F., and Cavallaro, R. A. (2003). Presenilin 1 gene silencing by S-adenosylmethionine: a treatment for Alzheimer disease? *FEBS Letters*, **541**, 145–148.

Shiio, Y. and Eisenman, R. N. (2003). Histone sumoylation is associated with transcriptional repression. *Proceedings of the National Academy of Sciences USA*, **100**, 13 225–13 230.

Shilatifard, A. (2006). Chromatin modifications by methylation and ubiquitination: implications in the regulation of gene expression. *Annual Review of Biochemistry*, **75**, 243–269.

Siegmund, K. D., Connor, C. M., Campan, M., *et al.* (2007). DNA methylation in the human cerebral cortex is dynamically regulated throughout the life span and involves differentiated neurons. *PLoS One*, **2**, e895.

Smith, Z. D., Gu, H., Bock, C., Gnirke, A., and Meissner, A. (2009). High-throughput bisulfite sequencing in mammalian genomes. *Methods*, **48**, 226–232.

Spalding, K. L., Bhardwaj, R. D., Buchholz, B. A., Druid, H., and Frisen, J. (2005). Retrospective birth dating of cells in humans. *Cell*, **122**, 133–143.

Suarez-Quian, C. A., Goldstein, S. R., Pohida, T., *et al.* (1999). Laser capture microdissection of single cells from complex tissues. *BioTechniques*, **26**, 328–335.

Turner, B. M. (2002). Cellular memory and the histone code. *Cell*, **111**, 285–291.

Weaver, I. C., Cervoni, N., Champagne, F. A., *et al.* (2004). Epigenetic programming by maternal behavior. *Nature Neuroscience*, **7**, 847–854.

Weber, M., Hellmann, I., Stadler, M. B., *et al.* (2007). Distribution, silencing potential and evolutionary impact of promoter DNA methylation in the human genome. *Nature Genetics*, **39**, 457–466.

Wynder, C., Hakimi, M. A., Epstein, J. A., Shilatifard, A., and Shiekhattar, R. (2005). Recruitment of MLL by HMG-domain protein iBRAF promotes neural differentiation. *Nature Cell Biology*, **7**, 1113–1117.

Xin, Y., Chanrion, B., Liu, M. M., *et al.* (2010). Genome-wide divergence of DNA methylation marks in cerebral and cerebellar cortices. *PLoS One*, **5**, e11357.

Yoder, J. A., Walsh, C. P., and Bestor, T. H. (1997). Cytosine methylation and the ecology of intragenomic parasites. *Trends in Genetics*, **13**, 335–340.

Yuan, E., Haghighi, F., White, S., *et al.* (2006). A single nucleotide polymorphism chip-based method for combined genetic and epigenetic profiling: validation in decitabine therapy and tumor/normal comparisons. *Cancer Research*, **66**, 3443–3451.

Zemach, A., McDaniel, I. E., Silva, P., and Zilberman, D. (2010). Genome-wide evolutionary analysis of eukaryotic DNA methylation. *Science*, **328**, 916–919.

Zhao, X., Ueba, T., Christie, B. R., *et al.* (2003). Mice lacking methyl-CpG binding protein 1 have deficits in adult neurogenesis and hippocampal function. *Proceedings of the National Academy of Sciences USA*, **100**, 6777–6782.

28 Epigenetic regulation in human neurodevelopmental disorders including autism, Rett syndrome, and epilepsy

Laura B. K. Herzing

28.1   Introduction

Phenotypic variability and overlapping symptomatology present considerable challenges to understanding the mechanisms underlying many human disorders, especially those involving neurodevelopmental and neuropsychiatric systems. It has long been understood that, for single-gene and especially for complex disorders, interplay between environmental and genetic factors contributes to such variability. More recently, it has become clear that epigenetics also plays a significant role in modifying phenotypic outcome, in addition to being causative in, e.g., disorders of genetic imprinting, wherein genes are preferentially or exclusively expressed from either the paternally or maternally inherited chromosome, as determined by allele-specific "imprints." These epigenetic modifications of chromatin, such as cytosine methylation (of DNA) or histone methylation or acetylation (in nucleosomes), may lead to modified nuclear localization or *trans*-interactions between genetic loci, and impact gene expression without altering the genetic sequence directly. Epigenetic "marks" may be susceptible to environmental modification, and may have subtle versus extreme, or local versus global effects on gene expression patterns. Thus, epigenetics provides a bridging mechanism between gene and environment, encompassing both phenotypic variability and overlap within and between disorders.

The study of epigenetics and its (dys)regulation continues to provide much insight into normal processes and the etiologies of many disorders. This chapter will focus on the role of epigenetics in neurodevelopmental disorders with overlapping autism spectrum and seizure phenotypes including Rett and Angelman syndromes, and will review data investigating the relevance of epigenetic modification of the chromosome 15q11–q13 locus in these disorders.

Epigenomics: From Chromatin Biology to Therapeutics, ed. K. Appasani. Published by Cambridge University Press. © Cambridge University Press 2012.

28.2 Altered epigenetic regulation in neurodevelopmental disorders

28.2.1 Mechanism

Epigenetic alterations in gene expression leading to abnormal phenotypic outcome may arise through multiple mechanisms. The simplest is a change in DNA or histone modification at a specific regulatory domain, most commonly an individual gene promoter. This can block gene expression, through increased chromatin condensation (heterochromatinization) or facilitating binding of a repressor, or, conversely, enhance gene expression through decreased chromatin condensation or activator binding. Similarly, altered chromatin modification may occur at enhancer elements, or at regulatory elements controlling large domains or gene clusters. Such changes may result from errors or inefficiencies in the erasing, establishment, or copying of epigenetic "marks" during germ cell development or somatic cell division, or consequential to DNA repair processes. Genetic deletions, duplications, or translocations of varying size may also encompass epigenetic regulatory regions, impacting the expression of genes not directly involved in the rearrangement.

Alterations in epigenetic marks at regulatory elements are routinely orchestrated by the recruitment of chromatin-remodeling enzymes such as histone deacetylases, DNA methyltransferases (commonly silencing), and others. Which enzyme complexes are recruited and the specific site modifications performed are dependent upon the initial state of the chromatin, and the make-up of the regulatory complexes that recognize it, which in turn are dependent upon the current "state" of the cell and the available components at any given time. In addition to protein components, non-coding RNAs (ncRNAs), including both long and microRNAs (miRNAs), can participate in these processes. Furthermore, "silencing" epigenetic modifications may occur as a result of overlapping "antisense" transcription, presumably as a consequence of altered chromatin state. Thus, any disruption in chromatin state, or in the level or functionality of the protein or RNA "recruiting" or "modifying" components themselves may lead to epigenetic disruption of gene expression. For long ncRNAs, the effect is most often limited to the same chromosome as the ncRNA expression (*cis*), whereas disruption of chromatin remodeling enzymes, which are often components of multiple regulatory complexes, may lead to wide-ranging or global disruption in gene expression.

Changes in epigenetic regulation may also arise consequential to variation in the availability of chemical substrates, such as methyl donors. Levels can be influenced by diet, and also by environmental exposure to agents that affect their availability. The latter can also affect functionality of proteins involved in chromatin remodeling directly.

28.2.2 Autism

A clinical diagnosis of autism involves recognition of deficits in social interaction and communication along with repetitive or stereotypical behaviors. There is considerable heterogeneity between patients with respect to severity and prevalence of symptoms. About 30% of patients will have undergone a period of relatively normal development, followed by loss of skills (regression). Some 25% of patients are without speech, 40% have moderate to severe levels of mental impairment, 25% will also develop seizures by early adolescence, and many suffer from anxiety, gastrointestinal complaints, and sleep cycle disruptions.

The prevalence of autism spectrum disorders (ASDs) is 1 : 120, with a 4 : 1 male-to-female ratio. Autism is a component of several syndromic neurodevelopmental disorders, including those described below, but most cases of autism occur apparently sporadically, with unknown cause (idiopathic). Family and twin studies suggest a strong genetic component to autism, and multiple loci throughout the genome have been suggested to be associated with autism. Of particular interest are associations to imprinted regions on chromosomes 7q and 15q; maternal duplications of the latter are associated with up to 3% of non-syndromic autism (described below). Patients with ASDs, along with those with other psychiatric disorders such as schizophrenia, have an increased incidence of novel small deletions or duplications throughout the genome (copy number variants, CNVs). A small number of these have been associated with known autism candidate genes, but most are of unknown consequence although they may disrupt common pathways (Pinto *et al.*, 2010). Genetic mutations have only been identified in a small handful of patients (rare variants), and promising risk alleles (common variants) have not yet been replicated across multiple studies (reviewed in State, 2010).

The variability in phenotypes between patients and especially between affected siblings, inconsistency in gene expression alterations between patients, and associations to known imprinted regions combined with a dearth of known major-effect gene loci together suggest that epigenetic alterations may play a role in the manifestation of autism. In support of this hypothesis, global profiling of lymphoblastoid DNA methylation differences between identical (MZ) twins discordant for a full autism diagnosis demonstrated differences in DNA methylation of two genes, *BCL-2* and *RORA*, as well as differences with unaffected siblings in multiple genes involved in pathways implicated in autism (Nguyen *et al.*, 2010). As for several of the genes involved in syndromic autism, described below, BCL-2 and RORA protein levels were decreased in postmortem brain tissue of some autism patients, suggesting that epigenetic misregulation of autism candidate genes may be a common contributing factor to the manifestation, or severity, of autism phenotypes.

28.2.3 Epilepsy

As for autism, epilepsy, or recurrent unprovoked seizures, is a common neurologic disorder that most often occurs sporadically, but may also be present in syndromic form. Epilepsy may be associated with cortical malformation or tumors, and can develop following trauma or infection. Although as a whole less heritable than autism, the study of familial epilepsies and clearly delineated seizure classifications have permitted the identification of many more causative genes in epilepsy (Andrade, 2009) than in autism. Both disorders, however, occur at increased prevalence in multiple neurological disorders, many of which have single, identified gene etiologies, including Angelman, Rett, Prader–Willi, and fragile X syndromes, as well as tuberous sclerosis. Similar alterations in epigenetic processes may underlie the manifestation of one or both disorders in these syndromes, depending on the unique profile of genes affected in each patient. Epilepsy is also associated with mutation of genes involved in chromatin remodeling, and examination of brain tissue from patients with epilepsy demonstrates abnormal genome-wide epigenetic modifications, including within candidate genes for epileptogenesis (reviewed in Qureshi and Mehler, 2010).

28.2.4   Fragile X syndrome

Fragile X syndrome (FRAXA) is the most common form of X-linked mental retardation (MR) in males. Up to 30% of patients meet full autism criteria and 14% report seizures, with a significant co-occurence of the two disorders (e.g., Berry-Kravis *et al.*, 2010) and a lower IQ in FRAXA patients with autism (Loesch *et al.*, 2007). In FRAXA, expansion of a CGG repeat (to >200 copies) in the promoter of the *FMR1* gene leads to epigenetic silencing of the locus and loss of expression of the FMRP protein, involved in regulating translation at the synapse. Pre-mutation carriers (55–200 copies) also have an increased incidence of autism (4%). The CGG repeats in premutation *FMR1* RNA can act as a "sink" for factors involved in transcription, splicing, mRNA trafficking, and translation, leading to tremor/ataxia (FXTAS) later in life. As both of these epigenetic-related mechanisms have genome-wide effects, it is unknown whether the comorbid autism in each case results from similar or separate gene (RNA) targets. A full discussion of the epigenetics of fragile X syndrome is provided elsewhere in this volume.

28.2.5   CHARGE

CHARGE (**c**oloboma, **h**eart defects, **a**tresia of the nasal choanae, growth/developmental **r**etardation, **g**enital and **e**ar abnormalities) is a relatively rare syndrome with broad clinical heterogeneity in which up to 50% of patients exhibit ASD and 10% exhibit seizures (Jongmans *et al.*, 2006). The CHARGE protein, CHD7, a chromodomain helicase with DNA binding, has also been identified within a CNV duplication in a patient with moderate MR (Montfort *et al.*, 2008). CHD7 binds to thousands of sites across the genome regulating gene expression and histone methylation in a cell- and developmental-specific manner, and is critical for neurogenesis and axon morphology (Melicharek *et al.*, 2010).

28.2.6   Rett syndrome

Rett syndrome (RTT) is the most common form of severe MR in females. It is the only other disorder besides idiopathic autism to exhibit regression, typically occurring between 6 and 18 months. This leads to characteristic loss of purposeful hand movements ("hand-wringing") and speech, and the development of ataxia, seizures, and respiratory and swallowing dysfunctions. Autistic features are common especially during the regressive period, and decreased neuronal size and disorganized, decreased dentritic branching have been reported in both

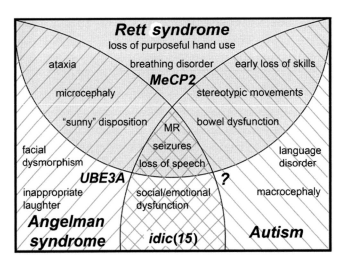

Figure 28.1 Phenotypic overlap between 15q11–q13-associated autism spectrum disorders (ASDs). A significant proportion of patients with Rett syndrome, Angelman syndrome, idiopathic autism, and maternal 15q duplication (idic 15) share multiple phenotypes including autistic features, mental retardation, seizures, and lack of functional speech. A small number of patients with clinical features of one class have also been found to carry the genetic etiology of another, highlighting the overlapping mechanisms underlying these disorders.

conditions (Figure 28.1). Rett syndrome is caused by mutation of the methyl-CpG binding protein 2 (*MECP2*) gene, located on the X chromosome. In females, X-chromosome inactivation leads to a mosaic state wherein cells expressing wild-type or mutant MeCP2 are co-mingled. Despite the presence of presumptively "wild-type" cells, in most brain regions all neurons exhibit a similar "immature" morphology. In males, the single mutant X chromosome leads to infantile encephalopathy and neonatal lethality, although atypical *MECP2* mutations have been described in patients with autism and in males with X-linked MR. ASD and MR are also associated with *MECP2* duplications, suggesting that any variation in the level of MeCP2 is disruptive to normal neuronal development and functioning, and predisposes to the development of ASD.

The MeCP2 protein was originally identified as a gene suppressor, binding to methylated DNA, recruiting histone deacetylases, and localizing to constitutive heterochromatic regions. Global expression profiling, however, did not identify large-effect enhancement of gene expression in RTT patients or animal models. Rather, MeCP2 has myriad binding sites across the genome (Yasui *et al.*, 2007) (Figure 28.2), and its loss leads to subtle expression changes, primarily down-regulation of multiple genes, including miRNAs (Urdinguio *et al.*, 2010), and to an increase in L1 retrotransposition, with the potential for variable mutational and epigenetic downstream effects (Muotri *et al.*, 2010). MeCP2, however, does play a role in the regulation of a subset of imprinted genes. Its regulation of the *Dlx5/6* locus, for example, has been associated with chromatin looping, which may be a general mechanism by which it affects chromatin states across the genome. Also, it interacts with ATRX to regulate developmental imprinted expression of *H19/*

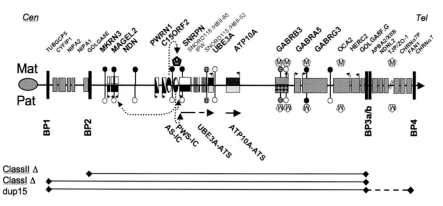

Figure 28.2 Schematic of the 15q11–q13 region. Maternal and paternal specific gene expression patterns, control loci, and epigenetic modifications are indicated for imprinted and biallelic gene clusters across the region. Common deletion and duplication endpoints (breakpoints, BP1–4) are indicated below the figure. Genes are demarcated by boxes with presence (black, gray) or absence (white) of expression from each allele as indicated; diagonal marking indicates putative random monoallelic expression; direction of transcription is shown by small arrows. The primarily paternal, long ncRNAs *Ube3a-ATS* and *Atp10a-ATS* arising from the SNRPN promoter region are indicated below the chromosome by a dashed, bold arrow. The AS- and PWS-IC regions are represented by ovals, with allele-specific activity indicated by black shading and sites of regulation/modification shown by dotted arrows. Sites of CpG methylation (black, full; pattern, partial; white, unmethylated) are indicated by lollipops (Sharp *et al.*, 2010), and sites of MeCP2 binding indicated by "M" pentagons.

Igf2 among others in an allele-specific manner (Kernohan *et al.*, 2010); and ATRX itself is mislocalized in RTT. ATRX, an SNF2-related chromatin remodeling protein, is mutated in ATR-X syndrome (**a**lpha-**t**halassemia, mental **r**etardation, **X**-linked) of which seizures, but not autism, are a strong component. ATRX binds to tandem repeat sequences in DNA, with expression patterns being skewed by variation in repeat length between alleles, even to the level of monoallelic expression (Law *et al.*, 2010). Thus, at the epigenomic level, the interplay between the levels or localization of MeCP2, ATRX, or other cofactors, combined with an individual's genetic sequence variation, could contribute to the phenotypic variability between patients, whereas stochastic effects on a smaller number of major-effect genes may contribute to the phenotypic overlap between syndromes.

28.3 15q11–q13-associated neurodevelopmental disorders

28.3.1 15q11–q13 locus
The proximal ~6-Mb region of chromosome 15 contains multiple genes involved in neuronal development and functioning, many of which fall within a complex domain of imprinting regulation (Figure 28.2). Parental-specific disruption of imprinted gene expression through deletion, duplication, or epigenetic misregulation gives rise to Prader–Willi syndrome (PWS), Angelman syndrome (AS), or ASD. In addition, linkage and association analyses have implicated this region in idiopathic autism as well as several other psychiatric and language-related disorders.

28.3.2   Imprint regulation

As diagrammed in Figure 28.2, 15q11–q13 contains clusters of paternally, maternally, and biallelically expressed genes. The synteny and general imprinted expression of the central 4-Mb region is conserved between human and mouse. Human patients carrying microdeletions disrupting the epigenetic regulation of the region have permitted identification of a bipartite imprinting control center (AS-IC, PWS-IC), further characterized using deletion and transgenic studies in mouse. These studies suggest a model in which the AS-IC, which carries activating histone modification marks exclusively on the maternal chromosome, promotes the establishment of silencing chromatin modifications on the PWS-IC during early development, removing its positive regulation, or enhancement, of flanking genes on this chromosome. In contrast, the PWS-IC remains active on the paternal chromosome, stimulating expression of the paternally expressed genes such as *NDN* and *SNRPN*. In somatic cells, the differentially modified ICs preferentially associate during interphase and replicate at different times during S phase, as is common for many imprinted regions. Furthermore, epigenetic modifications on the AS- and PWS-ICs remain intact throughout the process of induced pluripotent stem cell formation and epigenetic reprogramming, demonstrating the rigor of these regulatory mechanisms (Chamberlain *et al.*, 2010).

Also expressed from the *SNRPN* promoter region, primarily in neurons, is a long non-codingRNA, from which multiple small nucleolar RNAs are processed, associated with a general decondensation and "looping-out" of this region from the chromatin exclusively on the paternal locus (Leung *et al.*, 2009; Vitali *et al.*, 2010). The ncRNA is extensively spliced, and a proportion of the paternal transcripts extend across the *UBE3A* gene and, to a lesser extent, *ATP10A* (L.B.K. Herzing, unpublished data) (Figure 28.2). According to the standard model, the overlap of this paternal "antisense" transcript (*UBE3A-ATS*) inhibits paternal expression of *UBE3A*, likely through a stochastic interference mechanism as the *UBE3A* promoter remains unmethylated. Thus, in humans, *UBE3A* is also imprinted, with strong preferential maternal expression in neurons and to varying degrees in peripheral cells (Herzing *et al.*, 2002). The observed degree of "imprinting" in murine brain varies widely, depending on the technique and mouse model utilized (Gustin *et al.*, 2010).

In contrast, the imprint status of *ATP10A* is actively being debated, and the degree of preferential allelic expression appears largely dependent upon genetic background, tissue or cell-type specificity, developmental timing, and analysis technique. Strong preferential maternal expression, responsive to disruption of the AS-IC, has been shown in human brain (Herzing *et al.*, 2001), but Hogart *et al.* (2008) suggest that, alternatively, *ATP10A* is susceptible to gender-influenced *mono*allelic expression. As for *Ube3a*, the situation in mouse is even less straightforward. Several groups demonstrate strong exclusive or preferential maternal expression of *Atp10a* including in mouse olfactory bulb (OB), hippocampus (HP), and prefrontal cortex (PFC) (Gregg *et al.*, 2010; Herzing, unpublished data), but others find no deviance from biallelic expression or response to PWS-IC or *Ube3a-ATS* disruption (DuBose *et al.*, 2010).

The three GABA$_A$ receptor subunit genes, *GABRB3*, *GABRA5*, and *GABRG3*, along with the distal genes, are not imprinted and are biallelically expressed in human and mouse cortex (CX) under normal conditions, including, for *GABRB3/Gabrb3*, both the major and minor isoforms. However, murine *Gabrb3* is demarcated as a gene with paternal "complex imprinting" by Gregg *et al.* (2010), and additional underlying parental-specific expression and epigenetic regulatory biases may be unmasked upon loss of normal, region-wide or gene-specific biparental complement. For example, a paternal bias in total *GABRB3* expression has been detected in human CX using AS and PWS deletion and UPD patients (Hogart *et al.*, 2007), and epigenetic changes including intronic CpG island methylation and altered replication timing are also observed by this group. Thus, the localization of these genes adjacent to the PWS/AS imprinted locus may predispose them to epigenetic dysregulation.

28.3.3 Disorders associated with dysregulation of 15q11–q13

28.3.3.1 Maternal 15q11–q13 duplication

Maternal duplication of 15q11–q13 represents the most common recognized genetic etiology of non-syndromic ASD, occurring in up to 3% of patients. Interstitial maternal duplication may also result in language disorder without ASD, whereas patients carrying an extra, maternally derived isodicentric marker chromosome idic(15), usually containing two additional copies of the 15q11–q13 region, exhibit a more severe phenotype including seizures, motor delays, subtle dysmorphic facial features, behavioral problems, and variable MR. Paternal duplication, on the other hand, is only rarely associated with clinical phenotype, primarily involving complex rearrangements.

The association of maternal 15q duplication with ASD strongly suggests the involvement of the maternally expressed genes in phenotypic outcome. Maintenance of normal expression levels may be especially crucial for imprinted genes such as *UBE3A*, as evidenced by the severe phenotypes or lethality associated with cell-specific or global *dUbe3a* overexpression in *Drosophila* (Lu *et al.*, 2009). *UBE3A* expression arises from each duplicated allele in fibroblasts from patients with maternal 15q duplications, leading to an overall increase in expression (Herzing *et al.*, 2002), and *UBE3A* overexpression, along with overexpression of the *GABR$_A$* cluster genes, has been reported in CX from a typical patient with idic (15), but not in a patient with atypical, PWS-like features (Hogart *et al.*, 2009). A mouse model of 15q regional duplication, in contrast, yielded autistic-like, reduced social interaction only upon paternal inheritance of the duplication, despite overall, yet extremely variable, *Ube3a* and *Atp10a* overexpression in maternal duplication adult brain (Nakatani *et al.*, 2009). Another recent, gene-specific model demonstrated autistic-like features, including social, communication, and stereotypical defects only upon transgenic expression of three times the wild-type *Ube3a* protein levels (Smith *et al.*, 2011), together suggesting that phenotypic outcomes may differ between species.

In addition to alterations in expression of genes within 15q11–q13, microarray analysis of lymphoblastoid cells from patients with idic(15) has identified multiple downstream effects of 15q11-q13 duplication, with many genes overlapping those identified in non-duplication ASDs (Nishimura *et al.*, 2007). Thus, regional duplication and loss of biparental complement may lead to unexpected local epigenetic changes and alterations in gene expression, as evidenced by increased PWS-IC methylation and globally decreased paternal gene expression in the atypical (maternal) idic(15) patient described above, as well as have global consequences.

28.3.3.2 Prader–Willi syndrome

Prader–Willi syndrome (PWS) is characterized by hypotonia at birth, hypogonadism and small hands and feet, hyperphagia leading to obesity with associated complications, mild to moderate MR, tantrums, and obsessive–compulsive traits. PWS is caused by loss of the paternally expressed genes on 15q11–q12, through paternal regional deletion in 70% of patients mediated by large flanking repeat sequences ("breakpoints," BP) (Figure 28.2), the presence of two maternal chromosome 15s (maternal uniparental disomy, UPD) as a result of chromosomal non-dysjuction during meiosis, in ~26% of patients, and the remainder carrying altered epigenetic marks on or microdeletions of the PWS-IC. Recent identification of a patient with a microdeletion within the snoRNA cluster HBII-85 suggests that loss of these elements is sufficient to derive the primary PWS phenotype (Duker *et al.*, 2010). Other genes in the region contribute to the broader PWS phenotype, e.g., poor suckling (*Magel2*), which, in mouse, can be corrected by administration of oxytocin (Schaller *et al.*, 2010), itself an autism candidate gene. Within the broader PWS population, up to 25% of patients develop seizures, primarily those with deletions. It is hypothesized that the increased seizure prevalence in both PWS and AS deletion patients, below, is because these deletions also encompass the $GABR_A$ gene cluster leading to haplo-insufficiency for these genes, which are critical components of the receptor complex for the major inhibitory neurotransmitter, gamma-aminobutyric acid (GABA). Loss of individual $Gabr_A$ subunits in mouse leads to seizures, among other phenotypes, and downregulation of the minor isoform of the *GABRB3* receptor is associated with childhood absence epilepsy (CAE) and ASD (Delahantey *et al.*, 2011). In addition, about 20–40% of PWS patients also exhibit ASD, with the risk being highest in patients with maternal UPD, potentially arising from the predicted increase therein of maternally expressed genes, especially *UBE3A*, during development, as for 15q duplication ASD patients, above. Unlike for FRAXA, however, ASD does not correlate with level of mental impairment in PWS.

28.3.3.3 Angelman syndrome

Phenotypically, Angelman syndrome (AS) is much more severe than PWS, with profound MR, severe seizures, absent speech, and ataxic gait. As for PWS, 70% of AS patients carry a regional deletion, albeit of the maternal chromosome, with only 3–5% exhibiting paternal UPD and up to 9% with epigenetic modifications or deletions of the AS-IC. About 10% of AS patients have identified mutations in the *UBE3A* gene, which encodes a ubiquitin ligase, E6AP, involved in targeting specific proteins such as ECT2, PML, and ARC, for degradation. ECT2 plays a role in neuronal network formation (Reiter *et al.*, 2006), and ARC is involved in neurotransmitter receptor (AMPA) trafficking at the synapse (Greer *et al.*, 2010). The latter may provide an important link between the ASDs FRAXA and AS, as the FRM1 protein is also a key regulator of AMPA receptor trafficking. As mentioned above, seizures are most severe in AS deletion patients, although patients with *UBE3A* mutations exhibit seizures as well, findings which are replicated in mouse models of AS. The comorbidity of ASD with AS is difficult to confirm because of the severity of the mental impairment, although it has been suggested to be present in 40–100% of AS patients using standardized measures of ASD (reviewed in Grafodatskaya *et al.*, 2010).

In addition to the epigenetic regulation affected by the AS- and PWS-ICs and the *UBE3A-ATS* on *UBE3A* expression, an unexpected *trans* effect has been identified in a *Ube3a* heterozgous knockout mouse (Landers *et al.*, 2005). Here, maternal deletion of *Ube3a* exon 13 leads to approximately twofold upregulation of paternal *Ube3a-ATS* transcription in total brain and CB. The mechanism for this para-mutation-like phenomenon is not understood, although in this model the *Ube3a* deletion sequence overlaps that of a *Ube3a-ATS* splice variant, suggesting that the close physical association of maternal and paternal loci may permit sequence-specific crosstalk and co-regulation between alleles, or that *Ube3a* and *Ube3a-ATS* RNAs themselves interact, defining feedback regulation on the region. The increased *Ube3a-ATS* does not, however, appear to significantly impact paternal *Ube3a* expression (or allele-specific *Atp10a* expression), suggesting upregulation only occurs in cells with exclusive *Ube3a* maternal expression and not in brain cells with non-imprinted *Ube3a* expression normally lacking *Ube3a-ATS*.

A similar phenomenon is also observed in *Ube3a* exon 2 heterozygous knockout mice (Jiang, 1998). Here, we see greater variability in *Ube3a-ATS* overexpression between animals in CB and OB, including overexpression of *Atp10a-ATS* (Figure 28.3), but also a substantial decrease in total *Atp10a* expression and a small trend towards decreased *Gabrb3* expression. We suspect that the increased

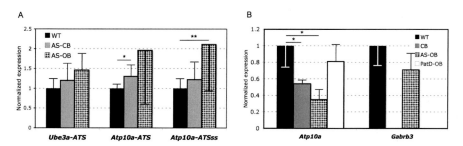

Figure 28.3 Dysregulation of imprinted expression in the *Ube3a* exon 2 knockout mouse brain. (A) Both *Ube3a-ATS* and *Atp10a-ATS* expression are upregulated in CB and OB of mice carrying a Ube3a exon 2 deletion. Expression was measured by real-time qRT-PCR using *Ube3a-ATS*-specific primers (Landers *et al.*, 2005) or semi-quantitatively using *Atp10a-ATS* intron 13 primers following standard or strand-specific (ss) RT-PCR (Herzing, unpublished data). (B) Total *Atp10a* expression is decreased in *Ube3a* exon 2 maternal deletion AS model mice, but not in paternal exon 2 deletion mice, as measured by real-time qRT-PCR. There is a trend towards decreased Gabrb3 expression, which did not reach significance. $n = 3-9$/group (*, $p < 0.05$; **, $p < 0.01$, Student's t-test).

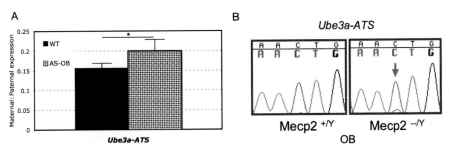

Figure 28.4 Increased maternal *Ube3a-ATS* expression is detected in OB from AS and RTT mice. (A) There is a 27% increase in the relative maternal:paternal ratio of *Ube3a-ATS* expression in OB of *Ube3a* exon 2 deletion mice, but not in CB or other tissues examined (not shown). Low baseline maternal *Ube3a-ATS* expression is observed by direct sequencing (not shown), exclusively in OB of these mixed-strain animals. (B) Maternal *Ube3a-ATS* expression can be detected exclusively in OB from *Mecp2* knockout male (−/Y) RTT model mice, but not in other tissues or in wild-type OB from this cross. $n = 4$ (*, $p < 0.05$; Student's *t*-test).

variability in *Ube3a-ATS* upregulation may be due to the fact that in these animals, the exon 2 deletion does not overlap a known spliced form of *Ube3a-ATS*, but only the primary transcript, thus reducing the efficacy of the response. Decreased total *Atp10a* expression, not quantified in exon 13 deletion mice, may be detected here because OB exhibits imprinted maternal expression of both *Ube3a* and *Atp10c* (Kashiwagi *et al.*, 2003), and also is a site of continuing neurogenesis, and thus may be susceptible to dysregulation of epigenetic mechanisms. Consistent with this scenario, we observe a slight upregulation of maternal *Ube3a-ATS* expression exclusively in OB in exon 2 deletion animals (Figure 28.4). The limited alteration in murine *Gabrb3* expression is consistent with the lack of significant decrease in GABR$_A$ ligand binding in brains of AS patients or mouse models, and may reflect the lower degree and increased variability of *Ube3a-ATS* response, and thus of regional effects outside direct *Ube3a-ATS* overlap. However, consistent with the direction of results presented here, Low and Chan (2010) report a 2.5× decrease in *Gabra5*, immediately proximal to *Gabrb3*, in adult female CB from *Ube3a* exon 2 deletion mice using Affymetrix (Santa Clara, CA, USA) Gene Chip Exon Array analysis. A second possibility for limited *Gabrb3* decrease may arise from a resistance of this locus, in mouse, to decreased total expression levels, as observed in mice with heterozygous *Gabrb3* deletion, which exhibit ~70% (rather than 50%) total expression levels.

28.3.3.4   Rett syndrome

There is considerable phenotypic overlap between RTT and AS (Figure 28.1), and some patients with clinically diagnosed AS have been found to carry mutations in *MECP2*. In turn, *UBE3A* expression, and to a lesser extent *GABRB3*, is decreased in brain from RTT patients. MeCP2 binds to several sequences across the 15q11–q13 region, most notably at the PWS-IC (Figure 28.2), and loss of this binding may lead to epigenetic modifications affecting the expression predominantly of maternally expressed genes. Several groups report decreased *Ube3a* and *Gabrb3*

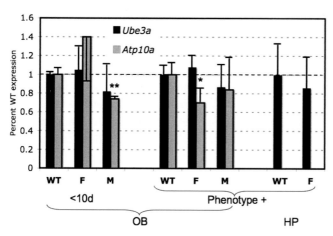

Figure 28.5 Variable decreased expression of maternally expressed genes in OB from *Mecp2*-knockout RTT model animals. Significant decreases in *Atp10a* expression are seen in OB from young male *Mecp2*-knockout (−/Y) and phenotype positive female *Mecp2* heterozygous knockout (+/−) animals on a mixed strain background as measured by real-time qRT-PCR. A trend towards decreased *Ube3a* is also observed in OB from male *Mecp2* −/Y animals, and in female *Mecp2* +/− HP, but does not reach significance due to the generally increased variance in expression observed in *Mecp2*-knockout animals. $n = 4$–8/group (*, $p < 0.05$; **, $p < 0.01$, Student's t-test). WT, wild-type; F, female; M, male.

expression in Rett syndrome mouse models, although expression variation may be shaped by strain and region-specific influences (Herzing, unpublished data) (Figure 28.5). Likewise, biallelic expression of *Ube3a-ATS* is detected in total brains and OB of Mecp2 knockout mice (Makedonski *et al.*, 2005) (Figure 28.4), but is exclusively paternal in CX and CB. Of note, loss of Mecp2 explicitly in GABA-ergic neurons recapitulates many features of RTT (Chao *et al.*, 2010), which effects may be compounded or mimicked by decreases in Gabr$_A$ receptors such as via epigenetic dysregulation within the 15q locus. Therefore, AS- and ASD-like phenotypes in RTT are likely be modulated by variable epigenetic modification of 15q11–q13 between patients, and may be especially sensitive to genetic background.

28.3.3.5 Idiopathic autism

In addition to the ASD associated with 15q11–q13 rearrangements, multiple lines of evidence suggest that genes within 15q11–q13 may contribute to idiopathic autism, as well. Linkage and association studies in many, but not all, autism samples have identified *GABRB3* as an autism candidate gene, with increased significance observed following division of patients into subgroups. Several high-density regional studies also implicate *UBE3A* and *ATP10A*. The function of *ATP10A* in brain is not known, although it is expressed in neurons and has homology to "flippases," which maintain the phospholipid profile of the inner and outer cell membrane. Disruption may result in a decrease in functionality of neuronal networks, or even apoptosis. Mutations in *UBE3A* have been identified in a handful of patients with autism, and maternal inheritance of the GABRB3 minor isoform variant associated with CAE has also recently been shown to increase the risk of ASD in the general population (Delahanty *et al.*, 2011). Disabling mutations in these genes are more likely to result in phenotypes of greater severity than autism. However, as in RTT and AS, decreased levels of E6AP, GABRB3, and GABRA5 (Fatemi *et al.*, 2010) have been described in CX of patients with idiopathic autism, as have monoallelic or skewed expression of genes within the $GABR_A$ cluster (Hogart *et al.*, 2007). In mouse, *Gabrb3* is one of several $Gabr_A$ subunits altered in select brain regions following genetic disruption of the ASD risk gene *PLAUR* (Eagleson *et al.*, 2010). Together, these findings suggest that epigenetic modifications of autism candidate gene expression in the general population may result in expression levels beyond thresholds required for normal neurologic development or functioning.

28.4   Therapeutic potential

The finding that restoration of wild-type Mecp2 expression levels in a mouse model of RTT can provide phentoypic rescue even after the onset of severe symptoms (Guy *et al.*, 2007) has demonstrated that, rather than being the result of irreversible developmental abnormalities, the deficits in ASDs may be reversible and thus amenable to therapeutic intervention. Whereas neuronal-specific gene replacement is currently unfeasible, multiple other approaches present themselves, including manipulating components of downstream pathways, or, in the case of genes that have been epigenetically modified, manipulation of the epigenetic "marks" themselves or of how they are interpreted. In the former case, glutamate receptor agonists, which have shown great promise in models and potentially in clinical trials in FRAXA, are being tested in other ASDs based on the premise that there may be an overlap in the underlying pathways between these disorders. In the latter case, dietary interventions such as methyl donor supplementation, and valproic acid, a histone deacetylase inhibitor, are being tested in AS and RTT patients or in RTT model systems, respectively, with the aim of restoring expression from the non-mutant, silenced allele (Peters *et al.*, 2010; Vecsler *et al.*, 2010). As our knowledge unfolds, combinations of these approaches may likely prove effective across multiple ASDs sharing common underlying pathways, whether disrupted through mutation or epigenetics.

28.5   Concluding remarks

The modulation of gene expression through epigenetic modification is proving to be a significant mechanism of gene regulation. At both the single gene and genome-wide level, epigenetics influences phenotypic outcome, and its disruption is a key component of multiple disorders including neurodevelopmental and autism spectrum disorder syndromes. Within these disorders, overlapping profiles of genes such as those on 15q11–q13 are disrupted to varying degrees, leading to phenotypic overlap between classes of patients but also to considerable variability within single diagnostic groups. Strategies to impact specific pathways and epigenetic modifications are being developed, which have the potential to be useful across these disorders regardless of underlying etiology.

REFERENCES

Andrade, D. M. (2009). Genetic basis in epilepsies caused by malformations of cortical development and in those with structurally normal brain. *Human Genetics*, **126**, 173–193.

Berry-Kravis, E., Raspa, M., Loggin-Hester, L., *et al.* (2010). Seizures in fragile X syndrome: characteristics and comorbid diagnoses. *American Journal of Intellectual and Developmental Disabilities*, **115**, 461–472.

Chamberlain, S. J., Chen, P.-F., Ng, K. Y., *et al.* (2010). Induced pluripotent stem cell models of the genomic imprinting disorders Angelman and Prader–Willi syndromes. *Proceedings of the National Academy of Sciences USA*, **107**, 17 668–17 673.

Chao, H. T., Chen, H., Samaco, R. C., *et al.* (2010). Dysfunction in GABA signalling mediates autism-like stereotypies and Rett syndrome phenotypes. *Nature*, **468**, 263–269.

Delahanty, R. J., Kang, J. Q., Brune, C. W., *et al.* (2011). Maternal transmission of a rare *GABRB3* signal peptide variant is associated with autism. *Molecular Psychiatry*, **16**, 86–96.

DuBose, A. J., Johnstone, K. A., Smith, E. Y., Hallett, R. A., and Resnick, J. L. (2010). *Atp10a*, a gene adjacent to the PWS/AS gene cluster, is not imprinted in mouse and is insensitive to the PWS-IC. *Neurogenetics*, **11**, 145–151.

Duker, A. L., Ballif, B. C., Bawle, E. V., *et al.* (2010). Paternally inherited microdeletion at 15q11.2 confirms a significant role for the *SNORD116* C/D box snoRNA cluster in Prader–Willi syndrome. *European Journal of Human Genetics*, **18**, 1196–1201.

Eagleson, K. L., Gravielle, M. C., Schlueter McFadyen-Detchum, L. J., *et al.* (2010). Genetic disruption of the autism spectrum disorder risk gene *PLAUR* induces GABA$_A$ receptor subunit changes. *Neuroscience*, **168**, 797–810.

Fatemi, S. H., Reutiman, T. J., Folsom, T. D., *et al.* (2010). mRNA and protein levels for GABA$_A$α4,α5,β1 and GABA$_B$R1 receptors are altered in brains from subjects with autism. *Journal of Autism and Developmental Disorders*, **40**, 743–750.

Grafodatskaya, D., Chung, B., Szatmari, P., and Weksberg, R. (2010). Autism spectrum disorders and epigenetics. *Journal of the American Academy of Child and Adolescent Psychiatry*, **49**, 794–809.

Greer, P. L., Hanayama, R., Bloodgood, B. L., *et al.* (2010). The Angelman syndrome protein Ube3A regulates synapse development by ubiquinating arc. *Cell*, **140**, 704–716.

Gregg, C., Zhang, J., Weissbourd, B., *et al.* (2010). High-resolution analysis of parent-of-origin allelic expression in the mouse brain. *Science*, **329**, 682–685.

Gustin, R. M., Bichell, T. J., Buber, M., *et al.* (2010). Tissue-specific variation of Ube3a protein expression in rodents and in a mouse model of Angelman syndrome. *Neurobiology of Disease*, **39**, 283–291.

Guy, J., Gan, J., Selfridge, J., Cobb, S., and Bird A. (2007). Reversal of neurological defects in a mouse model of Rett syndrome. *Science*, **315**, 1143–1147.

Herzing, L. B., Kim, S. J., Cook, E. H., Jr., and Ledbetter, D. H. (2001). The human aminophospholipid-transporting ATPase gene *ATP10C* maps adjacent to *UBE3A* and exhibits similar imprinted expression. *American Journal of Human Genetics*, **68**, 1501–1505.

Herzing, L. B. K., Cook, E. H., Jr., and Ledbetter, D. H. (2002). Allele-specific expression analysis by RNA-FISH demonstrates preferential maternal expression of *UBE3A* and determines imprint maintenance within 15q11–q13 duplications. *Human Molecular Genetics*, **11**, 1707–1718.

Hogart, A., Nagarajan, R. P., Patzel, K. A., Yasui, D. H., and Lasalle, J. M. (2007). 15q11–13 GABA$_A$ receptor genes are normally biallelically expressed in brain yet are subject to epigenetic dysregulation in autism-spectrum disorders. *Human Molecular Genetics*, **16**, 691–703.

Hogart, A., Patzel, K. A., and Lasalle, J. M. (2008). Gender influences monoallelic expression of *ATP10A* in human brain. *Human Genetics*, **124**, 235–242.

Hogart, A., Leung, K. N., Wang, N. J., *et al.* (2009). Chromosome 15q11–13 duplication syndrome brain reveals epigenetic alterations in gene expression not predicted from copy number. *Journal of Medical Genetics*, **46**, 86–93.

Hogart, A., Wu, D., LaSalle, J. M., and Schanen, N. C. (2010). The comorbidity of autism with the genomic disorders of chromosome 15q11.2–q13. *Neurobiology of Disease*, **38**, 181–191.

Jiang, Y.-H., Armstrong, D., Albrecht, U., *et al.* (1998). Mutation of Angelman ubiquitin ligase in mice causes increased cytoplasmic p 53 and deficits of contextual learning and long-term potentiation. *Neuron*, **21**, 799–811.

Jiang, Y.-H., Pan, Y., Zhu, L., *et al.* (2010). Altered ultrasonic vocalization and impaired learning and memory in Angelman syndrome mouse model with a large maternal deletion from *Ube3a* to *Gabrb3*. *PLoS One*, **5**, e12278.

Jongmans, M. C. J., Admiraal, R. J., van der Donk, K. P., *et al.* (2006). CHARGE syndrome: the phenotypic spectrum of mutations in the *CHD7* gene. *Journal of Medical Genetics*, **43**, 306–314.

Kashiwagi, A., Meguro, M., Hoshiya, H., *et al.* (2003). Predominant maternal expression of the mouse *Atp10a* in hippocampus and olfactory bulb. *Journal of Human Genetics*, **48**, 194–198.

Kernohan, K. D., Jiang, Y., Tremblay, D. C., *et al.* (2010). ATRX partners with cohesin and MeCP2 and contributes to developmental silencing of imprinted genes in the brain. *Developmental Cell*, **18**, 191–202.

Landers, M., Calciano, M. A., Colosi, D., *et al.* (2005). Maternal disruption of *Ube3a* leads to increased expression of *Ube3a-ATS* in *trans*. *Nucleic Acids Research*, **33**, 3976–3984.

Law, M. J., Lower, K. M., Voon, H. P. J., *et al.* (2010). ATR-X syndrome protein targets tandem repeats and influences allele-specific expression in a size-dependent manner. *Cell*, **143**, 367–378.

Leung, K. N., Vallero, R. O., DuBose, A. J., Resnick, J. L., and LaSalle, J. M. (2009). Imprinting regulates mammalian snoRNA-encoding chromatin decondensation and neuronal nucleolar size. *Human Molecular Genetics*, **18**, 4227–4238.

Loesch, D. Z., Bui, Q. M., Dissanayake, C., *et al.* (2007). Molecular and cognitive predictors of the continuum of autistic behaviours in fragile X. *Neuroscience and Biobehavioural Review*, **31**, 315–326.

Low, D. and Chen, K.-S. (2010). Genome-wide gene expression profiling of the Angelman syndrome mice with *Ube3a* mutation. *European Journal of Human Genetics*, **18**, 1228–1235.

Lu, Y., Wang, F., Li, Y., *et al.* (2009). The *Drosophila* homologue of the Angelman syndrome ubiquitin ligase regulates the formation of terminal dendritic branches. *Human Molecular Genetics* **18**, 454–462.

Makedonski, K., Abuhatzira L., Kaufman, Y., Razin, A., and Shemer, R. (2005). MeCP2 deficiency in Rett syndrome causes epigenetic aberrations at the PWS/AS imprinting center that affects UBE3A expression. *Human Molecular Genetics*, **14**, 1049–1058.

Melicharek, D. J., Ramirez, L. C., Singh, S., Thompson, R., and Marenda, D. R. (2010). Kismet/CHD7 regulates axon morphology, emory and locomotion in a *Drosophila* model of CHARGE syndrome. *Human Molecular Genetics*, **19**, 4253–4264.

Monfort, S., Roselio, M., Orellana, C., *et al.* (2008). Detection of known and novel genomic rearrangements by array-based comparative genomic hybridization: detection of *ZNF533* and duplication of the CHARGE syndrome genes. *Journal of Medical Genetics*, **45**, 432–437.

Muotri, A. R., Marchetto, M. C., Coufal, N. G., *et al.* (2010). L1 retrotransposition in neurons is modulated by MeCP2. *Nature*, **468**, 443–446.

Nakatani, J., Tamada, K., Hatanaka, F., *et al.* (2009). Abnormal behavior in a chromosome-engineered mouse model for human 15q11–13 duplication seen in autism. *Cell*, **137**, 1235–1246.

Nguyen, A T., Rauch, T. A., Pfeifer, G. P., and Hu, V. W. (2010). Global methylation profiling of lymphoblastoid cell lines reveals epigenetic contributions to autism spectrum disorders and a novel autism candidate gene, *RORA*, whose protein product is reduced in autistic brain. *Federation of the American Society for Experimental Biology Journal*, **24**, 3036–3051.

Nishimura, Y., Martin, C. L., Vazquez-Lopez, A., *et al.* (2007). Genome-wide expression profiling of lymphoblastoid cell lines distinguishes different forms of autism and reveals shared pathways. *Human Molecular Genetics*, **16**, 1682–1698.

Peters, S. U., Bird, L. M., Kimonis, V., *et al.* (2010). Double-blind therapeutic trial in Angelman syndrome using betaine and folic acid. *American Journal of Medical Genetics*, **152 Part A**, 1994–2001.

Pinto, D., Pagnamenta, A. T., Klei, L., *et al.* (2010). Functional impact of global rare copy number variation in autism spectrum disorders. *Nature*, **466**, 368–372.

Qureshi, I. A. and Mehler, M. F. (2010). Epigenetic mechanisms underlying human epileptic disorders and the process of epileptogenesis. *Neurobiology of Disease*, **39**, 53–60.

Reiter, L. T., Seagroves, T. N., Bowers, M., and Bier, E. (2006). Expression of the Rho-GEF Pbl/ECT2 is regulated by the UBE3A E3 ubiquitin ligase. *Human Molecular Genetics*, **15**, 2825–2835.

Schaller, F., Watrin, F., Sturny, R., *et al.* (2010). A single postnatal injection of oxytocin rescues the lethal feeding behaviour in mouse newborns deficient for the imprinted *Magel2* gene. *Human Molecular Genetics*, **19**, 4895–4905.

Sharp, A. J., Migliavacca, E., Dupre, Y., *et al.* (2010). Methylation profiling in individuals with uniparental disomy identifies novel differentially methylated regions on chromosome 15. *Genome Research*, **20**, 1271–1278.

Smith, S. E. P., Zhou, Y.-D., Zhang, G., *et al.* (2011). Increased gene dosage of *Ube3a* results in autism traits and decerased glutamate synaptic transmission in mice. *Science Translational Medicine*, **3**, 103ra97.

State, M. W. (2010). The genetics of child psychiatric disorders: focus on autism and Tourette syndrome. *Neuron*, **68**, 254–269.

Urdinguio, R. G., Fernandez, A. F., Lopez-Nieva, P., *et al.*, (2010). Disrupted microRNA expression caused by Mecp2 loss in a mouse model of Rett syndrome. *Epigenetics*, **5**, 656–663.

Vecsler, M., Simon, A. J., Amariglio, N., Rechavi, G., and Gak E. (2010). MeCP2 deficiency downregulates specific nuclear proteins that could be partially recovered by valproic acid in vitro. *Epigenetics*, **5**, 61–67.

Vitali, P., Royo, H., Marty, V., Bortolin-Cavaille, M-L., and Cavaille, J. (2010). Long nuclear-retained non-coding RNAs and allele-specific higher-order chromatin organization at imprinted snoRNA gene arrays. *Journal of Cell Science*, **123**, 70–83.

Yasui, D. H., Peddada, S., Bieda, M. C., *et al.* (2007). Integrated epigenomic analyses of neuronal MeCP2 reveals a role for long-range interaction with active genes. *Proceedings of the National Academy of Sciences USA*, **104**, 19 416–19 421.

29 The neurobiology of chromatin-associated mechanisms in the context of psychosis and mood spectrum disorders

Schahram Akbarian

29.1 Introduction

Epigenetic regulators of gene expression, including DNA methylation and various types of posttranslational histone modifications, could play a potential role in schizophrenia, depression, and other brain disorders associated with changes in gene expression and neuronal function. This chapter will provide a brief introduction to the concept of epigenetics as it pertains to the neurosciences and psychiatry. This includes brief remarks on the epigenetic heritability of disease, a critical discussion of methodological challenges when applying chromatin-based assays to brain tissue, and a summary of epigenetic alterations described to date in subjects diagnosed with psychosis or mood disorder. The chapter concludes with an outlook on epigenetic targets and novel treatments for cognitive and emotional brain disorders.

29.2 Schizophrenia and depression

Schizophrenia and depression are two major psychiatric diseases that lack consensus neuropathology and for a very large majority of affected individuals, the causative genetic, environmental, and other disease-relevant factors remain unclear. Schizophrenia affects 1% of the general population, often afflicts subjects in their late adolescence or early adult years but then takes a chronic course, with symptoms such as delusions, hallucinations, and cognitive dysfunction complemented by additional deficits in attention, motivation, and social behaviors (Ibrahim and Tamminga, 2010). Depression has a lifetime risk of 10–15% for the US general population, and like other disorders of mood and affect, shows significant overlap with schizophrenia in terms of genetic risk architecture. A number of rare DNA sequence alterations, including but not limited to the balanced translocation at the Disrupted-in-Schizophrenia 1 (*DISC-1*) locus (1q42) or the

Epigenomics: From Chromatin Biology to Therapeutics, ed. K. Appasani. Published by Cambridge University Press. © Cambridge University Press 2012.

22q11 deletion syndrome, could present in different individuals either as mood and anxiety disorder or psychosis (Porteous *et al.*, 2006; Green *et al.*, 2009). Moreover, genome-wide association studies identified a number of common single-nucleotide polymorphisms (SNPs) that confer genetic risk across a range of mood and psychosis spectrum disorders (Huang *et al.*, 2010; Domschke *et al.*, 2011; Liu *et al.*, 2011). As it pertains to treatment, conventional treatments for schizophrenia and depression mostly rely on drugs interfering with brain catecholamine (dopamine, norepinephrine) or indolamine (serotonin) signaling. Unfortunately, a substantial portion of patients show an incomplete or total lack of response to the currently available pharmacological options (Ibrahim and Tamminga, 2010; Krishnan and Nestler, 2010), and research on the underlying pathophysiology and/or novel treatments is a high priority for the National Institutes of Health and other sponsors (Insel, 2009).

29.3 Epigenetic and heritability of disease: (pre)clinical perspective

As discussed elsewhere in this book, chromatin is composed of the nucleosome core histones H2A, H2B, H3, and H4 (or their variant forms) which together form drum-like structures with 147 bp of genomic DNA wrapped around them (Wolffe, 1992; Hayes and Hansen, 2001), connected by linker histone and linker DNA which in vertebrates on average is limited to 30–45 bp. To date, more than 70 different types of epigenetic markings have been reported; these include histone lysine (K) acetylation, methylation and poly ADP-ribosylation, arginine (R) methylation, and serine (S), threonine (T), tyrosine (Y), and histidine (H) phosphorylation (Taverna *et al.*, 2007). In addition, a subset of the histone H2A, H2B, and H4 lysines are covalently linked to the small protein modifiers ubiquitin and small ubiquitin-like modifier (SUMO) (Shiio and Eisenman, 2003; Weake and Workman, 2008). Modifications of the genomic DNA include 5-methyl-cytosine and the related form 5-hydroxy-methyl-cytosine (Bonasio *et al.*, 2010). Many of these DNA and histone markings are discussed in more detail in other chapters of this book. Epigenetics, when more broadly defined, relates to the dynamic changes of these modifications under physiological or pathological conditions and implies that changes in chromatin structures serve as a key control point for gene expression regulation and genome organization in the absence of DNA sequence alterations (Bonasio *et al.*, 2010). From the perspective of the translational neurosciences, epigenetic dysregulation in psychiatric disorders is mostly discussed in the context of two different themes. The first refers to the topic of vertical transmission, or heritability, of epigenetic markings. In other words, is there a molecular signature associated with disease risk and that is not encoded in the DNA sequence but nonetheless potentially passed on from one generation to the next? This is an extremely interesting topic for hypotheses built on the so-called "missing heritability" of complex disorders such as schizophrenia and depression for which to date causative DNA sequence alterations remain unidentified for the majority of cases. However, some scholars have pointed out that "missing heritability" could still turn out to be an illusion pending further advancements in modeling of genetic risk architectures (Lander, 2011).

The second theme is centered on regulatory mechanisms governing chromatin structures and function in the animal and human brain. Most studies utilizing various preclinical model systems or diseased brain tissue are designed to test whether changes in expression of a gene of interest are accompanied by chromatin markings and epigenetic decorations at the corresponding locus in the genome. The underlying idea is that epigenetic markings serve as a type of cellular memory that regulates transcriptional activity at that locus in the context of a disease process, genetic polymorphism, pharmacological treatment, etc.

As far as it concerns the first of these two themes – heritability – conclusive evidence for or against a role of chromatin markings as a disease-associated factor transmissible from parent to offspring is still lacking. However, the existence of such mechanisms becomes increasingly plausible. Firstly, studies on human sperm imply that a significant portion of the genome, perhaps up to 4–5%, maintains nucleosomal organization including epigenetic decorations such as DNA methylation and some of the posttranslational histone modifications during germ-line transmission (Arpanahi et al., 2009; Hammoud et al., 2009). Additional investigations will be necessary to test whether or not these chromatin signals indeed could pass on heritable information. Secondly, a number of neurodevelopmental disorders, including Prader–Willi and Angelman syndromes, typically result from DNA variants affecting the parent-of-origin patterns of DNA and histone modifications at that locus that normally result in differential gene expression activities on the paternal and maternal chromosome (Nicholls and Knepper, 2001; Horsthemke and Wagstaff, 2008). Thirdly, strong (but not yet unambiguous) evidence for epigenetic inheritance has been uncovered in vertebrate animals, including laboratory mice: a frequently discussed example involves the dominant A^{vy} (agouti viable yellow) allele of the agouti gene (A) that affects coat color. A^{vy} is an insertion of the intracisternal A particle (IAP) retrotransposon upstream of A and drives expression of the gene by its cryptic promoter. Expression of A^{vy} results in yellow coat color, while the null allele a gives an agouti color. A^{vy} displays variable expressivity, so a spectrum of coat color is observed in the offspring of a A^{vy}/a mother. Epigenetic inheritance was evident by the observation that the distribution of different coat colors in the offspring is influenced by the phenotype of the A^{vy}/a mother (but not when the A^{vy} allele is contributed by the father), and maternal environment was ruled out as the underlying cause (Morgan et al., 1999). It was initially suggested that methylation status at the IAP retrotransposon is the basis of the inheritance but this was not supported in a subsequent study (Morgan et al., 1999; Blewitt et al., 2006). A more recent study reports that the fat content in the paternal diet exerts a moderate effect on DNA methylation patterns in offspring liver, including enhancers and proximal promoters of genes involved in fatty acid and cholesterol metabolism (Carone et al., 2010). The authors conclude that the murine epigenome is, at least to some extent, heritable and subject to environmental reprogramming (Carone et al., 2010). Interestingly, some of the aforementioned environmental influences may also operate in humans. For example, severe malnutrition in the periconceptual or early prenatal period not only increases the risk for schizophrenia and other psychiatric disease later in life, but also results in sustained alterations of DNA methylation pattern at the insulin growth factor 2 (*IGF2*) locus (Brown and Susser, 2008; Heijmans *et al.*, 2008; Bale *et al.*, 2010; Lumey *et al.*, 2011), which under normal (non-famine) conditions is a highly regulated imprinted region of the genome and a critical regulator of energy metabolism, brain weight and neural stem cell activities (Pidsley *et al.*, 2010; Lehtinen *et al.*, 2011).

29.4   Epigenetic dysregulation of gene expression in depression and schizophrenia

As already mentioned, depression and schizophrenia typically lack unambiguous neuropathology. In other words, a diseased brain often appears indistinguishable from controls when examined with conventional in vivo imaging or postmortem histological examination. Yet, at least on the cohort level, there are tell-tale signs that neural circuitries are compromised in the diseased cases. One exciting and more recent development involves the study of oscillations and synchronization, typically measured by electroencephalography and magnetencephalography. There is evidence that large cortical and subcortical neuronal networks operating in the beta (13–30 Hz) and gamma (30–200 Hz) range are abnormal in schizophrenia and related disease (Uhlhaas and Singer, 2010). These widespread deficits in synchronization are believed to underlie some of the cognitive deficits observed in the affected patients, including working memory and attention (Uhlhaas and Singer, 2010). There is a large body of literature suggesting that dysregulated gene transcription, indicative of compromised neural circuitry, contributes to defects in neural synchronization and other disordered brain function in psychosis and mood spectrum disorders (Large *et al.*, 2009; Krishnan and Nestler, 2010; Balu and Coyle, 2011). It is now generally accepted that alterations in RNA levels contribute to defects in GABAergic inhibitory neurotransmission and more generally, synapse organization and function, metabolism and mitochondrial functions, and oligodendrocyte pathology in such disorders (Lewis and Gonzalez-Burgos, 2006; Kim and Webster, 2010; Clay *et al.*, 2011; Luscher *et al.*, 2011). However, because no single gene transcript is unanimously affected, many investigators in the field view mood and psychosis spectrum disorders as maladies that, while undoubtedly biological in nature, are very difficult to study due to the absence of a unifying and unambiguous molecular and cellular pathology.

In this context, I envision that epigenetic approaches provide a real opportunity to move the field forward. Firstly, study of epigenetic markings at specific loci that encode the transcripts with altered levels in diseased brain could provide clues whether these changes involve transcriptional and posttranscriptional mechanisms (including initiation and elongation and splicing), and/or enhancer and promoter functions that are associated with a distinct chromatin signature on the level of DNA methylation and histone posttranslatioanl modifications (PTM) (Stadler *et al.*, 2005; Siegmund *et al.*, 2007; Mill *et al.*, 2008; Akbarian and Huang,

2009; Hon *et al.*, 2009; Cheung *et al.*, 2010; Luco *et al.*, 2010; Maunakea *et al.*, 2010; Bulger and Groudine, 2011). Secondly, given the inter-individual variability of the changes observed on the RNA (and protein) level, quantification of DNA methylation and histone PTM at the corresponding locus of the genome could provide additional evidence for or against a gene expression change in a particular disease case. Thus, combined analyses of epigenome and transcriptome will be extremely helpful to explore individual-specific gene expression signatures and offers additional layers of analyses beyond group or cohort-based differences which often are, at best, marginal in notoriously underpowered postmortem brain cohorts that typically are limited to (much) fewer than 100 samples for each study.

The following example should serve as a "proof of principle." Expression of the *GAD1* gene, encoding GAD67 glutamic acid decarboxylase which is a key enzyme for GABA synthesis, is frequently downregulated in inhibitory (GABAergic) neurons of cerebral cortex and hippocampus of schizophrenia and mood disorder cases (Akbarian *et al.*, 1995; Impagnatiello *et al.*, 1998; Volk *et al.*, 2000; Heckers *et al.*, 2002; Torrey *et al.*, 2005; Akbarian and Huang, 2006; Benes *et al.*, 2007; Straub *et al.*, 2007; Lisman *et al.*, 2008; Duncan *et al.*, 2010) and this alteration is viewed as a critical factor in a pathophysiological cascade thought to result in impaired GABAergic neurotransmission and desynchronization of neuronal networks in psychosis (Lewis *et al.*, 2005; Akbarian, 2008; Uhlhaas and Singer, 2010; Volk and Lewis, 2010). Furthermore, common SNPs surrounding the 5′ end of *GAD1* make a small but tractable contribution to genetic risk for childhood-onset schizophrenia, accelerated loss of cortical gray matter volumes during brain maturation and aging, and impaired activation of prefrontal cortex during cognitive tasks (Addington *et al.*, 2005; Straub *et al.*, 2007). Interestingly, subjects with schizophrenia that are biallelic for these at risk-SNPs show imbalances in histone methylation markings such as histone H3 trimethyl-lysine 4 (H3K4me3) and H3 trimethyl-lysine 27 (H3K27me3) which show differential enrichment around transcription start sites (TSS) and indicators of actual or potential transcription (H3K4me3) or repressive chromatin (H3K27me3), in conjunction with *deficits* in GAD1 RNA levels (Huang and Akbarian, 2007; Huang *et al.*, 2007) and DNA *hypo*methylation in portions of the *GAD1* gene body positioned further downstream from the *GAD1* TSS (Huang and Akbarian, 2007). The latter finding is in good agreement with the recent observation that many genes actively transcribed in brain show high levels of DNA cytosine methylation downstream from the primary TSS (Maunakea *et al.*, 2010). Together, these findings imply a subset of schizophrenia or mood disorder cases are affected by dysregulated *GAD1* expression, an alteration that in turn is brought about by a localized genetic risk architecture operating in tandem or sequentially with imbalances of multiple epigenetic markings in the surrounding chromatin (Huang and Akbarian, 2007; Huang *et al.*, 2007; Chen *et al.*, 2011). It is noteworthy that preliminary efforts aimed at "exporting" some of these findings from the aforementioned human postmortem and related animal brain studies (Dong *et al.*, 2005, 2010; Zhang *et al.*, 2010) to living human subjects were successful; for example, the mood-stabilzing drug valproate which is a broad inhibitor of class I/II histone deacetylase (HDAC), differentially affects *GAD1* expression and various histone PTMs in patient and control samples and furthermore, HDAC targets in the lymphocytes may perhaps also be overexpressed in diseased brain (Sharma *et al.*, 2008; Gavin *et al.*, 2009a, 2009b). It is not unrealistic to expect that these and related approaches could in the nearby future lead the way to a "pharmacoepigenomic" screen of selected patients (Abdolmaleky *et al.*, 2008), testing for correlations between the epigenetic assay and treatment response.

29.5 Chromatin assays on brain tissue: the problem of cellular heterogeneity

Many clinical and translational studies in neuroscience, psychiatry, and other fields of medicine exploring disease-related gene expression changes often use tissue homogenates to quantify DNA methylation and histone PTMs at proximal promoters and other sequences critical for transcriptional regulation (Li, 2002; Feinberg, 2004; Feng *et al.*, 2007; Maekawa and Watanabe, 2007). However, any disease-associated alterations in histones and DNA modifications at a specific gene or gene promoter do not necessarily offer clear inference about potential changes in underlying gene expression activity. This skepticism is based on the enormous degree of cellular heterogeneity of central nervous system tissues. Each brain region is uniquely defined by highly heterogeneous mixtures of glutamatergic and GABAergic neurons (which utilize either glutamate or gamma-amino-butyric acid as their major neurotransmitter), and various types of glia including astroglia, oligodendroglia, and microglia. To illustrate some of the underlying complexities, consider the GABAergic neurons of the cerebral cortex which comprise only 20–25% of the neuronal population in that structure, but this GABA neuron population is further subdivided into more than 15 different molecular phenotypes (Ascoli *et al.*, 2008; Burkhalter, 2008). There are also significant shifts in the overall proportion of neurons and glia during brain development and aging (Jiang *et al.*, 2008), and furthermore, microglia and other immune cells have the potential for large-scale invasion of brain tissue on the timescale of hours (Czigner *et al.*, 2007). Practically all of these cells package their genome into the core nucleosome organization described above; both the cells that express a specific RNA and those that do not may nonetheless carry the same type of covalent modification at a specific genomic locus. In contrast to in situ hybridization histochemistry and other histological techniques, virtually all chromatin assays designed to map DNA and histone modification patterns lack single-cell resolution. Instead, the majority of the preclinical and clinical studies focused on detection of histone or DNA modification changes in neuropsychiatric disease are based on assays utilizing brain tissue homogenate as input material, as opposed to specific cell populations (Jiang *et al.*, 2008; Matevossian and Akbarian, 2008). Therefore, most of these studies assume – but do not prove – that the specific cell population showing altered mRNA levels is also affected by the histone modification changes at the corresponding gene as observed in chromatin immunoprecipitates or some other assay from tissue homogenate. This lack of cell resolution in conventional chromatin assays is a significant limitation of the approach for tissue with a very high degree of cellular heterogeneity, including brain. Therefore, changes in chemical modifications of chromatin in tissue extracts from a particular brain specimen could, in principle, be brought about by alterations in the cellular composition of the tissue, and might not necessarily reflect a pathological change in transcriptional activity of the gene in question. These uncertainties in the correct interpretation of chromatin assays in brain tissue are rarely discussed in clinical studies on postmortem brain or even in preclinical work with animal brain. When neuronal and non-neuronal nuclei from human prefrontal cortex were collected separately by immunotagging and fluorescence-activated (cell) nuclei sorting, differential enrichment of the promoter-associated H3K4me3 marking was observed at thousands of loci and the difference in promoter-associated H3K4me3 between neuronal and non-neuronal chromatin was much higher than the changes in neuronal H3K4me3 occupancies during the transition from early infancy to old age (Cheung et al., 2010). The potential confound of cellular heterogeneity poses a particular problem for epigenetic studies in brain of subjects diagnosed with psychiatric disease because to date, studies exploring chromatin markings in clinical specimens and in the preclinical model systems mostly report changes that are quite subtle and highly variable between different postmortem brain cohorts for the same type of disorder (Abdolmaleky et al., 2005, 2006; Grayson et al., 2005; Mill et al., 2008). The challenges discussed here, however, are not insurmountable because protocols for cell-type-specific nuclei sorting from brain tissue are available (Spalding et al., 2005; Jiang et al., 2008; Matevossian and Akbarian, 2008) and several groups were able to profile epigenomes from defined cell populations of the adult brain (Cheung et al., 2010; Feng et al., 2010; Skene et al., 2010).

29.6   Epigenetic modifications as therapeutic targets in psychiatric disease

While the preceding sections of this chapter were mainly focused on how the study of epigenetic markings in clinical samples could shed more light on the underlying mechanisms of psychiatric disease, I will now focus the discussion on the therapeutic potential. There is no question that novel treatment options are urgently needed both for depression and schizophrenia. Currently prescribed antipsychotics are mainly aimed at dopaminergic and/or serotonergic receptor systems and exert therapeutic effects on psychotic symptoms in approximately 75% of patients. However, it is the cognitive impairment that is often the more disabling and persistent feature of schizophrenia and an important predictor for long-term outcome (Ibrahim and Tamminga, 2010). Unfortunately, there are no effective medical treatments for this symptom complex. Similarly, conventional antidepressant therapies primarily target monoamine metabolism and reuptake mechanisms at the terminals of serotonergic, noradrenergic, and dopaminergic neurons. However, up to 40% of cases show an insufficient response to these pharmacological treatments (Krishnan and Nestler, 2010).

There is ample evidence that some of the most frequently prescribed psychopharmacological agents are associated with chromatin remodeling events and

DNA and histone modification changes in brain cells, but whether these effects are required for therapeutic action or, alternatively, reflect some sort of epiphenomenon is presently unclear. The short-chain fatty-acid derivative and established histone deacetylase inhibitor (HDACi) valproic acid, widely prescribed for its mood-stabilizing and anticonvulsant effects, induces brain histone hyperacetylation at a select set of gene promoters when administered to animals at comparatively high doses (Dong *et al.*, 2005). There is additional evidence that pharmacological or genetic interference with histone acetylation in mouse or rat brain is associated with behavioral changes reminiscent of those elicited by conventional antidepressant drugs (Tsankova *et al.*, 2006; Renthal *et al.*, 2007; Schroeder *et al.*, 2007; Covington *et al.*, 2009; Zhu *et al.*, 2009; Grayson *et al.*, 2010; Hollis *et al.*, 2011). Therefore, new HDACi could emerge as promising therapeutic targets not only for cancer and other medical conditions (Marks, 2010), but for neuropsychiatric disorders as well. As it pertains to histone acetylation, future therapeutic targets may not only include the writers (acetyltransferases) or erasers (deacetylases) of this mark, but also some of its "reader" molecules. For example, bromodomains, which are present in many different types of nuclear proteins, bind to acetylated histones and small molecules interfering with some of these interactions recently emerged as powerful modulators of systemic inflammation (Nicodeme *et al.*, 2010). Histone acetylation is not the only type of histone PTM involved in the regulation of motivational and affective behaviors. For example, a robust antidepressant-like phenotype could be elicited by selective overexpression of the histone H3K9-specific methyltransferase, Set domain bifurcated 1 (Setdb1/ESET/KMT1C) in adult forebrain neurons (Jiang *et al.*, 2010). Interestingly, Setdb1 occupancy in neuronal chromatin was highly restricted, and may be confined to less than 0.75% of annotated genes (Jiang *et al.*, 2010). However, among these are several NMDA and other ionotropic glutamate receptor subunit genes, a finding of interest because mild to moderate inhibition of NMDA-receptor-mediated neurotransmission elicits a robust improvement of depressive symptoms in some mood disorder patients (Sanacora *et al.*, 2008). Indeed, Setdb1-mediated H3K9 methylation and repressive chromatin remodeling at NMDA receptor genes was associated antidepressant-like behavioral phenotypes in the Setdb1 transgenic mice (Jiang *et al.*, 2010). The molecular determinants that direct Setdb1 to its specific targets remain to be determined. The same molecule, Setdb1, recently emerged as an oncogene in a zebrafish melanoma model (Ceol *et al.*, 2011), where the mechanism of (tumorigenic) action involved selective dysregulation of a subset of *HOX* genes. Furthermore, monoamine oxidase inhibitors (MAOi) such as tranylcypromine or phenelzine – powerful antidepressants that exert their therapeutic effects mainly by elevating brain monoamine levels through inhibition of MAO-A/B – also function as weak blockers of LSD1-type histone demethylases (Mosammaparast and Shi, 2010). Recently several newer compounds were developed with much stronger activity against LSD1 (Mosammaparast and Shi, 2010) than the aforementioned clinical MAOi drugs, and it will be extremely interesting to explore these drugs in preclinical models for mood and psychosis spectrum disorders.

There is also early evidence for potential epigenetic targets in the treatment of psychosis spectrum disorders. For example, histone H3K4 methylation and expression levels and genomic occupancies of the H3K4-specific methyltransferase, mixed-lineage leukemia 1 (Mll1), were upregulated after treatment with the atypical antipsychotic clozapine (Huang *et al.*, 2007), a drug that improves working memory and other frontal-lobe-associated cognitive functions (Meltzer, 2004).

In addition, dopamine D2 receptor antagonists acting as typical antipsychotic drugs induce dynamic changes in histone acetylation and phosphorylation in the striatum (a forebrain structure intensely innervated by dopaminergic fibers and a key structure for the neuronal circuitry involved in psychosis) (Li *et al.*, 2004; Dong *et al.*, 2008; Bertran-Gonzalez *et al.*, 2009). In the same brain regions, various antipsychotic medications also induced DNA demethylation events at a subset of GABAergic gene promoters implicated in psychosis (Dong *et al.*, 2008). However, to the best of my knowledge, to date nearly all of the work aimed at potential epigenetic targets in psychosis is largely of a correlative nature. Therefore, more stringent approaches, including psychosis-related behavioral studies of genetically engineered mice with altered expression and activity of selected writer/eraser or reader proteins for specific epigenetic markings will be necessary in order to move the field forward.

ACKNOWLEDGMENTS

Work in the author's laboratory is supported by the International Mental Health Research Organization, Autism Speaks, the National Alliance for Research on Schizophrenia and Depression, and the National Institutes of Health.

REFERENCES

Abdolmaleky, H. M., Cheng, K. H., Russo, A., *et al.* (2005). Hypermethylation of the reelin (RELN) promoter in the brain of schizophrenic patients: a preliminary report. *American Journal of Medical Genetics, Part B, Neuropsychiatric Genetics*, **134**, 60–66.

Abdolmaleky, H. M., Cheng, K. H., Faraone, S. V., *et al.* (2006). Hypomethylation of MB-COMT promoter is a major risk factor for schizophrenia and bipolar disorder. *Human Molecular Genetics*, **15**, 3132–3145.

Abdolmaleky, H. M., Zhou, J. R., Thiagalingam, S., and Smith, C. L. (2008). Epigenetic and pharmacoepigenomic studies of major psychoses and potentials for therapeutics. *Pharmacogenomics*, **9**, 1809–1823.

Addington, A. M., Gornick, M., Duckworth, J., *et al.* (2005). GAD1 (2q31.1), which encodes glutamic acid decarboxylase (GAD67), is associated with childhood-onset schizophrenia and cortical gray matter volume loss. *Molecular Psychiatry*, **10**, 581–588.

Akbarian, S. (2008). Restoring GABAergic signaling and neuronal synchrony in schizophrenia. *American Journal of Psychiatry*, **165**, 1507–1509.

Akbarian, S. and Huang, H. S. (2006). Molecular and cellular mechanisms of altered GAD1/GAD67 expression in schizophrenia and related disorders. *Brain Research Reviews*, **52**, 293–304.

Akbarian, S. and Huang, H. S. (2009). Epigenetic regulation in human brain: focus on histone lysine methylation. *Biological Psychiatry*, **65**, 198–203.

Akbarian, S., Kim, J. J., Potkin, S. G., *et al.* (1995). Gene expression for glutamic acid decarboxylase is reduced without loss of neurons in prefrontal cortex of schizophrenics. *Archives of General Psychiatry*, **52**, 258–266.

Arpanahi, A., Brinkworth, M., Iles, D., *et al.* (2009). Endonuclease-sensitive regions of human spermatozoal chromatin are highly enriched in promoter and CTCF binding sequences. *Genome Research*, **19**, 1338–1349.

Ascoli, G. A., Alonso-Nanclares, L., Anderson, S. A., *et al.* (2008). Petilla terminology: nomenclature of features of GABAergic interneurons of the cerebral cortex. *Nature Reviews Neuroscience*, **9**, 557–568.

Bale, T. L., Baram, T. Z., Brown, A. S., *et al.* (2010). Early life programming and neurodevelopmental disorders. *Biological Psychiatry*, **68**, 314–319.

Balu, D. T. and Coyle, J. T. (2011). Neuroplasticity signaling pathways linked to the pathophysiology of schizophrenia. *Neuroscience and Biobehaviour Reviews*, **35**, 848–870.

Benes, F. M., Lim, B., Matzilevich, D., *et al.* (2007). Regulation of the GABA cell phenotype in hippocampus of schizophrenics and bipolars. *Proceedings of the National Academy of Sciences USA*, **104**, 10 164–10 169.

Bertran-Gonzalez, J., Hakansson, K., Borgkvist, A., *et al.* (2009). Histone H3 phosphorylation is under the opposite tonic control of dopamine D2 and adenosine A2A receptors in striatopallidal neurons. *Neuropsychopharmacology*, **34**, 1710–1720.

Blewitt, M. E., Vickaryous, N. K., Paldi, A., Koseki, H., and Whitelaw, E. (2006). Dynamic reprogramming of DNA methylation at an epigenetically sensitive allele in mice. *PLoS Genetics*, **2**, e49.

Bonasio, R., Tu, S., and Reinberg, D. (2010). Molecular signals of epigenetic states. *Science*, **330**, 612–616.

Brown, A. S. and Susser, E. S. (2008). Prenatal nutritional deficiency and risk of adult schizophrenia. *Schizophrenia Bulletin*, **34**, 1054–1063.

Bulger, M. and Groudine, M. (2011). Functional and mechanistic diversity of distal transcription enhancers. *Cell*, **144**, 327–339.

Burkhalter, A. (2008). Many specialists for suppressing cortical excitation. *Frontiers in Neurosciences*, **2**, 155–167.

Carone, B. R., Fauquier, L., Habib, N., *et al.* (2010). Paternally induced transgenerational environmental reprogramming of metabolic gene expression in mammals. *Cell*, **143**, 1084–1096.

Ceol, C. J., Houvras, Y., Jane-Valbuena, J., *et al.* (2011). The histone methyltransferase SETDB1 is recurrently amplified in melanoma and accelerates its onset. *Nature*, **471**, 513–517.

Chen, Y., Dong, E., and Grayson, D. R. (2011). Analysis of the GAD1 promoter: *trans*-acting factors and DNA methylation converge on the 5′ untranslated region. *Neuropharmacology*, **60**, 1075–1087.

Cheung, I., Shulha, H. P., Jiang, Y., *et al.* (2010). Developmental regulation and individual differences of neuronal H3K4me3 epigenomes in the prefrontal cortex. *Proceedings of the National Academy of Sciences USA*, **107**, 8824–8829.

Clay, H. B., Sillivan, S., and Konradil, C. (2011). Mitochondrial dysfunction and pathology in bipolar disorder and schizophrenia. *International Journal of Developmental Neurosciences*, **29**, 311–324.

Covington, H. E. III, Maze, I., LaPlant, Q. C., *et al.* (2009). Antidepressant actions of histone deacetylase inhibitors. *Journal of Neuroscience*, **29**, 11 451–11 460.

Czigner, A., Mihaly, A., Farkas, O., *et al.* (2007). Kinetics of the cellular immune response following closed head injury. *Acta Neurochirurgica*, **149**, 281–289.

Domschke, K., Lawford, B., Young, R., *et al.* (2011). Dysbindin (DTNBP1): a role in psychotic depression? *Journal of Psychiatric Research*, **45**, 588–595.

Dong, E., Agis-Balboa, R. C., Simonini, M. V., *et al.* (2005). Reelin and glutamic acid decarboxylase67 promoter remodeling in an epigenetic methionine-induced mouse model of schizophrenia. *Proceedings of the National Academy of Sciences USA*, **102**, 12 578–12 583.

Dong, E., Nelson, M., Grayson, D. R., Costa, E., and Guidotti, A. (2008). Clozapine and sulpiride but not haloperidol or olanzapine activate brain DNA demethylation. *Proceedings of the National Academy of Sciences USA*, **105**, 13 614–13 619.

Dong, E., Chen, Y., Gavin, D. P., Grayson, D. R., and Guidotti, A. (2010). Valproate induces DNA demethylation in nuclear extracts from adult mouse brain. *Epigenetics*, **5**, 730–735.

Duncan, C. E., Webster, M. J., Rothmond, D. A., *et al.* (2010). Prefrontal GABA(A) receptor alpha-subunit expression in normal postnatal human development and schizophrenia. *Journal of Psychiatric Research*, **44**, 673–681.

Feinberg, A. P. (2004). The epigenetics of cancer etiology. *Seminars in Cancer Biology*, **14**, 427–432.

Feng, J., Fouse, S., and Fan, G. (2007). Epigenetic regulation of neural gene expression and neuronal function. *Pediatric Research*, **61**, 58R–63R.

Feng, J., Zhou, Y., Campbell, S. L., *et al.* (2010). Dnmt1 and Dnmt3a maintain DNA methylation and regulate synaptic function in adult forebrain neurons. *Nature Neuroscience*, **13**, 423–430.

Gavin, D. P., Kartan, S., Chase, K., Jayaraman, S., and Sharma, R. P. (2009a). Histone deacetylase inhibitors and candidate gene expression: an in vivo and in vitro approach to studying chromatin remodeling in a clinical population. *Journal of Psychiatric Research*, **43**, 870–876.

Gavin, D. P., Rosen, C., Chase, K., *et al.* (2009b). Dimethylated lysine 9 of histone 3 is elevated in schizophrenia and exhibits a divergent response to histone deacetylase inhibitors in lymphocyte cultures. *Journal of Psychiatry and Neuroscience*, **34**, 232–237.

Grayson, D. R., Jia, X., Chen, Y., *et al.* (2005). Reelin promoter hypermethylation in schizophrenia. *Proceedings of the National Academy of Sciences USA*, **102**, 9341–9346.

Grayson, D. R., Kundakovic, M., and Sharma, R. P. (2010). Is there a future for histone deacetylase inhibitors in the pharmacotherapy of psychiatric disorders? *Molecular Pharmacology*, **77**, 126–135.

Green, T., Gothelf, D., Glaser, B., *et al.* (2009). Psychiatric disorders and intellectual functioning throughout development in velocardiofacial (22q11.2 deletion) syndrome. *Journal of the American Academy of Child and Adolescent Psychiatry*, **48**, 1060–1068.

Hammoud, S. S., Nix, D. A., Zhang, H., *et al.* (2009). Distinctive chromatin in human sperm packages genes for embryo development. *Nature*, **460**, 473–478.

Hayes, J. J. and Hansen, J. C. (2001). Nucleosomes and the chromatin fiber. *Current Opinions in Genetics and Development*, **11**, 124–129.

Heckers, S., Stone, D., Walsh, J., *et al.* (2002). Differential hippocampal expression of glutamic acid decarboxylase 65 and 67 messenger RNA in bipolar disorder and schizophrenia. *Archives of General Psychiatry*, **59**, 521–529.

Heijmans, B. T., Tobi, E. W., Stein, A. D., *et al.* (2008). Persistent epigenetic differences associated with prenatal exposure to famine in humans. *Proceedings of the National Academy of Sciences USA*, **105**, 17 046–17 049.

Hollis, F., Duclot, F., Gunjan, A., and Kabbaj, M. (2011). Individual differences in the effect of social defeat on anhedonia and histone acetylation in the rat hippocampus. *Hormonal Behaviour*, **59**, 331–337.

Hon, G., Wang, W., and Ren, B. (2009). Discovery and annotation of functional chromatin signatures in the human genome. *PLoS Computer Biology*, **5**, e1000566.

Horsthemke, B. and Wagstaff, J. (2008). Mechanisms of imprinting of the Prader–Willi/Angelman region. *American Journal of Medical Genetics, Part A*, **146**, 2041–2052.

Huang, H. S. and Akbarian, S. (2007). GAD1 mRNA expression and DNA methylation in prefrontal cortex of subjects with schizophrenia. *PLoS One*, **2**, e809.

Huang, H. S., Matevossian, A., Whittle, C., *et al.* (2007). Prefrontal dysfunction in schizophrenia involves mixed-lineage leukemia 1-regulated histone methylation at GABAergic gene promoters. *Journal of Neuroscience*, **27**, 11 254–11 262.

Huang, J., Perlis, R. H., Lee, P. H., *et al.* (2010). Cross-disorder genomewide analysis of schizophrenia, bipolar disorder, and depression. *American Journal of Psychiatry*, **167**, 1254–1263.

Ibrahim, H. M. and Tamminga, C. A. (2010). Schizophrenia: treatment targets beyond monoamine systems. *Annual Reviews of Pharmacology and Toxicology*, **51**, 189–209.

Impagnatiello, F., Guidotti, A. R., Pesold, C., *et al.* (1998). A decrease of reelin expression as a putative vulnerability factor in schizophrenia. *Proceedings of the National Academy of Sciences USA*, **95**, 15 718–15 723.

Insel, T. R. (2009). Disruptive insights in psychiatry: transforming a clinical discipline. *Journal of Clinical Investigation*, **119**, 700–705.

Jiang, Y., Matevossian, A., Huang, H. S., Straubhaar, J., and Akbarianl, S. (2008). Isolation of neuronal chromatin from brain tissue. *BMC Neuroscience*, **9**, 42.

Jiang, Y., Jakovcevski, M., Bharadwaj, R., *et al.* (2010). Setdb1 histone methyltransferase regulates mood-related behaviors and expression of the NMDA receptor subunit NR2B. *Journal of Neuroscience*, **30**, 7152–7167.

Kim, S. and Webster, M. J. (2010). Correlation analysis between genome-wide expression profiles and cytoarchitectural abnormalities in the prefrontal cortex of psychiatric disorders. *Molecular Psychiatry*, **15**, 326–336.

Krishnan, V. and Nestler, E. J. (2010). Linking molecules to mood: new insight into the biology of depression. *American Journal of Psychiatry*, **167**, 1305–1320.

Lander, E. S. (2011). Initial impact of the sequencing of the human genome. *Nature*, **470**, 187–197.

Large, C. H., Di Daniel, E., Li, X., and Georgel, M. S. (2009). Neural network dysfunction in bipolar depression: clues from the efficacy of lamotrigine. *Biochemical Society Transactions*, **37**, 1080–1084.

Lehtinen, M. K., Zappaterra, M. W., Chen, X., *et al.* (2011). The cerebrospinal fluid provides a proliferative niche for neural progenitor cells. *Neuron*, **69**, 893–905.

Lewis, D. A. and Gonzalez-Burgos, G. (2006). Pathophysiologically based treatment interventions in schizophrenia. *Nature Medicine*, **12**, 1016–1022.

Lewis, D. A., Hashimoto, T., and Volkl, D. W. (2005). Cortical inhibitory neurons and schizophrenia. *Nature Reviews in Neuroscience*, **6**, 312–324.

Li, E. (2002). Chromatin modification and epigenetic reprogramming in mammalian development. *Nature Reviews in Genetics*, **3**, 662–673.

Li, J., Guo, Y., Schroeder, F. A., *et al.* (2004). Dopamine D2-like antagonists induce chromatin remodeling in striatal neurons through cyclic AMP-protein kinase A and NMDA receptor signaling. *Journal of Neurochemistry*, **90**, 1117–1131.

Lisman, J. E., Coyle, J. T., Green, R. W., *et al.* (2008). Circuit-based framework for understanding neurotransmitter and risk gene interactions in schizophrenia. *Trends in Neurosciences*, **31**, 234–242.

Liu, J., Li, J., Li, T., *et al.* (2011). CTLA-4 confers a risk of recurrent schizophrenia, major depressive disorder and bipolar disorder in the Chinese Han population. *Brain, Behaviour and Immunity*, **25**, 429–433.

Luco, R. F., Pan, Q., Tominaga, K., *et al.* (2010). Regulation of alternative splicing by histone modifications. *Science*, **327**, 996–1000.

Lumey, L. H., Stein, A. D., and Susser, E. (2011). Prenatal famine and adult health. *Annual Reviews of Public Health*, **32**, 237–262.

Luscher, B., Shen, Q., and Sahirl, N. (2011). The GABAergic deficit hypothesis of major depressive disorder. *Molecular Psychiatry*, **16**, 383–406.

Maekawa, M. and Watanabe, Y. (2007). Epigenetics: relations to disease and laboratory findings. *Current Topics in Medicinal Chemistry*, **14**, 2642–2653.

Marks, P. A. (2010). The clinical development of histone deacetylase inhibitors as targeted anticancer drugs. *Expert Opinions in Investigative Drugs*, **19**, 1049–1066.

Matevossian, A. and Akbarian, S. (2008). Neuronal nuclei isolation from human postmortem brain tissue. *Journal of Visual Experiments*, **20**, 2–3.

Maunakea, A. K., Nagarajan, R. P., Bilenky, M., *et al.* (2010). Conserved role of intragenic DNA methylation in regulating alternative promoters. *Nature*, **466**, 253–257.

Meltzer, H. Y. (2004). What's atypical about atypical antipsychotic drugs? *Current Opinions in Pharmacology*, **4**, 53–57.

Mill, J., Tang, T., Kaminsky, Z., *et al.* (2008). Epigenomic profiling reveals DNA-methylation changes associated with major psychosis. *American Journal of Human Genetics*, **82**, 696–711.

Morgan, H. D., Sutherland, H. G., Martin, D. I., and Whitelaw, E. (1999). Epigenetic inheritance at the agouti locus in the mouse. *Nature Genetics*, **23**, 314–318.

Mosammaparast, N. and Shi, Y. (2010). Reversal of histone methylation: biochemical and molecular mechanisms of histone demethylases. *Annual Reviews of Biochemistry*, **79**, 155–179.

Nicholls, R. D. and Knepper, J. L. (2001). Genome organization, function, and imprinting in Prader–Willi and Angelman syndromes. *Annual Reviews of Genomics and Human Genetics*, **2**, 153–175.

Nicodeme, E., Jeffrey, K. L., Schaefer, U., *et al.* (2010). Suppression of inflammation by a synthetic histone mimic. *Nature*, **468**, 1119–1123.

Pidsley, R., Dempster, E. L., and Mill, J. (2010). Brain weight in males is correlated with DNA methylation at IGF2. *Molecular Psychiatry*, **15**, 880–881.

Porteous, D. J., Thomson, P., Brandon, N. J., and Millarl, J. K. (2006). The genetics and biology of DISC1: an emerging role in psychosis and cognition. *Biological Psychiatry*, **60**, 123–131.

Renthal, W., Maze, I., Krishnan, V., *et al.* (2007). Histone deacetylase 5 epigenetically controls behavioral adaptations to chronic emotional stimuli. *Neuron*, **56**, 517–529.

Sanacora, G., Zarate, C. A., Krystal, J. H., and Manjil, H. K. (2008). Targeting the glutamatergic system to develop novel, improved therapeutics for mood disorders. *Nature Reviews in Drug Discovery*, **7**, 426–437.

Schroeder, F. A., Lin, C. L., Crusio, W. E., and Akbarian, S. (2007). Antidepressant-like effects of the histone deacetylase inhibitor, sodium butyrate, in the mouse. *Biological Psychiatry*, **62**, 55–64.

Sharma, R. P., Grayson, D. R., and Gavinl, D. P. (2008). Histone deactylase 1 expression is increased in the prefrontal cortex of schizophrenia subjects: analysis of the National Brain Databank microarray collection. *Schizophrenia Research*, **98**, 111–117.

Shiio, Y. and Eisenman, R. N. (2003). Histone sumoylation is associated with transcriptional repression. *Proceedings of the National Academy of Sciences USA*, **100**, 13 225–13 230.

Siegmund, K. D., Connor, C. M., Campan, M., *et al.* (2007). DNA methylation in the human cerebral cortex is dynamically regulated throughout the life span and involves differentiated neurons. *PLoS One*, **2**, e895.

Skene, P. J., Illingworth, R. S., Webb, S., *et al.* (2010). Neuronal MeCP2 is expressed at near histone-octamer levels and globally alters the chromatin state. *Molecular Cell*, **37**, 457–468.

Spalding, K. L., Bhardwaj, R. D., Buchholz, B. A., Druid, H., and Frisenl, J. (2005). Retrospective birth dating of cells in humans. *Cell*, **122**, 133–143.

Stadler, F., Kolb, G., Rubusch, L., *et al.* (2005). Histone methylation at gene promoters is associated with developmental regulation and region-specific expression of ionotropic and metabotropic glutamate receptors in human brain. *Journal of Neurochemistry*, **94**, 324–336.

Straub, R. E., Lipska, B. K., Egan, M. F., *et al.* (2007). Allelic variation in *GAD1* (GAD67) is associated with schizophrenia and influences cortical function and gene expression. *Molecular Psychiatry*, **12**, 854–869.

Taverna, S. D., Li, H., Ruthenburg, A. J., Allis, C. D., and Patell, D. J. (2007). How chromatin-binding modules interpret histone modifications: lessons from professional pocket pickers. *Nature Structural and Molecular Biology*, **14**, 1025–1040.

Torrey, E. F., Barci, B. M., Webster, M. J., *et al.* (2005). Neurochemical markers for schizophrenia, bipolar disorder, and major depression in postmortem brains. *Biological Psychiatry*, **57**, 252–260.

Tsankova, N. M., Berton, O., Renthal, W., *et al.* (2006). Sustained hippocampal chromatin regulation in a mouse model of depression and antidepressant action. *Nature Neuroscience*, **9**, 519–525.

Uhlhaas, P. J. and Singer, W. (2010). Abnormal neural oscillations and synchrony in schizophrenia. *Nature Reviews in Neurosciences*, **11**, 100–113.

Volk, D. W. and Lewis, D. A. (2010). Prefrontal cortical circuits in schizophrenia. Current *Topics in Behavioural Neuroscience*, **4**, 485–508.

Volk, D. W., Austin, M. C., Pierri, J. N., Sampson, A. R., and Lewisl, D. A. (2000). Decreased glutamic acid decarboxylase67 messenger RNA expression in a subset of prefrontal cortical gamma-aminobutyric acid neurons in subjects with schizophrenia. *Archives of General Psychiatry*, **57**, 237–245.

Weake, V. M. and Workman, J. L. (2008). Histone ubiquitination: triggering gene activity. *Molecular Cell*, **29**, 653–663.

Wolffe, A. P. (1992). New insights into chromatin function in transcriptional control. *Federation of American Societies for Experimental Biologists*, **6**, 3354–3361.

Zhang, T. Y., Hellstrom, I. C., Bagot, R. C., *et al.* (2010). Maternal care and DNA methylation of a glutamic acid decarboxylase 1 promoter in rat hippocampus. *Journal of Neuroscience*, **30**, 13 130–13 137.

Zhu, H., Huang, Q., Xu, H., Niu, L., and Zhoul, J. N. (2009). Antidepressant-like effects of sodium butyrate in combination with estrogen in rat forced swimming test: involvement of 5-HT(1A) receptors. *Behavioural Brain Research*, **196**, 200–206.

30 Genome-wide DNA methylation analysis in patients with familial ATR-X mental retardation syndrome

Gemma Carvill* and Andrew Sharp

30.1 Introduction

Mental retardation (MR) is estimated to have a prevalence of 1–3% in the developed world, making it a common congenital disorder (Leonard and Wen, 2002). Patients afflicted with this condition exhibit impaired development of adaptive and cognitive abilities, which manifests before 18 years of age. Mental retardation is a genetically and clinically heterogeneous disorder with mutations in over 300 genes, resulting in both non-syndromic MR (MR occurs in isolation) and hundreds of MR syndromes (MR presents in conjunction with additional clinical features) (Chelly *et al.*, 2006; Chiurazzi *et al.*, 2008).

Mutations in these MR genes disrupt the organization of neuronal cells into complex networks as well as the ability of these networks to remodel in response to learning and experience. Thus, understanding the neurobiology that governs this pathophysiology requires knowledge of the molecular functions and pathways that are affected in MR pathogenesis. While a variety of metabolic, structural, and signal-transduction pathways are implicated in MR, this chapter will focus on the relatively recently elucidated pathogenic role of aberrant epigenetic mechanisms.

The role of epigenetic mechanisms in controlling neuronal development follows a number of lines of evidence. Firstly, neurogenesis is, at least in part, controlled by the protein restriction element 1 (RE1) silencing transcription factor (REST). In non-neuronal tissues, REST binds to DNA at short consensus sequences (RE1) and promotes protein–protein interaction with DNA methyltransferases (DNMTs), histone deacetylases (HDACs), methyl-binding domain proteins (MBDs), transcription factors, and chromatin remodelers and co-regulators (coREST). The formation of this REST-associated protein complex enforces a repressive epigenetic state which includes marks such as: DNA methylation, H3K9 di/trimethylation, and total H3K4

---

* Author to whom correspondence should be addressed.

Epigenomics: From Chromatin Biology to Therapeutics, ed. K. Appasani. Published by Cambridge University Press. © Cambridge University Press 2012.

demethylation at the target DNA region (Zheng *et al.*, 2009). In this way neuronal gene expression is restricted to the nervous system. Conversely, in neuronal progenitor cells the REST complex bound at neuronal promoters is generally associated with active chromatin modification marks. Rapid dissociation of REST from the RE1 site during neurogenesis is achieved through REST degradation or non-coding RNA (ncRNA)-mediated activation, hence permitting neuronal gene expression (Kuwabara *et al.*, 2004; Ballas and Mandel, 2005).

In addition to a role in neurogenesis, epigenetic mechanisms have been hypothesized to play a role in learning and memory. In order to develop long-term memories there is accumulating evidence that neuronal functioning makes use of epigenetic mechanisms in a manner similar to cell-state commitment during development (Swank and Sweatt, 2001; Levenson and Sweatt, 2005; Levenson *et al.*, 2006; Miller *et al.*, 2008). These epigenetic changes include both changes in DNA methylation and histone modifications, as well as interaction between the two. For example, it was demonstrated that acetylation of histone H3 occurs in rat hippocampal regions subsequent to contextual fear conditional training (Levenson *et al.*, 2004). Furthermore, DNMT inhibition prevented this memory-associated H3 acetylation, thus illustrating the necessity for interplay between these two epigenetic mechanisms for memory formation (Miller *et al.*, 2008). Finally, epigenetic mechanisms have been implicated in a variety of disorders, of which MR is often a core feature (Table 30.1).

Table 30.1 Disorders associated with aberrant epigenetic mechanisms for which MR is a core feature

Disorder	Manifestations	Etiology
Angelman syndrome	MR	Imprinting deregulation of *UBE3A* at 15q11–13 (maternal) (Bienvenu and Chelly, 2006)
Prader–Willi syndrome	MR	Imprinting deregulation of genes at 15q11–13 (paternal) (Bienvenu and Chelly, 2006)
ATR-X syndrome	MR	*ATRX* mutations → hypomethylation of specific repeat and satellite sequences (Gibbons *et al.*, 2000)
Coffin–Lowry syndrome	MR	*RSK2* mutations → histone phosphorylation (Harum *et al.*, 2001)
Rett syndrome	MR	*MECP2* mutations → recruits repressive and activating complexes to methylated DNA (Chahrour and Zoghbi, 2007; Yasui *et al.*, 2007)
Rubinstein–Taybi syndrome	MR	*CBP* mutations → histone acetylation (Roelfsema *et al.*, 2005)
ICF syndrome (immunodeficiency, centromeric region instability, and facial anomalies syndrome)	Chromosome instability, MR	*DNMT3B* mutations → DNA hypomethylation (Ehrlich *et al.*, 2008)

30.2   X-linked mental retardation and epigenetic mechanisms

It is estimated that 10–15% of MR is the result of mutations in genes located on the X chromosome, giving rise to the term X-linked mental retardation (XLMR) (Ropers and Hamel, 2005). To date, mutations in a total of 87 X-chromosomal genes have been shown to play a role in the pathogenesis of XLMR (Chiurazzi *et al.*, 2008; Cho *et al.* 2008; Molinari *et al.*, 2008; Tarpey *et al.*, 2009). These known XLMR genes are involved in a host of cellular processes (Figure 30.1). However, the largest number of XLMR genes are either known or postulated to play a role in transcription regulation (22%). This group of XLMR genes includes transcription activator/repressors and chromatin remodelers (e.g., *ATRX*). Interactions of these proteins with the two primary epigenetic marks, histone modifications and DNA methylation, induce a chromatin state that is applicable to the transcriptional state required. Thus, in this study it was hypothesized that mutations in XLMR genes involved in this transcription regulation pathway would lead to aberrant DNA methylation. In order to test this hypothesis we selected the candidate gene *ATRX* to investigate DNA methylation patterns in patients positive for mutations in this gene.

The *ATRX* gene is one of the 22% of XLMR genes that play a role transcriptional regulation. The participation of ATRX in gene regulation is supported by several lines of evidence, including:

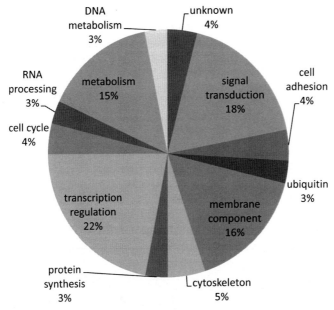

Figure 30.1 Pie chart illustrating the diverse protein functions of the 87 known XLMR genes. (Adapted from Chiurazzi *et al.*, 2008.)

- Association with chromatin remodeling complexes by interaction with proteins such as:
  - The repressive polycomb group(PcG) protein EZH2 (Cardoso *et al.*, 1998). EZH2 is involved in homeotic transcriptional repression during development
  - The heterochromatin protein HP1 (McDowell *et al.*, 1999)
  - Transcription/apoptosis cofactor, DAXX (Ishov *et al.*, 2004)
  - The MBD protein MeCP2 (Nan *et al.*, 2007)
- DNA hypomethylation at the acrocentric chromosomal regions which encompass the rDNA gene arrays in *ATRX* mutation positive patients. Conversely, DNA hypermethylation at Y-specific repeats (DYZ2) is also evident in these patients (Gibbons *et al.*, 2000)
- The protein's localization to promyelocytic leukemia nuclear bodies (PML-NBs), sites of transcription regulation (Xue *et al.*, 2003)
- The weak chromatin remodeling ability of ATRX, as demonstrated by mononucleosome and triple-helix DNA displacement activity (Xue *et al.*, 2003)
- Dysregulation of numerous genes involved in neurogenesis and neuronal development in conditional *Atrx* knockout mice (Levy *et al.*, 2008)
- The localization of ATRX to an H3K9me2 hotspot upstream of the *Xist* locus which controls X-inactivation in females (Baumann and De La Fuente, 2009).

Collectively, this evidence strongly argues for a chromatin remodeling function for ATRX. Also, the aforementioned observations have led to the hypothesis that ATRX forms part of a neuronal transcription repression complex which targets methylated promoter regions (Kramer and van Bokhoven, 2009).

While aberrations in DNA methylation have already been demonstrated in ATR-X patients at rDNA gene regions and Y-specific repeats (Gibbons *et al.*, 2000), no genome-wide investigations of DNA methylation status in ATR-X patients have been conducted to date. Furthermore, given the substantial evidence for the role of ATRX in gene regulation and the concomitant importance of DNA methylation in this process, it was hypothesized that aberrant DNA methylation was a pathogenic mechanism associated with *ATRX* mutations.

30.3   Methods

In order to test the hypothesis that aberrant DNA methylation analysis is part of the pathogenic spectrum associated with XLMR, we assessed genome-wide CpG island and promoter DNA methylation profile in an *ATRX*-mutation-positive family, XMR2. This family consisted of four MR patients with a known *ATRX* mutation (c.5987_6011del) and a number of unaffected individuals. DNA methylation profiles of the four affected individuals were compared to the profiles generated from three unaffected sibs. This genome-wide DNA methylation profiling (MeDIP-ChiP assay), statistical analysis, as well as the confirmatory DNA methylation investigations are described in the sections that follow.

30.3.1 Methylation-dependent immunoprecipitation(MeDIP)-ChiP assay

Methylation-dependent immunoprecipitation and array hybridization (MeDIP-ChiP) was employed to obtain the genome-wide promoter DNA methylation profiles of *ATRX* mutation positive patients and unaffected family members.

In order to enrich DNA samples for the methylated CpG portions of the genome, MeDIP was conducted by immunocapture of methylated cytosines using a 5-methylcytosine-specific antibody (Weber *et al.*, 2005). DNA samples were extracted from peripheral blood lymphocytes from the four *ATRX*-mutation-positive patients as well as three unaffected male sibs. DNA samples were sonicated to obtain fragment sizes of 200–800 bp (Branson 450D sonifier), denatured, and subjected to immunocapture by incubation with 10 µg of 5mc antibody (Diagenode, Liège, Belgium), followed by immunoprecipitation with Protein A sepharose beads (Life Technologies, Carlsbad, CA, USA) and finally a phenol–chloroform DNA extraction. Both immunoprecipitated and input DNA from each individual were labeled by random priming using Cy3- and Cy5-conjugated random nonamers (NimbleGen, Madison, WI, USA) respectively and hybridized to the HD2 deluxe CpG island and promoter array (NimbleGen). This commercially available array platform consists of 2.1 million 50–75mer probes which encompass all annotated promoters, CpG islands, and microRNA (miRNA) promoters, as well as manually selected ENCODE regions. These regions are covered at a mean resolution of 1 probe per 100 bp (NimbleGen).

All arrays were scanned on the G2565 Agilent scanner (Agilent Technologies, Santa Clara, CA, USA) at a 5-µm resolution according to manufacturer's recommendations. NimbleScan v. 2.5 (Nimblegen) software was used to analyze raw data and generate independent probe $\log_2$ fold changes. The SignalMap v. 1.9 (NimbleGen) software was used to visualize $\log_2$ fold changes according to genomic location of the probes. Duplicate array hybridizations and analyses were performed for each DNA sample.

30.3.2 Statistical analysis for differentially methylated region detection

In order to identify potential differentially methylated regions (DMRs), statistical analyses were performed to establish which loci harbored differential methylation ratios between the ATR-X patients and controls (unaffected family members). All statistical analyses were performed using software from the Bioconductor project (Gentleman *et al.*, 2004).

Prior to statistical analyses a quantile normalization was performed for each array, accounting for systematic bias incurred when comparing different arrays (e.g., DNA labeling, MeDIP enrichment) (Bolstad *et al.*, 2003). Also, outlier replacement (Sharp *et al.*, 2010) and a linear smoothing function was applied (Pelizzola *et al.*, 2008). Thereafter, a moderated *t*-score test was performed to identify potential DMRs, as employed in the LIMMA software package (Smyth,

2004), and a false-discovery rate (FDR) correction applied to account for multiple testing (Benjamini and Hochberg, 1995). Potential DMRs were identified by clusters of probes with significantly differential methylation fold changes at a relatively low stringency threshold. These clusters were defined as regions which encompassed a single probe with an FDR adjusted p-value ≤ 0.01 that were flanked by at least two additional probes with nominal p-values ≤ 0.01. These clusters were subject to further computational analyses and filtering methods.

30.3.3 Validation of putative differentially methylated regions by bisulfite DNA sequencing

In order to validate the results of the MeDIP-ChiP experiments, a number of DMRs were selected for bisulfite DNA sequencing, these included putative DMRs with the highest absolute value t-score statistic (four loci) and the highest average (for all differentially methylated probes in that region) absolute t-score statistic (four loci). These candidate DMRs were verified using Sanger DNA sequencing, subsequent to the treatment of genomic DNA with sodium bisulfite with the Epitect® bisulfite kit (QIAGEN, Hilden, Germany). Treatment of genomic DNA with sodium bisulfite results in the conversion of unmethylated cytosine residues to uracil by deamination, while methylated cytosines are protected from this conversion by the 5-methyl modification. PCR primers were designed to amplify each of the potential DMRs using Methyl Primer Express software v. 1.0 (Applied Biosystems, Foster City, CA, USA). Target DMRs were then amplified by PCR using the bisulfite-treated DNA as a template, amplicons were purified using exonuclease 1 and shrimp alkaline phosphatase (Fermentas, Waltham, MA, USA) and subject to DNA sequencing. Subsequently, unmethylated cytosine residues appear as thymine nucleotides on the resulting electropherogram, which were clearly distinguishable from the cytosine residues (which correspond to methylated Cs).

30.3.4 Quantification of DNA methylation at differentially methylated regions

Those loci that were found to harbor true-positive DMRs by bisulfite DNA sequencing were cloned and individual colonies sequenced in order to obtain a semi-quantitative measure of DNA methylation across the region in both the ATR-X patients and unaffected controls.

PCR was performed for each of the validated DMRs as above in the four affected males and the three unaffected sibs. PCR products were purified using the QIAGEN Gel Extraction Kit (QIAGEN) and subject to T/A cloning using the pGEM®-T Easy Vector System (Promega, Fitchburg, WI, USA) using chemically competent JM109 *E.coli* cells using a heat-shock protocol (Promega) and positive clone selection using blue–white screening. Twenty positive (white) colonies generated by T/A cloning in each patient and control individual of family XMR2 were selected for DNA sequencing. These colonies were PCR amplified using the plasmid-specific (pGEM®-T Easy Vector) forward and reverse primers and BiQ Analyzer was used for analysis (Bock *et al.*, 2005).

30.4   Results

30.4.1   Genome-wide CpG island and promoter DNA methylation array analysis

Comparison of genome-wide promoter DNA methylation profiles between four affected and three unaffected individuals of family XMR2 led to the identification of 112 unique putative DMRs. These putative DMRs were selected on the basis of one probe exhibiting a FDR-adjusted p-value ≤ 0.01, combined with at least two additional flanking probes with a nominal p-value ≤ 0.01. Subsequent to the removal of all DMRs containing overlapping regions with known copy-number variants (CNVs) and segmental duplications (SDs), a total of 107 unique putative DMRs were identified.

From these 107 putative DMRs, a total of eight loci were selected for validation. These candidate DMRs (Table 30.2) were selected as they possessed the highest absolute t-score value (rows 1–5, Table 30.2) or the highest absolute value average t-score (rows 1,4,6–8, Table 30.2) i.e., where the average t-score was calculated using t-score values for all probes exhibiting significant differentially methylation values.

30.4.2   Validation of putative differentially methylated regions

Bisulfite DNA sequence analysis of the eight putative DMRs prioritized for validation revealed one true-positive DMR, encompassing the central probe CHR16FS010387819 (Figure 30.2A). The DMR corresponds to an approximately 1-kb interval on chromosome 16 (16p13.13) (Figure 30.2B). This chromosomal interval encompasses a CpG island which corresponds to the promoter region of the Activating Transcription Factor 7 Interacting Protein 2 (ATF7IP2). The remaining seven putative DMRs prioritized for validation by bisulfite DNA sequencing showed no discernible difference between patients and controls at the CpG sites investigated.

Table 30.2 The eight DMRs selected for validation by bisulfite DNA sequencing; Highlighted in **bold** are the criteria upon which selection for validation was based

	Probe ID	t-score value	p-value	FDR[a] adjusted p-value	Average t-score value	Average FDR-adjusted p-value
1.	CHR16FS010387819	**9.1126**	2.49E-18	2.66E-12	**5.367544**	5.08E-02
2.	CHR11FS126375528	**−8.1971**	2.47E-15	4.77E-10	−4.34842	2.38E-01
3.	CHR17FS034606720	**−8.1907**	2.59E-15	4.77E-10	−4.43846	1.51E-01
4.	CHR05FS001853381	**7.8763**	2.45E-14	1.87E-09	**4.940175**	6.36E-02
5.	CHR02FS065138632	**6.8146**	2.98E-11	5.03E-07	4.176412	9.56E-02
6.	CHR19FS015436362	−6.8001	3.26E-11	5.43E-07	**−5.30631**	5.53E-03
7.	CHR18FS031177845	−7.0706	5.77E-12	1.58E-07	**−4.98466**	3.93E-02
8.	CHR05FS074359693	5.8915	7.40E-09	2.91E-05	**4.8544**	3.32E-03

[a] FDR, false-discovery rate.

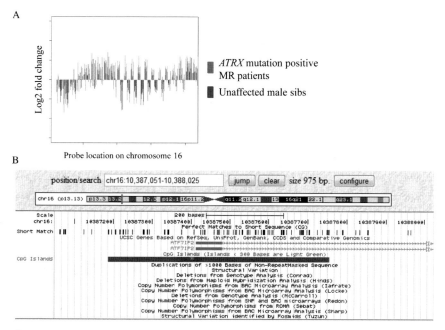

Figure 30.2 The DMR encompassing the probe CHR16FS010387819 identified by genome-wide promoter DNA methylation analysis in *ATRX*-mutation-positive patients and subsequent statistical analyses. (A) The output generated by statistical analysis also showing the hypermethylation of this 16p region in the ATR-X patients (red) as compared to the unaffected sibs (blue). (B) A UCSC exsert of the hypermethylated *ATF7IP2* locus in the ATR-X patients; this hypermethylation occurs in the CpG island upstream of *ATF7IP2*. No segmental duplications or CNVs are present in this region. See plate section for color version.

30.4.3 Quantification of DNA methylation at differentially methylated regions

The semi-quantitative DMR PCR cloning investigations showed variable methylation ratios across 23 CpG sites of between 8% and 34% in the ATR-X patients (XMR2.5/13/25/26) and 0–7% in the unaffected family members (XMR2.14/17/21) (Figure 30.3). In addition, the distribution of methylated cytosine residues demonstrated intercellular and inter-individual variability (Figure 30.3). These results show that *ATRX*-mutation-positive individuals have significantly increased levels of DNA methylation at the *ATF7IP2* promoter region as compared to the unaffected members of family XMR2.

30.5 Discussion

The comparative genome-wide DNA methylation analysis conducted between the four ATR-X patients and the three unaffected XMR2 family members identified a total of 112 putative DMRs. From the total of 112 potential DMRs, five regions (~4%) were excluded on the basis of the DNA sequence overlapping with known CNVs or SDs. These genomic regions are prone to rearrangements and/or are repetitive in nature. Furthermore, these changes in copy number are indistinguishable from methylation enrichment peaks in MeDIP-ChiP experiments (Vega

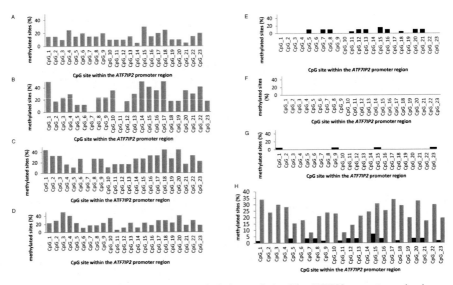

Figure 30.3 The semi-quantitative DNA methylation analysis of the *ATF7IP2* promoter region in *ATRX*-mutation-positive patients of family XMR2: (A) XMR2.5, (B) XMR2.13, (C) XMR2.25, and (D) XMR2.26, as well as the unaffected family members: (E) XMR2.14, (F) XMR2.27, and (G) XMR2.21. (H) For each of the 23 CpG sites located within this promoter region (*x*-axis) the corresponding proportion of cells with *ATF7IP2* methylation (*y*-axis) for the affected (gray) and unaffected (black) family members is shown. At all CpG sites the level of methylation was substantially higher (up to 34%) in *ATRX*-mutation-positive individuals, indicating hypermethylation of this *ATF7IP2* promoter region.

et al., 2009; A. Sharp, unpublished data). Thus these regions were excluded from any further analyses.

From the remaining 107 potential DMRs, the eight DMRs with the highest (nominal and average) absolute *t*-score values were selected for validation. Of these eight candidate DMRs, one locus, encompassing the *ATF7IP2* promoter region, was confirmed by bisulfite DNA sequencing. Not surprisingly, the true-positive *ATF7IP2* locus had the highest *t*-score statistic (9.1126, adjusted FDR *p*-value 2.49E-18) at a single probe, as well as the highest average *t*-score value (5.367544, adjusted FDR *p*-value 2.66E-12) across the region. The values of the seven candidate regions that were not verified using this technique had markedly lower *t*-score statistics (Table 30.2). These results suggest that a significantly high *t*-score value at a single probe should be accompanied by significantly high *t*-scores at flanking probes (therefore generating a high average *t*-score). While an attempt was made to account for the effects of a single deviant probe by selecting DMRs with at least two flanking probes with nominal *p*-values ≥0.01, these results suggest that future studies of this nature should employ stricter DMR selection criteria. In addition, the number of potential DMRs could be reduced by basing DMR selection criteria on a relatively high *t*-score as well *p*-value (as opposed to just the *p*-value). However, DNA methylation studies of this nature have not been performed before and therefore a relatively liberal statistical approach was originally employed in attempt to reduce false-negative rates.

The semi-quantitative analysis of the *ATF7IP2* promoter region using a cloning and colony-PCR technique showed variable methylation levels across the 23 CpG sites of this region. In ATR-X patients the proportion of CpG sites methylated was on average between 8% and 34%; conversely in unaffected family members this averaged methylation level was between 0% and 7%. This result clearly demonstrated the significant degree of *ATF7IP2* hypermethylation as a result of the *ATRX* mutation in family XMR2. Furthermore, aberrant DNA methylation at this locus has been confirmed by bisulfite sequencing in two additional patients from unrelated families harboring mutations in *ATRX* (data not shown). Taken together, these results support the presence of DNA hypermethylation in ATR-X patients at this locus. However, there was also significant intra-familial variability of methylation levels at this locus. For instance, there was a small level of methylation of CpG sites in the unaffected family members. Unaffected individual (XMR2.14) exhibited methylation at several CpG sites in up to 8% of cells; conversely individual (XMR2.17) did not demonstrate DNA methylation at any CpG sites. Furthermore, this intra-familial variability of methylation ratios at this locus was also apparent in the *ATRX*-mutation-positive patients, it is likely that this variability is an attribute of the *ATF7IP2* locus.

The *ATF7IP2* gene, also known as *MBD1-containing chromatin-associated factor 2* (*MCAF2*), is located at 16p13.13. *ATF7IP2* transcribes a 2046-bp mRNA that encodes a 681-amino-acid protein of approximately 100 kDa (Ichimura *et al.*, 2005). ATF7IP2 is known to bind to the transcription repression domain of the methylated-cytosine-binding protein MBD1 (Ichimura *et al.*, 2005). Furthermore, ATF7IP2 has also been shown to interact with the transcription activator Sp1 and to a lesser extent, the histone H3K9 methyltransferase SETDB1 (Ichimura *et al.*, 2005). The overlapping protein interactions of ATRX and ATF7IP2 suggest that these two proteins may form part of the same protein complex or repressive chromatin assembly pathway. This hypothesis is supported by several lines of empirical evidence. Firstly, ATF7IP2 forms a complex with SETDB1, responsible for the methylation of H3K9 (Ichimura *et al.*, 2005). In turn, this H3K9me3 mark has been shown to elicit ATRX localization at the *Xist* locus (during X-inactivation), Y-specific repeats (ATRX mutations elicit hypermethylation at this locus), and pericentric heterochromatin (Gibbons *et al.*, 2000; Baumann *et al.*, 2008; Baumann and De La Fuente, 2009). Finally, H3K9me3 serves as a mark for HP1, known to localize with ATRX at heterochromatic regions (McDowell *et al.*, 1999). These data suggest that ATRX targeting to these heterochromatic regions, which is dependent on H3K9me3, may be reliant on the histone methyltransferase activity of SETDB1 through cooperation with ATF7IP2.

There is also evidence that ATF7IP2 may have a transcription-activating role in addition to the aforementioned repressive function. ATF7IP2 associates with the transcription activator Sp1 (Ichimura *et al.*, 2005), known to localize to the PML-NBs, sites of active transcription (Vallian *et al.*, 1998). Furthermore, ATRX has been shown to interact with the transcription co-activator DAXX at the PML-NBs (Xue *et al.*, 2003). Once again these overlapping transcriptional regulatory roles of ATRX and ATF7IP2 suggest these proteins may participate in similar gene-regulation pathways. While the aforementioned empirical data provide evidence suggesting ATF7IP2 and ATRX have similar repressive/activating chromatin remodeling properties and potentially function in the same protein complex, this hypothesis requires further experimental validation. In addition, while alterations to the *ATF7IP2* transcript have not been previously associated with any disease phenotype, the overlapping function of ATF7IP2 and ATRX suggests that *ATF7IP2* is a good candidate MR gene.

The results presented here show that *ATRX* aberrations do not lead to widespread changes at CpG promoter regions. The microarray selected for this study targets only the genomic regions that are associated with CpG sites at promoter regions. Previously it was shown that *ATRX*-mutation-positive patients have DNA methylation aberrations at specific loci which do not correspond to promoter regions. These loci included hypomethylation at the rDNA gene arrays which localize to the short arms of acrocentric chromosomes, as well as hypermethylation of Y-specific repeats (DYZ2) (Gibbons *et al.*, 2000). The genome-wide CpG promoter HD2 array (NimbleGen) used in this study does not have probe coverage at these regions, most likely due to the repetitive nature of these regions.

The genome-wide promoter DNA methylation approach adopted in this study has been successfully employed for the detection of an aberrantly methylated gene, *ATF7IP2*. The results of these investigations have therefore provided a novel research direction for further investigations of a potential repressive chromatin complex including ATRX and ATF7IP2. These future studies will contribute to the understanding of the pathogenesis of ATR-X and related syndromes. Furthermore, *ATF7IP2* would present a good candidate for mutation screening in autosomal MR patient cohorts that, by future homozygosity mapping studies or CNV studies, are shown to be linked to the 16p13.13 locus. Finally, while the effects of ATRX ablation on promoter DNA methylation are not widespread, these investigations show that changes to cytosine methylation are a molecular feature of MR in at least one XLMR gene. Employing the methodology presented in this study for DNA methylation aberration investigations in additional XLMR "epigenetic" genes, will establish the extent of the role of this epigenetic mechanism in the pathogenesis of this disorder.

REFERENCES

Ballas, N. and Mandel, G. (2005). The many faces of REST oversee epigenetic programming of neuronal genes. *Current Opinion in Neurobiology*, **15**, 500–506.

Baumann, C. and De La Fuente, R. (2009). ATRX marks the inactive X chromosome (Xi) in somatic cells and during imprinted X chromosome inactivation in trophoblast stem cells. *Chromosoma*, **118**, 209–222.

Baumann, C., Schmidtmann, A., Muegge, K., and De La Fuente, R. (2008). Association of ATRX with pericentric heterochromatin and the Y chromosome of neonatal mouse spermatogonia. *BMC Molecular Biology*, **9**, 29.

Benjamini, Y. and Hochberg, Y. (1995). Controlling the false discovery rate: a practical and powerful approach to multiple testing. *Journal of the Royal Statistical Society, Series B, Methodological*, **57**, 289–300.

Bienvenu, T. and Chelly, J. (2006). Molecular genetics of Rett syndrome: when DNA methylation goes unrecognized. *Nature Reviews Genetics*, **7**, 415–426.

Bock, C., Reither, S., Mikeska, T., *et al.* (2005). BiQ Analyzer: visualization and quality control for DNA methylation data from bisulfite sequencing. *Bioinformatics*, **21**, 4067–4068.

Bolstad, B. M., Irizarry, R. A., Astrand, M., *et al.* (2003). A comparison of normalization methods for high-density oligonucleotide array data based on variance and bias. *Bioinformatics*, **19**, 185–193.

Cardoso, C., Timsit, S., Villard, L., *et al.* (1998). Specific interaction between the XNP/ATR-X gene product and the SET domain of the human EZH2 protein. *Human Molecular Genetics*, **7**, 679–684.

Chahrour, M. and Zoghbi, H. Y. (2007). The story of Rett syndrome: from clinic to neurobiology. *Neuron*, **56**, 422–437.

Chelly, J., Khelfaoui, M., Francis, F., *et al.* (2006). Genetics and pathophysiology of mental retardation. *European Journal of Human Genetics*, **14**, 701–713.

Chiurazzi, P., Schwartz, C. E., Gecz, J., *et al.* (2008). XLMR genes: update 2007. *European Journal of Human Genetics*, **16**, 422–434.

Cho, G., Bhat, S. S., Gao, J., *et al.* (2008). Evidence that *SIZN1* is a candidate X-linked mental retardation gene. *American Journal of Medical Genetics, Part A*, **146**, 2644–2650.

Ehrlich, M., Sanchez, C., Shao, C., *et al.* (2008). ICF, an immunodeficiency syndrome: DNA methyltransferase 3B involvement, chromosome anomalies, and gene dysregulation. *Autoimmunity*, **41**, 253–271.

Gentleman, R. C., Carey, V. J., Bates, D. M., *et al.* (2004). Bioconductor: open software development for computational biology and bioinformatics. *Genome Biology*, **5**, R80.

Gibbons, R. J., McDowell, T. L., Raman, S., *et al.* (2000). Mutations in *ATRX*, encoding a SWI/SNF-like protein, cause diverse changes in the pattern of DNA methylation. *Nature Genetics*, **24**, 368–371.

Harum, K. H., Alemi, L., and Johnston, M. V. (2001). Cognitive impairment in Coffin–Lowry syndrome correlates with reduced RSK2 activation. *Neurology*, **56**, 207–214.

Ichimura, T., Watanabe, S., Sakamoto, Y., *et al.* (2005). Transcriptional repression and heterochromatin formation by MBD1 and MCAF/AM family proteins. *Journal of Biological Chemistry*, **280**, 13928–13935.

Ishov, A. M., Vladimirova, O. V., and Maul, G. G. (2004). Heterochromatin and ND10 are cell-cycle regulated and phosphorylation-dependent alternate nuclear sites of the transcription repressor Daxx and SWI/SNF protein ATRX. *Journal of Cell Science*, **117**, 3807–3820.

Kramer, J. M. and van Bokhoven, H. (2009). Genetic and epigenetic defects in mental retardation. *International Journal of Biochemistry and Cell Biology*, **41**, 96–107.

Kuwabara, T., Hsieh, J., Nakashima, K., *et al.* (2004). A small modulatory dsRNA specifies the fate of adult neural stem cells. *Cell*, **116**, 779–793.

Leonard, H. and Wen, X. (2002). The epidemiology of mental retardation: challenges and opportunities in the new millennium. *Mental Retardation and Developmental Disability Research Reviews*, **8**, 117–134.

Levenson, J. M. and Sweatt, J. D. (2005). Epigenetic mechanisms in memory formation. *Nature Reviews in Neuroscience*, **6**, 108–118.

Levenson, J. M., O'Riordan, K. J., Brown, K. D., *et al.* (2004). Regulation of histone acetylation during memory formation in the hippocampus. *Journal of Biological Chemistry*, **279**, 40545–40559.

Levenson, J. M., Roth, T. L., Lubin, F. D., *et al.* (2006). Evidence that DNA (cytosine-5) methyltransferase regulates synaptic plasticity in the hippocampus. *Journal of Biological Chemistry*, **281**, 15763–15773.

Levy, M. A., Fernandes, A. D., Tremblay, D. C., *et al.* (2008). The SWI/SNF protein ATRX co-regulates pseudoautosomal genes that have translocated to autosomes in the mouse genome. *BMC Genomics*, **9**, 468.

McDowell, T. L., Gibbons, R. J., Sutherland, H., *et al.* (1999). Localization of a putative transcriptional regulator (ATRX) at pericentromeric heterochromatin and the short arms of acrocentric chromosomes. *Proceedings of the National Academy of Sciences USA*, **96**, 13983–13988.

Miller, C. A., Campbell, S. L., and Sweatt, J. D. (2008). DNA methylation and histone acetylation work in concert to regulate memory formation and synaptic plasticity. *Neurobiology of Learning and Memory*, **89**, 599–603.

Molinari, F., Foulquier, F., Tarpey, P. S., *et al.* (2008). Oligosaccharyltransferase-subunit mutations in nonsyndromic mental retardation. *American Journal of Human Genetics*, **82**, 1150–1157.

Nan, X., Hou, J., Maclean, A., *et al.* (2007). Interaction between chromatin proteins MECP2 and ATRX is disrupted by mutations that cause inherited mental retardation. *Proceedings of the National Academy of Sciences USA*, **104**, 2709–2714.

Pelizzola, M., Koga, Y., Urban, A. E., *et al.* (2008) MEDME: an experimental and analytical methodology for the estimation of DNA methylation levels based on microarray derived MeDIP-enrichment. *Genome Research*, **18**, 1652–1659.

Roelfsema, J. H., White, S. J., Ariyurek, Y., *et al.* (2005), Genetic heterogeneity in Rubinstein–Taybi syndrome: mutations in both the *CBP* and *EP300* genes cause disease. *American Journal of Human Genetics*, **76**, 572–580.

Ropers, H. H. & Hamel, B. C. (2005). X-linked mental retardation. *Nature Reviews Genetics*, **6**, 46–57.

Sharp, A. J., Migliavacca, E., Dupre, Y., *et al.* (2010). Methylation profiling in individuals with uniparental disomy identifies novel differentially methylated regions on chromosome 15. *Genome Research*, **20**, 1271–1278.

Smyth, G. K. (2004). Linear models and empirical Bayes methods for assessing differential expression in microarray experiments. *Statistical Applications in Genetics and Molecular Biology*, **3**, Article 3.

Swank, M. W. and Sweatt, J. D. (2001). Increased histone acetyltransferase and lysine acetyltransferase activity and biphasic activation of the ERK/RSK cascade in insular cortex during novel taste learning. *Journal of Neuroscience*, **21**, 3383–3391.

Tarpey, P. S., Smith, R., Pleasance, E., *et al.* (2009). A systematic, large-scale resequencing screen of X-chromosome coding exons in mental retardation. *Nature Genetics*, **41**, 535–543.

Vallian, S., Chin, K. V., and Chang, K. S. (1998) The promyelocytic leukemia protein interacts with Sp1 and inhibits its transactivation of the epidermal growth factor receptor promoter. *Molecular and Cellular Biology*, **18**, 7147–7156.

Vega, V. B., Cheung, E., Palanisamy, N., *et al.* (2009). Inherent signals in sequencing-based chromatin-immuno-precipitation control libraries. *PLoS One*, **4**, e5241.

Weber, M., Davies, J. J., Wittig, D., *et al.* (2005), Chromosome-wide and promoter-specific analyses identify sites of differential DNA methylation in normal and transformed human cells. *Nature Genetics*, **37**, 853–862.

Xue, Y., Gibbons, R., Yan, Z., *et al.* (2003). The ATRX syndrome protein forms a chromatin-remodeling complex with Daxx and localizes in promyelocytic leukemia nuclear bodies. *Proceedings of the National Academy of Sciences USA*, **100**, 10 635–10 640.

Yasui, D. H., Peddada, S., Bieda, M. C., *et al.* (2007). Integrated epigenomic analyses of neuronal MeCP2 reveal a role for long-range interaction with active genes. *Proceedings of the National Academy of Sciences USA*, **104**, 19 416–19 421.

Zheng, D., Zhao, K., and Mehler, M. F. (2009). Profiling RE1/REST-mediated histone modifications in the human genome. *Genome Biology*, **10**, R9.

31 Kinases and phosphatases in the epigenetic regulation of cognitive functions

Tamara B. Franklin and Isabelle M. Mansuy*

31.1 Introduction

Long-term memory formation is a complex process requiring *de novo* protein synthesis, which is often preceded by increased gene transcription rates. The role of transcription factors in the regulation of gene transcription and their consequent importance for long-term memory and reconsolidation is well known. However, recent evidence has suggested that changes in chromatin structure, also known to affect gene transcription, are likely to play a critical part in the establishment and maintenance of long-term memory. In particular, emphasis has been placed on histone posttranslational modifications (PTMs) that contribute to determining whether chromatin is in an "on" or "off" state, and whether genes are activated or silenced. These histone PTMs are multiple and varied, and co-occur on individual nucleosomes in what appears to be particular patterns. The presence of a "histone code," first proposed by David Allis, suggested that histone PTMs are established through complex crosstalk (Strahl and Allis, 2000; Baker *et al.*, 2008). In terms of memory, there may also be a particular histone code for memory formation, since particular types of memory are thought to be associated with specific histone modifications (Levenson *et al.*, 2005) (Figure 31.1). For example, contextual fear conditioning and latent inhibition involve two different memory systems. Contextual fear conditioning involves learning the association between a novel context and an aversive stimulus (Figure 31.2A). Latent inhibition trains the animal to form a spatial memory which blocks the formation of a contextual fear memory by pre-exposing an animal to a novel context for an extended period of time, such that they no longer form an association between the aversive stimulus and the novel context (Figure 31.2B). A clear dissociation between acetylation marks has emerged

---

* Author to whom correspondence should be addressed.

Epigenomics: From Chromatin Biology to Therapeutics, ed. K. Appasani. Published by Cambridge University Press. © Cambridge University Press 2012.

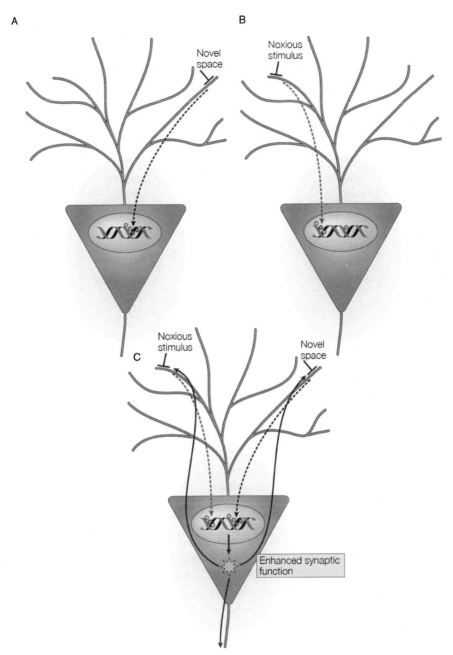

Figure 31.1 Specific epigenetic marks lead to particular memory formation – an example using contextual fear conditioning. In pyramidal neurons in the CA1 subregion of the hippocampus, (A) exposure of mice to a novel environment results in particular epigenetic marks that result in the formation of novel spatial memory, (B) exposure of mice to an aversive stimulus results in particular epigenetic marks leading to the formation of novel fear memories, and (C) temporally coupling the presentation of the novel environment and the aversive stimulus causes a combination of these epigenetic marks, leading to the formation of specific contextual fear memories. (From Levenson and Sweatt, 2005.)

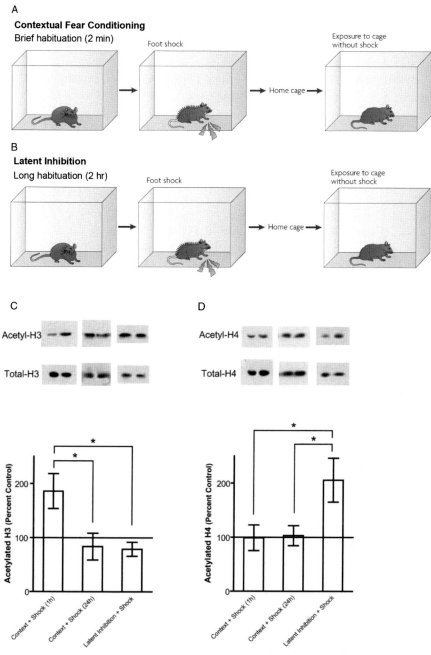

Figure 31.2 Memory-specific changes in histone acetylation in the CA1 subregion of the hippocampus. (A) Contextual fear conditioning involves a short habituation to the context followed by presentation of an aversive stimulus (foot shock) such that the animal associates the context with the aversive stimulus. (B) Latent inhibition blocks the formation of an associative contextual fear memory by pre-exposing the animal to the novel context for an extended period prior to presenting the aversive stimulus. Thus, the animal attains a spatial memory but not a fear memory. (C, D) Western blots showing that (C) contextual fear conditioning leads to increased H3 acetylation but not H4 acetylation, 1 hour after training, and (D) latent inhibition training leads to increased H4 acetylation, but not H3 acetylation. (Modified from Levenson *et al.*, 2004.)

between these two forms of memory; acetylation of H3 but not H4 is associated with contextual fear conditioning, while acetylation of H4 but not H3 is associated with latent inhibition in the CA1 subregion of the hippocampus (Levenson *et al.*, 2004) (Figure 31.2C and D).

The regulation of both acetylation and phosphorylation of histones has been implicated in long-term memory processes. In particular, protein kinases and protein phosphatases have been suggested to play a role in chromatin remodeling by directly regulating histone phosphorylation, but also indirectly by acting on proteins with histone acetyltransferase activities. The role of protein kinases and protein phosphatases on chromatin remodeling and long-term memory formation and maintenance are discussed below.

31.2   Role of kinases in epigenetic regulation of cognition: ERK/MAPK-cascade-regulated epigenetic marks required for long-term memory

31.2.1   ERK/MAPK-cascade-regulated epigenetic marks required for long-term memory formation

Long-term memories are thought to be the result of high levels of synaptic activity that lead to NMDA receptor activation, and consequently to an influx of calcium into the cytoplasm. The resulting heightened intracellular calcium concentration causes the activation of a variety of signaling pathways that lead to the activation of extracellular-signal-regulated kinase (ERK). Activation of ERK, and its subsequent effects on the activity of transcription factors important for memory formation are well known. However, recent findings have suggested that ERK activation may also play a role in another regulatory influence on gene transcription, chromatin remodeling, and more specifically on posttranslational modifications of histones including histone phosphorylation, but also histone acetylation. In vivo, contextual fear conditioning increases H3 acetylation and phosphorylation in the CA1 subregion of the hippocampus, dependent on NMDA receptor and ERK activation (Levenson *et al.*, 2004; Chwang *et al.*, 2006), and in vitro, ERK activation via either the protein kinase A (PKA) or protein kinase C (PKC) pathway both increase H3 acetylation and phosphorylation in hippocampal slices (Levenson *et al.*, 2004; Chwang *et al.*, 2006).

Two downstream targets of ERK, the protein kinase ribosomal S6 kinase 2 (RSK2) and mitogen- and stress-activated protein kinase 1 (MSK1), have been implicated to be important regulators of the epigenetic marks required for long-term memory formation. RSK2 is a serine/threonine protein kinase directly phosphorylated and activated by mitogen-activated protein kinase (MAPK) members of the ERK family on stimulation by growth factors, polypeptide hormones, neurotransmitters, and chemokines, amongst others (Hauge and Frodin, 2006). Interestingly, the activated form of RSK2 can act directly or indirectly to phosphorylate H3 (Sassone-Corsi *et al.*, 1999). This causes a change in the conformation of the chromatin structure which facilitates the binding of a protein with histone acetyltransferase (HAT) activity, cAMP response element binding protein (CREB) binding protein

(CBP), and consequently affects histone acetylation (Clayton *et al.*, 2000; Urdinguio *et al.*, 2009). In addition it has been suggested that in a quiescent cell, inactivated RSK2 binds CBP forming a complex in which both RSK2 kinase activity and CBP HAT activity are impaired (Merienne *et al.*, 2001). However, on stimulation by MAPK/ERK activation, phosphorylation of RSK2 results in the dissociation of the RSK2–CBP complex, thereby allowing heightened CBP HAT activity. Thus, RSK2 activation has two downstream effects which work in concert to promote gene transcription; it dissociates from CBP allowing for increased HAT activity by the protein, and it provides CBP access to H3 by phosphorylating H3 thereby favoring chromatin decondensation. This coupling of H3 phosphorylation and acetylation has been observed in the promoter region of an immediate early gene known to have increased expression following MAPK activation, *c-fos*, suggesting that a synergistic crosstalk exists between kinases and histone acetylation to promote gene transcription (Cheung *et al.*, 2000).

RSK2 has been suggested to play a role in cognition because a loss-of-function mutation in *Rsk2* (also known as *RPS6KA3*) results in Coffin–Lowry syndrome, a neurological disease characterized by skeletal abnormalities and mental retardation. In fibroblasts derived from patients with Coffin–Lowry syndrome, the impairment in RSK2 functioning results in deficits in H3 phosphorylation normally resulting from epidermal growth factor treatment (Sassone-Corsi *et al.*, 1999). However, it is still not yet known whether deficits in learning and memory observed in Coffin–Lowry patients are the result of altered epigenetic regulation or due to altered gene activation, since RSK2 also directly phosphorylates a number of proteins known to be involved in cognitive function (Urdinguio *et al.*, 2009). Nevertheless, due to the prevalence of altered epigenetic function in several forms of mental retardation (Franklin and Mansuy, 2011), it is likely that impaired epigenetic regulation by RSK2 is at least a contributing factor to the mental retardation present in Coffin–Lowry syndrome.

Similar to RSK2, MSK1 is also activated by MAPKs of the ERK family, but can also be activated by the p38 family MAPKs as a result of cellular stress stimuli and proinflammatory cytokines/factors (Hauge and Frodin, 2006). It is thought that MSK1 is important in long-term memory formation, since MSK1 knockout mice have impairments in performing associative fear conditioning and spatial memory tasks (Chwang *et al.*, 2007). Additionally, when normal hippocampal slices are stimulated with forskolin or phorbol 12,13-diacetate, drugs that lead to ERK activation by activating the cAMP/PKA pathway or PKC pathway, respectively, phosphorylation and acetylation of H3 increases (Chwang *et al.*, 2007). However, this is not seen in slices taken from MSK1 knockout mice. Following fear conditioning training, phosphorylation and acetylation of H3 are increased in the hippocampus in wild-type mice, but not in MSK1 knockout mice (Chwang *et al.*, 2007). These findings have led to the proposal of a model of MSK1 in chromatin remodeling involved in memory formation that suggests that activation of PKC or PKA leading to MSK1 activation mediates phosphorylation and acetylation of histones that result in increased gene transcription (Figure 31.3).

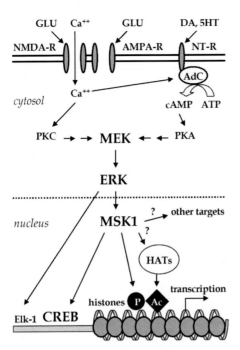

Figure 31.3 Mitogen- and stress-activated protein kinase 1 (MSK1) regulates chromatin structure during memory formation. Activation of cell surface receptors results in the activation of protein kinase C (PKC), protein kinase A (PKA), mitogen-activated protein kinase kinase (MEK), extracellular-signal-regulated kinase (ERK), and ultimately MSK1. In the nucleus, activated MSK1 leads to increased gene transcription through a variety of mechanisms. It can act on transcription factors, such as cAMP response element binding protein (CREB), and it can prompt histone phosphorylation and histone acetylation, perhaps through actions on histone acetyltransferases (HATs). (From Chwang *et al.*, 2007.)

31.2.2   Epigenetic regulation by inhibitor kappa B kinase α (IKKα) is required for memory reconsolidation

Memory reconsolidation refers to a process occurring during memory retrieval in which memories become labile, and therefore must be re-established and stabilized (Nader *et al.*, 2000). This process requires gene transcription and protein synthesis, and thus chromatin structure is likely to play a contributing factor (Nader, *et al.* 2000). Nuclear factor kappa B (NF-κB) is a transcription factor associated with the initiation of synaptic plasticity and the early stages of formation of long-term memory, but it has now also been implicated in regulating histone modifications during memory retrieval and reconsolidation. In vivo, NF-κB is tightly regulated by inhibitor kappa B (IκB) proteins. Binding of IκB causes NF-κB to remain in the cytoplasm, and thus its role as a transcriptional regulator is blocked. However, phosphorylation by the IκB kinase (IKK) complex marks IκB for degradation, an action that allows NF-κB to translocate to the nucleus and regulate gene transcription. An epigenetic role for NF-κB signaling has been suggested because inhibition of this pathway blocks the increase in phosphorylation and acetylation of H3 normally observed in the CA1 subregion of the hippocampus following re-exposure to a context previously associated with an aversive stimulus

(Lubin and Sweatt, 2007). Additionally, inhibition of NF-κB signaling impairs the reconsolidation of a contextual fear memory at a behavioral level. In particular, a catalytic subunit that forms part of the IKK complex, the IKKα subunit, is thought to be a key factor. Administration of sulfasalzine (SSZ), an inhibitor of IKKα, causes impairments in contextual fear memory reconsolidation, and blocks the increase in phosphorylation and acetylation of H3 normally observed in the hippocampus following reconsolidation, suggesting that the epigenetic effects mediated by this component of the NF-κB pathway is involved in memory reconsolidation in vivo (Lubin and Sweatt, 2007). The precise mechanism for how IKKα acts to increase both phosphorylation and acetylation of H3 is unknown, but is suggested to involve direct phosphorylation of H3 and interaction with CBP, a protein with histone acetyltransferase activity (Lubin and Sweatt, 2007) (Figure 31.4).

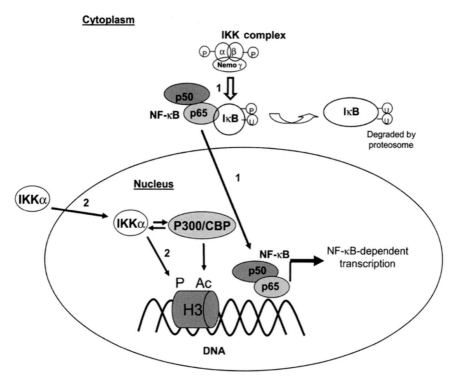

Figure 31.4 Inhibitor kappa B complex subunit α (IKKα) regulates chromatin structure during memory reconsolidation. Activation of IKKα kinase as a result of memory recall increases gene transcription in two different ways. (1) The DNA binding activity of the nuclear factor kappa B (NF-κB) complex is increased by the IKKα kinase protein activity in the IKK complex, thereby increasing NF-κB-dependent transcription. (2) Independent of the IKK complex, IKKα kinase activation increases H3 phosphorylation and, through its interaction with CBP, also H3 acetylation. This is thought to result in changes in chromatin structure towards more permissive states, allowing increased gene transcription of genes necessary for the restabilization of memory following memory retrieval. (From Lubin and Sweatt, 2007.)

31.3   Role of protein phosphatases in epigenetic regulation of cognition

In contrast to protein kinases, protein phosphatases regulate phosphorylation by removing a phosphate group from their substrate. In the context of learning and memory, protein phosphatases, such as calcineurin (also known as protein phosphatase 2B) and protein serine/threonine phosphatase 1 (PP1), decrease synaptic plasticity, limit learning and memory, and favor forgetting. These negative effects on learning and memory are potentially mediated by a variety of mechanisms including modulation of NMDA receptor currents and GTPase activity, in the case of calcineurin, and regulation of Ca^{2+}/calmodulin-dependent protein kinase II (CaMKII), the AMPA receptor subunit GluR1, and CREB-dependent gene expression, in the case of PP1 (Lee and Silva, 2009). However, at this point there is little known about the role that protein phosphatases play in regulating histone PTMs and how this relates to memory formation.

Several studies in our laboratory have begun to identify the epigenetic impact of PP1 and its association with the formation of long-term memory. Using a transgenic mouse in which PP1 is inhibited specifically in the nucleus, we have investigated the role that PP1 plays on histone PTMs in neurons and its downstream impact on gene transcription, with a focus on genes associated with long-term memory. PP1 binds to H3 both in vitro and in vivo in the adult mouse forebrain, and in vitro this results in dephosphorylation of serine 10 on H3 (Koshibu *et al.*, 2009). In addition to its direct role in histone phosphorylation, PP1 is thought to also indirectly regulate histone acetylation and methylation through interactions with other members of the epigenetic machinery in the nucleus. Inhibition of nuclear PP1 reduces the level of histone deacetylase 1 (HDAC1) bound to PP1, an effect that is associated with decreased HDAC activity in the nucleus, but not in the cytoplasm (Koshibu *et al.*, 2009). Additionally, PP1 inhibition is associated with reduced binding to a histone lysine demethylase JMJD2A, suggesting that its activity might also be regulated by PP1 (Koshibu *et al.*, 2009). Thus, PP1 may act to not only regulate H3 phosphorylation directly, but also to alter histone acetylation and methylation (Koshibu *et al.*, 2009). Consequently, inhibition of nuclear PP1 in mice results in increased phosphorylation of H3S10, increased acetylation of H3K14, H4K5, and H2B, and increased trimethylation of H3K36 (Koshibu *et al.*, 2009). These findings have been summarized in a model suggesting a PP1-dependent histone code regulating gene transcription (Figure 31.5A).

Interestingly, altered histone PTMs were also observed in the promoter region of several genes associated with long-term memory formation in the hippocampus of mice in which nuclear PP1 is inhibited. Histone PTMs which promote transcriptional activation were found to be increased in the promoter region of *creb*, and decreased in the promoter region of NF-κB (Koshibu *et al.*, 2009). Accordingly, RNA Pol II binding was increased at the *creb* promoter but decreased at the NF-κB promoter, *creb* expression was increased, and NF-κB expression was decreased (Koshibu *et al.*, 2009). In parallel, synaptic plasticity, as assessed by

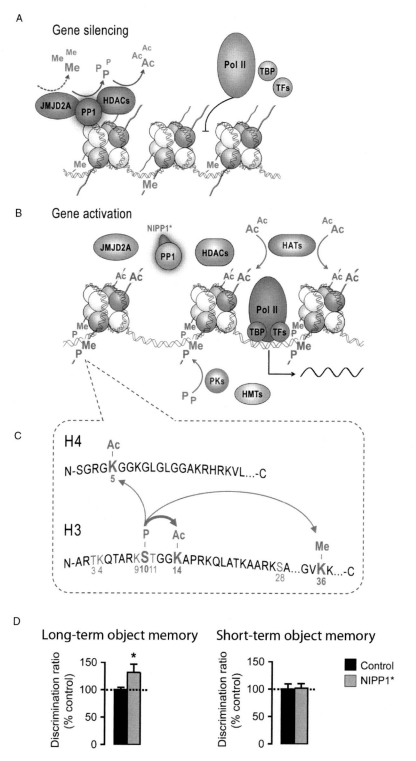

Figure 31.5 PP1-dependent histone code mediating gene transcription. (A) PP1 directly binds to chromatin resulting in histone dephosphorylation and deccreased histone acetylation and methylation through association with histone deacetylases (HDACs) and JMJD2A. This results in a repressive chromatin structure that prevents RNA polymerase II (Pol I), TATA box binding protein

hippocampal long-term potentiation (LTP), long-term, but not short-term object recognition memory (Figure 31.5D), and spatial memory, are all enhanced by nuclear inhibition of PP1 (Koshibu et al., 2009; Graff et al., 2010), suggesting that PP1 acts as a key regulator of epigenetic modifications contributing to long-term memory formation.

31.4   Conclusions

It is now becoming increasingly clear that chromatin incorporates a large number of signals that combine to affect transcriptional rates. Gene transcription is known to be an important part of the formation and maintenance of long-term memory, and thus chromatin remodeling is likely to play a key role in this process. Protein kinases and protein phosphatases are emerging as critical to the regulation of PTMs, not only in terms of histone phosphorylation, but also histone acetylation and histone methylation, through interaction with many other components of the epigenetic machinery. Thus, the decreased activity of protein kinases or increased activity of protein phosphatases are thought to result in memory impairments not only through aberrant transcription factor regulation, but also due to altered epigenetic marks resulting in less active chromatin structure. Further understanding of epigenetic regulation by protein kinases and protein phosphatases is required to fully understand the molecular processes involved in long-term memory formation. However, significant progress has been made in this emerging field towards elucidating the role of protein kinases and phosphatases in the epigenetic regulation of learning and memory.

REFERENCES

Baker, L. A., Allis, C. D., and Wang, G. G. (2008). PHD fingers in human diseases: disorders arising from misinterpreting epigenetic marks. *Mutation Research*, **647**, 3–12.

Caption for Figure 31.5 (cont.)

(TBP), and other transcription factors (TFs) from accessing the chromatin, resulting in gene silencing. (B) Nuclear inhibition of PP1 enhances histone phosphorylation on serine 10, mediated by protein kinases (PKs) such as MSK1, ERK1, or PKA. Nuclear inhibition of PP1 also reduces HDAC activity, thereby increasing H3K14 and H4K5 acetylation due to a dissociation of PP1 from HDACs. This is most likely also associated with histone acetyltransferase (HAT) activity. Lastly, nuclear inhibition of PP1 also results in reduced interaction between PP1 and JMJD2A, and consequently decreased JMJD2A activity, which leads to increased methylation of H3K36. Other histone methyltransferases (HMTs) are probably also involved. All these actions lead to a permissive chromatin structure, that allows transcriptional machinery access to the DNA, thereby promoting gene transcription. (C) In light gray, histone residues not differentially regulated by nuclear inhibition of PP1; in bold, histone residues with PTMs dependent on nuclear PP1. Thick arrows, well-established crosstalk between phosphorylation of H3S10 and acetylation of H3K14 known to be associated with memory formation. Thin arrows, potential crosstalk suggested by recent data. (D) Increased long-term, but not short-term, object discrimination ratio suggesting enhanced object recognition memory in mice in which nuclear PP1 is inhibited (NIPP1*). (Modified from Koshibu et al., 2009.)

Cheung, P., Tanner, K. G., Cheung, W. L., *et al.* (2000). Synergistic coupling of histone H3 phosphorylation and acetylation in response to epidermal growth factor stimulation. *Molecular Cell*, **5**, 905–915.

Chwang, W. B., O'Riordan, K. J., Levenson, J. M., *et al.* (2006). ERK/MAPK regulates hippocampal histone phosphorylation following contextual fear conditioning. *Learning and Memory*, **13**, 322–328.

Chwang, W. B., Arthur, J. S., Schumacher, A., *et al.* (2007). The nuclear kinase mitogen- and stress-activated protein kinase 1 regulates hippocampal chromatin remodeling in memory formation. *Journal of Neuroscience*, **27**, 12 732–12 742.

Clayton, A. L., Rose, S., Barratt, M. J., *et al.* (2000). Phosphoacetylation of histone H3 on *c-fos*- and *c-jun*-associated nucleosomes upon gene activation. *EMBO Journal*, **19**, 3714–3726.

Dantzer, R., O'Connor, J. C., Freund, G. G., *et al.* (2008). From inflammation to sickness and depression: when the immune system subjugates the brain. *Nature Reviews Neuroscience*, **9**, 46–56.

Franklin, T. B. and Mansuy, I. M. (2010). The prevalence of epigenetic mechanisms in the regulation of cognitive functions and behaviour. *Current Opinion in Neurobiology*, **20**, 441–449.

Graff, J., Koshibu, K., Jouvenceau, A., *et al.* (2010). Protein phosphatase 1-dependent transcriptional programs for long-term memory and plasticity. *Learning and Memory*, **17**, 355–363.

Hauge, C. and Frodin, M. (2006). RSK and MSK in MAP kinase signalling. *Journal of Cell Science*, **119**, 3021–3023.

Koshibu, K., Graff, J., Beullens, M., *et al.* (2009). Protein phosphatase 1 regulates the histone code for long-term memory. *Journal of Neuroscience*, **29**, 13 079–13 089.

Lee, Y. S. and Silva, A. J. (2009). The molecular and cellular biology of enhanced cognition. *Nature Reviews Neuroscience*, **10**, 126–140.

Levenson, J. M. and Sweatt, J. D. (2005). Epigenetic mechanisms in memory formation. *Nature Reviews Neuroscience*, **6**, 108–118.

Levenson, J. M., O'Riordan, K. J., Brown, K. D., *et al.* (2004). Regulation of histone acetylation during memory formation in the hippocampus. *Journal of Biological Chemistry*, **279**, 40 545–40 559.

Lubin, F. D. and Sweatt, J. D. (2007). The IkappaB kinase regulates chromatin structure during reconsolidation of conditioned fear memories. *Neuron*, **55**, 942–957.

Merienne, K., Pannetier, S., Harel-Bellan, A., *et al.* (2001). Mitogen-regulated RSK2–CBP interaction controls their kinase and acetylase activities. *Molecular and Cellular Biology*, **21**, 7089–7096.

Nader, K., Schafe, G. E., and Le Doux, J. E. (2000). Fear memories require protein synthesis in the amygdala for reconsolidation after retrieval. *Nature*, **406**, 722–726.

Sassone-Corsi, P., Mizzen, C. A., Cheung, P., *et al.* (1999). Requirement of Rsk-2 for epidermal growth factor-activated phosphorylation of histone H3. *Science*, **285**, 886–891.

Strahl, B. D. and Allis, C. D. (2000). The language of covalent histone modifications. *Nature*, **403**, 41–45.

Urdinguio, R. G., Sanchez-Mut, J. V., and Esteller, M. (2009). Epigenetic mechanisms in neurological diseases: genes, syndromes, and therapies. *Lancet Neurology*, **8**, 1056–1072.

Part VI

Epigenetic variation, polymorphism, and epidemiological perspectives

32 Epigenetic effects of childhood abuse on the human brain

Benoit Labonté and Gustavo Turecki*

32.1 Introduction

The environment in which we live, and especially the early life environment, shapes our behavior. Adversity during early life is strongly associated with problems in behavioral regulation and psychopathology in adulthood. Until recently, the mechanisms responsible for behavioral changes induced by early life adversity were not clear. However, recent evidence suggests that early life environment induces behavioral changes through epigenetic mechanisms controlling the expression of genes involved in the regulation of behavior. As such, the epigenome mediates the effects of environmental variability on behavioral, physiological, and pathological responses. Numerous findings in animals and humans support this view. This chapter will review the evidence suggesting that epigenetic changes are induced by the early environment and impact on the regulation of gene expression in the brain.

32.1.1 The burden of childhood adverse experience

With more than 3 million reports of child maltreatment in the USA in 2009, early life adversity is a major problem with an important social burden (Children's Bureau, 2010). From this number, parental neglect represents the major form of maltreatment followed by physical abuse and sexual abuse (Children's Bureau, 2010). From an epidemiological point of view, trauma exposure in children is estimated to range between 25% and 45%, although the rates reported vary considerably between studies and according to the definition of abuse types (Gorey and Leslie, 1997; McCauley *et al.*, 1997; Briere and Elliott, 2003; Scher *et al.*, 2004; Heim *et al.*, 2010). For instance, a prospective study reported prevalence of child maltreatment of 30% in women and 41% in men, from which 5% of

---

* Author to whom correspondence should be addressed.

Epigenomics: From Chromatin Biology to Therapeutics, ed. K. Appasani. Published by Cambridge University Press. © Cambridge University Press 2012.

cases were sexual abuse and 19% physical abuse (Briere and Elliott, 2003). Survey-based studies report rates varying between 12% and 32% for sexual abuse in females and between 5% and 14% in men (Gorey and Leslie, 1997; McCauley *et al.*, 1997; Briere and Elliott, 2003; Scher *et al.*, 2004; Heim *et al.*, 2010).

The economic burden of child maltreatment and trauma resides mainly in its impact on the development of psychopathology later during adulthood. Indeed, child trauma, and particularly child sexual and physical abuse (CSA and CPA), is associated with higher odds of psychiatric disorders including depression, anxiety, bipolar disorder, substance abuse, and suicide (Santa Mina and Gallop, 1998; Agid *et al.*, 1999; Kaplan and Klinetob, 2000; Kendler *et al.*, 2000; Heim and Nemeroff, 2001; Molnaret *et al.*, 2001; Kendler *et al.*, 2004; Evans *et al.*, 2005). Besides increasing the risk of psychiatric disorders, CSA and CPA also associate with earlier age of onset of psychopathology, chronic course, more severe outcomes, poorer recovery rate, and more importantly, with 12 times higher odds of suicidal behaviors (Brown and Moran, 1994; Bensley *et al.*, 1999; Dinwiddie *et al.*, 2000; Molnar *et al.*, 2001; Zlotnick *et al.*, 2001; Jaffee *et al.*, 2002; Gladstone *et al.*, 2004; Tanskanen *et al.*, 2004).

32.1.2   Epigenetic consequences of early life adversity on the brain

Early life adversity is frequently associated with maladaptive patterns of behavioral responses often leading to pervasive interpersonal difficulties, enhanced reactivity to stress, and increased risk of psychopathology. While substantial theoretical and empirical work supports the relationship between childhood adversity and development of negative mental health outcomes in adulthood, the critical question has been what molecular processes mediate these associations. In other words: "What long-lasting molecular mechanisms take place as a result of the adverse life experience that could be associated with increased risk for psychopathology?" Despite the complexity of this question, this chapter will review the evidence suggesting that molecular alterations result from variation in early life through epigenetic processes that modulate behaviors in animal models and increase risk for psychopathology in humans.

Epigenetics refers to the study of the epigenome, chemical and physical modifications taking place in or around the DNA molecule and altering the capacity of a gene to be activated and to produce the mRNA it encodes. Given the high complexity of the DNA organization, these modifications are expected to follow a defined pattern allowing the underlying molecular mechanisms to be performed correctly and to decode DNA in the context of chromatin. Epigenetic mechanisms refer to DNA methylation (Klose and Bird, 2006), histone modifications (Kouzarides, 2007), and more recently, posttranscriptional mechanisms such as microRNA (Schratt, 2009a, 2009b). Generally, DNA methylation has been suggested to direct transcriptional repression (Klose and Bird, 2006). However, recent evidence suggests that this may be particularly true for methylation found within a gene's promoter while intra- and intergenic methylation may be associated with the use of alternative promoters and active transcription (Maunakea *et al.*, 2010). At the chromatin level, high histone acetylation and low histone methylation levels have been associated with active transcription (Kouzarides, 2007). Epigenetic mechanisms are thought to be involved in the modification of gene expression induced by environmental factors. As such, it is possible to conceptualize the epigenome as an interface on which the environment can influence genetic processes, and through which behavior is regulated, at least in part, as a response to environmental adversity.

The following sections will review findings in gene systems that have been targeted by studies investigating epigenetic factors associated with the social environment. We will review animal studies using models of early life environmental variation, human studies investigating the effect of early life adversity, and studies focusing on suicide. While suicide is not a correlate of early life adversity – and in fact, most individuals who die by suicide do not have a history of abuse during childhood – a significant minority does, and in this subgroup, the association is very strong. Among the systems review in this chapter, we will focus in particular on gene systems coding for components of the hypothalamus–pituitary–adrenal (HPA) axis and related signaling hormones and molecules, neurotrophic factors and their receptors, as well as other signaling systems.

32.1.3 Alterations in the hypothalamus–pituitary–adrenal axis induced by early life adversity

Child abuse has been proposed to induce its long-term behavioral consequences partly by altering the neural circuits involved in the regulation of stress. The HPA axis is the main stress regulatory system (Pariante and Lightman, 2008). Under stressful conditions, corticotrophin-releasing factor (CRF) and vasopressin (AVP) are released from the hypothalamus. CRF and AVP induce the release of adreno-corticotropic hormone (ACTH) and pro-opiomelanocortin (POMC) from the pituitary gland to the blood which then travel to the adrenal cortex where they induce the release of glucocorticoids, cortisol in humans and corticosterone in rodents, to the blood. Glucocorticoids then act at each level of the HPA axis to decrease the release of CRF, AVP, POMC, and ACTH and regulate the stress response. While the HPA axis can be regulated at different levels, the main locus of regulation lies in the hippocampus where glucocorticoids bind glucocorticoid receptors (GR) and induce an inhibitory feedback on the activation of the HPA axis to bring back to basal levels the activity of the stress response.

From a structural point of view, childhood abuse and neglect have been associated with volume loss in hippocampus (Bremner et al., 1997; Stein et al., 1997; Driessen et al., 2000), altered cortical symmetry in frontal lobe (Carrion et al., 2001) and superior temporal gyrus (De Bellis et al., 2002), as well as a reduced neuronal density and/or neuronal integrity in the anterior cingulated gyrus (De Bellis et al., 2002). One study also reported poorer hippocampal activation on a memory task in patients with a history of childhood abuse (Bremner et al., 2003). Importantly, the structural consequences of child abuse are thought to be time-dependent, implying that particular brain regions may have unique windows of vulnerability to the effects of child abuse (Andersen et al., 2008).

From a molecular point of view, depressed patients with a history of child abuse have been reported to exhibit higher ACTH and cortisol levels following stress and dexamethasone (DEX) challenges (Heim *et al.*, 2000, 2008a). Interestingly, in these studies, both ACTH and cortisol levels did not differ significantly between depressed subjects without history of childhood abuse and controls (Heim *et al.*, 2000, 2008a). Childhood abuse, and particularly physical abuse, has also been shown to increase CRF levels (Carpenter *et al.*, 2004; Heim *et al.*, 2008b) and to decrease oxytocin levels in the cerebrospinal fluid (Heim *et al.*, 2009). More recently, low hippocampal GR levels have been reported in suicide completers with a history of childhood abuse (McGowan *et al.*, 2009) but not in non-abused suicide completers. Altogether, these alterations are believed to lead to important behavioral changes that may increase the predisposition toward suicidal behavior later in life.

This work is also substantiated with findings from animal work. For instance, the development of the HPA axis has been shown to be modulated by maternal behavior in rats. Depressive-like behaviors (Francis *et al.*, 1999) associated with altered HPA axis feedback (Liu *et al.*, 1997) and low GR mRNA hippocampal levels (Liu *et al.*, 1997) are common features in rats raised by mothers providing maternal care defined by low levels of licking and grooming (LG). A comprehensive model involving modulation at numerous levels, including hormonal, synaptic, and molecular changes, has been proposed in an attempt to characterize the molecular pathways involved in the modulatory effects of maternal behavior in rats. High maternal LG levels, which can also be mimicked by handling pups during early life, induce a physiological response involving the increase of thyroid hormone plasma levels. This increases serotonin (5-HT) activity in the raphe nuclei, and consequently, stimulates 5-HT turnover in the hippocampus and frontal cortex (Meaney *et al.*, 1987; Mitchell *et al.*, 1990; Smythe *et al.*, 1994). Via activation of the G-protein-coupled 5-HT$_7$ receptor (Laplante *et al.*, 2002), it is believed that 5-HT activates a c-AMP/PKA-dependent intracellular cascade increasing the expression of nerve growth factor 1-A (NGFI-A) and activator protein-2 (AP-2) in the hippocampus (Meaney *et al.*, 2000). NGFI-A and AP-2 are activating transcription factors with putative binding sites within the GR promoter region (McCormick *et al.*, 2000) that increase GR mRNA levels in the hippocampus of the offspring. This complex process is attenuated in rats raised by low-LG mothers according to the molecular and behavioral processes mentioned previously and resulting in relatively lower GR expression in the hippocampus. Interestingly, most of these regulatory changes are temporally stable and are maintained throughout adulthood. Moreover, cross-fostering studies report that these behavioral and molecular modifications are reversed when pups raised by low-LG mothers are transferred to high-LG mothers (Liu *et al.*, 1997; Francis *et al.*, 1999) within the first week of life.

32.2   The glucocorticoid receptor gene

As the HPA programming by maternal behavior is modified by cross-fostering and temporally stable, researchers hypothesized that the long-term effects of maternal behavior and early life environment variation on GR hippocampal expression could be due to epigenetic modifications. In rats, the GR gene is preceded by 10 non-coding exons and by 14 in humans (McCormick *et al.*, 2000; Turner and Muller, 2005). The expression of the non-coding exon 1_7 in rats and the human homolog 1_F has been shown to be specific to the hippocampus (Turner and Muller, 2005). Each of the untranslated exon 1 variants has its own promoter and multiple transcription factor binding sites, including NGFI-A (Meaney, 2001), have been identified in GR promoter sequences (Turner *et al.*, 2008). In offspring raised by low-LG rat mothers, CpG methylation levels in the exon 1_7 promoter region are significantly increased at almost all CpGs compared to offspring raised by high-LG mothers. More importantly, a CpG located in the 5' end of an NGFI-A binding site is methylated in almost 100% of offspring raised by low-LG mothers whereas it is almost not methylated in offspring from high-LG mothers (Weaver *et al.*, 2004).

These findings were recently translated to humans through studies investigating hippocampal tissue from individuals who died by suicide with and without a history of childhood adversity, as well as normal controls (McGowan *et al.*, 2009). In these studies, methylation levels in the exon 1_F promoter in abused suicide completers were higher than among non-abused suicides and normal, non-abused controls. In addition, similarly to what was found in rats, a significant hypermethylation in an NGFI-A binding site was found in abused suicide completers but not in the other groups. Through a series of cell functional assays, this epigenetic mark was shown to repress the binding of NGFI-A to its cognate DNA sequence and to decrease GR transcription (McGowan *et al.*, 2009). Altogether, this suggests that early life adversity may induce specific long-lasting epigenetic alterations affecting gene expression.

In a different study assessing the expression of several GR exon 1 variants expressed in the limbic system of depressed suicide completers, $GR1_F$ and $GR1_C$ hippocampal expression was significantly decreased in depressed suicide completers (Alt *et al.*, 2010). However, this was not associated with promoter hypermethylation. It should be noted that this study investigated methylation only in a limited region and promoter methylation levels reported were particularly low. On the other hand, NGFI-A protein levels in the hippocampus were significantly decreased in depressed suicide completers suggesting that the decrease in GR expression found in suicide completers may be mediated by different molecular pathways depending on the presence or the absence of early life adversity.

More recently, our group pushed further the investigation of the consequences of early life adversity on the epigenetic regulation of GR in the hippocampus of abused suicide completers. Our data indicated that the expression of the non-coding exons 1_C and 1_H is significantly decreased in suicide completers with a history of childhood abuse compared to non-abused suicides and controls. The assessment of methylation levels in the promoter of $GR1_C$ revealed methylation differences that are inversely correlated with $GR1_C$ expression in accordance with our previous finding on the 1_F variant. On the other hand, the $GR1_H$ promoter showed site-specific hypomethylation that was positively correlated with $GR1_H$ expression. In other words, less methylation significantly correlated with lower expression, which suggests that active demethylation is also a functional mechanism that may be affected by early life adversity. While this is a mechanism that has received less attention, more work is required in order to elucidate its potential implications in the context of early life adversity.

In addition to DNA methylation, chromatin changes have also been associated with poor maternal care. For instance, H3K9 acetylation, a marker of open euchromatin state (Kouzarides, 2007), was found to be lower in the GR1$_7$ promoter in low-LG raised rats. Pharmacological challenge with a histone deacetylase inhibitor, trichostatin A (TSA), restored methylation levels, increased NGFI-A binding to the promoter, and reinstated H3K9 acetylation and GR hippocampal levels (Weaver et al. 2004). Treated rats were also less reactive to stressful conditions. By decreasing H3K9 acetylation, DNA access to the transcriptional machinery and DNA binding proteins such as transcription factors and methylated DNA binding proteins is reduced. Functionally, these results suggest that variation in the early life environment in rats and early life adversity in human induces a coordinated remodeling of epigenetic mechanisms involving DNA methylation and chromatin modifications in the multiple promoters of GR leading to important changes in GR expression, and consequent regulation of the HPA axis.

32.3   The vasopressin and corticotrophin-releasing factor genes

Other components of the HPA axis have also been shown to be affected by early life stress. For instance, early life infant–maternal separation in mice, inducing stress-coping behavioral alterations in pups, has been shown to be associated with a long-lasting increase in corticosterone secretion and with an increased expression of POMC and AVP in the paraventricular nucleus (PVN) of the hypothalamus (Murgatroyd et al., 2009). The AVP gene in mice is composed of three coding exons and is oriented tail-to-tail with the *oxytocin* (Oxt) gene. Interestingly, the intergenic region between the AVP and the Oxt genes has been shown to include an enhancer modulating AVP expression (Gainer et al., 2001) and itself composed of a CpG island (Murgatroyd et al., 2009).

Methylation at multiple sites within the AVP enhancer was shown to be decreased in the PVN of stressed mice 6 weeks, 3 months, and 1 year following the stress regimen (Murgatroyd et al., 2009). Consistent with the repressive role of DNA methylation on expression, this was associated with overexpression of the AVP gene. The regulatory properties of this enhancer were defined by a deletion experiment. Deleting the first part of the enhancer partially reduced transcriptional activity, while removing the entire enhancer almost completely abolished the gene's activity. Furthermore, methylation of the enhancer also significantly reduced transcriptional activity. Interestingly, AVP expression was also significantly increased in stressed mice at 10 days, although no methylation differences were observed in the AVP enhancer. The AVP enhancer can putatively bind the methylated CpG binding protein MeCP2. However, because of the repressive role of MeCP2 on transcription, one would expect the opposite tendency concerning AVP expression. MeCP2 has nevertheless been shown to be susceptible to inactivation by neuronal-depolarization-induced phosphorylation, leading to its dissociation from putative targets (Chen *et al.*, 2003; Zhou *et al.*, 2006). Accordingly, higher neuronal activity-induced Ca^{2+}/calmodulin-dependent protein kinase II (CaMKII) immunoreactivity and phosphorylated MeCP2 levels have been reported in AVP-expressing neurons in the PVN of 10-day-old stressed mice. Altogether, these results suggest that in young stressed mice, methylation patterns in AVP enhancer allow the binding of MeCP2, which could then repress expression. However, since early life stress also increases neuronal activity in AVP-expressing neurons, MeCP2 gets phosphorylated and inactivated. Consequently, the repressive effect of MeCP2 on AVP expression is abolished. On the other hand, methylation levels in AVP enhancer decrease with time. This may decrease MeCP2 binding and allow AVP to be expressed at higher levels. Overall, these results nicely suggest that alterations in DNA methylation found outside of the promoter might also be involved in physiological and behavioral modifications induced by environmental factors.

The regulation of a related peptide, CRF, has been recently shown to be also associated with epigenetic regulation by the social environment. Accordingly, CRF expression in the PVN of chronically socially defeated mice was found to be increased (Elliott *et al.*, 2010). Interestingly, this effect was found only in animals susceptible to social stress and showing the normal subordinated behavior following chronic exposure to aggressive litter-mates as opposed to the resilient mice continuing to interact with their aggressor. This was associated with lower levels of methylation as reported by a reduced number of methylated clones in the susceptible group compared to control and resilient mice. A closer look at the methylation alteration induced by chronic social stress pointed to a single site of hypomethylation in the proximal promoter flanking the first exon and known to bind the CREB response element cAMP (Aguilera *et al.*, 2007). The importance of this site was further confirmed by luciferase assays showing that mutating a single base in the CRE binding site substantially reduced the cAMP-induced CRF promoter activity (Elliott *et al.*, 2010). These changes in methylation and expression were also accompanied by a significant decrease in the DNA methyltransferase 3b expression and by an increase in the expression of the demethylating candidate gadd45b. Interestingly, chronic treatment with the tricyclic imipramine attenuated the changes at the DNA methylation and expression levels induced by social stress (Elliott *et al.*, 2010). Consequently, these findings in both the AVP and the CRF genes strongly support the involvement of active demethylation in the long-term effects of early life adversity.

32.4   The brain-derived neurotrophic factor gene

Neurotrophic factors are important candidate molecules to understand the development of psychopathology because of their role in neuronal survival and plasticity, as well as their expression in brain regions from the limbic system, where emotions and related behaviors are processed. For instance, it is hypothesized that their alteration could partly underlie changes in plasticity observed in the brains of suicides as well as the mood symptoms observed in depressive patients. While the major neurotrophic factors include nerve growth factor (NGF), neurotrophin 3 and 4 (NT3/4), fibroblast growth factor (FGF), transforming growth factor (TGF), and brain-derived neurotrophic factor (BDNF), the latter has received most of the attention in neurobiological research of psychiatric conditions such as depressive disorders and suicide. For instance, low serum and brain BDNF expression has been reported in patients with major depression (Dwivedi *et al.*, 2003; Brunoni *et al.*, 2008; Pandey *et al.*, 2008) and these alterations were reversed by antidepressant treatment (Chen *et al.*, 2001; Sen *et al.*, 2008; Matrisciano *et al.*, 2009). In mice, BDNF depletion induces depressive-like behaviors (Chan *et al.*, 2006) while in rats, chronic stress and persistent pain reduces BDNF expression in the hippocampus (Duric and McCarson, 2005; Gronli *et al.*, 2006), and these effects are counteracted by antidepressant treatment (Rogoz *et al.*, 2005; Duric and McCarson, 2006; Xu *et al.*, 2006).

Epigenetic regulation of BDNF has recently been investigated in mice and rat models of depressive symptoms following stress-induced, negative social interactions (Tsankova *et al.*, 2006; Roth *et al.*, 2009), as well as in a rat model of exposure to traumatic events (Roth *et al.*, 2011). In both species, the BDNF gene contains nine 5′ non-coding first exons with their own promoter coding for the same protein (Aid *et al.*, 2007). The alternative splicing of these exons specifies the tissue in which BDNF is expressed (Aid *et al.*, 2007). In both species, epigenetic processes involved in the transcriptional control of BDNF have been shown to be altered by stress. For instance, chronic social stress in mice decreases the expression of two specific BDNF transcripts (III and IV) in the hippocampus (Tsankova *et al.*, 2006), while maternal maltreatment decreases prefrontal cortex BDNF mRNA expression in rats (Roth *et al.*, 2009). Although similar, these transcriptional alterations were shown to be induced by different epigenetic mechanisms. Indeed, chronic stress in mice raises H3K27 dimethylation levels in transcripts III and IV promoters (Tsankova *et al.*, 2006), while site-specific hypermethylation is found in transcripts IV and IX promoters of maltreated rats (Roth *et al.*, 2009). In the latter study, site-specific hypermethylation seems to follow a developmental pattern, exon IX promoter hypermethylation occurring immediately after the maltreatment regimen, while promoter IV methylation increases gradually to reach significantly altered levels only at adulthood. Surprisingly, in one of these studies (Tsankova *et al.*, 2006), no DNA methylation difference was found in association with histone modifications, while no histone modification was reported in association with DNA methylation alterations in the other study (Roth *et al.*, 2009). These findings illustrate that early life or chronic stressors may alter different epigenetic mechanisms with common transcriptional consequences, the latter leading to the compaction of chromatin in its heterochromatic state, and the former blocking the binding of transcription factors to DNA. On the other hand, these results may also highlight the heterogeneity of stress-induced epigenetic alterations between species.

Recently, the epigenetic regulation of BDNF has been shown to be altered in a rat model of post-traumatic stress disorder (PTSD) symptoms (Roth *et al.*, 2011).

Given the findings discussed above, the authors focused on exon IV. Stressed rats showed increased DNA methylation in the dorsal dentate gyrus and in the CA3 regions. Contrary to expected, a global hypomethylation was observed in the ventral CA1 region. These epigenetic alterations were accompanied by significant downregulation of BDNF exon IV expression in both the dorsal and ventral CA1 regions in the stressed rats relative to non-stressed rats. Interestingly, these alterations were restricted to the hippocampus since no alterations were found in the basolateral amygdala nor in the medial prefrontal cortex. These findings suggest that DNA methylation may be affected differently within the same structure depending on the function and the connections these regions have. Given that the expression of exon IV was decreased in the ventral CA1 without any significant changes in DNA methylation, these findings also suggests that, although DNA methylation may have an important role in the regulation of BDNF, other mechanisms are probably involved.

Pharmacological treatment with the tricyclic antidepressant imipramine was able to reverse the effect of chronic stress on BDNF transcription in mice (Tsankova et al., 2006). However, this reversal does not seem to be due to the reinstatement of altered histone modifications but rather to follow an indirect pathway. Indeed, chronic but not acute imipramine treatment did not reinstate H3K27 basal dimethylation levels, but rather decreased histone deacetylase 5 (HDAC5) levels in the hippocampus of chronically stressed mice leading to a global hyperacetylation in transcripts III and IV promoter regions. The importance of histone acetylation in the effect of antidepressant treatment has indeed been previously reported in animal models of stress-induced depression (Brown and Harris, 1978; Schroeder et al., 2007; Alt et al., 2010) Additionally, this hyperacetylation was associated with higher hippocampal levels of H3K4 dimethylation in the area of BDNF III and IV promoters with both modifications related to transcriptional activation. Consequently, these results suggest the existence of a compensatory mechanism in the reinstatement of basal BDNF levels by chronic imipramine treatment following chronic stress and emphasize the importance of chromatin hyperacetylation induced by antidepressant treatment.

Recently, the methylation state of BDNF was also assessed in postmortem brains from suicide completers (Keller et al., 2010). The human BDNF gene is also composed of 11 exons preceded by nine non-coding first exons regulating BDNF expression in different tissue (Pruunsild et al., 2007). In Keller and colleagues' study (Keller et al., 2010), three different methods were used to quantify methylation levels in a region encompassing part of non-coding exon IV and its promoter in the Wernicke area. Their results show that methylation in four CpGs located downstream to the promoter IV transcription initiation site was significantly increased in suicide completers compared to controls. These differences were specific to the BDNF promoter since the investigation of genome-wide methylation in these subjects did not reveal any significant difference between groups. In addition, BDNF expression in subjects with high methylation levels was significantly lower than in subjects with low and medium methylation levels, supporting the repressive effects of methylation within promoter on transcription.

32.5   The ribosomal RNA gene

Having the role to decode the mRNA into amino acids, the ribosomal RNA (rRNA) is a bottleneck structure for protein synthesis, allowing adequate cell function depending on the cell needs. The rRNA promoter is composed of two regulatory regions, namely the upstream control element (UCE) and the core promoter that binds the upstream binding factor (UBF) (Haltiner *et al.*, 1986; Learned *et al.*, 1986; Ghoshal *et al.*, 2004). The expression of rRNA genes has been shown to be epigenetically regulated in both mice (Santoro and Grummt, 2001) and humans (Ghoshal *et al.*, 2004; Brown and Szyf, 2007). In mice, the recruitment of transcription repressors has been suggested to induce chromatin modifications leading to methylation of a single CpG found within UBF binding sites in the UCE. This is thought to prevent UBF binding to its cognate sequence and to decrease rRNA expression (Santoro and Grummt, 2001). In humans, despite the fact that the CpG density in both promoter regions differs from that in mice (Santoro and Grummt, 2001; Ghoshal *et al.*, 2004), rRNA expression has nevertheless been shown to be epigenetically regulated (Brown and Szyf, 2007). Indeed, the active portion of the rRNA promoter associated with Pol I has been shown to be completely unmethylated while the inactive portion is almost fully methylated (Brown and Szyf, 2007).

The epigenetic control of rRNA gene expression has been shown to be dysregulated in the hippocampus of abused suicide completers (McGowan *et al.*, 2008). Abused suicide completers exhibited smaller rRNA expression levels associated with increased methylation in 21 out of 26 CpGs found within the rRNA core promoter and UCE compared to controls. From a mechanistic point of view, these results suggest that methylation represses the interaction of the UBF with the core promoter sequence and consequently decreases both the recruitment of transcriptional cofactors and the transcriptional activity of the RNA polymerase. Interestingly, these alterations seem to be specific to the hippocampus since no group difference in rRNA methylation pattern was found in the cerebellum. In addition, these results did not reflect global methylation differences, as genome-wide methylation levels did not reveal any methylation difference between abused suicides and controls.

32.6   The tropomyosin-related kinase B receptor gene

The transmembrane tropomyosin-related kinase B (TrkB) is the receptor for BDNF and has long been investigated in the neurobiology of mood and related disorders (Dwivedi *et al.*, 2003; Duman and Monteggia, 2006; Kim *et al.*, 2007; Dwivedi *et al.*, 2009). Expression microarray studies have reported lower TrkB expression in the prefrontal cortex of depressed subjects (Aston *et al.*, 2005; Nakatani *et al.*, 2006) and antidepressant treatment has been shown to increase its expression in cultured astrocytes (Mercier *et al.*, 2004).

The TrkB gene is found on chromosome 9 at locus q22.1 and has five splice variants. Splice variant T1 or TrkB-T1 is an astrocytic truncated form of TrkB lacking catalytic activity (Rose *et al.*, 2003). Recently, analysis of the methylation pattern in the promoter of a subset of suicide completers with low levels of TrkB-T1 expression revealed two sites where methylation levels were higher in suicide completers compared to controls (Ernst *et al.*, 2009a). The methylation pattern at those two sites was negatively correlated with the expression of TrkB-T1 in suicide completers, and this effect was specific to the prefrontal cortex, since no significant difference was found in the cerebellum. Such a pattern of expression and methylation is thought to increase predisposition to suicidal behaviors. In addition, suicide completers with low TrkB-T1 expression showed enrichment of H3K27 methylation in the TrkB promoter (Ernst *et al.*, 2009b), suggesting that the astrocytic variant of TrkB may be under the control of epigenetic mechanisms involving histone modifications and DNA methylation. Interestingly, recent data showed that mice overexpressing the TrkB-T1 variant are more susceptible to chronic social stress than wild-type mice by exhibiting consistent social avoidance (Razzoli *et al.*, 2011). Together, these data suggest that epigenetic changes in TrkB-T1 promoter inducing expression changes could define the vulnerability to chronic social stress and possibly to early-life adverse experience.

32.7   The GABAergic system

The GABAergic system has been the focus of many research studies in postmortem brain samples of psychiatric patients, and particularly individuals with histories of depression (Merali *et al.*, 2004; Torrey *et al.*, 2005; Klempan *et al.*, 2009), schizophrenia, or bipolar disorder, many of whom died by suicide (Akbarian *et al.*, 1995; Guidotti *et al.*, 2000; Volk *et al.*, 2000; Heckers *et al.*, 2002). For instance, reductions of reelin and glutamate decarboxylase 1 (GAD1) mRNA (Guidotti *et al.*, 2000) and an increase in DNA methyltransferase (DNMT1) expression (Veldic *et al.*, 2004; Kundakovic *et al.*, 2007) were previously reported in postmortem brains of schizophrenic and bipolar subjects who died by suicide. Consistently, promoter hypermethylation was reported for both genes in accordance with the methylating role of DNMT1 (Grayson *et al.*, 2005; Tamura *et al.*, 2007).

More recently, the hippocampal expression of GAD1 has been shown to be affected by maternal care in rats (Zhang *et al.*, 2010). Indeed, pups raised by mothers providing poor maternal care characterized by a low amount of LG have lower GAD1 hippocampal expression associated with promoter hypermethylation and lower levels of H3K9ac compared to pups raised by mothers providing high levels of maternal care. Interestingly, this is also associated with higher DNMT1 hippocampal levels. Functional assays revealed that the transcription factor NGFI-A binds the GAD1 promoter in order to increase GAD1 expression. Consequently, these results suggest that similarly to the regulation of GR in rat hippocampus, GAD1 expression is modulated by maternal behavior via epigenetic mechanisms involving DNA methylation interfering with the binding of activating transcription factors and by chromatin modifications (Zhang *et al.*, 2010).

These findings are in accordance with the study of Poulter and colleagues (Poulter *et al.*, 2008) which examined the expression of DNA methyltransferases as well as the GABA$_A$ receptor α1 subunit in the brain of suicide completers. Three hypermethylated CpG sites within α1 subunit promoter were identified in the prefrontal cortex of suicide completers and negatively correlated with DNMT3b protein expression. Besides DNMT3b, DNMT1 and DNMT3a levels have also been reported to be altered in the limbic system and brainstem of suicide completers. However, in this study there was no report of histories of early life adversity, and thus, one cannot assume that these effects would be similar in abused suicide completers.

32.8   Other epigenetic alterations in brains of suicide completers

In the light of what was discussed above, early life adversity seems to modify epigenetic control of gene expression. These changes can take place through histone modifications and/or DNA methylation. Moreover, epigenetic changes correlate with behavioral modifications in animals and humans, thus strongly suggesting that epigenetics may act as an interface mediating the effect of environment on the genome.

In addition to the findings discussed above, studies have focused on other functional systems, which had been implicated in depression and suicide. Among these systems, the polyamine and the serotonergic systems are noteworthy.

Polyamines are ubiquitous aliphatic molecules involved in cellular functions including growth, division, and signaling cascades (Gilad and Gilad, 2003; Mingue *et al.*, 2008). The polyamines also play a major role in the regulation of stress (Rhee *et al.*, 2007; Fiori and Turecki, 2008), being dependent on the activation of the HPA axis and the increased concentrations in circulating glucocorticoids (Gilad and Gilad, 2003). Furthermore, the emergence of the characteristic adult polyamine stress response correlates with the cessation of the hyporesponsive period of the HPA axis system (Gilad *et al.*, 1998). Previously, spermine synthase (SMS), spermidine/spermine N1-acetyltransferase (SAT1), and ornithine aminotransferase-like 1 (OATL1) expression has been shown to be altered in the limbic system of suicide completers with a history of depressive disorders (Sequeira *et al.*, 2006, 2007). However, follow-up studies revealed that epigenetic alterations in the promoter region of genes involved in polyamine synthesis do not account for these changes in expression. Up to date, a negative correlation between SAT1 expression and promoter methylation levels in BA 8/9 has been found (L. M. Fiori *et al.*, unpublished results).

The serotonergic system is a neurotransmitter system of great importance in psychiatry and has been extensively investigated in depression and suicide. Lower concentration, binding, neurotransmission, and reuptake of serotonin and its metabolites are risk markers for suicidality and major depression (Cronholm *et al.*, 1977; Bhagwagar and Cowen, 2008). Among the various serotonergic receptors, particular attention has been given to the 5-HT$_{2A}$ and its gene, as an important candidate in association studies of suicidal behavior (Turecki *et al.*, 1999;

Du *et al.*, 2001). One of the variants most commonly investigated was the 102 C/T polymorphism, located in exon 1 (Du *et al.*, 2000; De Luca *et al.*, 2007). Methylation in the C allele variant in this polymorphism has previously been associated with higher DNMT1 expression in the brain and leukocytes of healthy subjects (Polesskaya *et al.*, 2006). Although methylation was reported as increased in leukocytes from suicide ideators, a non-significant hypomethylation was reported in the prefrontal cortex of suicide completers carrying the C allele (De Luca *et al.*, 2009), suggesting that methylation levels may be different in individuals who committed suicide and those who are planning suicide. On the other hand, the functional significance of this hypermethylation in leukocytes remains to be explored, and since significance levels were not reached in brain tissue, further research is required.

32.9   Conclusion

In the light of what has been discussed, there is significant evidence suggesting that early life adversity affects molecular mechanisms involved in the regulation of behavior. These effects involve alterations in DNA methylation and histone modifications, which are believed to induce behavioral aberrations during development or later in life by affecting genes involved in crucial neuronal processes. Studies performed in postmortem brains from suicide completers with a history of childhood abuse highlighted several environmentally induced epigenetic alterations in the regulatory regions of genes involved in the regulation of stress (Table 32.1). Similarly, investigating the effect of variation in early life environment in animals has revealed useful information to expand our understanding of the molecular mechanisms involved in the effect of environmental stressors on the regulation of behavior (Table 32.2). Together, these findings suggest that epigenetics may act as a mechanism whereby environmental factors act on the modulation of long-term behavioral responses. In individuals with particular predispositions toward psychiatric disorders, these alterations may help trigger the expression of the illness. From a therapeutic point of view, it is tempting to speculate on the clinical potential these findings may provide. In the future, they could potentially lead to the development of tools for the identification of individuals at risk, and therefore, the possibility of preventive intervention. However, there are major challenges in their potential implementation, not the least of which are access to target tissue in living subjects, modification of epigenetic profiles, and appropriate delivery of such interventions. Given that this is a relatively new area of research, the current knowledge is significantly limited. Integrating genome-wide approaches will provide a more comprehensive view on the complexity of the relationship between early life adversity and the psychopathology of brain disorders. Furthermore, since these studies can provide information on the molecular nature of stress-induced psychopathologies, future work should assess whether similar alterations can be found in more accessible tissue.

Table 32.1 Summary of studies assessing epigenetic components in suicide

Study	Brain region	Gene	Findings
McGowan et al., 2009	Hippocampus	GR	↑ Methylation in NGFI-A binding site within GR promoter in the hippocampus of suicide completers with history of abuse
			↓ Expression of GR in the hippocampus of suicide completers with history of abuse
Alt et al., 2010	Amygdala	GR	↓GRα protein, ↑ expression of 1J, ↓ YY1 transcription factor
	Hippocampus		↓ Expression of GR1$_F$ and GR1$_C$, ↓ NGFI-A transcription factor
	Inferior frontal gyrus		↓GRβ protein, ↓ YY1 and Sp1 transcription factors
	Cingulate gyrus		↓GRα protein, ↑ expression of 1D, ↓ YY1, NGFI-A and Sp1 transcription factors
	Nucleus accumbens		↓ Expression of GR1$_B$, ↓ NGFI-A transcription factor; no methylation difference in promoters
McGowan et al., 2008	Hippocampus	rRNA	Overall hypermethylation of rRNA promoter in the hippocampus of suicide completers with history of abuse
			↓ Expression of rRNA gene in the hippocampus of suicide completers with history of abuse
Keller et al., 2010	Wernicke area	BDNF	Hypermethylation at four CpGs within promoter/exon IV in suicide completers; negative correlation between BDNF promoter methylation levels and expression
Ernst et al., 2009a	Frontal cortex	TrkB-T1	↑ Methylation in two sites within the promoter of TrkB-T1 in the frontal cortex of suicide completers
			↓ Expression of TrkB-T1 in the frontal cortex of suicide completers
Ernst et al., 2009b	Frontal cortex	TrkB-T1	↑ H3K27 methylation in the frontal cortex of suicide completers
			Negative correlation between H3K27 methylation levels and TrkB-T1 expression in the frontal cortex of suicide completers
Poulter et al., 2008	Frontopolar cortex	GABA$_A$ α1	↓ Expression of DNMT1 mRNA in suicide completers
			↑ expression of DNMT3b mRNA and protein levels in suicide completers
			↑ increased methylation at two sites in the promoter region of GABAA receptor subunit α1 in suicide completers
	Hippocampus, amygdala	DNMT1 DNMT3a DNMT3b	↓ Expression of DNMT1 and DNMT3b mRNA levels in suicide completers
	Brain stem		↓ Expression of DNMT3b mRNA in suicide completers
Fiori and Turecki, 2010	Prefrontal cortex	SMOX, SMS	No effects of promoter methylation in SMOX and SMS on expression levels
L.M. Fiori and G. Turecki (unpublished)	Prefrontal cortex	SAT1	Negative correlation between promoter methylation levels and expression of SAT1
De Luca et al., 2009	Prefrontal cortex	5-HT2A	↓ Methylation in the promoter region of 5-HT2A receptor associated with a C allele (trend) in the prefrontal cortex of suicide completers
			↑ Methylation in the promoter region of 5-HT2A receptor associated with a C allele in leukocytes of suicide attempters

Table 32.2 Summary of published studies assessing epigenetic components in animal models of stress-induced depressive symptoms

Study	Animal model	Brain region	Gene	Findings
Weaver et al., 2004	Low/high licking and grooming	Hippocampus	GR	In pups raised by LG mothers: Overall $GR1_7$ promoter hypermethylation ↓ H3K9 acetylation in $GR1_7$ promoter ↓ Binding of NGFIA in $GR1_7$ promoter
Tsankova et al., 2006	Intruder test	Hippocampus	BDNF	In stressed mice: ↓ Expression of transcript III and IV ↑ H3K27 dimethylation in transcript III and IV promoter No DNA methylation difference Chronic treatment with imipramine: ↑ H3 acetylation and H3K4 dimethylation in transcript III and IV promoter ↓ HDAC5 in hippocampus of stressed mice
Roth et al., 2009	Stressed mothers	Prefrontal cortex	BDNF	In maltreated rats: ↓ Expression of BDNF transcript IX from childhood to adulthood ↓ Expression of BDNF transcript IV at adulthood Overall hypermethylation in transcript IV promoter Transgenerational DNA methylation alterations in pups raised by abusive mothers
Roth et al., 2011	PTSD animal model		BDNF	In stressed rats:
		Hippocampus		Dorsal hippocampus: ↓ Expression of BDNF transcript IV ↑ Methylation in dentate gyrus ↑ Methylation in CA1 Ventral hippocampus: ↓ Expression of BDNF transcript IV ↓ Methylation in CA3
		Basolateral amygdala		No change in methylation pattern in transcript IV
		Prefrontal cortex		No change in methylation pattern in transcript IV
Murgatroyd et al., 2010	Maternal deprivation	Hypothalamic paraventricular nucleus	AVP	In stressed mice: ↑ Expression of AVP Site-specific hypomethylation in AVP intergenic region Phosphorylation of meCP2

Table 32.2 (cont.)

Study	Animal model	Brain region	Gene	Findings
Elliott *et al.*, 2010	Intruder test	Paraventricular nucleus	CRF	In stressed mice: ↑ Expression of CRF ↓ Methylation in CRF promoter (CRE) ↓ Expression of DNMT3b (acute), HDAC2 (long term) ↑ gadd45 (acute) In resilient mice: No differential CRF expression No differential methylation Chronic treatment with imipramine: Regulate behavior Regulate CRF expression Regulate CRF promoter methylation
Zhang *et al.*, 2010	Low/high licking and grooming	Hippocampus	GAD1	In pups raised by low-LG mothers: ↓ Expression of GAD1 ↑ Methylation within GAD1 promoter ↑ Expression of DNMT1 ↓ H3K9 acetylation in GAD1 promoter

REFERENCES

Agid, O., Shapira, B., Zislin, J., *et al.* (1999). Environment and vulnerability to major psychiatric illness: a case control study of early parental loss in major depression, bipolar disorder and schizophrenia. *Molecular Psychiatry*, **4**, 163–172.

Aguilera, G., Kiss, A., Liu, Y., *et al.* (2007). Negative regulation of corticotropin releasing factor expression and limitation of stress response. *Stress*, **10**, 153–161.

Aid, T., Kazantseva, A., Piirsoo, M., *et al.* (2007). Mouse and rat BDNF gene structure and expression revisited. *Journal of Neuroscience Research*, **85**, 525–535.

Akbarian, S., Kim, J. J., Potkin, S. G., *et al.* (1995). Gene expression for glutamic acid decarboxylase is reduced without loss of neurons in prefrontal cortex of schizophrenics. *Archives of General Psychiatry*, **52**, 258–266.

Alt, S. R., Turner, J. D., Klok, M. D., *et al.* (2010). Differential expression of glucocorticoid receptor transcripts in major depressive disorder is not epigenetically programmed. *Psychoneuroendocrinology*, **35**, 544–556.

Andersen, S. L., Tomada, A., Vincow, E. S., *et al.* (2008). Preliminary evidence for sensitive periods in the effect of childhood sexual abuse on regional brain development. *Journal of Neuropsychiatry Clinics and Neuroscience*, **20**, 292–301.

Aston, C., Jiang, L., and Sokolov, B. P. (2005). Transcriptional profiling reveals evidence for signaling and oligodendroglial abnormalities in the temporal cortex from patients with major depressive disorder. *Molecular Psychiatry*, **10**, 309–322.

Bensley, L. S., Van Eenwyk, J., Spieker, S. J., *et al.* (1999). Self-reported abuse history and adolescent problem behaviors. I. Antisocial and suicidal behaviors. *Journal of Adolescent Health*, **24**, 163–172.

Bhagwagar, Z. and Cowen, P. J. (2008). 'It's not over when it's over': persistent neurobiological abnormalities in recovered depressed patients. *Psychological Medicine*, **38**, 307–313.

Bremner, J. D., Randall, P., Vermetten, E., *et al.* (1997). Magnetic resonance imaging-based measurement of hippocampal volume in posttraumatic stress disorder related to childhood physical and sexual abuse: a preliminary report. *Biological Psychiatry*, **41**, 23–32.

Bremner, J. D., Vythilingam, M., Vermetten, E., *et al.* (2003). MRI and PET study of deficits in hippocampal structure and function in women with childhood sexual abuse and posttraumatic stress disorder. *American Journal of Psychiatry*, **160**, 924–932.

Briere, J. and Elliott, D. M. (2003). Prevalence and psychological sequelae of self-reported childhood physical and sexual abuse in a general population sample of men and women. *Child Abuse and Neglect*, **27**, 1205–1222.

Brown, G. W. and Harris, T. O. (1978). *Social Origins of Depression: A Study of Psychiatric Disorders in Women*. London: Tavistock.

Brown, G. W. and Moran, P. (1994). Clinical and psychosocial origins of chronic depressive episodes. I. A community survey. *British Journal of Psychiatry*, **165**, 447–456.

Brown, S. E. and Szyf, M. (2007). Epigenetic programming of the rRNA promoter by MBD3. *Molecular Cell Biology*, **27**, 4938–4952.

Brunoni, A. R., Lopes, M., and Fregni, F. (2008). A systematic review and meta-analysis of clinical studies on major depression and BDNF levels: implications for the role of neuroplasticity in depression. *International Journal of Neuropsychopharmacology*, **11**, 1169–1180.

Carpenter, L. L., Tyrka, A. R., McDougle, C. J., *et al.* (2004). Cerebrospinal fluid corticotropin-releasing factor and perceived early-life stress in depressed patients and healthy control subjects. *Neuropsychopharmacology*, **29**, 777–784.

Carrion, V. G., Weems, C. F., Eliez, S., *et al.* (2001). Attenuation of frontal asymmetry in pediatric posttraumatic stress disorder. *Biological Psychiatry*, **50**, 943–951.

Chan, J. P., Unger, T. J., Byrnes, J., *et al.* (2006). Examination of behavioral deficits triggered by targeting Bdnf in fetal or postnatal brains of mice. *Neuroscience*, **142**, 49–58.

Chen, B., Dowlatshahi, D., MacQueen, G. M., *et al.* (2001). Increased hippocampal BDNF immunoreactivity in subjects treated with antidepressant medication. *Biological Psychiatry*, **50**, 260–265.

Chen, W. G., Chang, Q., Lin, Y., *et al.* (2003). Derepression of BDNF transcription involves calcium-dependent phosphorylation of MeCP2. *Science*, **302**, 885–889.

Children's Bureau (2010). *Child Maltreatment 2009*. Washington DC: U.S. Department of Health and Human Services, Government Printing Office.

Cronholm, B., Asberg, M., Montgomery, S., and Schalling, D. (1977). Suicidal behaviour syndrome with low CSF 5-HIAA. *British Medical Journal*, **6063**, 776.

De Bellis, M. D., Keshavan, M. S., Frustaci, K., *et al.* (2002). Superior temporal gyrus volumes in maltreated children and adolescents with PTSD. *Biological Psychiatry*, **51**, 544–552.

De Luca, V., Likhodi, O., Kennedy, J. L., *et al.* (2007). Differential expression and parent-of-origin effect of the 5-HT2A receptor gene C102T polymorphism: analysis of suicidality in schizophrenia and bipolar disorder. *American Journal of Medical Genetics, Part B, Neuropsychiatric Genetics*, **144**, 370–374.

De Luca, V., Viggiano, E., Dhoot, R., *et al.* (2009). Methylation and QTDT analysis of the 5-HT2A receptor 102C allele: analysis of suicidality in major psychosis. *Journal of Psychiatric Research*, **43**, 532–537.

Dinwiddie, S., Heath, A. C., Dunne, M. P., *et al.* (2000). Early sexual abuse and lifetime psychopathology: a co-twin-control study. *Psychology and Medicine*, **30**, 41–52.

Driessen, M., Herrmann, J., Stahl, K., *et al.* (2000). Magnetic resonance imaging volumes of the hippocampus and the amygdala in women with borderline personality disorder and early traumatization. *Archives of General Psychiatry*, **57**, 1115–1122.

Du, L., Bakish, D., Lapierre, Y. D., *et al.* (2000). Association of polymorphism of serotonin 2A receptor gene with suicidal ideation in major depressive disorder. *American Journal of Medical Genetics*, **96**, 56–60.

Du, L., Faludi, G., Palkovits, M., *et al.* (2001). Serotonergic genes and suicidality. *Crisis*, **22**, 54–60.

Duman, R. S. and Monteggia, L. M. (2006). A neurotrophic model for stress-related mood disorders. *Biological Psychiatry*, **59**, 1116–1127.

Duric, V. and McCarson, K. E. (2005). Hippocampal neurokinin-1 receptor and brain-derived neurotrophic factor gene expression is decreased in rat models of pain and stress. *Neuroscience*, **133**, 999–1006.

Duric, V. and McCarson, K. E. (2006). Effects of analgesic or antidepressant drugs on pain- or stress-evoked hippocampal and spinal neurokinin-1 receptor and brain-derived neurotrophic factor gene expression in the rat. *Journal of Pharmacology and Experimental Therapy*, **319**, 1235–1243.

Dwivedi, Y., Rizavi, H. S., Conley, R. R., *et al.* (2003). Altered gene expression of brain-derived neurotrophic factor and receptor tyrosine kinase B in postmortem brain of suicide subjects. *Archives of General Psychiatry*, **60**, 804–815.

Dwivedi, Y., Rizavi, H. S., Zhang, H., *et al.* (2009). Neurotrophin receptor activation and expression in human postmortem brain: effect of suicide. *Biological Psychiatry*, **65**, 319–328.

Elliott, E., Ezra-Nevo, G., Regev, L., *et al.* (2010). Resilience to social stress coincides with functional DNA methylation of the *Crf* gene in adult mice. *Nature Neuroscience*, **13**, 1351–1353.

Ernst, C., Chen, E. S., and Turecki, G. (2009a). Histone methylation and decreased expression of TrkB-T1 in orbital frontal cortex of suicide completers. *Molecular Psychiatry*, **14**, 830–832.

Ernst, C., Deleva, V., Deng, X., *et al.* (2009b). Alternative splicing, methylation state, and expression profile of tropomyosin-related kinase B in the frontal cortex of suicide completers. *Archives of General Psychiatry*, **66**, 22–32.

Evans, E., Hawton, K., and Rodham, K. (2005). Suicidal phenomena and abuse in adolescents: a review of epidemiological studies. *Child Abuse and Neglect*, **29**, 45–58.

Fiori, L. M. and Turecki, G. (2008). Implication of the polyamine system in mental disorders. *Journal of Psychiatry and Neuroscience*, **33**, 102–110.

Francis, D., Diorio, J., Liu, D., *et al.* (1999). Nongenomic transmission across generations of maternal behavior and stress responses in the rat. *Science*, **286**, 1155–1158.

Gainer, H., Fields, R. L., and House, S. B. (2001). Vasopressin gene expression: experimental models and strategies. *Experimental Neurology*, **171**, 190–199.

Ghoshal, K., Majumder, S., Datta, J., *et al.* (2004). Role of human ribosomal RNA (rRNA) promoter methylation and of methyl-CpG-binding protein MBD2 in the suppression of rRNA gene expression. *Journal of Biological Chemistry*, **279**, 6783–6793.

Gilad, G. M. and Gilad, V. H. (2003). Overview of the brain polyamine-stress-response: regulation, development, and modulation by lithium and role in cell survival. *Cellular and Molecular Neurobiology*, **23**, 637–649.

Gilad, G. M., Gilad, V. H., Eliyayev, Y., *et al.* (1998). Developmental regulation of the brain polyamine-stress-response. *International Journal of Developmental Neuroscience*, **16**, 271–278.

Gladstone, G. L., Parker, G. B., Mitchell, P. B., *et al.* (2004). Implications of childhood trauma for depressed women: an analysis of pathways from childhood sexual abuse to deliberate self-harm and revictimization. *American Journal of Psychiatry*, **161**, 1417–1425.

Gorey, K. M. and Leslie, D. R. (1997). The prevalence of child sexual abuse: integrative review adjustment for potential response and measurement biases. *Child Abuse and Neglect*, **21**, 391–398.

Grayson, D. R., Jia, X., Chen, Y., *et al.* (2005). Reelin promoter hypermethylation in schizophrenia. *Proceedings of the National Academy of Sciences USA*, **102**, 9341–9346.

Gronli, J., Bramham, C., Murison, R., *et al.* (2006). Chronic mild stress inhibits BDNF protein expression and CREB activation in the dentate gyrus but not in the hippocampus proper. *Pharmacological and Biochemical Behaviour*, **85**, 842–849.

Guidotti, A., Auta, J., Davis, J. M., *et al.* (2000). Decrease in reelin and glutamic acid decarboxylase67 (GAD67) expression in schizophrenia and bipolar disorder: a postmortem brain study. *Archives of General Psychiatry*, **57**, 1061–1069.

Haltiner, M. M., Smale, S. T., and Tjian, R. (1986). Two distinct promoter elements in the human rRNA gene identified by linker scanning mutagenesis. *Molecular Cell Biology*, **6**, 227–235.

Heckers, S., Stone, D., Walsh, J., *et al.* (2002). Differential hippocampal expression of gluta-mic acid decarboxylase 65 and 67 messenger RNA in bipolar disorder and schizophre-nia. *Archives of General Psychiatry*, **59**, 521–529.

Heim, C. and Nemeroff, C. B. (2001). The role of childhood trauma in the neurobiology of mood and anxiety disorders: preclinical and clinical studies. *Biological Psychiatry*, **49**, 1023–1039.

Heim, C., Newport, D. J., Heit, S., *et al.* (2000). Pituitary–adrenal and autonomic responses to stress in women after sexual and physical abuse in childhood. *Journal of the American Medical Association*, **284**, 592–597.

Heim, C., Mletzko, T., Purselle, D., *et al.* (2008a). The dexamethasone/corticotropin-releasing factor test in men with major depression: role of childhood trauma. *Biological Psychiatry*, **63**, 398–405.

Heim, C., Newport, D. J., Mletzko, T., *et al.* (2008b). The link between childhood trauma and depression: insights from HPA axis studies in humans. *Psychoneuroendocrinology*, **33**, 693–710.

Heim, C., Young, L. J., Newport, D. J., *et al.* (2009). Lower CSF oxytocin concentrations in women with a history of childhood abuse. *Molecular Psychiatry*, **14**, 954–958.

Heim, C., Shugart, M., Craighead, W. E., *et al.* (2010). Neurobiological and psychiatric consequences of child abuse and neglect. *Developmental Psychobiology*, **52**, 671–690.

Jaffee, S. R., Moffitt, T. E., Caspi, A., *et al.* (2002). Differences in early childhood risk factors for juvenile-onset and adult-onset depression. *Archives of General Psychiatry*, **59**, 215–222.

Kaplan, M. J. and Klinetob, N. A. (2000). Childhood emotional trauma and chronic post-traumatic stress disorder in adult outpatients with treatment-resistant depression. *Journal of Nervous and Mental Disorders*, **188**, 596–601.

Keller, S., Sarchiapone, M., Zarrilli, F., *et al.* (2010). Increased BDNF promoter methylation in the Wernicke area of suicide subjects. *Archives of General Psychiatry*, **67**, 258–267.

Kendler, K. S., Bulik, C. M., Silberg, J., *et al.* (2000). Childhood sexual abuse and adult psychiatric and substance use disorders in women: an epidemiological and cotwin control analysis. *Archives of General Psychiatry*, **57**, 953–959.

Kendler, K. S., Kuhn, J. W., and Prescott, C. A. (2004). Childhood sexual abuse, stressful life events and risk for major depression in women. *Psychology and Medicine*, **34**, 1475–1482.

Kim, Y. K., Lee, H. P., Won, S. D., *et al.* (2007). Low plasma BDNF is associated with suicidal behavior in major depression. *Progress in Neuropsychopharmacology and Biological Psychiatry*, **31**, 78–85.

Klempan, T. A., Sequeira, A., Canetti, L., *et al.* (2009). Altered expression of genes involved in ATP biosynthesis and GABAergic neurotransmission in the ventral prefrontal cortex of suicides with and without major depression. *Molecular Psychiatry*, **14**, 175–189.

Klose, R. J. and Bird, A. P. (2006). Genomic DNA methylation: the mark and its mediators. *Trends in Biochemical Sciences*, **31**, 89–97.

Kouzarides, T. (2007). Chromatin modifications and their function. *Cell*, **128**, 693–705.

Kundakovic, M., Chen, Y., Costa, E., *et al.* (2007). DNA methyltransferase inhibitors coor-dinately induce expression of the human reelin and glutamic acid decarboxylase 67 genes. *Molecular Pharmacology*, **71**, 644–653.

Laplante, P., Diorio, J., and Meaney, M. J. (2002). Serotonin regulates hippocampal gluco-corticoid receptor expression via a 5-HT7 receptor. *Brain Research Developmental Brain Research*, **139**, 199–203.

Learned, R. M., Learned, T. K., Haltiner, M. M., *et al.* (1986). Human rRNA transcription is modulated by the coordinate binding of two factors to an upstream control element. *Cell*, **45**, 847–857.

Liu, D., Diorio, J., Tannenbaum, B., *et al.* (1997). Maternal care, hippocampal glucocorticoid receptors, and hypothalamic–pituitary–adrenal responses to stress. *Science*, **277**, 1659–1662.

Matrisciano, F., Bonaccorso, S., Ricciardi, A., *et al.* (2009). Changes in BDNF serum levels in patients with major depression disorder (MDD) after 6 months treatment with sertraline, escitalopram, or venlafaxine. *Journal of Psychiatric Research*, **43**, 247–254.

Maunakea, A. K., Nagarajan, R. P., Bilenky, M., *et al.* (2010). Conserved role of intragenic DNA methylation in regulating alternative promoters. *Nature*, **466**, 253–257.

McCauley, J., Kern, D. E., K. Kolodner, *et al.* (1997). Clinical characteristics of women with a history of childhood abuse: unhealed wounds. *Journal of the American Medical Association*, **277**, 1362–1368.

McCormick, J. A., Lyons, V., Jacobson, M. D., *et al.* (2000). 5'-heterogeneity of glucocorticoid receptor messenger RNA is tissue specific: differential regulation of variant transcripts by early-life events. *Molecular Endocrinology*, **14**, 506–517.

McGowan, P. O., Sasaki, A., Huang, T. C., *et al.* (2008). Promoter-wide hypermethylation of the ribosomal RNA gene promoter in the suicide brain. *PLoS One*, **3**, e2085.

McGowan, P. O., Sasaki, A., D'Alessio, A. C., *et al.* (2009). Epigenetic regulation of the glucocorticoid receptor in human brain associates with childhood abuse. *Nature Neuroscience*, **12**, 342–348.

Meaney, M. J. (2001). Maternal care, gene expression, and the transmission of individual differences in stress reactivity across generations. *Annual Reviews of Neuroscience*, **24**, 1161–1192.

Meaney, M. J., Aitken, D. H., and Sapolsky, R. M. (1987). Thyroid hormones influence the development of hippocampal glucocorticoid receptors in the rat: a mechanism for the effects of postnatal handling on the development of the adrenocortical stress response. *Neuroendocrinology*, **45**, 278–283.

Meaney, M. J., Diorio, J., Francis, D., *et al.* (2000). Postnatal handling increases the expression of cAMP-inducible transcription factors in the rat hippocampus: the effects of thyroid hormones and serotonin. *Journal of Neuroscience*, **20**, 3926–3935.

Merali, Z., Du, L., Hrdina, P., *et al.* (2004). Dysregulation in the suicide brain: mRNA expression of corticotropin-releasing hormone receptors and GABA(A) receptor subunits in frontal cortical brain region. *Journal of Neuroscience*, **24**, 1478–1485.

Mercier, G., Lennon, A. M., Renouf, B., *et al.* (2004). MAP kinase activation by fluoxetine and its relation to gene expression in cultured rat astrocytes. *Journal of Molecular Neuroscience*, **24**, 207–216.

Minguet, E. G., Vera-Sirera, F., Marina, A., *et al.* (2008). Evolutionary diversification in polyamine biosynthesis. *Molecular Biology and Evolution*, **25**, 2119–2128.

Mitchell, J. B., Iny, L. J., and Meaney, M. J. (1990). The role of serotonin in the development and environmental regulation of type II corticosteroid receptor binding in rat hippocampus. *Brain Research Developmental Brain Research*, **55**, 231–235.

Molnar, B. E., Berkman, L. F., and Buka, S. L. (2001). Psychopathology, childhood sexual abuse and other childhood adversities: relative links to subsequent suicidal behaviour in the US. *Psychological Medicine*, **31**, 965–977.

Murgatroyd, C., Patchev, A. V., Wu, Y., *et al.* (2009). Dynamic DNA methylation programs persistent adverse effects of early-life stress. *Nature Neuroscience*, **12**, 1559–1566.

Nakatani, N., Hattori, E., Ohnishi, T., *et al.* (2006). Genome-wide expression analysis detects eight genes with robust alterations specific to bipolar I disorder: relevance to neuronal network perturbation. *Human Molecular Genetics*, **15**, 1949–1962.

Pandey, G. N., Ren, X., Rizavi, H. S., *et al.* (2008). Brain-derived neurotrophic factor and tyrosine kinase B receptor signalling in post-mortem brain of teenage suicide victims. *International Journal of Neuropsychopharmacology*, **11**, 1047–1061.

Pariante, C. M. and Lightman, S. L. (2008). The HPA axis in major depression: classical theories and new developments. *Trends in Neurosciences*, **31**, 464–468.

Polesskaya, O. O., Aston, C., and Sokolov, B. P. (2006). Allele C-specific methylation of the 5-HT2A receptor gene: evidence for correlation with its expression and expression of DNA methylase DNMT1. *Journal of Neuroscience Research*, **83**, 362–373.

Poulter, M. O., Du, L., Weaver, I. C., *et al.* (2008). GABA$_A$ receptor promoter hypermethylation in suicide brain: implications for the involvement of epigenetic processes. *Biological Psychiatry*, **64**, 645–652.

Pruunsild, P., Kazantseva, A., Aid, T., *et al.* (2007). Dissecting the human BDNF locus: bidirectional transcription, complex splicing, and multiple promoters. *Genomics*, **90**, 397–406.

Razzoli, M., Domenici, E., Carboni, L., *et al.* (2011). A role for BDNF/TrkB signaling in behavioral and physiological consequences of social defeat stress. *Genes, Brain and Behavior*, **10**, 424–433.

Rhee, H. J., Kim, E. J., and Lee, J. K. (2007). Physiological polyamines: simple primordial stress molecules. *Journal of Cellular and Molecular Medicine*, **11**, 685–703.

Rogoz, Z., Skuza, G., and Legutko, B. (2005). Repeated treatment with mirtazepine induces brain-derived neurotrophic factor gene expression in rats. *Journal of Physiology and Pharmacology*, **56**, 661–671.

Rose, C. R., Blum, R., Pichler, B., *et al.* (2003). Truncated TrkB-T1 mediates neurotrophin-evoked calcium signalling in glia cells. *Nature*, **426**, 74–78.

Roth, T. L., Lubin, F. D., Funk, A. J., *et al.* (2009). Lasting epigenetic influence of early-life adversity on the BDNF gene. *Biological Psychiatry*, **65**, 760–769.

Roth, T. L., Zoladz, P. R., Sweatt, J. D., *et al.* (2011). Epigenetic modification of hippocampal Bdnf DNA in adult rats in an animal model of post-traumatic stress disorder. *Journal of Psychiatric Research*, **45**, 919–926.

Santa Mina, E. E. and Gallop, R. M. (1998). Childhood sexual and physical abuse and adult self-harm and suicidal behaviour: a literature review. *Canadian Journal of Psychiatry*, **43**, 793–800.

Santoro, R. and Grummt, I. (2001). Molecular mechanisms mediating methylation-dependent silencing of ribosomal gene transcription. *Molecular Cell*, **8**, 719–725.

Scher, C. D., Forde, D. R., McQuaid, J. R., *et al.* (2004). Prevalence and demographic correlates of childhood maltreatment in an adult community sample. *Child Abuse and Neglect*, **28**, 167–180.

Schratt, G. (2009a). Fine-tuning neural gene expression with microRNAs. *Current Opinion in Neurobiology*, **19**, 213–219.

Schratt, G. (2009b). MicroRNAs at the synapse. *Nature Reviews Neuroscience*, **10**, 842–849.

Schroeder, F. A., Lin, C. L., Crusio, W. E., *et al.* (2007). Antidepressant-like effects of the histone deacetylase inhibitor, sodium butyrate, in the mouse. *Biological Psychiatry*, **62**, 55–64.

Sen, S., Duman, R., and Sanacora, G. (2008). Serum brain-derived neurotrophic factor, depression, and antidepressant medications: meta-analyses and implications. *Biological Psychiatry*, **64**, 527–532.

Sequeira, A., Gwadry, F. G., Ffrench-Mullen, J. M., *et al.* (2006). Implication of SSAT by gene expression and genetic variation in suicide and major depression. *Archives of General Psychiatry*, **63**, 35–48.

Sequeira, A., Klempan, T., Canetti, L., *et al.* (2007). Patterns of gene expression in the limbic system of suicides with and without major depression. *Molecular Psychiatry*, **12**, 640–655.

Smythe, J. W., Rowe, W. B., and Meaney, M. J. (1994). Neonatal handling alters serotonin (5-HT) turnover and 5-HT2 receptor binding in selected brain regions: relationship to the handling effect on glucocorticoid receptor expression. *Brain Research Developmental Brain Research*, **80**, 183–189.

Stein, M. B., Koverola, C., Hanna, C., *et al.* (1997). Hippocampal volume in women victimized by childhood sexual abuse. *Psychological Medicine*, **27**, 951–959.

Tamura, Y., Kunugi, H., Ohashi, J., *et al.* (2007). Epigenetic aberration of the human *REELIN* gene in psychiatric disorders. *Molecular Psychiatry*, **12**, 519, 593–600.

Tanskanen, A., Hintikka, J., Honkalampi, K., *et al.* (2004). Impact of multiple traumatic experiences on the persistence of depressive symptoms: a population-based study. *Nordic Journal of Psychiatry*, **58**, 459–464.

Torrey, E. F., Barci, B. M., Webster, M. J., *et al.* (2005). Neurochemical markers for schizophrenia, bipolar disorder, and major depression in postmortem brains. *Biological Psychiatry*, **57**, 252–260.

Tsankova, N. M., Berton, O., Renthal, W., *et al.* (2006). Sustained hippocampal chromatin regulation in a mouse model of depression and antidepressant action. *Nature Neuroscience*, **9**, 519–525.

Turecki, G., Briere, R., Dewar, K., *et al.* (1999). Prediction of level of serotonin 2A receptor binding by serotonin receptor 2A genetic variation in postmortem brain samples from subjects who did or did not commit suicide. *American Journal of Psychiatry*, **156**, 1456–1458.

Turner, J. D. and Muller, C. P. (2005). Structure of the glucocorticoid receptor (NR3C1) gene 5′ untranslated region: identification, and tissue distribution of multiple new human exon 1. *Journal of Molecular Endocrinology*, **35**, 283–292.

Turner, J. D., Pelascini, L. P., Macedo, J. A., *et al.* (2008). Highly individual methylation patterns of alternative glucocorticoid receptor promoters suggest individualized epigenetic regulatory mechanisms. *Nucleic Acids Research*, **36**, 7207–7218.

Veldic, M., Caruncho, H. J., Liu, W. S., *et al.* (2004). DNA-methyltransferase 1 mRNA is selectively overexpressed in telencephalic GABAergic interneurons of schizophrenia brains. *Proceedings of the National Academy of Sciences USA*, **101**, 348–353.

Volk, D. W., Austin, M. C., Pierri, J. N., *et al.* (2000). Decreased glutamic acid decarboxylase67 messenger RNA expression in a subset of prefrontal cortical gamma-aminobutyric acid neurons in subjects with schizophrenia. *Archives of General Psychiatry*, **57**, 237–245.

Weaver, I. C., Cervoni, N., Champagne, F. A., *et al.* (2004). Epigenetic programming by maternal behavior. *Nature Neuroscience*, **7**, 847–854.

Xu, H., Chen, Z., He, J., *et al.* (2006). Synergetic effects of quetiapine and venlafaxine in preventing the chronic restraint stress-induced decrease in cell proliferation and BDNF expression in rat hippocampus. *Hippocampus*, **16**, 551–559.

Zhang, T. Y., Hellstrom, I. C., Bagot, R. C., *et al.* (2010). Maternal care and DNA methylation of a glutamic acid decarboxylase 1 promoter in rat hippocampus. *Journal of Neuroscience*, **30**, 13 130–13 137.

Zhou, Z., Hong, E. J., Cohen, S., *et al.* (2006). Brain-specific phosphorylation of MeCP2 regulates activity-dependent Bdnf transcription, dendritic growth, and spine maturation. *Neuron*, **52**, 255–269.

Zlotnick, C., Mattia, J. and Zimmerman, M. (2001). Clinical features of survivors of sexual abuse with major depression. *Child Abuse and Neglect*, **25**, 357–367.

33 X-linked expressed single nucleotide polymorphisms and dosage compensation

Lygia V. Pereira* and Joana C. Moreira de Mello

33.1 Introduction

In mammals dosage compensation of X-linked products between males and females is achieved by an extreme epigenetic process: the transcriptional inactivation of one of the two X chromosomes in female cells, a process called X-chromosome inactivation (XCI). XCI is primarily dependent on the expression of the *XIST/Xist* (inactive X-specific transcript) gene from the future inactive X, and accumulation in *cis* of its non-coding RNA. The interaction of Xist RNA with the future inactive X triggers a series of epigenetic changes in its chromatin, mainly DNA methylation and histone modifications, which determine its transcriptional silenced state. Here we discuss how XCI has been traditionally studied in humans, and how the use of contemporary genetic tools such as single nucleotide polymorphism (SNP) DNA/RNA genotyping to analyze allele-specific gene expression from the X chromosome can improve our understanding of this process.

In mammals, transcriptional silencing of all but one X chromosome in somatic cells was proposed by Mary Lyon in the early 1960s (Lyon, 1961, 1962), based on cytological evidence from feline and human cells – the condensed sex chromatin, or Barr body (Barr and Bertram, 1949) – and on variegated coat color in female mice and cats. She proposed that in different somatic cells of the same organism the sex chromatin could be either the paternal or the maternal X chromosome, and that the sex chromatin is actually a transcriptionally inactivated X chromosome (Xi). In humans the first report of XCI came from Beutler *et al.* (1962), who identified in females heterozygous for glucose-6-phosphate dehydrogenase (G6PD) deficiency two populations of red blood cells: one positive for the enzyme, and the other negative, according to the Xi in each cell.

* Author to whom correspondence should be addressed.

Epigenomics: From Chromatin Biology to Therapeutics, ed. K. Appasani. Published by Cambridge University Press. © Cambridge University Press 2012.

33.2   X-chromosome inactivation: counting, choice, initiation, and maintenance

X-chromosome inactivation (XCI) is an impressive example of epigenetic regulation. As it occurs very early in mammalian development, the initial events responsible for this process have been best studied in mice (reviewed in Payer and Lee, 2008). It is a complex multistep process, involving counting mechanisms (when the cell identifies how many X chromosomes are present per haploid genome), choosing the future Xi (which leads to a single active X – Xa), initiation and spreading of silencing. In addition, since the inactive state of the Xi is inherited throughout all subsequent mitoses, a mechanism responsible for the maintenance of the inactive state also exists.

One key locus for XCI is the X inactivation center (XIC), first described in mouse by Rastan and Cattanach (1983), and then in humans by Brown and colleagues (1991a). This locus is crucial for the "counting" and "choosing" mechanisms, and must be present in at least two copies for XCI to occur. Several regulatory elements required for the occurrence of X inactivation have been identified in the mouse, including *Xist*, *Tsix*, and *Xite* genes (Lee *et al.*, 1999), and the *DxPas34* locus (Ogawa and Lee, 2003). Before establishing XCI, the homologous Xs interact exclusively through their XIC regions in each cell of the embryo (Bacher *et al.*, 2006; Xu *et al.*, 2006). This interchromosomal pairing is transient but crucial for determining how many X chromosomes are present, and for choosing which one will remain active. However, all the precise events involved in these processes are not currently known.

The physical interaction between the two XICs leads to the upregulation of the *XIST/Xist* gene on the X chromosome to be inactivated. The *XIST/Xist* gene transcribes a non-coding RNA which remains in the nucleus, and is associated in *cis* with the future Xi (Brown *et al.*, 1991b, 1992; Brockdorff *et al.*, 1992; Clemson *et al.*, 1996). This in turn will lead to the modification of different histones on the Xi, and finally to the methylation of its DNA, leading to the typical heterochromatic state of that chromosome.

Once an X chromosome is inactivated, this state is stably maintained through mitosis as the result of a synergism between those multiple epigenetic modifications. They include methylation of CpG islands, hypoacetylation and methylation of specific residues in histones, the presence of the histone variant macroH2A, and the continuous expression in *cis* of *Xist* and its association with the Xi (Csankovszki *et al.*, 2001; Mietton *et al.*, 2009).

While most genes in the Xi are silenced, a subset of X-linked genes escapes XCI and is expressed from both Xa and Xi. Most of these escapee genes are located at the pseudoautosomal region, and thus have a Y-linked homolog, being biallelically expressed in both females and males. However, some X-linked genes outside the pseudoautosomal region and without a Y homolog also escape XCI. While in humans up to 15% of X-linked genes remain transcriptionally active in the Xi (Carrel and Willard, 2005), in mice only 3% escape XCI (Yang *et al.*, 2010). These genes, biallelically expressed exclusively in female cells, should explain the abnormal phenotypes associated with X chromosome aneuploidies in humans. In line with this, monosomy of the X chromosome leads to a much milder phenotype in mice than in humans (Lynn and Davies, 2007). Likewise, overexpression of escapee genes may be involved in the characteristic phenotypes of individuals with Klinefelter's syndrome (XXY), as recently shown in a mouse model for the disease (Werler *et al.*, 2011).

33.3   Imprinted and random X-chromosome inactivation

In mammals, there are two well-known patterns of XCI: imprinted, where the choice of the inactive X is dependent on its parental origin, and random, where maternal and paternal X are equally likely to be inactivated in each cell.

The imprinted pattern appears to be the primordial form of XCI, observed in marsupials, where the paternal X chromosome is preferentially inactivated in all cells of the embryo (Cooper *et al.*, 1971; Sharman, 1971; and reviewed in Deakin *et al.*, 2009). In mice, imprinted XCI is restricted to extra-embryonic lineages (Takagi and Sasaki, 1975; Wake *et al.*, 1976; West *et al.*, 1977), becoming evident as early as in the 4-cell stage (Huynh and Lee, 2003; Okamoto *et al.*, 2004; Patrat *et al.*, 2009), when expression of *Xist* exclusively from the paternal X results in its inactivation in every cell. At the blastocyst stage, cells from the epiblast reactivate the paternal Xi, and then go through a second round of XCI, this time randomly choosing the paternal or the maternal X as the inactive one, and leading to random XCI in the embryo proper (Mak *et al.*, 2004; Okamoto *et al.*, 2004). Interestingly, in humans both random and imprinted XCI have been reported in extra-embryonic tissues (reviewed in Moreira de Mello *et al.*, 2010), and this will be further discussed in the following sections.

33.4   Analysis of X-chromosome inactivation

In a given tissue or cell population, there is a specific ratio of cells with the maternal and the paternal X inactivated that can range from 100:0 to 50:50, depending on the pattern of XCI. Historically, two main X-linked polymorphisms have been most commonly used to determine XCI patterns in humans: the electrophoretically distinct variants for glucose-6-phosphate dehydrogenase protein (G6PD A and B), which migrate differently in starch gels, allowing the evaluation of the expression level of one isoform over the other, and thus the ratio of cells with the maternal or the paternal Xi; and a highly polymorphic trinucleotide (CAG) repeat in the first exon of the androgen receptor (*AR*) gene. A CpG island situated less than 100 bp upstream of this polymorphism is differentially methylated in the Xa and Xi. Assays with methylation-sensitive restriction enzymes followed by PCR amplification of that polymorphic trinucleotide enable the homologous X chromosomes to be distinguished from each other, and the quantification of the fraction of cells with one or the other X inactivated (Allen *et al.*, 1992).

For the study of genes escaping XCI, the most extensively used experimental system has been the rodent/human somatic cell hybrids containing the human Xi, which allows the identification of genes expressed from that chromosome. Using this model, an extensive chromosome-wide inactivation profile of the human X was generated, in which it was shown that up to 15% of the genes on the Xi escape inactivation, whereas another 10% show variable patterns of inactivation (Brown *et al.*, 1997; Carrel and Willard, 2005). Somatic cell hybrids were an important resource developed for mapping human genes, sequence-tagged sites, or biochemical markers to individual chromosomes or subchromosomal regions. However, for studying epigenetic characteristics of human cells, this may not be an adequate system. Cell fusion is capable of inducing global alterations that can culminate in a remarkable change in the epigenotype of the somatic nucleus, as observed by the reprogramming of adult cells into a pluripotent state when fused to embryonic stem cells (Tada *et al.*, 1997; Cowan *et al.*, 2005). Particularly with respect to the human X chromosome, there are also reports showing reactivation of genes on the Xi in mouse/human cell hybrids (Ellis *et al.*, 1987; Takagi, 1993; Yoshida *et al.*, 1997).

Therefore, the characterization of the epigenome of a given cell or tissue, and specifically of the epigenetic state of the X chromosome, should be performed in original samples of that cell or tissue. However, how is it possible to effectively distinguish the active from the inactive X in a given cell population?

The development of a large database of single nucleotide polymorphisms (SNPs) (Ross *et al.*, 2005) throughout the human genome enables investigation of allele-specific gene expression in normal human cells. SNPs are highly polymorphic markers spread throughout the genome at a frequency of approximately one in every 2000 base pairs. In particular, SNPs in expressed regions are extremely useful for the study of allele-specific gene expression (Vasques and Pereira, 2001). Once an informative SNP is found in a gene of interest, genotyping the SNP in mRNA from the tissue of interest will show which allele(s) is (are) being expressed (Figure 33.1). This approach has been used for the analysis of the inactivation pattern of two X-linked genes, *MAOA* and *GYG2*, which escape XCI in somatic cell hybrids (Stabellini *et al.*, 2009). Using a clonal population of normal female fibroblasts with the same Xi, it was shown that only one allele of those genes is expressed in the cell population, and thus, that in normal cells *MAOA* and *GYG2* are subjected to XCI, contrary to the findings in mouse/human hybrids.

Currently, according to the National Center for Biotechnology Information SNP database, there are more than 733 000 SNPs described in the X chromosome, almost 7000 of these in expressed regions, allowing a chromosome-wide analysis of XCI patterns. Using these resources, the issue of imprinted XCI in humans was recently revisited (Moreira de Mello *et al.*, 2010). While studies on human extra-embryonic tissues using one or two polymorphic markers on the X chromosome yield conflicting results (reviewed by Moreira de Mello *et al.*, 2010), the analysis of 27 SNPs in expressed regions of 22 X-linked genes, including *XIST*, revealed a complex pattern of XCI in the human term placenta (Figure 33.2). In this work, we were able to show that XCI in that human extra-embryonic tissue is random, and that the organ is composed of relatively large patches with the same inactive X, which may be interpreted as skewed XCI. Thus, the wider analysis of

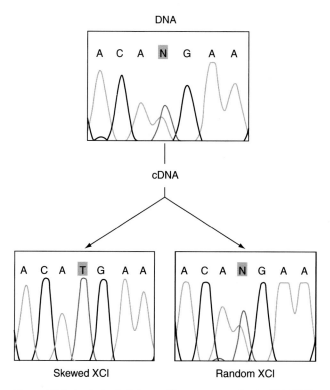

Figure 33.1 Analysis of allele-specific gene expression by DNA/RNA genotyping. An SNP present in the genomic DNA (upper panel) is analyzed in cDNA samples. Lower left panel: a sigle peak (T) indicates monoallelic expression of that gene. Lower right panel: the double peak indicates the presence of both kinds of transcripts, and thus of either biallelic expression, or random XCI.

allele-specific X-linked gene expression could finally explain the contradicting conclusions of the previous reports (Moreira de Mello *et al.*, 2010).

This approach has been adapted to a higher-throughput analysis of allele-specific gene expression using SNP-microarrays hybridized with cDNA, allowing the analysis of approximately 4000 genes in a single experiment (Gimelbrant *et al.*, 2007). This methodology was used to access the status of XCI in human embryonic stem cells, analyzing a microarray with 30 SNPs in expressed regions of the X chromosome (Lengner *et al.*, 2010).

33.5 Perspectives

Sequencing of the human genome has dramatically changed the way biomedical research is performed. In particular, the identification of common variations in the human genome has allowed the study of human phenotypic variability and migration, and the analysis of allele-specific gene expression – the functional manifestation of epigenetic modifications in the chromatin. However, the determination of the specific changes in chromatin that lead to an expression pattern is more complex. The combination of SNP analysis with chromatin immunoprecipitation using antibodies against different histone modifications and against

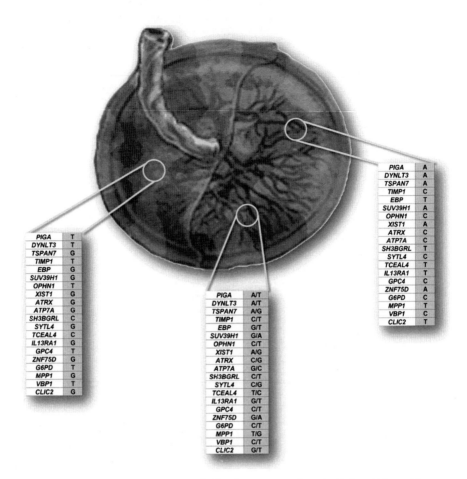

Figure 33.2 X-chromosome inactivation (XCI) in human term placenta. Pink and blue patches represent regions where the maternal and the paternal X is the active chromosome, respectively. Depending on the region analyzed, monoallelic or biallelic expression of the listed X-linked genes can be detected, suggesting imprinted or random XCI, respectively. See plate section for color version.

other chromatin-associated proteins can allow the construction of a clearer picture of the epigenetic states of the human genome, and in particular, of the two epigenetically distinct X chromosomes. That in turn can lead to the understanding of their association with gene expression and disease mechanisms.

REFERENCES

Allen, R. C., Zoghbi, H. Y., Moseley, A. B., Rosenblatt H. M., and Belmont, J. W. (1992). Methylation of HpaII and HhaI sites near the polymorphic CAG repeat in the human androgen-receptor gene correlates with X chromosome inactivation. *American Journal of Human Genetics*, **51**, 1229–1239.

Bacher, C. P., Guggiari, M., Brors, B., *et al.* (2006). Transient colocalization of X-inactivation centres accompanies the initiation of X inactivation. *Nature Cell Biology*, **8**, 293–299.

Barr, M. L. and Bertram, E. (1949). A morphological distinction between neurones of the male and female, and the behaviour of the nucleolar satellite during accelerated nucleoprotein synthesis. *Nature*, **163**, 676.

Beutler, E., Yeh, M., and Fairbanks, V. F. (1962). The normal human female as a mosaic of X-chromosome activity: studies using the gene for G-6-PD-deficiency as a marker. *Proceedings of the National Academy of Sciences USA*, **48**, 9–16.

Brockdorff, N., Ashworth, A., Kay, G. F., *et al.* (1992). The product of the mouse *Xist* gene is a 15 kb inactive X-specific transcript containing no conserved ORF and located in the nucleus. *Cell*, **71**, 515–526.

Brown, C. J., Lafreniere, R. G., Powers, V. E., *et al.* (1991a). Localization of the X inactivation centre on the human X chromosome in Xq13. *Nature*, **349**, 82–84.

Brown, C. J., Ballabio, A., Rupert J. L., *et al.* (1991b). A gene from the region of the human X inactivation centre is expressed exclusively from the inactive X chromosome. *Nature*, **349**, 38–44.

Brown, C. J., Hendrich, B. D., Rupert, J. L., *et al.* (1992). The human *XIST* gene: analysis of a 17 kb inactive X-specific RNA that contains conserved repeats and is highly localized within the nucleus. *Cell*, **71**, 527–542.

Brown, C. J., Carrel, L., and Willard, H. F. (1997). Expression of genes from the human active and inactive X chromosomes. *American Journal of Human Genetics*, **60**, 1333–1343.

Carrel, L. and Willard, H. F. (2005). X-inactivation profile reveals extensive variability in X-linked gene expression in females. *Nature*, **434**, 400–404.

Clemson, C. M., McNeil, J. A., Willard, H. F., and Lawrence, J. B. (1996). *XIST* RNA paints the inactive X chromosome at interphase: evidence for a novel RNA involved in nuclear/chromosome structure. *Journal of Cell Biology*, **132**, 259–275.

Cooper, D. W., VandeBerg, J. L., Sharman, G. B., and Poole, W. E. (1971). Phosphoglycerate kinase polymorphism in kangaroos provides further evidence for paternal X inactivation. *Nature New Biology*, **230**, 155–157.

Cowan, C. A., Atienza, J., Melton, D. A., and Eggan, K. (2005). Nuclear reprogramming of somatic cells after fusion with human embryonic stem cells. *Science*, **309**, 1369–1373.

Csankovszki, G., Nagy, A., and Jaenisch, R. (2001). Synergism of *Xist* RNA, DNA methylation, and histone hypoacetylation in maintaining X chromosome inactivation. *Journal of Cell Biology*, **153**, 773–784.

Deakin, J. E., Chaumeil, J., Hore, T. A., and Marshall Graves, J. A. (2009). Unravelling the evolutionary origins of X chromosome inactivation in mammals: insights from marsupials and monotremes. *Chromosome Research*, **17**, 671–685.

Ellis, N., Keitges, E., Gartler, S. M., and Rocchi, M. (1987). High-frequency reactivation of X-linked genes in Chinese hamster × human hybrid cells. *Somatic Cell and Molecular Genetics*, **13**, 191–204.

Gimelbrant, A., Hutchinson, J. N., Thompson, B. R., and Chess, A. (2007). Widespread monoallelic expression on human autosomes. *Science*, **318**, 1136–1140.

Huynh, K. D. and Lee, J. T. (2003). Inheritance of a pre-inactivated paternal X chromosome in early mouse embryos. *Nature*, **426**, 857–862.

Lee, J. T., Davidow, L. S., and Warshawsky, D. (1999). *Tsix*, a gene antisense to *Xist* at the X-inactivation centre. *Nature Genetics*, **21**, 400–404.

Lengner, C. J., Gimelbrant, A. A., Erwin, J. A., *et al.* (2010). Derivation of pre-X inactivation human embryonic stem cells under physiological oxygen concentrations. *Cell*, **141**, 872–883.

Lynn, P. M. and Davies, W. (2007). The 39, XO mouse as a model for the neurobiology of Turner syndrome and sex-biased neuropsychiatric disorders. *Behavioural Brain Research*, **179**, 173–182.

Lyon, M. F. (1961). Gene action in the X-chromosome of the mouse (*Mus musculus* L.). *Nature*, **190**, 372–373.

Lyon, M. F. (1962). Sex chromatin and gene action in the mammalian X-chromosome. *American Journal of Human Genetics*, **14**, 135–148.

Mak, W., Nesterova, T. B., de Napoles, M., *et al.* (2004). Reactivation of the paternal X chromosome in early mouse embryos. *Science*, **303**, 666–669.

Mietton, F., Sengupta, A. K., Molla, A., *et al.* (2009). Weak but uniform enrichment of the histone variant macroH2A1 along the inactive X chromosome. *Molecular and Cellular Biology*, **29**, 150–156.

Moreira de Mello, J. C., de Araujo, E. S., Stabellini, R., *et al.* (2010). Random X inactivation and extensive mosaicism in human placenta revealed by analysis of allele-specific gene expression along the X chromosome. *PLoS One*, **5**, e10947.

Ogawa, Y. and Lee, J. T. (2003). *Xite*, X-inactivation intergenic transcription elements that regulate the probability of choice. *Molecular Cell*, **11**, 731–743.

Okamoto, I., Otte, A. P., Allis, C. D., Reinberg, D., and Heard, E. (2004). Epigenetic dynamics of imprinted X inactivation during early mouse development. *Science*, **303**, 644–649.

Patrat, C., Okamoto, I., Diabangouaya, P., *et al.* (2009). Dynamic changes in paternal X-chromosome activity during imprinted X-chromosome inactivation in mice. *Proceedings of the National Academy of Sciences USA*, **106**, 5198–5203.

Payer, B. and Lee, J. T. (2008). X chromosome dosage compensation: how mammals keep the balance. *Annual Review of Genetics*, **42**, 733–772.

Rastan, S. and Cattanach, B. M. (1983). Interaction between the *Xce* locus and imprinting of the paternal X chromosome in mouse yolk-sac endoderm. *Nature*, **303**, 635–637.

Ross, M. T., Grafham, D. V., Coffey, A. J., *et al.* (2005). The DNA sequence of the human X chromosome. *Nature*, **434**, 325–337.

Sharman, G. B. (1971). Late DNA replication in the paternally derived X chromosome of female kangaroos. *Nature*, **230**, 231–232.

Stabellini, R., de Mello, J. C., Hernandes, L. M., and Pereira, L. V. (2009). *MAOA* and *GYG2* are submitted to X chromosome inactivation in human fibroblasts. *Epigenetics*, **4**, 388–393.

Tada, M., Tada, T., Lefebvre, L., Barton, S. C., and Surani, M. A. (1997). Embryonic germ cells induce epigenetic reprogramming of somatic nucleus in hybrid cells. *EMBO Journal*, **16**, 6510–6520.

Takagi, N. (1993). Variable X chromosome inactivation patterns in near-tetraploid murine EC × somatic cell hybrid cells differentiated in vitro. *Genetica*, **88**, 107–117.

Takagi, N. and Sasaki, M. (1975). Preferential inactivation of the paternally derived X chromosome in the extraembryonic membranes of the mouse. *Nature*, **256**, 640–642.

Vasques, L. R. and Pereira, L. V. (2001). Allele-specific X-linked gene activity in normal human cells assayed by expressed single nucleotide polymorphisms (cSNPs). *DNA Research*, **8**, 173–177.

Wake, N., Takagi, N., and Sasaki, M. (1976). Non-random inactivation of X chromosome in the rat yolk sac. *Nature*, **262**, 580–581.

Werler, S., Poplinski, A., Gromoll, J., and Wistuba, J. (2011). Expression of selected genes escaping from X inactivation in the 41, XXY* mouse model for Klinefelter's syndrome. *Acta Paediatrica*, **100**, 885–891.

West, J. D., Frels, W. I., Chapman, V. M., and Papaioannou, V. E. (1977). Preferential expression of the maternally derived X chromosome in the mouse yolk sac. *Cell*, **12**, 873–882.

Xu, N., Tsai, C. L., and Lee, J. T. (2006). Transient homologous chromosome pairing marks the onset of X inactivation. *Science*, **311**, 1149–1152.

Yang, F., Babak, T., Shendure, J., and Disteche, C. M. (2010). Global survey of escape from X inactivation by RNA-sequencing in mouse. *Genome Research*, **20**, 614–622.

Yoshida, I., Nishita, Y., Mohandas, T. K., and Takagi, N. (1997). Reactivation of an inactive human X chromosome introduced into mouse embryonal carcinoma cells by micro-cell fusion with persistent expression of *XIST*. *Experimental Cell Research*, **230**, 208–219.

Zhao, J., Sun, B. K., Erwin, J. A., Song, J. J., and Lee, J. T. (2008). Polycomb proteins targeted by a short repeat RNA to the mouse X chromosome. *Science*, **322**, 750–756.

34 Epigenomic diversity of colorectal cancer

Aditi Hazra* and Shuji Ogino

34.1 Introduction

34.1.1 Changes in DNA methylation in cancer

Colorectal cancer is the third most common cancer and the third leading cause of cancer mortality in the USA (Jemal *et al.*, 2010, 2011). Colorectal cancer is a complex multifactorial disease, resulting from germ-line and somatic genetic variations/alterations, epigenetic changes, and lifestyle and environmental risk factors (Ogino *et al.*, 2006a, 2006b, 2006c, 2006d, 2007a, 2007b, 2008b, 2008c, 2008d, 2009a, 2009b, 2009c, 2009d, 2009e; Chan *et al.*, 2007, 2009; Schernhammer *et al.*, 2008a, 2008b; Baba *et al.*, 2009; Meyerhardt *et al.*, 2009; Nosho *et al.*, 2009a, 2009b). The critical role of genetic alterations in colorectal carcinogenesis has been long recognized (Vogelstein *et al.*, 1988; Kinzler and Vogelstein, 1996). More recently, the epigenetic component in cancer development and progression has been identified (Jones and Baylin, 2007; Esteller, 2008). Epigenetic alterations including locus-specific DNA methylation of cytosine–guanine (CpG) dinucleotides and islands, global DNA methylation, histone modifications, and microRNA production have been reported to be altered in cancer. The most commonly studied epigenetic alteration in cancer is aberrant DNA methylation. A landmark study in 1983 reported that tumor cells showed global DNA hypomethylation compared to normal cells (Feinberg and Vogelstein, 1983). Specifically, DNA hypomethylation of repetitive DNA elements such as long interspersed nucleotide element-1 (LINE-1) may be involved in genomic instability by the reactivation of transposable DNA sequences (Ji *et al.*, 1997; Gaudet *et al.*, 2003), leading to colorectal carcinogenesis (Karpf and Matsui, 2005; Yamada *et al.*, 2005; Rodriguez *et al.*, 2006; Suzuki *et al.*, 2006; Estecio *et al.*, 2007). More recently, the role of DNA hypermethylation at promoter CpG

---

* Author to whom correspondence should be addressed.

Epigenomics: From Chromatin Biology to Therapeutics, ed. K. Appasani. Published by Cambridge University Press. © Cambridge University Press 2012.

dinucleotides and islands has also been demonstrated in cancer (Baylin *et al.*, 1986, 1987; Jones *et al.*, 1990). Aberrant DNA methylation of CpG promoter regions contributes to colorectal cancer by silencing key tumor suppressor genes, thus providing an additional mechanism of gene inactivation (Esteller, 2008; Grady and Carethers, 2008). In 1999, Toyota and colleagues characterized the CpG island methylator phenotype (CIMP) by the presence of widespread promoter CpG island methylation (Toyota *et al.*, 1999; Shen *et al.*, 2007b). This phenotype has been shown to be inversely associated with LINE-1 hypomethylation in colorectal cancer (Ogino *et al.*, 2008c). Colorectal tumors that are CIMP-high have a distinctive profile, including associations with older age at diagnosis, female gender, proximal tumor location, poor differentiation, microsatellite instability-high (MSI-high), and *BRAF* mutation (Kambara *et al.*, 2004; Samowitz *et al.*, 2005a; Weisenberger *et al.*, 2006; Kawasaki *et al.*, 2007; Slattery *et al.*, 2007; Nosho *et al.*, 2008; Curtin *et al.*, 2011). Recently, low-level CIMP, distinct from CIMP-high and CIMP-negative, has been shown to be associated with *KRAS* mutation (Ogino *et al.*, 2006e) in multiple studies (Barault *et al.*, 2008; Kim *et al.*, 2009; Dahlin *et al.*, 2010; Yagi *et al.*, 2010). CIMP-low shares common features with CIMP2 (Shen *et al.*, 2007b), and is variably termed intermediate methylation epigenotype (IME) (Yagi *et al.*, 2010). Other features of CIMP-low include associations with *MGMT* methylation and loss of expression (Ogino *et al.*, 2007d) low-level methylation in CIMP-high-specific CpG islands (Kawasaki *et al.*, 2008), and distinct CpG island methylation correlation structure (Tanaka *et al.*, 2010).

DNA methylation is reported to be reversible and thus can be a potential target for epigenetic therapy or chemoprevention (Jones and Baylin, 2007; Esteller, 2008). Given the importance of epigenetic aberrations in the development and progression of colorectal cancer, further cancer epigenetic epidemiologic research (Jablonka, 2004; Dumitrescu, 2009; Foley *et al.*, 2009; Kumar and Verma, 2009; Nise *et al.*, 2010; Relton and Davey Smith, 2010; Talens *et al.*, 2010; Ulrich and Grady, 2010; Ogino and Stampfer, 2010; Ogino *et al.*, 2011) will contribute to our understanding of colorectal cancer etiology and ultimately prevention of colorectal cancer.

34.1.2   Gobal DNA methylation level as measured in the LINE-1 repetitive element and risk of colorectal cancer and adenoma

Repetitive nucleotide elements such as LINE-1, Alu, and Satellite (Sat)-2 contain 33% of all CpG dinucleotides and comprise 30% of the human genome. Measurement of LINE-1 methylation has been emerged as a cost-effective and high-throughput method with a low DNA input to assess global DNA methylation status in population-based studies (Chalitchagorn *et al.*, 2004; Yang *et al.*, 2004; Matsuzaki *et al.*, 2005; Weisenberger *et al.*, 2005). The methylation of LINE-7 level has been shown to be a surrogate indicator of 5-methylcytosine content, i.e., cellular DNA methylation level (Weisenberger *et al.*, 2005; Yang *et al.*, 2006). In particular, LINE-1 methylation assay using pyrosequencing technology has been shown to be precise and highly reproducible (Yang *et al.*, 2004; Irahara *et al.*, 2009) and can be effectively implemented in large-scale epidemiologic studies using blood samples and paraffin-embedded normal colon and colon tumor tissue (Estecio *et al.*, 2007; Ogino *et al.*, 2008c, 2008d; Aparicio *et al.*, 2009; Figueiredo *et al.*, 2009; Nosho *et al.*, 2009a).

Sequences of LINE-1 correlate with increased chromosomal rearrangements (Feinberg and Tycko, 2004), influence transcriptional activity, and contribute to non-coding RNA expression (Peaston *et al.*, 2004; Faulkner *et al.*, 2009), and may provide alternative promoters (Speek, 2001) and correlate with chromatin modifications. Moreover, retrotransposons activated by DNA hypomethylation may transpose themselves throughout the genome, leading to gene disruptions (Han *et al.*, 2004) and possibly chromosomal instability (Yamada *et al.*, 2005; Howard *et al.*, 2008).

34.1.3 One-carbon nutrients, DNA methylation, and risk of colorectal cancer and adenoma

Waterland and colleagues demonstrated that exposing pregnant Agouti mice to methyl donors (e.g., folate, vitamin B12, and choline) altered the methylation state of metastable epialleles (Waterland and Jirtle, 2003, 2004). Environmental and lifestyle factors, such as diet and age, contribute to epigenetic variation in risk of human colorectal cancer (Liu *et al.*, 2006; Slattery *et al.*, 2007). Low dietary folate consumption prior to the presence of preneoplastic lesions, especially among alcohol consumers, is associated with increased risk of colorectal cancer (Giovannucci *et al.*, 1995, 1998; Boutron-Ruault *et al.*, 1996; Glynn *et al.*, 1996; Slattery *et al.*, 1997; Su and Arab, 2001; Giovannucci, 2004) and adenoma (Benito *et al.*, 1991; Giovannucci *et al.*, 1993; Bird *et al.*, 1995; Boutron-Ruault *et al.*, 1996; Tseng *et al.*, 1996). Folate from diet is directly linked to DNA methylation via the one-carbon metabolism pathway, where S-adenosylmethionine (SAM) (Alonso-Aperte *et al.*, 2008) is the universal methyl donor for several biological methylation reactions and for *de novo* deoxynucleoside triphosphate synthesis. The reduced availability of methyltetrahydrofolate (methyl-THF), the main circulating form of folate, decreases the biosynthesis of SAM, thus limiting the availability of methyl groups for methylation reactions (Friso *et al.*, 2002) and increases misincorporation of uracil for thymidine during DNA synthesis (Wickramasinghe and Fisa, 1994), leading to mutations and chromosomal breaks (Blount *et al.*, 1997). Notably, folate deficiency is associated with an increase in *de novo* DNA methyltransferase activity, suggesting that there may be an increase in site-specific CpG island methylation during states of methyl-group deficiency (Pogribny *et al.*, 1995, 1997). In a small, nested case–control study, colorectal cancer patients with low folate/high alcohol intake were more likely to have promoter hypermethylation than patients with high folate/low alcohol intake (van Engeland *et al.*, 2003). Moreover, the rs1801131 single nucleotide polymorphism (SNP) (codon 429, c.1286A>C, p.E429A) variant in methylenetetrahydrofolate reductase (*MTHFR*), a one-carbon metabolizing enzyme gene, has been associated with colon cancer with CpG island methylation (Curtin *et al.*, 2007; Hazra *et al.*, 2010). Thus, a balance between various metabolic intermediates of methyl-group (including THF, 5-methyl-THF, and 5,10-methylene-THF) seems to influence DNA methylation reaction. An animal study also supports a link between folate deficiency and global DNA hypomethylation in colon epithelium (Linhart *et al.*, 2009). In a randomized, double-blind, placebo-controlled study, folic acid supplementation was inversely associated with global DNA hypomethylation in normal colon mucosa (Pufulete *et al.*, 2005). In addition, folate supplementation has been shown to induce global DNA hypermethylation and limit the aggressiveness of glioma cells (Hervouet *et al.*, 2009). Notably, in colorectal cancer cell lines, global DNA hypomethylation was induced by folate depletion and reversed by folate supplementation (Wasson *et al.*, 2006). Of note, the Aspirin/Folate Polyp Prevention Trial has shown no significant association in LINE-1 methylation levels in normal colon mucosa and randomization to folic acid supplementation ($n = 388$) (Figueiredo *et al.*, 2009).

Several additional nutrients in the one-carbon metabolism pathway, including vitamin B_6, vitamin B_{12}, choline, and methionine, have been examined in relation to colorectal cancer risks, although further studies are needed (Giovannucci *et al.*, 1995, 1998; Boutron-Ruault *et al.*, 1996; Chen *et al.*, 1996; Glynn *et al.*, 1996; Ma *et al.*, 1997; Slattery *et al.*, 1997; Su and Arab, 2001; Wei *et al.*, 2005).

34.1.4   One-carbon nutrients, alcohol intake, and clinical outcome of cancer patients

In animal models, dietary and lifestyle factors significantly influence tumor progression and cancer survival (Yang *et al.*, 2001). With folic acid fortification having begun in 1997–8, and the increasing use of folic acid-containing supplements, the possible growth-promoting effect of folic acid on neoplasia lesions needs further study (Mason *et al.*, 2007; Ulrich and Potter, 2007) Moreover, cancer patients may consume more supplements than healthy individuals (Rock, 2007), possibly placing these patients at an increased risk for cancer recurrence and mortality. Thus, the investigation on the prognostic effect of folate on colorectal cancer patients remains a high-priority research area (Ulrich and Potter, 2007).

34.2   Methodology used for epigenetic assays in population-based studies

34.2.1   Quantitative analysis of promoter CpG island methylation in colorectal cancer

The combination of sodium bisulfite conversion and quantitative real-time PCR assay (MethyLight) (Eads *et al.*, 2000) technology offers scalable capability for epidemiologic studies using formalin-fixed paraffin embedded (FFPE) tissue (Ogino *et al.*, 2006e) and blood samples.

34.2.2   Measurement of LINE-1 methylation

Bisulfite-PCR followed by quantitative pyrosequencing is commonly used to measure LINE-1 methylation levels (Yang *et al.*, 2004). This LINE-1 methylation assay is precise, and has a coefficient of variation of 3–4% (Irahara *et al.*, 2009). The pyrosequencing assay measures LINE-1 methylation levels using a validated

146-bp amplicon representative of the genome (Ogino *et al.*, 2008a, 2008c; Irahara *et al.*, 2009; Baba *et al.*, 2010; Hazra *et al.*, 2010).

Alternative bisulfite-conversion-based methods for locus-specific analysis with a low sample requirement for FFPE tissue and blood samples include: pyrosequencing (Yang *et al.*, 2004; Choi *et al.*, 2007; Aparicio *et al.*, 2009; Figueiredo *et al.*, 2009; Wallace *et al.*, 2010) and EpiTyper (Mikeska *et al.*, 2007; Sepulveda *et al.*, 2009; Laird, 2010). Emerging genome-scale technologies using sodium bisulfite conversion, affinity enrichment, or enzyme digestion (Laird, 2010), including arrays and sequencing-based methods, need to be evaluated for reproducibility and reliability on various sample types including FFPE tissue in epidemiology and translational study settings.

34.3 Results on genomics and epigenetics in colorectal cancer

34.3.1 Global DNA (LINE-1) hypomethylation

Global DNA hypomethylation in colonic cells may lead to colorectal cancer by activating proto-oncogenes. In colorectal tumors, global DNA (LINE-1) methylation levels are reported to be highly variable (Estecio *et al.*, 2007; Ogino *et al.*, 2008c), and distributed approximately normally (Ogino *et al.*, 2008a). Recent data suggest that synchronous colorectal cancers (defined as two separate primary colorectal cancers in a single individual) exhibit a significant correlation of LINE-1 methylation levels (Nosho *et al.*, 2009a). This finding suggests the presence of a non-stochastic component and the possible contribution of genetic/environmental factors to the wide variability of LINE-1 methylation (Ogino *et al.*, 2008). Alternatively, the field effect of DNA hypomethylation may explain the observed concordant LINE-1 hypomethylation in synchronous colorectal cancers. In addition, global DNA (LINE-1) hypomethylation in colorectal cancer is associated positively with chromosomal instability (Matsuzaki *et al.*, 2005; Ogino *et al.*, 2008c) and inversely with microsatellite instability (MSI) (Estecio *et al.*, 2007; Ogino *et al.*, 2008c) and CIMP (Ogino *et al.*, 2008a). Moreover, studies have demonstrated that global DNA (LINE-1) hypomethylation in colon cancer is associated with high mortality (Frigola *et al.*, 2005; Ogino *et al.*, 2008c; Ahn *et al.*, 2011). Therefore, further studies on the potential etiologic factors for global DNA hypomethylation are needed.

34.3.2 CpG island methylator phenotype

Transcriptional silencing of tumor-suppressor genes by aberrant promoter CpG island methylation is a critical mechanism in tumor development and progression (Jones and Baylin, 2007; Esteller, 2008). Approximately 20% of colorectal cancers exhibit widespread CpG island methylation, termed the CpG island methylator phenotype (CIMP) (Toyota *et al.*, 1999, 2000; Van Rijnsoever *et al.*, 2002, 2003; Samowitz *et al.*, 2005a; Ogino *et al.*, 2006d). Inactivation of DNA mismatch repair gene *MLH1* by promoter DNA methylation, leading to microsatellite instability, is a common feature of colorectal cancers with a high degree of CIMP (CIMP-high) (Toyota *et al.*, 2000; Hawkins *et al.*, 2002; van Rijnsoever *et al.*, 2002; Samowitz *et al.*, 2005a; Ogino *et al.*, 2006d). CIMP-high colorectal cancers are associated with female gender, older age, proximal tumor location, poor differentiation, *BRAF* mutation, wild-type *TP53*, stable chromosomes, high-level LINE-1 methylation, and inactive β-catenin/Wnt (Toyota *et al.*, 1999, 2000; van Rijnsoever *et al.*, 2002, 2003; Samowitz *et al.*, 2005b), independent of MSI status (Samowitz *et al.*, 2005a; Ogino *et al.*, 2007c; Nosho *et al.*, 2008). Although the existence of CIMP as a distinct phenomenon has gained acceptance (Issa *et al.*, 2005; Samowitz *et al.*, 2005a; Curtin *et al.*, 2009), the cause and etiology of CIMP remain poorly understood. Recent work has shown that expression of *dnmt3b* (DNA methyltransferase 3B) in the mouse colon can induce colon tumors with specific CpG island methylation (Linhart *et al.*, 2007; Steine *et al.*, 2011). In human studies, DNMT3B expression is associated with CIMP-high in human colorectal cancer (Nosho *et al.*, 2009b) and CpG island methylation in adenoma (Ibrahim *et al.*, 2011). These data support a role of DNMT3B as an etiologic factor for CIMP-high colorectal tumors. However, the reports have been conflicting on the prognostic significance of CIMP in colorectal cancer. Since CIMP-high is strongly associated with both MSI-high (associated with superior survival: Popat and Houlston, 2005) and *BRAF* mutation (associated with reduced survival: Samowitz *et al.*, 2005b; Ferracin *et al.*, 2008; French *et al.*, 2008; Kim *et al.*, 2009; Ogino *et al.*, 2009c; Roth *et al.*, 2010), studies should include both factors as potential confounders in the statistical models (Ogino *et al.*, 2009c). Because MSI, CIMP, and *BRAF* mutations are interrelated, studies with a large sample size (e.g. $n > 500$ cases with 200 events) are needed to control for each confounding factor and have meaningful data. In other studies, CIMP-high has also been associated with worse patient survival among microsatellite stable cancers (Hawkins *et al.*, 2002; Ogino *et al.*, 2007e; Shen *et al.*, 2007a) which were likely due to the confounding effect of *BRAF* mutation; those earlier studies did not examine this effect of *BRAF* mutation (Hawkins *et al.*, 2002; van Rijnsoever *et al.*, 2003; Ogino *et al.*, 2007e; Shen *et al.*, 2007a). Utilizing the Nurses' Health Study (NHS) and Health Professional Follow-up Study (HPFS) cohorts, quantitative DNA methylation analysis (Eads *et al.*, 2000; Ogino *et al.*, 2006e) and a validated CIMP marker panel (Weisenberger *et al.*, 2006; Ogino *et al.*, 2007b; Nosho *et al.*, 2008) Ogino and colleagues have shown that CIMP-high is associated with superior survival after adjusting for tumor stage, MSI, and *BRAF* mutation (Ogino *et al.*, 2009c).

34.4   Summary

Epigenetic changes have been shown to be important in colorectal carcinogenesis and potentially reversible molecular targets for therapy or chemoprevention. Further epigenetic epidemiologic studies are needed to link dietary factors related to one-carbon metabolism, cellular epigenetic alterations, and hypothesized pathogenic mechanisms (Jablonka, 2004; Dumitrescu, 2009; Foley *et al.*, 2009; Relton and Davey Smith, 2010; Talens *et al.*, 2010).

REFERENCES

Ahn, J. B., Chung, W. B., Maeda, O., *et al.* (2011). DNA methylation predicts recurrence from resected stage III proximal colon cancer. *Cancer*, **117**, 1847–1854.

Alonso-Aperte, E., Gonzalez, M. P., Poo-Prieto, R., and Varela-Moreiras, G. (2008). Folate status and S-adenosylmethionine/S-adenosylhomocysteine ratio in colorectal adeno-carcinoma in humans. *European Journal of Clinical Nutrition*, **62**, 295–298.

Aparicio, A., North, B., Barske, L., *et al.* (2009). LINE-1 methylation in plasma DNA as a biomarker of activity of DNA methylation inhibitors in patients with solid tumors. *Epigenetics*, **4**, 176–184.

Baba, Y., Nosho, K., Shima, K., *et al.* (2009). Relationship of CDX2 loss with molecular features and prognosis in colorectal cancer. *Clinical Cancer Research*, **15**, 4665–4673.

Baba, Y., Huttenhower, C., Nosho, K., *et al.* (2010). Epigenomic diversity of colorectal cancer indicated by LINE-1 methylation in a database of 869 tumors. *Molecular Cancer*, **9**, 125.

Barault, L., Charon-Barra, C., Jooste, V., *et al.* (2008). Hypermethylator phenotype in spora-dic colon cancer: study on a population-based series of 582 cases. *Cancer Research*, **68**, 8541–8546.

Baylin, S. B., Hoppener, J. W., De Bustros, A., *et al.* (1986). DNA methylation patterns of the calcitonin gene in human lung cancers and lymphomas. *Cancer Research*, **46**, 2917–2922.

Baylin, S. B., Fearon, E. R., Vogelstein, B., *et al.* (1987). Hypermethylation of the 5' region of the calcitonin gene is a property of human lymphoid and acute myeloid malignan-cies. *Blood*, **70**, 412–417.

Benito, E., Stigglebout, A., Bosch, F., *et al.* (1991). Nutritional factors in colorectal cancer risk: a case-control study in Majorica. *International Journal of Cancer*, **49**, 161–167.

Bird, C. L., Swendseid, M. E., Witte, J. S., *et al.* (1995). Red cell and plasma folate, folate consumption, and the risk of colorectal adenomatous polyps. *Cancer Epidemiology, Biomarkers and Prevention*, **4**, 709–714.

Blount, B., Mack, M., Wehr, C., *et al.* (1997). Folate deficiency causes uracil misincorporation into human DNA and chromosome breakage: implications for cancer and neuronal damage. *Proceedings of the National Academy of Sciences USA*, **94**, 3290–3295.

Boutron-Ruault, M. C., Senesse, P., Faivre, J., Couillault, C., and Belghiti, C. (1996). Folate and alcohol intakes: related or independent roles in the adenoma-carcinoma sequence? *Nutrition and Cancer*, **26**, 337–346.

Chalitchagorn, K., Shuangshoti, S., Hourpai, N., *et al.* (2004). Distinctive pattern of LINE-1 methylation level in normal tissues and the association with carcinogenesis. *Oncogene*, **23**, 8841–8846.

Chan, A. T., Ogino, S., and Fuchs, C. S. (2007). Aspirin and the risk of colorectal cancer in relation to the expression of COX-2. *New England Journal of Medicine*, **356**, 2131–2142.

Chan, A. T., Ogino, S., and Fuchs, C. S. (2009). Aspirin use and survival after diagnosis of colorectal cancer. *Journal of the American Medical Association*, **302**, 649–658.

Chen, J., Giovannucci, E., Kelsey, K., *et al.* (1996). A methylenetetrahydrofolate reductase polymorphism and the risk of colorectal cancer. *Cancer Research*, **56**, 4862–4864.

Choi, I. S., Estecio, M. R., Nagano, Y., *et al.* (2007). Hypomethylation of LINE-1 and Alu in well-differentiated neuroendocrine tumors (pancreatic endocrine tumors and carci-noid tumors). *Modern Pathology*, **20**, 802–810.

Curtin, K., Slattery, M. L., Ulrich, C. M., *et al.* (2007). Genetic polymorphisms in one-carbon metabolism: associations with CpG island methylator phenotype (CIMP) in colon cancer and the modifying effects of diet. *Carcinogenesis*, **28**, 1672–1679.

Curtin, K., Samowitz, W. S., Wolff, R. K., *et al.* (2009). MSH6 G39E polymorphism and CpG island methylator phenotype in colon cancer. *Molecular Carcinogenesis*, **48**, 989–994.

Curtin, K., Slattery, M. L., and Samowitz, W. S. (2011). CpG Island methylation in colorectal cancer: past, present and future. *Pathology Research International*, **2011**, 902674.

Dahlin, A. M., Palmqvist, R., Henriksson, M. L., *et al.* (2010). The role of the CpG island methylator phenotype in colorectal cancer prognosis depends on microsatellite insta-bility screening status. *Clinical Cancer Research*, **16**, 1845–1855.

Dumitrescu, R. G. (2009). Epigenetic targets in cancer epidemiology. *Methods in Molecular Biology*, **471**, 457–467.

Eads, C. A., Danenberg, K. D., Kawakami, K., *et al.* (2000). MethyLight: a high-throughput assay to measure DNA methylation. *Nucleic Acids Research*, **28**, e32.

Estecio, M. R., Gharibyan, V., Shen, L., *et al.* (2007). LINE-1 hypomethylation in cancer is highly variable and inversely correlated with microsatellite instability. *PLoS One*, **2**, e399.

Esteller, M. (2008). Epigenetics in cancer. *New England Journal of Medicine*, **358**, 1148–1159.

Faulkner, G. J., Kimura, Y., Daub, C. O., *et al.* (2009). The regulated retrotransposon transcriptome of mammalian cells. *Nature Genetics*, **41**, 563–571.

Feinberg, A. P. and Tycko, B. (2004). The history of cancer epigenetics. *Nature Reviews Cancer*, **4**, 143–153.

Feinberg, A. P. and Vogelstein, B. (1983). Hypomethylation distinguishes genes of some human cancers from their normal counterparts. *Nature*, **301**, 89–92.

Ferracin, M., Gafa, R., Miotto, E., *et al.* (2008). The methylator phenotype in microsatellite stable colorectal cancers is characterized by a distinct gene expression profile. *Journal of Pathology*, **214**, 594–602.

Figueiredo, J. C., Grau, M. V., Wallace, K., *et al.* (2009). Global DNA hypomethylation (LINE-1) in the normal colon and lifestyle characteristics and dietary and genetic factors. *Cancer Epidemiology, Biomarkers and Prevention*, **18**, 1041–1049.

Foley, D. L., Craig, J. M., Morley, R., *et al.* (2009). Prospects for epigenetic epidemiology. *American Journal of Epidemiology*, **169**, 389–400.

French, A. J., Sargent, D. J., Burgart, L. J., *et al.* (2008). Prognostic significance of defective mismatch repair and *BRAF* V600E in patients with colon cancer. *Clinical Cancer Research*, **14**, 3408–3415.

Frigola, J., Sole, X., Paz, M. F., *et al.* (2005). Differential DNA hypermethylation and hypomethylation signatures in colorectal cancer. *Human Molecular Genetics*, **14**, 319–326.

Friso, S., Choi, S. W., Girelli, D., *et al.* (2002). A common mutation in the 5,10-methylenetetrahydrofolate reductase gene affects genomic DNA methylation through an interaction with folate status. *Proceedings of the National Academy of Sciences USA*, **99**, 5606–5611.

Gaudet, F., Hodgson, J. G., Eden, A., *et al.* (2003). Induction of tumors in mice by genomic hypomethylation. *Science*, **300**, 489–492.

Giovannucci, E. (2004). Alcohol, one-carbon metabolism, and colorectal cancer: recent insights from molecular studies. *Journal of Nutrition*, **134**, 2475S–2481S.

Giovannucci, E., Stampfer, M. J., Colditz, G. A., *et al.* (1993). Folate, methionine, and alcohol intake and risk of colorectal adenoma. *Journal of the National Cancer Institute*, **85**, 875–884.

Giovannucci, E., Rimm, E. B., Ascherio, A., *et al.* (1995). Alcohol, low-methionine–low-folate diets, and risk of colon cancer in men [see comments]. *Journal of the National Cancer Institute*, **87**, 265–273.

Giovannucci, E., Stampfer, M. J., Colditz, G. A., *et al.* (1998). Multivitamin use, folate, and colon cancer in women in the Nurses' Health Study. *Annals of Internal Medicine*, **129**, 517–524.

Glynn, S. A., Albanes, D., Pietinen, P., *et al.* (1996). Colorectal cancer and folate status: a nested case-control study among smokers. *Cancer Epidemiology, Biomarkers and Prevention*, **5**, 487–494.

Grady, W. M. and Carethers, J. M. (2008). Genomic and epigenetic instability in colorectal cancer pathogenesis. *Gastroenterology*, **135**, 1079–1099.

Han, J. S., Szak, S. T., and Boeke, J. D. (2004). Transcriptional disruption by the L1 retrotransposon and implications for mammalian transcriptomes. *Nature*, **429**, 268–274.

Hawkins, N., Norrie, M., Cheong, K., *et al.* (2002). CpG island methylation in sporadic colorectal cancers and its relationship to microsatellite instability. *Gastroenterology*, **122**, 1376–1387.

Hazra, A., Fuchs, C. S., Kawasaki, T., *et al.* (2010). Germline polymorphisms in the one-carbon metabolism pathway and DNA methylation in colorectal cancer. *Cancer Causes Control*, **21**, 331–345.

Hervouet, E., Debien, E., Campion, L., *et al.* (2009). Folate supplementation limits the aggressiveness of glioma via the remethylation of DNA repeats element and genes governing apoptosis and proliferation. *Clinical Cancer Research*, **15**, 3519–3529.

Howard, G., Eiges, R., Gaudet, F., Jaenisch, R., and Eden, A. (2008). Activation and transposition of endogenous retroviral elements in hypomethylation induced tumors in mice. *Oncogene*, **27**, 404–408.

Ibrahim, A. E., Arends, M. J., Silva, A. L., *et al.* (2011). Sequential DNA methylation changes are associated with DNMT3B overexpression in colorectal neoplastic progression. *Gut*, **60**, 499–508.

Irahara, N., Nosho, K., Baba, Y., *et al.* (2009). Precision of pyrosequencing assay to measure LINE-1 methylation in colon cancer, normal colonic mucosa, and peripheral blood cells. *Journal of Molecular Diagnostics*, **12**, 177–183.

Issa, J. P., Shen, L., and Toyota, M. (2005). CIMP, at last. *Gastroenterology*, **129**, 1121–1124.

Jablonka, E. (2004). Epigenetic epidemiology. *International Journal of Epidemiology*, **33**, 929–935.

Jemal, A., Siegel, R., Xu, J., and Ward, E. (2010). Cancer statistics, 2010. *CA Cancer Journal for Clinicians*, **60**, 277–300.

Jemal, A., Bray, F., Center, M. M., *et al.* (2011). Global cancer statistics. *CA Cancer Journal for Clinicians*, **61**, 69–90.

Ji, W., Hernandez, R., Zhang, X. Y., *et al.* (1997). DNA demethylation and pericentromeric rearrangements of chromosome 1. *Mutation Research*, **379**, 33–41.

Jones, P. A. and Baylin, S. B. (2007). The epigenomics of cancer. *Cell*, **128**, 683–692.

Jones, P. A., Wolkowicz, M. J., Rideout, W. M. III, *et al.* (1990). De novo methylation of the MyoD1 CpG island during the establishment of immortal cell lines. *Proceedings of the National Academy of Sciences USA*, **87**, 6117–6121.

Kambara, T., Simms, L. A., Whitehall, V. L., *et al.* (2004). *BRAF* mutation is associated with DNA methylation in serrated polyps and cancers of the colorectum. *Gut*, **53**, 1137–1144.

Karpf, A. R. and Matsui, S. (2005). Genetic disruption of cytosine DNA methyltransferase enzymes induces chromosomal instability in human cancer cells. *Cancer Research*, **65**, 8635–8639.

Kawasaki, T., Nosho, K., Ohnishi, M., *et al.* (2007). IGFBP3 promoter methylation in colorectal cancer: relationship with microsatellite instability, CpG island methylator phenotype, and p53. *Neoplasia*, **9**, 1091–1098.

Kawasaki, T., Ohnishi, M., Nosho, K., *et al.* (2008). CpG island methylator phenotype-low (CIMP-low) colorectal cancer shows not only few methylated CIMP-high-specific CpG islands, but also low-level methylation at individual loci. *Modern Pathology*, **21**, 245–255.

Kim, J. H., Shin, S. H., Kwon, H. J., Cho, N. Y., and Kang, G. H. (2009). Prognostic implications of CpG island hypermethylator phenotype in colorectal cancers. *Virchows Archive*, **455**, 485–494.

Kinzler, K. W. and Vogelstein, B. (1996). Lessons from hereditary colorectal cancer. *Cell*, **87**, 159–170.

Kumar, D. and Verma, M. (2009). Methods in cancer epigenetics and epidemiology. *Methods in Molecular Biology*, **471**, 273–288.

Laird, P. W. (2010). Principles and challenges of genomewide DNA methylation analysis. *Nature Reviews Genetics*, **11**, 191–203.

Linhart, H. G., Lin, H., Yamada, Y., *et al.* (2007). Dnmt3b promotes tumorigenesis in vivo by gene-specific de novo methylation and transcriptional silencing. *Genes and Development*, **21**, 3110–3122.

Linhart, H. G., Troen, A., Bell, G. W., *et al.* (2009). Folate deficiency induces genomic uracil misincorporation and hypomethylation but does not increase DNA point mutations. *Gastroenterology*, **136**, 227–235.

Liu, Y., Lan, Q., Siegfried, J. M., Luketich, J. D., and Keohavong, P. (2006). Aberrant promoter methylation of p16 and MGMT genes in lung tumors from smoking and never-smoking lung cancer patients. *Neoplasia*, **8**, 46–51.

Ma, J., Stampfer, M. J., Giovannucci, E., *et al.* (1997). Methylenetetrahydrofolate reductase polymorphism, dietary interactions and risk of colorectal cancer. *Cancer Research*, **57**, 1098–1102.

Mason, J. B., Dickstein, A., Jacques, P. F., *et al.* (2007). A temporal association between folic acid fortification and an increase in colorectal cancer rates may be illuminating important biological principles: a hypothesis. *Cancer Epidemiology, Biomarkers and Prevention*, **16**, 1325–1329.

Matsuzaki, K., Deng, G., Tanaka, H., *et al.* (2005). The relationship between global methylation level, loss of heterozygosity, and microsatellite instability in sporadic colorectal cancer. *Clinical Cancer Research*, **11**, 8564–8569.

Meyerhardt, J. A., Ogino, S., Kirkner, G. J., *et al.* (2009). Interaction of molecular markers and physical activity on mortality in patients with colon cancer. *Clinical Cancer Research*, **15**, 5931–5936.

Mikeska, T., Bock, C., El-Maarri, O., *et al.* (2007). Optimization of quantitative MGMT promoter methylation analysis using pyrosequencing and combined bisulfite restriction analysis. *Journal of Molecular Diagnostics*, **9**, 368–381.

Nise, M. S., Falaturi, P., and Erren, T. C. (2010). Epigenetics: origins and implications for cancer epidemiology. *Medical Hypotheses*, **74**, 377–382.

Nosho, K., Irahara, N., Shima, K., *et al.* (2008). Comprehensive biostatistical analysis of CpG island methylator phenotype in colorectal cancer using a large population-based sample. *PLoS One*, **3**, e3698.

Nosho, K., Kure, S., Irahara, N., *et al.* (2009a). A prospective cohort study shows unique epigenetic, genetic, and prognostic features of synchronous colorectal cancers. *Gastroenterology*, **137**, 1609–1620.

Nosho, K., Shima, K., Irahara, N., *et al.* (2009b). DNMT3B expression might contribute to CpG island methylator phenotype in colorectal cancer. *Clinical Cancer Research*, **15**, 3663–3671.

Ogino, S. and Stampfer, M. (2010). Lifestyle factors and microsatellite instability in colorectal cancer: the evolving field of molecular pathological epidemiology. *Journal of the National Cancer Institute*, **102**, 365–367.

Ogino, S., Brahmandam, M., Cantor, M., *et al.* (2006a). Distinct molecular features of colorectal carcinoma with signet ring cell component and colorectal carcinoma with mucinous component. *Modern Pathology*, **19**, 59–68.

Ogino, S., Brahmandam, M., Kawasaki, T., *et al.* (2006b). Combined analysis of COX-2 and p53 expressions reveals synergistic inverse correlations with microsatellite instability and CpG island methylator phenotype in colorectal cancer. *Neoplasia*, **8**, 458–464.

Ogino, S., Brahmandam, M., Kawasaki, T., *et al.* (2006c). Epigenetic profiling of synchronous colorectal neoplasias by quantitative DNA methylation analysis. *Modern Pathology*, **19**, 1083–1090.

Ogino, S., Cantor, M., Kawasaki, T., *et al.* (2006d). CpG island methylator phenotype (CIMP) of colorectal cancer is best characterised by quantitative DNA methylation analysis and prospective cohort studies. *Gut*, **55**, 1000–1006.

Ogino, S., Kawasaki, T., Brahmandam, M., *et al.* (2006e). Precision and performance characteristics of bisulfite conversion and real-time PCR (MethyLight) for quantitative DNA methylation analysis. *Journal of Molecular Diagnostics*, **8**, 209–217.

Ogino, S., Odze, R. D., Kawasaki, T., *et al.* (2006f). Correlation of pathologic features with CpG island methylator phenotype (CIMP) by quantitative DNA methylation analysis in colorectal carcinoma. *American Journal of Surgical Pathology*, **30**, 1175–1183.

Ogino, S., Hazra, A., Tranah, G. J., *et al.* (2007a). MGMT germline polymorphism is associated with somatic MGMT promoter methylation and gene silencing in colorectal cancer. *Carcinogenesis*, **28**, 1985–1990.

Ogino, S., Kawasaki, T., Kirkner, G.J., et al. (2007b). Evaluation of markers for CpG island methylator phenotype (CIMP) in colorectal cancer by a large population-based sample. *Journal of Molecular Diagnostics*, **9**, 305–314.

Ogino, S., Kawasaki, T., Kirkner, G.J., Ohnishi, M., and Fuchs, C.S. (2007c). 18q loss of heterozygosity in microsatellite-stable colorectal cancer is correlated with CpG island methylator phenotype-negative (CIMP-0) and inversely with CIMP-low and CIMP-high. *BMC Cancer*, **7**, 72.

Ogino, S., Kawasaki, T., Kirkner, G.J., et al. (2007d). Molecular correlates with MGMT promoter methylation and silencing support CpG island methylator phenotype-low (CIMP-low) in colorectal cancer. *Gut*, **56**, 1564–1571.

Ogino, S., Meyerhardt, J.A., Kawasaki, T., et al. (2007e). CpG island methylation, response to combination chemotherapy, and patient survival in advanced microsatellite stable colorectal carcinoma. *Virchows Archive*, **450**, 529–537.

Ogino, S., Kawasaki, T., Nosho, K., et al. (2008a). LINE-1 hypomethylation is inversely associated with microsatellite instability and CpG island methylator phenotype in colorectal cancer. *International Journal of Cancer*, **122**, 2767–2773.

Ogino, S., Kirkner, G.J., Nosho, K., et al. (2008b). Cyclooxygenase-2 expression is an independent predictor of poor prognosis in colon cancer. *Clinical Cancer Research*, **14**, 8221–8227.

Ogino, S., Nosho, K., Kirkner, G.J., et al. (2008c). A cohort study of tumoral LINE-1 hypomethylation and prognosis in colon cancer. *Journal of the National Cancer Institute*, **100**, 1734–1738.

Ogino, S., Nosho, K., Meyerhardt, J.A., et al. (2008d). Cohort study of fatty acid synthase expression and patient survival in colon cancer. *Journal of Clinical Oncology*, **26**, 5713–5720.

Ogino, S., Nosho, K., Baba, Y., et al. (2009a). A cohort study of STMN1 expression in colorectal cancer: body mass index and prognosis. *American Journal of Gastroenterology*, **104**, 2047–2056.

Ogino, S., Nosho, K., Irahara, N., et al. (2009b). A cohort study of cyclin d1 expression and prognosis in 602 colon cancer cases. *Clinical Cancer Research*, **15**, 4431–4438.

Ogino, S., Nosho, K., Kirkner, G.J., et al. (2009c). CpG island methylator phenotype, microsatellite instability, *BRAF* mutation and clinical outcome in colon cancer. *Gut*, **58**, 90–96.

Ogino, S., Nosho, K., Kirkner, G.J., et al. (2009d). *PIK3CA* mutation is associated with poor prognosis among patients with curatively resected colon cancer. *Journal of Clinical Oncology*, **27**, 1477–1484.

Ogino, S., Shima, K., Baba, Y., et al. (2009e). Colorectal cancer expression of peroxisome proliferator-activated receptor-gamma (PPARG, PPARgamma) is associated with good prognosis. *Gastroenterology*, **136**, 1242–1250.

Ogino, S., Shima, K., Nosho, K., et al. (2009f). A cohort study of p27 localization in colon cancer, body mass index, and patient survival. *Cancer Epidemiology, Biomarkers and Prevention*, **18**, 1849–1858.

Ogino, S., Chan, A.T., Fuchs, C.S., and Giovannucci, E. (2011). Molecular pathological epidemiology of colorectal neoplasia: an emerging transdisciplinary and interdisciplinary field. *Gut*, **60**, 397–411.

Peaston, A.E., Evsikov, A.V., Graber, J.H., et al. (2004). Retrotransposons regulate host genes in mouse oocytes and preimplantation embryos. *Developmental Cell*, **7**, 597–606.

Pogribny, I.P., Basnakian, A.G., Miller, B.J., et al. (1995). Breaks in genomic DNA and within the p53 gene are associated with hypomethylation in livers of folate/methyl-deficient rats. *Cancer Research*, **55**, 1894–1901. [Published erratum appears in *Cancer Research*, **55**, 2711.]

Pogribny, I., Miller, B., and James, S. (1997). Alterations in hepatic p53 gene methylation patterns during tumor progression with folate/methyl deficiency in the rat. *Cancer Letters*, **115**, 31–38.

Popat, S. and Houlston, R. S. (2005). A systematic review and meta-analysis of the relationship between chromosome 18q genotype, DCC status and colorectal cancer prognosis. *European Journal of Cancer*, **41**, 2060–2070.

Pufulete, M., Al-Ghnaniem, R., Khushal, A., *et al.* (2005). Effect of folic acid supplementation on genomic DNA methylation in patients with colorectal adenoma. *Gut*, **54**, 648–653.

Relton, C. L. and Davey Smith, G. (2010). Epigenetic epidemiology of common complex disease: prospects for prediction, prevention, and treatment. *PLoS Medicine*, **7**, e1000356.

Rock, C. L. (2007). Multivitamin–multimineral supplements: who uses them? *American Journal of Clinical Nutrition*, **85**, 277S–279S.

Rodriguez, J., Frigola, J., Vendrell, E., *et al.* (2006). Chromosomal instability correlates with genome-wide DNA demethylation in human primary colorectal cancers. *Cancer Research*, **66**, 8462–8468.

Roth, A. D., Tejpar, S., Delorenzi, M., *et al.* (2010). Prognostic role of *KRAS* and *BRAF* in stage II and III resected colon cancer: results of the translational study on the PETACC-3, EORTC 40993, SAKK 60–00 trial. *Journal of Clinical Oncology*, **28**, 466–474.

Samowitz, W., Albertsen, H., Herrick, J., *et al.* (2005a). Evaluation of a large, population-based sample supports a CpG island methylator phenotype in colon cancer. *Gastroenterology*, **129**, 837–845.

Samowitz, W. S., Sweeney, C., Herrick, J., *et al.* (2005b). Poor survival associated with the *BRAF* V600E mutation in microsatellite-stable colon cancers. *Cancer Research*, **65**, 6063–6069.

Schernhammer, E. S., Giovannuccci, E., Fuchs, C. S., and Ogino, S. (2008a). A prospective study of dietary folate and vitamin B and colon cancer according to microsatellite instability and *KRAS* mutational status. *Cancer Epidemiology, Biomarkers and Prevention*, **17**, 2895–2898.

Schernhammer, E. S., Ogino, S., and Fuchs, C. S. (2008b). Folate intake and risk of colon cancer in relation to p53 status. *Gastroenterology*, **135**, 770–780.

Sepulveda, A. R., Jones, D., Ogino, S., *et al.* (2009). CpG methylation analysis: current status of clinical assays and potential applications in molecular diagnostics – a report of the Association for Molecular Pathology. *Journal of Molecular Diagnostics*, **11**, 266–278.

Shen, L., Catalano, P. J., Benson, A. B. III, *et al.* (2007a). Association between DNA methylation and shortened survival in patients with advanced colorectal cancer treated with 5-fluorouracil-based chemotherapy. *Clinical Cancer Research*, **13**, 6093–6098.

Shen, L., Toyota, M., Kondo, Y., *et al.* (2007b). Integrated genetic and epigenetic analysis identifies three different subclasses of colon cancer. *Proceedings of the National Academy of Sciences USA*, **104**, 18 654–18 659.

Slattery, M. L., Schaffer, D., Edwards, S. L., Ma, K. N., and Potter, J. D. (1997). Are dietary factors involved in DNA methylation associated with colon cancer? *Nutrition and Cancer*, **28**, 52–62.

Slattery, M. L., Curtin, K., Sweeney, C., *et al.* (2007). Diet and lifestyle factor associations with CpG island methylator phenotype and *BRAF* mutations in colon cancer. *International Journal of Cancer*, **120**, 656–663.

Speek, M. (2001). Antisense promoter of human L1 retrotransposon drives transcription of adjacent cellular genes. *Molecular and Cellular Biology*, **21**, 1973–1985.

Steine, E. J., Ehrich, M., Bell, G. W., *et al.* (2011). Genes methylated by DNA methyltransferase 3b are similar in mouse intestine and human colon cancer. *Journal of Clinical Investigation*, **121**, 1748–1752.

Su, L. J. and Arab, L. (2001). Nutritional status of folate and colon cancer risk: evidence from NHANES I epidemiologic follow-up study. *Annals of Epidemiology*, **11**, 65–72.

Suzuki, K., Suzuki, I., Leodolter, A., *et al.* (2006). Global DNA demethylation in gastrointestinal cancer is age dependent and precedes genomic damage. *Cancer Cell*, **9**, 199–207.

Talens, R. P., Boomsma, D. I., Tobi, E. W., *et al.* (2010). Variation, patterns, and temporal stability of DNA methylation: considerations for epigenetic epidemiology. *FASEB Journal*, **24**, 3135–3144.

Tanaka, N., Huttenhower, C., Nosho, K., *et al.* (2010). Novel application of structural equation modeling to correlation structure analysis of CpG island methylation in colorectal cancer. *American Journal of Pathology*, **177**, 2731–2740.

Toyota, M., Ahuja, N., Ohe-Toyota, M., *et al.* (1999). CpG island methylator phenotype in colorectal cancer. *Proceedings of the National Academy of Sciences USA*, **96**, 8681–8686.

Toyota, M., Ohe-Toyota, M., Ahuja, N., and Issa, J. P. (2000). Distinct genetic profiles in colorectal tumors with or without the CpG island methylator phenotype. *Proceedings of the National Academy of Sciences USA*, **97**, 710–715.

Tseng, M., Murray, S. C., Kupper, L. L., and Sandler, R. S. (1996). Micronutrients and the risk of colorectal adenomas. *American Journal of Epidemiology*, **144**, 1005–1014.

Ulrich, C. M. and Grady, W. M. (2010). Linking epidemiology to epigenomics – where are we today? *Cancer Prevention Research*, **3**, 1505–1508.

Ulrich, C. M. and Potter, J. D. (2007). Folate and cancer: timing is everything. *Journal of the American Medical Association*, **297**, 2408–2409.

Van Engeland, M., Weijenberg, M. P., Roemen, G. M., *et al.* (2003). Effects of dietary folate and alcohol intake on promoter methylation in sporadic colorectal cancer: the Netherlands cohort study on diet and cancer. *Cancer Research*, **63**, 3133–3137.

Van Rijnsoever, M., Grieu, F., Elsaleh, H., Joseph, D., and Iacopetta, B. (2002). Characterisation of colorectal cancers showing hypermethylation at multiple CpG islands. *Gut*, **51**, 797–802.

Van Rijnsoever, M., Elsaleh, H., Joseph, D., Mccaul, K., and Iacopetta, B. (2003). CpG island methylator phenotype is an independent predictor of survival benefit from 5-fluorouracil in stage III colorectal cancer. *Clinical Cancer Research*, **9**, 2898–2903.

Vogelstein, B., Fearon, E. R., Hamilton, S. R., *et al.* (1988). Genetic alterations during colorectal tumor development. *New England Journal of Medicine*, **319**, 525–532.

Wallace, K., Grau, M. V., Levine, A. J., *et al.* (2010). Association between folate levels and CpG island hypermethylation in normal colorectal mucosa. *Cancer Prevention Research*, **3**, 1552–1564.

Wasson, G. R., McGlynn, A. P., McNulty, H., *et al.* (2006). Global DNA and p53 region-specific hypomethylation in human colonic cells is induced by folate depletion and reversed by folate supplementation. *Journal of Nutrition*, **136**, 2748–2753.

Waterland, R. A. and Jirtle, R. L. (2003). Transposable elements: targets for early nutritional effects on epigenetic gene regulation. *Molecular and Cellular Biology*, **23**, 5293–5300.

Waterland, R. A. and Jirtle, R. L. (2004). Early nutrition, epigenetic changes at transposons and imprinted genes, and enhanced susceptibility to adult chronic diseases. *Nutrition*, **20**, 63–68.

Wei, E. K., Giovannucci, E., Selhub, J., *et al.* (2005). Plasma vitamin B_6 and the risk of colorectal cancer and adenoma in women. *Journal of the National Cancer Institute*, **97**, 684–692.

Weisenberger, D. J., Campan, M., Long, T. I., *et al.* (2005). Analysis of repetitive element DNA methylation by MethyLight. *Nucleic Acids Research*, **33**, 6823–6836.

Weisenberger, D. J., Siegmund, K. D., Campan, M., *et al.* (2006). CpG island methylator phenotype underlies sporadic microsatellite instability and is tightly associated with *BRAF* mutation in colorectal cancer. *Nature Genetics*, **38**, 787–793.

Wickramasinghe, S. and Fisa, S. (1994). Bone marrow cells from vitamin B_{12}- and folate-deficient patients misincorporate uracil into DNA. *Blood*, **83**, 1656–1661.

Yagi, K., Akagi, K., Hayashi, H., *et al.* (2010). Three DNA methylation epigenotypes in human colorectal cancer. *Clinical Cancer Research*, **16**, 21–33.

Yamada, Y., Jackson-Grusby, L., Linhart, H., *et al.* (2005). Opposing effects of DNA hypomethylation on intestinal and liver carcinogenesis. *Proceedings of the National Academy of Sciences USA*, **102**, 13 580–13 585.

Yang, A. S., Estecio, M. R., Doshi, K., *et al.* (2004). A simple method for estimating global DNA methylation using bisulfite PCR of repetitive DNA elements. *Nucleic Acids Research*, **32**, e38.

Yang, A. S., Doshi, K. D., Choi, S. W., *et al.* (2006). DNA methylation changes after 5-aza-2′-deoxycytidine therapy in patients with leukemia. *Cancer Research*, **66**, 5495–5503.

Yang, W. C., Mathew, J., Velcich, A., *et al.* (2001). Targeted inactivation of the p21(WAF1/cip1) gene enhances Apc-initiated tumor formation and the tumor-promoting activity of a Western-style high-risk diet by altering cell maturation in the intestinal mucosal. *Cancer Research*, **61**, 565–569.

35 Epigenetic epidemiology: transgenerational responses to the environment

Lars Olov Bygren

35.1 Introduction

To some degree, the influence of epigenetic factors – environmental factors outside the gene such as a cell's exposure to chemical, physical, or biological agents modifying a gene, e.g., by methylation or phosphorylation – can be transmitted to an individual's descendants even if these factors are not present after fertilization or while programming takes place during gametogenesis (Petronis, 2010). We call this phenomenon *transgenerational epigenetic heritability* or *gametic epigenetic inheritance*. For example, methylation can stably segregate a parental epiallele of a plant gene over eight generations, producing two complex traits – flowering time and plant height (Johannes *et al.*, 2009). Such a phenomenon has been observed in zebrafish and in mammals (Silva and White, 1988; Macleod *et al.*, 1999; Ng *et al.*, 2010). In human beings, evidence links environmental exposures to long-lasting effects on the phenotype and to the transmission of sequels over generations. The evidence suggests that diseases with complex etiologies are not only the function of current risk factors in adult life but also of an individual's early life exposures and even experiences of an individual's ancestors.

This chapter discusses insights into complex etiologies, mainly of cardiovascular disease, related to epigenetic epidemiology. In the course of this discussion, several questions are addressed. Could pre-conception, prenatal, childhood, and life-course nutrition-related circumstances trigger and maintain systemic epigenetic responses that affect complex etiologies of disease such as cardiovascular disease? Could feast and famine leave markers during an individual's life, could they be maintained during cell divisions in somatic cells and leave markers in germ-line cells bridging the marks over to an individual's descendants?

Epigenomics: From Chromatin Biology to Therapeutics, ed. K. Appasani. Published by Cambridge University Press. © Cambridge University Press 2012.

35.2   A look back

One important study found a connection between early life exposure and adult myocardial infarction. Children raised in a poor region in Norway before and during World War II became subjects of a postwar welfare program, but also to an increased risk of adult myocardial infarction (Forsdahl, 1977). Similar studies in England concentrated on signs of intrauterine malnutrition experienced by newborns and found three patterns of adult cardiovascular risk factors depending on the trimester in which the embryo or fetus experienced intrauterine malnutrition (Barker and Osmond, 1986). In a Swedish study change of food availability during intrauterine life was the crucial agent. If food supply changed from plentiful to poor or from poor to plentiful the risk of adult stroke was doubled compared to stable exposure to poor or good availability of food (Bygren *et al.*, 2000). Finally, the caloric restrictions during the Dutch Hunger Winter 1944–5, when parts of the Netherlands experienced extreme food shortages, produced different adult disease responses depending on the intrauterine period of exposure. For example, embryonal and fetal exposure to the famine determined adult schizophrenia, diabetes, cardiovascular disease, and persistent methylation states influenced the insulin-like growth factor 2 (IGF2) gene and nine other loci (Heijmans *et al.*, 2009).

35.3   The primordial follicle and sperm

The doubled adult stroke risk after exposure to changed nutrition during the intrauterine life might have been due to changes in placenta size which brought the discussion over to epigenetics (placenta size is partly epigenetically determined [Barker *et al.*, 1990]). Could environments epigenetically affect not only the embryo and fetus but also the ovum and sperm during their continued development up to puberty? Could feast and famine during the high demand for food during the prepubescent growth spurt peak in growth velocity affect the germ-line? Could feast and famine result in opposite responses during the period of lowest total energy demand, i.e., during the slow growth in mid-childhood preceding the peak in growth late in childhood before the onset of puberty?

Good harvests and low food prices in the environment could be called feast. In the slow growth period (SGP) ancestors' feast is followed by high mortality among descendants. Famine – a time of crop failure and high food prices – during the SGP is associated with long life of the descendants (Bygren *et al.*, 2001) (Table 35.1). These transgenerational responses are sex-linked via the father and the sensitivities during development from embryo to puberty produce different signs. Famine during very early development, on the other hand, is followed by descendants' higher mortality, while famine during the SGP is followed by descendants' lower mortality, and famine during adolescence, again, is followed by ancestors' higher mortality from cardiovascular disease as shown in Figure 35.1. The epidemiological

Table 35.1 Probands' average longevity in years related to their ancestors' access to food during their own slow growth period (SGP) prior to the prepubertal peak

Variable[a]	Survival time		Probability
	Years	SE	p
Poor availability of food, any year during SGP			
Intercept	50.9	6.6	0.0001
Mother	−2.6	5.6	0.64
Father	5.5	5.7	0.34
Maternal grandmother	2.2	6.1	0.72
Maternal grandfather	−5.8	5.7	0.30
Paternal grandmother	10.8	6.3	0.09[b]
Paternal grandfather	15.8	5.8	0.01
F-value=2.945; *p*=0.01			
Adjusted R^2=0.12			
Good availability of food, any year during SGP			
Intercept	76.1	6.9	0.0001
Mother	−6.3	6.1	0.30
Father	−6.8	6.0	0.26
Maternal grandmother	8.7	5.6	0.12
Maternal grandfather	−1.1	6.0	0.85
Paternal grandmother	−6.2	6.0	0.31[c]
Paternal grandfather	−16.5	6.0	0.01
F-value=3.043; *p*=0.47			
Adjusted R^2=0.12			
Moderate availability of food during all SGP			
Intercept	62.2	4.3	0.0001
Mother	3.7	6.2	0.55
Father	4.0	7.6	0.59
Maternal grandmother	−16.4	8.1	0.05
Maternal grandfather	17.0	13.8	0.22
Paternal grandmother	−3.7	7.6	0.63
Paternal grandfather	4.3	10.6	0.68
F-value=0.943; *p*=0.47			
Adjusted R^2=−0.004			

[a] Ancestors are introduced into the model in this order: mother, father, maternal grandmother and grandfather, and paternal grandmother and grandfather. Availability of food classified from harvests and prices and other historical facts in poor (famines), good, and moderate availability.
[b] The step before the paternal grandfather was introduced in the model, *p* measured with a *t*-test was=0.02.
[c] The step before the paternal grandfather was introduced in the model, *p* measured with a *t*-test was =0.01.

findings indicate that epigenetic marks might link calorie-restriction-related circumstances during sensitive periods of development to adult cardiovascular disease in descendants. Again, to some degree, epigenetic factors can survive epigenetic changes after fertilization and after reprogramming during gametogenesis (Petronis, 2010). Calorie restriction in itself or the concomitant infections and stress in society might have led to metabolic signatures determining diseases of complex etiologies through epigenetic inheritance in cell generations and gametic epigenetic inheritance of human traits and diseases (Curley and Mashoodh, 2010; Ng *et al.*, 2010).

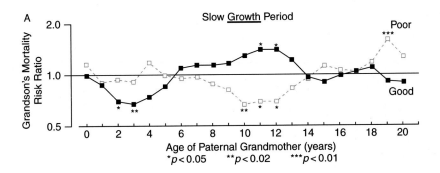

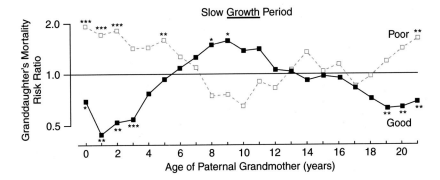

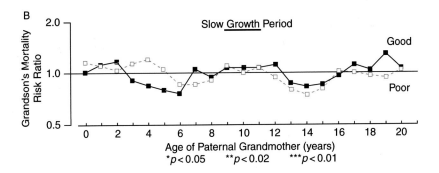

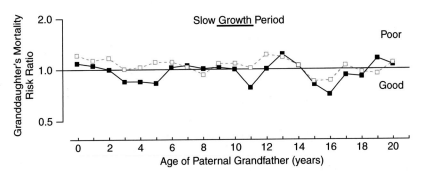

Figure 35.1 The effect of paternal grandparental food supply (good, filled squares; poor, open squares) at different times in their early life on the mortality rate of their grandchildren. Both (A) and (B) show on the y-axis the mortality risk ratio (RR) of the grandchildren separated by sex; first the grandsons' mortality RR and below this, the granddaughters' mortality RR. The age at which the paternal grandparent was exposed to good or poor food supply is given along the x-axis. The

35.4   A thought-provoking study

The Överkalix commune, a small commune in northeast Sweden, has proved to be of interest in the study of feast and famine in the nineteenth century. Using data collected from Överkalix, researchers studied transgenerational responses and possible epigenetic factors following ancestors of birth cohorts. The independent variables on feast and famine collected about ancestors were the availability of food during "strategic windows" in early life. To determine availability of food, the study used historical records. Data on harvest yields were gathered from agricultural statistical tables and data on food prices were gathered from pricing statistics. Additional information was gathered using general historical records. Selection was controlled by data on reproductive fitness and intergenerational change of standard deviations of outcomes. Socioeconomic circumstances included number of children in the family, the order in the sibship, professions, landownership, literacy, and marriages. These data provided information regarding the proband's socioeconomic circumstances throughout life.

Childhood environment had some influence on the mortality of the male probands. Their mother's death during their childhood was detrimental to their health as children or in adult life. The father's ownership of land and good literacy of the mother, on the other hand, protected the male proband. But, interestingly, ancestors' exposure to feast and famine in itself influenced the all-cause mortality of the descendants, the cardiovascular mortality, and the mortality from diabetes. In addition, the inheritance was largely via the grandfather and father (Bygren *et al.*, 2001; Pembrey *et al.*, 2006; Bygren, 2010).

35.5   Epigenetics and epidemiology

Combining human epigenetics and social medicine holds promise for the better understanding of complex diseases. Social medicine and epidemiology consider the person, the period, the environment, and the interactions between these. In the long run, we want to find the agents, present or absent, necessary for disease

---

Caption for Figure 35.1

grandchildren's mean mortality RR results are plotted for both those paternal grandparents who had a good food supply (solid line) and those who had a poor supply (dashed line) at the specified age on the x-axis. (A) First relates the paternal grandfather's exposure to his grandson's mortality RR and then below the paternal grandmothers' exposure to her granddaughters' mortality RR. (B) First relates paternal grandmother's exposure to her grandsons' mortality RR and then below the paternal grandfather's exposure to his granddaughters' mortality RR. The data points were obtained using a 3-year frame, advanced 1 year at a time, for grandparental age at exposure, to produce rolling means for the grandchild's (proband's) mortality RR for both "good" and "poor" ancestral food supply. Exposure to at least 1 year of surfeit of food or to at least 1 year of poor availability during a 3-year period denotes it as exposure to a "good" period or a "poor" period respectively. Food supply at age 0 years is the mean for the 33-month period from −267 days until the day before the second birthday and therefore includes fetal life. (Adapted from Pembrey *et al.* [2001], *European Journal of Human Genetics*, **14**, 163.)

and whether these agents are chemical, physical, mechanical, biological, or social. If the agents are well known, the pathophysiological mechanisms are sought. Contagiousness and resistance factors are studied directly in medical and other examinations and indirectly by looking at people's position in the social structure (age, gender, profession, educational status, etc.), their social environment (family, domicile, society, etc.), and their assets. The analyses are most often multivariable and discount variables causing blurring (confounders). In this discussion, the signals of interest are nutrition or lack of nutrition during times of famine and feast and concomitant infections and historic stresses in the family and in society.

Epigenetics is the study of changes in gene expression heritable in the cell line that occur without a change in DNA sequence. Epigenetic modifications include covalent modifications of DNA and histones as well as non-covalent changes regulating nucleosome positioning. It is well established that epigenetic regulations are critical components in the normal development and growth of cells. Increasing evidence indicates that dietary factors can modulate the epigenetic regulation of gene expression and induce permanent changes in metabolism and physiology, affecting the risk for adult-onset diseases. The reaction to the environment can modify DNA and histones and the epigenetic changes can be propagated during cell division and maintain the acquired phenotype. Heritability concerns somatic cells in one generation of individuals but includes germ-line cells and possibly gametic epigenetic heritability over generations. The erasure of epigenetic marks in the primordial germ cells might not be complete or mechanisms exist that partly restore parental epigenetic marks. Epigenetic stability or instability can largely replace the genetic and environmental components in traditional models, and inherited or acquired epigenetic regulation or misregulation can be a core unifying molecular mechanism of complex non-Mendelian inherited diseases (Petronis, 2010).

35.6   Early life development

In humans, early growth and development can be grossly divided into an intra-uterine–infancy–toddler growth phase, a slow growth phase (SGP), the prepubertal peak in growth velocity, and adolescence. Each phase has its own growth determinants such as nutrition, oxygen, growth hormone, insulin-like growth factor, and sex hormones. For the oocyte, the fetal–infancy period and the SGP are of special interest. Epigenetically competent oocytes are by-products of the germ cell's life history in the ovary. Coordinations, feedback through nutrition, hormones, growth factors, and circadian rhythm influence development depending on the stages of oogenesis. Primordial follicles mainly mature very early during development. For the spermatocyte the SGP is most influential since it is the period when the first viable pools of spermatocytes emerge and when some reprogramming of DNA methylation begins. Each sperm cell and oocyte has a unique DNA methylation profile and the variation in epigenetic marks greatly exceeds the variation in the DNA sequence profile, a different epigenetic starting point for the descendants (Flanagan *et al.*, 2006).

35.7 Possible mechanisms

The correlations between ancestors' early life exposures and descendants' cardio-vascular disease and other diseases indicate a gametic heritability in gene function without changes in DNA sequence and sociomedical inheritance. How could metabolic imprinting have influenced the development of the egg and sperm? In mammals, metabolic imprinting is complicated by erasure and waves of genome-wide epigenetic reprogramming. Despite these obstacles, gametic epigenetic inheritance sometimes occurs in mammals. Transmission, through the male germ-line in particular, suggests such epigenetic inheritance (Skinner, 2008). In the expression of three genes with major roles in development, gametic epigenetic heritability has been observed in mice by transcription initiated by RNAs related to the loci, either microRNAs or transcript fragments carried by the spermatozoon. Interestingly, human sperm contains a high load of RNA (Cuzin and Rasoulzadegan, 2010). Longitudinal studies have observed that in human beings DNA methylation clusters in families, again possibly indicating gametic epigenetic inheritance (Thorvaldsen and Bartolomei, 2007).

We know much about the epidemiological pattern regarding exposure to good or poor availability of food and the transgenerational responses in natural experiments, responses that might be understood in the light of epigenetics. Much remains, however, before the mechanisms behind the pedigree findings are understood with respect to genomic characterizations. One profile of response could be changes of DNA methylation in the descendants. Another mechanism could be seen in the correlation of microRNA with a genome-wide DNA methylation pattern. The obtained results would allow identification of specific genes subject to epigenetic modifications related to nutrition. The genes that regulate DNA methylation are differently expressed because of impairments or by DNA demethylase. Poor or good food availability during sensitive early life periods may modify histones in descendants. Possible mechanisms underlying these changes could be differential recruitment of some epigenetic components, such as histone deacetylase and methyl-CpG-binding proteins. The sex-linked deleterious influence via the father from overnutrition during windows in child development warrants analysis of the Y chromosome. For methylation, methyl donors are necessary and the fundamental supply of methyl radicals involves the folate cycle, the methionine cycle, the glycolysis cycle, and the urea cycle. The level of homocysteine is linked to epigenetics and has its own metabolism regulated by genes. Methyltransferases as well as demethylase are involved in the maintenance of DNA methylation and might have a key role in gametic epigenetic inheritance (Egger *et al.*, 2004).

35.8 Tissue-specific and general sensitivity

Tissue sensitivity is common in diseases with complex etiologies, but adult peripheral blood DNA can carry a methylation signature of early experience even if this is not the primary tissue of the developmental response. In studies of peripheral blood DNA periconceptional exposure to famine was associated with hypomethylation of the IGF2 gene 60 years later. Imprinted stem cells must have renewed the circulating blood. During their development from stem cells, blood cells and other cells captured promoters for epigenetic variation. The involvement of stem cells should indicate that germ cells might be affected with some epigenetic marks being transmitted to offspring. Significant correlations between tissues are found in the methylation part of the methylated CpG sites indicating a general sensitivity.

35.9   Selection and consanguinity, learning, mimicking, and material assets

Gametic epigenetic inheritance might be one of the factors behind the epidemiological findings. Selection has to be taken into account and can be detected as diminishing variance of a health outcome over generations. Another indication of selection is reproductive fitness – measured as age at first childbirth and number of children – in pedigrees with exposed ancestors. Sensitivity for consanguinity is done by rerunning the analyses with randomly withdrawing cousins. Material resources, the cultural environment, and the family environment in the nineteenth-century pedigrees can be understood using parental deaths during childhood up to 12 years, parents' literacy according to the clergy, the number of siblings, and the order within siblings (Bygren *et al.*, 2001; Pembrey *et al.*, 2006).

35.10   Concluding remarks: the near and distant future

Variation in the environment can affect gene expression and the expression can be inherited across generations. It can provide one source of heritable phenotypic variation that is not caused by hardwired genetics. In short, the epidemiological findings will have to be supplemented with profiles of DNA methylation, mRNA gene expression of genes regulating DNA methylation, histone modification, and Y-haplotypes and copy number changes. Inheritance by mimicking, learning, material, and cultural resources has to be taken into account. This combined epidemiological and epigenetic research field holds promise for a new understanding of complex etiologies of disease. If the epidemiological findings in humans indicating gametic epigenetic inheritance cannot be falsified, a door is opened to new human epidemiological and epigenetic research. Genetic anticipation that a complex disease in a family is more severe with every new generation because of the expansion of DNA repeats might sometimes be due to epigenetic marks being stably inherited. Unlike genetic damage, epigenetic changes can sometimes be reversed. Complete or substantial erasure of these epigenetic marks in the germ-line will halt the propagation of disease. This creates a great potential for the development of epigenetically based health promotion and prevention and even therapies such as inhibitors of enzymes that control epigenetic modifications (Egger *et al.*, 2004; Petronis, 2010).

REFERENCES

Barker, D. J. P. and Osmond, C. (1986). Infant mortality, childhood nutrition, and ischemic heart disease in England and Wales. *Lancet*, **1**, 1077–1081.

Barker, D. J. P., Bull, A. R., Osmond, C., and Simmonds, S. J. (1990). Fetal and placental size and risk of hypertension in adult life. *British Medical Journal*, **301**, 259–262.

Bygren, L. O. (2010). Epigenetic epidemiology and food availability. *UNSSC Nutrition News*, **37**, 13–14.

Bygren, L. O., Edvinsson, S., and Broström, G. (2000). Change in food availability during pregnancy: is it related to adult sudden death from cerebro- and cardiovascular disease in offspring? *American Journal of Human Biology*, **12**, 447–453.

Bygren, L. O., Kaati, G., and Edvinsson, S. (2001). Longevity determined by paternal ancestors' nutrition during their slow growth period. *Acta Biotheoretica*, **49**, 53–59.

Curley, J. P. and Mashoodh, R. (2010). Parent-of-origin and trans-generational germline influences on behavioral development: the interacting roles of mothers, fathers, and grandparents. *Developmental Psychobiology*, **523**, 12–30.

Cuzin, F. and Rasoulzadegan, M. (2010). Non-Mendelian epigenetic heredity: gametic RNAs as epigenetic regulators and transgenerational signal. *Essays in Biochemistry*, **48**, 101–106.

Egger, G., Liang, G., Aparicio, A., and Jones, P. A. (2004). Epigenetics in human disease and prospects for epigenetic therapy. *Nature*, **429**, 457–463.

Flanagan, J. M., Popendikyte, V., Pozdniakovaite, N., *et al.* (2006). Intra- and interindividual epigenetic variation in human germ cells. *American Journal of Human Genetics*, **79**, 67–84.

Forsdahl, A. (1977). Are poor living conditions in childhood and adolescents an important risk factor for arteriosclerotic heart disease? *British Journal of Preventive Social Medicine*, **31**, 91–95.

Johannes, F., Porcher, E., Teixeira, F. K., *et al.* (2009). Assessing the impact of transgenerational epigenetic variation on complex traits. *PLoS Genetics*, **26**, e1000530.

Lumey, L. H. (2008). Persistent epigenetic differences associated with prenatal exposure to famine in humans. *Proceedings of the National Academy of Sciences USA*, **105**, 1746–1749.

Macleod, D., Clark, V. H., and Bird, A. (1999). Absence of genome-wide changes in DNA methylation during development of the zebrafish. *Nature Genetics*, **23**, 139–140.

Ng, S. F., Lin, R. C., Laybutt, D. R., *et al.* (2010). Chronic high-fat diet in fathers programs β-cell dysfunction in female rat offspring. *Nature*, **467**, 963–966.

Pembrey, M. E., Bygren, L. O., Kaati, G., *et al.* (2006). Sex-specific, male-line transgenerational responses in humans. *European Journal of Human Genetics*, **14**, 159–166.

Petronis, A. (2010). Epigenetics as a unifying principle in the aetiology of complex traits and diseases. *Nature*, **465**, 721–727.

Silva, A. and White, R. (1988). Inheritance of blueprints for methylation patterns. *Cell*, **54**, 145–152.

Skinner, M. K. (2008). What is an epigenetic transgenerational phenotype? F3 or F2. *Reproductive Toxicology*, **25**, 2–6.

Thorvaldsen, J. L. and Bartolomei, M. S. (2007). Snapshot: imprinted genes cluster. *Cell*, **130**, 958.

Index

Printed in the United States
By Bookmasters